MyMathLab®:
Support You Need, When You Need It

MyMathLab is the world's leading online program in mathematics, integrating homework with support tools and tutorials in an easy-to-use format. MyMathLab helps you get up to speed on course material, visualize the content, and understand how math will play a role in your future career.

Review Prerequisite Skills

Integrated Review content identifies gaps in prerequisite skills and offers help for just those skills you need. With this targeted practice, you will be ready to learn new material.

Tutorial Videos

Tutorial videos are available for every section of the textbook and cover key examples from the text. These videos are especially handy if you miss a lecture or just need another explanation.

Interactive Figures

Interactive Figures illustrate key concepts and help you visualize the math. MyMathLab also includes assignable exercises that require use of Interactive Figures and instructional videos that explain the concept behind each figure.

www.mymathlab.com

fourteenth edition

Calculus

& ITS APPLICATIONS

Larry J. Goldstein
Goldstein Educational Technologies

David C. Lay
University of Maryland

David I. Schneider
University of Maryland

Nakhlé H. Asmar
University of Missouri

330 Hudson Street, NY, NY 10013

Director, Portfolio Management: *Deirdre Lynch*
Executive Editor: *Jeff Weidenaar*
Editorial Assistant: *Jennifer Snyder*
Content Producer: *Patty Bergin*
Managing Producer: *Karen Wernholm*
Producer, Production & Digital Studio: *Stephanie Green*
Manager, Content Development, Math: *Kristina Evans*
Product Marketing Manager: *Claire Kozar*
Marketing Assistant: *Jennifer Myers*
Senior Author Support/Technology Specialist: *Joe Vetere*
Manager, Rights and Permissions, Higher Education: *Gina Cheselka*
Manufacturing Buyer: *Carol Melville, LSC Communications*
Associate Director of Design: *Blair Brown*
Composition: *iEnergizer Aptara®, Inc.*
Text Design, Production Coordination, Composition, and Illustrations: *iEnergizer Aptara®, Inc.*
Cover Design: *Cenveo*
Cover Image: *Doug Chinnery/Getty Images*

Library of Congress Cataloging-in-Publication Data
Names: Goldstein, Larry Joel. | Lay, David C. | Schneider, David I. | Asmar, Nakhlé H.
Title: Calculus & its applications.
Other titles: Calculus and its applications
Description: Fourteenth edition / Larry J. Goldstein, Goldstein Educational Technologies, David C. Lay, University of Maryland, David I. Schneider, University of Maryland, Nakhlé H. Asmar, University of Missouri. | Boston : Pearson Education, [2018] | Includes indexes.
Identifiers: LCCN 2016030689| ISBN 9780134437774 (hardcover) | ISBN 0134437772 (hardcover)
Subjects: LCSH: Calculus—Textbooks.
Classification: LCC QA303.2 .G66 2018 | DDC 515—dc23
LC record available at https://lccn.loc.gov/2016030689
2   17

Student Edition ISBN-13: 978-0-134-43777-4
Student Edition ISBN-10:   0-134-43777-2

Contents

*These sections are optional in the sense that they are not prerequisites for later material.

Preface

This fourteenth edition of *Calculus & Its Applications*, and its Brief version, is written for either a one- or two-semester applied calculus course for students majoring in business or the social and biological sciences. Our goals for this revision reflect the original goals for the text, which include: to begin calculus as soon as possible; to present calculus in an intuitive yet intellectually satisfying way; and to integrate the many applications of calculus to business, life sciences, and social sciences. A primary goal of this revision was to make the calculus even more accessible and intuitive while maintaining the overall level and approach of previous editions.

The Series

This text is part of a highly successful series consisting of three texts: *Finite Mathematics & Its Applications*, *Calculus & Its Applications*, and *Calculus & Its Applications, Brief Version* (which contains Chapters 0–9 from the longer book). All three titles are available for purchase in a variety of formats, including as an eBook within the MyMathLab online course.

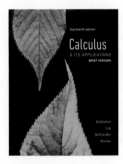

Content

The distinctive order of topics has proven over the years to be very successful. A strength of our approach is to build intuition with key concepts of calculus before the analytical material. For instance, the derivative is explained geometrically before the analytical material on limits is presented. And the definite integral is introduced intuitively via the notion of net change (before the material on Riemann sums).

The significant applications introduced early in the course serve to motivate students and make the mathematics more accessible. To allow you to reach the applications in Chapter 2 quickly, we present only the differentiation rules and curve sketching needed.

Several texts in applied calculus over the past decade or so have shifted coverage of exponential and logarithmic functions to earlier in the text. We have maintained our later treatment because (a) it allows for undistracted focus on these important functions and their applications, and (b) it provides a venue to review important concepts (such as the Chain Rule) in a new context. Chapter 5, which covers applications of exponential and logarithmic functions, stands apart from all other texts for this course with its straightforward and intuitive use of differential equations to model a variety of phenomena. A key focus of this revision was to bolster the strengths of our approach in Chapters 4 and 5, making them even more understandable.

Because most courses do not afford enough time to cover all of the topics in this text and because schools have different goals for the course(s), we have been strategic with the placement and organization of topics. To this end, the level of theoretical material may be adjusted to meet the needs of the students. For example, Section 1.4 "Limits and the Derivative" may be omitted entirely if the instructor does not wish to present the notion of limit beyond the required material that is contained in Section 1.3. In addition, sections considered optional are starred in the table of contents.

Prerequisites

Because students often enter this course with a variety of prerequisite skills, Chapter 0 is available to either cover in its entirety or as a source of remediation, depending on the pace of the course. In addition to being covered in Chapter 0, some important topics, such as the laws of exponents, are reviewed again when they are used in a later chapter.

The Prerequisite Skills Diagnostic Test prior to Chapter 0 may help you in evaluating student gaps in prerequisite skills. The answers to the diagnostic test are provided in the student edition along with references to where students can go for remediation. Diagnostic tests and remediation are also provided in MyMathLab—with the added benefits of auto-grading and personalized remediation for any skill-gaps that are identified.

New to This Edition

We are grateful for the many helpful suggestions made by reviewers and users of the text. We incorporated many of these into this new edition. We also analyzed aggregated student usage and performance data from MyMathLab for the previous edition of this text. The results of this analysis helped improve the quality and quantity of exercises that matter the most to instructors and students. Additionally, we made the following improvements in this edition:

- Revised Examples to more closely align with the exercise sets.
- Added blue "help text" next to steps within worked-out examples to help students understand key algebraic or numerical transitions.
- In cases where properties or theorems that were formerly numbered (e.g., Property 4) have a commonly used name (e.g., Power of a Quotient Rule), we used the name rather than the number. This allows for more intuitive explanations within examples and is better aligned to how concepts are explained in class.
- Updated graphing calculator screens to TI-84 Plus CE (color) format.
- Added *For Review* features in the side margin to either remind students of a concept that is needed at that point in the book or direct the student back to the section in which it was covered earlier in the book.
- Revised all 3-dimensional figures in the text using the latest software. The difference in most cases is stunning, as can be seen in the sample figure here. We took full advantage of these updates to make the figures more effective pedagogically.

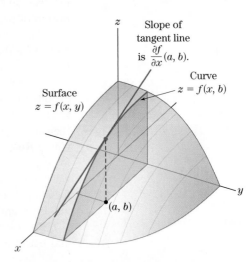

Figure 1 $\frac{\partial f}{\partial x}$ gives the slope of a curve formed by holding y constant at b.

- Updated data in examples and exercises. Many modern and student-centered applications have been added to help students see the relevance of the content.
- Added 225 new exercises and 30 new examples. The book now contains 4200 exercises and 520 worked-out examples.
- Updated examples and exercises to better reflect changes in culture and student population. And acknowledging that today's student population is more international than ever before, we changed the context of some problems to make them more understandable to students not born in the U.S.

New to MyMathLab

Many improvements have been made to the overall functionality of MyMathLab (MML) since the previous edition. However, beyond that, we have also increased and improved the content specific to this text.

- Instructors now have more exercises than ever to choose from in assigning homework. There are approximately 3300 assignable exercises in MML.
- The Conceptual Question Library provides additional conceptual questions for your use.
- We heard from users that the Annotated Instructor Edition for the previous edition required too much flipping of pages to find answers, so MML now contains a downloadable Instructor Answers document—*with all answers in one place*. (This augments the downloadable Instructor Solutions Manual, which contains all *solutions*.)
- Interactive Figures are now in HTML format (no plug-in required) and are supported by assignable exercises and tutorial videos.
- An Integrated Review version of the MML course contains pre-made quizzes to assess the prerequisite skills needed for each chapter, plus personalized remediation for any gaps in skills that are identified.
- New Setup & Solve exercises require students to show how they set up a problem as well as the solution, better mirroring what is required of students on tests.
- 97 new example-based videos were added.
- MathTalk videos highlight applications of the content of the course to business. The videos are supported by assignable exercises.
- Study skills modules help students with the life skills that can make the difference between passing and failing.
- Graphing Calculator and Excel Spreadsheet Manuals, specific to this course, are now downloadable from MML.

Content Changes

Chapter 0

- Added an example and exercises in 0.1 to illustrate the concept of piecewise-defined functions.
- Rewrote and simplified the introduction to 0.5 Exponents and Power Functions to make it more intuitive and easier to reference. Additionally, we added several examples to illustrate the rules of exponents.
- Added Examples 8 and 9 in 0.5 to illustrate the role of multiple factors in compound interest and investment accounts.
- Added Example 7 in 0.6 to illustrate various concepts from economics.
- Added four new exercises (45–48) in 0.6 to illustrate variations on the standard topic of compound interest.

Chapter 1

- Removed some of the proofs related to review material to simplify the presentation in 1.1.
- Added four new exercises (5–8) in 1.2 to illustrate the geometric meaning of the slope of a graph as the slope of the tangent line. Additionally, we modified two other exercises requiring reading and interpreting slopes of graphs.

- Simplified the discussion of limits in Examples 2 and 4 in 1.4.
- Included a discussion and a new Example 4 in 1.8 to illustrate the concepts of displacement and velocity.

Chapters 2 and 3

- Modified the Technology Exercises in 2.1 to make them more straightforward for students to answer.
- Improved and simplified the solutions within Example 4 in 2.4.
- Removed Example 4 in 2.6 which required more symbolic manipulation and use of constants than students would encounter in the exercises.
- Rewrote five examples in the Summary section of Chapter 2.
- Added ten new exercises in Chapter 3.

Chapter 4

- Revised Example 2 in 4.2 to better prepare students for the variety of exercises within the homework.
- Moved the material on the properties and graphs of exponential functions from 4.3 to 4.2.
- Replaced Examples 1, 2, and 3 from 4.3 with new examples that better build on the properties of derivatives introduced earlier. Example 3 introduces a new concept of combined returns to illustrate applications of linear combinations of exponential functions.
- Moved the material on differential equations in 4.3 to Chapter 5.
- Introduced forty new exercises in 4.3, including one on investment portfolios.
- Rewrote the introduction of 4.4 to better display and present the properties of logarithms.
- Changed Example 1 in 4.4 to better match the types of exercises in the homework.
- Used investment portfolios to illustrate applications involving solutions of equations with logarithms in Example 4 in 4.4.
- Rewrote the introduction of 4.5 to better display properties of the natural logarithm and its derivative.
- Added a new example on differentiation of the natural logarithm in 4.5.
- Modified and added over twenty exercises in 4.5.
- Rewrote the introduction of 4.6 to better present further properties of the logarithm.

Chapter 5

- Rewrote 5.1 to better show the applications of the derivative as a rate of change in setting up and solving differential equations. Changes to 5.1 also include:
  - Stated the solutions of the differential equation $y' = ky$ in a theorem.
  - Added an example on solving differential equations.
  - Stated the solutions of the initial value problem $y' = ky$, $y(0) = P_0$ in a theorem.
  - Illustrated the solutions of initial value problems with examples and a figure that a student can easily relate to.
  - Simplified the discussion on exponential decay and carbon dating.
  - Added twenty-three new exercises on differential equations and their applications.
- Rewrote and simplified the introduction to 5.2. Also in 5.2:
  - Introduced continuous compounding as a limit of the ordinary compounding from Chapter 0 and as a solution of a differential equation of the type that was discussed in 5.1.
  - Added narrative that compares continuous to ordinary compounding.
  - Added discussion, an example, and exercises on negative interest rates.
- Rewrote the introduction to 5.3 to better explain the concept of relative rate of change and how it relates to common pricing applications.
- Added an example in 5.3 on logarithmic derivatives.
- Included a summary of solutions of differential equations and their properties in the Chapter Summary.

Chapter 6

- Simplified Example 1 in 6.1 by adding more details to the solution.
- Rewrote the introduction to 6.2 and simplified the presentation by relating the new concept of definite integral to a common problem involving velocity and position.
- Simplified the numerical computations in Examples 3 and 5 in 6.2.
- Added nine new exercises to 6.2.
- Added a new example on online transportation in 6.4, as an application of area between two curves.

Chapter 7

- Revised all 3-dimensional figures in the chapter using the latest software.
- Added more help to Example 1 in 7.2 and added a 3-dimensional graph.
- In Example 4 in 7.3, added a new application of the second derivative test in two dimensions.
- Added twenty new exercises to 7.2, ten new exercises to 7.4, and two new exercises to 7.5.

Chapters 8–12

- Removed the application on population genetics from 10.6 (as this is now covered in Chapter 5).
- Revised exercises and examples to update real-world data.

Trusted Features

Though this edition has been improved in a variety of ways to reflect changing student needs, we have maintained the popular overall approach that has helped students be successful over the years.

Relevant and Varied Applications

We provide realistic applications that illustrate the uses of calculus in other disciplines and everyday life. The variety of applications is evident in the Index of Applications at the end of the text. Wherever possible, we attempt to use applications to motivate the mathematics. For example, our approach to the derivative in 1.3 is motivated by the slope formula and applications in 1.2, and applications of the net change functions in 6.2 motivate our approach to the definite integral in 6.2 and 6.3.

Plentiful Examples

The fourteenth edition includes 520 worked examples. Furthermore, we include computational details to enhance comprehension by students whose basic skills are weak. Knowing that students often refer back to examples for help, we built in fidelity between exercises and examples. In addition, students are given Now Try exercise references immediately following most examples to encourage them to check their understanding of the given example.

Exercises to Meet All Student Needs

The 4200 exercises are the most important part of the text, in our opinion. The exercises at the ends of the sections are typically arranged in the order in which the text proceeds, so that homework assignments may be made easily after only part of a section is discussed. Interesting applications and more challenging problems tend to be located near the ends of the exercise sets. Exercises have odd-even pairing, when appropriate. Chapter Review Exercises are designed to prepare students for end-of-chapter tests. Answers to the odd-numbered exercises, and all Chapter Review Exercises, are included at the back of the book.

Check Your Understanding Problems

The Check Your Understanding problems are a popular and useful feature of the book. They are carefully selected exercises located at the end of each section, just before the exercise set. Complete solutions follow the exercise set. These problems prepare students for the exercise sets beyond just covering simple examples. They give students a chance to think about the skills they are about to apply and reflect on what they've learned.

Use of Technology

As in previous editions, the use of graphing calculators is not required for the study of this text; however, for instructors who wish to incorporate them we have included instructions for their use in the Integrating Technology features within sections. In our discussions of graphing calculators, we specifically refer to the TI-84 Plus models. New to this edition, screen shots display the new color versions of the TI-84. Most exercise sets include problems at the end labeled Technology Exercises.

End-of-Chapter Study Aids

Near the end of each chapter is a set of problems entitled Fundamental Concept Check Exercises that help students recall key ideas of the chapter and focus on the relevance of these concepts as well as prepare for exams. Each chapter also contains a two-column grid giving a section-by-section summary of key terms and concepts with examples. Finally, each chapter has Chapter Review Exercises that provide more practice and preparation for chapter-level exams.

Technology and Supplements

MyMathLab® Online Course (access code required)

Built around Pearson's best-selling content, MyMathLab is an online homework, tutorial, and assessment program designed to work with this text to engage students and improve results. MyMathLab can be successfully implemented in any classroom environment—lab-based, hybrid, fully online, or traditional. **By addressing instructor and student needs, MyMathLab improves student learning.**

Used by more than 37 million students worldwide, MyMathLab delivers consistent, measurable gains in student learning outcomes, retention, and subsequent course success. Visit www.mymathlab.com/results to learn more.

Preparedness

One of the biggest challenges in applied calculus courses is making sure students are adequately prepared with the prerequisite skills needed to successfully complete their course work. Pearson offers a variety of content and course options to support students with just-in-time remediation and key-concept review.

▲ Chapter 4 Skills Check Quiz

▲ Chapter 4 Skills Review Homework

- **Integrated Review Courses** can be used for just-in-time prerequisite review. These courses provide additional prerequisite content on review topics, along with pre-made, assignable skill-check quizzes, personalized homework assignments, and videos integrated throughout the course content.

Motivation

Students are motivated to succeed when they're engaged in the learning experience and understand the relevance and power of mathematics. MyMathLab's online homework offers students immediate feedback and tutorial assistance that motivates them to do more, which means they retain more knowledge and improve their test scores.

- **Exercises with immediate feedback**—over 3300 assignable exercises—are based on the textbook exercises, and regenerate algorithmically to give students unlimited opportunity for practice and mastery. MyMathLab provides helpful feedback when students enter incorrect answers and includes optional learning aids including Help Me Solve This, View an Example, videos, and an eText.

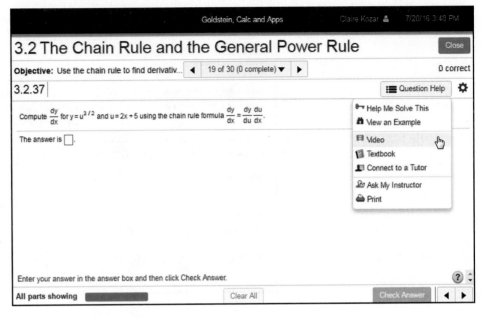

- **Setup and Solve Exercises** ask students to first describe how they will set up and approach the problem. This reinforces students' conceptual understanding of the process they are applying and promotes long-term retention of the skill.
- **Conceptual Question Library** exercises require students to apply their conceptual understanding to solve problems—mostly those involving applications.
- **MathTalk videos** connect the math to the real world (particularly business). The videos include assignable exercises to gauge students' understanding of video content.
- **Learning Catalytics™** is a student response tool that uses students' smartphones, tablets, or laptops to engage them in more interactive tasks and thinking. Learning Catalytics fosters student engagement and peer-to-peer learning with real-time analytics.

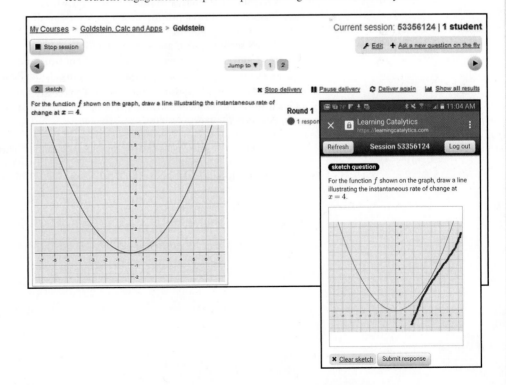

Learning and Teaching Tools

- **Interactive Figures** illustrate key concepts and allow manipulation for use as teaching and learning tools. Includes assignable exercises that require use of figures and instructional videos that explain the concept behind each figure.

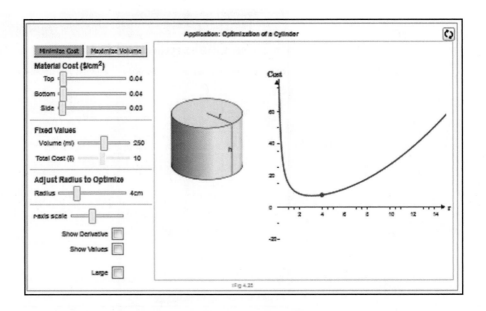

- **Instructional videos**—277 example-based videos—are available as learning aids within exercises and for self-study. The Guide to Video-Based Assignments makes it easy to assign videos for homework by showing which MyMathLab exercises correspond to each video.
- **Complete eText** is available to students through their MyMathLab courses for the lifetime of the edition, giving students unlimited access to the eText within any course using that edition of the textbook.
- **Skills for Success Modules** helps students with the life skills that can make the difference between passing and failing. Topics include "Time Management" and "Stress Management."
- **Excel Spreadsheet Manual,** specifically written for this course.
- **Graphing Calculator Manual,** specifically written for this course.
- **PowerPoint Presentations** are available for download for each section of the book.
- **Accessibility** and achievement go hand in hand. MyMathLab is compatible with the JAWS screen reader, and enables multiple-choice and free-response problem types to be read and interacted with via keyboard controls and math notation input. MyMathLab also works with screen enlargers, including ZoomText, MAGic, and SuperNova. And, all MyMathLab videos have closed-captioning. More information is available at http://mymathlab.com/accessibility.
- **A comprehensive gradebook** with enhanced reporting functionality allows you to efficiently manage your course.
  - **Item Analysis** tracks class-wide understanding of particular exercises so you can refine your class lectures or adjust the course/department syllabus. Just-in-time teaching has never been easier!

- **The Reporting Dashboard** provides insight to view, analyze, and report learning outcomes. Student performance data is presented at the class, section, and program levels in an accessible, visual manner so you'll have the information you need to keep your students on track.

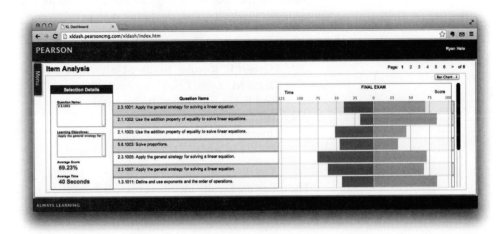

MyMathLab comes from an experienced partner with educational expertise and an eye on the future. Whether you are just getting started with MyMathLab, or have a question along the way, we're here to help you learn about our technologies and how to incorporate them into your course. To learn more about how MyMathLab helps students succeed, visit www.mymathlab.com or contact your Pearson rep.

MathXL® is the homework and assessment engine that runs MyMathLab. (MyMathLab is MathXL plus a learning management system.) MathXL access codes are also an option.

Student's Solutions Manual

ISBN-10: 0-134-46323-4 | ISBN-13: 978-0-134-46323-0
Contains fully worked-out solutions to odd-numbered exercises. Available in print and downloadable from within MyMathLab.

Instructor's Answers / Instructor's Solutions Manual (both downloadable)

ISBN-10: 0-134-46322-6 | ISBN-13: 978-0-134-46322-3
The Instructor's Answers document contains a list of answers to all student edition exercises. The Instructor's Solutions Manual contains solutions to all student edition exercises. Both are downloadable from the Pearson Instructor Resource Center www.pearsonhighered.com/irc, or from within MyMathLab.

TestGen (downloadable)

ISBN-10: 0-134-46324-2 | ISBN-13: 978-0-134-46324-7
TestGen enables instructors to build, edit, print, and administer tests using a bank of questions developed to cover all objectives in the text. TestGen is algorithmically based, allowing you to create multiple but equivalent versions of the same question or test. Instructors can also modify test bank questions or add new questions. The software and testbank are available to qualified instructors for download and installation from Pearson's online catalog www.pearsonhighered.com and from within MyMathLab.

PowerPoints

ISBN-10: 0-134-46333-1 | ISBN-13: 978-0-134-46333-9
Contains classroom presentation slides for this textbook featuring lecture content, worked-out examples, and key graphics from the text. Available to qualified instructors within MyMathLab or through the Pearson Instructor Resource Center www.pearsonhighered.com/irc.

Acknowledgments

While writing this book, we have received assistance from many people and our heartfelt thanks go out to them all. Especially, we should like to thank the following reviewers, who took the time and energy to share their ideas, preferences, and often their enthusiasm, with us during this revision:

Stephen Brick, University of South Alabama
Herbert E. Kasube, Bradley University
Diane M. Zych, Erie Community College
Paul E. Schwiegerling, Buffalo State College
Robert Sczech, Rutgers University
Omayra Ortega, Arizona State University
Marie Franzosa, Oregon State University

The following faculty members provided direction on the development of the MyMath-Lab course for this edition:

Henjin Chi, Indiana State University
Benjamin N. Levy, Northeastern University
Filix Maisch, Oregon State University
Shifat Mithila, Florida Atlantic University
Anthony Nelson, Tulsa Community College
Bill Schwedner, Pace University
Paul Schwiegerling, Buffalo State College
Salah Shalan, Schoolcraft College
David B. Williams, Clayton State University

We wish to thank the many people at Pearson who have contributed to the success of this book. We appreciate the efforts of the production, design, manufacturing, marketing, and sales departments. We are grateful to Lisa Collette for her thorough proofreading and Sarah Sponholz and Deana Richmond for their careful and thorough checking for accuracy. We are also grateful to Thomas Asmar (UC Berkeley) for his help in proofreading and updating the graphing calculator screens. Content Producer Patty Bergin did a fantastic job keeping the book on schedule. The authors wish to extend special thanks to editor Jeff Weidenaar.

If you have any comments or suggestions, we would like to hear from you. We hope you enjoy using this book as much as we have enjoyed writing it.

Larry J. Goldstein

David C. Lay

David I. Schneider
dis@math.umd.edu

Nakhlé H. Asmar
asmarn@missouri.edu

Prerequisite Skills Diagnostic Test

To the Student and the Instructor   Are you ready for calculus? This prerequisite skills diagnostic test evaluates basic mathematical skills that are required to begin the course. It is not intended to replace a placement test that your institution may already have. Each set of questions refers to a section in Chapter 0 of the text. If you miss several questions from one part, you may want to study the corresponding section from Chapter 0.

Calculate the given quantities using the laws of exponents. **(Section 0.5)**

1. $\dfrac{7^{3/2}}{49}\sqrt{7}$

2. $(3^{1/3}3^{1/2})^2\sqrt[3]{3}$

3. $\dfrac{\sqrt{5}}{\sqrt{15}\sqrt{3}}$

4. $2^{1/3}2^{1/2}2^{1/6}$

Simplify the given expressions. Your answer should not involve parentheses or negative exponents. **(Section 0.5)**

5. $\dfrac{x^2}{x^{-4}}$

6. $\dfrac{x^2(x^{-4}+1)}{x^{-2}}$

7. $\left(\dfrac{x}{x^2y^2}\right)^3 y^8$

8. $\left(\dfrac{1}{xy}\right)^{-2}\left(\dfrac{x}{y}\right)^2$

Given $f(x) = \dfrac{x}{x+1}$ and $g(x) = x+1$, express the following as rational functions. **(Section 0.3)**

9. $f(x) + g(x)$

10. $f(x)g(x)$

11. $\dfrac{g(x)}{f(x)}$

12. $f(x) - \dfrac{g(x)}{x+1}$

Given $f(t) = t^2$ and $g(t) = \dfrac{t}{t+1}$, calculate the following functions. Simplify your answer as much as possible. **(Section 0.3)**

13. $f(g(t))$

14. $g(f(t))$

15. $f(f(g(t)))$

16. $f(g(t+1))$

For the given $f(x)$, find $\dfrac{f(x+h)-f(x)}{h}$ and simplify your answer as much as possible. **(Section 0.3)**

17. $f(x) = x^2 + 2x$

18. $f(x) = \dfrac{1}{x}$

19. $f(x) = \sqrt{x}$ (*Hint:* Rationalize the numerator.)

20. $f(x) = x^3 - 1$

Graph the following functions. Determine clearly the intercepts of the graphs. **(Section 0.2)**

21. $f(x) = 2x - 1$

22. $f(x) = -x$

23. $f(x) = -\dfrac{x-1}{2}$

24. $f(x) = 3$

Find the points of intersection (if any) of the pairs of curves. **(Section 0.4)**

25. $y = 3x + 1, y = -x - 2$

26. $y = \dfrac{x}{2}, y = 3$

27. $y = 4x - 7, y = 0$

28. $y = 2x + 3, y = 2x - 2$

Factor the given polynomials. **(Section 0.4)**

29. $x^2 + 5x - 14$

30. $x^2 + 5x + 4$

31. $x^3 + x^2 - 2x$

32. $x^3 - 2x^2 - 3x$

Solve the given equations by factoring first. **(Section 0.4)**

33. $x^2 - 144 = 0$

34. $x^2 + 4x + 4 = 0$

35. $x^3 + 8x^2 + 15x = 0$

36. $6x^3 + 11x^2 + 3x = 0$

Solve using the quadratic formula. **(Section 0.4)**

37. $2x^2 + 3x - 1 = 0$

38. $x^2 + x - 1 = 0$

39. $-3x^2 + 2x - 4 = 0$

40. $x^2 + 4x - 4 = 0$

Solve the given equations. **(Section 0.4)**

41. $x^2 - 3x = 4x - 10$

42. $4x^2 + 2x = -2x + 3$

43. $\dfrac{1}{x + 1} = x + 1$

44. $\dfrac{x^3}{x^2 + 2x - 1} = x - 1$

Introduction

Often, it is possible to give a succinct and revealing description of a situation by drawing a graph. For example, Fig. 1 describes the amount of money in a bank account drawing 5% interest, compounded daily. The graph shows that, as time passes, the amount of money in the account grows. Figure 2 depicts the weekly sales of a breakfast cereal at various times after advertising has ceased. The graph shows that the longer the time since the last advertisement, the fewer the sales. Figure 3 shows the size of a bacteria culture at various times. The culture grows larger as time passes. But there is a maximum size that the culture cannot exceed. This maximum size reflects the restrictions imposed by food supply, space, and similar factors. The graph in Fig. 4 describes the decay of the radioactive isotope iodine 131. As time passes, less and less of the original radioactive iodine remains.

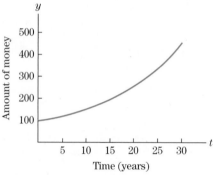

Figure 1 Growth of money in a savings account.

Figure 2 Decrease in sales of breakfast cereal.

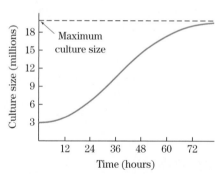

Figure 3 Growth of a bacteria culture.

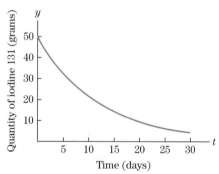

Figure 4 Decay of radioactive iodine.

Each graph in Figs. 1 to 4 describes a change that is taking place. The amount of money in the bank is changing, as are the sales of cereal, the size of the bacteria culture, and the amount of iodine. Calculus provides mathematical tools to study each change in a quantitative way.

ach graph in Figures 1 through 4 of the introduction depicts a relationship between two quantities. For example, Fig. 4 illustrates the relationship between the quantity of iodine (measured in grams) and time (measured in days). The basic quantitative tool for describing such relationships is a **function**. In this preliminary chapter, we develop the concept of a function and review important algebraic operations on functions used later in the text.

0.1   Functions and Their Graphs

Real Numbers

Most applications of mathematics use real numbers. For purposes of such applications (and the discussions in this text), it suffices to think of a real number as a decimal. A **rational number** is one that may be written as a fraction, and consequently as a finite or infinitely repeating decimal, such as

$$-\frac{5}{2} = -2.5, \quad 1, \quad \frac{13}{3} = 4.333\ldots \quad \text{Rational numbers.}$$

An **irrational number** cannot be represented as a fraction, but instead has an infinite decimal representation whose digits form no repeating pattern, such as

$$-\sqrt{2} = -1.414213\ldots, \quad \pi = 3.14159\ldots \quad \text{Irrational numbers.}$$

The real numbers are described geometrically by a **number line**, as in Fig. 1. Each number corresponds to one point on the line, and each point determines one real number.

Figure 1 The real number line.

We use four types of inequalities to compare real numbers.

$x < y$ x is less than y

$x \leq y$ x is less than or equal to y

$x > y$ x is greater than y

$x \geq y$ x is greater than or equal to y

The double inequality $a < b < c$ is shorthand for the pair of inequalities $a < b$ and $b < c$. Similar meanings are assigned to other double inequalities, such as $a \leq b < c$. The three numbers in a double inequality, such as $1 < 3 < 4$ or $4 > 3 > 1$, should have the same relative positions on the number line as in the inequality. Thus $3 < 4 > 1$ is never written because the numbers are "out of order."

Geometrically, the inequality $x \leq b$ means that either x equals b or x lies to the left of b on the number line. The set of real numbers x that satisfies the double inequality $a \leq x \leq b$ corresponds to the line segment between a and b, including the endpoints. This set is sometimes denoted by $[a, b]$ and is called the **closed interval** from a to b. If a and b are removed from the set, the set is written as (a, b) and is called the **open interval** from a to b. The notation for various line segments is listed in Table 1. Note that an interval endpoint that is included (e.g., both endpoints of $[a, b]$) is drawn as a solid circle, whereas an endpoint not included (e.g., the endpoint a in (a, b)) is drawn as an unfilled circle.

Table 1   **Intervals on the Number Line**

Inequality	Geometric Description	Interval Notation
$a \leq x \leq b$	●———● a b	$[a, b]$
$a < x < b$	○———○ a b	(a, b)
$a \leq x < b$	●———○ a b	$[a, b)$
$a < x \leq b$	○———● a b	$(a, b]$
$a \leq x$	●——— a	$[a, \infty)$
$a < x$	○——— a	(a, ∞)
$x \leq b$	———● b	$(-\infty, b]$
$x < b$	———○ b	$(-\infty, b)$

The symbols ∞ ("infinity") and $-\infty$ ("minus infinity") do not represent actual real numbers. Rather, they indicate that the corresponding line segment extends infinitely far to the right or left. An inequality that describes such an infinite interval may be written in two ways. For instance, $a \leq x$ is equivalent to $x \geq a$.

EXAMPLE 1   **Graphing Intervals** Describe each of the following intervals both graphically and in terms of inequalities.

(a) $(-1, 2)$ (b) $[-2, \pi]$ (c) $(2, \infty)$ (d) $(-\infty, \sqrt{2}]$

SOLUTION   The line segments and the inequalities corresponding to the intervals are shown in Fig. 2(a) through (d).

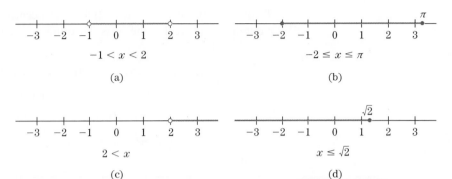

Figure 2 Intervals on the Number Line

> **Now Try Exercise 7**

EXAMPLE 2   **Using Inequalities** Let the variable x describe the profit that a company anticipates to earn in the current fiscal year. The business plan calls for a profit of at least \$5 million. Describe this aspect of the business plan in the language of intervals.

SOLUTION   The phrase "at least" means "greater than or equal to." The business plan requires that $x \geq 5$ (where the units are millions of dollars). This is equivalent to saying that x lies in the infinite interval $[5, \infty)$.   «

Functions

A **function** of a variable x is a rule f that assigns each value of x to a unique number, $f(x)$, called the **value of the function at** x. (We read "$f(x)$" as "f of x".) The variable of a function is called the **independent variable**. The set of values that the independent variable is allowed to assume is called the **domain** of the function. The domain of a function may be explicitly specified as part of the definition of a function, or it may be understood from context, as you will see in Examples 5 and 7. The **range** of a function is the set of values that the function assumes.

The functions we shall meet in this book will usually be defined by algebraic formulas. For example, the domain of the function

$$f(x) = 3x - 1$$

consists of all real numbers x. This function is the rule that takes a number, multiplies it by 3, and then subtracts 1. If we specify a value of x—say, $x = 2$—then we find the value of the function at 2 by substituting 2 for x in the formula:

$$f(2) = 3(2) - 1 = 5.$$

EXAMPLE 3   **Evaluating a Function** Let f be the function with domain all real numbers x and defined by the formula

$$f(x) = 3x^3 - 4x^2 - 3x + 7.$$

Find $f(2)$ and $f(-2)$.

SOLUTION To find $f(2)$, we substitute 2 for every occurrence of x in the formula for $f(x)$:

$$f(2) = 3(2)^3 - 4(2)^2 - 3(2) + 7 \qquad \text{Substitute 2 for } x.$$
$$= 3(8) - 4(4) - 3(2) + 7 \qquad \text{Evaluate exponents.}$$
$$= 24 - 16 - 6 + 7 \qquad \text{Multiply.}$$
$$= 9. \qquad \text{Add and subtract.}$$

To find $f(-2)$, we substitute (-2) for each occurrence of x in the formula for $f(x)$. The parentheses ensure that the -2 is substituted correctly. For instance, x^2 must be replaced by $(-2)^2$, not -2^2:

$$f(-2) = 3(-2)^3 - 4(-2)^2 - 3(-2) + 7 \quad \text{Substitute } (-2) \text{ for } x.$$
$$= 3(-8) - 4(4) - 3(-2) + 7 \qquad \text{Evaluate exponents.}$$
$$= -24 - 16 + 6 + 7 \qquad \text{Multiply.}$$
$$= -27. \qquad \text{Add and subtract.}$$

>> *Now Try Exercise 13*

EXAMPLE 4 **Temperature Scales** If x represents the temperature of an object in degrees Celsius, then the temperature in degrees Fahrenheit is a function of x, given by $f(x) = \frac{9}{5}x + 32$.
(a) Water freezes at 0°C (C = Celsius) and boils at 100°C. What are the corresponding temperatures in degrees Fahrenheit (F = Fahrenheit)?
(b) Aluminum melts at 660°C. What is its melting point in degrees Fahrenheit?

SOLUTION (a) $f(0) = \frac{9}{5}(0) + 32 = 32$. Water freezes at 32°F.
$f(100) = \frac{9}{5}(100) + 32 = 180 + 32 = 212$. Water boils at 212°F.
(b) $f(660) = \frac{9}{5}(660) + 32 = 1188 + 32 = 1220$. Aluminum melts at 1220°F.

>> *Now Try Exercise 19*

In the preceding examples, the functions had domains consisting of all real numbers or an interval. For some functions, the domain may consist of several intervals, with a different formula defining the function on each interval. Here is an illustration of this phenomenon.

EXAMPLE 5 **A Function Defined by Multiple Formulas** A company that runs an online movie streaming website charges users $3 per movie. If a user streams 10 or more movies in one month, the company adds a 10% overusage fee to the original cost. Let x denote the number of movies a user streams a month and let $f(x)$ be the amount of dollars the company charges the user as a function of x.
(a) Describe $f(x)$.
(b) Find $f(5)$ and $f(15)$.

SOLUTION (a) The formula $f(x)$ depends on whether $0 \le x < 10$ or $10 \le x$. When $0 \le x < 10$, the charge is $3x$ dollars. When $10 \le x$, the charge is $3x$ plus 10% of $3x$ or $3x + 0.1(3x) = 3.3x$ dollars. The domain consists of the values of x in one of the two intervals $[0, 10)$ and $[10, \infty)$. In each of these intervals, the function is defined by a separate formula:

$$f(x) = \begin{cases} 3x & \text{for } 0 \le x < 10 \\ 3.3x & \text{for } 10 \le x. \end{cases}$$

(b) Since $x = 5$ satisfies $0 \le x < 10$, we use the first formula for $f(x)$: $f(5) = 3(5) = 15$. So it costs $15 to stream 5 movies in one month. Since $x = 15$ satisfies $10 \le x$, we use the second formula for $f(x)$: $f(15) = 3.3(15) = 49.5$. So it costs $49.50 to stream 15 movies in one month.

>> *Now Try Exercise 57*

In calculus, it is often necessary to substitute an algebraic expression for x and simplify the result, as illustrated in the following example.

EXAMPLE 6	**Evaluating a Function** If $f(x) = (4 - x)/(x^2 + 3)$, what is (a) $f(a)$? (b) $f(a + 1)$?

SOLUTION

(a) Here, a represents some number. To find $f(a)$, we substitute a for x wherever x appears in the formula defining $f(x)$:

$$f(a) = \frac{4 - a}{a^2 + 3}.$$

(b) To evaluate $f(a + 1)$, substitute $a + 1$ for each occurrence of x in the formula for $f(x)$:

$$f(a + 1) = \frac{4 - (a + 1)}{(a + 1)^2 + 3}.$$

We can simplify the expression for $f(a + 1)$ using the fact that $(a + 1)^2 = (a + 1)(a + 1) = a^2 + 2a + 1$:

$$f(a + 1) = \frac{4 - (a + 1)}{(a + 1)^2 + 3} \overset{\text{Expand}}{=} \frac{4 - a - 1}{a^2 + 2a + 1 + 3} \overset{\text{Add and Subtract}}{=} \frac{3 - a}{a^2 + 2a + 4}.$$ ≪

More about the Domain of a Function   When defining a function, it is necessary to specify the domain of the function, which is the set of acceptable values of the variable. In the preceding examples, we explicitly specified the domains of the functions considered. However, throughout the remainder of the text, we will usually mention functions without specifying domains. In such circumstances, we will understand the intended domain to consist of all numbers for which the defining formula(s) makes sense. For example, consider the function

$$f(x) = x^2 - x + 1.$$

The expression on the right may be evaluated for any value of x. So, in the absence of any explicit restrictions on x, the domain is understood to consist of all numbers. As a second example, consider the function

$$f(x) = \frac{1}{x}.$$

Here x may be any number except zero (division by zero is not permissible.) So the domain intended is the set of nonzero numbers. Similarly, when we write

$$f(x) = \sqrt{x},$$

we understand the domain of $f(x)$ to be the set of all nonnegative numbers, since the square root of a real number x is defined if and only if $x \geq 0$.

EXAMPLE 7	**Domains of Functions** Find the domains of the following functions:

(a) $f(x) = \sqrt{4 + x}$ (b) $g(x) = \dfrac{1}{\sqrt{1 + 2x}}$ (c) $h(x) = \sqrt{1 + x} - \sqrt{1 - x}$

SOLUTION

(a) Since we cannot take the square root of a negative number, we must have $4 + x \geq 0$, or equivalently, $x \geq -4$. So the domain of f is $[-4, \infty)$.

(b) Here, the domain consists of all x for which

$$1 + 2x > 0$$
$$2x > -1 \qquad \text{Subtract 1 from both sides.}$$
$$x > -\frac{1}{2} \qquad \text{Divide both sides by 2.}$$

The domain is the open interval $\left(-\frac{1}{2}, \infty\right)$.

(c) In order to be able to evaluate both square roots that appear in the expression of $h(x)$, we must have

$$1 + x \geq 0 \quad \text{and} \quad 1 - x \geq 0.$$

The first inequality is equivalent to $x \geq -1$, and the second inequality to $x \leq 1$. Since x must satisfy both inequalities, it follows that the domain of h consists of the closed interval $[-1, 1]$. » *Now Try Exercise 23*

Graphs of Functions Often it is helpful to describe a function f geometrically, using a rectangular xy-coordinate system. Given any x in the domain of f, we can plot the point $(x, f(x))$. This is the point in the xy-plane whose y-coordinate is the value of the function at x. The set of *all* such points $(x, f(x))$ usually forms a curve in the xy-plane and is called the **graph** of the function $f(x)$.

It is possible to approximate the graph of $f(x)$ by plotting the points $(x, f(x))$ for a representative set of values of x and joining them by a smooth curve. (See Fig. 3.) The more closely spaced the values of x, the closer the approximation.

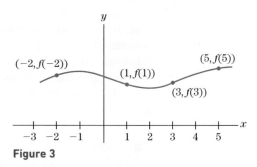

Figure 3

EXAMPLE 8 **Sketching a Graph by Plotting Points** Sketch the graph of the function $f(x) = x^3$.

SOLUTION The domain consists of all real numbers x. We choose some representative values of x and tabulate the corresponding values of $f(x)$. We then plot the points $(x, f(x))$ and draw a smooth curve through the points. (See Fig. 4.)

x	$f(x) = x^3$
0	0
1	1
2	8
3	27
−1	−1
−2	−8
−3	−27

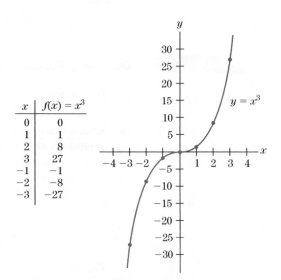

Figure 4 Graph of $f(x) = x^3$.

«

EXAMPLE 9

A Graph with a Restricted Domain Sketch the graph of the function $f(x) = 1/x$.

SOLUTION

The domain of the function consists of all numbers except zero. The table in Fig. 5 lists some representative values of x and the corresponding values of $f(x)$. A function often has interesting behavior for x near a number not in the domain. So, when we chose representative values of x from the domain, we included some values close to zero. The points $(x, f(x))$ are plotted and the graph sketched in Fig. 5.

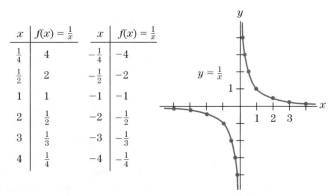

x	$f(x) = \frac{1}{x}$	x	$f(x) = \frac{1}{x}$
$\frac{1}{4}$	4	$-\frac{1}{4}$	-4
$\frac{1}{2}$	2	$-\frac{1}{2}$	-2
1	1	-1	-1
2	$\frac{1}{2}$	-2	$-\frac{1}{2}$
3	$\frac{1}{3}$	-3	$-\frac{1}{3}$
4	$\frac{1}{4}$	-4	$-\frac{1}{4}$

Figure 5 Graph of $f(x) = \frac{1}{x}$.

>> *Now Try Exercise 27*

Now that graphing calculators and computer graphing programs are widely available, we seldom need to sketch graphs by hand-plotting a large number of points on graph paper. However, to use such a calculator or program effectively, we must know in advance which part of a curve to display. Critical features of a graph may be missed or misinterpreted if, for instance, the scale on the x- or y-axis is inappropriate.

An important use of calculus is to identify key features of a function that should appear in its graph. In many cases, only a few points need be plotted, and the general shape of the graph is easy to sketch by hand. For more complicated functions, a graphing program is helpful. Even then, calculus provides a way of checking that the graph on the computer screen has the correct shape. Algebraic calculations are usually part of the analysis. The appropriate algebraic skills are reviewed in this chapter.

Note also that analytic solutions to problems typically provide more precise information than graphing calculators and can provide insight into the behavior of the functions involved in the solution.

The connection between a function and its graph is explored in this section and in Section 0.6.

EXAMPLE 10

Reading a Graph Suppose that f is the function whose graph is given in Fig. 6. Notice that the point $(x, y) = (3, 2)$ is on the graph of f.
(a) What is the value of the function when $x = 3$?
(b) Find $f(-2)$.
(c) What is the domain of f? What is its range?

SOLUTION

(a) Since $(3, 2)$ is on the graph of f, the y-coordinate 2 must be the value of f at the x-coordinate 3. That is, $f(3) = 2$.
(b) To find $f(-2)$, we look at the y-coordinate of the point on the graph where $x = -2$. From Fig. 6, we see that $(-2, 1)$ is on the graph of f. Thus, $f(-2) = 1$.
(c) The points on the graph of $f(x)$ all have x-coordinates between -3 and 5 inclusive; and for each value of x between -3 and 5, there is a point $(x, f(x))$ on the graph. So the domain consists of those x for which $-3 \leq x \leq 5$. From Fig. 6, the function assumes all values between .2 and 2.5. Thus, the range of f is [.2, 2.5].

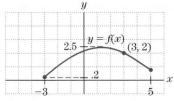

Figure 6

>> *Now Try Exercise 37*

As we saw in Example 10, the graph of f can be used to picture the domain of f on the x-axis and its range on the y-axis. The general situation is illustrated in Fig. 7.

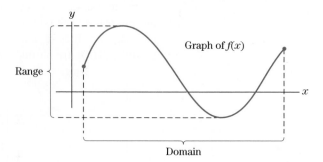

Figure 7 Domain and range.

To every x in the domain, a function assigns one and only one value of y, that is, the function value $f(x)$. This implies, among other things, that not every curve is the graph of a function. To see this, refer first to the curve in Fig. 6, which *is* the graph of a function. It has the following important property: For each x between -3 and 5 inclusive, there is a *unique* y such that (x, y) is on the curve. The variable y is called the **dependent variable**, since its value depends on the value of the independent variable x. Refer to the curve in Fig. 8. It cannot be the graph of a function because a function f must assign to each x in its domain a *unique* value $f(x)$. However, for the curve of Fig. 8, there corresponds to $x = 3$ (for example) more than one y-value: $y = 1$ and $y = 4$.

The essential difference between the curves in Figs. 6 and 8 leads us to the following test.

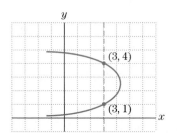

Figure 8 A curve that is *not* the graph of a function.

> **The Vertical Line Test** A curve in the xy-plane is the graph of a function if and only if each vertical line cuts or touches the curve at no more than one point.

EXAMPLE 11 **Vertical Line Test** Which of the curves in Fig. 9 are graphs of functions?

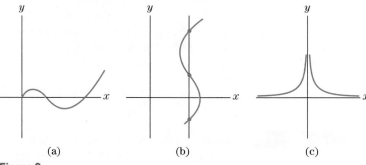

(a) (b) (c)

Figure 9

SOLUTION The curve in (a) is the graph of a function. It appears that vertical lines to the left of the y-axis do not touch the curve at all. This simply means that the function represented in (a) is defined only for $x \geq 0$. The curve in (b) is *not* the graph of a function because some vertical lines cut the curve in three places. The curve in (c) is the graph of a function whose domain is all nonzero x. (There is no point on the curve in (c) whose x-coordinate is 0.)

» Now Try Exercise 29

There is another notation for functions that we will find useful. Suppose that $f(x)$ is a function. When $f(x)$ is graphed on an xy-coordinate system, the values of $f(x)$ give the y-coordinates of points of the graph. For this reason, the function is often abbreviated by the letter y, and we find it convenient to speak of "the function $y = f(x)$." For example, the function $y = 2x^2 + 1$ refers to the function $f(x)$ for which $f(x) = 2x^2 + 1$. The graph of a function $f(x)$ is often called the **graph of the equation** $y = f(x)$.

INCORPORATING

TECHNOLOGY

Graphing Functions The use of graphing calculators is not required for the study of this text; however, graphing calculators are very useful tools that can be used to simplify computations, draw graphs, and sometimes enhance understanding of the fundamental topics of calculus. Helpful information about the use of calculators will appear at the end of some sections in subsections titled "Incorporating Technology." The examples in this text use TI-83/84 calculators, and the keystrokes will be identical on every available version of these calculators. Other models and brands of graphing calculators should function similarly. (The designation TI-83/84 refers to the TI-83, TI-84, TI-83+, TI-84+, TI-83+ Silver Edition and TI-84+ Silver Edition calculators.)

Consider the function $f(x) = x^3 - 2$. To graph $f(x)$, follow these steps:

Step 1 Press ⟨Y=⟩. We will define our function in the calculator as Y_1. Move the cursor up if necessary, so that it is placed directly after the expression "\ Y_1 =." Press ⟨CLEAR⟩ to ensure that no formulas are entered for Y_1.

Step 2 Enter $X^3 - 2$. The variable X may be entered using the ⟨X,T,Θ,n⟩ key. (See Fig. 10(a).)

Step 3 Press ⟨GRAPH⟩. (See Fig. 10(b).)

(a)

(a)

(b)

Figure 10

To change the parameters for the viewing window, follow these steps:

Step 1 Press ⟨WINDOW⟩.

Step 2 Change the parameters to their desired values.

One of the most important tasks in using a graphing calculator is to determine the viewing window that shows the features of interest. For this example, we will simply set the view to $[-3, 3]$ by $[-29, 29]$. We will also set the value of *Yscl* to 5. The parameter *Yscl* and its sibling, *Xscl*, set the distance between tick marks on their respective axes. To do this, set the window parameters on your calculator to match those in Fig. 11(a). The value of *Xres* sets the screen resolution, and we will leave it at its default setting.

(b)

Figure 11

Step 3 Press ⟨GRAPH⟩ to see the results. (See Fig. 11(b).)

Check Your Understanding 0.1

Solutions can be found following the section exercises.

1. Is the point $(3, 12)$ on the graph of the function $g(x) = x^2 + 5x - 10$?

2. Sketch the graph of the function $h(t) = t^2 - 2$.

EXERCISES 0.1

Draw the following intervals on the number line.

1. $[-1, 4]$

2. $(4, 3\pi)$

3. $[-2, \sqrt{2}]$

4. $[1, \frac{3}{2}]$

5. $(-\infty, 3)$

6. $(4, \infty)$

Use intervals to describe the real numbers satisfying the inequalities in Exercises 7–12.

7. $2 \leq x < 3$

8. $-1 < x < \frac{3}{2}$

9. $x < 0$ and $x \geq -1$

10. $x \geq -1$ and $x < 8$

11. $x < 3$

12. $x \geq \sqrt{2}$

13. If $f(x) = x^2 - 3x$, find $f(0)$, $f(5)$, $f(3)$, and $f(-7)$.

14. If $f(x) = x^3 + x^2 - x - 1$, find $f(1)$, $f(-1)$, $f(\frac{1}{2})$, and $f(a)$.

15. If $f(x) = x^2 - 2x$, find $f(a + 1)$ and $f(a + 2)$.

16. If $h(s) = s/(1 + s)$, find $h(\frac{1}{2})$, $h(-\frac{3}{2})$, and $h(a + 1)$.

17. If $f(x) = 3x + 2$ and $h \neq 0$, find $\dfrac{f(3 + h) - f(3)}{h}$. Simplify your answer as much as possible.

18. If $f(x) = x^2$ and $h \neq 0$, find $\dfrac{f(1 + h) - f(1)}{h}$. Simplify your answer as much as possible.

19. Temperature Scales The boiling point of tungsten is approximately 5933 Kelvin.
   (a) Find the boiling point of tungsten in degrees Celsius, given that the equation to convert $x°C$ to Kelvin is $k(x) = x + 273$.
   (b) Find the boiling point of tungsten in degrees Fahrenheit. (Tungsten, also known as wolfram, has the highest boiling point out of all known metals.)

20. Computer Sales An office supply firm finds that the number of laptop computers sold in year x is given approximately by the function $f(x) = 150 + 2x + x^2$, where $x = 0$ corresponds to 2015.
   (a) What does $f(0)$ represent?
   (b) Find the number of laptops sold in 2020.

In Exercises 21–24, describe the domain of the function.

21. $f(x) = \dfrac{8x}{(x - 1)(x - 2)}$

22. $f(t) = \dfrac{1}{\sqrt{t}}$

23. $g(x) = \dfrac{1}{\sqrt{3 - x}}$

24. $g(x) = \dfrac{4}{x(x + 2)}$

In Exercises 25–28, sketch the graph of the function.

25. $f(x) = x^2 + 1$

26. $f(x) = 2x^2 - 1$

27. $f(x) = \sqrt{x + 1}$

28. $f(x) = x + 1$

In Exercises 29–34, decide which curves are graphs of functions.

29.

30.

31.

32.

33.

34.

Exercises 35–42 relate to the function whose graph is sketched in Fig. 12.

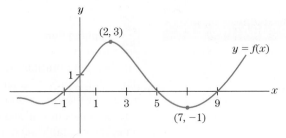

Figure 12

35. Find $f(0)$ and $f(7)$.

36. Find $f(2)$ and $f(-1)$.

37. Is $f(4)$ positive or negative?

38. Is $f(6)$ positive or negative?

39. What is the range of f?

40. For what values of x does $f(x) = 0$?

41. For what values of x is $f(x) \leq 0$?

42. For what values of x is $f(x) \geq 0$?

Exercises 43–46 relate to Fig. 13. When a drug is injected into a person's muscle tissue, the concentration y of the drug in the blood is a function of the time elapsed since the injection. The graph of a typical time–concentration function f is given in Fig. 13, where $t = 0$ corresponds to the time of the injection.

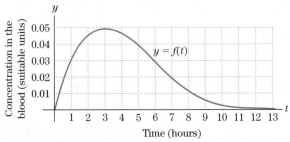

Figure 13 Drug time-concentration curve.

43. What is the concentration of the drug when $t = 1$ and $t = 5$?

44. What is the value of the time–concentration function f when $t = 6$?

45. What is the range of f?

46. At what time does $f(t)$ attain its largest value?

47. Is the point $(3, 12)$ on the graph of the function $f(x) = (x - \frac{1}{2})(x + 2)$?

48. Is the point $(-2, 12)$ on the graph of the function $f(x) = x(5 + x)(4 - x)$?

49. Is the point $(1, 1)$ on the graph of the function $g(x) = (3x - 1)/(x^2 + 1)$?

50. Is the point $(4, \frac{1}{4})$ on the graph of the function $g(x) = (x^2 + 4)/(x + 2)$?

51. Find the y-coordinate of the point $(a + 1, ?)$ if this point lies on the graph of the function $f(x) = x^3$.

52. Find the y-coordinate of the point $(2 + h, ?)$ if this point lies on the graph of the function $f(x) = (5/x) - x$.

In Exercises 53–56, compute $f(1)$, $f(2)$, and $f(3)$.

53. $f(x) = \begin{cases} \sqrt{x} & \text{for } 0 \leq x < 2 \\ 1 + x & \text{for } 2 \leq x \leq 5 \end{cases}$

54. $f(x) = \begin{cases} 1/x & \text{for } 1 \leq x \leq 2 \\ x^2 & \text{for } 2 < x \end{cases}$

55. $f(x) = \begin{cases} \pi x^2 & \text{for } x < 2 \\ 1 + x & \text{for } 2 \leq x \leq 2.5 \\ 4x & \text{for } 2.5 < x \end{cases}$

56. $f(x) = \begin{cases} 3/(4 - x) & \text{for } x < 2 \\ 2x & \text{for } 2 \leq x < 3 \\ \sqrt{x^2 - 5} & \text{for } 3 \leq x \end{cases}$

57. Commission on Gold Purchases A brokerage firm charges 6% commission on gold purchases in amounts from \$50 to \$3000. For purchases exceeding \$3000, the firm charges 2% of the amount purchased plus a \$15 handling fee. Let x denote the amount of gold purchased (in dollars), and let $f(x)$ be the commission charge as a function of x. (a) Describe $f(x)$. (b) Find $f(3000)$ and $f(4500)$.

58. Figure 14(a) shows the number 2 on the x-axis and the graph of a function. Let h represent a positive number and label a possible location for the number $2 + h$. Plot the point on the graph whose first coordinate is $2 + h$, and label the point with its coordinates.

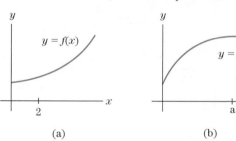

(a)

(b)

Figure 14

59. Figure 14(b) shows the number a on the x-axis and the graph of a function. Let h represent a negative number and label a possible location for the number $a + h$. Plot the point on the graph whose first coordinate is $a + h$, and label the point with its coordinates.

TECHNOLOGY EXERCISES

60. Response of a Muscle When a solution of acetylcholine is introduced into the heart muscle of a frog, it diminishes the force with which the muscle contracts. The data from experiments of the biologist A. J. Clark are closely approximated by a function of the form

$$R(x) = \frac{100x}{b + x}, \quad x \geq 0,$$

where x is the concentration of acetylcholine (in appropriate units), b is a positive constant that depends on the particular frog, and $R(x)$ is the response of the muscle to the acetylcholine, expressed as a percentage of the maximum possible effect of the drug.

(a) Suppose $b = 15$. Find the response of the muscle when $x = 30$.

(b) Determine the value of b if $R(50) = 30$; that is, if a concentration of $x = 50$ units produces a 30% response.

61. What is wrong with entering the function $f(x) = \dfrac{1}{x + 1}$ into a graphing utility as $\mathbf{Y_1 = 1/X + 1}$?

62. What is wrong with entering the function $f(x) = x^{3/4}$ into a graphing utility as $\mathbf{Y_1 = X\^3/4}$?

In Exercises 63–64, graph the function with the specified viewing window setting.

63. $f(x) = -x^2 + 2x + 2;\ [-2, 4]$ by $[-8, 5]$

64. $f(x) = \dfrac{1}{x^2 + 1};\ [-4, 4]$ by $[-.5, 1.5]$

Solutions to Check Your Understanding 0.1

1. If $(3, 12)$ is on the graph of $g(x) = x^2 + 5x - 10$, we must have $g(3) = 12$. This is not the case, however, because

$$g(3) = 3^2 + 5(3) - 10$$
$$= 9 + 15 - 10 = 14.$$

Thus $(3, 12)$ is *not* on the graph of $g(x)$.

2. Choose some representative values for t, say, $t = 0,\ \pm 1,\ \pm 2,\ \pm 3$. For each value of t, calculate $h(t)$ and plot the point $(t, h(t))$. See Fig. 15.

t	$h(t) = t^2 - 2$
0	−2
1	−1
2	2
3	7
−1	−1
−2	2
−3	7

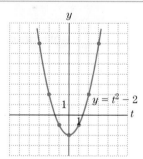

Figure 15 Graph of $h(t) = t^2 - 2$.

0.2 Some Important Functions

In this section, we introduce some of the functions that will play a prominent role in our discussion of calculus.

Linear Functions

As we shall see in Chapter 1, a knowledge of the algebraic and geometric properties of straight lines is essential for the study of calculus. Every straight line is the graph of a linear equation of the form

$$Ax + By = C,$$

where A, B, and C are given constants, with A and B not both zero. If $B \neq 0$, then we may solve the equation for y to obtain an equation of the form

$$y = mx + b, \tag{1}$$

for appropriate numbers m and b. If $B = 0$, then we may solve the equation for x to obtain an equation of the form

$$x = a, \tag{2}$$

for an appropriate number a. So every straight line is the graph of an equation of type (1) or (2). The graph of an equation of the form (1) is a nonvertical line [Fig. 1(a)], whereas the graph of (2) is a vertical line [Fig. 1(b)]. Note that equation (2) does not define a function. Its graph fails the vertical line test.

The straight line of Fig. 1(a) is the graph of the function $f(x) = mx + b$. Such a function, which is defined for all x, is called a **linear function**. Note that the straight line of Fig. 1(b) is not the graph of a function, since the vertical line test is violated.

An important special case of a linear function occurs if the value of m is zero; that is, $f(x) = b$ for some number b. In this case, $f(x)$ is called a **constant function**, since it assigns the same number b to every value of x. Its graph is the horizontal line whose equation is $y = b$. (See Fig. 2.)

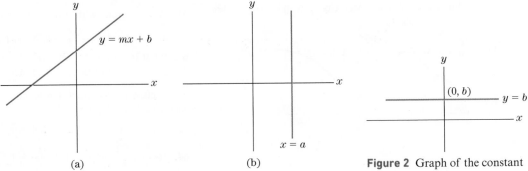

(a)

(b)

Figure 1

Figure 2 Graph of the constant function $f(x) = b$.

EXAMPLE 1   **Sketching a Linear Function** Sketch the graph of $3x - y = 2$.

SOLUTION   Since the equation is linear, its graph is a straight line. (See Fig. 3.) To simplify finding points on the line, we solve for y first and get

$$y = 3x - 2.$$

Even though only two points are needed to identify and graph the line, in the following figure we compute three points to verify that it is correct.

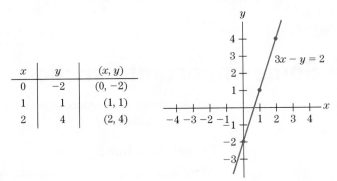

x	y	(x, y)
0	-2	$(0, -2)$
1	1	$(1, 1)$
2	4	$(2, 4)$

Figure 3

≫ Now Try Exercise 9

Linear functions often arise in real-life situations, as the next two examples show.

EXAMPLE 2

A Model for an EPA Fine When the U.S. Environmental Protection Agency found a certain company emitting ozone-depleting substances from refrigeration equipment, it fined the company $600,000, plus $50,000 per month until the company complied with federal emission regulations. Express the total fine as a function of the number of months, x, the company continued to violate the federal regulations.

SOLUTION

The variable fine for x months of pollution, at $50,000 per month, is $50,000x$ dollars. The total fine is therefore given by the function

$$f(x) = 600,000 + 50,000x. \qquad \textbf{» Now Try Exercise 17}$$

Since the graph of a linear function is a line, we may sketch it by locating any two points on the graph and drawing the line through them. For example, to sketch the graph of the function $f(x) = -\frac{1}{2}x + 3$, we may select two convenient values of x—say, 0 and 4—and compute $f(0) = -\frac{1}{2}(0) + 3 = 3$ and $f(4) = -\frac{1}{2}(4) + 3 = 1$. The line through the points $(0, 3)$ and $(4, 1)$ is the graph of the function. (See Fig. 4.)

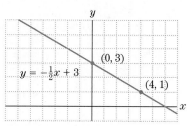

Figure 4

A simple cost function for a business consists of two parts: the **fixed costs**, such as rent, insurance, and business loans, which must be paid no matter how many items of a product are produced, and the **variable costs**, which depend on the number of items produced.

EXAMPLE 3

Cost Suppose that a computer software company produces and sells a new operating system at a cost of $55 per copy, and the company has fixed costs of $15,000 per month. Express the total monthly cost as a function of the number of copies sold, x, and compute the cost when $x = 1000$.

SOLUTION

The monthly variable cost is $55x$ dollars. Thus,

$$[\text{total cost}] = [\text{fixed costs}] + [\text{variable costs}]$$
$$C(x) = 15,000 + 55x.$$

When sales are at 1000 copies per month, the cost is

$$C(1000) = 15,000 + 55(1000) = 70,000 \text{ (dollars)}.$$

(See Fig. 5.)

Figure 5 A linear cost function.   **» Now Try Exercise 19**

The point at which the graph of a linear function intersects the y-axis is called the *y-intercept* of the graph. The point at which the graph intersects the x-axis is called the *x-intercept*. The next example shows how to determine the intercepts of a linear function.

EXAMPLE 4   **Intercepts**   Determine the intercepts of the graph of the linear function $f(x) = 2x + 5$.

SOLUTION   Since the y-intercept is on the y-axis, its x-coordinate is 0. The point on the line with x-coordinate zero has y-coordinate

$$f(0) = 2(0) + 5 = 5.$$

So, the y-intercept is $(0, 5)$. Since the x-intercept is on the x-axis, its y-coordinate is 0. Since $f(x)$ gives the y-coordinate, we must have

$$2x + 5 = 0 \qquad \text{Set } f(x) \text{ equal to 0.}$$
$$2x = -5 \qquad \text{Subtract 5 from each side.}$$
$$x = -\tfrac{5}{2}. \qquad \text{Divide each side by 2.}$$

So, $\left(-\tfrac{5}{2}, 0\right)$ is the x-intercept. (See Fig. 6.)

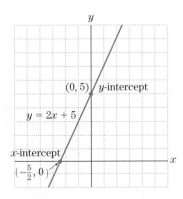

Figure 6 Graph of $f(x) = 2x + 5$.

» *Now Try Exercise 15*

Piecewise-Defined Functions

The function in the next example is described by two expressions. Functions described by more than one expression are said to be **piecewise defined**.

EXAMPLE 5   **A Piecewise-Defined Function**   Sketch the graph of the following function.

$$f(x) = \begin{cases} \frac{5}{2}x - \frac{1}{2} & \text{for } -1 \le x \le 1 \\ \frac{1}{2}x - 2 & \text{for } x > 1 \end{cases}$$

SOLUTION   The formula for f depends on whether $-1 \le x \le 1$ or $x > 1$. So we expect the graph to consist of two portions. To draw the portion when $-1 \le x \le 1$, we use the formula $f(x) = \frac{5}{2}x - \frac{1}{2}$, which is a linear function. If $x = -1$, then $f(-1) = \frac{5}{2}(-1) - \frac{1}{2} = -\frac{5}{2} - \frac{1}{2} = -\frac{6}{2} = -3$. If $x = 1$, then $f(1) = \frac{5}{2}(1) - \frac{1}{2} = \frac{5}{2} - \frac{1}{2} = \frac{4}{2} = 2$. So the points $(-1, -3)$ and $(1, 2)$ are on the graph. The portion of the graph for $-1 \le x \le 1$ is the line segment obtained by connecting these two points (Fig. 7).

To draw the portion of the graph when $x > 1$, we use the formula $f(x) = \frac{1}{2}x - 2$, which is also a linear function. Choose two values of x with $x > 1$; say, $x = 2$ and $x = 4$. For $x = 2$, $f(2) = \frac{1}{2}(2) - 2 = 1 - 2 = -1$; and for $x = 4$, $f(4) = \frac{1}{2}(4) - 2 = 2 - 2 = 0$. Thus the points $(2, -1)$ and $(4, 0)$ are on the graph. Plotting these two points and connecting them with a straight line, we obtain the graph for $x > 1$ (Fig. 7). Note that the point $(2, -1)$ is not on this portion.

The graph of f consists of the two portions shown in Fig. 7.

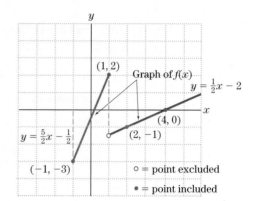

Figure 7 Graph of a function specified by two expressions.

○ = point excluded
● = point included

>> *Now Try Exercise 33*

Quadratic Functions

Economists utilize average cost curves that relate the average unit cost of manufacturing a commodity to the number of units to be produced. (See Fig. 8.) Ecologists use curves that relate the net primary production of nutrients in a plant to the surface area of the foliage. (See Fig. 9.) Each curve is bowl-shaped, opening either up or down. The simplest functions whose graphs resemble these curves are the quadratic functions.

Figure 8 Average cost curve.

Figure 9 Production of nutrients.

A **quadratic function** is a function of the form

$$f(x) = ax^2 + bx + c,$$

where a, b, and c are constants and $a \neq 0$. The domain of such a function consists of all numbers. The graph of a quadratic function is called a **parabola**. Two typical parabolas are drawn in Figures 10 and 11.

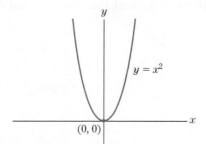

Figure 10 Graph of $f(x) = x^2$.

Figure 11 Graph of $f(x) = -x^2 + 4x + 5$.

Note how the parabola in Fig. 10 opens up and has its lowest point (vertex) at $(0, 0)$; while the parabola in Fig. 11 opens down and has its highest point (vertex) at $(2, 9)$.

Indeed, you may recall that a parabola opens up if $a > 0$ and down if $a < 0$, and that the vertex is located at

$$\left(-\frac{b}{2a}, f\left(-\frac{b}{2a}\right)\right).$$

We shall derive these facts and develop advanced techniques for sketching graphs of quadratic functions after we have some calculus at our disposal.

Polynomial and Rational Functions

A **polynomial function** $f(x)$ is of the form

$$f(x) = a_n x^n + a_{n-1}x^{n-1} + \cdots + a_0,$$

where n is a nonnegative integer and $a_0, a_1, \ldots, a_n$ are given numbers. Some examples of polynomial functions are

$$f(x) = 5x^3 - 3x^2 - 2x + 4$$
$$g(x) = x^4 - x + 1.$$

Of course, linear and quadratic functions are special cases of polynomial functions. The domain of a polynomial function consists of all numbers.

A function expressed as the quotient of two polynomials is called a **rational function**. Some examples are

$$h(x) = \frac{x^2 + 1}{x}$$
$$k(x) = \frac{x + 3}{x^2 - 4}.$$

The domain of a rational function excludes all values of x for which the denominator is zero. For example, the domain of $h(x)$ excludes $x = 0$, whereas the domain of $k(x)$ excludes $x = 2$ and $x = -2$. As we shall see, both polynomial and rational functions arise in applications of calculus.

Rational functions are used in environmental studies as **cost–benefit** models. The cost of removing a pollutant from the atmosphere is estimated as a function of the percentage of the pollutant removed. The higher the percentage removed, the greater the "benefit" to the people who breathe that air. The issues here are complex, of course, and the definition of "cost" is debatable. The cost to remove a small percentage of pollutant may be fairly low. But the removal cost of the final 5% of the pollutant, for example, may be terribly expensive.

EXAMPLE 6   **A Cost–Benefit Model** Suppose that a cost–benefit function is given by

$$f(x) = \frac{50x}{105 - x}, \quad 0 \le x \le 100,$$

where x is the percentage of some pollutant to be removed and $f(x)$ is the associated cost (in millions of dollars). (See Fig. 12.) Find the cost to remove 70%, 95%, and 100% of the pollutant.

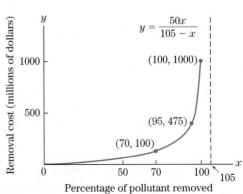

Figure 12 A cost-benefit model.

SOLUTION   The cost to remove 70% is

$$f(70) = \frac{50(70)}{105 - 70} = 100 \text{ (million dollars).}$$

Similar calculations show that

$$f(95) = 475 \quad \text{and} \quad f(100) = 1000.$$

Observe that the cost to remove the last 5% of the pollutant is $f(100) - f(95) = 1000 - 475 = 525$ million dollars. This is more than five times the cost to remove the first 70% of the pollutant!   **» Now Try Exercise 21**

Power Functions

Functions of the form $f(x) = x^r$ are called **power functions**. The meaning of x^r is clear when $r = n$ is a positive integer; in this case, the function $f(x) = x^n$ is a polynomial. However, the power function $f(x) = x^r$ may be defined for any number r. We delay until Section 0.5 a discussion of power functions, where we will review the meaning of x^r in the case when r is a rational number.

Absolute Value Functions

The absolute value of a number is the distance of x from 0. It is denoted by $|x|$ and is defined by

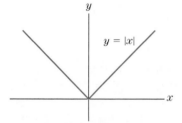

$y = |x|$

$$|x| = \begin{cases} x & \text{if } x \text{ is positive or zero.} \\ -x & \text{if } x \text{ is negative.} \end{cases}$$

For example, $|5| = 5$, $|0| = 0$, and $|-3| = -(-3) = 3$.

The function defined for all numbers real x by

$$f(x) = |x|$$

Figure 13 Graph of the absolute value function.

is called the **absolute value function**. Its graph coincides with the graph of the equation $y = x$ for $x \geq 0$ and with the graph of the equation $y = -x$ for $x < 0$. (See Fig. 13.)

INCORPORATING TECHNOLOGY

Evaluating a Function Consider the quadratic function $f(x) = -x^2 + 4x + 5$. To use your graphing calculator to compute the value of $f(-5)$, follow these steps:

Step 1   Press $\boxed{\text{Y=}}$, enter the expression $-X^2 + 4X + 5$ for Y_1, and return to the home screen. (Recall that the variable X may be entered with the $\boxed{\text{X,T,}\Theta,n}$ key; to enter the expression $-X^2$, use the following key sequence: $\boxed{\text{(-)}}$ $\boxed{\text{X,T,}\Theta,n}$ $\boxed{x^2}$.)

Step 2   From the home screen, press $\boxed{\text{VARS}}$ to access the variables menu, and then press $\boxed{\triangleright}$ to access the Y-VARS submenu. Next press $\boxed{1}$. You will be presented with a list of the y-variables Y_1, Y_2, and so on. Select Y_1.

Step 3   Now press $\boxed{\text{(}}$, -5, $\boxed{\text{)}}$, $\boxed{\text{ENTER}}$.

The output (see Fig. 14) shows that $f(-5) = -40$. As an alternative, you can first assign the value -5 to X and then ask for the value of Y_1 (see Fig. 15). To assign the value -5 to X, use $\boxed{\text{(-)}}$ $\boxed{5}$ $\boxed{\text{STO}\triangleright}$ $\boxed{\text{X,T,}\Theta,n}$.

Figure 14

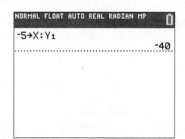

Figure 15

Check Your Understanding 0.2

Solutions can be found following the section exercises.

1. A photocopy service has a fixed cost of $2000 per month (for rent, depreciation of equipment, etc.) and variable costs of $0.04 for each page it reproduces for customers. Express its total cost as a (linear) function of the number of pages copied per month.

2. Determine the intercepts of the graph of

$$f(x) = -\tfrac{3}{8}x + 6.$$

EXERCISES 0.2

Graph the following equations.

1. $y = 2x - 1$
2. $y = 3$
3. $y = 3x + 1$
4. $y = -\tfrac{1}{2}x - 4$
5. $y = -2x + 3$
6. $y = 0$
7. $x - y = 0$
8. $3x + 2y = -1$
9. $x = 2y - 1$
10. $x = -4$

Determine the intercepts of the graphs of the following equations.

11. $f(x) = 9x + 3$
12. $f(x) = -\tfrac{1}{2}x - 1$
13. $f(x) = 5$
14. $f(x) = 14$
15. $x - 5y = 0$
16. $2 + 3x = 2y$

17. **Cost of Car Rentals** In some cities, you can rent a car for $24 per day and $.45 per mile.
    (a) Find the cost of renting the car for one day and driving 200 miles.
    (b) If the car is to be rented for one day, express the total rental expense as a function of the number x of miles driven. (Assume that for each fraction of a mile driven the same fraction of $.45 is charged.)

18. **Right to Drill** A gas company will pay a property owner $5000 for the right to drill on his land for natural gas and $.10 for each thousand cubic feet of gas extracted from the land. Express the amount of money the landowner will receive as a function of the amount of gas extracted from the land.

19. **Medical Expense** In 2010, a patient paid $700 per day for a semiprivate hospital room and $1900 for an appendectomy operation. Express the total amount paid for an appendectomy as a function of the number of days of hospital confinement.

20. **Velocity of a Baseball** When a baseball thrown at 85 miles per hour is hit by a bat swung at x miles per hour, the ball travels $6x - 40$ feet. (*Source: The Physics of Baseball.*) (This formula assumes that $50 \leq x \leq 90$ and that the bat is 35 inches long, weighs 32 ounces, and strikes a waist-high pitch so that the plane of the swing lies at $35°$ from the horizontal.) How fast must the bat be swung for the ball to travel 350 feet?

21. **Cost–Benefit** Let $f(x)$ be the cost–benefit function from Example 6. If 70% of the pollutant has been removed, what is the added cost to remove another 5%? How does this compare with the cost to remove the final 5% of the pollutant? (See Example 6.)

22. **Cost of Cleaning a Pollutant** Suppose that the cost (in millions of dollars) to remove x percent of a certain pollutant is given by the cost–benefit function

$$f(x) = \frac{20x}{102 - x} \quad \text{for } 0 \leq x \leq 100.$$

    (a) Find the cost to remove 85% of the pollutant.
    (b) Find the cost to remove the final 5% of the pollutant.

23. **Enzyme Kinetics** In biochemistry, such as in the study of enzyme kinetics, we encounter a linear function of the form $f(x) = (K/V)x + 1/V$, where K and V are constants.
    (a) If $f(x) = .2x + 50$, find K and V so that $f(x)$ may be written in the form $f(x) = (K/V)x + 1/V$.
    (b) Find the x-intercept and y-intercept of the line $y = (K/V)x + 1/V$ (in terms of K and V).

24. The constants K and V in Exercise 23 are often determined from experimental data. Suppose that a line is drawn through data points and has x-intercept $(-500, 0)$ and y-intercept $(0, 60)$. Determine K and V so that the line is the graph of the function $f(x) = (K/V)x + 1/V$. [*Hint*: Use Exercise 23(b).]

Each quadratic function in Exercises 25–30 has the form $y = ax^2 + bx + c$. Identify a, b, and c.

25. $y = 3x^2 - 4x$
26. $y = \dfrac{x^2 - 6x + 2}{3}$
27. $y = 3x - 2x^2 + 1$
28. $y = 3 - 2x + 4x^2$
29. $y = 1 - x^2$
30. $y = \tfrac{1}{2}x^2 + \sqrt{3}x - \pi$

Sketch the graphs of the following functions.

31. Sketch the parabola $f(x) = 2x^2 - 4x$.
32. Sketch the parabola $g(t) = -t^2 + 4t - 3$.

33. $f(x) = \begin{cases} 3 & \text{for } x < 2 \\ 2x + 1 & \text{for } x \geq 2 \end{cases}$

34. $f(x) = \begin{cases} \tfrac{1}{2}x & \text{for } 0 \leq x < 4 \\ 2x - 3 & \text{for } 4 \leq x \leq 5 \end{cases}$

35. $f(x) = \begin{cases} 4 - x & \text{for } 0 \leq x < 2 \\ 2x - 2 & \text{for } 2 \leq x < 3 \\ x + 1 & \text{for } x \geq 3 \end{cases}$

36. $f(x) = \begin{cases} 4x & \text{for } 0 \leq x < 1 \\ 8 - 4x & \text{for } 1 \leq x < 2 \\ 2x - 4 & \text{for } x \geq 2 \end{cases}$

Evaluate each of the functions in Exercises 37–42 at the given value of x.

37. $f(x) = x^{100}, x = -1$
38. $f(x) = x^5, x = \tfrac{1}{2}$
39. $f(x) = |x|, x = 10^{-2}$
40. $f(x) = |x|, x = \pi$
41. $f(x) = |x|, x = -2.5$
42. $f(x) = |x|, x = -\tfrac{2}{3}$

TECHNOLOGY EXERCISES

In Exercises 43–46, use your graphing calculator to find the value of the given function at the indicated values of x.

43. $f(x) = 3x^3 + 8; x = -11, x = 10$
44. $f(x) = x^4 + 2x^3 + x - 5; x = -\tfrac{1}{2}, x = 3$
45. $f(x) = \tfrac{1}{2}x^2 + \sqrt{3}x - \pi; x = -2, x = 20$
46. $f(x) = \dfrac{2x - 1}{x^3 + 3x^2 + 4x + 1}; x = 2, x = 6$

Solutions to Check Your Understanding 0.2

1. If x represents the number of pages copied per month, then the variable cost is $.04x$ dollars. Now, [total cost] = fixed cost] + [variable cost]. If we define

$$f(x) = 2000 + .04x,$$

then $f(x)$ gives the total cost per month.

2. To find the y-intercept, evaluate $f(x)$ at $x = 0$:

$$f(0) = -\tfrac{3}{8}(0) + 6 = 0 + 6 = 6.$$

To find the x-intercept, set $f(x) = 0$ and solve for x:

$$-\tfrac{3}{8}x + 6 = 0$$
$$\tfrac{3}{8}x = 6$$
$$x = \tfrac{8}{3} \cdot 6 = 16.$$

Therefore, the y-intercept is $(0, 6)$ and the x-intercept is $(16, 0)$.

0.3 The Algebra of Functions

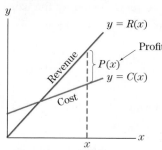

Figure 1 Profit equals revenue minus cost.

Many functions encountered later in the text can be viewed as combinations of other functions. For example, let $P(x)$ represent the profit a company makes on the sale of x units of some commodity. If $R(x)$ denotes the revenue received from the sale of x units, and if $C(x)$ is the cost of producing x units, then

$$\boxed{\begin{array}{c} P(x) = \quad R(x) \quad - C(x) \\ \text{[profit]} = \text{[revenue]} - \text{[cost]}. \end{array}}$$

Writing the profit function in this way makes it possible to predict the behavior of $P(x)$ from properties of $R(x)$ and $C(x)$. For instance, we can determine when the profit $P(x)$ is positive by observing whether $R(x)$ is greater than $C(x)$. (See Fig. 1.)

Operations on Functions

The next four examples review the algebraic techniques needed to combine functions by addition, subtraction, multiplication, and division.

Basic Properties of Fractions   The following basic properties of fractions are useful when you are dealing with rational expressions. (All letters occurring in a denominator are nonzero.)

1. Multiplication $a \cdot \dfrac{b}{c} = \dfrac{ab}{c}$

2. $\dfrac{a}{b} \cdot \dfrac{c}{d} = \dfrac{ac}{bd}$

3. Simplifying common factors $\dfrac{ac}{bc} = \dfrac{a}{b}$

4. Division $\dfrac{a}{b} \div \dfrac{c}{d} = \dfrac{a}{b} \cdot \dfrac{d}{c} = \dfrac{ad}{bc}$

5. Addition $\dfrac{a}{b} + \dfrac{c}{d} = \dfrac{ad + bc}{bd}$

EXAMPLE 1   **Operations on Functions** Let $f(x) = 2x + 4$ and $g(x) = 2x - 6$. Find

(a) $f(x) + g(x)$,   (b) $f(x) - g(x)$,   (c) $\dfrac{f(x)}{g(x)}$,   and   (d) $f(x)g(x)$.

SOLUTION   (a) $f(x) + g(x) = (2x + 4) + (2x - 6) = 2x + 4 + 2x - 6 = 4x - 2$

(b) $f(x) - g(x) = (2x + 4) - (2x - 6) = 2x + 4 - 2x + 6 = 10.$

(c) $\dfrac{f(x)}{g(x)} = \dfrac{2x + 4}{2x - 6}$

To simplify the expression for $\dfrac{f(x)}{g(x)}$, you must first find a common factor. (You cannot cross out $2x$ from the numerator and denominator. You must factor first.)

$$\frac{2x + 4}{2x - 6} = \frac{2(x + 2)}{2(x - 3)} \qquad \text{Factor first.}$$

$$= \frac{x + 2}{x - 3} \qquad \text{Cancel common factors.}$$

This expression is in its simplest form.

(d) To simplify the expression for $f(x)g(x)$, we substitute the values of $f(x)$ and $g(x)$, then carry out the multiplication indicated in $(2x + 4)(2x - 6)$. We must be careful to multiply each term of $2x + 4$ by each term of $2x - 6$. A common order for multiplying such expressions is (1) the First terms, (2) the Outer terms, (3) the Inner terms, and (4) the Last terms. (This procedure may be remembered by the word FOIL.)

$$f(x)g(x) = (2x + 4)(2x - 6) \qquad \text{FOIL method illustrated}$$

$$= 4x^2 - 12x + 8x - 24 \qquad \text{Multiply terms.}$$

$$= 4x^2 - 4x - 24. \qquad \text{Combine like terms.}$$

》》 Now Try Exercise 3

EXAMPLE 2   **Adding Rational Functions** Express $g(x) + h(x)$ as a rational function, where

$$g(x) = \frac{2}{x} \quad \text{and} \quad h(x) = \frac{3}{x - 1}.$$

SOLUTION   First, we have

$$g(x) + h(x) = \frac{2}{x} + \frac{3}{x - 1}, \quad x \neq 0, 1.$$

The restriction $x \neq 0, 1$ comes from the fact that $g(x)$ is defined only for $x \neq 0$ and $h(x)$ is defined only for $x \neq 1$. (A rational function is not defined for values of the variable for which the denominator is 0.) To add two fractions, their denominators must be the same. A common denominator for $\dfrac{2}{x}$ and $\dfrac{3}{x - 1}$ is $x(x - 1)$. If we multiply $\dfrac{2}{x}$ by $\dfrac{x - 1}{x - 1}$, we obtain an equivalent expression whose denominator is $x(x - 1)$. Similarly, if we multiply $\dfrac{3}{x - 1}$ by $\dfrac{x}{x}$, we obtain an equivalent expression whose denominator is $x(x - 1)$. Thus,

$$\frac{2}{x} + \frac{3}{x - 1} = \frac{2}{x} \cdot \frac{x - 1}{x - 1} + \frac{3}{x - 1} \cdot \frac{x}{x} \qquad \begin{array}{l}\text{Find common}\\ \text{denominator.}\end{array}$$

$$= \frac{2(x - 1)}{x(x - 1)} + \frac{3x}{x(x - 1)} \qquad \text{Multiply.}$$

$$= \frac{2(x - 1) + 3x}{x(x - 1)} \qquad \text{Combine.}$$

$$= \frac{5x - 2}{x(x - 1)}. \qquad \text{Simplify.}$$

So,

$$g(x) + h(x) = \frac{5x - 2}{x(x - 1)}.$$

》》 Now Try Exercise 9

EXAMPLE 3	**Multiplying Rational Functions** Find $f(t)g(t)$, where

$$f(t) = \frac{t}{t-1} \quad \text{and} \quad g(t) = \frac{t+2}{t+1}.$$

SOLUTION To multiply rational functions, multiply numerator by numerator and denominator by denominator:

$$f(t)g(t) = \frac{t}{t-1} \cdot \frac{t+2}{t+1} = \frac{t(t+2)}{(t-1)(t+1)}.$$

An alternative way of expressing $f(t)g(t)$ is obtained by carrying out the indicated multiplications:

$$f(t)g(t) = \frac{t^2 + 2t}{t^2 + t - t - 1} = \frac{t^2 + 2t}{t^2 - 1}.$$

The choice of which expression to use for $f(t)g(t)$ depends on the particular application.   **» Now Try Exercise 15**

EXAMPLE 4	**Quotient of Rational Functions** Find

$$\frac{f(x)}{g(x)}, \quad \text{where } f(x) = \frac{x}{x-3} \text{ and } g(x) = \frac{x+1}{x-5}.$$

SOLUTION The function $f(x)$ is defined only for $x \neq 3$, and $g(x)$ is defined only for $x \neq 5$. The quotient $f(x)/g(x)$ is therefore not defined for $x = 3, 5$. Moreover, the quotient is not defined for values of x for which $g(x)$ is equal to 0; that is, $x = -1$. Thus, the quotient is defined for $x \neq -1, 3, 5$. To divide $f(x)$ by $g(x)$, we multiply $f(x)$ by the reciprocal of $g(x)$. The reciprocal of $g(x) = \frac{x+1}{x-5}$ is the function $\frac{x-5}{x+1}$, and so

$$\frac{f(x)}{g(x)} = \frac{x}{x-3} \cdot \frac{x-5}{x+1} \qquad \text{Multiply by the reciprocal of } g.$$

$$= \frac{x(x-5)}{(x-3)(x+1)} \qquad \text{Multiply numerator by numerator, denominator by denominator.}$$

$$= \frac{x^2 - 5x}{x^2 + x - 3x - 3} \qquad \text{Expand, FOIL.}$$

$$= \frac{x^2 - 5x}{x^2 - 2x - 3} \qquad \text{Simplify.}$$

» Now Try Exercise 17

Composition of Functions

Another important way of combining two functions $f(x)$ and $g(x)$ is to substitute the function $g(x)$ for every occurrence of the variable x in $f(x)$. The resulting function is called the **composition** (or **composite**) of $f(x)$ and $g(x)$ and is denoted by $f(g(x))$.

EXAMPLE 5	**Composition of Functions** Let $f(x) = x^2 + 3x + 1$ and $g(x) = x - 5$. Find (a) $f(g(x))$ and (b) $g(f(x))$.

SOLUTION (a) $f(g(x)) = [g(x)]^2 + 3g(x) + 1$ $\qquad$ Substitute $g(x)$ for x in the formula for $f(x)$.

$\qquad = (x-5)^2 + 3(x-5) + 1$ $\qquad$ Substitute $g(x)$ by its formula.

$\qquad = (x^2 - 10x + 25) + (3x - 15) + 1$ $\qquad$ FOIL and expand.

$\qquad = x^2 - 7x + 11.$ $\qquad$ Simplify.

(b) $g(f(x)) = f(x) - 5$ Substitute $f(x)$ for x in $g(x)$.

$\qquad\qquad = x^2 + 3x + 1 - 5$ Substitute $f(x)$ by its formula.

$\qquad\qquad = x^2 + 3x - 4.$ Simplify.

» Now Try Exercise 29

Note: You can see from this example that, in general, $f(g(x)) \neq g(f(x))$.

    Later in the text we shall need to study expressions of the form $f(x + h)$, where $f(x)$ is a given function and h represents some number. The meaning of $f(x + h)$ is that $x + h$ is to be substituted for each occurrence of x in the formula for $f(x)$. In fact, $f(x + h)$ is just a special case of $f(g(x))$, where $g(x) = x + h$.

EXAMPLE 6

Evaluating a Function If $f(x) = x^3$, find $\frac{f(x + h) - f(x)}{h}$, where $h \neq 0$:

SOLUTION

Let's start by evaluating $f(x + h)$.

$$f(x + h) = (x + h)^3 = x^3 + 3x^2h + 3xh^2 + h^3$$

Substitute $x + h$ for x and expand.

So the numerator has the value

$$f(x + h) - f(x) = (x^3 + 3x^2h + 3xh^2 + h^3) - x^3$$
$$= 3x^2h + 3xh^2 + h^3.$$ Simplify.

So,

$$\frac{f(x + h) - f(x)}{h} = \frac{3x^2h + 3xh^2 + h^3}{h}$$

$$= \frac{(3x^2 + 3xh + h^2)\cancel{h}}{\cancel{h}}$$ Factor h in numerator.

$$= 3x^2 + 3xh + h^2.$$ Cancel the common factor h.

We were able to divide the numerator and denominator by h because $h \neq 0$.

» Now Try Exercise 33

EXAMPLE 7

Freshwater Biology In a certain lake, the bass feed primarily on minnows, and the minnows feed on plankton. If the size of the bass population is a function $f(n)$ of the number n of minnows in the lake, and the number of minnows is a function $g(x)$ of the amount x of plankton in the lake, express the size of the bass population as a function of the amount of plankton, if $f(n) = 50 + \sqrt{n/150}$ and $g(x) = 4x + 3$.

SOLUTION

The number of minnows is equal to both n and $g(x)$, so we have $n = g(x)$. Substituting $g(x)$ for n in $f(n)$, we find that the size of the bass population is given by

$$f(g(x)) = 50 + \sqrt{\frac{g(x)}{150}} = 50 + \sqrt{\frac{4x + 3}{150}}.$$

» Now Try Exercise 37

INCORPORATING TECHNOLOGY

Graphing Compositions of Functions To use your graphing calculator to graph the function $f(g(x))$, where $f(x) = x^2$ and $g(x) = x + 3$, follow these steps:

Step 1 Press $\boxed{Y=}$, and set $Y_1 = X^2$ and $Y_2 = X + 3$.

Step 2 Set $Y_3 = Y_1(Y_2)$ by accessing the Y-VARS submenu of $\boxed{VARS}$.

Step 3 To graph our composite function $Y_3 = Y_1(Y_2)$ without also graphing Y_1 and Y_2, we must first deselect the functions Y_1 and Y_2. To deselect Y_1, place your cursor on top of the equal sign after $\setminus Y_1$ and press $\boxed{ENTER}$. Similarly, deselect Y_2; your screen should look like Fig. 2(a). Finally, press $\boxed{GRAPH}$. [See Fig. 2(b).]

(a)   (b)

Figure 2

Check Your Understanding 0.3

Solutions can be found following the section exercises.

1. Let $f(x) = x^5$, $g(x) = x^3 - 4x^2 + x - 8$.
  (a) Find $f(g(x))$.   **(b)** Find $g(f(x))$.

2. Let $f(x) = x^2$. Calculate $\dfrac{f(1 + h) - f(1)}{h}$ and simplify.

EXERCISES 0.3

Let $f(x) = x^2 + 1$, $g(x) = 9x$, and $h(x) = 5 - 2x^2$. Calculate the following functions.

1. $f(x) + g(x)$

2. $f(x) - h(x)$

3. $f(x)g(x)$

4. $g(x)h(x)$

5. $\dfrac{f(t)}{g(t)}$

6. $\dfrac{g(t)}{h(t)}$

In Exercises 7–12, express $f(x) + g(x)$ as a rational function. Carry out all multiplications.

7. $f(x) = \dfrac{2}{x - 3}$, $g(x) = \dfrac{1}{x + 2}$

8. $f(x) = \dfrac{3}{x - 6}$, $g(x) = \dfrac{-2}{x - 2}$

9. $f(x) = \dfrac{x}{x - 8}$, $g(x) = \dfrac{-x}{x - 4}$

10. $f(x) = \dfrac{-x}{x + 3}$, $g(x) = \dfrac{x}{x + 5}$

11. $f(x) = \dfrac{x + 5}{x - 10}$, $g(x) = \dfrac{x}{x + 10}$

12. $f(x) = \dfrac{x + 6}{x - 6}$, $g(x) = \dfrac{x - 6}{x + 6}$

Let $f(x) = \dfrac{x}{x - 2}$, $g(x) = \dfrac{5 - x}{5 + x}$, and $h(x) = \dfrac{x + 1}{3x - 1}$. Express the following as rational functions.

13. $f(x) - g(x)$

14. $f(t) - h(t)$

15. $f(x)g(x)$

16. $g(x)h(x)$

17. $\dfrac{f(x)}{g(x)}$

18. $\dfrac{h(s)}{f(s)}$

19. $f(x + 1)g(x + 1)$

20. $f(x + 2) + g(x + 2)$

21. $\dfrac{g(x + 5)}{f(x + 5)}$

22. $f\left(\dfrac{1}{t}\right)$

23. $g\left(\dfrac{1}{u}\right)$

24. $h\left(\dfrac{1}{x^2}\right)$

Let $f(x) = x^6$, $g(x) = \dfrac{x}{1 - x}$, and $h(x) = x^3 - 5x^2 + 1$. Calculate the following functions.

25. $f(g(x))$

26. $h(f(t))$

27. $h(g(x))$

28. $g(f(x))$

29. $g(h(t))$

30. $f(h(x))$

31. If $f(x) = x^2$, find $f(x + h) - f(x)$ and simplify.

32. If $f(x) = 1/x$, find $f(x + h) - f(x)$ and simplify.

33. If $g(t) = 4t - t^2$, find $\dfrac{g(t + h) - g(t)}{h}$ and simplify.

34. If $g(t) = t^3 + 5$, find $\dfrac{g(t + h) - g(t)}{h}$ and simplify.

35. **Cost** After t hours of operation, an assembly line has assembled $A(t) = 20t - \frac{1}{2}t^2$ power lawn mowers, $0 \le t \le 10$. Suppose that the factory's cost of manufacturing x units is $C(x)$ dollars, where $C(x) = 3000 + 80x$.
  (a) Express the factory's cost as a (composite) function of the number of hours of operation of the assembly line.
  (b) What is the cost of the first 2 hours of operation?

36. **Cost** During the first $\frac{1}{2}$ hour, the employees of a machine shop prepare the work area for the day's work. After that, they turn out 10 precision machine parts per hour, so the output after t hours is $f(t)$ machine parts, where $f(t) = 10(t - \frac{1}{2}) = 10t - 5$, $\frac{1}{2} \le t \le 8$. The total cost of producing x machine parts is $C(x)$ dollars, where $C(x) = .1x^2 + 25x + 200$.
  (a) Express the total cost as a (composite) function of t.
  (b) What is the cost of the first 4 hours of operation?

37. **Conversion Scales** Table 1 shows a conversion table for men's hat sizes for three countries. The function $g(x) = 8x + 1$ converts from British sizes to French sizes, and the function $f(x) = \frac{1}{8}x$ converts from French sizes to U.S. sizes. Determine the function $h(x) = f(g(x))$ and give its interpretation.

Table 1 Conversion Table for Men's Hat Sizes

Britain	$6\frac{1}{2}$	$6\frac{5}{8}$	$6\frac{3}{4}$	$6\frac{7}{8}$	7	$7\frac{1}{8}$	$7\frac{1}{4}$	$7\frac{3}{8}$
France	53	54	55	56	57	58	59	60
U.S.	$6\frac{5}{8}$	$6\frac{3}{4}$	$6\frac{7}{8}$	7	$7\frac{1}{8}$	$7\frac{1}{4}$	$7\frac{3}{8}$	$7\frac{1}{2}$

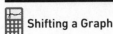
38. Shifting a Graph Let $f(x) = x^2$. Graph the functions $f(x + 1)$, $f(x - 1)$, $f(x + 2)$, and $f(x - 2)$. Make a guess about the relationship between the graph of a general function $f(x)$ and the graph of $f(g(x))$, where $g(x) = x + a$ for some constant a. Test your guess on the functions $f(x) = x^3$ and $f(x) = \sqrt{x}$.

39. Shifting a Graph Let $f(x) = x^2$. Graph the functions $f(x) + 1$, $f(x) - 1$, $f(x) + 2$, and $f(x) - 2$. Make a guess about the relationship between the graph of a general function $f(x)$ and the graph of $f(x) + c$ for some constant c. Test your guess on the functions $f(x) = x^3$ and $f(x) = \sqrt{x}$.

40. Based on the results of Exercises 38 and 39, sketch the graph of $f(x) = (x - 1)^2 + 2$ without using a graphing calculator. Check your result with a graphing calculator.

41. Based on the results of Exercises 38 and 39, sketch the graph of $f(x) = (x + 2)^2 - 1$ without using a graphing calculator. Check your result with a graphing calculator.

42. Let $f(x) = x^2 + 3x + 1$ and let $g(x) = x^2 - 3x - 1$. Graph the two functions $f(g(x))$ and $g(f(x))$ together in the window $[-4, 4]$ by $[-10, 10]$ and determine if they are the same function.

43. Let $f(x) = \dfrac{x}{x - 1}$ and graph the function $f(f(x))$ in the window $[-15, 15]$ by $[-10, 10]$. Trace to examine the coordinates of several points on the graph and then determine the formula for $f(f(x))$.

Solutions to Check Your Understanding 0.3

1. (a) $f(g(x)) = [g(x)]^5 = (x^3 - 4x^2 + x - 8)^5$

(b) $g(f(x)) = [f(x)]^3 - 4[f(x)]^2 + f(x) - 8$

$= (x^5)^3 - 4(x^5)^2 + x^5 - 8$

$= x^{15} - 4x^{10} + x^5 - 8$

2. $\dfrac{f(1 + h) - f(1)}{h} = \dfrac{(1 + h)^2 - 1^2}{h}$

$= \dfrac{1 + 2h + h^2 - 1}{h}$

$= \dfrac{2h + h^2}{h} = 2 + h$

0.4   Zeros of Functions—The Quadratic Formula and Factoring

A **zero** of a function $f(x)$ is a value of x for which $f(x) = 0$. For instance, the function $f(x)$ whose graph is shown in Fig. 1 has $x = -3$, $x = 3$, and $x = 7$ as zeros. To solve many application problems, we shall need to determine zeros of functions or, what amounts to the same thing, to solve the equation $f(x) = 0$.

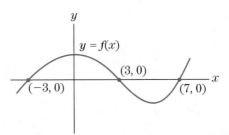

Figure 1 Zeros of a function.

In Section 0.2, we found zeros of linear functions. In this section, our emphasis is on zeros of quadratic functions.

The Quadratic Formula

Consider the quadratic function $f(x) = ax^2 + bx + c$, $a \neq 0$. The zeros of this function are precisely the solutions of the quadratic equation

$$ax^2 + bx + c = 0.$$

One way of solving such an equation is by using the **quadratic formula**.

> **Quadratic Formula**   The solutions of the equation $ax^2 + bx + c = 0$ are
> $$x = \frac{-b \pm \sqrt{b^2 - 4ac}}{2a}.$$

The $\pm$ sign tells us to form two expressions, one with $+$ and one with $-$. The quadratic formula implies that a quadratic equation has at most two roots. It will have none if the expression $b^2 - 4ac$ is negative and one if $b^2 - 4ac$ equals 0. The quadratic formula is derived at the end of this section.

EXAMPLE 1   **Using the Quadratic Formula** Solve the quadratic equation $3x^2 - 6x + 2 = 0$.

SOLUTION   Here, $a = 3$, $b = -6$, and $c = 2$. To determine whether a solution exists, we start by computing $b^2 - 4ac$:

$$b^2 - 4ac = (-6)^2 - 4(3)(2) = 36 - 24 = 12.$$

Since $b^2 - 4ac > 0$, we have two distinct solutions given by

$$x = \frac{-b \pm \sqrt{b^2 - 4ac}}{2a} \qquad \text{Quadratic formula.}$$

$$= \frac{-(-6) \pm \sqrt{12}}{2(3)} \qquad \text{Substitute the values of } a, b, \text{ and } c.$$

$$= \frac{6 \pm 2\sqrt{3}}{2(3)} \qquad \text{Note: } \sqrt{12} = \sqrt{(4)(3)} = \sqrt{4}\sqrt{3} = 2\sqrt{3}$$

$$= \frac{2(3 \pm \sqrt{3})}{(2)(3)} \qquad \text{Factor 2 from the numerator.}$$

$$= \frac{3 \pm \sqrt{3}}{3} \qquad \text{Cancel out the common factor.}$$

To simplify further, we note a useful fact about the addition of fractions with a common denominator: If A, B, and $C \neq 0$ are any real numbers, then

$$\frac{A + B}{C} = \frac{A}{C} + \frac{B}{C}.$$

So

$$x = \frac{3 \pm \sqrt{3}}{3} = \frac{3}{3} \pm \frac{\sqrt{3}}{3} = 1 \pm \frac{\sqrt{3}}{3}.$$

The solutions of the equation are $1 + \sqrt{3}/3$ and $1 - \sqrt{3}/3$. (See Fig. 2.)

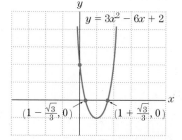

Figure 2

» *Now Try Exercise 1*

EXAMPLE 2   **Finding Zeros of Functions** Find the zeros of the following quadratic functions.
(a) $f(x) = 4x^2 - 4x + 1$ **(b)** $f(x) = \frac{1}{2}x^2 - 3x + 5$

SOLUTION   **(a)** We must solve $4x^2 - 4x + 1 = 0$. Here, $a = 4$, $b = -4$, and $c = 1$, so

$$\sqrt{b^2 - 4ac} = \sqrt{(-4)^2 - 4(4)(1)} = \sqrt{0} = 0.$$

Thus, there is only one zero:

$$x = \frac{-(-4) \pm 0}{2(4)} = \frac{4}{8} = \frac{1}{2}.$$

The graph of $f(x)$ is sketched in Fig. 3.

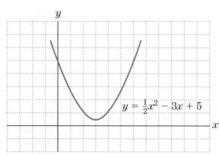

Figure 3   **Figure 4**

(b) We must solve $\frac{1}{2}x^2 - 3x + 5 = 0$. Here, $a = \frac{1}{2}$, $b = -3$, and $c = 5$, so,

$$\sqrt{b^2 - 4ac} = \sqrt{(-3)^2 - 4(\tfrac{1}{2})(5)} = \sqrt{9 - 10} = \sqrt{-1}.$$

The square root of a negative number is undefined, so we conclude that $f(x)$ has no zeros. The reason is clear from Fig. 4. The graph of $f(x)$ lies entirely above the x-axis and has no x-intercepts.   **» Now Try Exercise 5**

The common problem of finding where two curves intersect amounts to finding the zeros of a function.

EXAMPLE 3   **Intersection of Graphs** Find the points of intersection of the graphs of the functions $y = x^2 + 1$ and $y = 4x$. (See Fig. 5.)

SOLUTION   If a point (x, y) is on both graphs, then its coordinates must satisfy both equations. That is, x and y must satisfy $y = x^2 + 1$ and $y = 4x$. Equating the two expressions for y, we have

$$x^2 + 1 = 4x.$$

To use the quadratic formula, we rewrite the equation in the form

$$x^2 - 4x + 1 = 0.$$

From the quadratic formula,

$$x = \frac{4 \pm \sqrt{16 - 4}}{2} = \frac{4 \pm \sqrt{12}}{2} = \frac{4 \pm 2\sqrt{3}}{2} = 2 \pm \sqrt{3}.$$

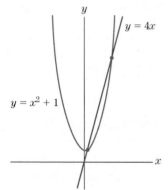

Figure 5 Points of intersection of two graphs.

Thus, the x-coordinates of the points of intersection are $2 + \sqrt{3}$ and $2 - \sqrt{3}$. To find the y-coordinates, we substitute these values of x into either equation, $y = x^2 + 1$ or $y = 4x$. The second equation is simpler. We obtain $y = 4(2 + \sqrt{3}) = 8 + 4\sqrt{3}$ and $y = 4(2 - \sqrt{3}) = 8 - 4\sqrt{3}$. Thus, the points of intersection are $(2 + \sqrt{3}, 8 + 4\sqrt{3})$ and $(2 - \sqrt{3}, 8 - 4\sqrt{3})$.   **» Now Try Exercise 33**

EXAMPLE 4   **Profit and Breakeven Points** A cable television company estimates that with x thousand subscribers its monthly revenue and cost (in thousands of dollars) are

$$R(x) = 32x - .21x^2$$
$$C(x) = 195 + 12x.$$

Determine the company's **breakeven points**; that is, find the number of subscribers at which profit equals zero, that is, the revenue equals the cost. (See Fig. 6.)

Figure 6 Breakeven points.

SOLUTION   Let $P(x)$ be the profit function:

$$P(x) = R(x) - C(x)$$
$$= (32x - .21x^2) - (195 + 12x)$$
$$= -.21x^2 + 20x - 195.$$

The breakeven points occur where the profit is zero. Thus, we must solve

$$-.21x^2 + 20x - 195 = 0.$$

From the quadratic formula,

$$x = \frac{-20 \pm \sqrt{20^2 - 4(-.21)(-195)}}{2(-.21)} = \frac{-20 \pm \sqrt{236.2}}{-.42}$$
$$\approx 47.62 \pm 36.59 = 11.03 \text{ and } 84.21.$$

The breakeven points occur where the company has 11,030 or 84,210 subscribers. Between those two levels, the company will be profitable.   **» Now Try Exercise 45**

Factoring

If $f(x)$ is a polynomial, we can often write $f(x)$ as a product of linear factors (i.e., factors of the form $ax + b$). If this can be done, then we can determine the zeros of $f(x)$ by setting each linear factor equal to zero and solving for x. (The reason is that the product of numbers can be zero only when one of the numbers is zero.)

EXAMPLE 5   **Factoring Quadratic Polynomials** Factor the following quadratic polynomials.
(a) $x^2 + 7x + 12$ (b) $x^2 - 13x + 12$
(c) $x^2 - 4x - 12$ (d) $x^2 + 4x - 12$

SOLUTION   Note, first, that, for any numbers c and d,

$$(x + c)(x + d) = x^2 + (c + d)x + cd.$$

In the quadratic on the right, the constant term is the product cd, whereas the coefficient of x is the sum $c + d$.

(a) Think of all integers c and d such that $cd = 12$. Then, choose the pair that satisfies $c + d = 7$; that is, take $c = 3$, $d = 4$. Thus,

$$x^2 + 7x + 12 = (x + 3)(x + 4).$$

(b) We want $cd = 12$. Since 12 is positive, c and d must be both positive or both negative. We must also have $c + d = -13$. These facts lead us to

$$x^2 - 13x + 12 = (x - 12)(x - 1).$$

(c) We want $cd = -12$. Since -12 is negative, c and d must have opposite signs. Also, they must sum to give -4. We find that

$$x^2 - 4x - 12 = (x - 6)(x + 2).$$

(d) This is almost the same as part (c).

$$x^2 + 4x - 12 = (x + 6)(x - 2).$$ **» Now Try Exercise 13**

EXAMPLE 6   **Factoring Quadratic Polynomials**   Factor the following polynomials.
(a) $x^2 - 6x + 9$ (b) $x^2 - 25$
(c) $3x^2 - 21x + 30$ (d) $20 + 8x - x^2$

SOLUTION   (a) We look for $cd = 9$ and $c + d = -6$. The solution is $c = d = -3$, and

$$x^2 - 6x + 9 = (x - 3)(x - 3) = (x - 3)^2.$$

In general,

$$x^2 - 2cx + c^2 = (x - c)(x - c) = (x - c)^2.$$

(b) We use the **difference of squares identity**:

$$x^2 - c^2 = (x + c)(x - c).$$

Hence,

$$x^2 - 25 = (x + 5)(x - 5).$$

(c) We first factor out a common factor of 3 and then use the method of Example 5:

$$3x^2 - 21x + 30 = 3(x^2 - 7x + 10)$$
$$= 3(x - 5)(x - 2).$$

(d) We first factor out -1 to make the coefficient of x^2 equal to $+1$:

$$20 + 8x - x^2 = (-1)(x^2 - 8x - 20)$$
$$= (-1)(x - 10)(x + 2).$$ **» Now Try Exercise 17**

EXAMPLE 7   **Factoring Higher Degree Polynomials**   Factor the following polynomials.
(a) $x^2 - 8x$ (b) $x^3 + 3x^2 - 18x$ (c) $x^3 - 10x$

SOLUTION   In each case, we first factor out a common factor of x:

(a) $x^2 - 8x = x(x - 8)$.
(b) $x^3 + 3x^2 - 18x = x(x^2 + 3x - 18) = x(x + 6)(x - 3)$.
(c) $x^3 - 10x = x(x^2 - 10)$. To factor $x^2 - 10$, we use the difference of squares identity $x^2 - c^2 = (x + c)(x - c)$, where $c^2 = 10$ and $c = \sqrt{10}$. Thus,

$$x^3 - 10x = x(x^2 - 10) = x(x + \sqrt{10})(x - \sqrt{10}).$$
» Now Try Exercise 23

EXAMPLE 8   **Solving Equations**   Solve the following equations.

(a) $x^2 - 2x - 15 = 0$ (b) $x^2 - 20 = x$ (c) $\dfrac{x^2 + 10x + 25}{x + 1} = 0$

SOLUTION   (a) The equation $x^2 - 2x - 15 = 0$ may be written in the form

$$(x - 5)(x + 3) = 0.$$

The product of two numbers is zero if one or the other of the numbers (or both) is zero. Hence,

$$x - 5 = 0 \quad \text{or} \quad x + 3 = 0.$$

That is,

$$x = 5 \quad \text{or} \quad x = -3.$$

(b) First, we must rewrite the equation $x^2 - 20 = x$ in the form $ax^2 + bx + c = 0$; that is,

$$x^2 - x - 20 = 0$$
$$(x - 5)(x + 4) = 0.$$

We conclude that

$$x - 5 = 0 \quad \text{or} \quad x + 4 = 0;$$

that is,

$$x = 5 \quad \text{or} \quad x = -4.$$

(c) A rational function will be zero only if the numerator is zero. Thus,

$$x^2 + 10x + 25 = 0$$
$$(x + 5)^2 = 0$$
$$x + 5 = 0.$$

That is,

$$x = -5.$$

Since the denominator is not 0 at $x = -5$, we conclude that $x = -5$ is the solution.

≫ Now Try Exercise 35

EXAMPLE 9   **Solving a Rational Equation**

Solve $-\dfrac{3}{x} + \dfrac{8}{x - 1} = -1$

SOLUTION   Note at the outset that the denominators appearing in the equation cannot be zero. So $x \neq 0$ and $x - 1 \neq 0$ or $x \neq 1$. We can get rid of the denominators, x and $x - 1$, by multiplying both sides of the equation by $x(x - 1)$. We have

$$x(x - 1)\left[-\frac{3}{x} + \frac{8}{x - 1}\right] = -x(x - 1) \quad \text{Multiply both sides by } x(x - 1).$$

$$-\frac{3x(x - 1)}{x} + \frac{8x(x - 1)}{x - 1} = -x^2 + x \quad \text{Distribute.}$$

$$-3(x - 1) + 8x = -x^2 + x \quad \text{Cancel the common factors } x \text{ and } x - 1.$$

$$-3x + 3 + 8x = -x^2 + x \quad \text{Distribute.}$$

$$x^2 + 4x + 3 = 0 \quad \text{Move all terms to one side and simplify.}$$

$$(x + 1)(x + 3) = 0 \quad \text{Factor.}$$

$$x + 1 = 0 \quad \text{or} \quad x + 3 = 0$$

$$x = -1 \quad \text{or} \quad x = -3 \quad \text{Solve.}$$

Since none of these values are excluded (remember $x \neq 0, 1$), we can verify that they are indeed solutions by substituting back into the equation. For example, substituting $x = -1$ in the left side of the equation yields

$$\frac{-3}{(-1)} + \frac{8}{(-1) - 1} = 3 + \frac{8}{(-2)} = 3 - 4 = -1,$$

implying that $x = -1$ is a solution. Similarly, you can verify that $x = -3$ is also a solution.

>> *Now Try Exercise 39*

In Example 6(b), we used the difference of squares identity to factor. Let us recall some algebraic identities that can be useful when factoring.

Identities for Factoring

Difference of squares	$A^2 - B^2 = (A - B)(A + B)$
Perfect square	$A^2 + 2AB + B^2 = (A + B)^2$
Perfect square	$A^2 - 2AB + B^2 = (A - B)^2$
Difference of cubes	$A^3 - B^3 = (A - B)(A^2 + AB + B^2)$
Sum of cubes	$A^3 + B^3 = (A + B)(A^2 - AB + B^2)$

EXAMPLE 10   **Factoring Sums and Differences of Cubes**

Factor **(a)** $x^3 + 27$ **(b)** $x^3 + 1$ **(c)** $x^3 - 8$.

SOLUTION
   (a) $x^3 + 27 = x^3 + 3^3 = (x + 3)(x^2 - 3x + 9)$ Sum of cubes.
   (b) $x^3 + 1 = x^3 + 1^3 = (x + 1)(x^2 - x + 1)$ Sum of cubes.
   (c) $x^3 - 8 = x^3 - 2^3 = (x - 2)(x^2 + 2x + 4)$ Difference of cubes.

>> *Now Try Exercise 27*

Derivation of the Quadratic Formula

$$ax^2 + bx + c = 0 \; (a \neq 0)$$
$$ax^2 + bx = -c$$
$$4a^2x^2 + 4abx = -4ac \qquad \text{Both sides multiplied by } 4a.$$
$$4a^2x^2 + 4abx + b^2 = b^2 - 4ac \qquad b^2 \text{ added to both sides to complete the square.}$$

Now, note that $4a^2x^2 + 4abx + b^2 = (2ax + b)^2$. To check this, simply multiply out the right side:

$$(2ax + b)^2 = b^2 - 4ac$$
$$2ax + b = \pm\sqrt{b^2 - 4ac}$$
$$2ax = -b \pm \sqrt{b^2 - 4ac}$$
$$x = \frac{-b \pm \sqrt{b^2 - 4ac}}{2a}.$$

INCORPORATING TECHNOLOGY

NORMAL FLOAT AUTO REAL RADIAN CL
CALC ZERO

Zero
X=-1.618034   Y=0

Figure 7

Computing Zeros of a Function To find the values of x at which $x^3 - 2x + 1 = 0$, follow those steps:

Step 1   Press [Y=] and enter the formula for the polynomial for Y_1. Press [GRAPH] to graph the function. The zeros of this polynomial appear to occur in the interval $[-2, 2]$, so press [WINDOW] and set $X\,min = -2$ and $X\,max = 2$.

Step 2   Now, press [2nd] [CALC] and then [2] to begin the process of finding the zeros. When prompted, enter -2 for the left bound, -1 for the right bound, and -1.5 for the guess. We can enter any value between our left and right bounds for the guess. The calculator then computes for us that $x^3 - 2x + 1 = 0$ when $x = -1.618034$. (See Fig. 7.)

We can now repeat this process with different bounds to find the other value of x at which $x^3 - 2x + 1 = 0$. Finally, notice that the menu behind [2nd] [CALC] includes a number of useful routines. Of particular interest for this section is option **5: intersect**, which will compute for us the intersection points of two functions, Y_1 and Y_2.

One problem in graphing a function is to find a domain (Xmin to Xmax) that contains all the zeros. For a polynomial function, there is an easy solution to this problem. Write the polynomial in the form $c(x^n + a_{n-1}x^{n-1} + \cdots + a_0)$, and let M be the number that is one more than the largest magnitude of the coefficients $a_{n-1}, \ldots, a_0$. Then, the interval $[-M, M]$ will contain all the zeros of the polynomial. For instance, all the zeros of the polynomial $x^3 - 10x^2 + 9x + 8$ lie in the interval $[-11, 11]$. After you view the polynomial on the domain $[-M, M]$, you can usually find a smaller domain that also contains all the zeros.

Check Your Understanding 0.4

Solutions can be found following the section exercises.

1. Solve the equation $x - \dfrac{14}{x} = 5$.

2. Use the quadratic formula to solve $7x^2 - 35x + 35 = 0$.

EXERCISES 0.4

Use the quadratic formula to find the zeros of the functions in Exercises 1–6.

1. $f(x) = 2x^2 - 7x + 6$

2. $f(x) = 3x^2 + 2x - 1$

3. $f(t) = 4t^2 - 12t + 9$

4. $f(x) = \frac{1}{4}x^2 + x + 1$

5. $f(x) = -2x^2 + 3x - 4$

6. $f(a) = 11a^2 - 7a + 1$

Use the quadratic formula to solve the equations in Exercises 7–12.

7. $5x^2 - 4x - 1 = 0$

8. $x^2 - 4x + 5 = 0$

9. $15x^2 - 135x + 300 = 0$

10. $z^2 - \sqrt{2}z - \frac{5}{4} = 0$

11. $\frac{3}{2}x^2 - 6x + 5 = 0$

12. $9x^2 - 12x + 4 = 0$

Factor the polynomials in Exercises 13–30.

13. $x^2 + 8x + 15$

14. $x^2 - 10x + 16$

15. $x^2 - 16$

16. $x^2 - 1$

17. $3x^2 + 12x + 12$

18. $2x^2 - 12x + 18$

19. $30 - 4x - 2x^2$

20. $15 + 12x - 3x^2$

21. $3x - x^2$

22. $4x^2 - 1$

23. $6x - 2x^3$

24. $16x + 6x^2 - x^3$

25. $x^3 - 1$

26. $x^3 + 125$

27. $8x^3 + 27$

28. $x^3 - \frac{1}{8}$

29. $x^2 - 14x + 49$

30. $x^2 + x + \frac{1}{4}$

Find the points of intersection of the pairs of curves in Exercises 31–38.

31. $y = 2x^2 - 5x - 6$, $y = 3x + 4$

32. $y = x^2 - 10x + 9$, $y = x - 9$

33. $y = x^2 - 4x + 4$, $y = 12 + 2x - x^2$

34. $y = 3x^2 + 9$, $y = 2x^2 - 5x + 3$

35. $y = x^3 - 3x^2 + x$, $y = x^2 - 3x$

36. $y = \frac{1}{2}x^3 - 2x^2$, $y = 2x$

37. $y = \frac{1}{2}x^3 + x^2 + 5$, $y = 3x^2 - \frac{1}{2}x + 5$

38. $y = 30x^3 - 3x^2$, $y = 16x^3 + 25x^2$

Solve the equations in Exercises 39–44.

39. $\dfrac{21}{x} - x = 4$

40. $x + \dfrac{2}{x - 6} = 3$

41. $x + \dfrac{14}{x + 4} = 5$

42. $1 = \dfrac{5}{x} + \dfrac{6}{x^2}$

43. $\dfrac{x^2 + 14x + 49}{x^2 + 1} = 0$

44. $\dfrac{x^2 - 8x + 16}{1 + \sqrt{x}} = 0$

45. Breakeven Points Suppose that the cable television company's cost function in Example 4 changes to $C(x) = 275 + 12x$. Determine the new breakeven points.

46. Velocity When a car is moving at x miles per hour and the driver decides to slam on the brakes, the car will travel $x + \frac{1}{20}x^2$ feet. (The general formula is $f(x) = ax + bx^2$, where the constant a depends on the driver's reaction time and the constant b depends on the weight of the car and the type of tires.) If a car travels 175 feet after the driver decides to stop, how fast was the car moving? (*Source: Applying Mathematics: A Course in Mathematical Modelling.*)

TECHNOLOGY EXERCISES

In Exercises 47–50, find the zeros of the function. (Use the specified viewing window.)

47. $f(x) = x^2 - x - 2$; $[-4, 5]$ $[-4, 10]$

48. $f(x) = x^3 - 3x + 2$; $[-3, 3]$ $[-10, 10]$

49. $f(x) = \sqrt{x + 2} - x + 2$; $[-2, 7]$ by $[-2, 4]$

50. $f(x) = \dfrac{x}{x + 2} - x^2 + 1$; $[-1.5, 2]$ by $[-2, 3]$

In Exercises 51–54, find the points of intersection of the graphs of the functions. (Use the specified viewing window.)

51. $f(x) = 2x - 1$; $g(x) = x^2 - 2$; $[-4, 4]$ by $[-6, 10]$

52. $f(x) = -x - 2$; $g(x) = -4x^2 + x + 1$; $[-2, 2]$ by $[-5, 2]$

53. $f(x) = 3x^4 - 14x^3 + 24x - 3$; $g(x) = 2x - 30$; $[-3, 5]$ by $[-80, 30]$

54. $f(x) = \dfrac{1}{x}$; $g(x) = \sqrt{x^2 - 1}$; $[0, 4]$ by $[-1, 3]$

In Exercises 55–58, find a good window setting for the graph of the function. The graph should show all the zeros of the polynomial.

55. $f(x) = x^3 - 22x^2 + 17x + 19$

56. $f(x) = x^4 - 200x^3 - 100x^2$

57. $f(x) = 3x^3 + 52x^2 - 12x - 12$

58. $f(x) = 2x^5 - 24x^4 - 24x + 2$

Solutions to Check Your Understanding 0.4

1. Multiply both sides of the equation by x. Then, $x^2 - 14 = 5x$. Now, take the term $5x$ to the left side of the equation and solve by factoring:

$$x^2 - 5x - 14 = 0$$
$$(x - 7)(x + 2) = 0$$
$$x = 7 \quad \text{or} \quad x = -2$$

2. In this case, each coefficient is a multiple of 7. To simplify the arithmetic, we divide both sides of the equation by 7 before using the quadratic formula:

$$x^2 - 5x + 5 = 0$$
$$\sqrt{b^2 - 4ac} = \sqrt{(-5)^2 - 4(1)(5)} = \sqrt{5}$$
$$x = \frac{-b \pm \sqrt{b^2 - 4ac}}{2a} = \frac{5 \pm \sqrt{5}}{2 \cdot 1} = \frac{5}{2} \pm \frac{1}{2}\sqrt{5}.$$

0.5   Exponents and Power Functions

In this section, we review the operations with exponents that occur frequently throughout the text. We begin with the definition b^n for any nonzero number b and any nonnegative integer n.

Definition	Examples
$b^n = \underbrace{b \cdot b \cdot \,\cdots\, \cdot b}_{n \text{ times}}$	$2^4 = 2 \cdot 2 \cdot 2 \cdot 2 = 16$ $(-5)^3 = (-5)(-5)(-5) = -125$
$b^{-n} = \dfrac{1}{b^n}$	$2^{-4} = \dfrac{1}{2^4} = \dfrac{1}{16}$ $(-5)^{-3} = \dfrac{1}{(-5)^3} = \dfrac{1}{-125} = -\dfrac{1}{125}$
$b^0 = 1$	$2^0 = 1, \quad (-5)^0 = 1$

Next, we consider numbers of the form $b^{1/n}$, where n is a positive integer. For instance,

$2^{1/2}$ is the positive number whose square is 2: $2^{1/2} = \sqrt{2}$

$2^{1/3}$ is the positive number whose cube is 2: $2^{1/3} = \sqrt[3]{2}$

$2^{1/4}$ is the positive number whose fourth power is 2: $2^{1/4} = \sqrt[4]{2}$

and so on. In general, when b is nonnegative, $b^{1/n}$ is zero or the positive number whose nth power is b.

If n is even, there is no number whose nth power is b if b is negative. Therefore, $b^{1/n}$ is not defined if n is even and $b < 0$. When n is odd, we may permit b to be negative as well as positive. For example, $(-8)^{1/3}$ is the number whose cube is -8; that is, $(-8)^{1/3} = -2$.

Thus, when b is negative and n is odd, we again define $b^{1/n}$ to be the number whose nth power is b.

Let's summarize this discussion and introduce fractional powers, such as $b^{m/n}$, where b is any number, m and n are positive integers. We will further assume that m/n is in lowest terms.

Definition	Examples
$b^{\frac{1}{n}}$ or $\sqrt[n]{b}$ denotes the nth root of b. If n is even, then b must be ≥ 0.	$9^{1/2} = \sqrt{9} = 3$ $8^{1/3} = \sqrt[3]{8} = 2$ $(-8)^{1/3} = \sqrt[3]{-8} = -2$
$b^{-\frac{1}{n}} = \dfrac{1}{b^{\frac{1}{n}}} = \dfrac{1}{\sqrt[n]{b}}$ whenever $b^{\frac{1}{n}}$ is defined and $\neq 0$.	$2^{-\frac{1}{2}} = \dfrac{1}{2^{\frac{1}{2}}} = \dfrac{1}{\sqrt{2}}$ $4^{-\frac{1}{2}} = \dfrac{1}{4^{\frac{1}{2}}} = \dfrac{1}{2}$
$b^{m/n} = (b^{1/n})^m$, whenever $b^{1/n}$ is defined	$25^{3/2} = (25^{1/2})^3 = 5^3 = 125$ $(-8)^{2/3} = ((-8)^{1/3})^2 = (-2)^2 = 4$
$b^{-m/n} = \frac{1}{b^{m/n}}$, whenever $b^{m/n}$ is defined and $\neq 0$	$8^{-5/3} = \dfrac{1}{8^{5/3}} = \dfrac{1}{2^5} = \dfrac{1}{32}$

Exponents may be manipulated algebraically according to the following rules:

Laws of Exponents

1. $b^r b^s = b^{r+s}$ Product rule.
2. $b^{-r} = \dfrac{1}{b^r}$ Changing sign of exponents.
3. $\dfrac{b^r}{b^s} = b^r \cdot b^{-s} = b^{r-s}$ Quotient rule.
4. $(b^r)^s = b^{rs}$ Power of a power.
5. $(ab)^r = a^r b^r$ Power of a product.
6. $\left(\dfrac{a}{b}\right)^r = \dfrac{a^r}{b^r}$ Power of a quotient.

EXAMPLE 1   **Laws of Exponents**   Use the laws of exponents to calculate the following quantities.

   (a) $2^{1/2}\, 50^{1/2}$ **(b)** $(2^{1/2}\, 2^{1/3})^6$ **(c)** $\dfrac{5^{3/2}}{\sqrt{5}}$

SOLUTION   **(a)** $2^{1/2}\, 50^{1/2} = (2 \cdot 50)^{1/2}$ Power of a product.

$\qquad\qquad\qquad\quad = \sqrt{100}$

$\qquad\qquad\qquad\quad = 10$

   (b) $(2^{1/2}\, 2^{1/3})^6 = (2^{(1/2)+(1/3)})^6$ Product rule.

$\qquad\qquad\qquad\quad = (2^{5/6})^6$ Add the exponents: $\frac{1}{2} + \frac{1}{3} = \frac{5}{6}$.

$\qquad\qquad\qquad\quad = 2^{(5/6)6}$ Power of a power.

$\qquad\qquad\qquad\quad = 2^5$ Multiply exponent.

$\qquad\qquad\qquad\quad = 32$

   (c) $\dfrac{-5^{3/2}}{\sqrt{5}} = \dfrac{-5^{3/2}}{5^{1/2}}$ Change radical into exponent.

$\qquad\qquad\quad = -5^{(3/2)-(1/2)}$ Quotient rule.

$\qquad\qquad\quad = -5^1$ Subtract exponents.

$\qquad\qquad\quad = -5$

≫ Now Try Exercise 31

In words, the changing sign of exponents rule (Law 2) states that we can move a number raised to a power from denominator to numerator by changing the sign of the power. This law can be combined with other laws of exponents to yield useful formulas.

Law	Examples	In words
7. $\dfrac{a^{-r}}{b^{-s}} = \dfrac{b^s}{a^r}$	$\dfrac{2^{-3}}{5^{-2}} = \dfrac{5^2}{2^3} = \dfrac{25}{8}$	To move a number raised to a power, from numerator to denominator or from denominator to numerator change the sign of its exponents.
8. $\left(\dfrac{a}{b}\right)^{-r} = \left(\dfrac{b}{a}\right)^{r}$	$\left(\dfrac{2}{3}\right)^{-2} = \left(\dfrac{3}{2}\right)^{2}$ $\left(\dfrac{2}{3}\right)^{-1/2} = \left(\dfrac{3}{2}\right)^{1/2} = \sqrt{\dfrac{3}{2}}$	To raise a fraction to a negative power, invert the fraction and change the sign of the exponent.

Verification of Law 7

$$\frac{a^{-r}}{b^{-s}} = \frac{1/a^r}{1/b^s} \qquad \text{Changing sign of exponent law 2.}$$

$$= \frac{1}{a^r} \cdot \frac{b^s}{1} \qquad \text{To divide by } 1/b^s, \text{ we multiply by its inverse.}$$

$$= \frac{b^s}{a^r}$$

Law 8 can be verified similarly. Laws of exponents apply to algebraic expressions as well. In what follows, we use them with expressions such as x^r, where r is a rational number.

EXAMPLE 2   **Simplifying Algebraic Expressions**   Simplify the following expressions.

(a) $\dfrac{1}{x^{-4}}$ (b) $\dfrac{x^2}{x^5}$ (c) $\sqrt{x}\,(x^{3/2} + 3\sqrt{x})$

SOLUTION   (a) $\dfrac{1}{x^{-4}} = x^{-(-4)}$ Changing sign of exponents with $r = -4$.

$= x^4$

(b) $\dfrac{x^2}{x^5} = x^{2-5}$ Quotient rule.

$= x^{-3}$ Subtract exponents.

It is also correct to write this answer as $\dfrac{1}{x^3}$.

(c) $\sqrt{x}\,(x^{3/2} + 3\sqrt{x}) = x^{1/2}(x^{3/2} + 3x^{1/2})$ Change radical into rational exponent $^1/_2$.

$= x^{1/2}x^{3/2} + 3x^{1/2}x^{1/2}$ Distribute.

$= x^{(1/2)+(3/2)} + 3x^{(1/2)+(1/2)}$ Product rule.

$= x^2 + 3x$ Simplify.

» Now Try Exercise 47

Recall from Section 0.2 that a **power function** is a function of the form

$$f(x) = x^r$$

for some number r.

EXAMPLE 3   **Operations with Laws of Exponents** Let $f(x)$ and $g(x)$ be the power functions

$$f(x) = x^{-1} \quad \text{and} \quad g(x) = x^{1/2}.$$

Determine the following functions.

(a) $\dfrac{f(x)}{g(x)}$ (b) $f(x)g(x)$ (c) $\dfrac{g(x)}{f(x)}$

SOLUTION   (a) $\dfrac{f(x)}{g(x)} = \dfrac{x^{-1}}{x^{1/2}} = x^{-1-(1/2)} = x^{-3/2} = \dfrac{1}{x^{3/2}}$

(b) $f(x)g(x) = x^{-1}x^{1/2} = x^{-1+(1/2)} = x^{-1/2} = \dfrac{1}{x^{1/2}} = \dfrac{1}{\sqrt{x}}$

(c) $\dfrac{g(x)}{f(x)} = \dfrac{x^{1/2}}{x^{-1}} = x^{(1/2)-(-1)} = x^{3/2}$

》 Now Try Exercise 75

When factoring fractional powers of x, factor out the smallest exponent and keep in mind that factoring is like dividing: You need to divide each factor by the common factor to find the missing factor.

EXAMPLE 4   **Factoring Fractional Powers** Factor the given expression by factoring out the smallest power of x:

(a) $x^{-\frac{1}{3}} + 2x^{\frac{2}{3}}$ (b) $x^{-\frac{5}{3}} + \dfrac{1}{x^2}$.

SOLUTION   (a) Since the smaller power of x is $-\frac{1}{3}$, we have

$$x^{-\frac{1}{3}} + 2x^{\frac{2}{3}} = x^{-\frac{1}{3}}\left(1 + \frac{2x^{\frac{2}{3}}}{x^{-\frac{1}{3}}}\right)$$ Divide by the common factor to find the missing factor.

$$= x^{-\frac{1}{3}}\left(1 + 2x^{\frac{2}{3}} \cdot x^{\frac{1}{3}}\right)$$ Change sign of exponent.

$$= x^{-\frac{1}{3}}\left(1 + 2x^{\frac{2}{3}+\frac{1}{3}}\right)$$ Product rule.

$$= x^{-\frac{1}{3}}(1 + 2x).$$

(b) $$x^{-\frac{5}{3}} + \frac{1}{x^2} = x^{-\frac{5}{3}} + x^{-2}$$ Write $\frac{1}{x^2}$ as x^{-2}.

$$= x^{-2}\left(\frac{x^{-\frac{5}{3}}}{x^{-2}} + 1\right)$$ Factor out x^{-2}, the smaller of the two exponents.

$$= x^{-2}\left(x^{-\frac{5}{3}} \cdot x^2 + 1\right)$$ Write $\frac{1}{x^{-2}}$ as x^2.

$$= x^{-2}\left(x^{-\frac{5}{3}+2} + 1\right)$$ Product rule.

$$= x^{-2}\left(x^{\frac{1}{3}} + 1\right)$$ Add exponents.

》 Now Try Exercise 83

Compound Interest

The subject of compound interest provides a significant application of exponents. We introduce this topic at this point with a view toward using it as a source of applied problems throughout the book.

When money is deposited in a savings account, interest is paid at stated intervals. If this interest is added to the account and thereafter earns interest itself, then the interest is called **compound interest**. The original amount deposited is called the **principal amount**. The principal amount plus the compound interest is called the **compound amount**. The interval between interest payments is referred to as the **interest period**. In formulas for compound interest, the interest rate is expressed as a decimal rather than a percentage. Thus, 6% is written as .06.

If $1000 is deposited at 6% annual interest, compounded annually, the compound amount at the end of the first year will be

$$A_1 = \underset{\text{principal}}{1000} + \underset{\text{interest}}{1000(.06)} = 1000(1 + .06).$$

At the end of the second year, the compound amount will be

$$A_2 = \underset{\substack{\text{compound}\\\text{amount}}}{A_1} + \underset{\text{interest}}{A_1(.06)} = A_1(1 + .06)$$

$$= [1000(1 + .06)](1 + .06) = 1000(1 + .06)^2.$$

At the end of 3 years,

$$A_3 = A_2 + A_2(.06) = A_2(1 + .06)$$

$$= [1000(1 + .06)^2](1 + .06) = 1000(1 + .06)^3.$$

After n years, the compound amount will be

$$A = 1000(1 + .06)^n.$$

In this example, the interest period was 1 year. The important point to note, however, is that at the end of each interest period, the amount on deposit grew by a factor of $(1 + .06)$. In general, if the interest rate is i instead of .06, the compound amount will grow by a factor of $(1 + i)$ at the end of each interest period.

Simple Compound Interest If a principal amount P is invested at a compound interest rate i per interest period, for a total of n interest periods, the compound amount A at the end of the nth period will be

$$A = P(1 + i)^n. \tag{1}$$

EXAMPLE 5 **Compound Interest** If $5000 is invested at 8% per year, with interest compounded annually, what is the compound amount after 3 years?

SOLUTION Substituting $P = 5000$, $i = .08$, and $n = 3$ into the compound interest formula, we have

$$A = 5000(1 + .08)^3 = 5000(1.08)^3$$

$$= 5000(1.259712) = 6298.56 \text{ dollars.} \qquad \blacktriangleleft\blacktriangleleft$$

It is common practice to state the interest rate as a percentage per year ("per annum"), even though each interest period is often shorter than 1 year. If the annual rate is r and if interest is paid and compounded m times per year, the interest rate i for each period is given by

$$[\text{rate per period}] = i = \frac{r}{m} = \frac{[\text{annual interest rate}]}{[\text{periods per year}]}.$$

Many banks pay interest quarterly. If the stated annual rate is 5%, then $i = .05/4 = .0125$.

> **Compound Interest with Multiple Interest Periods**   If interest is compounded for
> t years, with m interest periods each year, there will be a total of mt interest periods.
> If in formula (1) we replace n by mt and replace i by r/m, we obtain the following
> formula for the compound amount:
>
> $$A = P\left(1 + \frac{r}{m}\right)^{mt}, \tag{2}$$
>
> where P = principal amount
> $\quad r$ = interest rate per annum
> $\quad m$ = number of interest periods per year
> $\quad t$ = number of years.

EXAMPLE 6   **Compound Interest**   Suppose that \$1000 is deposited in a savings account that pays 6%
per annum, compounded quarterly. If no additional deposits or withdrawals are made,
how much will be in the account at the end of 1 year?

SOLUTION   We use the compound interest formula (2) with $P = 1000$, $r = .06$, $m = 4$, and $t = 1$:

$$A = 1000\left(1 + \frac{.06}{4}\right)^4 = 1000(1.015)^4$$

$$\approx 1000(1.06136355) \approx 1061.36 \text{ dollars.}$$

>> *Now Try Exercise 99*

Note that the \$1000 in Example 6 earned a total of \$61.36 in (compound) interest.
This is 6.136% of \$1000. Savings institutions sometimes advertise this rate as the **effective
annual interest rate**. That is, the savings institutions mean that, *if* they paid interest only
once a year, they would have to pay a rate of 6.136% to produce the same earnings as their
6% rate compounded quarterly. The stated rate of 6% is often called the **nominal rate**.

The effective annual rate increases if the interest is compounded more often. Some
savings institutions compound interest monthly or even daily.

EXAMPLE 7   **Monthly Compound**   If the interest in Example 6 were compounded monthly, how much
would be in the account at the end of 1 year? What about the case when 6% annual
interest is compounded daily?

SOLUTION   For monthly compounding, $m = 12$. From the compound interest formula (2), we have

$$A = 1000\left(1 + \frac{.06}{12}\right)^{12} = 1000(1.005)^{12} \approx 1061.68 \text{ dollars.}$$

The effective rate in this case is 6.168%.
A "bank year" usually consists of 360 days (to simplify calculations). So, for daily
compounding, we take $m = 360$. Then,

$$A = 1000\left(1 + \frac{.06}{360}\right)^{360} \approx 1000(1.00016667)^{360}$$

$$\approx 1000(1.06183251) \approx 1061.83 \text{ dollars.}$$

With daily compounding, the effective rate is 6.183%.   «

EXAMPLE 8   **Account Balance as a Function of Time**   Suppose the \$1000 invested in Example 6 is
invested for t years.
(a) Express the account balance $A(t)$ as a function t, the number of years that the principal has been in the account.
(b) Calculate the account balance at the end of 1 year and at the end of 3 years.

SOLUTION (a) We use the compound interest formula (2) with $P = 1000$, $r = .06$, and $m = 4$. We then leave t as an independent variable and express the account balance as $A(t)$, a function of t:

$$A(t) = 1000\left(1 + \frac{.06}{4}\right)^{4t}$$

(b) We use $A(t)$ in part (a) to calculate $A(1)$ and $A(3)$:

$$A(1) = 1000\left(1 + \frac{.06}{4}\right)^{4(1)} \approx \$1061.36;$$

$$A(3) = 1000\left(1 + \frac{.06}{4}\right)^{4(3)} \approx \$1195.62$$

Thus, the balance after one year is \$1061.36 and after 3 years it is \$1195.62. «

In the preceding examples, we assumed that the interest rate, r, was fixed. In fact, r can be thought of as the expected rate of "return" on any investment. For a savings account, this rate may be fixed. For an investment in stock, this rate may be highly variable, even negative.

EXAMPLE 9 **Account Balance as a Function of Return** Suppose the \$5000 in Example 5 is invested for 3 years at rate r per annum, compounded annually.
(a) Express the account balance $A(r)$ as a function of r.
(b) Calculate the account balance for $r = .08$ and $r = .15$

SOLUTION (a) We use formula (2) with $P = 5000$, $m = 1$ and $t = 3$. We leave r as an independent variable and express the account balance $A(r)$ as a function of r:

$$A(r) = 5000(1 + r)^3$$

(b) We use $A(r)$ in part (a) to calculate $A(.08)$ and $A(.15)$:

$$A(.08) = 5000(1 + .08)^3 = \$6298.56$$
$$A(.15) = 5000(1 + .15)^3 \approx \$7604.38$$ «

EXAMPLE 10 **A Zero Coupon Bond** A corporation issues a bond costing \$200 and paying interest compounded monthly. The interest is accumulated until the bond reaches maturity. (A security of this sort is called a **zero coupon bond**.) If after 5 years the bond is worth \$500, what is the annual interest rate?

SOLUTION Let r denote the annual interest rate. The value A of the bond after 5 years $= 60$ months is given by the compound interest formula:

$$A = 200\left(1 + \frac{r}{12}\right)^{60}.$$

We must find r that satisfies

$$500 = 200\left(1 + \frac{r}{12}\right)^{60} \qquad \text{Substitute 500 for } A.$$

$$2.5 = \left(1 + \frac{r}{12}\right)^{60}. \qquad \text{Divide both sides by 200.}$$

Raise both sides to the power $\frac{1}{60}$, and apply the laws of exponents to obtain

$$(2.5)^{1/60} = \left[\left(1 + \frac{r}{12}\right)^{60}\right]^{\frac{1}{60}} = \left(1 + \frac{r}{12}\right)^{60 \cdot \frac{1}{60}} = 1 + \frac{r}{12}$$

$$r = 12 \cdot ((2.5)^{1/60} - 1). \qquad \text{Rearrange terms to isolate } r \text{ on one side.}$$

Using a calculator, we see that $r \approx .18466$. That is, the annual interest rate is 18.466%. (A bond paying a rate of interest this high is generally called a **junk bond**.) «

INCORPORATING TECHNOLOGY

Scientific Notation By default, your TI-83/84-type calculator will display the result of a calculation in 10 digits. For example, if asked to calculate 1/3, your calculator will return .3333333333, and 7/3 returns 2.333333333. In each case, 10 digits are given in the answer.

However, if the answer cannot display 10 digits (or if the absolute value is less than .001), your calculator will express the answer in scientific notation. Scientific notation expresses numbers in two parts. The significant digits display with one digit to the left of the decimal. The appropriate power of 10 displays to the right of E, as in **2.5E-4**. This stands for 2.5×10^{-4} or 0.00025. Similarly, **1E12** stands for 1×10^{12} or 1,000,000,000,000. (*Note:* Multiplying a number by 10^{-4} moves the decimal point four places to the left, and multiplying by 10^{12} moves the decimal point 12 places to the right.)

Check Your Understanding 0.5

Solutions can be found following the section exercises.

1. Compute the following.

(a) -5^2 (b) $16^{0.75}$

2. Simplify the following.

(a) $(4x^3)^2$ (b) $\dfrac{\sqrt[3]{x}}{x^2}$ (c) $\dfrac{2 \cdot (x+5)^6}{x^2 + 10x + 25}$

EXERCISES 0.5

In Exercises 1–28, compute the numbers.

1. 3^3

2. $(-2)^3$

3. 1^{100}

4. 0^{25}

5. $(.1)^4$

6. $(100)^4$

7. -4^2

8. $(.01)^3$

9. $(16)^{1/2}$

10. $(27)^{1/3}$

11. $(.000001)^{1/3}$

12. $\left(\dfrac{1}{125}\right)^{1/3}$

13. 6^{-1}

14. $\left(\dfrac{1}{2}\right)^{-1}$

15. $(.01)^{-1}$

16. $(-5)^{-1}$

17. $8^{4/3}$

18. $16^{3/4}$

19. $(25)^{3/2}$

20. $(27)^{2/3}$

21. $(1.8)^0$

22. $9^{1.5}$

23. $16^{0.5}$

24. $(81)^{0.75}$

25. $4^{-1/2}$

26. $\left(\dfrac{1}{8}\right)^{-2/3}$

27. $(.01)^{-1.5}$

28. $1^{-1.2}$

In Exercises 29–40, use the laws of exponents to compute the numbers.

29. $5^{1/3} \cdot 200^{1/3}$

30. $(3^{1/3} \cdot 3^{1/6})^6$

31. $6^{1/3} \cdot 6^{2/3}$

32. $(9^{4/5})^{5/8}$

33. $\dfrac{10^4}{5^4}$

34. $\dfrac{3^{5/2}}{3^{1/2}}$

35. $(2^{1/3} \cdot 3^{2/3})^3$

36. $20^{0.5} \cdot 5^{0.5}$

37. $\left(\dfrac{8}{27}\right)^{2/3}$

38. $(125 \cdot 27)^{1/3}$

39. $\dfrac{7^{4/3}}{7^{1/3}}$

40. $(6^{1/2})^0$

In Exercises 41–70, use the laws of exponents to simplify the algebraic expressions. Your answer should not involve parentheses or negative exponents.

41. $(xy)^6$

42. $(x^{1/3})^6$

43. $\dfrac{x^4 \cdot y^5}{xy^2}$

44. $\dfrac{1}{x^{-3}}$

45. $x^{-1/2}$

46. $(x^3 \cdot y^6)^{1/3}$

47. $\left(\dfrac{x^4}{y^2}\right)^3$

48. $\left(\dfrac{x}{y}\right)^{-2}$

49. $(x^3 y^5)^4$

50. $\sqrt{1+x}(1+x)^{3/2}$

51. $x^5 \cdot \left(\dfrac{y^2}{x}\right)^3$

52. $x^{-3} \cdot x^7$

53. $(2x)^4$

54. $\dfrac{-3x}{15x^4}$

55. $\dfrac{-x^3 y}{-xy}$

56. $\dfrac{x^3}{y^{-2}}$

57. $\dfrac{x^{-4}}{x^3}$

58. $(-3x)^3$

59. $\sqrt[3]{x} \cdot \sqrt[3]{x^2}$

60. $(9x)^{-1/2}$

61. $\left(\dfrac{3x^2}{2y}\right)^3$

62. $\dfrac{x^2}{x^5 y}$

63. $\dfrac{2x}{\sqrt{x}}$

64. $\dfrac{1}{yx^{-5}}$

65. $(16x^8)^{-3/4}$

66. $(-8y^9)^{2/3}$

67. $\sqrt{x}\left(\dfrac{1}{4x}\right)^{5/2}$

68. $\dfrac{(25xy)^{3/2}}{x^2 y}$

69. $\dfrac{(-27x^5)^{2/3}}{\sqrt[3]{x}}$

70. $(-32y^{-5})^{3/5}$

Let $f(x) = \sqrt[3]{x}$ and $g(x) = \dfrac{1}{x^2}$. Calculate the following functions. Take $x > 0$.

71. $f(x)g(x)$

72. $\dfrac{f(x)}{g(x)}$

73. $\dfrac{g(x)}{f(x)}$

74. $[f(x)]^3 g(x)$

75. $[f(x)g(x)]^3$

76. $\sqrt{\dfrac{f(x)}{g(x)}}$

77. $\sqrt{f(x)g(x)}$

78. $\sqrt[3]{f(x)g(x)}$

79. $f(g(x))$

80. $g(f(x))$

81. $f(f(x))$

82. $g(g(x))$

The expressions in Exercises 83–88 may be factored as shown. Find the missing factors.

83. $\sqrt{x} - \dfrac{1}{\sqrt{x}} = \dfrac{1}{\sqrt{x}}(\quad)$

84. $2x^{2/3} - x^{-1/3} = x^{-1/3}(\quad)$

85. $x^{-1/4} + 6x^{1/4} = x^{-1/4}(\quad)$

86. $\sqrt{\dfrac{x}{y}} - \sqrt{\dfrac{y}{x}} = \sqrt{xy}(\quad)$

87. Explain why $\sqrt{a} \cdot \sqrt{b} = \sqrt{ab}$.

88. Explain why $\sqrt{a}/\sqrt{b} = \sqrt{a/b}$.

In Exercises 89–96, evaluate $f(4)$.

89. $f(x) = x^2$

90. $f(x) = x^3$

91. $f(x) = x^{-1}$

92. $f(x) = x^{1/2}$

93. $f(x) = x^{3/2}$

94. $f(x) = x^{-1/2}$

95. $f(x) = x^{-5/2}$

96. $f(x) = x^0$

Calculate the compound amount from the given data in Exercises 97–104.

97. principal = \$500, compounded annually, 6 years, annual rate = 6%

98. principal = \$700, compounded annually, 8 years, annual rate = 8%

99. principal = \$50,000, compounded quarterly, 10 years, annual rate = 9.5%

100. principal = \$20,000, compounded quarterly, 3 years, annual rate = 12%

101. principal = \$100, compounded monthly, 10 years, annual rate = 5%

102. principal = \$500, compounded monthly, 1 year, annual rate = 4.5%

103. principal = \$1500, compounded daily, 1 year, annual rate = 6%

104. principal = \$1500, compounded daily, 3 years, annual rate = 6%

105. Annual Compound Assume that a couple invests \$1000 upon the birth of their daughter. Assume that the investment earns 6.8% compounded annually. What will the investment be worth on the daughter's 18th birthday?

106. Annual Compound with Deposits Assume that a couple invests \$4000 each year for 4 years in an investment that earns 8% compounded annually. What will the value of the investment be 8 years after the first amount is invested?

107. Quarterly Compound Assume that a \$500 investment earns interest compounded quarterly. Express the value of the investment after 1 year as a polynomial in the annual rate of interest r.

108. Semiannual Compound Assume that a \$1000 investment earns interest compounded semiannually. Express the value of the investment after 2 years as a polynomial in the annual rate of interest r.

109. Velocity When a car's brakes are slammed on at a speed of x miles per hour, the stopping distance is $\frac{1}{20}x^2$ feet. Show that when the speed is doubled the stopping distance increases fourfold.

TECHNOLOGY EXERCISES

In Exercises 110–113, convert the numbers from graphing calculator form to standard form (that is, without E).

110. `5E-5`

111. `8.103E-4`

112. `1.35E13`

113. `8.23E-6`

Solutions to Check Your Understanding 0.5

1. (a) $-5^2 = -25$. [Note that -5^2 is the same as $-(5^2)$. This number is different from $(-5)^2$, which equals 25. Whenever there are no parentheses, apply the exponent first and then apply the other operations.]

(b) Since $.75 = \frac{3}{4}$, $16^{0.75} = 16^{3/4} = (\sqrt[4]{16})^3 = 2^3 = 8$.

2. (a) Apply Law 5 with $a = 4$ and $b = x^3$. Then, use Law 4.

$$(4x^3)^2 = 4^2 \cdot (x^3)^2 = 16 \cdot x^6$$

[A common error is to forget to square the 4. If that had been our intent, we would have asked for $4(x^3)^2$.]

(b) $\dfrac{\sqrt[3]{x}}{x^3} = \dfrac{x^{1/3}}{x^3} = x^{(1/3)-3} = x^{-8/3}$

[The answer can also be given as $1/x^{8/3}$.] When simplifying expressions involving radicals, it is usually a good idea to first convert the radicals to exponents.

(c) $\dfrac{2(x+5)^6}{x^2 + 10x + 25} = \dfrac{2 \cdot (x+5)^6}{(x+5)^2} = 2(x+5)^{6-2}$

$$= 2(x+5)^4$$

[Here, the third law of exponents was applied to $(x + 5)$. The laws of exponents apply to any algebraic expression.]

0.6 Functions and Graphs in Applications

The key step in solving many applied problems in this text is to construct appropriate functions or equations. Once this is done, the remaining mathematical steps are usually straightforward. This section focuses on representative applied problems and reviews skills needed to set up and analyze functions, equations, and their graphs.

Geometric Problems

Many applications involve dimensions, areas, or volumes of objects similar to those in Fig. 1. When a problem involves a plane figure, such as a rectangle or circle, we must distinguish between the *perimeter* and the *area* of the figure. The perimeter of a figure, or "distance around" the figure, is a *length* or *sum of lengths*. Typical units, if specified, are inches, feet, centimeters, meters, and so on. Area involves the *product of two lengths*, and the units are *square* inches, *square* feet, *square* centimeters, and so on.

Rectangle   Rectangular box   Cylinder

Figure 1 Geometric figures.

EXAMPLE 1   **Cost** Suppose that the longer side of the rectangle in Fig. 1 has twice the length of the shorter side, and let x denote the length of the shorter side.
(a) Express the perimeter of the rectangle as a function of x.
(b) Express the area of the rectangle as a function of x.
(c) If the rectangle represents a kitchen countertop to be constructed of a durable material costing \$25 per square foot, write a function $C(x)$ that expresses the cost of the material as a function of x, where lengths are in feet.

SOLUTION   (a) The rectangle is shown in Fig. 2. The length of the longer side is $2x$. If the perimeter is denoted by P, then P is the sum of the lengths of the four sides of the rectangle: $x + 2x + x + 2x$. That is, $P = 6x$.

Figure 2

(b) The area A of the rectangle is the product of the lengths of two adjacent sides. That is, $A = x \cdot 2x = 2x^2$.
(c) Here, the area is measured in square feet. The basic principle for this part is

$$\begin{bmatrix} \text{cost of} \\ \text{materials} \end{bmatrix} = \begin{bmatrix} \text{cost per} \\ \text{square foot} \end{bmatrix} \cdot \begin{bmatrix} \text{number of} \\ \text{square feet} \end{bmatrix}$$

$$C(x) \quad = \quad 25 \quad \cdot \quad 2x^2$$

$$= 50x^2 \text{ (dollars).}$$ **» Now Try Exercise 3**

When a problem involves a three-dimensional object, such as a box or cylinder, we must distinguish between the *surface area* of the object and the *volume* of the object. Surface area is an area, of course, so it is measured in *square* units. Typically, the surface area is a *sum of areas* (each area is a product of two lengths). The volume of an object is often a *product of three lengths* and is measured in *cubic* units.

EXAMPLE 2   **Surface Area** A rectangular box has a square copper base, wooden sides, and a wooden top. The copper costs \$21 per square foot and the wood costs \$2 per square foot.
(a) Write an expression giving the surface area (that is, the sum of the areas of the bottom, the top, and the four sides of the box) in terms of the dimensions of the box. Also, write an expression giving the volume of the box.
(b) Write an expression giving the total cost of the materials used to make the box in terms of the dimensions.

SOLUTION

Figure 3 Closed box.

(a) The first step is to assign variables to the dimensions of the box. Denote the length of one (and therefore every) side of the square base by x, and denote the height of the box by h. (See Fig. 3.)

The top and bottom each have area x^2, and each of the four sides has area xh. Therefore, the surface area is $2x^2 + 4xh$. The volume of the box is the product of the length, width, and height. Because the base is square, the volume is x^2h.

(b) When the various surfaces of the box have different costs per square foot, the cost of each is computed separately:

$$[\text{cost of bottom}] = [\text{cost per sq. ft.}] \cdot [\text{area of bottom}] = 21x^2;$$

$$[\text{cost of top}] = [\text{cost per sq. ft.}] \cdot [\text{area of top}] = 2x^2;$$

$$[\text{cost of one side}] = [\text{cost per sq. ft.}] \cdot [\text{area of one side}] = 2xh.$$

The total cost is

$$C = [\text{cost of bottom}] + [\text{cost of top}] + 4 \cdot [\text{cost of one side}]$$
$$= 21x^2 + 2x^2 + 4 \cdot 2xh = 23x^2 + 8xh.$$ ≪

Business Problems

Many business applications in this text involve cost, revenue, and profit functions.

EXAMPLE 3

Cost A toy manufacturer has fixed costs of $3000 (such as rent, insurance, and business loans) that must be paid no matter how many toys are produced. In addition, there are variable costs of $2 per toy. At a production level of x toys, the variable costs are $2 \cdot x$ (dollars) and the total cost is

$$C(x) = 3000 + 2x \quad \text{(dollars)}.$$

(a) Find the cost of producing 2000 toys.
(b) What additional cost is incurred if the production level is raised from 2000 toys to 2200 toys?
(c) To answer the question "How many toys may be produced at a cost of $5000?" should you compute $C(5000)$ or should you solve the equation $C(x) = 5000$?

SOLUTION

(a) $C(2000) = 3000 + 2(2000) = 7000$ (dollars).
(b) The total cost when $x = 2200$ is $C(2200) = 3000 + 2(2200) = 7400$ (dollars). So, the *increase* in cost when production is raised from 2000 to 2200 toys is

$$C(2200) - C(2000) = 7400 - 7000 = 400 \text{ (dollars)}.$$

(c) This is an important type of question. The phrase "how many toys" implies that the quantity x is unknown. Therefore, we find the answer by solving $C(x) = 5000$ for x:

$$3000 + 2x = 5000$$
$$2x = 2000$$
$$x = 1000 \text{ (toys)}.$$

Another way to analyze this problem is to look at the types of units involved. The input x of the cost function is the *quantity* of toys, and the output of the cost function is the *cost*, measured in dollars. Since the question involves 5000 *dollars*, it is the *output* that is specified. The input x is unknown. **≫ Now Try Exercise 21**

EXAMPLE 4

Cost, Profit, and Revenue The toys in Example 3 sell for $10 apiece. When x toys are sold, the revenue (amount of money received) $R(x)$ is $10x$ dollars. Given the same cost function, $C(x) = 3000 + 2x$, the profit (or loss) $P(x)$ generated by the x toys will be

$$P(x) = R(x) - C(x)$$
$$= 10x - (3000 + 2x) = 8x - 3000.$$

(a) To determine the revenue generated by 8000 toys, should you compute $R(8000)$ or should you solve the equation $R(x) = 8000$?

(b) If the revenue from the production and sale of some toys is \$7000, what is the corresponding profit?

SOLUTION

(a) The revenue is unknown, but the input to the revenue function is known. So compute $R(8000)$ to find the revenue.

(b) The profit is unknown, so we want to compute the value of $P(x)$. Unfortunately, we don't know the value of x. However, the fact that the revenue is \$7000 enables us to solve for x. Thus, the solution has two steps:

(i) Solve $R(x) = 7000$ to find x.

$$10x = 7000$$
$$x = 700 \quad \text{(toys)}$$

(ii) Compute $P(x)$ when $x = 700$.

$$P(x) = 8(700) - 3000$$
$$= 2600 \quad \text{(dollars)} \quad \text{«}$$

Functions and Graphs

When a function arises in an applied problem, the graph of the function provides useful information. Every statement or task involving a function corresponds to a feature or task involving its graph. This "graphical" point of view will broaden your understanding of functions and strengthen your ability to work with them.

Modern graphing calculators and calculus computer software provide excellent tools for thinking geometrically about functions. Most popular graphing calculators and programs provide a **cursor** or **cross hairs** that may be moved to any point on the screen, with the x- and y-coordinates of the cursor displayed somewhere on the screen. The next example shows how geometric calculations with the graph of a function correspond to the more familiar numerical computations. This example is worth reading even if a computer (or calculator) is unavailable.

EXAMPLE 5

Cost To plan for future growth, a company analyzes production costs for one of its products and estimates that the cost (in dollars) of operating at a production level of x units per hour is given by the function

$$C(x) = 150 + 59x - 1.8x^2 + .02x^3.$$

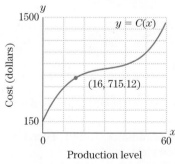

Figure 4 Graph of a cost function.

Suppose that the graph of this function is available, either displayed on the screen of a graphing utility or perhaps printed on graph paper in a company report. (See Fig. 4.)

(a) The point $(16, 715.12)$ is on the graph. What does that say about the cost function $C(x)$?

(b) We can solve the equation $C(x) = 900$ graphically by finding a certain point on the graph and reading its x- and y-coordinates. Describe how to locate the point. How do the coordinates of the point provide the solution to the equation $C(x) = 900$?

(c) You may complete the task "Find $C(45)$" graphically by finding a point on the graph. Describe how to locate the point. How do the coordinates of the point provide the value of $C(45)$?

SOLUTION

(a) The fact that $(16, 715.12)$ is on the graph of $C(x)$ means that $C(16) = 715.12$. That is, if the production level is 16 units per hour, the cost is \$715.12.

(b) To solve $C(x) = 900$ graphically, locate 900 on the y-axis and move to the right until you reach the point $(?, 900)$ on the graph of $C(x)$. (See Fig. 5.) The x-coordinate of the point is the solution of $C(x) = 900$. Estimate x graphically ($x \approx 39.04$).

Figure 5

Figure 6

(c) To find $C(45)$ graphically, locate 45 on the x-axis and move up until you reach the point $(45, ?)$ on the graph of $C(x)$. (See Fig. 6.) The y-coordinate of the point is the value of $C(45)$. [In fact, $C(45) = 982.50$.]   **» Now Try Exercise 41**

The two final examples illustrate how to extract information about a function by examining its graph.

EXAMPLE 6   **Height of a Ball** A ball is thrown straight up into the air from the top of a 64-foot tower. The function $h(t)$, the height of the ball (in feet) after t seconds, has the graph shown in Fig. 7. (*Note*: This graph is not a picture of the physical path of the ball; the ball is thrown vertically into the air.)
(a) What is the height of the ball after 1 second?
(b) After how many seconds does the ball reach its greatest height, and what is this height?
(c) After how many seconds does the ball hit the ground?
(d) When is the height 64 feet?

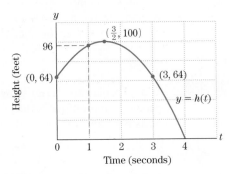

Figure 7 Graph of a height function.

SOLUTION   (a) Since the point $(1, 96)$ is on the graph of $h(t)$, $h(1) = 96$, the height of the ball after 1 second is 96 feet.
(b) The highest point on the graph of the function has coordinates $(\frac{3}{2}, 100)$. Therefore, after $\frac{3}{2}$ seconds the ball achieves its greatest height, 100 feet.
(c) The ball hits the ground when the height is 0. This occurs after 4 seconds.
(d) The height of 64 feet occurs twice, at times $t = 0$ and $t = 3$ seconds.
   » Now Try Exercise 47

EXAMPLE 7   **Revenue Function** Figure 8 shows the graph of the function $R(x)$, the revenue obtained from selling x bicycles.
(a) Estimate the number of bicycles the manufacturer must sell to achieve a revenue of $102,000.

Figure 8 Graph of revenue from bicycle sales.

(b) What is the approximate revenue from the sale of 1000 bicycles? 1100 bicycles?

(c) What additional revenue is derived from the sale of 100 more bicycles if the current sales level is 1000 bicycles?

SOLUTION

(a) Find the approximate location of 102,000 on the y-axis; then scan horizontally to find the point on $R(x)$ that has this y-coordinate (see Fig. 9). From this point, scan vertically to find the x-coordinate, which is approximately 600. The point on the graph is (600, 102,000), meaning that approximately 600 bicycles must be sold to achieve a revenue of $102,000.

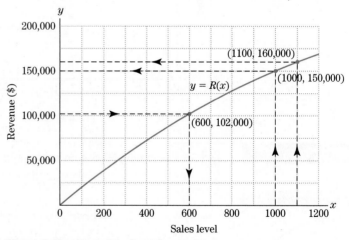

Figure 9 Reading values from the graph.

(b) Find the location of 1000 on the x-axis; then scan vertically to find the point on $R(x)$ that has this x-coordinate (see Fig. 9). From this point, scan horizontally to find the y-coordinate, which is approximately 150,000. The point on the graph is (1000, 150,000), meaning that the approximate revenue from the sale of 1000 bicycles is $150,000. Use a similar process to determine that the approximate revenue from the sale of 1100 bicycles is $160,000.

(c) When the value of x increases from 1000 to 1100, the revenue increases from $150,000 to $160,000. Therefore, the additional revenue increase is $10,000. ◀◀

Table 1 summarizes most of the concepts in Examples 3 through 7. Although stated here for a profit function, the concepts will arise later for many other types of functions as well. Each statement about the profit is translated into a statement about $f(x)$ and a statement about the graph of $f(x)$. The graph in Fig. 10 illustrates each statement.

Table 1   Using Graphs in Applied Problems

Assume that $f(x)$ is the profit in dollars at production level x.

Applied Problem	Function	Graph
When production is at 2 units, the profit is $7.	$f(2) = 7$.	The point $(2, 7)$ is on the graph in Fig. 10.
Determine the number of units that generate a profit of $12.	Solve $f(x) = 12$ for x.	Find the x-coordinate(s) of the point(s) on the graph whose y-coordinate is 12.
Determine the profit when the production level is 4 units.	Evaluate $f(4)$.	Find the y-coordinate of the point on the graph whose x-coordinate is 4.
Find the production level that maximizes the profit.	Find x such that $f(x)$ is as large as possible.	Find the x-coordinate of the highest point, M, on the graph.
Determine the maximum profit.	Find the maximum value of $f(x)$.	Find the y-coordinate of the highest point on the graph.
Determine the change in profit when the production level is changed from 6 to 7 units.	Find $f(7) - f(6)$.	Determine the difference in the y-coordinates of the points with x-coordinates 7 and 6.
The profit decreases when the production level is changed from 6 to 7 units.	The function value decreases when x changes from 6 to 7.	The point on the graph with x-coordinate 6 is higher than the point with x-coordinate 7.

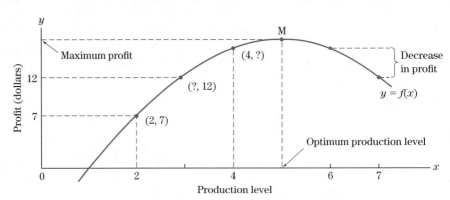

Figure 10 Graph of a profit function.

Figure 11

Figure 12

Approximating Maximum or Minimum Values For a certain commodity, it is found that when x thousands of units are sold, the revenue (in thousands of dollars) is given by

$$R(x) = -x^2 + 6.233x \quad \text{for} \quad 0 \le x \le 6.233.$$

To approximate the number of units that must be sold to maximize revenue, and to find the approximate maximum revenue, we begin by entering the function into the calculator. Press $\boxed{Y=}$ and enter the formula for $R(x)$ for Y_1. Press $\boxed{\text{WINDOW}}$ and set $X\,min = 0$ and $X\,max = 6.233$.

There are two methods of approximating the maximum value. The first method uses the $\boxed{\text{TRACE}}$ command. After entering the function into the calculator, press $\boxed{\text{TRACE}}$. We use the $\boxed{\triangleleft}$ and $\boxed{\triangleright}$ to trace the cursor along the graph. As we do so, the values of X and Y are updated; and by observing the value of X, we conclude that maximum revenue of approximately $9713 is achieved when approximately 3117 units are sold. (See Fig. 11.)

The second method uses the built-in **maximum routine**. We begin by pressing $\boxed{\text{2nd}}$ [CALC] and then $\boxed{4}$ to start the maximum routine. Enter 2 for the left bound, 4 for the right bound, and 3 for the guess. The result is shown in Fig. 12 and also indicates that maximum revenue of approximately $9713 is achieved when approximately 3117 units are sold.

Similarly, we can use $\boxed{\text{TRACE}}$ or **minimum** to find the minimum value of a function.

Check Your Understanding 0.6

Solutions can be found following the section exercises.

Consider the cylinder shown in Fig. 13.

1. Assign variables to the dimensions of the cylinder.

2. The girth of the cylinder is the circumference of the colored circle in the figure. Express the girth in terms of the dimensions of the cylinder.

3. What is the area of the bottom (or top) of the cylinder?

4. What is the surface area of the side of the cylinder? (*Hint*: Imagine cutting the side of the cylinder and unrolling the cylinder to form a rectangle.)

Figure 13

EXERCISES 0.6

In Exercises 1–6, assign variables to the dimensions of the geometric object.

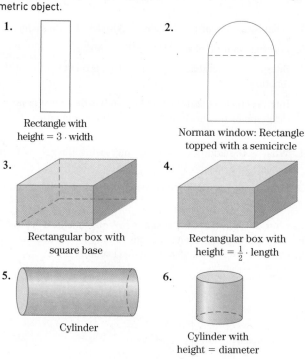

1.

Rectangle with height = 3 · width

2.

Norman window: Rectangle topped with a semicircle

3.

Rectangular box with square base

4.

Rectangular box with height = $\frac{1}{2}$ · length

5.

Cylinder

6.

Cylinder with height = diameter

Exercises 7–14 refer to the letters assigned to the figures in Exercises 1–6.

7. **Perimeter, Area** Consider the rectangle in Exercise 1. Write an expression for the perimeter. If the area is 25 square feet, write this fact as an equation.

8. **Perimeter, Area** Consider the rectangle in Exercise 1. Write an expression for the area. Write an equation expressing the fact that the perimeter is 30 centimeters.

9. **Area, Circumference** Consider a circle of radius r. Write an expression for the area. Write an equation expressing the fact that the circumference is 15 centimeters.

10. **Perimeter, Area** Consider the Norman window of Exercise 2. Write an expression for the perimeter. Write an equation expressing the fact that the area is 2.5 square meters.

11. **Volume, Surface Area** Consider the rectangular box in Exercise 3, and suppose that it has no top. Write an expression for the volume. Write an equation expressing the fact that the surface area is 65 square inches.

12. **Surface Area, Volume** Consider the closed rectangular box in Exercise 4. Write an expression for the surface area. Write an equation expressing the fact that the volume is 10 cubic feet.

13. **Volume, Surface Area, Cost** Consider the cylinder of Exercise 5. Write an equation expressing the fact that the volume is 100 cubic inches. Suppose that the material to construct the left end costs $5 per square inch, the material to construct the right end costs $6 per square inch, and the material to construct the side costs $7 per square inch. Write an expression for the total cost of material for the cylinder.

14. **Surface Area, Volume** Consider the cylinder of Exercise 6. Write an equation expressing the fact that the surface area is 30π square inches. Write an expression for the volume.

15. **Fencing a Rectangular Corral** Consider a rectangular corral with a partition down the middle, as shown in Fig. 14. Assign letters to the outside dimensions of the corral. Write an equation expressing the fact that 5000 feet of fencing is needed to construct the corral (including the partition). Write an expression for the total area of the corral.

Figure 14

16. **Fencing a Rectangular Corral** Consider a rectangular corral with two partitions, as in Fig. 15. Assign letters to the outside dimensions of the corral. Write an equation expressing the fact that the corral has a total area of 2500 square feet. Write an expression for the amount of fencing needed to construct the corral (including both partitions).

Figure 15

17. **Cost of Fencing** Consider the corral of Exercise 16. If the fencing for the boundary of the corral costs $10 per foot and the fencing for the inner partitions costs $8 per foot, write an expression for the total cost of the fencing.

18. Cost of an Open Box Consider the rectangular box of Exercise 3. Assume that the box has no top, the material needed to construct the base costs $5 per square foot, and the material needed to construct the sides costs $4 per square foot. Write an equation expressing the fact that the total cost of materials is $150. (Use the dimensions assigned in Exercise 11.)

19. If the rectangle in Exercise 1 has a perimeter of 40 cm, find the area of the rectangle.

20. If the cylinder in Exercise 6 has a volume of 54π cubic inches, find the surface area of the cylinder.

21. Cost A specialty shop prints custom slogans and designs on T-shirts. The shop's total cost at a daily sales level of x T-shirts is $C(x) = 73 + 4x$ dollars.
(a) At what sales level will the cost be $225?
(b) If the sales level is at 40 T-shirts, how much will the cost rise if the sales level changes to 50 T-shirts?

22. Cost, Revenue, Profit A college student earns income by typing term papers on a computer, which she leases (along with a printer). The student charges $4 per page for her work, and she estimates that her monthly cost when typing x pages is $C(x) = .10x + 75$ dollars.
(a) What is the student's profit if she types 100 pages in 1 month?
(b) Determine the change in profit when the typing business rises from 100 to 101 pages per month.

23. Profit A frozen yogurt stand makes a profit of $P(x) = .40x - 80$ dollars when selling x scoops of yogurt per day.
(a) Find the breakeven sales level, that is, the level at which $P(x) = 0$.
(b) What sales level generates a daily profit of $30?
(c) How many more scoops of yogurt will have to be sold to raise the daily profit from $30 to $40?

24. Profit A cellular telephone company estimates that, if it has x thousand subscribers, its monthly profit is $P(x)$ thousand dollars, where $P(x) = 12x - 200$.
(a) How many subscribers are needed for a monthly profit of 160 thousand dollars?
(b) How many new subscribers would be needed to raise the monthly profit from 160 to 166 thousand dollars?

25. Cost, Revenue, Profit An average sale at a small florist shop is $21, so the shop's weekly revenue function is $R(x) = 21x$, where x is the number of sales in 1 week. The corresponding weekly cost is $C(x) = 9x + 800$ dollars.
(a) What is the florist shop's weekly profit function?
(b) How much profit is made when sales are at 120 per week?
(c) If the profit is $1000 for 1 week, what is the revenue for the week?

26. Cost, Revenue, Profit A catering company estimates that, if it has x customers in a typical week, its expenses will be approximately $C(x) = 550x + 6500$ dollars, and its revenue will be approximately $R(x) = 1200x$ dollars.
(a) How much profit will the company earn in 1 week when it has 12 customers?
(b) How much profit is the company making each week if the weekly costs are running at a level of $14,750?

Exercises 27–32 refer to the function $f(r)$, which gives the cost (in cents) of constructing a 100-cubic-inch cylinder of radius r inches. The graph of $f(r)$ is shown in Fig. 16.

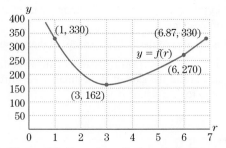

Figure 16 Cost of a cylinder.

27. What is the cost of constructing a cylinder of radius 6 inches?

28. For what value(s) of r is the cost 330 cents?

29. Interpret the fact that the point (3, 162) is on the graph of the function.

30. Interpret the fact that the point (3, 162) is the lowest point on the graph of the function. What does this say in terms of cost versus radius?

31. What is the additional cost of increasing the radius from 3 inches to 6 inches?

32. How much is saved by increasing the radius from 1 inch to 3 inches?

Exercises 33–36 refer to the cost and revenue functions in Fig. 17. The cost of producing x units of goods is $C(x)$ dollars and the revenue from selling x units of goods is $R(x)$ dollars.

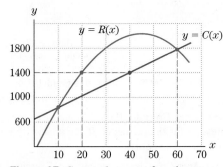

Figure 17 Cost and revenue functions.

33. What are the revenue and cost from the production and sale of 30 units of goods?

34. At what level of production is the revenue $1400?

35. At what level of production is the cost $1400?

36. What is the profit from the manufacture and sale of 30 units of goods?

Exercises 37–40 refer to the cost function in Fig. 18.

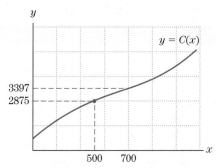

Figure 18 A cost function.

37. The point (1000, 4000) is on the graph of the function. Restate this fact in terms of the function $C(x)$.

38. Translate the task "solve $C(x) = 3500$ for x" into a task involving the graph of the function.

39. Translate the task "find $C(400)$" into a task involving the graph.

40. If 500 units of goods are produced, estimate the cost of producing 100 more units of goods.

Exercises 41–44 refer to the profit function in Fig. 19.

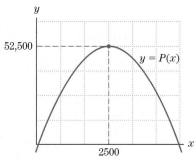

Figure 19 A profit function.

41. The point (2500, 52,500) is the highest point on the graph of the function. What does this say in terms of profit versus quantity?

42. The point (1500, 42,500) is on the graph of the function. Restate this fact in terms of the function $P(x)$.

43. Translate the task "solve $P(x) = 30,000$" into a task involving the graph of the function.

44. Translate the task "find $P(2000)$" into a task involving the graph.

Height of a Ball A ball is thrown straight up into the air. The function $h(t)$ gives the height of the ball (in feet) after t seconds. In Exercises 45–50, translate the task into both a statement involving the function and a statement involving the graph of the function.

45. Find the height of the ball after 3 seconds.

46. Find the time at which the ball attains its greatest height.

47. Find the greatest height attained by the ball.

48. Determine when the ball will hit the ground.

49. Determine when the height of the ball is 100 feet.

50. Find the height of the ball when it is first released.

TECHNOLOGY EXERCISES

51. Height of a Ball A ball thrown straight up into the air has height $-16x^2 + 80x$ feet after x seconds.
   (a) Graph the function in the window

   $$[0, 6] \text{ by } [-30, 120].$$

   (b) What is the height of the ball after 3 seconds?
   (c) At what times will the height be 64 feet?
   (d) At what time will the ball hit the ground?
   (e) When will the ball reach its greatest height? What is that height?

52. Cost The daily cost (in dollars) of producing x units of a certain product is given by the function

   $$C(x) = 225 + 36.5x - .9x^2 + .01x^3.$$

   (a) Graph $C(x)$ in the window $[0, 70]$ by $[-400, 2000]$.
   (b) What is the cost of producing 50 units of goods?
   (c) Consider the situation as in part (b). What is the additional cost of producing one more unit of goods?
   (d) At what production level will the daily cost be $510?

53. Revenue from Sales A store estimates that the total revenue (in dollars) from the sale of x bicycles per year is given by the function $R(x) = 250x - .2x^2$.
   (a) Graph $R(x)$ in the window

   $$[200, 500] \text{ by } [42000, 75000].$$

   (b) What sales level produces a revenue of $63,000?
   (c) What revenue is received from the sale of 400 bicycles?
   (d) Consider the situation of part (c). If the sales level were to decrease by 50 bicycles, by how much would revenue fall?
   (e) The store believes that, if it spends $5000 in advertising, it can raise the total sales from 400 to 450 bicycles next year. Should it spend the $5000? Explain your conclusion.

Solutions to Check Your Understanding 0.6

1. Let r be the radius of the circular base and let h be the height of the cylinder.

2. Girth $= 2\pi r$ (the circumference of the circle).

3. Area of the bottom $= \pi r^2$.

4. The cylinder is a rolled-up rectangle of height h and base $2\pi r$ (the circumference of the circle). The area is $2\pi rh$. (See Fig. 20.)

Figure 20 Unrolled side of a cylinder.

CHAPTER 0 Summary

KEY TERMS AND CONCEPTS	EXAMPLES

0.1 Functions and Their Graphs

A **rational number** is one that may be represented as a repeating decimal or as a fraction.

An **irrational number** has an infinite decimal representation with no repeating pattern.

Inequalities can be used to describe intervals on the real **number line**.

1. Rational numbers: $\frac{3}{4} = .75$, $\frac{1}{3} = .333\ldots$

Irrational numbers:

$\sqrt{3} = 1.73205\ldots$, $\sqrt{\pi} = 1.77245\ldots$

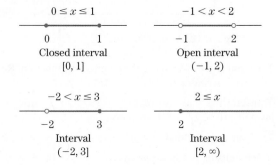

$0 \le x \le 1$	$-1 < x < 2$
Closed interval [0, 1]	Open interval (−1, 2)
$-2 < x \le 3$	$2 \le x$
Interval (−2, 3]	Interval [2, ∞)

A **function** is a rule that assigns to each member of one set (called the **domain**) exactly one member of another set (called the **range**).

Given a function $y = f(x)$, for x in the domain of f, you can plot the point $(x, f(x))$. The set of all such points $(x, f(x))$ is called the **graph of the function** $f(x)$.

2. The domain of $f(x) = \dfrac{1}{\sqrt{x-1}}$ consists of all x such that $x - 1 > 0$ or $x > 1$. The domain is $(1, \infty)$.

3. To graph the parabola $f(x) = x^2 + 4x - 1$,

   (a) Make a table of values $\left(\text{include the vertex at } x = -\dfrac{b}{2a}\right)$.

   (b) Plot the points (x, y) from your table.

   (c) Connect the points with a smooth graph.

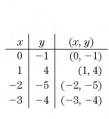

x	y	(x, y)
0	−1	(0, −1)
1	4	(1, 4)
−2	−5	(−2, −5)
−3	−4	(−3, −4)

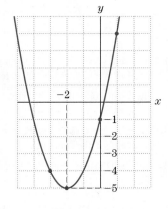

The **vertical line test** tells us that a graph is the graph of a function if, and only if, each vertical line touches the graph at no more than one point.

4.

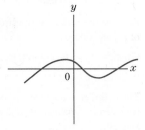

Not the graph of a function   Graph of a function

KEY TERMS AND CONCEPTS	EXAMPLES

0.2 Some Important Functions

The **graph of a linear function** is a straight line written in the form $Ax + By = C$.

When $B \neq 0$, you can solve for y and put the equation in the form $y = mx + b$.

1. The linear equation $3x + 2y = 7$ can be put in the form $y = mx + b$ as follows:

$$3x + 2y = 7$$

$$y = -\frac{3}{2}x + \frac{7}{2}.$$

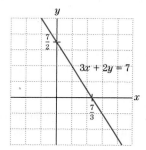

Piecewise-defined functions are defined by more than one formula.

2. The function $y = \begin{cases} 2 - x & \text{if } x > 1 \\ 3 + 2x & \text{if } x \leq 1 \end{cases}$

is defined piecewise. To plot it, we use the formula $2 - x$ on the interval $x > 1$, and the formula $3 + 2x$ on the interval $x \leq 1$.

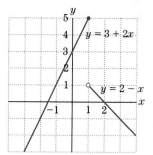

The graph of a **quadratic function** $f(x) = ax^2 + bx + c \ (a \neq 0)$ is a parabola with vertex at $\left(-\frac{b}{2a}, f\left(-\frac{b}{2a}\right)\right)$. The parabola opens upward if $a > 0$ and downward if $a < 0$.

3. The parabola $f(x) = x^2 - 3x + 1$ opens upward ($a = 1 > 0$) and has its vertex at $\left(-\frac{b}{2a}, f\left(-\frac{b}{2a}\right)\right) = \left(\frac{3}{2}, f\left(\frac{3}{2}\right)\right) = \left(\frac{3}{2}, -\frac{5}{4}\right)$. The parabola $g(x) = -2x^2 + 4x$ opens downward ($a = -2 < 0$) and has its vertex at $\left(-\frac{b}{2a}, f\left(-\frac{b}{2a}\right)\right) = (1, f(1)) = (1, 2)$.

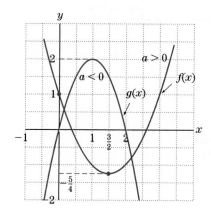

KEY TERMS AND CONCEPTS	EXAMPLES

A function expressed as the quotient of two polynomials is called a **rational function**. The domain of the function is restricted to those values for which the denominator is not 0.

4. The domain of the function $f(x) = \frac{x}{x-2}$ consists of all $x \neq 2$.

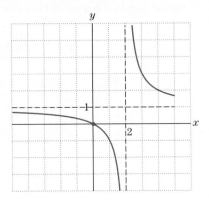

A **power function** is of the form x^r, where $r \neq 0$ is a real number. The domain of the function is restricted to those values for which the formula can be evaluated.

5. The domain of the function $f(x) = x^{1/2} = \sqrt{x}$ is restricted to all $x \geq 0$, because we cannot take the square root of a negative number. The function $g(x) = x^{1/3} = \sqrt[3]{x}$ is defined for all x.

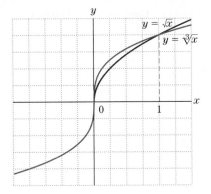

An **absolute value function** is described by means of the absolute value sign.

6. The absolute value functions $f(x) = |x|$ and $g(x) = |x + 2|$.

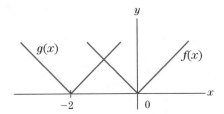

0.3 The Algebra of Functions

Given two functions $f(x)$ and $g(x)$ and constant k, you can perform **algebraic operations on functions** as illustrated by the following:

Sum: $f(x) + g(x)$

Difference: $f(x) - g(x)$

Product: $f(x)g(x)$

Quotient: $\dfrac{f(x)}{g(x)}$

For the functions $f(x) = 3x - 2$ and $g(x) = \dfrac{1}{2x - 1}$:

$$f(x) - g(x) = (3x - 2) - \frac{1}{2x - 1}$$
$$= \frac{(3x - 2)(2x - 1)}{2x - 1}$$
$$ - \frac{1}{2x - 1}$$
$$= \frac{6x^2 - 3x - 4x + 2 - 1}{2x - 1}$$
$$= \frac{6x^2 - 7x + 1}{2x - 1}$$

$$f(x)g(x) = (3x - 2) \cdot \frac{1}{2x - 1}$$
$$= \frac{3x - 2}{2x - 1}$$

$$\frac{f(x)}{g(x)} = (3x - 2) \div \frac{1}{2x - 1}$$
$$= (3x - 2) \cdot \frac{2x - 1}{1}$$
$$= (3x - 2)(2x - 1)$$
$$= 6x^2 - 7x + 2$$

KEY TERMS AND CONCEPTS	EXAMPLES

In addition to algebraic operations, you can perform the following important operation:

Composition: $f(g(x))$ or $g(f(x))$

$$f(g(x)) = 3[g(x)] - 2 \qquad g(f(x)) = \frac{1}{2f(x) - 1}$$

$$= \frac{3}{2x - 1} - 2 \qquad = \frac{1}{2(3x - 2) - 1}$$

$$= \frac{3}{2x - 1} - 2\frac{(2x - 1)}{2x - 1} \qquad = \frac{1}{6x - 5}.$$

$$= \frac{-4x + 5}{2x - 1}$$

0.4 Zeros of Functions—The Quadratic Formula and Factoring

A **zero of a function** is a value of x for which $f(x) = 0$. On a graph, the zeros of $f(x)$ correspond to an x-intercept.

If $f(x)$ is a polynomial of the second degree, you can find the zeros of f by applying the quadratic formula: The solutions of

$$ax^2 + bx + c = 0 \qquad (a \neq 0)$$

are given by $x = \dfrac{-b \pm \sqrt{b^2 - 4ac}}{2a}$.

1. The parabola $f(x) = x^2 + x - 1$ has its zeros at

$$x = \frac{-1 \pm \sqrt{1 - 4(1)(-1)}}{2} = \frac{-1 \pm \sqrt{5}}{2}$$

$$x_1 = \frac{-1 + \sqrt{5}}{2} \approx .62; \quad x_2 = \frac{-1 - \sqrt{5}}{2} \approx -1.62.$$

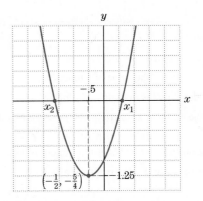

For a polynomial of degree two or higher, we can find the zeros by factoring and using the fact that $x = a$ is a zero if, and only if, $(x - a)$ is a factor of the polynomial.

2. To find the zeros of $f(x) = x^3 - 7x + 6$, we first note that $x = 1$ is a root of the equation $x^3 - 7x + 6 = 0$. This implies that $(x - 1)$ is a factor of $x^3 - 7x + 6$. Factoring, we find

$$x^3 - 7x + 6 = (x - 1)(x^2 + x - 6)$$
$$= (x - 1)(x - 2)(x + 3).$$

Thus, the zeros of f are $x = 1, 2, -3$.

0.5 Exponents and Power Functions

To perform operations on **power functions**, we use the laws of exponents and simplify.

Let $f(x) = \sqrt{x}$, $g(x) = 2\sqrt[3]{x}$, and $h(x) = \dfrac{1}{x^6}$. In terms of powers of x, we have $f(x) = x^{1/2}$, $g(x) = 2x^{1/3}$, and $h(x) = x^{-6}$. We use the laws of exponents to simplify the following operations:

$$f(x)g(x) = x^{1/2}2x^{1/3}$$
$$= 2x^{\frac{1}{2} + \frac{1}{3}} = 2x^{\frac{5}{6}}$$

$$\frac{f(x)}{g(x)} = \frac{x^{1/2}}{2x^{1/3}} \qquad f(g(x)) = [g(x)]^{1/2}$$

$$= \frac{1}{2}x^{1/2}x^{-1/3} \qquad = (2x^{1/3})^{1/2}$$

$$= \frac{1}{2}x^{\frac{1}{2} - \frac{1}{3}} \qquad = 2^{1/2}(x^{1/3})^{1/2}$$

$$= \frac{1}{2}x^{\frac{1}{6}} \qquad = 2^{1/2}x^{\frac{1}{2} \cdot \frac{1}{3}}$$

$$= 2^{1/2}x^{\frac{1}{6}}$$

KEY TERMS AND CONCEPTS	EXAMPLES
0.6 Functions and Graphs in Applications	
The key step in solving many applied problems is to construct appropriate functions or equations that model the quantity under consideration.	If the problem involves a geometric object, you can model the problem by following the outlined steps:

    (a) Draw a figure that represents the geometric object.

    (b) Label the figure with the given data and assign letters to quantities that may vary.

    (c) Derive an equation that represents the quantity of interest in terms of the variables in step 2.

CHAPTER 0   Fundamental Concept Check Exercises

1. Explain the relationships and differences among real numbers, rational numbers, and irrational numbers.

2. What are the four types of inequalities, and what do they each mean?

3. What is the difference between an open interval and a closed interval from a to b?

4. What is a function?

5. What is meant by "the value of a function at x"?

6. What is meant by the domain and range of a function?

7. What is the graph of a function, and how is it related to vertical lines?

8. What is a linear function? Constant function? Give examples.

9. What are the x- and y-intercepts of a function, and how are they found?

10. What is a quadratic function? What shape does its graph have?

11. Define and give an example of each of the following types of functions.
    (a) quadratic function   (b) polynomial function
    (c) rational function   (d) power function

12. What is meant by the absolute value of a number?

13. What five operations on functions are discussed in this chapter? Give an example of each.

14. What is a zero of a function?

15. Give two methods for finding the zeros of a quadratic function.

16. State the six laws of exponents.

17. In the formula $A = P(1 + i)^n$, what do A, P, i, and n represent?

18. Explain how to solve $f(x) = b$ geometrically from the graph of $y = f(x)$.

19. Explain how to find $f(a)$ geometrically from the graph of $y = f(x)$.

CHAPTER 0   Review Exercises

1. Let $f(x) = x^3 + \dfrac{1}{x}$. Evaluate $f(1)$, $f(3)$, $f(-1)$, $f(-\tfrac{1}{2})$, and $f(\sqrt{2})$.

2. Let $f(x) = 2x + 3x^2$. Evaluate $f(0)$, $f(-\tfrac{1}{4})$, and $f(1/\sqrt{2})$.

3. Let $f(x) = x^2 - 2$. Evaluate $f(a - 2)$.

4. Let $f(x) = [1/(x + 1)] - x^2$. Evaluate $f(a + 1)$.

Determine the domains of the following functions.

5. $f(x) = \dfrac{1}{x(x + 3)}$

6. $f(x) = \sqrt{x - 1}$

7. $f(x) = \sqrt{x^2 + 1}$

8. $f(x) = \dfrac{1}{\sqrt{3x}}$

9. Is the point $\left(\tfrac{1}{2}, -\tfrac{3}{5}\right)$ on the graph of the function $h(x) = (x^2 - 1)/(x^2 + 1)$?

10. Is the point $(1, -2)$ on the graph of the function $k(x) = x^2 + (2/x)$?

Factor the polynomials in Exercises 11–14.

11. $5x^3 + 15x^2 - 20x$

12. $3x^2 - 3x - 60$

13. $18 + 3x - x^2$

14. $x^5 - x^4 - 2x^3$

15. Find the zeros of the quadratic function $y = 5x^2 - 3x - 2$.

16. Find the zeros of the quadratic function $y = -2x^2 - x + 2$.

17. Find the points of intersection of the curves $y = 5x^2 - 3x - 2$ and $y = 2x - 1$.

18. Find the points of intersection of the curves $y = -x^2 + x + 1$ and $y = x - 5$.

Let $f(x) = x^2 - 2x$, $g(x) = 3x - 1$, and $h(x) = \sqrt{x}$. Find the following functions.

19. $f(x) + g(x)$

20. $f(x) - g(x)$

21. $f(x)h(x)$

22. $f(x)g(x)$

23. $f(x)/h(x)$

24. $g(x)h(x)$

Let $f(x) = x/(x^2 - 1)$, $g(x) = (1 - x)/(1 + x)$, and $h(x) = 2/(3x + 1)$. Express the following as rational functions.

25. $f(x) - g(x)$

26. $f(x) - g(x + 1)$

27. $g(x) - h(x)$

28. $f(x) + h(x)$

29. $g(x) - h(x - 3)$

30. $f(x) + g(x)$

Let $f(x) = x^2 - 2x + 4$, $g(x) = 1/x^2$, and $h(x) = 1/(\sqrt{x} - 1)$. Determine the following functions.

31. $f(g(x))$

32. $g(f(x))$

33. $g(h(x))$

34. $h(g(x))$

35. $f(h(x))$

36. $h(f(x))$

37. Simplify $(81)^{3/4}$, $8^{5/3}$, and $(.25)^{-1}$.

38. Simplify $(100)^{3/2}$ and $(.001)^{1/3}$.

39. Carbon Monoxide Levels The population of a city is estimated to be $750 + 25t + .1t^2$ thousand people t years from the present. Ecologists estimate that the average level of carbon monoxide in the air above the city will be $1 + .4x$ ppm (parts per million) when the population is x thousand people. Express the carbon monoxide level as a function of the time t.

40. Advertising The revenue $R(x)$ (in thousands of dollars) that a company receives from the sale of x thousand units is given by $R(x) = 5x - x^2$. The sales level x is in turn a function $f(d)$ of the number d of dollars spent on advertising, where

$$f(d) = 6\left(1 - \frac{200}{d + 200}\right).$$

Express the revenue as a function of the amount spent on advertising.

In Exercises 41–44, use the laws of exponents to simplify the algebraic expressions.

41. $(\sqrt{x + 1})^4$

42. $\dfrac{xy^3}{x^{-5}y^6}$

43. $\dfrac{x^{3/2}}{\sqrt{x}}$

44. $\sqrt[3]{x}\,(8x^{2/3})$

45. Monthly Compound Suppose that $15,000 is deposited in a savings account that pays 4% per annum, compounded monthly, for t years.
 (a) Express the account balance $A(t)$ as a function of t, the number of years that the principal has been in the account.
 (b) Calculate the account balance at the end of 2 years and at the end of 5 years.

46. Biannual Compound Suppose that $7000 is deposited in a savings account that pays 9% per annum, compounded biannually, for t years.
 (a) Express the account balance $A(t)$ as a function of t, the number of years that the principal has been in the account.
 (b) Calculate the account balance at the end of 10 years and at the end of 20 years.

47. Varying the Rate of Change Suppose that $15,000 is deposited in a savings account that pays a rate of interest r per annum, compounded annually, for 10 years.
 (a) Express the account balance $A(r)$ as a function of r.
 (b) Calculate the account balance for $r = 0.04$ and $r = 0.06$.

48. Varying the Rate of Change Suppose that $7000 is deposited in a savings account that pays a rate of interest r per annum, compounded annually, for 20 years.
 (a) Express the account balance $A(r)$ as a function of r.
 (b) Calculate the account balance for $r = .07$ and $r = .12$.

The Derivative

We are all familiar with the notion of slope of a line and its importance when analyzing linear functions. Geometrically, the slope tells us whether a line is rising or falling. In Fig. 1, L_1 has a positive slope, while L_2 has a negative slope. The slope measures steepness and tells us how fast a line is rising or falling. In Fig. 2, L_1 has a smaller slope than L_2.

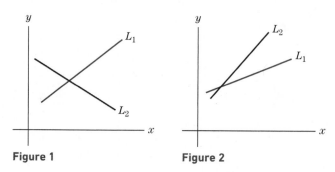

Figure 1   **Figure 2**

In applied problems, the slope represents the rate of change of a linear function. If $C(x) = 3x + 12$ is the cost in dollars of manufacturing x items of a certain product, then the slope 3 tells us that cost rises at the rate of 3 dollars per item (Fig. 3).

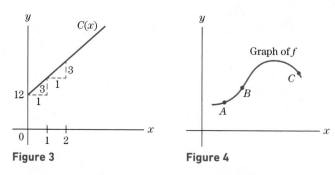

Figure 3   **Figure 4**

How can we measure the rate of change when the function is not linear? In fact, we already have an intuitive understanding of what the rate of change should be. Look at the graph of the function f in Fig. 4. It is geometrically clear that the graph is rising at the points A and B and falling at the point C. It is also clear that the graph is rising faster, or is steeper, at B than at A. With this concept of steepness in mind, reading the graph of f suggests that the rate of change of f is positive at A and B and negative at C, and that the rate of change is larger at B than at A. In this chapter, we explore an important idea of calculus, which establishes a correspondence between the steepness of the graph and the rate of change of the function. To measure the steepness of the graph at a particular point, we will introduce the *derivative*, which is a fundamental tool of calculus. By studying the derivative, we will be able to deal numerically with rates of change in applied problems.

1.1   The Slope of a Straight Line

As we shall see later, the study of straight lines is crucial for the study of the steepness of curves. So, this section is devoted to a discussion of the geometric and algebraic properties of straight lines. Since a vertical line is not the graph of a function (it fails the vertical line test), we focus our discussion on nonvertical lines.

Equations of Nonvertical Lines   A nonvertical line L has an equation of the form

$$y = mx + b. \tag{1}$$

The number m is called the **slope** of L and the point $(0, b)$ is called the **y-intercept**. The equation (1) is called the **slope–intercept equation of L**.

If we set $x = 0$, we see that $y = b$, so $(0, b)$ is on the line L. Thus, the y-intercept tells us where the line L crosses the y-axis. The slope measures the steepness of the line. In Fig. 1, we give three examples of lines with slope $m = 2$. In Fig. 2, we give three examples of lines with slope $m = -2$.

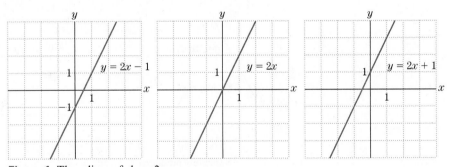

Figure 1 Three lines of slope 2.

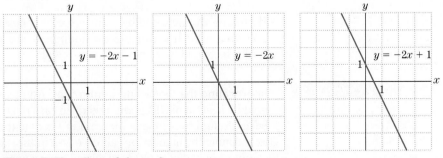

Figure 2 Three lines of slope −2.

To conceptualize the meaning of slope, think of walking along a line from left to right. On lines of positive slope, we will be walking uphill; the greater the slope, the steeper the ascent. On lines of negative slope, we will be walking downhill; the more negative the slope, the steeper the descent. Walking on lines of zero slope corresponds to walking on level ground. In Fig. 3, we have graphed lines with $m = 3, 1, \frac{1}{3}, 0, -\frac{1}{3}, -1, -3$, all having $b = 0$. The reader can readily verify our conceptualization of slope for these lines.

The slope and y-intercept of a straight line often have physical interpretations, as the following three examples illustrate.

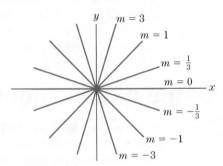

Figure 3

EXAMPLE 1   **Fixed and Marginal Costs** A manufacturer finds that the total cost of producing x units of a commodity is $2x + 1000$ dollars. What is the economic significance of the y-intercept and the slope of the line $y = 2x + 1000$? (See Fig. 4.)

SOLUTION   The y-intercept is $(0, 1000)$. In other words, when $x = 0$ (no units produced), the cost is still $y = 1000$ dollars. The number 1000 represents the *fixed costs* of the manufacturer—those overhead costs, such as rent and insurance, that must be paid no matter how many items are produced.

The slope of the line is 2. This number represents the cost of producing each additional unit. To see this, we can calculate some typical costs.

Quantity Produced	Total Cost
$x = 1500$	$y = 2(1500) + 1000 = 4000$
$x = 1501$	$y = 2(1501) + 1000 = 4002$
$x = 1502$	$y = 2(1502) + 1000 = 4004$

Figure 4 A cost function.

Each time x is increased by 1, the value of y increases by 2. The number 2 is called the **marginal cost**. It is equal to the additional cost that is incurred when the production level is increased by 1 unit, from x to $x + 1$. ≪

EXAMPLE 2   **Interpreting the Slope and y-Intercept** An apartment complex has a storage tank to hold its heating oil. The tank was filled on January 1, but no more deliveries of oil will be made until sometime in March. Let t denote the number of days after January 1, and let y denote the number of gallons of oil in the tank. Current records of the apartment complex show that y and t are related approximately by the equation

$$y = 30,000 - 400t.$$

What interpretation can be given to the y-intercept and slope of this line? (See Fig. 5.)

SOLUTION

The y-intercept is (0, 30,000). This value of y corresponds to $t = 0$, so there were 30,000 gallons of oil in the tank on January 1. Let us examine how fast the oil is removed from the tank.

Days after January 1	Gallons of Oil in the Tank
$t = 0$	$y = 30{,}000 - 400(0) = 30{,}000$
$t = 1$	$y = 30{,}000 - 400(1) = 29{,}600$
$t = 2$	$y = 30{,}000 - 400(2) = 29{,}200$
$t = 3$	$y = 30{,}000 - 400(3) = 28{,}800$
$\vdots$	$\vdots$

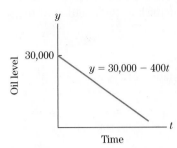

Figure 5 Amount of heating oil in a tank.

The oil level in the tank drops by 400 gallons each day; that is, the oil is being used at the rate of 400 gallons per day. The slope of the line is -400. Thus, the slope gives the rate at which the level of oil in the tank is changing. The negative sign on the -400 indicates that the oil level is decreasing rather than increasing. **≪**

EXAMPLE 3

Depreciation of Equipment For tax purposes, businesses are allowed to regard equipment as decreasing in value (or depreciating) each year. The amount of depreciation may be taken as an income tax deduction. If the value y of a piece of equipment x years after its purchase is given by

$$y = 500{,}000 - 50{,}000x,$$

interpret the y-intercept and the slope of the graph.

SOLUTION

The y-intercept is (0, 500,000) and corresponds to the value of y when $x = 0$. That is, the y-intercept gives the original value, $500,000, of the equipment. The slope indicates the rate at which the equipment is changing in value. Thus, the value of the equipment is decreasing at the rate of $50,000 per year. **≫ Now Try Exercise 47**

Properties of the Slope of a Line

Let us now review several useful properties of the slope of a straight line.

> **Plotting a Line Using the Slope and a Point** If we start at a point on a line of slope m and move 1 unit to the right, then we must move m units in the y-direction in order to return to the line.
>
>

> **Slope Formula** We can compute the slope of a line by knowing two points on the line. If (x_1, y_1) and (x_2, y_2) are on the line, the slope of the line is
>
> $$m = \frac{y_2 - y_1}{x_2 - x_1}.$$

Point–Slope Form The equation of a line can be obtained if we know the slope and one point on the line. If the slope is m and if (x_1, y_1) is on the line, the equation of the line is

$$y - y_1 = m(x - x_1).$$

This equation is called the **point–slope form** of the equation of the line.

Parallel Lines Distinct lines of the same slope are parallel. Conversely, if two non-vertical lines are parallel, they have the same slope.

Perpendicular Lines When two lines are perpendicular, excluding the case of the vertical and horizontal line, the product of their slopes is -1.

FOR REVIEW

Two lines with slopes m_1 and m_2 are perpendicular if and only if $m_1 \cdot m_2 = -1$, or $m_2 = -\dfrac{1}{m_1}$. Also a horizontal line (slope 0) is perpendicular to a vertical line (no slope).

Calculations Involving Slope of a Line

EXAMPLE 4 **Finding the Slope and y-Intercept** Find the slope and the y-intercept of the line whose equation is $2x + 3y = 6$.

SOLUTION We put the equation into slope–intercept form by solving for y in terms of x.

$$3y = -2x + 6$$

$$y = -\frac{2}{3}x + 2$$

The slope is $-\frac{2}{3}$ and the y-intercept is $(0, 2)$.   **» Now Try Exercise 5**

EXAMPLE 5 **Sketching a Line Given a Point and a Slope** Sketch the graph of the line
(a) passing through $(2, -1)$ with slope 3,
(b) passing through $(2, 3)$ with slope $-\frac{1}{2}$.

SOLUTION We follow the method for plotting a line using the slope. (See Fig. 6.) In each case, we begin at the given point, move 1 unit to the right, and then move m units in the y-direction (upward for positive m, downward for negative m). The new point reached will also be on the line. Draw the straight line through these two points.   **» Now Try Exercise 29**

  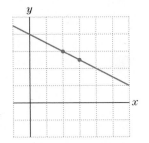

Figure 6

EXAMPLE 6	**Slope of Line through Two Points** Find the slope of the line passing through the points $(6, -2)$ and $(9, 4)$.

SOLUTION   We use the slope formula with $(x_1, y_1) = (6, -2)$ and $(x_2, y_2) = (9, 4)$. Then,

$$\frac{y_2 - y_1}{x_2 - x_1} = \frac{4 - (-2)}{9 - 6} = \frac{6}{3} = 2.$$

Thus, the slope is 2. [We would have reached the same answer if we had let $(x_1, y_1) = (9, 4)$ and $(x_2, y_2) = (6, -2)$.] The slope is just the difference of the y-coordinates divided by the difference of the x-coordinates, with each difference formed in the same order.   **≪**

EXAMPLE 7	**Equation of Line Given a Point and a Slope** Find an equation of the line passing through $(-1, 2)$ with slope 3.

SOLUTION   We let $(x_1, y_1) = (-1, 2)$ and $m = 3$, and we use the point–slope form. The equation of the line is

$$y - 2 = 3[x - (-1)]$$
$$y - 2 = 3(x + 1).$$ **≫ Now Try Exercise 19**

EXAMPLE 8	**Equation of Line through Two Points** Find an equation of the line passing through the points $(1, -2)$ and $(2, -3)$.

SOLUTION   By the slope formula, the slope of the line is

$$\frac{-3 - (-2)}{2 - 1} = \frac{-3 + 2}{1} = -1.$$

Since $(1, -2)$ is on the line, we can use the point–slope form to get the equation of the line:

$$y - (-2) = (-1)(x - 1)$$
$$y + 2 = -x + 1$$
$$y = -x - 1.$$ **≫ Now Try Exercise 11**

EXAMPLE 9	**Equation of Line Parallel to a Given Line** Find an equation of the line passing through $(5, 3)$ parallel to the line $2x + 5y = 7$.

SOLUTION   We first find the slope of the line $2x + 5y = 7$.

$$2x + 5y = 7$$
$$5y = 7 - 2x$$
$$y = -\frac{2}{5}x + \frac{7}{5}$$

The slope of this line is $-\frac{2}{5}$. By the parallel lines property, any line parallel to this line will also have slope $-\frac{2}{5}$. Using the given point $(5, 3)$ and the point–slope form, we get the desired equation:

$$y - 3 = -\frac{2}{5}(x - 5).$$

This equation can also be written as

$$y = -\frac{2}{5}x + 5.$$ **≫ Now Try Exercise 23**

The Slope as a Rate of Change

Given a linear function $y = L(x) = mx + b$, as x varies from x_1 to x_2, y varies from $y_1 = L(x_1)$ to $y_2 = L(x_2)$. The *rate of change* of $y = L(x)$ over the interval from x_1 to x_2 is the change in y divided by the change in x:

$$\left[\begin{array}{c} \text{rate of change of } y = L(x) \\ \text{over the interval from } x_1 \text{ to } x_2 \end{array} \right] = \frac{[\text{Change in } y]}{[\text{Change in } x]}$$

$$= \frac{y_2 - y_1}{x_2 - x_1} = m,$$

by the slope formula. Thus, the rate of change of a linear function over any interval is constant and equal to the slope m. As we did in Example 2, we will refer to the slope as *the* rate of change of the function, without mentioning an underlying interval. The fact that the rate of change of a linear function is constant characterizes linear functions (see Exercise 65) and has many important applications.

EXAMPLE 10 **Temperature in a City** As a cold front moved through a midwestern city, the temperature dropped at the rate of 3°F per hour between noon and 8 P.M. Express the temperature as a function of time and find the temperature at noon, given that at 1 P.M. the temperature was 47°F.

SOLUTION Let t denote time measured in hours since noon and $T(t)$ the temperature of the town at time t. The fact that the temperature was dropping at a (constant) rate of 3°F per hour tells us that the temperature is a linear function whose graph is falling. Thus, its slope is $m = -3$, and so $T(t) = -3t + b$. To determine b, we use the fact that at 1 P.M. the temperature was 47°F or $T(1) = 47$:

$$(-3)(1) + b = 47$$
$$b = 50.$$

So $T(t) = -3t + 50$. The graph of the temperature is shown in Fig. 7. To determine the temperature at noon, we set $t = 0$ and get $T(0) = 50°F$. **≫ Now Try Exercise 61**

Figure 7 Temperature in a city.

Note the importance of the sign of the slope. An increasing linear function has a positive rate of change and thus a positive slope (Example 1). A decreasing linear function has a negative rate of change and thus a negative slope (Example 10).

INCORPORATING TECHNOLOGY

At the end of each section in Chapter 0, we reviewed in detail many very useful techniques for graphing functions using a TI-83/84 type of calculator. All those techniques can be used to analyze the graphs of nonlinear functions. We encourage you to review those sections before attempting to solve the Technology Exercises for this section.

Check Your Understanding 1.1

Solutions can be found following the section exercises.

Find the slopes of the following lines.

1. The line whose equation is $x = 3y - 7$

2. The line going through the points $(2, 5)$ and $(2, 8)$

EXERCISES 1.1

Find the slopes and y-intercepts of the following lines.

1. $y = 3 - 7x$

2. $y = \dfrac{3x + 1}{5}$

3. $x = 2y - 3$

4. $y = 6$

5. $y = \dfrac{x}{7} - 5$

6. $4x + 9y = -1$

Find an equation of the given line.

7. Slope is -1; $(7, 1)$ on line

8. Slope is 2; $(1, -2)$ on line

9. Slope is $\frac{1}{2}$; $(2, 1)$ on line

10. Slope is $\frac{7}{3}$; $(\frac{1}{4}, -\frac{2}{5})$ on line

11. $(\frac{5}{7}, 5)$ and $(-\frac{5}{7}, -4)$ on line

12. $(\frac{1}{2}, 1)$ and $(1, 4)$ on line

13. $(0, 0)$ and $(1, 0)$ on line

14. $(-\frac{1}{2}, -\frac{1}{7})$ and $(\frac{2}{3}, 1)$ on line

15. Horizontal through $(2, 9)$

16. x-intercept is 1; y-intercept is -3

17. x-intercept is $-\pi$; y-intercept is 1

18. Slope is 2; x-intercept is -3

19. Slope is -2; x-intercept is -2

20. Horizontal through $(\sqrt{7}, 2)$

21. Parallel to $y = x$; $(2, 0)$ on line

22. Parallel to $x + 2y = 0$; $(1, 2)$ on line

23. Parallel to $y = 3x + 7$; x-intercept is 2

24. Parallel to $y - x = 13$; y-intercept is 0

25. Perpendicular to $y + x = 0$; $(2, 0)$ on line

26. Perpendicular to $y = -5x + 1$; $(1, 5)$ on line

In Exercises 27–30, we specify a line by giving the slope and one point on the line. Start at the given point and use plotting a line using a slope and a point to sketch the graph of the line.

27. $m = 1$, $(1, 0)$ on line

28. $m = \frac{1}{2}$, $(-1, 1)$ on line

29. $m = -\frac{1}{3}$, $(1, -1)$ on line

30. $m = 0$, $(0, 2)$ on line

31. Each of lines (A), (B), (C), and (D) in the figure is the graph of one of the equations (a), (b), (c), and (d). Match each equation with its graph.

(a) $x + y = 1$

(b) $x - y = 1$

(c) $x + y = -1$

(d) $x - y = -1$

(A)   (B)

(C)   (D)

32. The line through the points $(-1, 2)$ and $(3, b)$ is parallel to $x + 2y = 0$. Find b.

In Exercises 33–36, refer to a line of slope m. If you begin at a point on the line and move h units in the x-direction, how many units must you move in the y-direction to return to the line?

33. $m = \frac{1}{3}$, $h = 3$

34. $m = 2$, $h = \frac{1}{2}$

35. $m = -3$, $h = .25$

36. $m = \frac{2}{3}$, $h = \frac{1}{2}$

In Exercises 37 and 38, we specify a line by giving the slope and one point on the line. We give the first coordinate of some points on the line. Without deriving the equation of the line, find the second coordinate of each point.

37. Slope is 2, $(1, 3)$ on line; $(2, \ )$; $(3, \ )$; $(0, \ )$.

38. Slope is -3, $(2, 2)$ on line; $(3, \ )$; $(4, \ )$; $(1, \ )$.

39. If $f(x)$ is a linear function, $f(1) = 0$, and $f(2) = 1$, what is $f(3)$?

40. Is the line through the points (3, 4) and (−1, 2) parallel to the line $2x + 3y = 0$? Justify your answer.

For each pair of lines in the following figures, determine the one with the greater slope.

41.

42.

Find the equation and sketch the graph of the following lines.

43. With slope −2 and y-intercept (0, −1)

44. With slope $\frac{1}{3}$ and y-intercept (0, 1)

In Exercises 45 and 46, two lines intersect the graph of a function $y = f(x)$ as shown in the figure. Find a and $f(a)$.

45.

46.

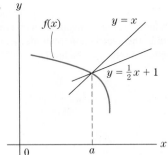

47. Marginal Cost Let $C(x) = 12x + 1100$ denote the total cost (in dollars) of manufacturing x units of a certain commodity per day.
  (a) What is the total cost if the production is set at 10 units per day?

(b) What is the marginal cost?
  (c) Use (b) to determine the additional cost of raising the daily production level from 10 to 11 units.

48. Refer to Exercise 47. Use the formula for $C(x)$ to show directly that $C(x + 1) − C(x) = 12$. Interpret your result as it pertains to the marginal cost.

49. Price of Gasoline The price of 1 gallon of unleaded gasoline at the pump dropped to \$2.19 on January 1, 2015, and continued to fall at the rate of 4 cents per month for the next 9 months. Express the price of 1 gallon of unleaded gasoline as a function of time for the period starting January 1, 2015. What was the price of 15 gallons of gasoline on April 1, 2015? On September 1, 2015?

50. Impact of Mad Cow Disease on Canadian Beef Exports The discovery of one case of bovine spongiform encephalopathy, or mad cow disease, in May 2003 in Canada led to an immediate ban on all Canadian beef exports. At the beginning of September 2003, the ban was lifted, and exports of Canadian beef rose at a steady rate of \$42.5 million per month. Express the value of the monthly exports of Canadian beef as a function of time for the period starting the first day of September 2003. What was the value of the monthly exports at the end of December 2003, when presumably the exports regained their normal level? (*Source: International Trade Division Statistics, Canada.*)

51. Cost of Shipping and Handling An online bookstore charges \$5 plus 3% of the purchase price of books for shipping and handling. Find a function $C(x)$ that expresses the shipping and handling charge for a book order that costs x dollars.

52. Quit Ratio In industry, the relationship between wages and the quit ratio of employees is defined to be the percentage of employees that quit within 1 year of employment. The quit ratio of a large restaurant chain that paid its employees the minimum hourly wage (\$7.25 per hour) was .2 or 20 employees per 100. When the company raised the hourly wage to \$8, the quit ratio dropped to .18, or 18 employees per 100.
  (a) Assuming a linear relationship between the quit ratio $Q(x)$ and the hourly wage x, find an expression for $Q(x)$.
  (b) What should the hourly wage be for the quit ratio to drop to 10 employees per 100?

53. Price Affects Sales When the owner of a gas station sets the price of 1 gallon of unleaded gasoline at \$2.10, she can sell approximately 1500 gallons per day. When she sets the price at \$2.25 per gallon, she can sell approximately 1250 gallons per day. Let $G(x)$ denote the number of gallons of unleaded gasoline sold per day when the price is x dollars. Assume that $G(x)$ is a linear function of x. Approximately how many gallons will be sold per day if the price is set at \$2.34 per gallon?

54. Refer to Exercise 53. Where should the owner set the price if she wants to sell 2200 gallons per day?

55. Marginal Cost Analysis A company manufactures and sells fishing rods. The company has a fixed cost of \$1500 per day and a total cost of \$2200 per day when the production is set at 100 rods per day. Assume that the total cost $C(x)$ is linearly related to the daily production level x.

(a) Express the total cost as a function of the daily production level.

(b) What is the marginal cost at production level $x = 100$?

(c) What is the additional cost of raising the daily production level from 100 to 101 rods? Answer this question in two different ways: (1) by using the marginal cost and (2) by computing $C(101) - C(100)$.

56. **Interpreting the Slope and y-Intercept** A salesperson's weekly pay depends on the volume of sales. If she sells x units of goods, her pay is $y = 5x + 60$ dollars. Give an interpretation of the slope and the y-intercept of this straight line.

57. **Interpreting the Slope and y-Intercept** The demand equation for a manufacturer is $y = -.02x + 7$, where x is the number of units produced and y is the price. That is, to sell x units of goods, the price must be $y = -.02x + \$7$. Interpret the slope and y-intercept of this line.

58. **Converting Fahrenheit to Celsius** Temperatures of 32°F and 212°F correspond to temperatures of 0°C and 100°C. The linear equation $y = mx + b$ converts Fahrenheit temperatures to Celsius temperatures. Find m and b. What is the Celsius equivalent of 98.6°F?

59. **Intravenous Injection** A drug is administered to a patient through an IV (intravenous) injection at the rate of 6 milliliters (mL) per minute. Assuming that the patient's body already contained 1.5 mL of this drug at the beginning of the infusion, find an expression for the amount of the drug in the body x minutes from the start of the infusion.

60. Refer to Exercise 59. If the patient's body eliminates the drug at the rate of 2 mL per hour, find an expression for the amount of the drug in the body x minutes from the start of the infusion.

61. **Diver's Ascent** After inspecting a sunken ship at a depth of 212 feet, a diver starts her slow ascent to the surface of the ocean, rising at the rate of 2 feet per second. Find $y(t)$, the depth of the diver, measured in feet from the ocean's surface, as a function of time t (in seconds).

62. **Diver's Ascent** The diver in the previous exercise is supposed to stop for 5 minutes and decompress at 150 feet depth. Assuming that the diver will continue her ascent, after decompressing, at the same rate of 2 feet per second, find $y(t)$ in this case and determine how long it will take the diver to reach the surface of the ocean.

63. **Sale of T-Shirts** A T-shirt shop owner has a fixed cost of $230 and a marginal cost of $7 per T-shirt to manufacture x T-shirts per day. Let $C(x)$ denote the cost to manufacture x T-shirts per day.

(a) Find $C(x)$.

(b) If the shop owner decides to sell the T-shirts at $12 each, find $R(x)$, the total revenue from selling x T-shirts per day.

64. **Breakeven** In order for a business to break even, revenue has to equal cost. Determine the minimum number of T-shirts that should be sold in the previous exercise to break even.

65. If, for some constant m,

$$\frac{f(x_2) - f(x_1)}{x_2 - x_1} = m$$

for all $x_1 \neq x_2$, show that $f(x) = mx + b$, where b is some constant. [*Hint:* Fix x_1 and take $x = x_2$; then, solve for $f(x)$.]

66. (a) Draw the graph of any function $f(x)$ that passes through the point (3, 2).

(b) Choose a point to the right of $x = 3$ on the x-axis and label it $3 + h$.

(c) Draw the straight line through the points $(3, f(3))$ and $(3 + h, f(3 + h))$.

(d) What is the slope of this straight line (in terms of h)?

67. **Urban World Population** Let y denote the percentage of the world population that is urban x years after 2014. According to data from the United Nations, 54 percent of the world's population was urban in 2014, and projections show that this percentage will increase to 66 percent by 2050. Assume that y is a linear function of x since 2014.

(a) Determine y as a function of x.

(b) Interpret the slope as a rate of change.

(c) Find the percentage of the world's population that is urban in 2020.

(d) Determine the year in which 72% of the world's population will be urban.

TECHNOLOGY EXERCISES

68. Let y denote the average amount claimed for itemized deductions on a tax return reporting x dollars of income. According to Internal Revenue Service data, y is a linear function of x. Moreover, in a recent year income tax returns reporting $20,000 of income averaged $729 in itemized deductions, while returns reporting $50,000 averaged $1380.

(a) Determine y as a function of x.

(b) Graph this function in the window [0, 75000] by [0, 2000].

(c) Give an interpretation of the slope in applied terms.

(d) Determine graphically the average amount of itemized deductions on a return reporting $75,000.

(e) Determine graphically the income level at which the average itemized deductions are $1600.

(f) If the income level increases by $15,000, by how much do the average itemized deductions increase?

Solutions to Check Your Understanding 1.1

1. We solve for y in terms of x.

$$y = \frac{1}{3}x + \frac{7}{3}$$

The slope of the line is the coefficient of x, that is, $\frac{1}{3}$.

2. The line passing through these two points is a vertical line; therefore, its slope is undefined.

1.2   **The Slope of a Curve at a Point**

To extend the concept of slope from straight lines to more general curves, we must first discuss the notion of the tangent line to a curve at a point.

We have a clear idea of what is meant by the tangent line to a circle at a point P. It is the straight line that touches the circle at just the one point P. Let us focus on the region near P, designated by the dashed rectangle shown in Fig. 1. The enlarged portion of the circle looks almost straight, and the straight line that it resembles is the tangent line. Further enlargements would make the circle near P look even straighter and have an even closer resemblance to the tangent line. In this sense, the tangent line to the circle at the point P is the straight line through P that *best approximates* the circle near P. In particular, the tangent line at P reflects the steepness of the circle at P. Thus, it seems reasonable to define the *slope* of the circle at P to be the slope of the tangent line at P.

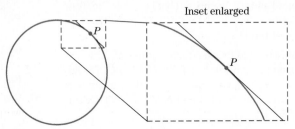

Figure 1 Enlarged portion of a circle.

Similar reasoning leads us to a suitable definition of slope for an arbitrary curve at a point P. Consider the three curves drawn in Fig. 2. We have drawn an enlarged version of the dashed box around each point P. Notice that the portion of each curve lying in the boxed region looks almost straight. If we further magnify the curve near P, it would appear even straighter. Indeed, if we apply higher and higher magnification, the portion of the curve near P would approach a certain straight line more and more exactly. (See Fig. 3.) This straight line is called the *tangent line to the curve at P*. This line best approximates the curve near P and leads us to the following definition.

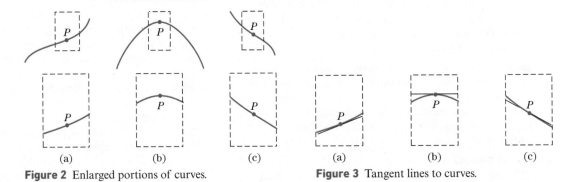

(a)   (b)   (c)   (a)   (b)   (c)

Figure 2 Enlarged portions of curves.   **Figure 3** Tangent lines to curves.

> **DEFINITION Slope of a Curve**   The **slope of a curve at a point** P is defined to be the slope of the tangent line to the curve at P.

Let us consider some examples and applications involving slopes of curves and tangent lines. Later, in the next section, we shall describe a process for constructing the tangent line and computing its slope.

EXAMPLE 1   **Slope of a Graph**   The graph of $f(x) = x^2$ and the tangent line at the point $P = (1, 1)$ are shown in Fig. 4. Find the slope of the graph at P.

SOLUTION

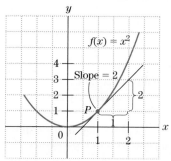

Figure 4 Tangent line and slope at the point P.

The slope of the graph at a point is by definition the slope of the tangent line at that point. Figure 4 shows that the tangent line at P rises 2 units for each unit change in x. Thus, the slope of the tangent line at $P = (1, 1)$ is

$$[\text{slope of tangent line at } P] = \frac{[\text{change in } y]}{[\text{change in } x]} = \frac{2}{1} = 2$$

(see Fig. 4), and hence the slope of the graph at P is 2.   **» Now Try Exercise 3**

Slope of a Graph as a Rate of Change

We have just defined the slope of a curve at a point P to be the slope of the tangent line to the curve at P. Now, the portion of the curve near P can be, at least within an approximation, replaced by the tangent line at P. Since the slope of a linear function measures its rate of change, we obtain the following important interpretation of the slope of a curve.

> **Slope of a Curve as a Rate of Change**   The slope of a curve at a point P—that is, the slope of the tangent line at P—measures the rate of change of the curve as it passes through P.

EXAMPLE 2   **Slope as a Rate of Change**   In Example 2 of Section 1.1, the apartment complex used approximately 400 gallons of oil per day. Suppose that we keep a continuous record of the oil level in the storage tank. The graph for a typical 2-day period appears in Fig. 5. What is the physical significance of the slope of the graph at the point P?

SOLUTION   The curve near P is closely approximated by its tangent line. So, think of the curve as being replaced by its tangent line near P. Then, the slope at P is just the rate of decrease of the oil level at 7 A.M. on March 5.

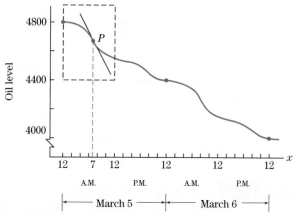

Figure 5 Oil level in a storage tank.   **«**

Notice that during the entire day of March 5, the graph in Fig. 5 seems to be the steepest at 7 A.M. That is, the oil level is falling the fastest at that time. This corresponds to the fact that most people awaken around 7 A.M., turn up their thermostats, take showers, and so on. You can estimate the rate of consumption of oil at 7 A.M. by estimating the slope of the tangent line at P in Fig. 5.

EXAMPLE 3   **Slope of a Graph**   In Fig. 6, we show an enlarged version of a portion of the graph designated by a dashed rectangle in Fig. 5. Estimate the slope of the graph at the point P and give an interpretation of your result.

SOLUTION

The slope of the graph at P is by definition the slope of the tangent line at P. From Fig. 6 we see that the tangent line at P falls approximately 250 units for a 4-unit change in x. So the slope of the graph at P is

$$[\text{slope of tangent line at } P] = \frac{[\text{change in } y]}{[\text{change in } x]} = \frac{-250}{4} = -62.5.$$

Thus, the oil level at 7 A.M. is decreasing at the rate of 62.5 gallons per hour.

≫ Now Try Exercise 27

Figure 6 Oil level in a storage tank around 7 A.M.

Examples 2 and 3 provide typical illustrations of the manner in which slopes can be interpreted as rates of change. We shall return to this important idea in later sections.

Slope Formulas

We know that the slope of a straight line is constant and does not depend on where we are on the line. This is clearly not the case with other curves. In general, the slope of a curve at a point will depend on the point. We saw in Example 1 that the slope of the tangent line to the graph of $y = x^2$ at the point $(1, 1)$ is 2. In fact, we can show that the slope of the tangent line to the graph of $y = x^2$ at the point $(3, 9)$ is 6; and the slope at $\left(-\frac{5}{2}, \frac{25}{4}\right)$ is -5. The various tangent lines are shown in Fig. 7. As we will demonstrate in the next section, in calculus we can usually compute the slopes by using formulas. For the parabola $y = x^2$, we notice that the slope at each point is two times the x-coordinate of the point. This is a general fact for this graph, which is expressed by the following *slope formula*:

$$[\text{slope of the graph of } y = x^2 \text{ at the point } (x, y)] = 2x$$

(see Fig. 8). This simple formula will be derived in the next section. For now, we will use it to get further insight into the topics of slopes of curves and tangent lines.

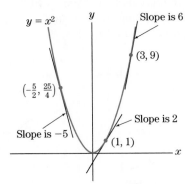

Figure 7 Graph of $y = x^2$.

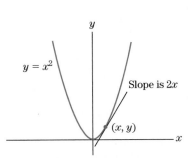

Figure 8 Slope of tangent line to $y = x^2$.

EXAMPLE 4

Using a Slope Formula
(a) What is the slope of the graph of $y = x^2$ at the point $\left(\frac{3}{4}, \frac{9}{16}\right)$?
(b) Write the equation of the tangent line to the graph of $y = x^2$ at the point $\left(\frac{3}{4}, \frac{9}{16}\right)$.

SOLUTION

(a) The x-coordinate of $\left(\frac{3}{4}, \frac{9}{16}\right)$ is $\frac{3}{4}$, so the slope of $y = x^2$ at this point is $2\left(\frac{3}{4}\right) = \frac{3}{2}$.
(b) We shall write the equation of the tangent line in the point–slope form. The point is $\left(\frac{3}{4}, \frac{9}{16}\right)$ and the slope is $\frac{3}{2}$ by part (a). Hence, the equation is

$$y - \frac{9}{16} = \frac{3}{2}\left(x - \frac{3}{4}\right).$$

≫ Now Try Exercise 13

INCORPORATING
TECHNOLOGY

▦ **Approximating the Slope of a Graph at a Point** The slope of a graph at a point can be approximated by zooming in on the graph. In Fig. 9 we have the graph of the curve $y = 3^x$, with window setting ZDecimal, obtained with ZOOM 4. We now zoom in

Figure 9

on the point (0, 1) by hitting ⌐TRACE⌐ to find the point (0, 1) and then ⌐ZOOM⌐ ⌐2⌐ ⌐ENTER⌐ to zoom in. If we zoom in close enough (Fig. 10), the graph starts to look like a straight line. We approximate the slope of this line by measuring the ratio of rise over run in the window in Fig. 10(b), and find

$$\frac{3^{0.1} - 3^{-0.1}}{.1 - (-.1)} = \frac{1.116 - 0.896}{.2} = \frac{.22}{.2} = 1.1.$$

(a)

(b)

Figure 10

Hence, the slope of the curve at the point (0, 1) is approximately 1.1. (We will learn how to calculate this slope in Chapter 4. To find a point on the graph, enter ⌐TRACE⌐ and then enter the *x*-value.)

Check Your Understanding 1.2

Solutions can be found following the section exercises.

1. Refer to Fig. 11.
   (a) What is the slope of the curve at (3, 4)?
   (b) What is the equation of the tangent line at the point where *x* = 3?

2. What is the equation of the tangent line to the graph of $y = \frac{1}{2}x + 1$ at the point (4, 3)?

Figure 11

EXERCISES 1.2

Estimate the slope of each of the following curves at the designated point *P*.

1.

2.

3.

4.

5.

6.

7.

8.
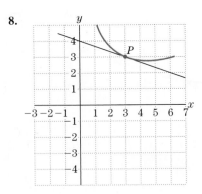

Exercises 9–12 refer to the points in Fig. 12. Assign one of the following descriptors to each point: large positive slope, small positive slope, zero slope, small negative slope, large negative slope.

9. A and B **10.** C and D **11.** E and F

12. Assign a value from the set $\{-6, -\frac{1}{2}, 0, 1, 8\}$ to the slope of the graph at each point in Figure 12.

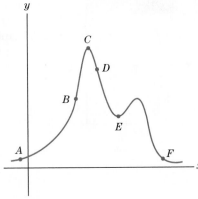

Figure 12

In Exercises 13–20, find the slope of the tangent line to the graph of $y = x^2$ at the point indicated and then write the corresponding equation of the tangent line.

13. $(-.4, .16)$ **14.** $(-2, 4)$ **15.** $\left(\frac{1}{3}, \frac{1}{9}\right)$ **16.** $(-1.5, 2.25)$

17. Find the slope of the tangent line to the graph of $y = x^2$ at the point where $x = -\frac{1}{4}$.

18. Find the slope of the tangent line to the graph of $y = x^2$ where $x = -.2$.

19. Write the equation of the tangent line to the graph of $y = x^2$ at the point where $x = 2.5$.

20. Find the equation of the tangent line to $y = x^2$ at the point where $x = 2.1$.

21. Find the point on the graph of $y = x^2$ where the curve has slope $\frac{7}{2}$.

22. Find the point on the graph of $y = x^2$ where the curve has slope -6.

23. Find the point on the graph of $y = x^2$ where the tangent line is parallel to the line $2x + 3y = 4$.

24. Find the point on the graph of $y = x^2$ where the tangent line is parallel to the line $3x - 2y = 2$.

25. Price of Crude Oil Figure 13 shows the price of 1 barrel of crude oil on the New York Stock Exchange from March 1, 2015, to March 1, 2016. Determine the price decrease from March 1, 2015, to January 1, 2016. Also determine whether the price was rising, falling, or holding steady on these days.

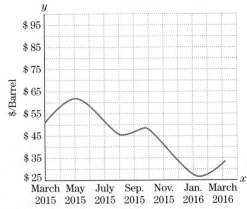

Figure 13 Price of crude oil, March 1, 2015, to March 1, 2016.

26. Refer to Fig. 13. Do you agree with the statement that the price of one barrel of crude oil fell on July 1, 2015, and on December 1, 2015, at approximately the same rate? Justify your answer.

27. Refer to Fig. 14, which shows an enlarged version of one portion of the curve in Fig. 13. Estimate the price of one barrel of crude oil on January 20, 2016, and the rate at which the price was rising on that day. (Your answer for the rate should be in dollars per day.)

Figure 14 Price of crude oil.

28. Refer to Fig. 14. Estimate the price of one barrel of crude oil on January 12, 2015, and the rate at which it was rising on that day.

In the next section, we shall see that the tangent line to the graph of $y = x^3$ at the point (x, y) has slope $3x^2$. See Fig. 15. Using this result, find the slope of the curve at the points in Exercises 29–31.

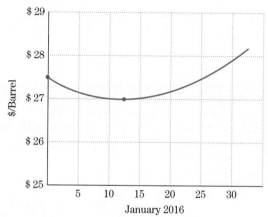

Figure 15 Slope of tangent line to $y = x^3$.

29. $(2, 8)$

30. $\left(\frac{3}{2}, \frac{27}{8}\right)$

31. $\left(-\frac{1}{2}, -\frac{1}{8}\right)$

32. Write the equation of the line tangent to the graph of $y = x^3$ at the point where $x = -1$.

In Exercises 33 and 34, you are shown the tangent line to the graph of $f(x) = x^2$ at the point $(a, f(a))$. Find a, $f(a)$, and the slope of the parabola at $(a, f(a))$.

33.

34.

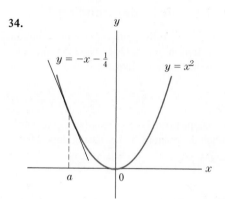

35. Find the point(s) on the graph in Fig. 15 where the slope is equal to $\frac{3}{2}$.

36. Find the points on the graph in Fig. 15 where the tangent line is parallel to $y = 2x$.

37. Let l be the line through the points P and Q in Fig. 16.
  (a) If $P = (2, 4)$ and $Q = (5, 13)$, find the slope of the line l and the length of the line segment d.
  (b) Suppose that the point Q moves toward P along the graph. Does the slope of the line l increase or decrease?

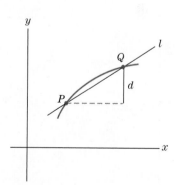

Figure 16 Line through P and Q.

38. In Fig. 17, h represents a positive number, and $3 + h$ is the number h units to the right of 3. Draw line segments on the graph having the following lengths.
  (a) $f(3)$ **(b)** $f(3 + h)$
  (c) $f(3 + h) - f(3)$ **(d)** h
  (e) Draw a line of slope $\dfrac{f(3 + h) - f(3)}{h}$.

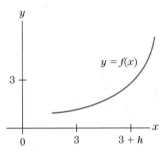

Figure 17 Geometric representation of values.

In Exercises 39–42, you are given a function and a point on the graph of the function. Zoom in on the graph at the given point until it starts to look like a straight line. Estimate the slope of the graph at the point indicated.

39. $f(x) = 2x^2 - 3x + 2$, $(0, 2)$.

40. $f(x) = \dfrac{x - 1}{x + 1}$, $(1, 0)$.

41. $f(x) = \sqrt{x + 3}$, $(1, 2)$.

42. $f(x) = \sqrt[3]{x + 6}$, $(2, 2)$.

Solutions to Check Your Understanding 1.2

1. (a) The slope of the curve at the point $(3, 4)$ is, by definition, the slope of the tangent line at $(3, 4)$. Note that the point $(4, 6)$ is also on the line. Therefore, the slope is

$$\frac{6 - 4}{4 - 3} = \frac{2}{1} = 2.$$

(b) Use the point–slope formula. The equation of the line passing through the point $(3, 4)$ and having slope 2 is

$$y - 4 = 2(x - 3) \quad \text{or} \quad y = 2x - 2.$$

2. The tangent line at $(4, 3)$ is, by definition, the line that best approximates the curve at $(4, 3)$. Since the "curve" in this case is itself a line, the curve and its tangent line at $(4, 3)$ (and at every other point) must be the same. Therefore, the equation is $y = \frac{1}{2}x + 1$.

1.3 # The Derivative and Limits

Suppose that a curve is the graph of a function $f(x)$ and that we have a tangent line at every point on the graph. We saw in the previous section that it is often possible to find a formula that gives the slope of the curve $y = f(x)$ at any point. This slope formula is called the **derivative** of $f(x)$ and is written $f'(x)$. For each value of x, $f'(x)$ gives the slope of the curve $y = f(x)$ at the point with first coordinate x. (See Fig. 1.) (As we shall see, some curves do not have tangent lines at every point. At values of x corresponding to such points, the derivative $f'(x)$ is not defined. For the sake of the current discussion, which is designed to develop an intuitive feeling for the derivative, let us assume that the graph of $f(x)$ has a tangent line for each x in the domain of f.)

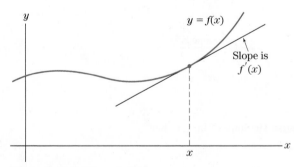

Figure 1 Definition of $f'(x)$.

In the previous section, we considered some interesting examples of slope formulas. Here, we add a few more examples and state them in the terminology of derivatives. At the end of the section, we use the fact that the derivative is a slope formula for the tangent line to describe a geometric construction of the tangent line and hence of the derivative. This construction will lead us naturally into the topic of limits, which will be treated in detail in the following section. The process of computing $f'(x)$ for a given function $f(x)$ is called **differentiation**.

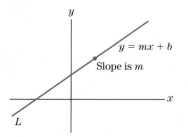

Figure 2 Derivative of a linear function.

Examples of Derivatives: The Power Rule

The case of a linear function $f(x) = mx + b$ is particularly simple. The graph of $y = mx + b$ is a straight line L of slope m. Recall that, by definition, the tangent line to a curve at a point is the line that best approximates the curve at that point. Since the curve L is itself a line, the tangent line to L at any point is the line L itself. So the slope of the graph is m at every point. (See Fig. 2.) In other words, the value of the derivative $f'(x)$ is always equal to m. We summarize this fact as follows:

Derivative of Linear Function If $f(x) = mx + b$, then we have

$$f'(x) = m. \tag{1}$$

Set $m = 0$ in $f(x) = mx + b$. Then, the function becomes $f(x) = b$, which has the value b for each value of x. The graph is a horizontal line of slope 0, so $f'(x) = 0$ for all x. (See Fig. 3.) Thus, we have the following fact:

Figure 3 Derivative of a constant function.

Constant Rule The derivative of a constant function $f(x) = b$ is zero. That is,

$$f'(x) = 0. \tag{2}$$

Next, consider the function $f(x) = x^2$. As we stated in Section 1.2 (and will prove at the end of this section), the slope of the graph of $y = x^2$ at the point (x, y) is equal to $2x$. That is, the value of the derivative $f'(x)$ is $2x$:

If $f(x) = x^2$, then its derivative is the function $2x$. That is,

$$f'(x) = 2x. \tag{3}$$

In Exercises 29–32 of Section 1.2, we made use of the fact that the slope of the graph of $y = x^3$ at the point (x, y) is $3x^2$. This can be restated in terms of derivatives as follows:

If $f(x) = x^3$, then the derivative is $3x^2$. That is,

$$f'(x) = 3x^2. \tag{4}$$

One reason that calculus is so useful is that it provides general techniques that can be easily used to determine derivatives. One such general rule, which contains formulas (3) and (4) as special cases, is the power rule.

Power Rule Let r be any number, and let $f(x) = x^r$. Then $f'(x) = rx^{r-1}$.

Indeed, if $r = 2$, then $f(x) = x^2$ and $f'(x) = 2x^{2-1} = 2x$, which is formula (3). If $r = 3$, then $f(x) = x^3$ and $f'(x) = 3x^{3-1} = 3x^2$, which is (4). We shall prove the power rule in Chapter 4.

EXAMPLE 1   **Applying the Power Rule**   Let $f(x) = \sqrt{x}$. What is $f'(x)$?

SOLUTION   Recall that $\sqrt{x} = x^{1/2}$. We may apply the power rule with $r = \frac{1}{2}$.

$$f(x) = x^{1/2}$$

$$f'(x) = \frac{1}{2}x^{1/2-1} = \frac{1}{2}x^{-1/2}$$

$$= \frac{1}{2} \cdot \frac{1}{x^{1/2}} = \frac{1}{2\sqrt{x}}$$ **» Now Try Exercise 13**

FOR REVIEW

Rational exponents are discussed in Sec. 0.5, p. 33. Examples:

$$\sqrt{x} = x^{\frac{1}{2}}; \frac{1}{x} = x^{-1};$$

$$\frac{1}{\sqrt{x}} = x^{-\frac{1}{2}}.$$

Another important special case of the power rule occurs for $r = -1$, corresponding to $f(x) = x^{-1}$. In this case, $f'(x) = (-1)x^{-1-1} = -x^{-2}$. However, since $x^{-1} = 1/x$ and $x^{-2} = 1/x^2$, the power rule for $r = -1$ may also be written as follows:

$$\text{If } f(x) = \frac{1}{x}, \text{ then } f'(x) = -\frac{1}{x^2} \quad (x \neq 0). \tag{5}$$

(The formula gives $f'(x)$ for $x \neq 0$. The derivative of $f(x)$ is not defined at $x = 0$ since $f(x)$ itself is not defined there.)

Geometric Meaning of the Derivative and Equation of the Tangent Line

We should, at this stage at least, keep the geometric meaning of the derivative clearly in mind. Figure 4 shows the graphs of x^2 and x^3 together with the interpretations of formulas (3) and (4) in terms of slope.

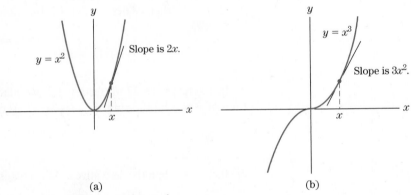

(a)   (b)

Figure 4 Derivatives of x^2 and x^3.

In general, the derivative of $f(x)$ at $x = a$ is the slope of the graph of $f(x)$ at the point $(a, f(a))$. Since the slope of the graph at a point is by definition the slope of the tangent line at that point, we have

---

Geometric Meaning of the Derivative

$$f'(a) = \text{slope of the graph of } f(x) \text{ at } (a, f(a))$$

$$= \text{slope of the tangent line to the graph of } f(x) \text{ at } (a, f(a)).$$

This geometric interpretation of the derivative will be used on many occasions in this book.

EXAMPLE 2

Evaluating the Derivative at One Point Find the slope of the curve $y = 1/x$ at $\left(2, \frac{1}{2}\right)$.

SOLUTION

Set $f(x) = 1/x$. The point $\left(2, \frac{1}{2}\right)$ corresponds to $x = 2$, so to find the slope at this point, we compute $f'(2)$. From (5), we have $f'(x) = \frac{-1}{x^2}$, and so

$$f'(2) = -\frac{1}{2^2} = -\frac{1}{4}.$$

Thus, the slope of $y = 1/x$ at the point $\left(2, \frac{1}{2}\right)$ is $-\frac{1}{4}$. (See Fig. 5.)

>> *Now Try Exercise 23*

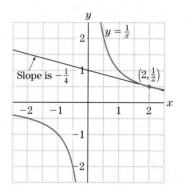

Figure 5 Derivative of $\frac{1}{x}$.

WARNING

Do not confuse $f'(2)$, the value of the derivative at 2, with $f(2)$, the value of the y-coordinate at the point on the graph at which $x = 2$. In Example 2, we have $f'(2) = -\frac{1}{4}$, whereas $f(2) = \frac{1}{2}$. The number $f'(2)$ gives the *slope* of the graph at $x = 2$; the number $f(2)$ gives the *height* of the graph at $x = 2$. «

Using the derivative, we can find the slope of the tangent line, and this, in turn, allows us to find the equation of the tangent line at a given point on the graph.

EXAMPLE 3

Finding an Equation of the Tangent Line at a Given Point Find an equation of the tangent line to the graph of $f(x) = \frac{1}{x}$ at the point $\left(2, \frac{1}{2}\right)$.

SOLUTION

We found in Example 2 that the slope of the tangent line at $\left(2, \frac{1}{2}\right)$ is $f'(2) = -\frac{1}{4}$. To be able to find the equation of the tangent line, we need one point on this line. Clearly, the point $\left(2, \frac{1}{2}\right)$ is on the tangent line (it is the point of contact of the tangent line with the graph). So, the point–slope equation of the tangent line is

$$y - \frac{1}{2} = -\frac{1}{4}(x - 2).$$

In slope–intercept form, the equation is

$$y = -\frac{1}{4}x + \frac{1}{2} + \frac{1}{2} \quad \text{or} \quad y = -\frac{1}{4}x + 1.$$

The tangent line is shown in Fig. 5.

>> *Now Try Exercise 33*

EXAMPLE 4

Finding the Equation of the Tangent Line at a Given x Find the point–slope equation of the tangent line to the graph of $f(x) = \frac{1}{x^2}$ at $x = 2$.

SOLUTION

In this problem, we are not given a point on the tangent line, but only its first coordinate $x = 2$. Since the point is on the graph of $f(x) = \frac{1}{x^2}$, we get the second coordinate by plugging the x-value into $f(x)$:

$$y = f(2) = \frac{1}{2^2} = \frac{1}{4}.$$

Thus, $\left(2, \frac{1}{4}\right)$ is the point on the graph *and* the tangent line. Next, we find the slope of the tangent line. For this purpose, we compute $f'(x)$ by using the power rule:

$$f(x) = \frac{1}{x^2} = x^{-2}$$

$$f'(x) = (-2)x^{-2-1} = (-2)x^{-3} = \frac{-2}{x^3}.$$

The slope of the tangent line when $x = 2$ is

$$f'(2) = \frac{-2}{2^3} = -\frac{1}{4}.$$

In point–slope form, the equation of the tangent line is $y - \frac{1}{4} = -\frac{1}{4}(x - 2)$.

≫ Now Try Exercise 39

In general, to find the point–slope equation of the tangent line to the graph of $y = f(x)$ at the point with first coordinate $x = a$, proceed as follows:

Step 1   Find the point of contact of the graph and the tangent line by evaluating $f(x)$ at $x = a$. This yields the point $(a, f(a))$.

Step 2   Find the slope of the tangent line by evaluating the derivative $f'(x)$ at $x = a$. This yields the slope $m = f'(a)$.

Using the point $(a, f(a))$ and the slope $m = f'(a)$, we obtain the equation of the tangent line:

Equation of the Tangent Line

$$y - f(a) = f'(a)(x - a). \tag{6}$$

You do not need to memorize formula (6), but you should be able to derive it in a given situation, as we did in Example 4.

EXAMPLE 5   **Equation of the Tangent Line at a Given Point** Find the equation of the tangent line to the graph of $f(x) = \sqrt{x}$ at the point $(1, 1)$.

SOLUTION   From Example 1, we have $f'(x) = \frac{1}{2\sqrt{x}}$. Thus, the slope of the tangent line at the point $(1, 1)$ is $f'(1) = \frac{1}{2\sqrt{1}} = \frac{1}{2}$. Since the tangent line passes through the point $(1, 1)$, its equation in point-slope form is

$$y - 1 = \frac{1}{2}(x - 1).$$

Note how this follows from (6) with $a = 1$, $f(a) = 1$, and $f'(a) = f'(1) = \frac{1}{2}$. Figure 6 shows the graph of $y = \sqrt{x}$ and the tangent line at the point $(1, 1)$. (To plot the tangent line, start at the point $(1, 1)$; move over 1 unit to the right and then up $\frac{1}{2}$ unit.)

Figure 6 Tangent to $f(x) = \sqrt{x}$ at the point $(1, 1)$. Slope of tangent line is $f'(1) = \frac{1}{2\sqrt{1}} = \frac{1}{2}$.

≫ Now Try Exercise 41

EXAMPLE 6

Using the Derivative to Solve Tangent Problems The line $y = -\frac{1}{4}x + b$ is tangent to the graph of $y = \frac{1}{x}$ at the point $P = \left(a, \frac{1}{a}\right)$ $(a > 0)$. Find P and determine b (Fig. 7).

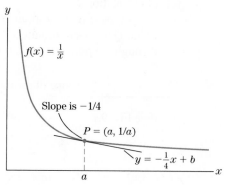

Figure 7

SOLUTION

The problem is illustrated in Fig. 7, where the tangent line $y = -\frac{1}{4}x + b$ is drawn at the point $P = \left(a, \frac{1}{a}\right)$. The tangent line has slope $-\frac{1}{4}$, and this slope must equal $f'(a)$. Since $f'(x) = -\frac{1}{x^2}$, we conclude that $-\frac{1}{4} = f'(a) = -\frac{1}{a^2}$. Solving for a, we find

$$-\frac{1}{4} = -\frac{1}{a^2} \qquad \text{Given equation.}$$

$$a^2 = 4 \qquad \text{Cross-multiply.}$$

$$a = \pm 2. \qquad \text{Take square root.}$$

Since $a > 0$, we take $a = 2$, and so $P = \left(2, \frac{1}{2}\right)$. To determine b, we use the fact that P is a point on the tangent line. Plugging the coordinates of P into the equation of the tangent line and solving for b, we get

$$\frac{1}{2} = -\frac{1}{4}(2) + b = -\frac{1}{2} + b$$

$$b = 1.$$

Hence, the equation of the tangent line is $y = -\frac{1}{4}x + 1$.   **» Now Try Exercise 43**

Notation   The operation of forming a derivative $f'(x)$ from a function $f(x)$ is also indicated by the symbol $\dfrac{d}{dx}$ (read "the derivative with respect to x"). Thus,

$$\frac{d}{dx}f(x) = f'(x).$$

For example,

$$\frac{d}{dx}(x^6) = 6x^5, \qquad \frac{d}{dx}(x^{5/3}) = \frac{5}{3}x^{2/3}, \qquad \frac{d}{dx}\left(\frac{1}{x}\right) = -\frac{1}{x^2}.$$

When working with an equation of the form $y = f(x)$, we often write $\dfrac{dy}{dx}$ as a symbol for the derivative $f'(x)$. For example, if $y = x^6$, we may write

$$\frac{dy}{dx} = 6x^5.$$

Limits and the Secant-Line Calculation of the Derivative

So far, we have said nothing about how to derive differentiation formulas such as (3), (4), or (5). Let us remedy that omission now. The derivative gives the slope of the tangent line, so we must describe a procedure for calculating that slope.

The fundamental idea for calculating the slope of the tangent line at a point P is to approximate the tangent line very closely by *secant lines*. A secant line at P is a straight line passing through P and a nearby point Q on the curve. (See Fig. 8.) By moving Q very close to P, we can make the slope of the secant line approximate the slope of the tangent line to any desired degree of accuracy. Let us see what this amounts to in terms of calculations.

Suppose that the point P is $(x, f(x))$. Also, Q is h horizontal units away from P. Then, Q has x-coordinate $x + h$ and y-coordinate $f(x + h)$. The slope of the secant line through the points $P = (x, f(x))$ and $Q = (x + h, f(x + h))$ is simply

$$[\text{slope of secant line}] = \frac{f(x + h) - f(x)}{(x + h) - x} = \frac{f(x + h) - f(x)}{h}.$$

(See Fig. 9.)

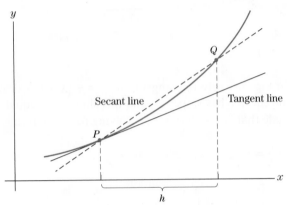

Figure 8 A secant-line approximation to a tangent line.

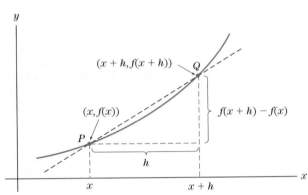

Figure 9 Computing the slope of a secant line.

To move Q close to P along the curve, we let h approach zero. Then, the secant line approaches the tangent line, and so,

$$[\text{slope of secant line}] \quad \text{approaches} \quad [\text{slope of tangent line}];$$

that is,

$$\frac{f(x + h) - f(x)}{h} \quad \text{approaches} \quad f'(x).$$

Let us summarize the process as follows:

Calculating the Derivative Informally   To calculate $f'(x)$:

1. Write the difference quotient $\dfrac{f(x + h) - f(x)}{h}$ for $h \neq 0$.
2. Simplify this difference quotient.
3. Let h approach zero. The quantity $\dfrac{f(x + h) - f(x)}{h}$ will approach $f'(x)$.

Another way of stating Step 3 is to say that the *limit* of $\frac{f(x + h) - f(x)}{h}$ is $f'(x)$, as h approaches 0. The concept of limit and the three-step method just outlined will be the subject of our next section.

EXAMPLE 7   **Verification of the Power Rule for $r = 2$** Apply the three-step method to show that the derivative of $f(x) = x^2$ is $f'(x) = 2x$.

SOLUTION   Here, $f(x) = x^2$, so the slope of the secant line is

$$\frac{f(x + h) - f(x)}{h} = \frac{(x + h)^2 - x^2}{h} \quad (h \neq 0).$$

By multiplying out, we have $(x + h)^2 = x^2 + 2xh + h^2$, so,

$$\frac{f(x + h) - f(x)}{h} = \frac{x^2 + 2xh + h^2 - x^2}{h} = \frac{(2x + h)\cancel{h}}{\cancel{h}}$$

$$= 2x + h.$$

(*Note:* Since $h \neq 0$, we were able to divide both numerator and denominator by h.)

As h approaches 0 (that is, as the secant line approaches the tangent line), the quantity $2x + h$ approaches $2x$. Thus, we have

$$f'(x) = 2x.$$ **» Now Try Exercise 71**

In the next section, we formulate the three-step method that we just described using limits and derive the *limit definition* of the derivative.

INCORPORATING TECHNOLOGY

Numerical Derivatives and Tangent Lines TI-83/84 calculators have a numerical derivative routine, **nDeriv**, accessed by pressing MATH 8. For example, to compute the derivative of $f(x) = \sqrt{x}$ at $x = 3$, we proceed as indicated in Fig. 10. We can also graph the tangent line to a function at a point. For example, to graph the tangent line $f(x) = \sqrt{x}$ at $x = 3$, we proceed as follows: First, enter $Y_1 = \sqrt{X}$ and press GRAPH. Now, from the graph window, press 2nd [DRAW] 5 to select **Tangent**. Enter 3 to set $x = 3$ and press ENTER. The result is shown in Fig. 11. The equation of the tangent line is also shown in Fig. 11. Note how the slope is equal to the numerical derivative that we obtained in Fig. 10.

Figure 10

Figure 11

Check Your Understanding 1.3

Solutions can be found following the section exercises.

1. Consider the curve $y = f(x)$ in Fig. 12.
   (a) Find $f(5)$.   (b) Find $f'(5)$.

2. Let $f(x) = 1/x^4$.
   (a) Find its derivative.   (b) Find $f'(2)$.

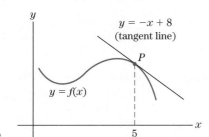

$y = -x + 8$
(tangent line)

$y = f(x)$

Figure 12

EXERCISES 1.3

Use formulas (1) and (2) and the power rule to find the derivatives of the following functions.

1. $f(x) = 3x + 7$

2. $f(x) = -2x$

3. $f(x) = \dfrac{3x}{4} - 2$

4. $f(x) = \dfrac{2x - 6}{7}$

5. $f(x) = x^7$

6. $f(x) = x^{-2}$

7. $f(x) = x^{2/3}$

8. $f(x) = x^{-1/2}$

9. $f(x) = \dfrac{1}{\sqrt{x^5}}$

10. $f(x) = \dfrac{1}{x^3}$

11. $f(x) = \sqrt[3]{x}$

12. $f(x) = \dfrac{1}{\sqrt[5]{x}}$

13. $f(x) = \dfrac{1}{x^{-2}}$

14. $f(x) = \sqrt[7]{x^2}$

15. $f(x) = 4^2$

16. $f(x) = \pi$

In Exercises 17–24, find the derivative of $f(x)$ at the designated value of x.

17. $f(x) = x^3$ at $x = \frac{1}{2}$

18. $f(x) = x^5$ at $x = \frac{3}{2}$

19. $f(x) = \dfrac{1}{x}$ at $x = \frac{2}{3}$

20. $f(x) = \frac{1}{3}$ at $x = 2$

21. $f(x) = x + 11$ at $x = 0$

22. $f(x) = x^{1/3}$ at $x = 8$

23. $f(x) = \sqrt{x}$ at $x = \frac{1}{16}$

24. $f(x) = \dfrac{1}{\sqrt[5]{x^2}}$ at $x = 32$

25. Find the slope of the curve $y = x^4$ at $x = 2$.

26. Find the slope of the curve $y = x^5$ at $x = \frac{1}{3}$.

27. If $f(x) = x^3$, compute $f(-5)$ and $f'(-5)$.

28. If $f(x) = 2x + 6$, compute $f(0)$ and $f'(0)$.

29. If $f(x) = x^{1/3}$, compute $f(8)$ and $f'(8)$.

30. If $f(x) = 1/x^2$, compute $f(1)$ and $f'(1)$.

31. If $f(x) = 1/x^5$, compute $f(-2)$ and $f'(-2)$.

32. If $f(x) = x^{3/2}$, compute $f(16)$ and $f'(16)$.

In Exercises 33–40, find an equation of the tangent line to the graph of $y = f(x)$ at the given x. Do not apply formula (6), but proceed as we did in Example 4.

33. $f(x) = x^3$, $x = -2$

34. $f(x) = x^2$, $x = -\frac{1}{2}$

35. $f(x) = 3x + 1$, $x = 4$

36. $f(x) = 5$, $x = -2$

37. $f(x) = \sqrt{x}$, $x = \frac{1}{9}$

38. $f(x) = \dfrac{1}{x}$, $x = .01$

39. $f(x) = \dfrac{1}{\sqrt{x}}$, $x = 1$

40. $f(x) = \dfrac{1}{x^3}$, $x = 3$

41. The point–slope form of the equation of the tangent line to the graph of $y = x^4$ at the point $(1, 1)$ is $y - 1 = 4(x - 1)$. Explain how this equation follows from formula (6).

42. The tangent to the graph of $y = \frac{1}{x}$ at the point $P = (a, \frac{1}{a})$, where $a > 0$, is perpendicular to the line $y = 4x + 1$. Find P.

43. The line $y = 2x + b$ is tangent to the graph of $y = \sqrt{x}$ at the point $P = (a, \sqrt{a})$. Find P and determine b.

44. The line $y = ax + b$ is tangent to the graph of $y = x^3$ at the point $P = (-3, -27)$. Find a and b.

45. (a) Find the point on the curve $y = \sqrt{x}$ where the tangent line is parallel to the line $y = \frac{x}{8}$.
(b) On the same axes, plot the curve $y = \sqrt{x}$, the line $y = \frac{x}{8}$, and the tangent line to $y = \sqrt{x}$ that is parallel to $y = \frac{x}{8}$.

46. There are two points on the graph of $y = x^3$ where the tangent lines are parallel to $y = x$. Find these points.

47. Is there any point on the graph of $y = x^3$ where the tangent line is perpendicular to $y = x$? Justify your answer.

48. The graph of $y = f(x)$ goes through the point $(2, 3)$ and the equation of the tangent line at that point is $y = -2x + 7$. Find $f(2)$ and $f'(2)$.

In Exercises 49–56, find the indicated derivative.

49. $\dfrac{d}{dx}(x^8)$

50. $\dfrac{d}{dx}(x^{-3})$

51. $\dfrac{d}{dx}(x^{3/4})$

52. $\dfrac{d}{dx}(x^{-1/3})$

53. $\dfrac{dy}{dx}$ if $y = 1$

54. $\dfrac{dy}{dx}$ if $y = x^{-4}$

55. $\dfrac{dy}{dx}$ if $y = x^{1/5}$

56. $\dfrac{dy}{dx}$ if $y = \dfrac{x - 1}{3}$

57. Consider the curve $y = f(x)$ in Fig. 13. Find $f(6)$ and $f'(6)$.

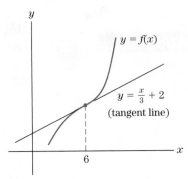

Figure 13

58. Consider the curve $y = f(x)$ in Fig. 14. Find $f(1)$ and $f'(1)$.

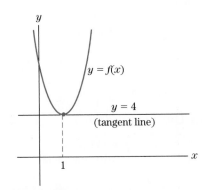

Figure 14

59. In Fig. 15, the straight line $y = \frac{1}{4}x + b$ is tangent to the graph of $f(x) = \sqrt{x}$. Find the values of a and b.

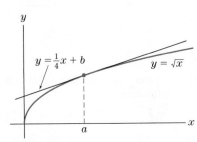

Figure 15

60. In Fig. 16, the straight line is tangent to the graph of $f(x) = 1/x$. Find the value of a.

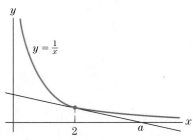

Figure 16

61. Consider the curve $y = f(x)$ in Fig. 17. Find a and $f(a)$. Estimate $f'(a)$.

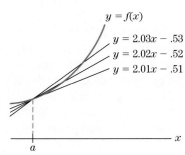

Figure 17

62. Consider the curve $y = f(x)$ in Fig. 18. Estimate $f'(1)$.

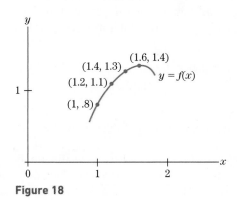

Figure 18

63. In Fig. 19, find the equation of the tangent line to $f(x)$ at the point A.

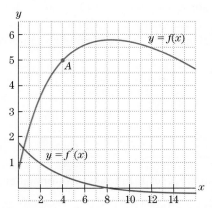

Figure 19

64. In Fig. 20, find the equation of the tangent line to $f(x)$ at the point P.

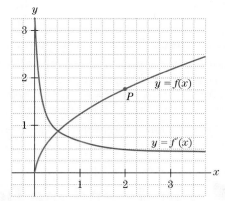

Figure 20

In Exercises 65–70, compute the difference quotient

$$\frac{f(x + h) - f(x)}{h}.$$

Simplify your answer as much as possible.

65. $f(x) = 2x^2$ **66.** $f(x) = x^2 - 7$

67. $f(x) = -x^2 + 2x$ **68.** $f(x) = -2x^2 + x + 3$

69. $f(x) = x^3$ [*Hint:* $(a + b)^3 = a^3 + 3a^2b + 3ab^2 + b^3$.]

70. $f(x) = 2x^3 + x^2$

In Exercises 71–76, apply the three-step method to compute the derivative of the given function.

71. $f(x) = -x^2$ **72.** $f(x) = 3x^2 - 2$

73. $f(x) = 7x^2 + x - 1$ **74.** $f(x) = x + 3$

75. $f(x) = x^3$ **76.** $f(x) = 2x^3 - x$

77. (a) Draw two graphs of your choice that represent a function $y = f(x)$ and its vertical shift $y = f(x) + 3$.
   (b) Pick a value of x and consider the points $(x, f(x))$ and $(x, f(x) + 3)$. Draw the tangent lines to the curves at these points and describe what you observe about the tangent lines.
   (c) Based on your observation in part (b), explain why

   $$\frac{d}{dx}f(x) = \frac{d}{dx}(f(x) + 3).$$

78. Use the approach of Exercise 77 to show that

$$\frac{d}{dx}f(x) = \frac{d}{dx}(f(x) + c)$$

for any constant c. [*Hint:* Compare the tangent lines to the graphs of $f(x)$ and $f(x) + c$.]

TECHNOLOGY EXERCISES

In Exercises 79–84, use a derivative routine to obtain the value of the derivative. Give the value to 5 decimal places.

79. $f'(0)$, where $f(x) = 2^x$

80. $f'(1)$, where $f(x) = \dfrac{1}{1 + x^2}$

81. $f'(1)$, where $f(x) = \sqrt{1 + x^2}$

82. $f'(3)$, where $f(x) = \sqrt{25 - x^2}$

83. $f'(2)$, where $f(x) = \dfrac{x}{1 + x}$

84. $f'(0)$, where $f(x) = 10^{1+x}$

Solutions to Check Your Understanding 1.3

1. (a) The number $f(5)$ is the y-coordinate of the point P. Since the tangent line passes through P, the coordinates of P satisfy the equation $y = -x + 8$. Since its x-coordinate is 5, its y-coordinate is $-5 + 8 = 3$. Therefore, $f(5) = 3$.

(b) The number $f'(5)$ is the slope of the tangent line at P, $y = -x + 8$. So $f'(5) = -1$.

2. (a) The function $1/x^4$ can be written as the power function x^{-4}. Here, $r = -4$. Therefore,

$$f'(x) = (-4)x^{(-4)-1} = -4x^{-5} = \frac{-4}{x^5}.$$

(b) $f'(2) = -4/2^5 = -4/32 = -\frac{1}{8}.$

1.4   Limits and the Derivative

The notion of a limit is one of the fundamental ideas of calculus. Indeed, any "theoretical" development of calculus rests on an extensive use of the theory of limits. Even in this book, where we have adopted an intuitive viewpoint, limit arguments are used occasionally (although in an informal way). In the previous section, limits were used to define the derivative. They arose naturally from our geometric construction of the tangent line as a *limit* of secant lines. In this section, we continue our brief introduction to limits and state some additional properties that are required in our development of the derivative and several other areas of calculus. We start with a definition.

> **DEFINITION   Limit of a Function**   Let $g(x)$ be a function and a a number. We say that the number L *is* the **limit** of $g(x)$ as x approaches a, provided that $g(x)$ can be made arbitrarily close to L for all x sufficiently close (but not equal) to a. In this case, we write
>
> $$\lim_{x \to a} g(x) = L.$$

In other words, the limit of $g(x)$ is L as x approaches a, if the values of $g(x)$ approach L as x approaches a.

If, as x approaches a, the values $g(x)$ do *not* approach a specific number, we say that the limit of $g(x)$ as x approaches a *does not exist*. We saw several examples of limits in the previous section. Let us give some further basic examples before turning to the properties of limits.

EXAMPLE 1   **Computing a Limit Using a Table of Values**   Determine $\lim\limits_{x \to 2}(3x - 5)$.

SOLUTION   We make a table of values of x approaching 2 and the corresponding values of $3x - 5$:

x	$3x - 5$	x	$3x - 5$
2.1	1.3	1.9	.7
2.01	1.03	1.99	.97
2.001	1.003	1.999	.997
2.0001	1.0003	1.9999	.9997

As x approaches 2, we see that $3x - 5$ approaches 1. In terms of our notation,

$$\lim_{x \to 2}(3x - 5) = 1.$$

$\ll$

EXAMPLE 2 **Computing Limits Using Graphs** For each of the functions in Fig. 1, determine if $\lim_{x \to 2} g(x)$ exists. (The circles drawn on the graphs are meant to represent breaks in the graph, indicating that the functions under consideration are not defined at $x = 2$.)

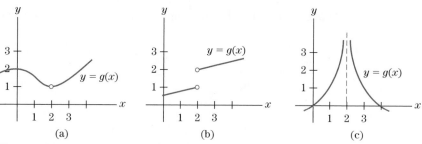

Figure 1 Functions with an undefined point.

SOLUTION (a) We can see that as x gets closer and closer to 2, the values of $g(x)$ get closer and closer to 1. This is true for values of x to both the right and the left of 2. So $\lim_{x \to 2} g(x) = 1$.

(b) As x approaches 2 from the right, $g(x)$ approaches 2. However, as x approaches 2 from the left, $g(x)$ approaches 1. For a limit to exist, the function must approach the *same* value from each direction. So $\lim_{x \to 2} g(x)$ does not exist.

(c) As x approaches 2, the values of $g(x)$ become larger and larger and do not approach a fixed number. So $\lim_{x \to 2} g(x)$ does not exist.   **>> Now Try Exercise 5**

EXAMPLE 3 **Two Basic Limits** Let a and C be any real numbers. Then
(a) $\lim_{x \to a} C = C$ Limit of a constant.
(b) $\lim_{x \to a} x = a$ Limit of $g(x) = x$.

SOLUTION (a) Because the constant function $g(x) = C$ always has the value C, its limit as x approaches any number a is C. Thus, $\lim_{x \to a} C = C$.

(b) As x approaches a, the function $g(x) = x$ clearly approaches a. In terms of our notation, $\lim_{x \to a} x = a$.   **«**

The following limit theorems, which we cite without proof, allow us to reduce the computation of limits for combinations of functions to computations of limits involving the component functions.

Limit Theorems   If $\lim_{x \to a} f(x)$ and $\lim_{x \to a} g(x)$ both exist, we have the following results.

(I) If k is a constant, then $\lim_{x \to a} k \cdot f(x) = k \cdot \lim_{x \to a} f(x)$.   Constant multiple

(II) If r is a positive constant and $[f(x)]^r$ is defined for $x \neq a$, then $\lim_{x \to a} [f(x)]^r = \left[\lim_{x \to a} f(x) \right]^r$.   Powers of a limit

(III) $\lim_{x \to a} [f(x) + g(x)] = \lim_{x \to a} f(x) + \lim_{x \to a} g(x)$.   Limit of a sum

(IV) $\lim_{x \to a} [f(x) - g(x)] = \lim_{x \to a} f(x) - \lim_{x \to a} g(x)$.   Limit of a difference

(V) $\lim_{x \to a} [f(x) \cdot g(x)] = \left[\lim_{x \to a} f(x) \right] \cdot \left[\lim_{x \to a} g(x) \right]$.   Limit of a product

(VI) If $\lim_{x \to a} g(x) \neq 0$, then $\lim_{x \to a} \dfrac{f(x)}{g(x)} = \dfrac{\lim_{x \to a} f(x)}{\lim_{x \to a} g(x)}$.   Limit of a quotient

EXAMPLE 4 **Using Properties of Limits** Use the limit theorems to compute the following limits.

(a) $\lim\limits_{x \to 2} x^3$ (b) $\lim\limits_{x \to 2} 5x^3$ (c) $\lim\limits_{x \to 2} (5x^3 - 15)$

(d) $\lim\limits_{x \to 2} \sqrt{5x^3 - 15}$ (e) $\lim\limits_{x \to 2} (\sqrt{5x^3 - 15}/x^5)$

SOLUTION (a) From Example 3(b), $\lim\limits_{x \to 2} x = 2$. By the powers of a limit,

$$\lim\limits_{x \to 2} x^3 = \left(\lim\limits_{x \to 2} x\right)^3 = 2^3 = 8.$$

(b) $\lim\limits_{x \to 2} 5x^3 = 5 \lim\limits_{x \to 2} x^3$ Constant multiple

$\qquad\qquad = 5 \cdot 8$ By part (a)

$\qquad\qquad = 40$

(c) $\lim\limits_{x \to 2} (5x^3 - 15) = \lim\limits_{x \to 2} 5x^3 - \lim\limits_{x \to 2} 15$ Limit of a difference

By Example 3(a) $\lim\limits_{x \to 2} 15 = 15$. By part (b), $\lim\limits_{x \to 2} 5x^3 = 40$. Thus,

$$\lim\limits_{x \to 2} (5x^3 - 15) = 40 - 15 = 25.$$

(d) $\lim\limits_{x \to 2} \sqrt{5x^3 - 15} = \lim\limits_{x \to 2} (5x^3 - 15)^{1/2}$

$\qquad\qquad = \left[\lim\limits_{x \to 2} (5x^3 - 15)\right]^{1/2}$ Powers of a limit with $r = \frac{1}{2}$, $f(x) = 5x^3 - 15$

$\qquad\qquad = 25^{1/2}$ By part (c)

$\qquad\qquad = 5$

(e) The limit of the denominator is $\lim\limits_{x \to 2} x^5$, which is $2^5 = 32$, a nonzero number. So by limit of a quotient, we have

$$\lim\limits_{x \to 2} \frac{\sqrt{5x^3 - 15}}{x^5} = \frac{\lim\limits_{x \to 2} \sqrt{5x^3 - 15}}{\lim\limits_{x \to 2} x^5}$$

$$= \frac{5}{32}$$ By part (d).

» Now Try Exercise 13

The following facts, which may be deduced by repeated applications of the various limit theorems, are extremely handy in evaluating limits:

Limit Theorems (*continued*)

(VII) **Limit of a Polynomial Function** Let $p(x)$ be a polynomial function, a any number. Then,

$$\lim\limits_{x \to a} p(x) = p(a).$$

(VIII) **Limit of a Rational Function** Let $r(x) = p(x)/q(x)$ be a rational function, where $p(x)$ and $q(x)$ are polynomials. Let a be a number such that $q(a) \neq 0$. Then,

$$\lim\limits_{x \to a} r(x) = r(a).$$

In other words, to determine a limit of a polynomial or a rational function, simply evaluate the function at $x = a$, provided, of course, that the function is defined at $x = a$. For instance, we can rework the solution to Example 4(c) as follows:

$$\lim\limits_{x \to 2} (5x^3 - 15) = 5(2)^3 - 15 = 25.$$

Many situations require algebraic simplifications before the limit theorems can be applied.

EXAMPLE 5   **Factoring Techniques** Compute the following limit.

$$\lim_{x \to 3} \frac{x^2 - 9}{x - 3}$$

SOLUTION   The function $\dfrac{x^2 - 9}{x - 3}$ is not defined when $x = 3$, since

$$\frac{3^2 - 9}{3 - 3} = \frac{0}{0},$$

which is undefined. This causes no difficulty, since the limit as x approaches 3 depends only on the values of x *near* 3 and excludes considerations of the value at $x = 3$ itself. Since the denominator approaches 0 when taking the limit, we may not apply Limit Theorem VI directly. To evaluate the limit, note that $x^2 - 9 = (x - 3)(x + 3)$. So, for $x \neq 3$,

$$\frac{x^2 - 9}{x - 3} = \frac{(x - 3)(x + 3)}{x - 3} = x + 3.$$

As x approaches 3, $x + 3$ approaches 6. Therefore,

$$\lim_{x \to 3} \frac{x^2 - 9}{x - 3} = 6.$$

 » Now Try Exercise 21

FOR REVIEW

If $p(x)$ is a polynomial and $p(a) = 0$, then $(x - a)$ is a factor of $p(x)$. See pp. 27–28.

Limit Definition of the Derivative

Our discussion of the derivative in the previous section was based on an intuitive geometric concept of the tangent line. This geometric approach led us to the three-step method for computing the derivative as a limit of a difference quotient. This process may be considered independently of its geometric interpretation and used to *define* $f'(x)$. In fact, we say that f is **differentiable** at x if

$$\frac{f(x + h) - f(x)}{h}$$

approaches some number as h approaches 0, and we denote this limiting number by $f'(x)$; in symbols, using the limit notation,

DEFINITION **Derivative as a Limit**

$$f'(x) = \lim_{h \to 0} \frac{f(x + h) - f(x)}{h} \tag{1}$$

If the **difference quotient**

$$\frac{f(x + h) - f(x)}{h}$$

does not approach any specific number as h approaches 0, we say that f is *nondifferentiable* at x. Essentially, all the functions in this text are differentiable at all points in their domain. A few exceptions are described in Section 1.5.

Using Limits to Calculate $f'(a)$

Step 1 Write the difference quotient $\dfrac{f(a + h) - f(a)}{h}$.

Step 2 Simplify the difference quotient.

Step 3 Find the limit as $h \to 0$. This limit is $f'(a)$.

EXAMPLE 6   **Computing a Derivative from the Limit Definition**   Use limits to compute the derivative $f'(5)$ for the following functions.

(a) $f(x) = 15 - x^2$ (b) $f(x) = \dfrac{1}{2x - 3}$

SOLUTION   (a) $\dfrac{f(5 + h) - f(5)}{h} = \dfrac{[15 - (5 + h)^2] - (15 - 5^2)}{h}$ Step 1

$= \dfrac{15 - (25 + 10h + h^2) - (15 - 25)}{h}$ Step 2

$= \dfrac{-10h - h^2}{h} = -10 - h.$

Therefore,

$$f'(5) = \lim_{h \to 0} (-10 - h) = -10.$$ Step 3

(b) $\dfrac{f(5 + h) - f(5)}{h} = \dfrac{\dfrac{1}{2(5 + h) - 3} - \dfrac{1}{2(5) - 3}}{h}$ Step 1

$= \dfrac{\dfrac{1}{7 + 2h} - \dfrac{1}{7}}{h} = \dfrac{\dfrac{7 - (7 + 2h)}{(7 + 2h)7}}{h}$ Step 2

$= \dfrac{-2h}{(7 + 2h)7 \cdot h} = \dfrac{-2}{(7 + 2h)7} = \dfrac{-2}{49 + 14h}.$

Therefore,

$$f'(5) = \lim_{h \to 0} \dfrac{-2}{49 + 14h} = -\dfrac{2}{49}.$$ Step 3

»» Now Try Exercise 29

NOTE   When computing the limits in Example 6, we considered only values of h near 0 (and not $h = 0$ itself). Therefore, we were freely able to divide both numerator and denominator by h. **«**

It is important to distinguish between the derivative at a specific point (as in Example 6) and a derivative formula. A derivative at a specific point is a number, while a derivative formula is a function. The three-step method that we outlined can be used to find derivative formulas, as we now illustrate.

EXAMPLE 7   **Computing Derivative from the Definition**   Find the derivative of $f(x) = x^2 + 2x + 2$.

SOLUTION   We have

$$\dfrac{f(x + h) - f(x)}{h} = \dfrac{(x + h)^2 + 2(x + h) + 2 - (x^2 + 2x + 2)}{h}$$ Step 1

$= \dfrac{x^2 + 2xh + h^2 + 2x + 2h + 2 - x^2 - 2x - 2}{h}$ Step 2

$= \dfrac{2xh + h^2 + 2h}{h} = \dfrac{(2x + h + 2)\not h}{\not h}$

$= 2x + 2 + h.$

Therefore,

$$f'(x) = \lim_{h \to 0} (2x + 2 + h) = 2x + 2.$$ Step 3

»» Now Try Exercise 33

Let us go back to the power rule in the previous section and verify it in two special cases: $r = -1$ and $r = \frac{1}{2}$.

EXAMPLE 8 **Derivative of a Rational Function** Find the derivative of $f(x) = \frac{1}{x}$, $x \neq 0$.

SOLUTION We have

$$\frac{f(x+h) - f(x)}{h} = \frac{\frac{1}{x+h} - \frac{1}{x}}{h} \qquad \text{Step 1}$$

$$= \frac{1}{h}\left[\frac{1}{x+h} - \frac{1}{x}\right] \qquad \text{Step 2}$$

$$= \frac{1}{h}\left[\frac{x - (x+h)}{(x+h)x}\right] = \frac{1}{\not{h}}\left[\frac{-\not{h}}{(x+h)x}\right]$$

$$= \frac{-1}{(x+h)x}.$$

Step 3 Using Limit Theorem VIII, we find that, as h approaches 0, $\frac{-1}{(x+h)x}$ approaches $\frac{-1}{x^2}$. Hence,

$$f'(x) = \frac{-1}{x^2}.$$

>> **Now Try Exercise 41**

In evaluating the next limit, we use a useful rationalization technique.

EXAMPLE 9 **Derivative of a Function with a Radical** Find the derivative of $f(x) = \sqrt{x}$, $x > 0$.

SOLUTION Recall the algebraic identity $(a + b)(a - b) = a^2 - b^2$. So,

$$(\sqrt{x+h} - \sqrt{x})(\sqrt{x+h} + \sqrt{x}) = (\sqrt{x+h})^2 - (\sqrt{x})^2 = x + h - x = h.$$

We have

$$\frac{f(x+h) - f(x)}{h} = \frac{\sqrt{x+h} - \sqrt{x}}{h} \qquad \text{Step 1}$$

$$= \frac{\sqrt{x+h} - \sqrt{x}}{h} \cdot \frac{\sqrt{x+h} + \sqrt{x}}{\sqrt{x+h} + \sqrt{x}} \qquad \text{Step 2}$$

$$= \frac{x + h - x}{h(\sqrt{x+h} + \sqrt{x})} = \frac{\not{h}}{\not{h}(\sqrt{x+h} + \sqrt{x})}$$

$$= \frac{1}{\sqrt{x+h} + \sqrt{x}}.$$

Therefore,

$$f'(x) = \lim_{h \to 0} \frac{1}{\sqrt{x+h} + \sqrt{x}} = \frac{1}{\sqrt{x} + \sqrt{x}} = \frac{1}{2\sqrt{x}}. \qquad \text{Step 3}$$

>> **Now Try Exercise 45**

Infinity and Limits Consider the function $f(x)$ whose graph is sketched in Fig. 2. As x grows large, the value of $f(x)$ approaches 2. In this circumstance, we say that 2 is the *limit of $f(x)$ as x approaches infinity*. Infinity is denoted by the symbol ∞. The preceding limit statement is expressed in the following notation:

$$\lim_{x \to \infty} f(x) = 2.$$

In a similar vein, consider the function whose graph is sketched in Fig. 3. As x grows large in the negative direction, the value of $f(x)$ approaches 0. In this circumstance, we say that 0 is the *limit of $f(x)$ as x approaches minus infinity*. In symbols,

$$\lim_{x \to -\infty} f(x) = 0.$$

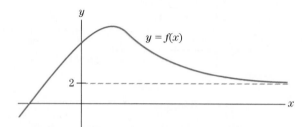

Figure 2 Function with a limit as x approaches infinity.

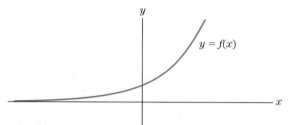

Figure 3 Function with a limit as x approaches minus infinity.

EXAMPLE 10   **Limits at Infinity** Calculate the following limits.

(a) $\lim_{x \to \infty} \dfrac{1}{x^2 + 1}$ (b) $\lim_{x \to \infty} \dfrac{6x - 1}{2x + 1}$

SOLUTION   (a) As x increases without bound, so does $x^2 + 1$. Therefore, $1/(x^2 + 1)$ approaches 0 as x approaches ∞.

(b) Both $6x - 1$ and $2x + 1$ increase without bound as x does. To determine the limit of their quotient, we employ an algebraic trick. Divide both numerator and denominator by x to obtain

$$\lim_{x \to \infty} \frac{6x - 1}{2x + 1} = \lim_{x \to \infty} \frac{6 - \dfrac{1}{x}}{2 + \dfrac{1}{x}}.$$

As x increases without bound, $1/x$ approaches 0, so $6 - (1/x)$ approaches 6 and $2 + (1/x)$ approaches 2. Thus, the desired limit is $6/2 = 3$.

≫ Now Try Exercise 57

The limit theorems that we discussed earlier apply in the context of limits at infinity as our next example illustrates.

EXAMPLE 11   **Limits at Infinity** Refer to Fig. 2 to compute the limits $\lim_{x \to \infty} (1 - 3f(x))$ and $\lim_{x \to \infty} [f(x)]^2$.

SOLUTION   From Fig. 2, we see that $\lim_{x \to \infty} f(x) = 2$. Now, using the limit theorems, we have

$$\lim_{x \to \infty} (1 - 3f(x)) = \lim_{x \to \infty} (1) + \lim_{x \to \infty} (-3f(x)) \quad \text{Limit of a difference}$$

$$= 1 - 3 \lim_{x \to \infty} f(x) \qquad\qquad \text{Constant multiple}$$

$$= 1 - 3(2) = -5$$

and

$$\lim_{x\to\infty} [f(x)]^2 = \lim_{x\to\infty} [f(x)\cdot f(x)]$$
$$= [\lim_{x\to\infty} f(x)]\cdot[\lim_{x\to\infty} f(x)] \quad \text{Limit of a product}$$
$$= (2)(2) = 4.$$

» Now Try Exercise 65

INCORPORATING

TECHNOLOGY

Using Tables to Find Limits Consider the function

$$y = \frac{x^2 - 9}{x - 3}$$

from Example 5. This function is undefined at $x = 3$, but if you examine its values near $x = 3$, as in Fig. 4, you will be convinced that

$$\lim_{x\to 3} \frac{x^2 - 9}{x - 3} = 6.$$

To generate the tables in Fig. 4, first press $\boxed{\text{Y=}}$ and assign the function $\frac{x^2-9}{x-3}$ to Y_1. Press $\boxed{\text{2nd}}$ [TBLSET] and set **Indpnt** to **Ask**, leaving the other settings at their default values. Finally, press $\boxed{\text{2nd}}$ [TABLE] and enter the values shown for X.

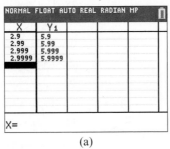

(a)   (b)

Figure 4

Check Your Understanding 1.4

Solutions can be found following the section exercises.

Determine which of the following limits exist. Compute the limits that exist.

1. $\displaystyle\lim_{x\to 6} \frac{x^2 - 4x - 12}{x - 6}$

2. $\displaystyle\lim_{x\to 6} \frac{4x + 12}{x - 6}$

EXERCISES 1.4

For each of the following functions $g(x)$, determine whether $\displaystyle\lim_{x\to 3} g(x)$ exists. If so, give the limit.

1.

2.

3.

4.

5.

6.

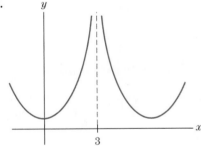

Determine which of the following limits exist. Compute the limits that exist.

7. $\lim_{x \to 1} (1 - 6x)$

8. $\lim_{x \to 2} \dfrac{x}{x - 2}$

9. $\lim_{x \to 3} \sqrt{x^2 + 16}$

10. $\lim_{x \to 4} (x^3 - 7)$

11. $\lim_{x \to 5} \dfrac{x^2 + 1}{5 + x}$

12. $\lim_{x \to 6} \left(\sqrt{6x} + 3x - \dfrac{1}{x} \right)(x^2 - 4)$

13. $\lim_{x \to 7} (x + \sqrt{x - 6})(x^2 - 2x + 1)$

14. $\lim_{x \to 8} \dfrac{\sqrt{5x - 4} - 1}{3x^2 + 2}$

15. $\lim_{x \to -5} \dfrac{\sqrt{x^2 - 5x - 36}}{8 - 3x}$

16. $\lim_{x \to 10} (2x^2 - 15x - 50)^{20}$

17. $\lim_{x \to 0} \dfrac{x^2 + 3x}{x}$

18. $\lim_{x \to 1} \dfrac{x^2 - 1}{x - 1}$

19. $\lim_{x \to 2} \dfrac{-2x^2 + 4x}{x - 2}$

20. $\lim_{x \to 3} \dfrac{x^2 - x - 6}{x - 3}$

21. $\lim_{x \to 4} \dfrac{x^2 - 16}{4 - x}$

22. $\lim_{x \to 5} \dfrac{2x - 10}{x^2 - 25}$

23. $\lim_{x \to 6} \dfrac{x^2 - 6x}{x^2 - 5x - 6}$

24. $\lim_{x \to 7} \dfrac{x^3 - 2x^2 + 3x}{x^2}$

25. $\lim_{x \to 8} \dfrac{x^2 + 64}{x - 8}$

26. $\lim_{x \to 9} \dfrac{1}{(x - 9)^2}$

27. Compute the limits that exist, given that

$$\lim_{x \to 0} f(x) = -\frac{1}{2} \quad \text{and} \quad \lim_{x \to 0} g(x) = \frac{1}{2}.$$

(a) $\lim_{x \to 0} (f(x) + g(x))$

(b) $\lim_{x \to 0} (f(x) - 2g(x))$

(c) $\lim_{x \to 0} f(x) \cdot g(x)$

(d) $\lim_{x \to 0} \dfrac{f(x)}{g(x)}$

28. Use the limit definition of the derivative to show that if $f(x) = mx + b$, then $f'(x) = m$.

Use limits to compute the following derivatives.

29. $f'(3)$, where $f(x) = x^2 + 1$

30. $f'(2)$, where $f(x) = x^3$

31. $f'(0)$, where $f(x) = x^3 + 3x + 1$

32. $f'(0)$, where $f(x) = x^2 + 2x + 2$

In Exercises 33–36, apply the three-step method to compute $f'(x)$ for the given function. Follow the steps that we used in Example 6. Make sure to simplify the difference quotient as much as possible before taking limits.

33. $f(x) = x^2 + 1$

34. $f(x) = -x^2 + 2$

35. $f(x) = x^3 - 1$

36. $f(x) = -3x^2 + 1$

In Exercises 37–48, use limits to compute $f'(x)$. [*Hint*: In Exercises 45–48, use the rationalization trick of Example 8.]

37. $f(x) = 3x + 1$

38. $f(x) = -x + 11$

39. $f(x) = x + \dfrac{1}{x}$

40. $f(x) = \dfrac{1}{x^2}$

41. $f(x) = \dfrac{x}{x + 1}$

42. $f(x) = -1 + \dfrac{2}{x - 2}$

43. $f(x) = \dfrac{1}{x^2 + 1}$

44. $f(x) = \dfrac{x}{x + 2}$

45. $f(x) = \sqrt{x + 2}$

46. $f(x) = \sqrt{x^2 + 1}$

47. $f(x) = \dfrac{1}{\sqrt{x}}$

48. $f(x) = x\sqrt{x}$

Each limit in Exercises 49–54 is a definition of $f'(a)$. Determine the function $f(x)$ and the value of a.

49. $\lim_{h \to 0} \dfrac{(1 + h)^2 - 1}{h}$

50. $\lim_{h \to 0} \dfrac{(2 + h)^3 - 8}{h}$

51. $\lim\limits_{h \to 0} \dfrac{\dfrac{1}{10 + h} - .1}{h}$

52. $\lim\limits_{h \to 0} \dfrac{(64 + h)^{1/3} - 4}{h}$

53. $\lim\limits_{h \to 0} \dfrac{\sqrt{9 + h} - 3}{h}$

54. $\lim\limits_{h \to 0} \dfrac{(1 + h)^{-1/2} - 1}{h}$

Compute the following limits.

55. $\lim\limits_{x \to \infty} \dfrac{1}{x^2}$

56. $\lim\limits_{x \to -\infty} \dfrac{1}{x^2}$

57. $\lim\limits_{x \to \infty} \dfrac{5x + 3}{3x - 2}$

58. $\lim\limits_{x \to \infty} \dfrac{1}{x - 8}$

59. $\lim\limits_{x \to \infty} \dfrac{10x + 100}{x^2 - 30}$

60. $\lim\limits_{x \to \infty} \dfrac{x^2 + x}{x^2 - 1}$

In Exercises 61–66, refer to Fig. 5 to find the given limit.

61. $\lim\limits_{x \to 0} f(x)$

62. $\lim\limits_{x \to \infty} f(x)$

63. $\lim\limits_{x \to 0} x f(x)$

64. $\lim\limits_{x \to \infty} (1 + 2f(x))$

65. $\lim\limits_{x \to \infty} (1 - f(x))$

66. $\lim\limits_{x \to 0} [f(x)]^2$

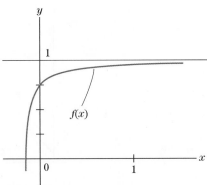

Figure 5

TECHNOLOGY EXERCISES

Examine the graph of the function and evaluate the function-at-large values of x to guess the value of the limit.

67. $\lim\limits_{x \to \infty} \sqrt{25 + x} - \sqrt{x}$

68. $\lim\limits_{x \to \infty} \dfrac{x^2}{2^x}$

69. $\lim\limits_{x \to \infty} \dfrac{x^2 - 2x + 3}{2x^2 + 1}$

70. $\lim\limits_{x \to \infty} \dfrac{-8x^2 + 1}{x^2 + 1}$

Solutions to Check Your Understanding 1.4

1. The function under consideration is a rational function. Since the denominator has value 0 at $x = 6$, we cannot immediately determine the limit by just evaluating the function at $x = 6$. Also,

$$\lim\limits_{x \to 6} (x - 6) = 0.$$

Since the function in the denominator has the limit 0, we cannot apply Limit Theorem VI. However, since the definition of limit considers only values of x different from 6, the quotient can be simplified by factoring and canceling:

$$\frac{x^2 - 4x - 12}{x - 6} = \frac{(x + 2)(x - 6)}{(x - 6)} = x + 2 \quad \text{for } x \neq 6.$$

Now, $\lim\limits_{x \to 6} (x + 2) = 8$. Therefore,

$$\lim\limits_{x \to 6} \frac{x^2 - 4x - 12}{x - 6} = 8.$$

2. No limit exists. It is easily seen that $\lim\limits_{x \to 6} (4x + 12) = 36$ and $\lim\limits_{x \to 6} (x - 6) = 0$. As x approaches 6, the denominator gets very small and the numerator approaches 36. For example, if $x = 6.00001$, the numerator is 36.00004 and the denominator is .00001. The quotient is 3,600,004. As x approaches 6 even more closely, the quotient gets arbitrarily large and cannot possibly approach a limit.

1.5 Differentiability and Continuity

In the preceding section, we defined differentiability of $f(x)$ at $x = a$ in terms of a limit. If this limit does not exist, then we say that $f(x)$ is *nondifferentiable* at $x = a$. Geometrically, the nondifferentiability of $f(x)$ at $x = a$ can manifest itself in several different ways. First, the graph of $f(x)$ could have no tangent line at $x = a$. Second, the graph could have a vertical tangent line at $x = a$. (Recall that slope is not defined for vertical lines.) Some of the various geometric possibilities are illustrated in Fig. 1.

The following example illustrates how functions that are nondifferentiable at a specific point can arise in practice.

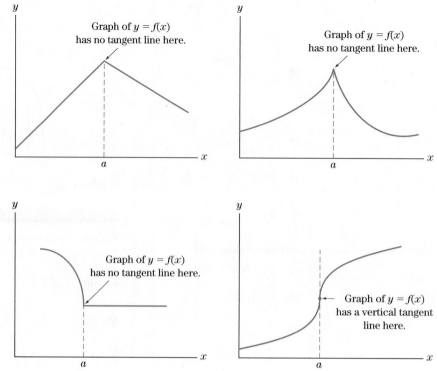

Figure 1 Functions that are nondifferentiable at $x = a$.

EXAMPLE 1

Hauling Cost A railroad company charges $10 per mile to haul a boxcar up to 200 miles and $8 per mile for each mile exceeding 200. In addition, the railroad charges a $1000 handling charge per boxcar. Graph the cost of sending a boxcar x miles.

SOLUTION

If x is at most 200 miles, then the cost $C(x)$ is given by $C(x) = 1000 + 10x$ dollars. The cost for 200 miles is $C(200) = 1000 + 2000 = 3000$ dollars. If x exceeds 200 miles, then the total cost will be

$$C(x) = \underbrace{3000}_{\substack{\text{cost of first} \\ \text{200 miles}}} + \underbrace{8(x - 200)}_{\substack{\text{cost of miles in} \\ \text{excess of 200}}} = 1400 + 8x.$$

Thus,

$$C(x) = \begin{cases} 1000 + 10x, & \text{for } 0 < x \le 200, \\ 1400 + 8x, & \text{for } x > 200. \end{cases}$$

Figure 2 Cost of hauling a boxcar.

The graph of $C(x)$ is sketched in Fig. 2. Note that $C(x)$ is not differentiable at $x = 200$.   ≪

Closely related to the concept of differentiability is that of continuity. We say that a function $f(x)$ is **continuous** at $x = a$ provided that, roughly speaking, its graph has no breaks (or gaps) as it passes through the point $(a, f(a))$. That is, $f(x)$ is continuous at $x = a$ provided that we can draw the graph through $(a, f(a))$ without lifting our pencil from the paper. The functions whose graphs are drawn in Figs. 1 and 2 are continuous for all values of x. By contrast, however, the function whose graph is drawn in Fig. 3(a) is not continuous (we say it is **discontinuous**) at $x = 1$ and $x = 2$, since the graph has breaks there. Similarly, the function whose graph is drawn in Fig. 3(b) is discontinuous at $x = 2$.

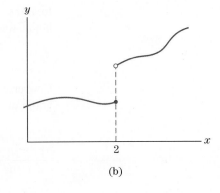

Figure 3 Functions with discontinuities.

Discontinuous functions can occur in applications, as the following example shows.

EXAMPLE 2

Manufacturing Cost A manufacturing plant is capable of producing 15,000 units in one shift of 8 hours. For each shift worked, there is a fixed cost of $2000 (for light, heat, and the like). If the variable cost (the cost of labor and raw materials) is $2 per unit, graph the cost $C(x)$ of manufacturing x units.

SOLUTION   If $x \leq 15,000$, a single shift will suffice, so

$$C(x) = 2000 + 2x, \qquad 0 \leq x \leq 15,000.$$

If x is between 15,000 and 30,000, one extra shift will be required, and

$$C(x) = 4000 + 2x, \qquad 15,000 < x \leq 30,000.$$

If x is between 30,000 and 45,000, the plant will need to work three shifts, and

$$C(x) = 6000 + 2x, \qquad 30,000 < x \leq 45,000.$$

The graph of $C(x)$ for $0 \leq x \leq 45,000$ is drawn in Fig. 4. Note that the graph has breaks at two points.

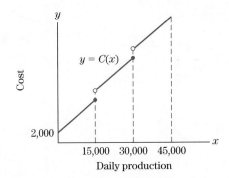

Figure 4 Cost function of a manufacturing plant.

The relationship between differentiability and continuity is this:

Theorem I   If $f(x)$ is differentiable at $x = a$, then $f(x)$ is continuous at $x = a$.

Note, however, that the converse statement is definitely false: A function may be continuous at $x = a$ but still not be differentiable there. The functions whose graphs are drawn in Fig. 1 provide examples of this phenomenon. We prove Theorem I at the end of the section.

Just as with differentiability, the notion of continuity can be phrased in terms of limits. For $f(x)$ to be continuous at $x = a$, the values of $f(x)$ for all x near a must be close to $f(a)$ (otherwise, the graph would have a break at $x = a$). In fact, the closer x is to a, the closer $f(x)$ must be to $f(a)$ (again, to avoid a break in the graph). In terms of limits, we must therefore have

$$\lim_{x \to a} f(x) = f(a).$$

Conversely, an intuitive argument shows that, if the preceding limit relation holds, the graph of $y = f(x)$ has no break at $x = a$.

DEFINITION Continuity at a Point   A function $f(x)$ is continuous at $x = a$ provided that the following limit relation holds:

$$\lim_{x \to a} f(x) = f(a). \tag{1}$$

For formula (1) to hold, three conditions must be fulfilled:

1. $f(x)$ must be defined at $x = a$.
2. $\lim\limits_{x \to a} f(x)$ must exist.
3. The limit $\lim\limits_{x \to a} f(x)$ must have the value $f(a)$.

A function will fail to be continuous at $x = a$ when any one of these conditions fails to hold. The various possibilities are illustrated in the next example.

EXAMPLE 3   **Continuity at a Point** Determine whether the functions whose graphs are drawn in Fig. 5 are continuous at $x = 3$. Use the limit definition.

(a)

(b)

(c)

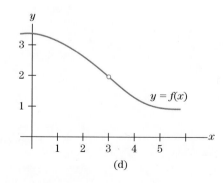

(d)

Figure 5

SOLUTION   (a) Here, $\lim\limits_{x \to 3} f(x) = 2$. However, $f(3) = 4$. So,

$$\lim_{x \to 3} f(x) \neq f(3)$$

and $f(x)$ is not continuous at $x = 3$. (Geometrically, this is clear. The graph has a break at $x = 3$.)

(b) $\lim_{x \to 3} g(x)$ does not exist, so $g(x)$ is not continuous at $x = 3$.

(c) $\lim_{x \to 3} h(x)$ does not exist, so $h(x)$ is not continuous at $x = 3$.

(d) $f(x)$ is not defined at $x = 3$, so $f(x)$ is not continuous at $x = 3$.

» Now Try Exercise 5

Using our result on the limit of a polynomial function (Section 1.4), we see that

$$p(x) = a_0 + a_1 x + \cdots + a_n x^n, \quad a_0, \ldots, a_n \text{ constants,}$$

is continuous for all x. Similarly, a rational function

$$\frac{p(x)}{q(x)}, \quad p(x), q(x) \text{ polynomials,}$$

is continuous at all x for which $q(x) \neq 0$.

Proof of Theorem I Suppose that f is differentiable at $x = a$, and let us show that f is continuous at $x = a$. It is enough to show that

$$\lim_{x \to a} (f(x) - f(a)) = 0. \tag{2}$$

For $x \neq a$, write $x = a + h$, where $h \neq 0$, and note that x approaches a if, and only if, h approaches 0. Then,

$$\lim_{x \to a} (f(x) - f(a)) = \lim_{h \to 0} \left[\overbrace{\frac{f(a+h) - f(a)}{h}}^{f(x)} \cdot h \right]$$

$$= \lim_{h \to 0} \left[\frac{f(a+h) - f(a)}{h} \right] \cdot \lim_{h \to 0} h,$$

by Limit Theorem V of the previous section. To evaluate the product of these limits, note that $\lim_{h \to 0} h = 0$ and, since f is differentiable at a,

$$\lim_{h \to 0} \frac{f(a+h) - f(a)}{h} = f'(a).$$

So the product of the limits is $f'(a) \cdot 0 = 0$, and hence formula (2) holds.

INCORPORATING TECHNOLOGY

Piecewise-Defined Functions Let us graph the function

$$f(x) = \begin{cases} (5/2)x - 1/2 & \text{for } -1 \leq x \leq 1, \\ (1/2)x - 2 & \text{for } x > 1. \end{cases}$$

To do so, we will enter the function into our calculator as

$$Y_1 = (-1 \leq X) * (X \leq 1) * ((5/2)X - (1/2)) + (X > 1) * ((1/2)X - 2)$$

and then press $\boxed{\text{GRAPH}}$. The result is shown in Fig. 6 with window setting **ZDecimal** and graph mode set to **Dot**. The graph mode can be changed from the $\boxed{\text{MODE}}$ menu.

To understand this expression, let us look closely at the first term. The inequality relations can be accessed and entered into the formula from the menu under $\boxed{\text{2nd}}$ [TEST]. The calculator gives the expression $-1 \leq X$ the value 1, when the value of X is greater than or equal to -1, and it gives the expression the value 0 otherwise. Similarly, the calculator gives the expression $X \leq 1$ the value 1, when the value of X is less than or equal to 1, and it gives the expression the value 0 otherwise. Therefore, $(-1 \leq X) * (X \leq 1)$ has

Figure 6

value 1 when both inequalities are true, and it has value 0 otherwise. And, in that case, the inequality $X > 1$ will have value 0. Hence, our formula for Y_1 takes on the value of

$$(5/2)X - (1/2)$$

when $-1 \le X \le 1$ and, following the same rules, our formula takes on the value of $(1/2)X - 2$ when $X > 1$.

Check Your Understanding 1.5

Solutions can be found following the section exercises.

Let $f(x) = \begin{cases} \dfrac{x^2 - x - 6}{x - 3} & \text{for } x \ne 3 \\ 4 & \text{for } x = 3. \end{cases}$

1. Is $f(x)$ continuous at $x = 3$?
2. Is $f(x)$ differentiable at $x = 3$?

EXERCISES 1.5

Is the function, whose graph is drawn in Fig. 7, continuous at the following values of x?

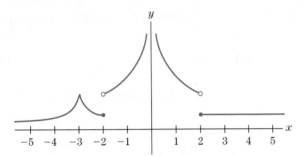

Figure 7

1. $x = 0$	**2.** $x = -3$	**3.** $x = 3$
4. $x = .001$	**5.** $x = -2$	**6.** $x = 2$

Is the function, whose graph is drawn in Fig. 7, differentiable at the following values of x?

7. $x = 0$	**8.** $x = -3$	**9.** $x = 3$
10. $x = .001$	**11.** $x = -2$	**12.** $x = 2$

Determine whether each of the following functions is continuous and/or differentiable at $x = 1$.

13. $f(x) = x^2$

14. $f(x) = \dfrac{1}{x}$

15. $f(x) = \begin{cases} x + 2 & \text{for } -1 \le x \le 1 \\ 3x & \text{for } 1 < x \le 5 \end{cases}$

16. $f(x) = \begin{cases} x^3 & \text{for } 0 \le x < 1 \\ x & \text{for } 1 \le x \le 2 \end{cases}$

17. $f(x) = \begin{cases} 2x - 1 & \text{for } 0 \le x \le 1 \\ 1 & \text{for } 1 < x \end{cases}$

18. $f(x) = \begin{cases} x & \text{for } x \ne 1 \\ 2 & \text{for } x = 1 \end{cases}$

19. $f(x) = \begin{cases} \dfrac{1}{x - 1} & \text{for } x \ne 1 \\ 0 & \text{for } x = 1 \end{cases}$

20. $f(x) = \begin{cases} x - 1 & \text{for } 0 \le x < 1 \\ 1 & \text{for } x = 1 \\ 2x - 2 & \text{for } x > 1 \end{cases}$

The functions in Exercises 21–26 are defined for all x except for one value of x. If possible, define $f(x)$ at the exceptional point in a way that makes $f(x)$ continuous for all x.

21. $f(x) = \dfrac{x^2 - 7x + 10}{x - 5}, x \ne 5$

22. $f(x) = \dfrac{x^2 + x - 12}{x + 4}, x \ne -4$

23. $f(x) = \dfrac{x^3 - 5x^2 + 4}{x^2}, x \ne 0$

24. $f(x) = \dfrac{x^2 + 25}{x - 5}, x \ne 5$

25. $f(x) = \dfrac{(6 + x)^2 - 36}{x}, x \ne 0$

26. $f(x) = \dfrac{\sqrt{9 + x} - \sqrt{9}}{x}, x \ne 0$

27. Computing Income Tax The tax that you pay to the federal government is a percentage of your *taxable income*, which is what remains of your gross income after you subtract your allowed deductions. In a recent year, there were five rates or brackets for a single taxpayer, as shown in Table 1.

Table 1　Single Taxpayer Rates

Amount Over	But Not Over	Tax Rate
$0	$27,050	15%
$27,050	$65,550	27.5%
$65,550	$136,750	30.5%
$136,750	$297,350	35.5%
$297,350	· · ·	39.1%

So, if you are single and your taxable income was less than $27,050, your tax is your taxable income times 15% (.15). The maximum amount of tax that you will pay on your income in this first bracket is 15% of $27,050, or (.15) × 27,050 = 4057.50 dollars. If your taxable income is more than $27,050 but less than $65,550, your tax is $4057.50 plus 27.5% of the amount in excess of $27,050. So, for example, if your taxable income is $50,000, your tax is 4057.5 + .275(50,000 − 27,050) = 4057.5 + .275 × 22,950 = $10,368.75. Let x denote your taxable income and $T(x)$ your tax.

(a) Find a formula for $T(x)$ if x is not over \$136,750.
(b) Plot the graph of $T(x)$ for $0 \le x \le 136{,}750$.
(c) Find the maximum amount of tax that you will pay on the portion of your income in the second tax bracket. Express this amount as a difference between two values of $T(x)$.

28. Refer to Exercise 27.
(a) Find a formula for $T(x)$ for all taxable income x.
(b) Plot $T(x)$.
(c) Determine the maximum amount of tax that you will pay on the portion of your income in the fourth tax bracket.

29. Revenue from Sales The owner of a photocopy store charges 7 cents per copy for the first 100 copies and 4 cents per copy for each copy exceeding 100. In addition, there is a setup fee of \$2.50 for each photocopying job.
(a) Determine $R(x)$, the revenue from selling x copies.
(b) If it costs the store owner 3 cents per copy, what is the profit from selling x copies? (Recall that profit is revenue minus cost.)

30. Do Exercise 29 if it costs 10 cents per copy for the first 50 copies and 5 cents per copy for each copy exceeding 50, and there is no setup fee.

31. Department Store Sales The graph in Fig. 8 shows the total sales in thousands of dollars in a department store during a typical 24-hour period.
(a) Estimate the rate of sales during the period between 8 A.M. and 10 A.M.
(b) Which 2-hour interval in the day sees the highest rate of sales and what is this rate?

32. Refer to Exercise 31.
(a) From midnight to noon, which 2-hour time intervals have the same rate of sales and what is this rate?
(b) What is the total amount of sales between midnight and 8 A.M.? Compare this amount to the total sales in the period between 8 A.M. and 10 A.M.

In Exercises 33 and 34, determine the value of a that makes the function $f(x)$ continuous at $x = 0$.

33. $f(x) = \begin{cases} 1 & \text{for } x \ge 0 \\ x + a & \text{for } x < 0 \end{cases}$

34. $f(x) = \begin{cases} 2(x - a) & \text{for } x \ge 0 \\ x^2 + 1 & \text{for } x < 0 \end{cases}$

Figure 8

Solutions to Check Your Understanding 1.5

1. The function $f(x)$ is defined at $x = 3$, that is, $f(3) = 4$. When computing $\lim_{x \to 3} f(x)$, we exclude consideration of $x = 3$; therefore, we can simplify the expression for $f(x)$ as follows:

$$f(x) = \frac{x^2 - x - 6}{x - 3} = \frac{(x - 3)(x + 2)}{x - 3} = x + 2$$

$$\lim_{x \to 3} f(x) = \lim_{x \to 3} (x + 2) = 5.$$

Since $\lim_{x \to 3} f(x) = 5 \ne 4 = f(3)$, $f(x)$ is not continuous at $x = 3$.

2. There is no need to compute any limits to answer this question. By Theorem I, since $f(x)$ is not continuous at $x = 3$, it cannot possibly be differentiable there.

1.6 Some Rules for Differentiation

Three additional rules of differentiation greatly extend the number of functions that we can differentiate.

1. Constant-multiple rule: $\dfrac{d}{dx} [k \cdot f(x)] = k \cdot \dfrac{d}{dx} [f(x)]$, k a constant.

2. Sum rule: $\dfrac{d}{dx} [f(x) + g(x)] = \dfrac{d}{dx} [f(x)] + \dfrac{d}{dx} [g(x)]$.

3. General power rule: $\dfrac{d}{dx} ([g(x)]^r) = r \cdot [g(x)]^{r-1} \cdot \dfrac{d}{dx} [g(x)]$.

We shall discuss these rules, present some examples, and then prove the first two rules at the end of the section.

The Constant-Multiple Rule Starting with a function $f(x)$, we can multiply it by a constant number k to obtain a new function $k \cdot f(x)$. For instance, if $f(x) = x^2 - 4x + 1$ and $k = 2$, then,

$$2f(x) = 2(x^2 - 4x + 1) = 2x^2 - 8x + 2.$$

The constant-multiple rule says that the derivative of the new function $k \cdot f(x)$ is just k times the derivative of the original function. In other words, when faced with the differentiation of a constant times a function, simply carry along the constant and differentiate the function.

EXAMPLE 1 **Using the Constant-Multiple Rule** Calculate.

(a) $\dfrac{d}{dx}(2x^5)$ (b) $\dfrac{d}{dx}\left(\dfrac{x^3}{4}\right)$ (c) $\dfrac{d}{dx}\left(-\dfrac{3}{x}\right)$ (d) $\dfrac{d}{dx}(5\sqrt{x})$

SOLUTION (a) With $k = 2$ and $f(x) = x^5$, we have

$$\frac{d}{dx}(2 \cdot x^5) = 2 \cdot \frac{d}{dx}(x^5) = 2(5x^4) = 10x^4. \quad \text{Constant-multiple rule.}$$

(b) Write $\dfrac{x^3}{4}$ in the form $\dfrac{1}{4} \cdot x^3$. Then,

$$\frac{d}{dx}\left(\frac{x^3}{4}\right) = \frac{1}{4} \cdot \frac{d}{dx}(x^3) = \frac{1}{4}(3x^2) = \frac{3}{4}x^2. \quad \text{Constant-multiple and power rules.}$$

(c) Write $-\dfrac{3}{x}$ in the form $(-3) \cdot \dfrac{1}{x}$. Then,

$$\frac{d}{dx}\left(-\frac{3}{x}\right) = (-3) \cdot \frac{d}{dx}\left(\frac{1}{x}\right) = (-3) \cdot \frac{-1}{x^2} = \frac{3}{x^2}.$$

(d) $\dfrac{d}{dx}(5\sqrt{x}) = 5\dfrac{d}{dx}(\sqrt{x}) = 5\dfrac{d}{dx}(x^{1/2}) = \dfrac{5}{2}x^{-1/2}.$ Power rule: $\dfrac{d}{dx}(x^{1/2}) = x^{\frac{1}{2}-1} = x^{-\frac{1}{2}}.$

This answer may also be written in the form $\dfrac{5}{2\sqrt{x}}$.   **» Now Try Exercise 3**

The Sum Rule To differentiate a sum of functions, differentiate each function individually and add together the derivatives. Another way of saying this is "the derivative of a sum of functions is the sum of the derivatives."

EXAMPLE 2 **Using the Sum Rule** Find each of the following.

(a) $\dfrac{d}{dx}(x^3 + 5x)$ (b) $\dfrac{d}{dx}\left(x^4 - \dfrac{3}{x^2}\right)$ (c) $\dfrac{d}{dx}(2x^7 - x^5 + 8)$

SOLUTION (a) Let $f(x) = x^3$ and $g(x) = 5x$. Then,

$$\frac{d}{dx}(x^3 + 5x) = \frac{d}{dx}(x^3) + \frac{d}{dx}(5x) \quad \text{Sum rule}$$

$$= 3x^2 + 5. \quad \text{Power rule}$$

(b) The sum rule applies to differences as well as sums (see Exercise 46). Indeed, by the sum rule,

$$\frac{d}{dx}\left(x^4 - \frac{3}{x^2}\right) = \frac{d}{dx}(x^4) + \frac{d}{dx}\left(-\frac{3}{x^2}\right) \quad \text{Sum rule}$$

$$= \frac{d}{dx}(x^4) - 3\frac{d}{dx}(x^{-2}) \quad \text{Constant-multiple rule}$$

$$= 4x^3 - 3(-2x^{-3}) \quad \text{Power rule}$$

$$= 4x^3 + 6x^{-3}.$$

After some practice, we usually omit most or all of the intermediate steps and simply write

$$\frac{d}{dx}\left(x^4 - \frac{3}{x^2}\right) = 4x^3 + 6x^{-3}.$$

(c) We apply the sum rule repeatedly and use the fact that the derivative of a constant function is 0:

$$\frac{d}{dx}(2x^7 - x^5 + 8) = \frac{d}{dx}(2x^7) - \frac{d}{dx}(x^5) + \frac{d}{dx}(8)$$

$$= 2(7x^6) - 5x^4 + 0$$

$$= 14x^6 - 5x^4. \qquad \text{➤➤ Now Try Exercise 7}$$

NOTE ▶ The differentiation of a function *plus* a constant is different from the differentiation of a constant *times* a function. Figure 1 shows the graphs of $f(x)$, $f(x) + 2$, and $2 \cdot f(x)$, where $f(x) = x^3 - \frac{3}{2}x^2$. For each x, the graphs of $f(x)$ and $f(x) + 2$ have the same slope. In contrast, for each x, the slope of the graph of $2 \cdot f(x)$ is twice the slope of the graph of $f(x)$. Upon differentiation, an added constant disappears, whereas a constant that multiplies a function is carried along.

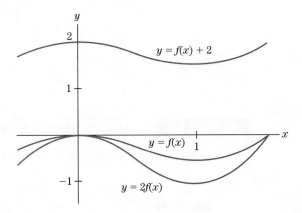

Figure 1 Two effects of a constant on the graph of $f(x)$.   ◀◀

The General Power Rule   Frequently, we will encounter expressions of the form $[g(x)]^r$, for instance, $(x^3 + 5)^2$, where $g(x) = x^3 + 5$ and $r = 2$. The general power rule says that, to differentiate $[g(x)]^r$, we must first treat $g(x)$ as if it were simply an x, form $r[g(x)]^{r-1}$, and then multiply it by a "correction factor" $g'(x)$. Thus,

$$\frac{d}{dx}(x^3 + 5)^2 = 2(x^3 + 5)^1 \cdot \frac{d}{dx}(x^3 + 5)$$

$$= 2(x^3 + 5) \cdot (3x^2)$$

$$= 6x^2(x^3 + 5).$$

In this special case, it is easy to verify that the general power rule gives the correct answer. We first expand $(x^3 + 5)^2$ and then differentiate.

$$(x^3 + 5)^2 = (x^3 + 5)(x^3 + 5) = x^6 + 10x^3 + 25$$

From the constant-multiple rule and the sum rule, we have

$$\frac{d}{dx}(x^3 + 5)^2 = \frac{d}{dx}(x^6 + 10x^3 + 25)$$

$$= 6x^5 + 30x^2 + 0$$

$$= 6x^2(x^3 + 5).$$

The two methods give the same answer. We state the rule for ease of reference.

General Power Rule

$$\frac{d}{dx}[g(x)]^r = r[g(x)]^{r-1} \cdot g'(x)$$

Note that if we set $g(x) = x$ in the general power rule, we recover the power rule. So, the general power rule contains the power rule as a special case.

EXAMPLE 3 **Differentiating a Radical** Differentiate $\sqrt{1 - x^2}$.

SOLUTION

$$\frac{d}{dx}(\sqrt{1 - x^2}) = \frac{d}{dx}[(1 - x^2)^{1/2}] \qquad \text{Change the radical to a power } \frac{1}{2}.$$

$$= \frac{1}{2}(1 - x^2)^{-1/2} \cdot \frac{d}{dx}(1 - x^2) \qquad \text{General power rule.}$$

$$= \frac{1}{2}(1 - x^2)^{-1/2} \cdot (-2x)$$

$$= \frac{-x}{(1 - x^2)^{1/2}} = \frac{-x}{\sqrt{1 - x^2}} \qquad \begin{array}{l}\text{Change the negative power to a} \\ \text{radical and simplify.}\end{array}$$

>> *Now Try Exercise 15*

EXAMPLE 4 **Using the General Power Rule** Differentiate

$$y = \frac{1}{x^3 + 4x}.$$

SOLUTION

$$y = \frac{1}{x^3 + 4x} = (x^3 + 4x)^{-1} \qquad \begin{array}{l}\text{Change denominator to a} \\ (-1) \text{ power.}\end{array}$$

$$\frac{dy}{dx} = (-1)(x^3 + 4x)^{-2} \cdot \frac{d}{dx}(x^3 + 4x) \qquad \text{General power rule}$$

$$= \frac{-1}{(x^3 + 4x)^2}(3x^2 + 4)$$

$$= -\frac{3x^2 + 4}{(x^3 + 4x)^2}.$$

>> *Now Try Exercise 21*

Some derivatives require more than one differentiation rule.

EXAMPLE 5   **Using the General Power Rule** Differentiate $5\sqrt[3]{1 + x^3}$.

SOLUTION
$$\frac{d}{dx}\left(5\sqrt[3]{1 + x^3}\right) = 5\frac{d}{dx}\left[\sqrt[3]{1 + x^3}\right] \qquad \text{Constant-multiple rule}$$

$$= 5\frac{d}{dx}\left[(1 + x^3)^{1/3}\right] \qquad \text{Change radical to a power.}$$

$$= 5\left(\frac{1}{3}\right)(1 + x^3)^{\frac{1}{3}-1}\frac{d}{dx}\left[(1 + x^3)\right] \qquad \text{General power rule}$$

$$= \frac{5}{3}(1 + x^3)^{-\frac{2}{3}}(3x^2) = 5x^2(1 + x^3)^{-\frac{2}{3}}.$$

» Now Try Exercise 25

Example 5 illustrates how the differentiation rules can be combined. For example, as a by-product of the constant-multiple rule and the general power rule, we have

$$\frac{d}{dx}\left(k \cdot [g(x)]^r\right) = kr \cdot [g(x)]^{r-1}g'(x).$$

Let us verify both rules when x has the value a. Recall that if $f(x)$ is differentiable at $x = a$, then its derivative is the limit

$$\lim_{h \to 0}\frac{f(a + h) - f(a)}{h}.$$

APPENDIX Proofs of the Constant-Multiple and Sum Rules

Constant-Multiple Rule   We assume that $f(x)$ is differentiable at $x = a$. We must prove that $k \cdot f(x)$ is differentiable at $x = a$ and that its derivative is $k \cdot f'(x)$. This amounts to showing that the limit

$$\lim_{h \to 0}\frac{k \cdot f(a + h) - k \cdot f(a)}{h}$$

exists and has the value $k \cdot f'(a)$. However,

$$\lim_{h \to 0}\frac{k \cdot f(a + h) - k \cdot f(a)}{h}$$

$$= \lim_{h \to 0} k\left[\frac{f(a + h) - f(a)}{h}\right]$$

$$= k \cdot \lim_{h \to 0}\frac{f(a + h) - f(a)}{h} \qquad \text{Constant-multiple Limit Theorem I, Sec. 1.4.}$$

$$= k \cdot \lim_{h \to 0}\frac{f(a + h) - f(a)}{h} = kf'(a) \qquad \text{Since } f(x) \text{ is differentiable at } x = a.$$

which is what we desired to show.

Sum Rule   We assume that both $f(x)$ and $g(x)$ are differentiable at $x = a$. We must prove that $f(x) + g(x)$ is differentiable at $x = a$ and that its derivative is $f'(a) + g'(a)$. That is, we must show that the limit

$$\lim_{h \to 0}\frac{[f(a + h) + g(a + h)] - [f(a) + g(a)]}{h}$$

exists and equals $f'(a) + g'(a)$. Using Limit of a sum, Theorem III, Sec. 1.4, and the fact that $f(x)$ and $g(x)$ are differentiable at $x = a$, we have

$$\lim_{h \to 0} \frac{[f(a + h) + g(a + h)] - [f(a) + g(a)]}{h}$$

$$= \lim_{h \to 0} \left[\frac{f(a + h) - f(a)}{h} + \frac{g(a + h) - g(a)}{h} \right]$$

$$= \lim_{h \to 0} \frac{f(a + h) - f(a)}{h} + \lim_{h \to 0} \frac{g(a + h) - g(a)}{h} \qquad \text{Limit of a sum.}$$

$$= f'(a) + g'(a).$$

In Chapter 3, the general power rule will be proved as a special case of the chain rule.

Check Your Understanding 1.6

Solutions can be found following the section exercises.

1. Find the derivative $\frac{d}{dx}(x)$.

2. Differentiate the function $y = \dfrac{x + (x^5 + 1)^{10}}{3}$.

EXERCISES 1.6

Differentiate.

1. $y = 6x^3$

2. $y = 3x^4$

3. $y = 3\sqrt[3]{x}$

4. $y = \dfrac{1}{3x^3}$

5. $y = \dfrac{x}{2} - \dfrac{2}{x}$

6. $f(x) = 12 + \dfrac{1}{7^3}$

7. $f(x) = x^4 + x^3 + x$

8. $y = 4x^3 - 2x^2 + x + 1$

9. $y = (2x + 4)^3$

10. $y = (x^2 - 1)^3$

11. $y = (x^3 + x^2 + 1)^7$

12. $y = (x^2 + x)^{-2}$

13. $y = \dfrac{4}{x^2}$

14. $y = 4(x^2 - 6)^{-3}$

15. $y = 3\sqrt[3]{2x^2 + 1}$

16. $y = 2\sqrt{x + 1}$

17. $y = 2x + (x + 2)^3$

18. $y = (x - 1)^3 + (x + 2)^4$

19. $y = \dfrac{1}{5x^5}$

20. $y = (x^2 + 1)^2 + 3(x^2 - 1)^2$

21. $y = \dfrac{1}{x^3 + 1}$

22. $y = \dfrac{2}{x + 1}$

23. $y = x + \dfrac{1}{x + 1}$

24. $y = 2\sqrt[4]{x^2 + 1}$

25. $f(x) = 5\sqrt{3x^3 + x}$

26. $y = \dfrac{1}{x^3 + x + 1}$

27. $y = 3x + \pi^3$

28. $y = \sqrt{1 + x^2}$

29. $y = \sqrt{1 + x + x^2}$

30. $y = \dfrac{1}{2x + 5}$

31. $y = \dfrac{2}{1 - 5x}$

32. $y = \dfrac{7}{\sqrt{1 + x}}$

33. $y = \dfrac{45}{1 + x + \sqrt{x}}$

34. $y = (1 + x + x^2)^{11}$

35. $y = x + 1 + \sqrt{x + 1}$

36. $y = \pi^2 x$

37. $f(x) = \left(\dfrac{\sqrt{x}}{2} + 1 \right)^{3/2}$

38. $y = \left(x - \dfrac{1}{x} \right)^{-1}$

In Exercises 39 and 40, find the slope of the graph of $y = f(x)$ at the designated point.

39. $f(x) = 3x^2 - 2x + 1,\ (1, 2)$

40. $f(x) = x^{10} + 1 + \sqrt{1 - x},\ (0, 2)$

41. Find the slope of the tangent line to the curve $y = x^3 + 3x - 8$ at (2, 6).

42. Write the equation of the tangent line to the curve $y = x^3 + 3x - 8$ at (2, 6).

43. Find the slope of the tangent line to the curve $y = (x^2 - 15)^6$ at $x = 4$. Then write the equation of this tangent line.

44. Find the equation of the tangent line to the curve $y = \dfrac{8}{x^2 + x + 2}$ at $x = 2$.

45. Differentiate the function $f(x) = (3x^2 + x - 2)^2$ in two ways.
    (a) Use the general power rule.
    (b) Multiply $3x^2 + x - 2$ by itself and then differentiate the resulting polynomial.

46. Using the sum rule and the constant-multiple rule, show that for any functions $f(x)$ and $g(x)$
$$\frac{d}{dx}[f(x) - g(x)] = \frac{d}{dx}f(x) - \frac{d}{dx}g(x).$$

47. Figure 2 contains the curves $y = f(x)$ and $y = g(x)$ and the tangent line to $y = f(x)$ at $x = 1$, with $g(x) = 3 \cdot f(x)$. Find $g(1)$ and $g'(1)$.

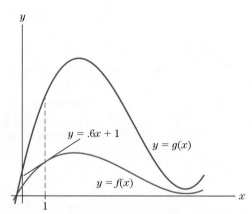

Figure 2 Graphs of $f(x)$ and $g(x) = 3f(x)$.

48. Figure 3 contains the curves $y = f(x)$, $y = g(x)$, and $y = h(x)$ and the tangent lines to $y = f(x)$ and $y = g(x)$ at $x = 1$, with $h(x) = f(x) + g(x)$. Find $h(1)$ and $h'(1)$.

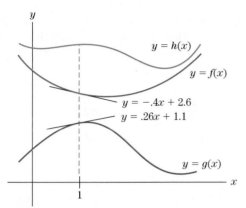

Figure 3 Graphs of $f(x)$, $g(x)$, and $h(x) = f(x) + g(x)$.

49. If $f(5) = 2$, $f'(5) = 3$, $g(5) = 4$, and $g'(5) = 1$, find $h(5)$ and $h'(5)$, where $h(x) = 3f(x) + 2g(x)$.

50. If $g(3) = 2$ and $g'(3) = 4$, find $f(3)$ and $f'(3)$, where $f(x) = 2 \cdot [g(x)]^3$.

51. If $g(1) = 4$ and $g'(1) = 3$, find $f(1)$ and $f'(1)$, where $f(x) = 5 \cdot \sqrt{g(x)}$.

52. If $h(x) = [f(x)]^2 + \sqrt{g(x)}$, determine $h(1)$ and $h'(1)$, given that $f(1) = 1$, $g(1) = 4$, $f'(1) = -1$, and $g'(1) = 4$.

53. The tangent line to the curve $y = \frac{1}{3}x^3 - 4x^2 + 18x + 22$ is parallel to the line $6x - 2y = 1$ at two points on the curve. Find the two points.

54. The tangent line to the curve $y = x^3 - 6x^2 - 34x - 9$ has slope 2 at two points on the curve. Find the two points.

55. The straight line in the figure is tangent to the graph of $f(x)$. Find $f(4)$ and $f'(4)$.

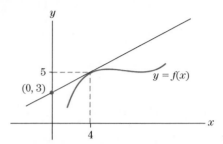

56. The straight line in the figure is tangent to the parabola. Find the value of b.

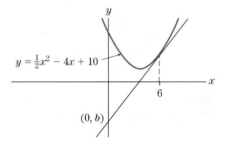

Solutions to Check Your Understanding 1.6

1. The problem asks for the derivative of the function $y = x$, a straight line of slope 1. Therefore,

$$\frac{d}{dx}(x) = 1.$$

The result can also be obtained from the power rule with $r = 1$. If $f(x) = x^1$, then,

$$\frac{d}{dx}(f(x)) = 1 \cdot x^{1-1} = x^0 = 1.$$

See the figure

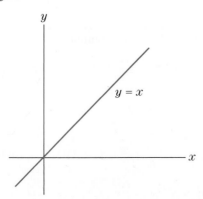

2. All three rules are required to differentiate this function.

$$\frac{dy}{dx} = \frac{d}{dx}\frac{1}{3} \cdot [x + (x^5 + 1)^{10}]$$

$$= \frac{1}{3}\frac{d}{dx}[x + (x^5 + 1)^{10}] \qquad \text{Constant-multiple rule.}$$

$$= \frac{1}{3}\left[\frac{d}{dx}(x) + \frac{d}{dx}(x^5 + 1)^{10}\right] \qquad \text{Sum rule.}$$

$$= \frac{1}{3}[1 + 10(x^5 + 1)^9 \cdot (5x^4)] \qquad \text{General power rule.}$$

$$= \frac{1}{3}[1 + 50x^4(x^5 + 1)^9]$$

1.7 More about Derivatives

In many applications, it is convenient to use variables other than x and y. One might, for instance, study the function $f(t) = t^2$ instead of writing $f(x) = x^2$. In this case, the notation for the derivative involves t rather than x, but the concept of the derivative as a slope formula is unaffected. (See Fig. 1.) When the independent variable is t instead of x, we write $\dfrac{d}{dt}$ in place of $\dfrac{d}{dx}$. For instance,

$$\frac{d}{dt}(t^3) = 3t^2, \qquad \frac{d}{dt}(2t^2 + 3t) = 4t + 3.$$

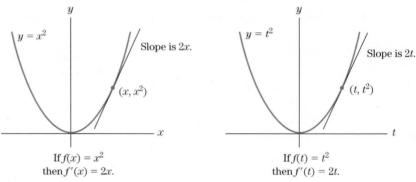

Figure 1 The same function but different variables.

Recall that if y is a function of x—say, $y = f(x)$—we may write $\dfrac{dy}{dx}$ in place of $f'(x)$. We sometimes call $\dfrac{dy}{dx}$ "the derivative of y with respect to x." Similarly, if v is a function of t, the derivative of v with respect to t is written as $\dfrac{dv}{dt}$. For example, if $v = 4t^2$, then $\dfrac{dv}{dt} = 8t$.

Of course, other letters can be used to denote variables. The formulas

$$\frac{d}{dP}(P^3) = 3P^2, \qquad \frac{d}{ds}(s^3) = 3s^2, \qquad \frac{d}{dz}(z^3) = 3z^2$$

all express the same basic fact that the slope formula for the cubic curve $y = x^3$ is given by $3x^2$.

EXAMPLE 1 **Differentiating with Respect to a Specific Variable** Compute.

(a) $\dfrac{ds}{dp}$ if $s = 3(p^2 + 5p + 1)^{10}$

(b) $\dfrac{d}{dt}(at^2 + St^{-1} + S^2)$

SOLUTION (a) $\dfrac{d}{dp}3(p^2 + 5p + 1)^{10} = 30(p^2 + 5p + 1)^9 \cdot \dfrac{d}{dp}(p^2 + 5p + 1)$ Constant-multiple and general power rules

$$= 30(p^2 + 5p + 1)^9(2p + 5).$$

(b) Although the expression $at^2 + St^{-1} + S^2$ contains several letters, the notation $\dfrac{d}{dt}$ indicates that, for purposes of calculating the derivative with respect to t, all letters except t are to be considered constants. Hence,

$$\frac{d}{dt}(at^2 + St^{-1} + S^2) = \frac{d}{dt}(at^2) + \frac{d}{dt}(St^{-1}) + \frac{d}{dt}(S^2) \quad \text{Sum rule.}$$

$$= a \cdot \frac{d}{dt}(t^2) + S \cdot \frac{d}{dt}(t^{-1}) + 0 \quad \text{Constant-multiple rule.}$$

$$= 2at - St^{-2}. \quad \text{Power rule.}$$

$$\left[\text{The derivative } \frac{d}{dt}(S^2) \text{ is zero because } S^2 \text{ is a constant.} \right] \qquad \text{➤➤ \textit{Now Try Exercise 7}}$$

The Second Derivative When we differentiate a function $f(x)$, we obtain a new function $f'(x)$ that is a formula for the slope of the curve $y = f(x)$. If we differentiate the function $f'(x)$, we obtain what is called the **second derivative** of $f(x)$, denoted by $f''(x)$. That is,

$$\frac{d}{dx}f'(x) = f''(x).$$

EXAMPLE 2 **Second Derivatives** Find the second derivatives of the following functions.
  (a) $f(x) = x^3 + (1/x)$ **(b)** $f(x) = 2x + 1$ **(c)** $f(t) = t^{1/2} + t^{-1/2}$

SOLUTION **(a)** $f(x) = x^3 + (1/x) = x^3 + x^{-1}$

  $f'(x) = 3x^2 - x^{-2}$

  $f''(x) = 6x + 2x^{-3}.$

  (b) $f(x) = 2x + 1$

  $f'(x) = 2$ A constant function whose value is 2

  $f''(x) = 0$ The derivative of a constant function is zero.

  (c) $f(t) = t^{1/2} + t^{-1/2}$

  $f'(t) = \frac{1}{2}t^{-1/2} - \frac{1}{2}t^{-3/2}$

  $f''(t) = -\frac{1}{4}t^{-3/2} + \frac{3}{4}t^{-5/2}$ ➤➤ *Now Try Exercise 15*

The first derivative of a function $f(x)$ gives the slope of the graph of $f(x)$ at any point. The second derivative of $f(x)$ gives important additional information about the shape of the curve near any point. We shall examine this subject carefully in the next chapter.

Other Notation for Derivatives Unfortunately, the process of differentiation does not have a standardized notation. Consequently, it is important to become familiar with alternative terminology.

If y is a function of x—say, $y = f(x)$—then we may denote the first and second derivatives of this function in several ways.

Prime Notation	$\dfrac{d}{dx}$ Notation
$f'(x)$	$\dfrac{d}{dx}f(x)$
y'	$\dfrac{dy}{dx}$
$f''(x)$	$\dfrac{d^2}{dx^2}f(x)$
y''	$\dfrac{d^2y}{dx^2}$

The notation $\dfrac{d^2}{dx^2}$ is purely symbolic $\left(\dfrac{d^2}{dx^2}\text{ does NOT mean "}d\text{ squared over }d\text{ times}\right.$ x squared"$\Big)$. It reminds us that the second derivative is obtained by differentiating $\dfrac{d}{dx}f(x)$; that is,

$$f'(x) = \frac{d}{dx}f(x)$$

$$f''(x) = \frac{d}{dx}\left[\frac{d}{dx}f(x)\right].$$

If we evaluate the derivative $f'(x)$ at a specific value of x—say, $x = a$—we get a number $f'(a)$ that gives the slope of the curve $y = f(x)$ at the point $(a, f(a))$. Another way of writing $f'(a)$ is

$$\left.\frac{dy}{dx}\right|_{x=a}.$$

If we have a second derivative $f''(x)$, then its value when $x = a$ is written

$$f''(a) \quad \text{or} \quad \left.\frac{d^2y}{dx^2}\right|_{x=a}.$$

| EXAMPLE 3 | Evaluating a Second Derivative |

If $y = x^4 - 5x^3 + 7$, find $\left.\dfrac{d^2y}{dx^2}\right|_{x=3}$.

SOLUTION

$$\frac{dy}{dx} = \frac{d}{dx}(x^4 - 5x^3 + 7) = 4x^3 - 15x^2$$

$$\frac{d^2y}{dx^2} = \frac{d}{dx}(4x^3 - 15x^2) = 12x^2 - 30x$$

$$\left.\frac{d^2y}{dx^2}\right|_{x=3} = 12(3)^2 - 30(3) = 108 - 90 = 18.$$

≫ Now Try Exercise 25

| EXAMPLE 4 | Evaluating First and Second Derivatives |

If $s = t^3 - 2t^2 + 3t$, find $\left.\dfrac{ds}{dt}\right|_{t=-2}$ and $\left.\dfrac{d^2s}{dt^2}\right|_{t=-2}$.

SOLUTION

$$\frac{ds}{dt} = \frac{d}{dt}(t^3 - 2t^2 + 3t) = 3t^2 - 4t + 3$$

$$\left.\frac{ds}{dt}\right|_{t=-2} = 3(-2)^2 - 4(-2) + 3 = 12 + 8 + 3 = 23.$$

To find the value of the second derivative at $t = -2$, we must first differentiate $\dfrac{ds}{dt}$.

$$\frac{d^2s}{dt^2} = \frac{d}{dt}(3t^2 - 4t + 3) = 6t - 4$$

$$\left.\frac{d^2s}{dt^2}\right|_{t=-2} = 6(-2) - 4 = -12 - 4 = -16.$$

≫ Now Try Exercise 29

The Derivative as a Rate of Change

Suppose that $y = f(x)$ is a function and $P = (a, f(a))$ a point on its graph (Fig. 2). Recall from Section 1.2 that the slope of the graph at P (the slope of the tangent line at P) measures the rate of change of the graph at P. Since the slope of the graph is $f'(a)$, we have the following very useful interpretation of the derivative:

> **Derivative as a Rate of Change**
>
> $$f'(a) = \text{rate of change of } f(x) \text{ at } x = a. \tag{1}$$

That is, as the graph of $f(x)$ is going through the point P, it is changing at the rate of $f'(a)$ units in the y-direction for every 1-unit change in x. From Fig. 2, we see that on the tangent line, the change in y, for 1 unit change in x, is equal to the slope $f'(a)$. On the graph of $f(x)$, the change in $f(x)$, for 1 unit change in x, is approximately equal to the change on the tangent line. Thus,

> **Derivative as Approximate Rate of Change for Increase of 1 Unit**
>
> $$f(a + 1) - f(a) \approx f'(a); \tag{2}$$

equivalently,

$$f(a + 1) \approx f(a) + f'(a). \tag{3}$$

(The symbol $\approx$ is used to indicate that, in general, we have an approximation and not an equality.)

This formula allows us to approximate the value of $f(x + 1)$ using the values of $f(x)$ and $f'(x)$. A generalization of this formula will be derived in the next section.

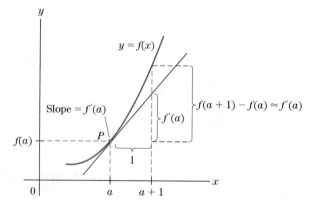

Figure 2 The slope of the graph of $y = f(x)$ at P is the slope of the tangent line at P. This slope measures the rate of change of f as its graph goes through P.

EXAMPLE 5　**Declining Sales** At the end of the holiday season in January, the sales at a department store are expected to fall (Fig. 3). It is estimated that for the x day of January the sales will be

$$S(x) = 3 + \frac{9}{(x + 1)^2} \text{ thousand dollars.}$$

(a) Compute $S(2)$ and $S'(2)$ and interpret your results.
(b) Estimate the value of sales on January 3 and compare your result with the exact value $S(3)$.

SOLUTION　(a) We have

$$S(2) = 3 + \frac{9}{(2 + 1)^2} = 4 \text{ thousand dollars.}$$

Figure 3 The sales decrease starting the first day of January.

To compute $S'(2)$, write $\frac{9}{(x+1)^2}$ as $9 \cdot (x+1)^{-2}$; then,

$$S'(x) = \frac{d}{dx}[S(x)] = \overbrace{\frac{d}{dx}[3]}^{=0} + \frac{d}{dx}(9 \cdot (x+1)^{-2})$$

$$= 9 \cdot \frac{d}{dx}((x+1)^{-2})$$

$$= 9(-2)(x+1)^{-3}\frac{d}{dx}(x+1) = -18(x+1)^{-3}(1) = \frac{-18}{(x+1)^3};$$

$$S'(2) = \frac{-18}{(2+1)^3} = -\frac{2}{3} \approx -.667.$$

The equations $S(2) = 4$ and $S'(2) = -.667$ tell us that on January 2 the sales are $4000 and are falling at the rate of .667 thousand dollars per day, or $667 per day.

(b) To estimate $S(3)$, we use (3):

$$S(3) \approx S(2) + S'(2).$$

Thus, our estimate of the sales on January 3 is $4000 - 667 = \$3333$. To compare with the exact value of sales on January 3, we compute $S(3)$ from the formula:

$$S(3) = 3 + \frac{9}{(3+1)^2} = 3 + \frac{9}{16} = \frac{57}{16} = 3.5625 \text{ thousand dollars, or } \$3562.5.$$

This is close to our estimate of $3333.   **»** *Now Try Exercise 45*

The Marginal Concept in Economics

For the sake of this discussion, suppose that $C(x)$ is a cost function (the cost of producing x units of a commodity), measured in dollars. A problem of interest to economists is to approximate the quantity $C(a + 1) - C(a)$, which is the additional cost that is incurred if the production level is increased by 1 unit from $x = a$ to $x = a + 1$. Note that $C(a + 1) - C(a)$ is also the cost of producing the $(a + 1)$ unit. Taking $f(x) = C(x)$ in (2), we find that

$$\text{additional cost} = C(a + 1) - C(a) \approx C'(a).$$

Economists refer to the derivative $C'(a)$ as the **marginal cost at production level** a or the **marginal cost** of producing a units of a commodity.

DEFINITION   **Marginal Cost**   If $C(x)$ is a cost function, then the **marginal cost function** is $C'(x)$. The marginal cost of producing a units, $C'(a)$, is approximately equal to $C(a + 1) - C(a)$, which is the additional cost that is incurred when the production level is increased by 1 unit from a to $a + 1$.

Before we give an example, let us note that if $C(x)$ is measured in dollars, where x is the number of items, then $C'(x)$, being a rate of change, is measured in dollars per item.

EXAMPLE 6 **Marginal Cost** Suppose that the cost of producing x items is $C(x) = .005x^3 - .5x^2 + 28x + 300$ dollars, and daily production is 50 items.
(a) What is the extra cost of increasing daily production from 50 to 51 items?
(b) What is the marginal cost when $x = 50$?

SOLUTION (a) The change in cost when daily production is raised from 50 to 51 items is $C(51) - C(50)$, which equals

$$[.005(51)^3 - .5(51)^2 + 28(51) + 300] - [.005(50)^3 - .5(50)^2 + 28(50) + 300]$$
$$= 1090.755 - 1075$$
$$= 15.755.$$

(b) The marginal cost at production level 50 is $C'(50)$.

$$C'(x) = .015x^2 - x + 28$$
$$C'(50) = 15.5.$$

Notice that 15.5 is close to the actual cost in part (a) of increasing production by one item. **» Now Try Exercise 41**

Our discussion of cost and marginal cost applies as well to other economic quantities such as profit and revenue. In fact, in economics, the derivatives are often described by the adjective *marginal*. Here are two more definitions of marginal functions and their interpretations.

> **DEFINITION** **Marginal Revenue and Marginal Profit** If $R(x)$ is the revenue generated from the production of x units of a certain commodity, $C(x)$ is the cost function, and $P(x)$ is the corresponding profit, the **marginal revenue function** is $R'(x)$ and the **marginal profit function** is $P'(x)$. (Recall that $P(x) = R(x) - C(x)$.)

The marginal revenue of producing a units, $R'(a)$, is an approximation of the additional revenue that results from increasing the production level by 1 unit, from a to $a + 1$:

$$R(a + 1) - R(a) \approx R'(a).$$

Similarly, for the marginal profit, we have

$$P(a + 1) - P(a) \approx P'(a).$$

The following example illustrates how marginal functions help in the decision-making process in economics.

EXAMPLE 7 **Predicting Profits** Let $R(x)$ denote the revenue (in thousands of dollars) generated from the production of x units of a certain commodity.
(a) Given that $R(4) = 7$ and $R'(4) = -.5$, estimate the additional revenue that results from increasing the production level by 1 unit from $x = 4$ to $x = 5$.
(b) Estimate the revenue generated from the production of 5 units.
(c) Is it profitable to raise the production to 5 units if the cost (in thousands of dollars) to produce x units is given by $C(x) = x + \frac{4}{x+1}$?

SOLUTION   (a) Using formula (2) with $a = 4$, we find

$$R(5) - R(4) \approx R'(4) = -.5 \text{ thousand dollars.}$$

Thus, if the production level is raised to 5 units, the revenue will drop by about $500 dollars.

(b) The revenue that results from the production of 4 units is $R(4) = 7$, or $7000. By part (a), this revenue will drop by about $500 if we raise the production from 4 to 5 units. Thus, at production level $x = 5$, the revenue will be approximately $7000 - 500 = \$6500$.

(c) Let $P(x)$ denote the profit (in thousands of dollars) that results from the production of x units. We have $P(x) = R(x) - C(x)$. At production level $x = 5$, we have

$$C(5) = 5 + \frac{4}{5 + 1} = \frac{17}{3} \approx 5.667 \text{ thousand dollars or } \$5667;$$

also, from part (b), the revenue is $R(5) \approx \$6500$. So, the profit at production level $x = 5$ is $P(5) \approx 6500 - 5667 = \$833$. So, it is still profitable to raise the production level to $x = 5$, even though the cost will rise and the revenue will fall.

» Now Try Exercise 47

INCORPORATING TECHNOLOGY

Although functions can be specified (and differentiated) in graphing calculators with letters other than X, only functions in X can be graphed. Therefore, we will always use X as the variable.

Figure 4 contains the graphs of $f(x) = x^2$ and its first and second derivatives. These three graphs can be obtained with any of the function editor settings in Figs. 5(a) and 5(b).

Figure 4 Graph of x^2 and its first two derivatives.

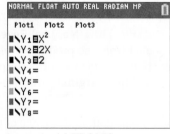

(a) TI-84 Plus

Figure 5

(b) TI-84 Plus

Check Your Understanding 1.7

Solutions can be found following the section exercises.

1. Let $f(t) = t + (1/t)$. Find $f''(2)$.

2. Differentiate $g(r) = 2\pi rh$.

EXERCISES 1.7

Find the first derivatives.

1. $f(t) = (t^2 + 1)^5$

2. $f(P) = P^3 + 3P^2 - 7P + 2$

3. $v(t) = 4t^2 + 11\sqrt{t} + 1$

4. $g(y) = y^2 - 2y + 4$

5. $y = T^5 - 4T^4 + 3T^2 - T - 1$

6. $x = 16t^2 + 45t + 10$

7. Find $\dfrac{d}{dP}\left(3P^2 - \tfrac{1}{2}P + 1\right)$.   8. Find $\dfrac{d}{ds}\sqrt{s^2 + 1}$.

9. Find $\dfrac{d}{dt}(a^2t^2 + b^2t + c^2)$.   10. Find $\dfrac{d}{dP}(T^2 + 3P)^3$.

In Exercises 11–20, find the first and second derivatives.

11. $y = x + 1$

12. $y = (x + 12)^3$

13. $y = \sqrt{x}$

14. $y = 100$

15. $y = \sqrt{x + 1}$

16. $v = 2t^2 + 3t + 11$

17. $f(r) = \pi r^2$

18. $y = \pi^2 + 3x^2$

19. $f(P) = (3P + 1)^5$

20. $T = (1 + 2t)^2 + t^3$

Compute the following.

21. $\dfrac{d}{dx}(2x + 7)^2\Big|_{x=1}$

22. $\dfrac{d}{dt}\left(t^2 + \dfrac{1}{t + 1}\right)\Big|_{t=0}$

23. $\dfrac{d}{dz}(z^2 + 2z + 1)^7\Big|_{z=-1}$

24. $\dfrac{d^2}{dx^2}(3x^4 + 4x^2)\Big|_{x=2}$

25. $\left.\dfrac{d^2}{dx^2}(3x^3 - x^2 + 7x - 1)\right|_{x=2}$

26. $\left.\dfrac{d}{dx}\left(\dfrac{dy}{dx}\right)\right|_{x=1}$, where $y = x^3 + 2x - 11$

27. $f'(1)$ and $f''(1)$, when $f(t) = \dfrac{1}{2+t}$

28. $g'(0)$ and $g''(0)$, when $g(T) = (T+2)^3$

29. $\left.\dfrac{d}{dt}\left(\dfrac{dv}{dt}\right)\right|_{t=2}$, where $v(t) = 3t^3 + \dfrac{4}{t}$

30. $\dfrac{d}{dt}\left(\dfrac{dv}{dt}\right)$, where $v = 2t^2 + \dfrac{1}{t+1}$

31. **A Revenue Function** A company finds that the revenue R generated by spending x dollars on advertising is given by $R = 1000 + 80x - .02x^2$, for $0 \le x \le 2000$. Find $\left.\dfrac{dR}{dx}\right|_{x=1500}$.

32. **Daily Volume of Business** A supermarket finds that its average daily volume of business, V (in thousands of dollars), and the number of hours, t, that the store is open for business each day are approximately related by the formula

$$V = 20\left(1 - \dfrac{100}{100 + t^2}\right), \qquad 0 \le t \le 24.$$

Find $\left.\dfrac{dV}{dt}\right|_{t=10}$.

33. If $s = PT$, find

(a) $\dfrac{ds}{dP}$, (b) $\dfrac{ds}{dT}$.

34. If $s = P^2 T$, find

(a) $\dfrac{d^2 s}{dP^2}$, (b) $\dfrac{d^2 s}{dT^2}$.

35. If $s = Tx^2 + 3xP + T^2$, find:

(a) $\dfrac{ds}{dx}$ (b) $\dfrac{ds}{dP}$ (c) $\dfrac{ds}{dT}$

36. If $s = 7x^2 y\sqrt{z}$, find:

(a) $\dfrac{d^2 s}{dx^2}$ (b) $\dfrac{d^2 s}{dy^2}$ (c) $\dfrac{ds}{dz}$

37. **Manufacturing Cost** Let $C(x)$ be the cost (in dollars) of manufacturing x bicycles per day in a certain factory. Interpret $C(50) = 5000$ and $C'(50) = 45$.

38. Estimate the cost of manufacturing 51 bicycles per day in Exercise 37.

39. **A Revenue Function** The revenue from producing (and selling) x units of a product is given by $R(x) = 3x - .01x^2$ dollars.
(a) Find the marginal revenue at a production level of 20.
(b) Find the production levels where the revenue is $200.

40. **Profit and Marginal Profit** Let $P(x)$ be the profit from producing (and selling) x units of goods. Match each question with the proper solution.

Questions
A. What is the profit from producing 1000 units of goods?
B. At what level of production will the marginal profit be 1000 dollars?

C. What is the marginal profit from producing 1000 units of goods?
D. For what level of production will the profit be 1000 dollars?

Solutions
(a) Compute $P'(1000)$.
(b) Find a value of a for which $P'(a) = 1000$.
(c) Set $P(x) = 1000$ and solve for x.
(d) Compute $P(1000)$.

41. **Revenue and Marginal Revenue** Let $R(x)$ denote the revenue (in thousands of dollars) generated from the production of x units of computer chips per day, where each unit consists of 100 chips.
(a) Represent the following statement by equations involving R or R': When 1200 chips are produced per day, the revenue is $22,000 and the marginal revenue is $.75 per chip.
(b) If the marginal cost of producing 1200 chips is $1.5 per chip, what is the marginal profit at this production level?

42. Refer to Exercise 41. Is it profitable to produce 1300 chips per day if the cost of producing 1200 chips per day is $14,000?

43. **Sales at a Department Store** Let $S(x)$ represent the total sales (in thousands of dollars) for the month x in the year 2005 at a certain department store. Represent each following statement by an equation involving S or S'.
(a) The sales at the end of January reached $120,560 and were rising at the rate of $1500 per month.
(b) At the end of March, the sales for this month dropped to $80,000 and were falling by about $200 a day. (Use 1 month = 30 days.)

44. **Comparing Rates of Change**
(a) In Example 5, find the total sales for January 10, and determine the rate at which sales are falling on that day.
(b) Compare the rate of change of sales on January 2 (Example 5) to the rate on January 10. What can you infer about the rate of change of sales?

45. **Predicting Sales** Refer to Example 5.
(a) Compute $S(10)$ and $S'(10)$.
(b) Use the data in part (a) to estimate the total sales on January 11. Compare your estimate to the actual value given by $S(11)$.

46. **Correcting a Prediction** The financial analysts at the store in Example 5 corrected their projections and are now expecting the total sales for the x day of January to be

$$T(x) = \dfrac{24}{5} + \dfrac{36}{5(3x+1)^2} \text{ thousand dollars.}$$

(a) Let $S(x)$ be as in Example 5. Compute $T(1)$, $T'(1)$, $S(1)$, and $S'(1)$.
(b) Compare and interpret the data in part (a) as they pertain to the sales on January 1.

47. **Sales of Computers**
(a) Let $A(x)$ denote the number (in hundreds) of computers sold when x thousand dollars is spent on advertising. Represent the following statement by equations involving A or A': When $8000 is spent on advertising, the number of computers sold is 1200 and is rising at the rate of 50 computers for each $1000 spent on advertising.
(b) Estimate the number of computers that will be sold if $9000 is spent on advertising.

48. Estimating Sales of Toys A toy company introduces a new video game on the market. Let $S(x)$ denote the number of videos sold on the day, x, since the item was introduced. Let n be a positive integer. Interpret $S(n)$, $S'(n)$, and $S(n) + S'(n)$.

49. Third Derivative The *third derivative* of a function $f(x)$ is the derivative of the second derivative $f''(x)$ and is denoted by $f'''(x)$. Compute $f'''(x)$ for the following functions:

(a) $f(x) = x^5 - x^4 + 3x$

(b) $f(x) = 4x^{5/2}$

50. Compute the third derivatives of the following functions:

(a) $f(t) = t^{10}$ (b) $f(z) = \dfrac{1}{z + 5}$

TECHNOLOGY EXERCISES

51. For the given function, simultaneously graph the functions $f(x)$, $f'(x)$, and $f''(x)$ with the specified window setting. *Note:* Since we have not yet learned how to differentiate the given function, you must use your graphing utility's differentiation command to define the derivatives.

$$f(x) = \frac{x}{1 + x^2}, [-4, 4] \text{ by } [-2, 2].$$

52. Consider the cost function of Example 6.

(a) Graph $C(x)$ in the window $[0, 60]$ by $[-300, 1260]$.

(b) For what level of production will the cost be \$535?

(c) For what level of production will the marginal cost be \$14?

Solutions to Check Your Understanding 1.7

1. $f(t) = t + t^{-1}$

$f'(t) = 1 + (-1)t^{(-1)-1} = 1 - t^{-2}$

$f''(t) = -(-2)t^{(-2)-1} = 2t^{-3} = \dfrac{2}{t^3}$

Therefore,

$$f''(2) = \frac{2}{2^3} = \frac{1}{4}.$$

[*Note:* It is essential first to compute the function $f''(t)$ and *then* to evaluate the function at $t = 2$.]

2. The expression $2\pi rh$ contains two numbers, 2 and π, and two letters, r and h. The notation $g(r)$ tells us that the expression $2\pi rh$ is to be regarded as a function of r. Therefore, h, and hence $2\pi h$, is to be treated as a constant, and differentiation is done with respect to the variable r. That is,

$$g(r) = (2\pi h)r$$
$$g'(r) = 2\pi h.$$

1.8   The Derivative as a Rate of Change

As we saw in previous sections, an important interpretation of the slope of a function at a point is as a rate of change. In this section, we reexamine this interpretation and discuss some additional applications in which it proves useful. The first step is to understand what is meant by *average rate of change* of a function $f(x)$.

Consider a function $y = f(x)$ defined on the interval $a \leq x \leq b$. The average rate of change of $f(x)$ over this interval is the change in $f(x)$ divided by the length of the interval:

AVERAGE RATE OF CHANGE

$$\begin{bmatrix} \text{average rate of change of } f(x) \\ \text{over the interval } a \leq x \leq b \end{bmatrix} = \frac{f(b) - f(a)}{b - a}.$$

EXAMPLE 1   **Average Rates of Change** If $f(x) = x^2$, calculate the average rate of change of $f(x)$ over the following intervals:

(a) $1 \leq x \leq 2$ (b) $1 \leq x \leq 1.1$ (c) $1 \leq x \leq 1.01$

SOLUTION   (a) The average rate of change over the interval $1 \leq x \leq 2$ is

$$\frac{2^2 - 1^2}{2 - 1} = \frac{3}{1} = 3.$$

(b) The average rate of change over the interval $1 \leq x \leq 1.1$ is

$$\frac{1.1^2 - 1^2}{1.1 - 1} = \frac{.21}{.1} = 2.1.$$

(c) The average rate of change over the interval $1 \le x \le 1.01$ is

$$\frac{1.01^2 - 1^2}{1.01 - 1} = \frac{.0201}{.01} = 2.01.$$

>> *Now Try Exercise 1*

FOR REVIEW

The difference quotient was introduced in Sec. 1.3 and used in Sec. 1.4 to compute the derivative. See pp. 78 and 85.

In the special case where b is $a + h$, the value of $b - a$ is $(a + h) - a$, or h, and the average rate of change of the function over the interval is the familiar difference quotient

$$\frac{f(a + h) - f(a)}{h}.$$

Geometrically, this quotient is the slope of the secant line in Fig. 1. Recall that as h approaches 0, the slope of the secant line approaches the slope of the tangent line. Thus, the average rate of change approaches $f'(a)$. For this reason, $f'(a)$ is called the (**instantaneous**) **rate of change** of $f(x)$ exactly at the point where $x = a$.

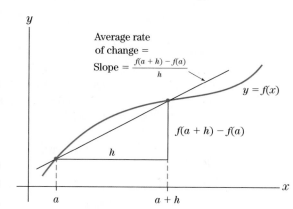

Figure 1 Average rate of change is the slope of the secant line.

Derivative as Instantaneous Rate of Change   The derivative $f'(a)$ measures the instantaneous rate of change of $f(x)$ at $x = a$.

From now on, unless we explicitly use the word *average* when we refer to the rate of change of a function, we mean the *instantaneous* rate of change.

EXAMPLE 2   **Comparing the Average and Instantaneous Rates**   Consider the function $f(x) = x^2$ of Example 1. Calculate the rate of change of $f(x)$ at $x = 1$, and compare your results to those in Example 1.

SOLUTION   The rate of change of $f(x)$ at $x = 1$ is equal to $f'(1)$. We have

$$f'(x) = 2x$$
$$f'(1) = 2 \cdot 1 = 2.$$

That is, the rate of change is 2 units per unit change in x. Notice how the average rates of change in Example 1 approach the (instantaneous) rate of change as the intervals beginning at $x = 1$ shrink.

>> *Now Try Exercise 3*

EXAMPLE 3   **Rates of Change of a Population Model**   The function $f(t)$ in Fig. 2 gives the population of the United States t years after the beginning of 1800. The figure also shows the tangent line through the point $(40, 17)$.

(a) What was the average rate of growth of the United States population from 1840 to 1870?

(b) How fast was the population growing in 1840?

(c) Was the population growing faster in 1810 or 1880?

Figure 2 U.S. population from 1800 to 1900.

SOLUTION

(a) Since 1870 is 70 years after 1800 and 1840 is 40 years after 1800, the question asks for the average rate of change of $f(t)$ over the interval $40 \leq t \leq 70$. This value is

$$\frac{f(70) - f(40)}{70 - 40} = \frac{40 - 17}{30} = \frac{23}{30} \approx .77.$$

Therefore, from 1840 to 1870, the population grew at the average rate of about .77 million, or 770,000 people per year.

(b) The rate of growth of $f(t)$ at $t = 40$ is $f'(40)$, that is, the slope of the tangent line at $t = 40$. Since $(40, 17)$ and $(90, 42)$ are two points on the tangent line, the slope of the tangent line is

$$\frac{42 - 17}{90 - 40} = \frac{25}{50} = .5.$$

Therefore, in 1840 the population was growing at the rate of .5 million, or 500,000 people per year.

(c) The graph is clearly steeper in 1880 than in 1810. Therefore, the population was growing faster in 1880 than in 1810.   **» Now Try Exercise 5**

Velocity and Acceleration

An everyday illustration of rate of change is given by the velocity of a moving object. Suppose that we are driving a car along a straight road and at each time t we let $s(t)$ be our position on the road, measured from some convenient reference point. See Fig. 3, where distances are positive to the right of the car. For the moment we shall assume that we are proceeding in only the positive direction along the road.

Figure 3 Position of a car traveling on a straight road.

At any instant, the car's speedometer tells us how fast we are moving, that is, how fast our position $s(t)$ is changing. To show how the speedometer reading is related to our calculus concept of a derivative, let us examine what is happening at a specific time, say, $t = 2$. Consider a short time interval of duration h from $t = 2$ to $t = 2 + h$. Our car will move from position $s(2)$ to position $s(2 + h)$, a distance of $s(2 + h) - s(2)$. Thus, the **average velocity from** $t = 2$ **to** $t = 2 + h$ is

$$\frac{[\text{distance traveled}]}{[\text{time elapsed}]} = \frac{s(2 + h) - s(2)}{h}. \tag{1}$$

If the car is traveling at a steady speed during this time period, the speedometer reading will equal the average velocity in formula (1).

From our discussion in Section 1.3, the ratio (1) approaches the derivative $s'(2)$ as h approaches zero. For this reason we call $s'(2)$ **the (instantaneous) velocity at** $t = 2$. This number will agree with the speedometer reading at $t = 2$ because, when h is very small, the car's speed will be nearly steady over the time interval from $t = 2$ to $t = 2 + h$; so the average velocity over this time interval will be nearly the same as the speedometer reading at $t = 2$.

The reasoning used for $t = 2$ holds for an arbitrary t as well. Thus, the following definition makes sense:

> **DEFINITION Velocity**   If $s(t)$ denotes the position function of an object moving in a straight line, then the velocity $v(t)$ of the object at time t is given by
>
> $$v(t) = s'(t).$$

In our discussion, we assumed that the car moved in the positive direction. If the car moves in the opposite direction, the ratio (1) and the limiting value $s'(2)$ will be negative. So, we interpret negative velocity as movement in the negative direction along the road.

The derivative of the velocity function $v(t)$ is called the **acceleration function** and is often written as $a(t)$:

> **Acceleration as First Derivative of Velocity**
>
> $$a(t) = v'(t).$$

Because $v'(t)$ measures the rate of change of the velocity $v(t)$, this use of the word *acceleration* agrees with our common usage in connection with automobiles. Note that since $v(t) = s'(t)$, the acceleration is actually the second derivative of the position function $s(t)$,

> **Acceleration as Second Derivative of Position**
>
> $$a(t) = s''(t).$$

The discussion that we presented using the example of a car moving in a straight line applies to any object moving in a straight line. For ease of reference, we set the notation in this more general situation. Suppose that an object is moving in a straight line. Denote by $s(t)$ its position at time t, measured from some reference point. The **displacement** of the object as time varies from a to b is given by $s(b) - s(a)$. It is the net change in its position from $t = a$ to $t = b$. The **average velocity** of the object during the time interval $[a, b]$ is the average rate of change of the position function and is given by

$$[\text{average velocity}] = \frac{[\text{displacement}]}{[\text{time elapsed}]} = \frac{s(b) - s(a)}{b - a}.$$

The **instantaneous velocity** or simply **velocity** of the object at time t is the instantaneous rate of change of the position function and is given by $v(t) = s'(t)$.

| EXAMPLE 4 | **Velocity and Average Velocities** When a ball is thrown straight up into the air, its position may be measured as the vertical distance from the ground. Regard "up" as the positive direction, and let $s(t)$ be the height of the ball in feet after t seconds. Suppose that $s(t) = -16t^2 + 128t + 5$. Figure 4 contains a graph of the function $s(t)$. *Note:* The graph does *not* show the path of the ball. The ball is rising *straight* up and then falling *straight* down. |

(a) Find the displacement of the ball in the intervals of time [1, 2], [1, 1.5] and [1, 1.1].

(b) Find the average velocity of the ball during the intervals of time [1, 2], [1, 1.5] and [1, 1.1].

(c) What is the velocity of the ball at $t = 1$? How is this value related to the average velocities found in (b)?

SOLUTION

(a) Let us first compute the position of the ball at $t = 1$, 1.1, 1.5, and 2:

$$s(1) = -16(1)^2 + 128(1) + 5 = 117 \text{ feet,}$$
$$s(1.1) = -16(1.1)^2 + 128(1.1) + 5 = 126.44 \text{ feet,}$$
$$s(1.5) = -16(1.5)^2 + 128(1.5) + 5 = 161 \text{ feet,}$$
$$s(2) = -16(2)^2 + 128(2) + 5 = 197 \text{ feet.}$$

With the help of these values, we compute the displacements: $s(2) - s(1) = 197 - 117 = 80$ feet; $s(1.5) - s(1) = 161 - 117 = 44$ feet; $s(1.1) - s(1) = 126.44 - 117 = 9.44$ feet.

(b) Average velocity in [1, 2] is

$$\frac{[\text{displacement}]}{[\text{time elapsed}]} = \frac{s(2) - s(1)}{2 - 1} = \frac{80}{1} = 80 \text{ feet per second.}$$

In [1, 1.5], the average velocity is

$$\frac{s(1.5) - s(1)}{1.5 - 1} = \frac{44}{.5} = 88 \text{ feet per second,}$$

and in [1, 1.1], the average velocity is

$$\frac{s(1.1) - s(1)}{1.1 - 1} = \frac{9.44}{.1} = 94.4 \text{ feet per second.}$$

(c) The velocity is the rate of change of the position function, so

$$v(t) = s'(t) = -32t + 128.$$

When $t = 1$, the velocity is $v(1) = -32(1) + 128 = 96$ feet per second. The average velocities (80, 88, 94.4 in feet per second) are approaching the instantaneous velocity (96 feet per second) at time $t = 1$ as the length of the time interval approaches 0.

» Now Try Exercise 5

Figure 4 Height of ball in the air.

| EXAMPLE 5 | **Position, Velocity, and Acceleration** With the notation of Example 4, let $s(t) = -16t^2 + 128t + 5$. |

(a) What is the velocity after 2 seconds?

(b) What is the acceleration after 2 seconds?

(c) At what time is the velocity -32 feet per second? (The negative sign indicates that the ball's height is decreasing; that is, the ball is falling.)

(d) When is the ball at a height of 117 feet?

SOLUTION

(a) From the previous example, $v(t) = s'(t) = -32t + 128$. The velocity when $t = 2$ is $v(2) = -32(2) + 128 = 64$ feet per second.

(b) The acceleration is the rate of change (or derivative) of the velocity: $a(t) = v'(t) = -32$. The acceleration is -32 feet per second per second for all t. This constant acceleration is due to the downward (and therefore negative) force of gravity.

(c) Since the velocity is given and the time is unknown, we set $v(t) = -32$ and solve for t:

$$-32t + 128 = -32$$
$$-32t = -160$$
$$t = 5.$$

The velocity is -32 feet per second when t is 5 seconds.

(d) The question here involves the height function, not the velocity. Since the height is given and the time is unknown, we set $s(t) = 117$ and solve for t:

$$-16t^2 + 128t + 5 = 117 \quad \text{Subtract 117 from both sides then factor } -16.$$
$$-16(t^2 - 8t + 7) = 0 \quad \text{Factor.}$$
$$-16(t - 1)(t - 7) = 0.$$

The ball is at a height of 117 feet once on the way up (when $t = 1$ second) and once on the way down (when $t = 7$ seconds).   **▸▸ Now Try Exercise 13**

Approximating the Change in a Function

Consider the function $f(x)$ near $x = a$. As we just saw, the average rate of change of f over a small interval of length h is approximately equal to the instantaneous change at an endpoint of the interval. That is,

$$\frac{f(a + h) - f(a)}{h} \approx f'(a).$$

Multiplying both sides of this approximation by h, we have

Approximating Change with the Derivative

$$f(a + h) - f(a) \approx f'(a) \cdot h. \tag{2}$$

Note that, when $h = 1$, we obtain formula (2) of the previous section.

If x changes from a to $a + h$, then the change in the value of the function $f(x)$ is approximately $f'(a)$ times the change h in the value of x. This result holds for both positive and negative values of h. In applications, the right side of formula (2) is calculated and used to estimate the left side.

Figure 5 contains a geometric interpretation of (2). Given the small change h in x, the quantity $f'(a) \cdot h$ gives the corresponding change in y *along the tangent line* at $(a, f(a))$. In contrast, the quantity $f(a + h) - f(a)$ gives the change in y *along the curve* $y = f(x)$. When h is small, $f'(a) \cdot h$ is a good approximation to the change in $f(x)$.

Figure 5 Change in y along the tangent line and along the graph of $y = f(x)$.

Before we consider an application, it is instructive to present an alternative derivation of (2). From Section 1.3, the equation of the tangent line at P is

$$y - f(a) = f'(a)(x - a). \tag{3}$$

Now the tangent line approximates the graph of f near P, so if in formula (3) we replace y (the value of the tangent line) by $f(x)$, we get

$$f(x) - f(a) \approx f'(a)(x - a), \tag{4}$$

and (2) follows from (4) upon setting $h = x - a$ and $x = a + h$.

| EXAMPLE 6 | **Approximating a Production Function** Let the production function $p(x)$ give the number of units of goods produced when x units of labor are employed. If 5000 units of labor are currently employed, $p(5000) = 300$, and $p'(5000) = 2$. |

(a) Interpret $p(5000) = 300$.
(b) Interpret $p'(5000) = 2$.
(c) Estimate the number of *additional* units of goods produced when x is increased from 5000 to $5000\frac{1}{2}$ units of labor.
(d) Estimate the *change* in the number of units of goods produced when x is decreased from 5000 to 4999 units of labor.

SOLUTION
(a) When 5000 units of labor are employed, 300 units of goods will be produced.
(b) If 5000 units of labor are currently employed and we consider adding more labor, productivity will increase at approximately the rate of 2 units of goods for each additional unit of labor.
(c) Here, $h = \frac{1}{2}$. By formula (2), the change in $p(x)$ will be approximately

$$p'(5000) \cdot \tfrac{1}{2} = 2 \cdot \tfrac{1}{2} = 1.$$

About 1 additional unit will be produced. Therefore, about 301 units of goods will be produced when $5000\frac{1}{2}$ units of labor are employed.
(d) Here, $h = -1$, since the amount of labor is reduced. The change in $p(x)$ will be approximately

$$p'(5000) \cdot (-1) = 2 \cdot (-1) = -2.$$

About two fewer units (that is, 298) of goods will be produced.

>> Now Try Exercise 19

To simplify the use of formula (2) in marginal cost analysis, let us rewrite it with a cost function, as follows:

$$C(a + h) - C(a) \approx C'(a) \cdot h. \tag{5}$$

Taking $h = 1$, we obtain the familiar formula of the previous section,

$$C(a + 1) - C(a) \approx C'(a). \tag{6}$$

Thus, for a whole unit increase in production ($h = 1$), the change in cost is approximately equal to the marginal cost, but for an increase by h units, we must modify the marginal cost proportionally.

| EXAMPLE 7 | **Marginal Cost** The total cost in thousands of dollars of producing x units of a certain commodity is $C(x) = 6x^2 + 2x + 10$. |

(a) Find the marginal cost function.
(b) Find the cost and marginal cost when 10 units are produced.
(c) Use marginal cost to approximate the cost of producing the 11th unit.
(d) Use marginal cost to estimate the additional cost incurred if the production level is increased from 10 to 10.5 units.

SOLUTION
(a) The marginal cost function is the derivative of the cost function:

$$C'(x) = 12x + 2 \text{ (thousand dollars per unit).}$$

(b) When 10 units are produced, the cost is

$$C(10) = 6(10^2) + 2(10) + 10 = 630 \text{ thousand dollars,}$$

and the marginal cost is $C'(10) = 12(10) + 2 = 122$ thousand dollars per unit.

(c) The cost of producing the 11th unit is the difference in cost as x varies from 10 to 11 or $C(11) - C(10)$. By formula (6), this difference can be approximated by the marginal cost $C'(10) = 122$. Hence, to produce the 11th unit, it will cost approximately 122 thousand dollars.

(d) The additional cost when we raise the production level by half a unit from 10 to 10.5 is $C(10.5) - C(10)$. According to formula (5), with $a = 10$ and $h = .5$,

$$C(10.5) - C(10) \approx C'(10) \cdot (.5) = (122)(.5) = 61 \text{ thousand dollars.}$$

>> *Now Try Exercise 29*

Units for Rate of Change   Table 1 shows the units appearing in several examples from this section. In general,

$$[\text{unit of measure for } f'(x)]$$
$$= [\text{unit of measure for } f(x)] \text{ per } [\text{unit of measure for } x].$$

Table 1 Units of Measure

Example	Unit for $f(t)$ or $f(x)$	Unit for t or x	Unit for $f'(t)$ or $f'(x)$
U.S. population	millions of people	year	millions of people per year
Ball in air	feet	second	feet per second
Cost function	dollars	item	dollars per item

Check Your Understanding 1.8

Solutions can be found following the section exercises.

Let $f(t)$ be the temperature (in degrees Celsius) of a liquid at time t (in hours). The rate of temperature change at time a has the value $f'(a)$. Listed next are typical questions about $f(t)$ and its slope at various points. Match each question with the proper method of solution.

Questions

1. What is the temperature of the liquid after 6 hours (that is, when $t = 6$)?

2. When is the temperature rising at the rate of 6 degrees per hour?

3. By how many degrees did the temperature rise during the first 6 hours?

4. When is the liquid's temperature only 6 degrees?

5. How fast is the temperature of the liquid changing after 6 hours?

6. What is the average rate of increase in the temperature during the first 6 hours?

Methods of Solution

(a) Compute $f(6)$.
(b) Set $f(t) = 6$ and solve for t.
(c) Compute $[f(6) - f(0)]/6$.
(d) Compute $f'(6)$.
(e) Find a value of a for which $f'(a) = 6$.
(f) Compute $f(6) - f(0)$.

EXERCISES 1.8

1. If $f(x) = x^2 + 3x$, calculate the average rate of change of $f(x)$ over the following intervals
   (a) $1 \le x \le 2$,
   (b) $1 \le x \le 1.5$,
   (c) $1 \le x \le 1.1$.

2. If $f(x) = 3x^2 + 2$, calculate the average rate of change of $f(x)$ over the following intervals
   (a) $0 \le x \le .5$,
   (b) $0 \le x \le .1$,
   (c) $0 \le x \le .0.01$.

3. **Average and Instantaneous Rates of Change** Suppose that $f(x) = 4x^2$.
   (a) What is the average rate of change of $f(x)$ over each of the intervals 1 to 2, 1 to 1.5, and 1 to 1.1?
   (b) What is the (instantaneous) rate of change of $f(x)$ when $x = 1$?

4. **Average and Instantaneous Rates of Change** Suppose that $f(x) = -6/x$.
   (a) What is the average rate of change of $f(x)$ over each of the intervals 1 to 2, 1 to 1.5, and 1 to 1.2?
   (b) What is the (instantaneous) rate of change of $f(x)$ when $x = 1$?

5. **Average and Instantaneous Rates of Change** Suppose that $f(t) = t^2 + 3t - 7$.
   (a) What is the average rate of change of $f(t)$ over the interval 5 to 6?
   (b) What is the (instantaneous) rate of change of $f(t)$ when $t = 5$?

6. **Average and Instantaneous Rates of Change** Suppose that $f(t) = 3t + 2 - \dfrac{12}{t}$.
   (a) What is the average rate of change of $f(t)$ over the interval 2 to 3?
   (b) What is the (instantaneous) rate of change of $f(t)$ when $t = 2$?

7. **Motion of an Object** An object moving in a straight line travels $s(t)$ kilometers in t hours, where $s(t) = 2t^2 + 4t$.
   (a) What is the object's velocity when $t = 6$?
   (b) How far has the object traveled in 6 hours?
   (c) When is the object traveling at the rate of 6 kilometers per hour?

8. **Effect of Advertising on Sales** After an advertising campaign, the sales of a product often increase and then decrease. Suppose that t days after the end of the advertising, the daily sales are $f(t) = -3t^2 + 32t + 100$ units. What is the average rate of growth in sales during the fourth day, that is, from time $t = 3$ to $t = 4$? At what (instantaneous) rate are the sales changing when $t = 2$?

9. **Average Daily Output** An analysis of the daily output of a factory assembly line shows that about $60t + t^2 - \frac{1}{12}t^3$ units are produced after t hours of work, $0 \le t \le 8$. What is the rate of production (in units per hour) when $t = 2$?

10. **Rate of Change of Volume after Liquid** Liquid is pouring into a large vat. After t hours, there are $5t + \sqrt{t}$ gallons in the vat. At what rate is the liquid flowing into the vat (in gallons per hour) when $t = 4$?

11. **Maximum Height** A toy rocket is fired straight up into the air. Let $s(t) = -6t^2 + 72t$ denote its position in feet after t seconds.
    (a) Find the velocity after t seconds.
    (b) Find the acceleration after t seconds.
    (c) When does the rocket reach its maximum height? [*Hint:* What happens to the velocity when the rocket reaches its maximum height?]
    (d) What is the maximum height reached by the rocket?

12. **Analysis of a Moving Particle** Refer to Fig. 6, where $s(t)$ represents the position of a car moving in a straight line.

    (a) Was the car going faster at A or at B?
    (b) Is the velocity increasing or decreasing at B? What does this say about the acceleration at B?
    (c) What happened to the car's velocity at C?
    (d) In which direction was the car moving at D?
    (e) What happened at E?
    (f) What happened after F?

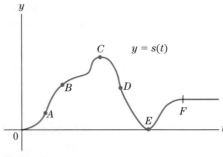

Figure 6

13. **Position of a Toy Rocket** A toy rocket fired straight up into the air has height $s(t) = 160t - 16t^2$ feet after t seconds.
    (a) What is the rocket's initial velocity (when $t = 0$)?
    (b) What is the velocity after 2 seconds?
    (c) What is the acceleration when $t = 3$?
    (d) At what time will the rocket hit the ground?
    (e) At what velocity will the rocket be traveling just as it smashes into the ground?

14. **Height of a Helicopter** A helicopter is rising straight up in the air. Its distance from the ground t seconds after takeoff is $s(t)$ feet, where $s(t) = t^2 + t$.
    (a) How long will it take for the helicopter to rise 20 feet?
    (b) Find the velocity and the acceleration of the helicopter when it is 20 feet above the ground.

15. **Height of a Ball** Let $s(t)$ be the height (in feet) after t seconds of a ball thrown straight up into the air. Match each question with the proper solution.

Questions
A. What is the velocity of the ball after 3 seconds?
B. When is the velocity 3 feet per second?
C. What is the average velocity during the first 3 seconds?
D. When is the ball 3 feet above the ground?
E. When does the ball hit the ground?
F. How high is the ball after 3 seconds?
G. How far did the ball travel during the first 3 seconds?

Solutions
a. Set $s(t) = 0$ and solve for t.
b. Compute $s'(3)$.
c. Compute $s(3)$.
d. Set $s'(t) = 3$ and solve for t.
e. Find a value of a for which $s(a) = 3$.
f. Compute $[s(3) - s(0)]/3$.
g. Compute $s(3) - s(0)$.

16. **Average Speed** Table 2 gives a car's trip odometer reading (in miles) at 1 hour into a trip and at several nearby times. What is the average speed during the time interval from 1 to 1.05 hours? Estimate the speed at time 1 hour into the trip.

Table 2 Trip Odometer Readings at Several Times

Time	Trip Meter
.96	43.2
.97	43.7
.98	44.2
.99	44.6
1.00	45.0
1.01	45.4
1.02	45.8
1.03	46.3
1.04	46.8
1.05	47.4

17. Velocity and Position A particle is moving in a straight line in such a way that its position at time t (in seconds) is $s(t) = t^2 + 3t + 2$ feet to the right of a reference point, for $t \geq 0$.
   (a) What is the velocity of the object when the time is 6 seconds?
   (b) Is the object moving toward the reference point when $t = 6$? Explain your answer.
   (c) What is the object's velocity when the object is 6 feet from the reference point?

18. Interpreting Rates of Change on a Graph A car is traveling from New York to Boston and is partway between the two cities. Let $s(t)$ be the distance from New York during the next minute. Match each behavior with the corresponding graph of $s(t)$ in Fig. 7.
   (a) The car travels at a positive steady speed.
   (b) The car is stopped.
   (c) The car is backing up.
   (d) The car is accelerating.
   (e) The car is decelerating.

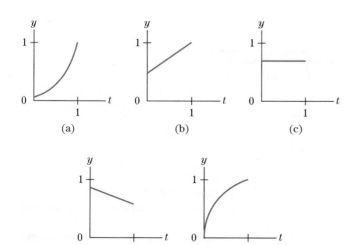

Figure 7 Possible graphs of $s(t)$.

19. Estimating the Values of a Function If $f(100) = 5000$ and $f'(100) = 10$, estimate each of the following.
   (a) $f(101)$ **(b)** $f(100.5)$
   (c) $f(99)$ **(d)** $f(98)$
   (e) $f(99.75)$

20. Estimating the Values of a Function If $f(25) = 10$ and $f'(25) = -2$, estimate each of the following.
   (a) $f(27)$ **(b)** $f(26)$
   (c) $f(25.25)$ **(d)** $f(24)$
   (e) $f(23.5)$

21. Temperature of a Cup of Coffee Let $f(t)$ be the temperature of a cup of coffee t minutes after it has been poured. Interpret $f(4) = 120$ and $f'(4) = -5$. Estimate the temperature of the coffee after 4 minutes and 6 seconds, that is, after 4.1 minutes.

22. Rate of Elimination of a Drug Suppose that 5 mg of a drug is injected into the bloodstream. Let $f(t)$ be the amount present in the bloodstream after t hours. Interpret $f(3) = 2$ and $f'(3) = -.5$. Estimate the number of milligrams of the drug in the bloodstream after $3\frac{1}{2}$ hours.

23. Price Affects Sales Let $f(p)$ be the number of cars sold when the price is p dollars per car. Interpret the statements $f(10,000) = 200,000$ and $f'(10,000) = -3$.

24. Advertising Affects Sales Let $f(x)$ be the number of toys sold when x dollars are spent on advertising. Interpret the statements $f(100,000) = 3,000,000$ and $f'(100,000) = 30$.

25. Rate of Sales Let $f(x)$ be the number (in thousands) of computers sold when the price is x hundred dollars per computer. Interpret the statements $f(12) = 60$ and $f'(12) = -2$. Then, estimate the number of computers sold if the price is set at \$1250 per computer.

26. Marginal Cost Let $C(x)$ be the cost (in dollars) of manufacturing x items. Interpret the statements $C(2000) = 50,000$ and $C'(2000) = 10$. Estimate the cost of manufacturing 1998 items.

27. Marginal Profit Let $P(x)$ be the profit (in dollars) from manufacturing and selling x cars. Interpret $P(100) = 90,000$ and $P'(100) = 1200$. Estimate the profit from manufacturing and selling 99 cars.

28. Price of a Company's Stock Let $f(x)$ be the value in dollars of one share of a company x days since the company went public.
   (a) Interpret the statements $f(100) = 16$ and $f'(100) = .25$.
   (b) Estimate the value of one share on the 101st day since the company went public.

29. Marginal Cost Analysis Consider the cost function $C(x) = 6x^2 + 14x + 18$ (thousand dollars).
   (a) What is the marginal cost at production level $x = 5$?
   (b) Estimate the cost of raising the production level from $x = 5$ to $x = 5.25$.
   (c) Let $R(x) = -x^2 + 37x + 38$ denote the revenue in thousands of dollars generated from the production of x units. What is the breakeven point? (Recall that the breakeven point is when revenue is equal to cost.)
   (d) Compute and compare the marginal revenue and marginal cost at the breakeven point. Should the company

increase production beyond the breakeven point? Justify your answer using marginals.

30. Estimate how much the function

$$f(x) = \frac{1}{1 + x^2}$$

will change if x decreases from 1 to .9.

31. Health Expenditures National health expenditures (in billions of dollars) from 1980 to 1998 are given by the function $f(t)$ in Fig. 8.
   (a) How much money was spent in 1987?
   (b) Approximately how fast were expenditures rising in 1987?
   (c) When did expenditures reach $1 trillion?
   (d) When were expenditures rising at the rate of $100 billion per year?

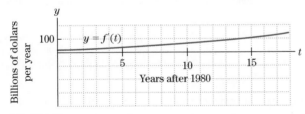

Figure 8 National health expenditures

32. Velocity and Acceleration In an 8-second test run, a vehicle accelerates for several seconds and then decelerates. The function $s(t)$ gives the number of feet traveled after t seconds and is graphed in Fig. 9.
   (a) How far has the vehicle traveled after 3.5 seconds?
   (b) What is the velocity after 2 seconds?
   (c) What is the acceleration after 1 second?
   (d) When will the vehicle have traveled 120 feet?
   (e) When, during the second part of the test run, will the vehicle be traveling at the rate of 20 feet per second?
   (f) What is the greatest velocity? At what time is this greatest velocity reached? How far has the vehicle traveled at this time?

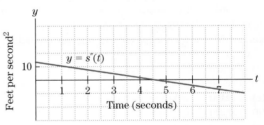

Figure 9

TECHNOLOGY EXERCISES

33. Judgment Time In a psychology experiment, people improved their ability to recognize common verbal and semantic information with practice. Their judgment time after t days of practice was $f(t) = .36 + .77(t - .5)^{-0.36}$ seconds. (*Source: American Journal of Psychology.*)
   (a) Display the graphs of $f(t)$ and $f'(t)$ in the window [.5, 6] by [−3, 3]. Use these graphs to answer the following questions.
   (b) What was the judgment time after 4 days of practice?
   (c) After how many days of practice was the judgment time about .8 second?
   (d) After 4 days of practice, at what rate was judgment time changing with respect to days of practice?
   (e) After how many days was judgment time changing at the rate of −.08 second per day of practice?

34. Position of a Ball A ball thrown straight up into the air has height $s(t) = 102t - 16t^2$ feet after t seconds.
 (a) Display the graphs of $s(t)$ and $s'(t)$ in the window $[0, 7]$ by $[-100, 200]$. Use these graphs to answer the remaining questions.
 (b) How high is the ball after 2 seconds?
 (c) When, during descent, is the height 110 feet?
 (d) What is the velocity after 6 seconds?
 (e) When is the velocity 70 feet per second?
 (f) How fast is the ball traveling when it hits the ground?

Solutions to Check Your Understanding 1.8

1. Method (a). The question involves $f(t)$, the temperature at time t. Since the time is given, compute $f(6)$.

2. Method (e). The question involves $f'(t)$, the rate of change of temperature. The interrogative *when* indicates that the time is unknown. Set $f'(t) = 6$ and solve for t.

3. Method (f). The question asks for the change in the value of the function from time 0 to time 6, $f(6) - f(0)$.

4. Method (b). The question involves $f(t)$, and the time is unknown. Set $f(t) = 6$ and solve for t.

5. Method (d). The question involves $f'(t)$, and the time is given. Compute $f'(6)$.

6. Method (c). The question asks for the average rate of change of the function during the time interval 0 to 6, $[f(6) - f(0)]/6$.

| CHAPTER 1 | **Summary** |

| **KEY TERMS AND CONCEPTS** | **EXAMPLES** |

1.1 The Slope of a Straight Line

The **slope–intercept form** of the equation of a nonvertical line is $y = mx + b$, where m is the slope and b is the y-intercept. The **point–slope form** of the equation of a nonvertical line is

$$(y - y_1) = m(x - x_1),$$

where m is the slope and (x_1, y_1) is a point on the line.

1. The line $y = 3x - 1$ has slope $m = 3$ and y-intercept -1.

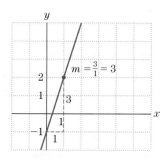

The slope tells us that every 1-unit change in x corresponds a 3-unit change in y, as illustrated.

2. To find the point–slope equation of the line through the points $(-1, 2)$ and $(0, 1)$, we first find the slope

$$m = \frac{y_2 - y_1}{x_2 - x_1} = \frac{1 - 2}{0 - (-1)} = -1.$$

Given any two distinct points on the line (x_1, y_1) and (x_2, y_2), then,

$$m = \frac{y_2 - y_1}{x_2 - x_1} = \frac{\text{change in } y}{\text{change in } x} = \frac{\text{Rise}}{\text{Run}}.$$

Thus, $m = -1$, and the equation of the line is

$$(y - 1) = -1(x - 0) \text{ or } y = -x + 1.$$

KEY TERMS AND CONCEPTS	**EXAMPLES**

The equation of a horizontal line is of the form $y = c$. The slope is 0.

The equation of a vertical line is of the form $x = a$. The slope of a vertical line is not defined.

Parallel lines have equal slopes.

Perpendicular lines have opposite reciprocal slopes.

Parallel lines   Perpendicular lines

3. The lines $y = x + 1$ and $y = x - 1$ are parallel. They have equal slope 1.

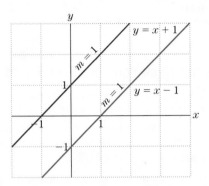

4. The lines $y = -2x + 1$ and $y = \frac{1}{2}x$ are perpendicular. They have opposite reciprocal slopes, -2 and $-\frac{1}{-2} = \frac{1}{2}$.

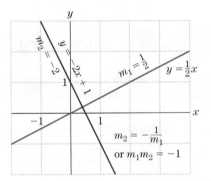

1.2 The Slope of a Curve at a Point

The figure shows the graph of $y = f(x)$ along with the tangent line at a point P on the graph. The **slope of the graph** at P is the slope of the tangent line at P.

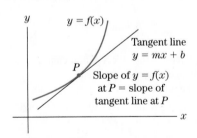

1. The graph of $y = f(x)$ and its tangent line at the point $P = (1, 2)$ are shown. To find the slope of the graph at the point P, we note that the slope of the tangent line at P is $\frac{1}{2}$.

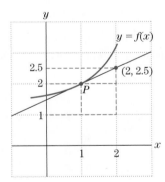

Since the slope of the graph is the slope of the tangent line, we conclude that that the graph at P has slope $\frac{1}{2}$.

KEY TERMS AND CONCEPTS	EXAMPLES
The slope of the tangent line is usually given by a formula. For example, the slope formula for the function $f(x) = x^2$ is $2x$.	**2.** Consider the point $P = (1.2, 1.44)$ on the parabola $y = x^2$. To find the slope m of the tangent line to the parabola at P, we evaluate the slope formula $m = 2x$ at the first coordinate $x = 1.2$ of P and get $m = 2(1.2) = 2.4$. Thus, the tangent line at the point P has slope 2.4.

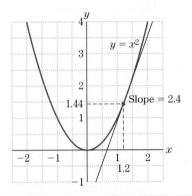

1.3 The Derivative and Limits

The *derivative* of $y = f(x)$ is the slope formula that gives the slope of the tangent line to the graph at a point (x, y) on the graph of $y = f(x)$. The derivative is denoted by $f'(x)$ or $\frac{d}{dx}[f(x)]$.

Derivative formulas:
If $f(x) = mx + b$, then $f'(x) = m$.
(Derivative of a linear function is its slope.)

If $f(x) = c$, then $f'(x) = 0$.
(Derivative of a constant function is 0.)

The Power Rule: If $f(x) = x^n$, then $f'(x) = nx^{n-1}$.

1. The following are examples that we obtain by applying the derivative formulas:

(a) $\dfrac{d}{dx}(-7x + 4) = -7$

(b) $\dfrac{d}{dx}(3^2) = 0$ Derivative of a constant.

(c) $\dfrac{d}{dx}(x^3) = 3x^2$ Power rule with $n = 3$.

(d) $\dfrac{d}{dx}\left[\dfrac{1}{\sqrt{x}}\right] = \dfrac{d}{dx}\left[x^{-1/2}\right]$

$$= -\frac{1}{2}x^{-\frac{1}{2}-1} \qquad \text{Power rule with } n = -\tfrac{1}{2}.$$

$$= -\frac{1}{2}x^{-\frac{3}{2}} = -\frac{1}{2x^{\frac{3}{2}}}$$

(e) $\dfrac{d}{dx}\left[\dfrac{1}{x^2}\right] = \dfrac{d}{dx}\left[x^{-2}\right] = -2x^{-3} = \dfrac{-2}{x^3}$ Power rule with $n = -2$.

2. To find the equation of the tangent line to the graph of $f(x) = x^5$ at the point $P = (-1, -1)$, we first find the derivative (slope formula) $f'(x) = 5x^4$. Evaluate the derivative at $x = -1$: $f'(-1) = 5(-1)^4 = 5$. Thus, the slope of the tangent line at P is 5 and the point–slope equation of the tangent line is $y - (-1) = 5(x - (-1))$.

| **KEY TERMS AND CONCEPTS** | **EXAMPLES** |

1.4 Limits and the Derivative

Let $g(x)$ be a function and a a number on the x-axis that may or may not be in the domain of g. We say that the *limit* of $g(x)$ as x approaches a is L if the values of $g(x)$ can be made arbitrarily close to a number L by taking x sufficiently close (but not equal) to a. In symbols, we write

$$\lim_{x \to a} g(x) = L.$$

Limits arise naturally when we are studying derivatives. The **limit definition of the derivative** is

$$f'(x) = \lim_{h \to 0} \frac{f(x + h) - f(x)}{h}.$$

When the limit $f'(x)$ exists, we say that f is *differentiable* at x.

Examples of limits:

(a) $\lim\limits_{x \to 0} 2 = 2$

(b) $\lim\limits_{x \to -1} x^3 = (-1)^3 = -1$

(c) $\lim\limits_{x \to 4} (3x - 2) = 3(4) - 2 = 10$

(d) $\lim\limits_{x \to 2} \dfrac{x^2 - 4}{x - 2} = \lim\limits_{x \to 2} \dfrac{(x - 2)(x + 2)}{x - 2}$ Factor.

$\qquad\qquad\quad = \lim\limits_{x \to 2} (x + 2) = 4$

(e) $\lim\limits_{x \to 2} \dfrac{x^2 + 4}{x - 2} = \dfrac{8}{0}$ Limit does not exist.

Use the limit definition to find $f'(1)$, where $f(x) = \sqrt{x}$.

$$f'(1) = \lim_{h \to 0} \frac{f(1 + h) - f(1)}{h}$$

$$= \lim_{h \to 0} \frac{\sqrt{1 + h} - 1}{h}$$

$$= \lim_{h \to 0} \frac{(\sqrt{1 + h} - 1)(\sqrt{1 + h} + 1)}{h(\sqrt{1 + h} + 1)} \quad \begin{array}{l}\text{Rationalize the}\\\text{numerator.}\end{array}$$

$$= \lim_{h \to 0} \frac{(\sqrt{1 + h})^2 - 1^2}{h(\sqrt{1 + h} + 1)}$$

$$= \lim_{h \to 0} \frac{(1 + h) - 1}{h(\sqrt{1 + h} + 1)}$$

$$= \lim_{h \to 0} \frac{\not h}{\not h(\sqrt{1 + h} + 1)}$$

$$= \lim_{h \to 0} \frac{1}{\sqrt{1 + h} + 1} = \frac{1}{\sqrt{1} + 1} = \frac{1}{2}.$$

1.5 Differentiability and Continuity

The function $f(x)$ is **differentiable** at $x = a$, if $f'(a)$ exists. If $f'(a)$ does not exist, we say that f is *nondifferentiable* at $x = a$.

Geometrically, if f is differentiable at $x = a$, the graph has a tangent line at the point $(a, f(a))$ and the slope of the tangent line is $f'(a)$. In particular, if f is differentiable at $x = a$, then the graph is "smooth" at $x = a$.

A function $f(x)$ is **continuous** at $x = a$ if

- $f(a)$ is defined.

- $\lim\limits_{x \to a} f(x) = f(a)$.

Geometrically, if f is continuous at $x = a$, then you should be able to trace the graph of $f(x)$, going through both sides of the point $(a, f(a))$, without lifting your pencil from the paper.

If a function is not continuous at $x = a$, we say that it is **discontinuous** at $x = a$.

A basic theorem states that if f is differentiable at $x = a$, then f is continuous at $x = a$.

The converse of the result is not true: A function may be continuous at a point without being differentiable at that point.

A polynomial $p(x)$ is continuous for all x.

A rational function $\frac{p(x)}{q(x)}$, where p and q are polynomials, is continuous for all x such that $q(x) \neq 0$.

Consider the functions depicted in the figures.

The function $f(x)$ is discontinuous at $x = 1$, because $\lim_{x \to 1} f(x)$ does not exist.

The function $g(x)$ is discontinuous at $x = 1$, because $g(1)$ is not defined.

The function $h(x)$ is continuous for all x but is not differentiable at $x = 1$ because the graph does not have a tangent line at $x = 1$.

KEY TERMS AND CONCEPTS	EXAMPLES

1.6 Some Rules for Differentiation

In computing derivatives, we can use basic rules for differentiation such as the following:

- Constant-multiple rule:

$$\frac{d}{dx}[k \cdot f(x)] = k \cdot \frac{d}{dx}[f(x)], \quad k \text{ a constant.}$$

- Sum rule:

$$\frac{d}{dx}[f(x) + g(x)] = \frac{d}{dx}[f(x)] + \frac{d}{dx}[g(x)].$$

- General power rule:

$$\frac{d}{dx}([g(x)]^r) = r \cdot [g(x)]^{r-1} \cdot \frac{d}{dx}[g(x)].$$

Using properties of derivatives, compute the following derivatives:

(a) $\dfrac{d}{dx}[-7x^2] = -7\dfrac{d}{dx}[x^2]$ Constant-multiple rule.

$= (-7)2x = -14x.$ Power rule.

(b) $\dfrac{d}{dx}[\sqrt{x} - 3x] = \dfrac{d}{dx}[\sqrt{x}] + \dfrac{d}{dx}[-3x]$ Sum rule.

$= \dfrac{d}{dx}[x^{1/2}] - 3\dfrac{d}{dx}[x]$ Constant-multiple rule.

$= \dfrac{1}{2}(x^{-1/2}) - 3 = \dfrac{1}{2\sqrt{x}} - 3.$

(c) $\dfrac{d}{dx}\left[\dfrac{1}{\sqrt{x^2 + 1}}\right] = \dfrac{d}{dx}[(x^2 + 1)^{-1/2}]$

$= -\dfrac{1}{2}(x^2 + 1)^{-\frac{1}{2}-1}\dfrac{d}{dx}[(x^2 + 1)]$ General power rule.

$= -\dfrac{1}{2}(x^2 + 1)^{-\frac{3}{2}}(2x)$

$= -\dfrac{x}{(x^2 + 1)^{\frac{3}{2}}}.$

1.7 More about Derivatives

To denote the derivative of $y = f(x)$, we can use any one of the symbols y', $f'(x)$, or $\frac{d}{dx}[f(x)]$. The symbol $\frac{d}{dx}$ tells us to take the derivative with respect to x. If you want to take the derivative with respect to a variable other than x, then use that variable in the derivative symbol. For example, if you want to differentiate with respect to t, then use the symbol $\frac{d}{dt}$. In the presence of the symbol $\frac{d}{dt}$, we treat t as the independent variable and all other letters are treated as constants. So,

$$\frac{d}{dx}[x^3] = 3x^2, \quad \text{but} \quad \frac{d}{dt}[x^3] = 0$$

because the derivative of a constant (in this case, x^3) is 0.

When we differentiate a function $f(x)$, we obtain a new function $f'(x)$. Since $f'(x)$ is itself a function, we can take its derivative and obtain what is called the *second derivative* of $f(x)$, denoted by $f''(x)$, is defined by:

$$f''(x) = \frac{d}{dx}[f'(x)].$$

You can keep on differentiating the derivatives. For example, the third derivative is $f'''(x) = \frac{d}{dx}[f''(x)]$, and so on. Another convenient notation for the *higher-order derivatives* of $y = f(x)$ is illustrated by the following:

$$f''(x) = \frac{d^2}{dx^2}[f(x)]; \quad f'''(x) = \frac{d^3}{dx^3}[f(x)].$$

Compute the given derivatives.

(a) $\dfrac{d}{dt}[x^2 + 3t - t^2] = \dfrac{d}{dt}[x^2] + 3\dfrac{d}{dt}[t] - \dfrac{d}{dt}[t^2]$

Sum rule and constant-multiple rule.

$= 0 + 3 - 2t = -2t + 3$

Treat x as a constant.

(b) $\dfrac{d}{dx}[xt + t^2 - 7x^3] = \dfrac{d}{dx}[xt] + \dfrac{d}{dx}[t^2] - 7\dfrac{d}{dx}[x^3]$

$= t\dfrac{d}{dx}[x] + 0 - 7(3x^2) = t - 21x^2$

Treat t as a constant.

Compute the given second-order derivative.

(c) $\dfrac{d^2}{dx^2}[(4x + 1)^3] = \dfrac{d}{dx}\left[\dfrac{d}{dx}[(4x + 1)^3]\right]$

Derivative of the derivative.

$= \dfrac{d}{dx}\left[3(4x + 1)^2\dfrac{d}{dx}[(4x + 1)]\right]$

General power rule.

$= \dfrac{d}{dx}[3(4x + 1)^2(4)] = \dfrac{d}{dx}[12(4x + 1)^2]$

$= 12\dfrac{d}{dx}[(4x + 1)^2]$

$= 12(2)(4x + 1)\dfrac{d}{dx}[4x + 1]$

$= 24(4x + 1)(4)$

$= 96(4x + 1).$

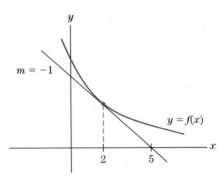

Figure 1

66. In Fig. 2, the straight line is tangent to the graph of $f(x) = x^3$. Find the value of a.

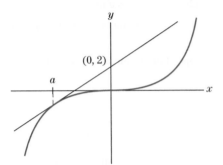

Figure 2

67. Height of a Helicopter A helicopter is rising at a rate of 32 feet per second. At a height of 128 feet the pilot drops a pair of binoculars. After t seconds, the binoculars have height $s(t) = -16t^2 + 32t + 128$ feet from the ground. How fast will they be falling when they hit the ground?

68. Rate of Output of a Coal Mine Each day the total output of a coal mine after t hours of operation is approximately $40t + t^2 - \frac{1}{15}t^3$ tons, $0 \le t \le 12$. What is the rate of output (in tons of coal per hour) at $t = 5$ hours?

Exercises 69–72 refer to Fig. 3, where $s(t)$ is the number of feet traveled by a person after t seconds of walking along a straight path.

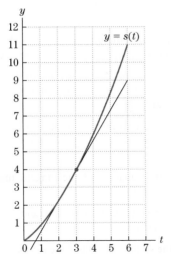

Figure 3 Walker's progress.

69. How far has the person traveled after 6 seconds?

70. What is the person's average velocity from time $t = 1$ to $t = 4$?

71. What is the person's velocity at time $t = 3$?

72. Without calculating velocities, determine whether the person is traveling faster at $t = 5$ or at $t = 6$.

73. Marginal Cost A manufacturer estimates that the hourly cost of producing x units of a product on an assembly line is $C(x) = .1x^3 - 6x^2 + 136x + 200$ dollars.
  (a) Compute $C(21) - C(20)$, the extra cost of raising the production from 20 to 21 units.
  (b) Find the marginal cost when the production level is 20 units.

74. Number of Subway Passengers The number of people riding the subway daily from Silver Spring, Maryland, to Washington's Metro Center is a function $f(x)$ of the fare, x cents. If $f(235) = 4600$ and $f'(235) = -100$, approximate the daily number of riders for each of the following costs:
  (a) 237 cents   **(b)** 234 cents
  (c) 240 cents   **(d)** 232 cents

75. Height of a Child Let $h(t)$ be a boy's height (in inches) after t years. If $h'(12) = 1.5$, how much will his height increase (approximately) between ages 12 and $12\frac{1}{2}$?

76. Compound Interest If you deposit \$100 in a savings account at the end of each month for 2 years, the balance will be a function $f(r)$ of the interest rate, $r\%$. At 7% interest (compounded monthly), $f(7) = 2568.10$ and $f'(7) = 25.06$. Approximately how much additional money would you earn if the bank paid $7\frac{1}{2}\%$ interest?

Determine whether the following limits exist. If so, compute the limit.

77. $\lim\limits_{x \to 2} \dfrac{x^2 - 4}{x - 2}$

78. $\lim\limits_{x \to 3} \dfrac{1}{x^2 - 4x + 3}$

79. $\lim\limits_{x \to 4} \dfrac{x - 4}{x^2 - 8x + 16}$

80. $\lim\limits_{x \to 5} \dfrac{x - 5}{x^2 - 7x + 2}$

Use limits to compute the following derivatives.

81. $f'(5)$, where $f(x) = 1/(2x)$.

82. $f'(3)$, where $f(x) = x^2 - 2x + 1$.

83. What geometric interpretation may be given to
$$\frac{(3 + h)^2 - 3^2}{h}$$
in connection with the graph of $f(x) = x^2$?

84. As h approaches 0, what value is approached by
$$\frac{\dfrac{1}{2 + h} - \dfrac{1}{2}}{h}?$$

KEY TERMS AND CONCEPTS	EXAMPLES

1.6 Some Rules for Differentiation

In computing derivatives, we can use basic rules for differentiation such as the following:

• Constant-multiple rule:

$$\frac{d}{dx}[k \cdot f(x)] = k \cdot \frac{d}{dx}[f(x)], \quad k \text{ a constant.}$$

• Sum rule:

$$\frac{d}{dx}[f(x) + g(x)] = \frac{d}{dx}[f(x)] + \frac{d}{dx}[g(x)].$$

• General power rule:

$$\frac{d}{dx}([g(x)]^r) = r \cdot [g(x)]^{r-1} \cdot \frac{d}{dx}[g(x)].$$

Using properties of derivatives, compute the following derivatives:

(a) $\dfrac{d}{dx}[-7x^2] = -7\dfrac{d}{dx}[x^2]$ Constant-multiple rule.

$\qquad = (-7)2x = -14x.$ Power rule.

(b) $\dfrac{d}{dx}[\sqrt{x} - 3x] = \dfrac{d}{dx}[\sqrt{x}] + \dfrac{d}{dx}[-3x]$ Sum rule.

$\qquad = \dfrac{d}{dx}[x^{1/2}] - 3\dfrac{d}{dx}[x]$ Constant-multiple rule.

$\qquad = \dfrac{1}{2}(x^{-1/2}) - 3 = \dfrac{1}{2\sqrt{x}} - 3.$

(c) $\dfrac{d}{dx}\left[\dfrac{1}{\sqrt{x^2+1}}\right] = \dfrac{d}{dx}[(x^2+1)^{-1/2}]$

$\qquad = -\dfrac{1}{2}(x^2+1)^{-\frac{1}{2}-1}\dfrac{d}{dx}[(x^2+1)]$ General power rule.

$\qquad = -\dfrac{1}{2}(x^2+1)^{-\frac{3}{2}}(2x)$

$\qquad = -\dfrac{x}{(x^2+1)^{\frac{3}{2}}}.$

1.7 More about Derivatives

To denote the derivative of $y = f(x)$, we can use any one of the symbols y', $f'(x)$, or $\frac{d}{dx}[f(x)]$. The symbol $\frac{d}{dx}$ tells us to take the derivative with respect to x. If you want to take the derivative with respect to a variable other than x, then use that variable in the derivative symbol. For example, if you want to differentiate with respect to t, then use the symbol $\frac{d}{dt}$. In the presence of the symbol $\frac{d}{dt}$, we treat t as the independent variable and all other letters are treated as constants. So,

$$\frac{d}{dx}[x^3] = 3x^2, \quad \text{but} \quad \frac{d}{dt}[x^3] = 0$$

because the derivative of a constant (in this case, x^3) is 0.

When we differentiate a function $f(x)$, we obtain a new function $f'(x)$. Since $f'(x)$ is itself a function, we can take its derivative and obtain what is called the *second derivative* of $f(x)$, denoted by $f''(x)$, is defined by:

$$f''(x) = \frac{d}{dx}[f'(x)].$$

You can keep on differentiating the derivatives. For example, the third derivative is $f'''(x) = \frac{d}{dx}[f''(x)]$, and so on. Another convenient notation for the *higher-order derivatives* of $y = f(x)$ is illustrated by the following:

$$f''(x) = \frac{d^2}{dx^2}[f(x)]; \quad f'''(x) = \frac{d^3}{dx^3}[f(x)].$$

Compute the given derivatives.

(a) $\dfrac{d}{dt}[x^2 + 3t - t^2] = \dfrac{d}{dt}[x^2] + 3\dfrac{d}{dt}[t] - \dfrac{d}{dt}[t^2]$

$\qquad$ Sum rule and constant-multiple rule.

$\qquad = 0 + 3 - 2t = -2t + 3$

$\qquad$ Treat x as a constant.

(b) $\dfrac{d}{dx}[xt + t^2 - 7x^3] = \dfrac{d}{dx}[xt] + \dfrac{d}{dx}[t^2] - 7\dfrac{d}{dx}[x^3]$

$\qquad = t\dfrac{d}{dx}[x] + 0 - 7(3x^2) = t - 21x^2$

$\qquad$ Treat t as a constant.

Compute the given second-order derivative.

(c) $\dfrac{d^2}{dx^2}[(4x+1)^3] = \dfrac{d}{dx}\left[\dfrac{d}{dx}[(4x+1)^3]\right]$

$\qquad$ Derivative of the derivative.

$\qquad = \dfrac{d}{dx}\left[3(4x+1)^2\dfrac{d}{dx}[(4x+1)]\right]$

$\qquad$ General power rule.

$\qquad = \dfrac{d}{dx}[3(4x+1)^2(4)] = \dfrac{d}{dx}[12(4x+1)^2]$

$\qquad = 12\dfrac{d}{dx}[(4x+1)^2]$

$\qquad = 12(2)(4x+1)\dfrac{d}{dx}[4x+1]$

$\qquad = 24(4x+1)(4)$

$\qquad = 96(4x+1).$

KEY TERMS AND CONCEPTS	EXAMPLES

1.8 The Derivative as a Rate of Change

Let $y = f(x)$ be a function defined on the interval $a \le x \le b$. Then,

$$\left[\begin{array}{c} \text{average rate of change of } f(x) \\ \text{over the interval } a \le x \le b \end{array}\right] = \frac{f(b) - f(a)}{b - a}.$$

The **(instantaneous) rate of change** of $f(x)$ at the point where $x = a$ is $f'(a)$.

If $s(t)$ denotes the position function of an object moving in a straight line, then the **velocity** $v(t)$ of the object at time t is given by

$$v(t) = s'(t).$$

The derivative of the velocity function $v(t)$ is called the **acceleration**:

$$a(t) = v'(t) \quad \text{or} \quad a(t) = s''(t).$$

To approximate the change in a function, we can use the formula
$$f(a + h) - f(a) \approx f'(a) \cdot h.$$

If we let $a + h = x$, then $h = x - a$, and the formula becomes
$$f(x) - f(a) \approx f'(a) \cdot (x - a).$$
Equivalently,
$$f(x) \approx f(a) + f'(a) \cdot (x - a).$$

1. The average rate of change of $f(x) = \dfrac{1}{x}$ on the interval $1 \le x \le 3$ is

$$\frac{f(3) - f(1)}{3 - 1} = \frac{\frac{1}{3} - 1}{2} = \frac{-\frac{2}{3}}{2} = -\frac{2}{6} = -\frac{1}{3}.$$

2. The instantaneous rate of change of $f(x) = \dfrac{1}{x}$ at $x = 3$ is $f'(3)$:

$$f'(x) = -\frac{1}{x^2}; \quad f'(3) = -\frac{1}{9}.$$

3. The position of a ball thrown up into the air is given by

$$s(t) = -16t^2 + 64t + 5,$$

where t is in seconds and $s(t)$ in feet. To find the time when the velocity is 0, we first find the velocity at any time t, then set it equal to 0 and solve for t:

$$v(t) = s'(t) = -32t + 64$$

$$-32t + 64 = 0$$

$$t = \frac{64}{32} = 2.$$

Thus, the velocity is 0 when $t = 2$ seconds.

4. Given $C(2) = 3.5$ and $C'(2) = .7$, we can approximate $C(1.8)$ as follows: Take $a = 2$, $x = 1.8$; then,

$$C(1.8) \approx C(2) + C'(2)(x - a)$$

$$= 3.5 + (.7)(1.8 - 2)$$

$$= 3.5 - .14 = 3.36.$$

So, $C(1.8) \approx 3.36$.

CHAPTER 1   Fundamental Concept Check Exercises

1. Define the slope of a nonvertical line and give a physical description.

2. What is the point–slope form of the equation of a line?

3. Describe how to find an equation for a line when you know the coordinates of two points on the graph of a line.

4. What can you say about the slopes of parallel lines? Perpendicular lines?

5. Give a physical description of what is meant by the slope of $f(x)$ at the point $(2, f(2))$.

6. What does $f'(2)$ represent?

7. Explain why the derivative of a constant function is 0.

8. State the power rule, the constant-multiple rule, and the sum rule, and give an example of each.

9. Explain how to calculate $f'(2)$ as the limit of slopes of secant lines through the point $(2, f(2))$.

10. In your own words, explain the meaning of $\lim_{x \to 2} f(x) = 3$. Give an example of a function with this property.

11. Give the limit definition of $f'(2)$, that is, the slope of $f(x)$ at the point $(2, f(2))$.

12. In your own words, explain the meaning of $\lim_{x \to \infty} f(x) = 3$. Give an example of such a function $f(x)$. Do the same for $\lim_{x \to -\infty} f(x) = 3$.

13. In your own words, explain the meaning of "$f(x)$ is continuous at $x = 2$." Give an example of a function $f(x)$ that is not continuous at $x = 2$.

14. In your own words, explain the meaning of "$f(x)$ is differentiable at $x = 2$." Give an example of a function $f(x)$ that is not differentiable at $x = 2$.

15. State the general power rule and give an example.

16. Give two different notations for the first derivative of $f(x)$ at $x = 2$. Do the same for the second derivative.

17. What is meant by the average rate of change of a function over an interval?

18. How is an (instantaneous) rate of change related to average rates of change?

19. Explain the relationship between derivatives and velocity and acceleration.

20. What expression involving a derivative gives an approximation to $f(a + h) - f(a)$?

21. Describe marginal cost in your own words.

22. How do you determine the proper units for a rate of change? Give an example.

CHAPTER 1   Review Exercises

Find the equation and sketch the graph of the following lines.

1. With slope -2, y-intercept $(0, 3)$.

2. With slope $\frac{3}{4}$, y-intercept $(0, -1)$.

3. Through $(2, 0)$, with slope 5.

4. Through $(1, 4)$, with slope $-\frac{1}{3}$.

5. Parallel to $y = -2x$, passing through $(3, 5)$.

6. Parallel to $-2x + 3y = 6$, passing through $(0, 1)$.

7. Through $(-1, 4)$ and $(3, 7)$.

8. Through $(2, 1)$ and $(5, 1)$.

9. Perpendicular to $y = 3x + 4$, passing through $(1, 2)$.

10. Perpendicular to $3x + 4y = 5$, passing through $(6, 7)$.

11. Horizontal with height 3 units above the x–axis.

12. Vertical and 4 units to the right of the y-axis.

13. The y-axis.

14. The x-axis.

Differentiate.

15. $y = x^7 + x^3$

16. $y = 5x^8$

17. $y = 6\sqrt{x}$

18. $y = x^7 + 3x^5 + 1$

19. $y = 3/x$

20. $y = x^4 - \dfrac{4}{x}$

21. $y = (3x^2 - 1)^8$

22. $y = \frac{3}{4}x^{4/3} + \frac{4}{3}x^{3/4}$

23. $y = \dfrac{1}{5x - 1}$

24. $y = (x^3 + x^2 + 1)^5$

25. $y = \sqrt{x^2 + 1}$

26. $y = \dfrac{5}{7x^2 + 1}$

27. $f(x) = 1/\sqrt[4]{x}$

28. $f(x) = (2x + 1)^3$

29. $f(x) = 5$

30. $f(x) = \dfrac{5x}{2} - \dfrac{2}{5x}$

31. $f(x) = [x^5 - (x - 1)^5]^{10}$

32. $f(t) = t^{10} - 10t^9$

33. $g(t) = 3\sqrt{t} - \dfrac{3}{\sqrt{t}}$

34. $h(t) = 3\sqrt{2}$

35. $f(t) = \dfrac{2}{t - 3t^3}$

36. $g(P) = 4P^{0.7}$

37. $h(x) = \frac{3}{2}x^{3/2} - 6x^{2/3}$

38. $f(x) = \sqrt{x + \sqrt{x}}$

39. If $f(t) = 3t^3 - 2t^2$, find $f'(2)$.

40. If $V(r) = 15\pi r^2$, find $V'(\frac{1}{3})$.

41. If $g(u) = 3u - 1$, find $g(5)$ and $g'(5)$.

42. If $h(x) = -\frac{1}{2}$, find $h(-2)$ and $h'(-2)$.

43. If $f(x) = x^{5/2}$, what is $f''(4)$?

44. If $g(t) = \frac{1}{4}(2t - 7)^4$, what is $g''(3)$?

45. Find the slope of the graph of $y = (3x - 1)^3 - 4(3x - 1)^2$ at $x = 0$.

46. Find the slope of the graph of $y = (4 - x)^5$ at $x = 5$.

Compute.

47. $\dfrac{d}{dx}(x^4 - 2x^2)$

48. $\dfrac{d}{dt}(t^{5/2} + 2t^{3/2} - t^{1/2})$

49. $\dfrac{d}{dP}(\sqrt{1 - 3P})$

50. $\dfrac{d}{dn}(n^{-5})$

51. $\dfrac{d}{dz}(z^3 - 4z^2 + z - 3)\Big|_{z=-2}$

52. $\dfrac{d}{dx}(4x - 10)^5\Big|_{x=3}$

53. $\dfrac{d^2}{dx^2}(5x + 1)^4$

54. $\dfrac{d^2}{dt^2}(2\sqrt{t})$

55. $\dfrac{d^2}{dt^2}(t^3 + 2t^2 - t)\Big|_{t=-1}$

56. $\dfrac{d^2}{dP^2}(3P + 2)\Big|_{P=4}$

57. $\dfrac{d^2y}{dx^2}$, where $y = 4x^{3/2}$

58. $\dfrac{d}{dt}\left(\dfrac{dy}{dt}\right)$, where $y = \dfrac{1}{3t}$.

59. What is the slope of the graph of $f(x) = x^3 - 4x^2 + 6$ at $x = 2$? Write the equation of the line tangent to the graph of $f(x)$ at $x = 2$.

60. What is the slope of the curve $y = 1/(3x - 5)$ at $x = 1$? Write the equation of the line tangent to this curve at $x = 1$.

61. Find the equation of the tangent line to the curve $y = x^2$ at the point $(\frac{3}{2}, \frac{9}{4})$. Sketch the graph of $y = x^2$ and sketch the tangent line at $(\frac{3}{2}, \frac{9}{4})$.

62. Find the equation of the tangent line to the curve $y = x^2$ at the point $(-2, 4)$. Sketch the graph of $y = x^2$ and sketch the tangent line at $(-2, 4)$.

63. Determine the equation of the tangent line to the curve $y = 3x^3 - 5x^2 + x + 3$ at $x = 1$.

64. Determine the equation of the tangent line to the curve $y = (2x^2 - 3x)^3$ at $x = 2$.

65. In Fig. 1, the straight line has slope -1 and is tangent to the graph of $f(x)$. Find $f(2)$ and $f'(2)$.

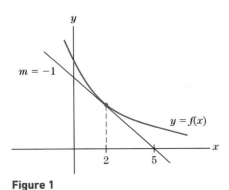

Figure 1

66. In Fig. 2, the straight line is tangent to the graph of $f(x) = x^3$. Find the value of a.

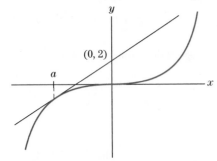

Figure 2

67. **Height of a Helicopter** A helicopter is rising at a rate of 32 feet per second. At a height of 128 feet the pilot drops a pair of binoculars. After t seconds, the binoculars have height $s(t) = -16t^2 + 32t + 128$ feet from the ground. How fast will they be falling when they hit the ground?

68. **Rate of Output of a Coal Mine** Each day the total output of a coal mine after t hours of operation is approximately $40t + t^2 - \frac{1}{15}t^3$ tons, $0 \le t \le 12$. What is the rate of output (in tons of coal per hour) at $t = 5$ hours?

Exercises 69–72 refer to Fig. 3, where $s(t)$ is the number of feet traveled by a person after t seconds of walking along a straight path.

Figure 3 Walker's progress.

69. How far has the person traveled after 6 seconds?

70. What is the person's average velocity from time $t = 1$ to $t = 4$?

71. What is the person's velocity at time $t = 3$?

72. Without calculating velocities, determine whether the person is traveling faster at $t = 5$ or at $t = 6$.

73. **Marginal Cost** A manufacturer estimates that the hourly cost of producing x units of a product on an assembly line is $C(x) = .1x^3 - 6x^2 + 136x + 200$ dollars.
    (a) Compute $C(21) - C(20)$, the extra cost of raising the production from 20 to 21 units.
    (b) Find the marginal cost when the production level is 20 units.

74. **Number of Subway Passengers** The number of people riding the subway daily from Silver Spring, Maryland, to Washington's Metro Center is a function $f(x)$ of the fare, x cents. If $f(235) = 4600$ and $f'(235) = -100$, approximate the daily number of riders for each of the following costs:
    (a) 237 cents   (b) 234 cents
    (c) 240 cents   (d) 232 cents

75. **Height of a Child** Let $h(t)$ be a boy's height (in inches) after t years. If $h'(12) = 1.5$, how much will his height increase (approximately) between ages 12 and $12\frac{1}{2}$?

76. **Compound Interest** If you deposit $100 in a savings account at the end of each month for 2 years, the balance will be a function $f(r)$ of the interest rate, $r\%$. At 7% interest (compounded monthly), $f(7) = 2568.10$ and $f'(7) = 25.06$. Approximately how much additional money would you earn if the bank paid $7\frac{1}{2}\%$ interest?

Determine whether the following limits exist. If so, compute the limit.

77. $\lim\limits_{x \to 2} \dfrac{x^2 - 4}{x - 2}$

78. $\lim\limits_{x \to 3} \dfrac{1}{x^2 - 4x + 3}$

79. $\lim\limits_{x \to 4} \dfrac{x - 4}{x^2 - 8x + 16}$

80. $\lim\limits_{x \to 5} \dfrac{x - 5}{x^2 - 7x + 2}$

Use limits to compute the following derivatives.

81. $f'(5)$, where $f(x) = 1/(2x)$.

82. $f'(3)$, where $f(x) = x^2 - 2x + 1$.

83. What geometric interpretation may be given to
$$\frac{(3 + h)^2 - 3^2}{h}$$
in connection with the graph of $f(x) = x^2$?

84. As h approaches 0, what value is approached by
$$\frac{\dfrac{1}{2 + h} - \dfrac{1}{2}}{h}?$$

2

Applications of the Derivative

Calculus techniques can be applied to a wide variety of problems in real life. We consider many examples in this chapter. In each case, we construct a function as a mathematical model of some problem and then analyze the function and its derivatives to gain information about the original problem. Our principal method for analyzing a function will be to sketch its graph. For this reason, we devote the first part of the chapter to curve sketching and understanding the behavior of functions.

2.1   Describing Graphs of Functions

Figure 1 An increasing function.

Let's examine the graph of a typical function, such as the one shown in Fig. 1, and introduce some terminology to describe its behavior. First, observe that the graph is either rising or falling, depending on whether we look at it from left to right or from right to left. To avoid confusion, we shall always follow the accepted practice of reading a graph from left to right.

Let's now examine the behavior of a function $f(x)$ in an interval throughout which it is defined.

DEFINITION   **Increasing and Decreasing Functions**   We say that a function $f(x)$ is **increasing in an interval** if the graph continuously rises as x goes from left to right through the interval. That is, whenever x_1 and x_2 are in the interval with $x_1 < x_2$, we have $f(x_1) < f(x_2)$. We say that $f(x)$ is **increasing at** $x = c$ provided that $f(x)$ is increasing in some open interval on the x-axis that contains the point c.

We say that a function $f(x)$ is **decreasing in an interval** provided that the graph continuously falls as x goes from left to right through the interval. That is, whenever x_1 and x_2 are in the interval with $x_1 < x_2$, we have $f(x_1) > f(x_2)$. We say that $f(x)$ is **decreasing at** $x = c$ provided that $f(x)$ is decreasing in some open interval that contains the point c.

Figure 2 shows graphs that are increasing and decreasing at $x = c$. Observe in Fig. 2(d) that when $f(c)$ is negative and $f(x)$ is decreasing, the values of $f(x)$ become *more* negative. When $f(c)$ is negative and $f(x)$ is increasing, as in Fig. 2(e), the values of $f(x)$ become *less* negative.

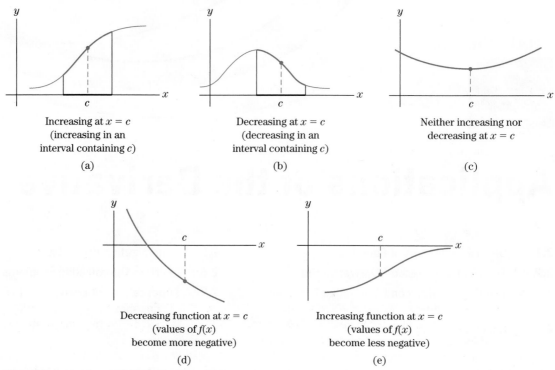

Increasing at $x = c$
(increasing in an
interval containing c)

(a)

Decreasing at $x = c$
(decreasing in an
interval containing c)

(b)

Neither increasing nor
decreasing at $x = c$

(c)

Decreasing function at $x = c$
(values of $f(x)$
become more negative)

(d)

Increasing function at $x = c$
(values of $f(x)$
become less negative)

(e)

Figure 2 Increasing and decreasing functions at $x = c$.

Extreme Points

A **relative extreme point** or an **extremum** of a function is a point at which its graph changes from increasing to decreasing, or vice versa. We distinguish the two possibilities in an obvious way.

> **DEFINITION**   **Relative Maximum and Minimum**   A **relative maximum point** is a point at which the graph changes from increasing to decreasing; a **relative minimum point** is a point at which the graph changes from decreasing to increasing. (See Fig. 3.)

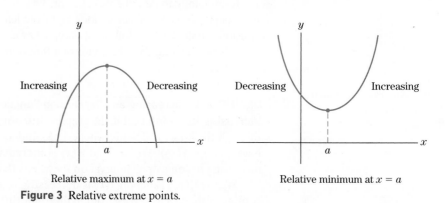

Increasing   Decreasing

Decreasing   Increasing

Relative maximum at $x = a$

Relative minimum at $x = a$

Figure 3 Relative extreme points.

The adjective *relative* in these definitions indicates that a point is maximal or minimal relative only to nearby points on the graph. The adjective *local* is also used in place of *relative*.

> **DEFINITION   Absolute Maximum and Minimum**   The **maximum value** (or **absolute maximum value**) of a function is the largest value that the function assumes on its domain. The **minimum value** (or **absolute minimum value**) of a function is the smallest value that the function assumes on its domain.

Functions may or may not have maximum or minimum values. (See Fig. 4.) However, it can be shown that a continuous function whose domain is an interval of the form $a \leq x \leq b$ has both a maximum and a minimum value.

FOR REVIEW

A dot at the end of a graph indicates that the graph ends there.

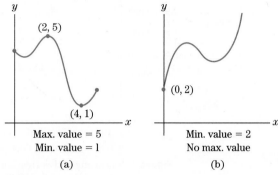

Figure 4

Maximum values and minimum values of functions usually occur at relative maximum points and relative minimum points, as in Fig. 4(a). However, they can occur at endpoints of the domain, as in Fig. 4(b). If so, we say that the function has an **endpoint extreme value** (or **endpoint extremum**).

Relative maximum points and endpoint maximum points are higher than any nearby points. The maximum value of a function is the y-coordinate of the highest point on its graph. (The highest point is called the *absolute maximum point*.) Similar considerations apply to minima. (*Minima* and *maxima* are the plural forms of *minimum* and *maximum*.)

EXAMPLE 1   **Concentration of a Drug in the Bloodstream**   When a drug is injected intramuscularly (into a muscle), the concentration of the drug in the blood has the time–concentration curve shown in Fig. 5. Describe this graph using the terms introduced previously.

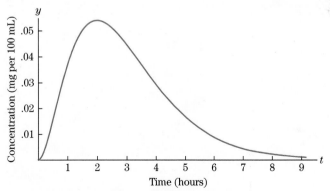

Figure 5 A drug time–concentration curve.

SOLUTION   Initially (when $t = 0$), there is no drug in the blood, so 0 is the minimum value of the graph. When the drug is injected into the muscle, it begins to diffuse into the bloodstream. The drug concentration in the blood increases until it reaches its maximum value at $t = 2$. After this time, the concentration begins to decrease as the body's metabolic processes remove the drug from the blood. Eventually, the drug concentration decreases to a level so small that, for all practical purposes, it is zero.

≫ Now Try Exercise 1

FOR REVIEW

The slope of the graph at a point P is the slope of the tangent line at P.

Changing Slope

An important but subtle feature of a graph is the way the graph's slope *changes* (as we look from left to right). The graphs in Fig. 6 are both increasing, but there is a fundamental difference in the way they are increasing. Graph I, which describes the U.S. gross federal debt per person, is steeper for 1990 than for 1960. That is, the *slope* of graph I *increases* as we move from left to right. A newspaper description of graph I might read: U.S. gross federal debt per capita rose at an increasing rate during the years 1960 to 1990.

 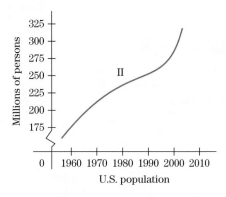

Figure 6 Increasing and decreasing slopes.

In contrast, the *slope* of graph II *decreases* as we move from left to right from 1960 to 1990. Although the U.S. population is rising each year, the rate of increase declines throughout the years from 1960 to 1990. That is, the slope becomes less positive. The media might say: During the 1960s, 1970s, and 1980s, U.S. population rose at a decreasing rate.

EXAMPLE 2 **Daylight Hours in a City** The daily number of hours of sunlight in Washington, D.C., increased from 9.45 hours on December 21 to 12 hours on March 21 and then increased to 14.9 hours on June 21. From December 22 to March 21, the daily increase was greater than the previous daily increase, and from March 22 to June 21, the daily increase was less than the previous daily increase. Draw a possible graph of the number of hours of daylight as a function of time.

SOLUTION Let $f(t)$ be the number of hours of daylight t months after December 21. See Fig. 7. The first part of the graph, December 21 to March 21, is increasing at an increasing rate. The second part of the graph, March 21 to June 21, is increasing at a decreasing rate.

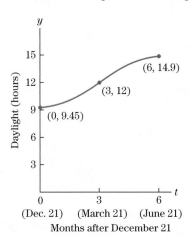

Figure 7 Hours of daylight in Washington, D.C.

CAUTION Recall that when a negative quantity decreases, it becomes more negative. (Think about the temperature outside when it is below zero and the temperature is falling.) So if the slope of a graph is negative and the slope is decreasing, the slope is becoming more negative, as in Fig. 8. This technical use of the term *decreasing* runs counter to our intuition, because in popular discourse, *decrease* often means to become smaller in size. «

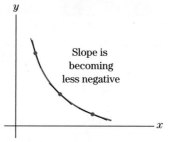

Figure 8 Slope is decreasing.   **Figure 9** Slope is increasing.

It is true that the curve in Fig. 9 is becoming "less steep" in a nontechnical sense (since steepness, if it were defined, would probably refer to the magnitude of the slope). However, the slope of the curve in Fig. 9 is increasing because it is becoming less negative. The popular press would probably describe the curve in Fig. 9 as decreasing at a decreasing rate, because the rate of fall tends to taper off. Since this terminology is potentially confusing, we shall not use it.

Concavity

The U.S. debt and population graphs in Fig. 6 may also be described in geometric terms: Between 1960 and 1990, graph I opens up and lies above its tangent line at each point, whereas graph II opens down and lies below its tangent line at each point between 1960 and 1990. (See Fig. 10.)

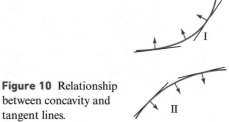

Figure 10 Relationship between concavity and tangent lines.

> **DEFINITION Concavity** We say that a function $f(x)$ is **concave up** at $x = a$ if there is an open interval on the x-axis containing a, throughout which the graph of $f(x)$ lies above its tangent line. Equivalently, $f(x)$ is concave up at $x = a$ if the slope of the graph increases as we move from left to right through $(a, f(a))$.

   Graph I in Fig. 10 is an example of a function that is concave up at each point.
   Similarly, we say that a function $f(x)$ is **concave down** at $x = a$ if there is an open interval on the x-axis containing a, throughout which the graph of $f(x)$ lies below its tangent line. Equivalently, $f(x)$ is concave down at $x = a$ if the slope of the graph decreases as we move from left to right through $(a, f(a))$. Graph II in Fig. 10 is concave down at each point.

> **DEFINITION Inflection Point** An **inflection point** is a point on the graph of a function at which the function is continuous and at which the graph changes from concave up to concave down or vice versa.

   At such a point, the graph crosses its tangent line. (See Fig. 11.) The continuity condition means that the graph cannot break at an inflection point.

Figure 11 Inflection points.

EXAMPLE 3   **Describing a Graph** Use the terms defined earlier to describe the graph shown in Fig. 12.

Figure 12

SOLUTION   (a) For $x < 3$, $f(x)$ is increasing and concave down.

(b) Relative maximum point at $x = 3$.

(c) For $3 < x < 4$, $f(x)$ is decreasing and concave down.

(d) Inflection point at $x = 4$.

(e) For $4 < x < 5$, $f(x)$ is decreasing and concave up.

(f) Relative minimum point at $x = 5$.

(g) For $x > 5$, $f(x)$ is increasing and concave up.   **>> Now Try Exercise 7**

Intercepts, Undefined Points, and Asymptotes

A point at which a graph crosses the y-axis is called a **y-intercept**, and a point at which it crosses the x-axis is called an **x-intercept**. A function can have, at most, one y-intercept. Otherwise, its graph would violate the vertical line test for a function. Note, however, that a function can have any number of x-intercepts (possibly none). The x-coordinate of an x-intercept is sometimes called a **zero** of the function, since the function has the value zero there. (See Fig. 13.)

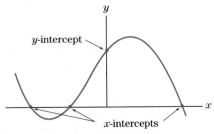

Figure 13 Intercepts of a graph.

Some functions are not defined for all values of x. For instance, $f(x) = 1/x$ is not defined for $x = 0$, and $f(x) = \sqrt{x}$ is not defined for $x < 0$. (See Fig. 14.) Many functions that arise in applications are defined only for $x \geq 0$. A properly drawn graph should leave no doubt as to the values of x for which the function is defined.

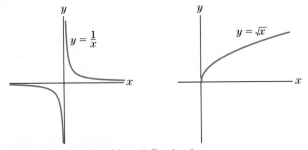

Figure 14 Graphs with undefined points.

Graphs in applied problems sometimes straighten out and approach some straight line as x gets large (Fig. 15). Such a straight line is called an **asymptote** of the curve. The most common asymptotes are horizontal, as in (a) and (b) of Fig. 15. In Example 1, the t-axis is an asymptote of the drug time–concentration curve (Fig. 5).

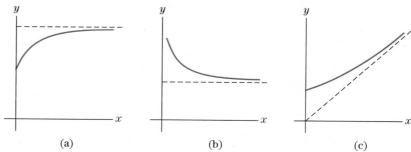

(a)   (b)   (c)

Figure 15 Graphs that approach asymptotes as x gets large.

Some asymptotes are slanted, as illustrated in Fig. 15(c).

The horizontal asymptotes of a graph may be determined by calculation of the limits

$$\lim_{x \to \infty} f(x) \quad \text{and} \quad \lim_{x \to -\infty} f(x).$$

If either limit exists, the value of the limit determines a horizontal asymptote.

Occasionally, a graph will approach a vertical line as x approaches some fixed value, as in Fig. 16. Such a line is a **vertical asymptote**. Most often, we expect a vertical asymptote at a value x that would result in division by zero in the definition of $f(x)$. For example, $f(x) = 1/(x - 3)$ has a vertical asymptote $x = 3$.

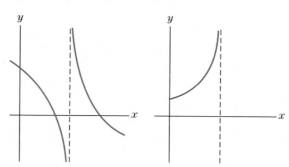

Figure 16 Examples of vertical asymptotes.

We now have six categories for describing the graph of a function.

Describing a Graph

1. Intervals in which the function is increasing (or decreasing), relative maximum points, relative minimum points
2. Maximum value, minimum value
3. Intervals in which the function is concave up (or concave down), inflection points
4. x-intercepts, y-intercept
5. Undefined points
6. Asymptotes

For us, the first three categories will be the most important. However, the last three categories should not be forgotten.

Check Your Understanding 2.1

Solutions can be found following the section exercises.

1. Does the slope of the curve in Fig. 17 increase or decrease as *x* increases?

Figure 17

2. At which labeled point on the graph in Fig. 18 is the slope the least?

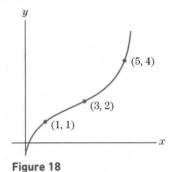

Figure 18

EXERCISES 2.1

Exercises 1–4 refer to graphs (a)–(f) in Fig. 19.

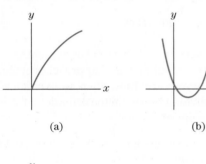

(a)　(b)

(c)　(d)

(e)　(f)

Figure 19

1. Which functions are increasing for all *x*?

2. Which functions are decreasing for all *x*?

3. Which functions have the property that the slope always increases as *x* increases?

4. Which functions have the property that the slope always decreases as *x* increases?

Describe each of the following graphs. Your descriptions should include each of the six categories mentioned previously.

5.

6.

7.

8.

9.

10.

11.

12.

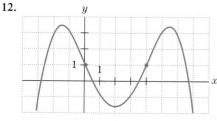

13. Describe the way the *slope* changes as you move along the graph (from left to right) in Exercise 5.

14. Describe the way the *slope* changes on the graph in Exercise 6.

15. Describe the way the *slope* changes on the graph in Exercise 8.

16. Describe the way the *slope* changes on the graph in Exercise 10.

Exercises 17 and 18 refer to the graph in Fig. 20.

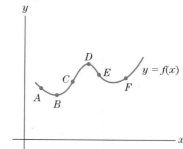

Figure 20

17. **(a)** At which labeled points is the function increasing?
   (b) At which labeled points is the graph concave up?
   (c) Which labeled point has the most positive slope?

18. **(a)** At which labeled points is the function decreasing?
   (b) At which labeled points is the graph concave down?

(c) Which labeled point has the most negative slope (that is, negative and with the greatest magnitude)?

In Exercises 19–22, draw the graph of a function $y = f(x)$ with the stated properties.

19. Both the function and the slope increase as x increases.

20. The function increases and the slope decreases as x increases.

21. The function decreases and the slope increases as x increases. [*Note:* The slope is negative but becomes less negative.]

22. Both the function and the slope decrease as x increases. [*Note:* The slope is negative and becomes more negative.]

23. **Annual World Consumption of Oil** The annual world consumption of oil rises each year. Furthermore, the amount of the annual *increase* in oil consumption is also rising each year. Sketch a graph that could represent the annual world consumption of oil.

24. **Average Annual Income** In certain professions, the average annual income has been rising at an increasing rate. Let $f(T)$ denote the average annual income at year T for persons in one of these professions and sketch a graph that could represent $f(T)$.

25. **A Patient's Temperature** At noon, a child's temperature is 101°F and is rising at an increasing rate. At 1 P.M. the child is given medicine. After 2 P.M. the temperature is still increasing but at a decreasing rate. The temperature reaches a peak of 103° at 3 P.M. and decreases to 100° by 5 P.M. Draw a possible graph of the function $T(t)$, the child's temperature at time t.

26. **A Cost Function** Let $C(x)$ denote the total cost of manufacturing x units of some product. Then $C(x)$ is an increasing function for all x. For small values of x, the rate of increase of $C(x)$ decreases (because of the savings that are possible with "mass production"). Eventually, however, for large values of x, the cost $C(x)$ increases at an increasing rate. (This happens when production facilities are strained and become less efficient.) Sketch a graph that could represent $C(x)$.

27. **Blood Flow through the Brain** One method of determining the level of blood flow through the brain requires the person to inhale air containing a fixed concentration of N_2O, nitrous oxide. During the first minute, the concentration of N_2O in the jugular vein grows at an increasing rate to a level of .25%. Thereafter, it grows at a decreasing rate and reaches a concentration of about 4% after 10 minutes. Draw a possible graph of the concentration of N_2O in the vein as a function of time.

28. **Pollution** Suppose that some organic waste products are dumped into a lake at time $t = 0$ and that the oxygen content of the lake at time t is given by the graph in Fig. 21. Describe the graph in physical terms. Indicate the significance of the inflection point at $t = b$.

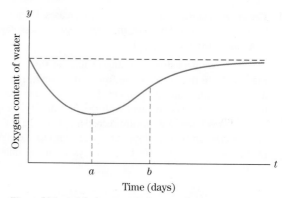

Figure 21 A lake's recovery from pollution.

29. Number of U.S. Farms Figure 22 gives the number of U.S. farms in millions from 1920 ($t = 20$) to 2000 ($t = 100$). In what year was the number of farms decreasing most rapidly?

Figure 22 Number of U.S. farms.

30. Consumer Price Index Figure 23 shows the graph of the consumer price index for the years 1983 ($t = 0$) through 2002 ($t = 19$). This index measures how much a basket of commodities that costs $100 in the beginning of 1983 would cost at any given time. In what year was the rate of increase of the index greatest? The least?

Figure 23 Consumer price index.

31. Velocity of a Parachutist Let $s(t)$ be the distance (in feet) traveled by a parachutist after t seconds from the time of opening the chute, and suppose that $s(t)$ has the line $y = -15t + 10$

as an asymptote. What does this imply about the velocity of the parachutist? [*Note:* Distance traveled downward is given a negative value.]

32. Let $P(t)$ be the population of a bacteria culture after t days and suppose that $P(t)$ has the line $y = 25,000,000$ as an asymptote. What does this imply about the size of the population?

In Exercises 33–36, sketch the graph of a function having the given properties.

33. Defined for $0 \leq x \leq 10$; relative maximum point at $x = 3$; absolute maximum value at $x = 10$

34. Relative maximum points at $x = 1$ and $x = 5$; relative minimum point at $x = 3$; inflection points at $x = 2$ and $x = 4$

35. Defined and increasing for all $x \geq 0$; inflection point at $x = 5$; asymptotic to the line $y = (\frac{3}{4})x + 5$

36. Defined for $x \geq 0$; absolute minimum value at $x = 0$; relative maximum point at $x = 4$; asymptotic to the line $y = (\frac{x}{2}) + 1$

37. Consider a smooth curve with no undefined points.
   (a) If it has two relative maximum points, must it have a relative minimum point?
   (b) If it has two relative extreme points, must it have an inflection point?

38. If the function $f(x)$ has a relative minimum at $x = a$ and a relative maximum at $x = b$, must $f(a)$ be less than $f(b)$?

TECHNOLOGY EXERCISES

39. Graph the function

$$f(x) = \frac{1}{x^3 - 2x^2 + x - 2}$$

in the window $[0, 4]$ by $[-15, 15]$. For what value of x does $f(x)$ have a vertical asymptote?

40. The graph of the function

$$f(x) = \frac{2x^2 - 1}{.5x^2 + 6}$$

has a horizontal asymptote of the form $y = c$. Estimate the value of c by graphing $f(x)$ in the window $[0, 50]$ by $[-1, 6]$.

41. Simultaneously graph the functions

$$y = \frac{1}{x} + x \quad \text{and} \quad y = x$$

in the window $[-6, 6]$ by $[-6, 6]$. Describe the asymptote of the first function.

Solutions to Check Your Understanding 2.1

1. The curve is concave up, so the slope increases. Even though the curve itself is decreasing, the slope becomes less negative as we move from left to right.

2. At $x = 3$. We have drawn in tangent lines at three points in Fig. 24. Note that as we move from left to right, the slopes decrease steadily until the point (3, 2), at which time they start to increase. This is consistent with the fact that the graph is concave down (hence slopes are decreasing) to the left of (3, 2) and concave up (hence, slopes are increasing) to the right of (3, 2). Extreme values of slopes always occur at inflection points.

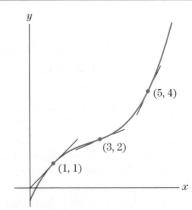

Figure 24

2.2 The First- and Second-Derivative Rules

$f'(a) > 0$

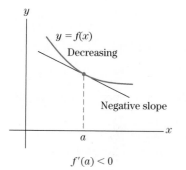

$f'(a) < 0$

Figure 1 Illustration of the first-derivative rule.

We shall now show how properties of the graph of a function $f(x)$ are determined by properties of the derivatives, $f'(x)$ and $f''(x)$. These relationships will provide the key to the curve sketching and optimization discussed in the rest of the chapter. Throughout this chapter, we assume that we are dealing with functions that are not "too badly behaved." More precisely, it suffices to assume that all our functions have continuous first and second derivatives in the interval(s) (in x) where we are considering their graphs.

We begin with a discussion of the first derivative of a function $f(x)$. Suppose that for some value of x, say $x = a$, the derivative $f'(a)$ is positive. Then the tangent line at $(a, f(a))$ has positive slope and is a rising line (moving from left to right, of course). Since the graph of $f(x)$ near $(a, f(a))$ resembles its tangent line, the function must be increasing at $x = a$. Similarly, when $f'(a) < 0$, the function is decreasing at $x = a$. (See Fig. 1.)

Thus, we have the following useful result.

> **First-Derivative Rule** If $f'(a) > 0$, then $f(x)$ is increasing at $x = a$. If $f'(a) < 0$, then $f(x)$ is decreasing at $x = a$.

In other words, a function is increasing whenever the value of its derivative is positive; a function is decreasing whenever the value of its derivative is negative. The first-derivative rule says nothing about the case when the derivative of a function is zero. If $f'(a) = 0$, the function might be increasing or decreasing or have a relative extreme point at $x = a$.

EXAMPLE 1 **Graphing Using Properties of the Derivative** Sketch the graph of a function $f(x)$ that has all the following properties.
(a) $f(3) = 4$
(b) $f'(x) > 0$ for $x < 3$, $f'(3) = 0$, and $f'(x) < 0$ for $x > 3$.

SOLUTION The only specific point on the graph is $(3, 4)$ [property (a)]. We plot this point and then use the fact that $f'(3) = 0$ to sketch the tangent line at $x = 3$. (See Fig. 2.)

From property (b) and the first-derivative rule, we know that $f(x)$ must be increasing for x less than 3 and decreasing for x greater than 3. A graph with these properties might look like the curve in Fig. 3.

Figure 2

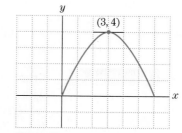

Figure 3

>> *Now Try Exercise 7*

The second derivative of a function $f(x)$ gives useful information about the concavity of the graph of $f(x)$. Suppose that $f''(a)$ is negative. Then, since $f''(x)$ is the derivative of $f'(x)$, we conclude that $f'(x)$ has a negative derivative at $x = a$. In this case, $f'(x)$ must be a decreasing function at $x = a$; that is, the slope of the graph of $f(x)$ is decreasing as we move from left to right on the graph near $(a, f(a))$. (See Fig. 4.) This means that the graph of $f(x)$ is concave down at $x = a$. A similar analysis shows that if $f''(a)$ is positive, then $f(x)$ is concave up at $x = a$. Thus, we have the following rule.

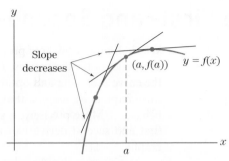

Figure 4 Illustration of the second-derivative rule.

The second derivative, $f''(x)$, is the derivative of the derivative $f'(x)$ (Sec. 1.7, p. 105).

Second-Derivative Rule   If $f''(a) > 0$, then $f(x)$ is concave up at $x = a$. If $f''(a) < 0$, then $f(x)$ is concave down at $x = a$.

CAUTION   When $f''(a) = 0$, the second-derivative rule gives no information. In this case, the function might be concave up, concave down, or neither at $x = a$. ≪

The following table shows how a graph may combine the properties of increasing, decreasing, concave up, and concave down.

Conditions on the Derivatives	Description of $f(x)$ at $x = a$	Graph of $y = f(x)$ near $x = a$
1. $f'(a)$ positive $f''(a)$ positive	$f(x)$ increasing $f(x)$ concave up	
2. $f'(a)$ positive $f''(a)$ negative	$f(x)$ increasing $f(x)$ concave down	
3. $f'(a)$ negative $f''(a)$ positive	$f(x)$ decreasing $f(x)$ concave up	
4. $f'(a)$ negative $f''(a)$ negative	$f(x)$ decreasing $f(x)$ concave down	

EXAMPLE 2   **Graphing Using First and Second Derivatives** Sketch the graph of a function $f(x)$ with all the following properties.
(a) $(2, 3)$, $(4, 5)$, and $(6, 7)$ are on the graph.
(b) $f'(6) = 0$ and $f'(2) = 0$.
(c) $f''(x) > 0$ for $x < 4$, $f''(4) = 0$, and $f''(x) < 0$ for $x > 4$.

SOLUTION   First, we plot the three points from property (a) and then sketch two tangent lines, using the information from property (b). (See Fig. 5.) From property (c) and the second-derivative rule, we know that $f(x)$ is concave up for $x < 4$. In particular, $f(x)$ is concave up at $(2, 3)$. Also, $f(x)$ is concave down for $x > 4$, in particular, at $(6, 7)$. Note that $f(x)$ must have an inflection point at $x = 4$ because the concavity changes there. We now sketch small portions of the curve near $(2, 3)$ and $(6, 7)$. (See Fig. 6.) We can now complete the sketch (Fig. 7), taking care to make the curve concave up for $x < 4$ and concave down for $x > 4$.

Figure 5

Figure 6

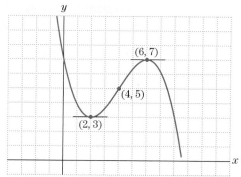

Figure 7

» Now Try Exercise 11

Connections between the Graphs of $f(x)$ and $f'(x)$

Think of the derivative of $f(x)$ as a "slope function" for $f(x)$. The "y-values" on the graph of $y = f'(x)$ are the *slopes* of the corresponding points on the original graph $y = f(x)$. This important connection is illustrated in the next three examples.

EXAMPLE 3   **Connections between the Graphs of f and f'** The function $f(x) = 8x - x^2$ is graphed in Fig. 8 along with the slope at several points. How is the slope changing on the graph? Compare the slopes on the graph with the y-coordinates of the points on the graph of $f'(x)$ in Fig. 9.

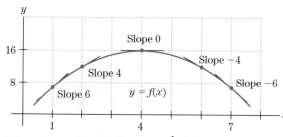

Figure 8 Graph of $f(x) = 8x - x^2$.

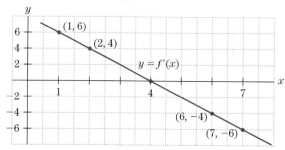

Figure 9 Graph of the derivative of the function in Fig. 8.

SOLUTION   The slopes are decreasing (as we move from left to right). That is, $f'(x)$ is a decreasing function. Observe that the y-values of $f'(x)$ decrease to zero at $x = 4$ and then continue to decrease for x-values greater than 4.   **«**

The graph in Fig. 8 is the shape of a typical revenue curve for a manufacturer. In this case, the graph of $f'(x)$ in Fig. 9 would be the *marginal revenue curve*. The graph in the next example has the shape of a typical cost curve. Its derivative produces a *marginal cost curve*.

EXAMPLE 4   **Location of the Inflection Point on the Graph** The function $f(x) = \frac{1}{3}x^3 - 4x^2 + 18x + 10$ is graphed in Fig. 10. The slope decreases at first and then increases. Use the graph of $f'(x)$ to verify that the slope in Fig. 10 is minimum at the inflection point, where $x = 4$.

Figure 10 Graph of $f(x)$.

SOLUTION   Several slopes are marked on the graph of $f(x)$. These values are y-coordinates of points on the graph of the derivative, $f'(x) = x^2 - 8x + 18$. Observe in Fig. 11 that the y-values on the graph of $f'(x)$ decrease at first and then begin to increase. The minimum value of $f'(x)$ occurs at $x = 4$, which is the first coordinate of the inflection point on the graph of $f(x)$ in Fig. 10.

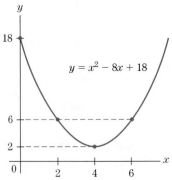

Figure 11 Graph of $f'(x)$.   **» Now Try Exercise 23**

EXAMPLE 5 **Properties of $f(x)$ Deduced from the Graph of $f'(x)$** Figure 12 shows the graph of $y = f'(x)$, the derivative of a function $f(x)$.
(a) What is the slope of the graph of $f(x)$ when $x = 1$?
(b) Describe how the values of $f'(x)$ change on the interval $1 \le x \le 2$.
(c) Describe the shape of the graph of $f(x)$ on the interval $1 \le x \le 2$.
(d) For what values of x does the graph of $f(x)$ have a horizontal tangent line?
(e) Explain why $f(x)$ has a relative maximum at $x = 3$.

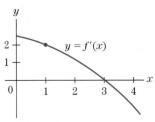

Figure 12

SOLUTION
(a) Since $f'(1)$ is 2, $f(x)$ has slope 2 when $x = 1$.
(b) The values of $f'(x)$ are positive and decreasing as x increases from 1 to 2.
(c) On the interval $1 \le x \le 2$, the slope of the graph of $f(x)$ is positive and is decreasing as x increases. Therefore, the graph of $f(x)$ is increasing and concave down.
(d) The graph of $f(x)$ has a horizontal tangent line when the slope is 0, that is, when $f'(x)$ is 0. This occurs at $x = 3$.
(e) Since $f'(x)$ is positive to the left of $x = 3$ and negative to the right of $x = 3$, the graph of $f(x)$ changes from increasing to decreasing at $x = 3$. Therefore, $f(x)$ has a relative maximum at $x = 3$.   **» Now Try Exercise 29**

Figure 13 shows the graph of a function $f(x)$ whose derivative has the shape shown in Fig. 12. Reread Example 5 and its solution while referring to Fig. 13.

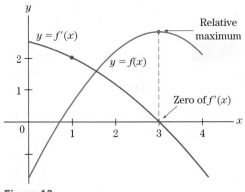

Figure 13

INCORPORATING TECHNOLOGY

Graphing Derivatives Once we have defined a function Y_1 in our calculator, we can graph both the function and its derivative by setting $Y_2 = \mathbf{nDeriv}(Y_1, X, X)$. In Fig. 14, we do this for the function $f(x) = 8x - x^2$ from Example 3. The graph is displayed in the window $[-1, 10]$ and $[-10, 20]$.

(a)   (b)

Figure 14

Check Your Understanding 2.2

Solutions can be found following the section exercises.

1. Make a good sketch of the function $f(x)$ near the point where $x = 2$, given that $f(2) = 5$, $f'(2) = 1$, and $f''(2) = -3$.

2. The graph of $f(x) = x^3$ is shown in Fig. 15.

Figure 15

(a) Is the function increasing at $x = 0$?

(b) Compute $f'(0)$.

(c) Reconcile your answers to parts (a) and (b) with the first derivative rule.

3. The graph of $y = f'(x)$ is shown in Fig. 16. Explain why $f(x)$ must have a relative minimum point at $x = 3$.

Figure 16

EXERCISES 2.2

Exercises 1–4 refer to the functions whose graphs are given in Fig. 17.

1. Which functions have a positive first derivative for all x?

2. Which functions have a negative first derivative for all x?

3. Which functions have a positive second derivative for all x?

4. Which functions have a negative second derivative for all x?

5. Which one of the graphs in Fig. 18 could represent a function $f(x)$ for which $f(a) > 0$, $f'(a) = 0$, and $f''(a) < 0$?

6. Which one of the graphs in Fig. 18 could represent a function $f(x)$ for which $f(a) = 0$, $f'(a) < 0$, and $f''(a) > 0$?

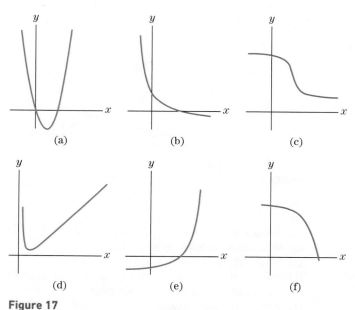

(a)   (b)   (c)

(d)   (e)   (f)

Figure 17

(a)   (b)

(c)   (d)

Figure 18

In Exercises 7–12, sketch the graph of a function that has the properties described.

7. $f(2) = 1$; $f'(2) = 0$; concave up for all x.

8. $f(-1) = 0$; $f'(x) < 0$ for $x < -1$, $f'(-1) = 0$ and $f'(x) > 0$ for $x > -1$.

9. $f(3) = 5$; $f'(x) > 0$ for $x < 3$, $f'(3) = 0$ and $f'(x) > 0$ for $x > 3$.

10. $(-2, -1)$ and $(2, 5)$ are on the graph; $f'(-2) = 0$ and $f'(2) = 0$; $f''(x) > 0$ for $x < 0$, $f''(0) = 0$, $f''(x) < 0$ for $x > 0$.

11. $(0, 6)$, $(2, 3)$, and $(4, 0)$ are on the graph; $f'(0) = 0$ and $f'(4) = 0$; $f''(x) < 0$ for $x < 2$, $f''(2) = 0$, $f''(x) > 0$ for $x > 2$.

12. $f(x)$ defined only for $x \geq 0$; $(0, 0)$ and $(5, 6)$ are on the graph; $f'(x) > 0$ for $x \geq 0$; $f''(x) < 0$ for $x < 5$, $f''(5) = 0$, $f''(x) > 0$ for $x > 5$.

In Exercises 13–18, use the given information to make a good sketch of the function $f(x)$ near $x = 3$.

13. $f(3) = 4$, $f'(3) = -\frac{1}{2}$, $f''(3) = 5$

14. $f(3) = -2$, $f'(3) = 0$, $f''(3) = 1$

15. $f(3) = 1$, $f'(3) = 0$, inflection point at $x = 3$, $f'(x) > 0$ for $x > 3$

16. $f(3) = 4$, $f'(3) = -\frac{3}{2}$, $f''(3) = -2$

17. $f(3) = -2$, $f'(3) = 2$, $f''(3) = 3$

18. $f(3) = 3$, $f'(3) = 1$, inflection point at $x = 3$, $f''(x) < 0$ for $x > 3$

19. Refer to the graph in Fig. 19. Fill in each box of the grid with either POS, NEG, or 0.

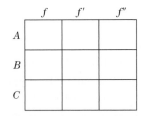

Figure 19

20. The first and second derivatives of the function $f(x)$ have the values given in Table 1.
   (a) Find the x-coordinates of all relative extreme points.
   (b) Find the x-coordinates of all inflection points.

Table 1   Values of the First Two Derivatives of a Function

x	$f'(x)$	$f''(x)$
$0 \leq x < 2$	Positive	Negative
2	0	Negative
$2 < x < 3$	Negative	Negative
3	Negative	0
$3 < x < 4$	Negative	Positive
4	0	0
$4 < x \leq 6$	Negative	Negative

21. Suppose that Fig. 20 contains the graph of $y = s(t)$, the distance traveled by a car after t hours. Is the car going faster at $t = 1$ or $t = 2$?

22. Suppose that Fig. 20 contains the graph of $y = v(t)$, the velocity of a car after t hours. Is the car going faster at $t = 1$ or $t = 2$?

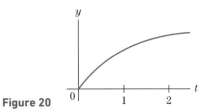

Figure 20

23. Refer to Fig. 21.

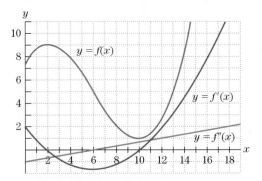

Figure 21

   (a) Looking at the graph of $f'(x)$, determine whether $f(x)$ is increasing or decreasing at $x = 9$. Look at the graph of $f(x)$ to confirm your answer.
   (b) Looking at the values of $f'(x)$ for $1 \leq x < 2$ and $2 < x \leq 3$, explain why the graph of $f(x)$ must have a relative maximum at $x = 2$. What are the coordinates of the relative maximum point?
   (c) Looking at the values of $f'(x)$ for x close to 10, explain why the graph of $f(x)$ has a relative minimum at $x = 10$.
   (d) Looking at the graph of $f''(x)$, determine whether $f(x)$ is concave up or concave down at $x = 2$. Look at the graph of $f(x)$ to confirm your answer.
   (e) Looking at the graph of $f''(x)$, determine where $f(x)$ has an inflection point. Look at the graph of $f(x)$ to confirm your answer. What are the coordinates of the inflection point?
   (f) Find the x-coordinate of the point on the graph of $f(x)$ at which $f(x)$ is increasing at the rate of 6 units per unit change in x.

24. In Fig. 22, the t-axis represents time in minutes.

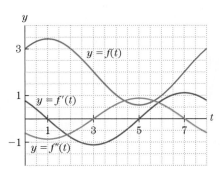

Figure 22

   (a) What is $f(2)$?
   (b) Solve $f(t) = 1$.
   (c) When does $f(t)$ attain its greatest value?
   (d) When does $f(t)$ attain its least value?
   (e) What is the rate of change of $f(t)$ at $t = 7.5$?
   (f) When is $f(t)$ decreasing at the rate of 1 unit per minute? That is, when is the rate of change equal to -1?
   (g) When is $f(t)$ decreasing at the greatest rate?
   (h) When is $f(t)$ increasing at the greatest rate?

Exercises 25–36 refer to Fig. 23, which contains the graph of $f'(x)$, the derivative of the function $f(x)$.

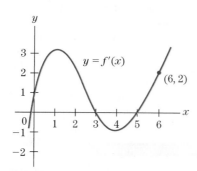

Figure 23

25. Explain why $f(x)$ must be increasing at $x = 6$.

26. Explain why $f(x)$ must be decreasing at $x = 4$.

27. Explain why $f(x)$ has a relative maximum at $x = 3$.

28. Explain why $f(x)$ has a relative minimum at $x = 5$.

29. Explain why $f(x)$ must be concave up at $x = 0$.

30. Explain why $f(x)$ must be concave down at $x = 2$.

31. Explain why $f(x)$ has an inflection point at $x = 1$.

32. Explain why $f(x)$ has an inflection point at $x = 4$.

33. If $f(6) = 3$, what is the equation of the tangent line to the graph of $y = f(x)$ at $x = 6$?

34. If $f(6) = 8$, what is an approximate value of $f(6.5)$?

35. If $f(0) = 3$, what is an approximate value of $f(.25)$?

36. If $f(0) = 3$, what is the equation of the tangent line to the graph of $y = f(x)$ at $x = 0$?

37. **Level of Water from Melting Snow** Melting snow causes a river to overflow its banks. Let $h(t)$ denote the number of inches of water on Main Street t hours after the melting begins.
   (a) If $h'(100) = \frac{1}{3}$, by approximately how much will the water level change during the next half hour?
   (b) Which of the following two conditions is the better news?
      (i) $h(100) = 3$, $h'(100) = 2$, $h''(100) = -5$
      (ii) $h(100) = 3$, $h'(100) = -2$, $h''(100) = 5$

38. **Changes in Temperature** $T(t)$ is the temperature on a hot summer day at time t hours.
   (a) If $T'(10) = 4$, by approximately how much will the temperature rise from 10:00 to 10:45?
   (b) Which of the following two conditions is the better news if you do not like hot weather? Explain your answer.
      (i) $T(10) = 95$, $T'(10) = 4$, $T''(10) = -3$
      (ii) $T(10) = 95$, $T'(10) = -4$, $T''(10) = 3$

39. Decide which of the curves in Fig. 24 could *not* be the graph of $f(x) = (3x^2 + 1)^4$ for $x \geq 0$. Decide which of the curves in Fig. 24 could *not* be the graph of $f(x) = (3x^2 + 1)^4$ for $x \geq 0$ by considering the derivative of $f(x)$. Explain your answer.

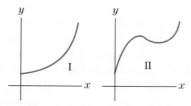

Figure 24

40. By looking at the first derivative, decide which of the curves in Fig. 24 could *not* be the graph of $f(x) = x^3 - 9x^2 + 24x + 1$ for $x \geq 0$. [*Hint:* Factor the formula for $f'(x)$.]

41. By looking at the second derivative, decide which of the curves in Fig. 25 could be the graph of $f(x) = x^{5/2}$.

Figure 25

42. Match each observation (a)–(e) with a conclusion (A)–(E).
   Observations
   (a) The point $(3, 4)$ is on the graph of $f'(x)$.
   (b) The point $(3, 4)$ is on the graph of $f(x)$.
   (c) The point $(3, 4)$ is on the graph of $f''(x)$.
   (d) The point $(3, 0)$ is on the graph of $f'(x)$, and the point $(3, 4)$ is on the graph of $f''(x)$.
   (e) The point $(3, 0)$ is on the graph of $f'(x)$, and the point $(3, -4)$ is on the graph of $f''(x)$.

   Conclusions
   (A) $f(x)$ has a relative minimum point at $x = 3$.
   (B) When $x = 3$, the graph of $f(x)$ is concave up.
   (C) When $x = 3$, the tangent line to the graph of $y = f(x)$ has slope 4.
   (D) When $x = 3$, the value of $f(x)$ is 4.
   (E) $f(x)$ has a relative maximum point at $x = 3$.

43. **Number of Farms in the U.S.** The number of farms in the United States t years after 1925 is $f(t)$ million, where f is the function graphed in Fig. 26(a). [The graphs of $f'(t)$ and $f''(t)$ are shown in Fig. 26(b).]

(1925)   (a)

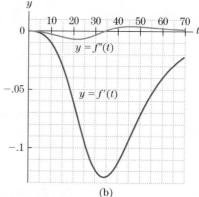

(b)

Figure 26

(a) Approximately how many farms were there in 1990?

(b) At what rate was the number of farms declining in 1990?

(c) In what year were there about 6 million farms?

(d) When was the number of farms declining at the rate of 60,000 farms per year?

(e) When was the number of farms declining fastest?

44. Drug Diffusion in the Bloodstream After a drug is taken orally, the amount of the drug in the bloodstream after t hours is $f(t)$ units. Figure 27 shows partial graphs of $f'(t)$ and $f''(t)$.

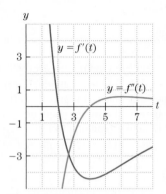

Figure 27

(a) Is the amount of the drug in the bloodstream increasing or decreasing at $t = 5$ hours?

(b) Is the graph of $f(t)$ concave up or concave down at $t = 5$ hours?

(c) When is the level of the drug in the bloodstream decreasing the fastest?

(d) At what time is the greatest level of drug in the bloodstream reached?

(e) When is the level of the drug in the bloodstream decreasing at the rate of 3 units per hour?

In Exercises 45 and 46, display the graph of the *derivative* of $f(x)$ in the specified window. Then use the graph of $f'(x)$ to determine the approximate values of x at which the graph of $f(x)$ has relative extreme points and inflection points. Then check your conclusions by displaying the graph of $f(x)$.

45. $f(x) = (x^2)3x^5 - 20x^3 - 120x$; $[-4, 4]$ by $[-325, 325]$

46. $f(x) = (x^2)x^4 - x^2$; $[-1.5, 1.5]$ by $[-.75, 1]$

Solutions to Check Your Understanding 2.2

1. Since $f(2) = 5$, the point $(2, 5)$ is on the graph. Since $f'(2) = 1$, the tangent line at the point $(2, 5)$ has slope 1. Draw in the tangent line. [See Fig. 28(a).] Near the point $(2, 5)$, the graph looks approximately like the tangent line. Since $f''(2) = -3$, a negative number, the graph is concave down at the point $(2, 5)$. Now we are ready to sketch the graph. [See Fig. 28(b).]

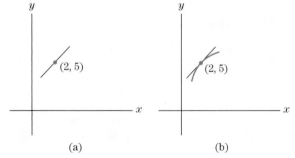

Figure 28

2. (a) Yes. The graph is steadily increasing as we pass through the point $(0, 0)$.

(b) Since $f'(x) = 3x^2$, $f'(0) = 3 \cdot 0^2 = 0$.

(c) There is no contradiction here. The first-derivative rule says that if the derivative is positive the function is increasing. However, it does not say that this is the only condition under which a function is increasing. As we have just seen, sometimes we can have the first derivative zero and the function still increasing.

3. Since $f'(x)$, the derivative of $f(x)$, is negative to the left of $x = 3$ and positive to the right of $x = 3$, $f(x)$ is decreasing to the left of $x = 3$ and increasing to the right of $x = 3$. Therefore, by the definition of a relative minimum point, $f(x)$ has a relative minimum point at $x = 3$.

2.3 The First- and Second-Derivative Tests and Curve Sketching

In this section and the next, we develop our ability to sketch the graphs of functions. There are two important reasons for doing so. First, a geometric "picture" of a function is often easier to comprehend than its abstract formula. Second, the material in this section will provide a foundation for the applications in Sections 2.5 through 2.7.

A "sketch" of the graph of a function $f(x)$ should convey the general shape of the graph; it should show where $f(x)$ is defined and where it is increasing and decreasing, and it should indicate, insofar as possible, where $f(x)$ is concave up and concave down. In addition, one or more key points should be accurately located on the graph. These points usually include extreme points, inflection points, and *x*- and *y*-intercepts. Other features of a graph may be important, too, but we shall discuss them as they arise in examples and applications.

Our general approach to curve sketching will involve four main steps:

1. Starting with $f(x)$, we compute $f'(x)$ and $f''(x)$.
2. Next, we locate all relative maximum and relative minimum points and make a partial sketch.
3. We study the concavity of $f(x)$ and locate all inflection points.
4. We consider other properties of the graph, such as the intercepts, and complete the sketch.

The first step was the main subject of Chapter 1. We discuss the second and third steps in this section and then present several completely worked examples that combine all four steps in the next section. For the sake of our discussion, unless otherwise stated, we will assume that the function $f(x)$ has continuous first and second derivatives.

Locating Relative Extreme Points

It is clear from the graph of $f(x)$ that the tangent line at a relative maximum or a relative minimum point has zero slope. Indeed, at a relative extreme point, the derivative changes sign, because the graph is increasing when $f'(x) > 0$ and decreasing when $f'(x) < 0$, and this can only happen when $f'(x)$ is zero. Thus, we may state the following useful rule.

> Look for possible relative extreme points of
> $f(x)$ by setting $f'(x) = 0$ and solving for x. 　　　(1)

A number a in the domain of f such that $f'(a) = 0$ is called a **critical number** or **critical value** of f. [We also call a value a in the domain of f a critical value if $f'(a)$ does not exist.] If a is a critical value, the point $(a, f(a))$ is called a **critical point**. Thus, if f has a relative extreme value at $x = a$, then a must be a critical value of f. But not every critical value of f yields a relative extreme point. The condition $f'(a) = 0$ tells us only that the tangent line is horizontal at a. In Fig. 1, we show four cases that can occur when $f'(a) = 0$. We can see from the figure that we have an extreme point when the first

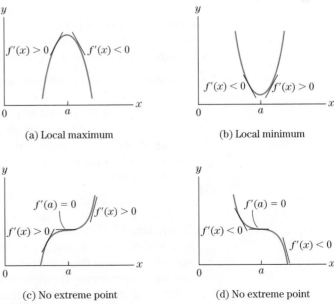

(a) Local maximum　　　　　　　　(b) Local minimum

(c) No extreme point　　　　　　　　(d) No extreme point

Figure 1

derivative, $f'(x)$, changes sign at $x = a$. But if the sign of the first derivative does not change, we do not have an extreme point.

Based on these observations and the first-derivative rule of the previous section, we obtain the following useful test for extreme points.

The First-Derivative Test (for local extreme points)   Suppose that $f'(a) = 0$.

(a) If f' changes from positive to negative at $x = a$, then f has a local maximum at a. [See Fig. 1(a).]

(b) If f' changes from negative to positive at $x = a$, then f has a local minimum at a. [See Fig. 1(b).]

(c) If f' does not change sign at a (that is, f' is either positive on both sides of a [Fig. 1(c)] or negative on both sides of a [Fig. 1(d)]), then f has no local extreme point at a.

EXAMPLE 1   **Applying the First-Derivative Test** Find the local maximum and minimum points of $f(x) = \frac{1}{3}x^3 - 2x^2 + 3x + 1$.

SOLUTION   First, we find the critical values and critical points of f:

$$f'(x) = \frac{1}{3}(3)x^2 - 2(2)x + 3$$

$$= x^2 - 4x + 3$$

$$= (x - 1)(x - 3).$$

The first derivative $f'(x) = 0$ if $x - 1 = 0$ or $x - 3 = 0$. Thus, the critical values are

$$x = 1 \quad \text{and} \quad x = 3.$$

Substitute the critical values into the expression of f:

$$f(1) = \frac{1}{3}(1)^3 - 2(1)^2 + 3(1) + 1 = \frac{1}{3} + 2 = \frac{7}{3};$$

$$f(3) = \frac{1}{3}(3)^3 - 2(3)^2 + 3(3) + 1 = 1.$$

Thus, the critical points are $\left(1, \frac{7}{3}\right)$ and $(3, 1)$. To determine whether we have a relative maximum, minimum, or neither at a critical point, we apply the first derivative test. This requires a careful study of the sign of $f'(x)$, which can be facilitated with the help of a chart. Here is how you can set up the chart:

- Divide the real line into intervals with the critical values as endpoints.
- Since the sign of f' depends on the signs of its two factors $x - 1$ and $x - 3$, determine the signs of the factors of f' over each interval. Usually, we do this by testing the sign of a factor at points selected from each interval.
- In each interval, use a plus sign if the factor is positive or a minus sign if it is negative. Then determine the sign of f' over each interval by multiplying the signs of the factors and using

$$(+) \cdot (+) = +; \quad (+) \cdot (-) = -; \quad (-) \cdot (+) = -; \quad (-) \cdot (-) = +.$$

- A plus sign of f' corresponds to an increasing portion of the graph of f and a minus sign to a decreasing portion. Denote an increasing portion with an upward arrow and a decreasing portion with a downward arrow. The sequence of arrows should convey the general shape of the graph and, in particular, tell you whether or not your critical values correspond to extreme points. Here is the chart:

Critical Values		1		3	
Intervals	$x < 1$		$1 < x < 3$		$3 < x$
$f'(x)$	$+$	0	$-$	0	$+$
$f(x)$	Increasing on $(-\infty, 1)$	$\frac{7}{3}$	Decreasing on $(1, 3)$	1	Increasing on $(3, \infty)$

Local maximum $\left(1, \frac{7}{3}\right)$ Local minimum $(3, 1)$

You can see from the chart that the sign of f' varies from positive to negative at $x = 1$. Thus, according to the first-derivative test, f has a local maximum at $x = 1$. Also, the sign of f' varies from negative to positive at $x = 3$; so f has a local minimum at $x = 3$. You can confirm these assertions by following the direction of the arrows in the last row of the chart. In conclusion, f has a local maximum at $\left(1, \frac{7}{3}\right)$ and a local minimum at $(3, 1)$. (See Fig. 2.)

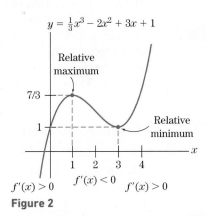

$y = \frac{1}{3}x^3 - 2x^2 + 3x + 1$

Relative maximum

7/3

1

Relative minimum

$f'(x) > 0$ $f'(x) < 0$ $f'(x) > 0$

Figure 2

>> *Now Try Exercise 5*

The previous example illustrates cases (a) and (b) in Fig. 1, which correspond to local extreme points on the graph at each critical value. The following example illustrates cases (c) and (d) in Fig. 1.

EXAMPLE 2   **Applying the First-Derivative Test** Find the local maximum and minimum points of $f(x) = (3x - 1)^3$.

SOLUTION   We begin by finding the first derivative and the critical values. Using the general power rule, we have

$$f'(x) = 3(3x - 1)^2(3) = 9(3x - 1)^2$$
$$f'(x) = 0 \Rightarrow 3x - 1 = 0$$
$$\Rightarrow x = \tfrac{1}{3}.$$

Plugging this value into the original expression for f, we find

$$f\left(\tfrac{1}{3}\right) = \left(3 \cdot \tfrac{1}{3} - 1\right)^3 = 0.$$

Thus, $\left(\tfrac{1}{3}, 0\right)$ is the only critical point. To determine whether it is a local maximum, minimum, or neither, we use the first derivative test. Look at $9(3x - 1)^2$. This expression is always nonnegative, because a square is always ≥ 0 (there is no need for a sign chart in this case). Consequently, f' is positive on both sides of $x = \tfrac{1}{3}$; so, since f' does not change sign at $x = \tfrac{1}{3}$, we conclude from part (c) of the first derivative test that there is no relative extreme point. Indeed, since $f'(x) \geq 0$ for all x, the graph is always increasing. (See Fig. 3.)

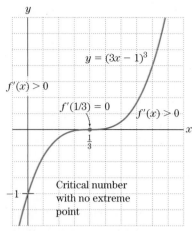

Figure 3

» Now Try Exercise 7

Using Concavity to Determine Extreme Points

Studying the variation of the graph of f and determining the extreme points can be tedious tasks. Any shortcut that helps us avoid excessive computations is always appreciated. Let us look back at Fig. 1 and give an alternative description of the extreme points using concavity. At $x = a$ the concavity is downward in Fig. 1(a), upward in Fig. 1(b), and neither in Figs. 1(c) and (d). These observations, together with the second-derivative rule from the previous section, are the basis for the following important test.

The Second-Derivative Test (for local extreme points)

(a) If $f'(a) = 0$ and $f''(a) < 0$, then f has a local maximum at a. [See Fig. 1(a).]

(b) If $f'(a) = 0$ and $f''(a) > 0$, then f has a local minimum at a. [See Fig. 1(b).]

EXAMPLE 3   **Applying the Second-Derivative Test** The graph of the quadratic function $f(x) = \frac{1}{4}x^2 - x + 2$ is a parabola and so has one relative extreme point. Find it and sketch the graph.

SOLUTION   We begin by computing the first and second derivatives of $f(x)$:

$$f(x) = \frac{1}{4}x^2 - x + 2$$

$$f'(x) = \frac{1}{2}x - 1$$

$$f''(x) = \frac{1}{2}.$$

Setting $f'(x) = 0$, we have $\frac{1}{2}x - 1 = 0$, so that $x = 2$ is the only critical value. Thus, $f'(2) = 0$. Geometrically, this means that the graph of $f(x)$ will have a horizontal tangent line at the point where $x = 2$. To plot this point, we substitute the value 2 for x in the original expression for $f(x)$:

$$f(2) = \frac{1}{4}(2)^2 - (2) + 2 = 1.$$

Figure 4 shows the point $(2, 1)$ together with the horizontal tangent line. Is $(2, 1)$ a relative extreme point? To decide, we look at $f''(x)$. Since $f''(x) = \frac{1}{2}$, which is positive, the graph of $f(x)$ is concave up at $x = 2$, so $(2, 1)$ is a local minimum by the second-derivative test. A partial sketch of the graph near $(2, 1)$ should look something like Fig. 5.

Figure 4

Figure 5

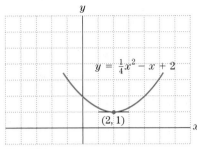

Figure 6

We see that (2, 1) is a relative minimum point. In fact, it is the only relative extreme point, for there is no other place where the tangent line is horizontal. Since the graph has no other "turning points," it must be decreasing before it gets to (2, 1) and then increasing to the right of (2, 1). Note that since $f''(x)$ is positive (and equal to $\frac{1}{2}$) for all x, the graph is concave up at each point. A completed sketch is given in Fig. 6.

>> **Now Try Exercise 13**

Related to Example 3, observe that if $f(x) = ax^2 + bx + c$, then $f(x)$ has a relative minimum point when $a > 0$ and a relative maximum point when $a < 0$.

EXAMPLE 4   **Applying the Second-Derivative Test** Locate all possible relative extreme points on the graph of the function $f(x) = x^3 - 3x^2 + 5$. Check the concavity at these points and use this information to sketch the graph of $f(x)$:

SOLUTION   We have

$$f(x) = x^3 - 3x^2 + 5$$
$$f'(x) = 3x^2 - 6x$$
$$f''(x) = 6x - 6.$$

To find the critical values, let us factor the expression for $f'(x)$:

$$3x^2 - 6x = 3x(x - 2).$$

From this factorization, it is clear that $f'(x)$ will be zero if and only if $x = 0$ or $x = 2$. In other words, the graph will have horizontal tangent lines when $x = 0$ and $x = 2$, and nowhere else.

To plot the points on the graph where $x = 0$ and $x = 2$, we substitute these values back into the original expression for $f(x)$. That is, we compute

$$f(0) = (0)^3 - 3(0)^2 + 5 = 5$$
$$f(2) = (2)^3 - 3(2)^2 + 5 = 1.$$

Figure 7 shows the points (0, 5) and (2, 1), along with the corresponding tangent lines.

Next, we check the sign of $f''(x)$ at $x = 0$ and $x = 2$ and apply the second-derivative test:

$$f''(0) = 6(0) - 6 = -6 < 0 \quad \text{Local maximum}$$
$$f''(2) = 6(2) - 6 = 6 > 0. \quad \text{Local minimum}$$

Since $f''(0)$ is negative, the graph is concave down at $x = 0$; since $f''(2)$ is positive, the graph is concave up at $x = 2$. A partial sketch of the graph is given in Fig. 8.

It is clear from Fig. 8 that (0, 5) is a relative maximum point and (2, 1) is a relative minimum point. Since they are the only turning points, the graph must be increasing before it gets to (0, 5), decreasing from (0, 5) to (2, 1), and then increasing again to the right of (2, 1). A sketch incorporating these properties appears in Fig. 9.

Figure 7

Figure 8

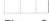

Figure 9

» Now Try Exercise 19

The facts that we used to sketch Fig. 9 could equally well be used to produce the graph in Fig. 10. Which graph really corresponds to $f(x) = x^3 - 3x^2 + 5$? The answer will be clear when we find the inflection points on the graph of $f(x)$.

Locating Inflection Points

Under the assumption that $f''(x)$ is continuous, an inflection point of a function $f(x)$ can occur only at a value of x for which $f''(x)$ is zero, because the curve is concave up where $f''(x)$ is positive and concave down where $f''(x)$ is negative. Thus, we have the following rule:

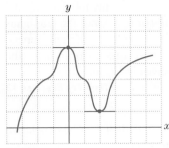

Figure 10

> Look for possible inflection points by setting $f''(x) = 0$ and solving for x. (2)

Once we have a value of x where the second derivative is zero—say, at $x = b$,—we must check the concavity of $f(x)$ at nearby points to see if the concavity really changes at $x = b$.

EXAMPLE 5   **Locating an Inflection Point** Find the inflection points of the function $f(x) = x^3 - 3x^2 + 5$, and explain why the graph in Fig. 9 has the correct shape.

SOLUTION   From Example 4, we have $f''(x) = 6x - 6 = 6(x - 1)$. Clearly, $f''(x) = 0$ if and only if $x = 1$. We will want to plot the corresponding point on the graph, so we compute

$$f(1) = (1)^3 - 3(1)^2 + 5 = 3.$$

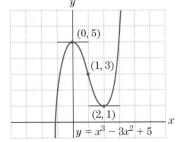

Figure 11

Therefore, the only possible inflection point is $(1, 3)$.

Now look back at Fig. 8, where we indicated the concavity of the graph at the relative extreme points. Since $f(x)$ is concave down at $(0, 5)$ and concave up at $(2, 1)$, the concavity must reverse somewhere between these points. Hence, $(1, 3)$ must be an inflection point. Furthermore, since the concavity of $f(x)$ reverses nowhere else, the concavity at all points to the left of $(1, 3)$ must be the same (that is, concave down). Similarly, the concavity of all points to the right of $(1, 3)$ must be the same (concave up). Thus, the graph in Fig. 9 has the correct shape. The graph in Fig. 10 has too many "wiggles," caused by frequent changes in concavity; that is, there are too many inflection points. A correct sketch showing the one inflection point at $(1, 3)$ is given in Fig. 11.

» Now Try Exercise 25

EXAMPLE 6   **Graphing Using Properties of f' and f''** Sketch the graph of $y = -\frac{1}{3}x^3 + 3x^2 - 5x$.

SOLUTION   Let

$$f(x) = -\frac{1}{3}x^3 + 3x^2 - 5x.$$

Then,

$$f'(x) = -x^2 + 6x - 5$$
$$f''(x) = -2x + 6.$$

Step 1 Finding the critical points
We set $f'(x) = 0$ and solve for x:

$$-(x^2 - 6x + 5) = 0$$
$$-(x - 1)(x - 5) = 0$$
$$x = 1 \quad \text{or} \quad x = 5 \quad \text{Critical values}$$

Substituting these values of x back into $f(x)$, we find that

$$f(1) = -\tfrac{1}{3}(1)^3 + 3(1)^2 - 5(1) = -\tfrac{7}{3}$$
$$f(5) = -\tfrac{1}{3}(5)^3 + 3(5)^2 - 5(5) = \tfrac{25}{3}.$$

Step 2 Determining the extreme points
The information we have so far is given in Fig. 12(a). We obtain the sketch in Fig. 12(b) by computing

$$f''(1) = -2(1) + 6 = 4 > 0 \quad \text{Local minimum}$$
$$f''(5) = -2(5) + 6 = -4 < 0 \quad \text{Local maximum}$$

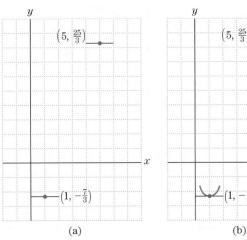

(a)   (b)

Figure 12

The curve is concave up at $x = 1$ because $f''(1)$ is positive (local minimum), and the curve is concave down at $x = 5$ because $f''(5)$ is negative (local maximum).

Step 3 Concavity and inflection points
Since the concavity reverses somewhere between $x = 0$ and $x = 5$, there must be at least one inflection point. If we set $f''(x) = 0$, we find that

$$-2x + 6 = 0$$
$$x = 3.$$

So the inflection point must occur at $x = 3$. To plot the inflection point, we compute

$$f(3) = -\tfrac{1}{3}(3)^3 + 3(3)^2 - 5(3) = 3.$$

The final sketch of the graph is given in Fig. 13.   **» Now Try Exercise 29**

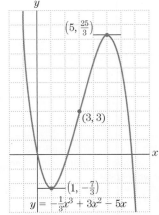

Figure 13

CAUTION The argument in Example 6, that there must be an inflection point because concavity reverses, is valid whenever $f(x)$ is a polynomial. However, it does not always apply to a function whose graph has a break in it. For example, the function $f(x) = 1/x$ is concave down at $x = -1$ and concave up at $x = 1$, but there is no inflection point in between. «

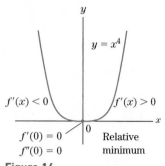

Figure 14

In Examples 1 and 2 we used the first derivative test to locate the extreme points; but in the remaining examples we used the second-derivative test. In general, which test should we use? If $f''(x)$ is straightforward to compute (for example, when $f(x)$ is a polynomial), you should try the second-derivative test first. If the second derivative is tedious to compute or if $f''(a) = 0$, use the first-derivative test. Remember that when $f''(a) = 0$, the second-derivative test is inconclusive. Keep the following examples in mind. The function $f(x) = x^3$ is always increasing and has no local maximum or minimum, even though $f'(0) = 0$ and $f''(0) = 0$. The function $f(x) = x^4$ is shown in Fig. 14. We have $f'(x) = 4x^3$ and $f''(x) = 12x^2$. So $f'(0) = 0$ and $f''(0) = 0$, but, as you can see from Fig. 14, you have a local minimum at $x = 0$. The reason follows from the first-derivative test, since f' changes sign at $x = 0$ and goes from negative to positive.

Example 6 here should serve as a summary of the techniques introduced so far.

Check Your Understanding 2.3

Solutions can be found following the section exercises.

1. Which of the curves in Fig. 15 could possibly be the graph of a function of the form $f(x) = ax^2 + bx + c$, where $a \neq 0$?

2. Which of the curves in Fig. 16 could be the graph of a function of the form $f(x) = ax^3 + bx^2 + cx + d$, where $a \neq 0$?

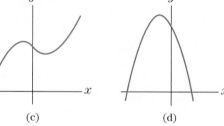

(a)   (b)   (c)   (d)

Figure 15

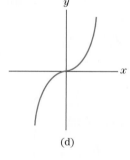

(a)   (b)   (c)   (d)

Figure 16

EXERCISES 2.3

Each of the graphs of the functions in Exercises 1–8 has one relative maximum and one relative minimum point. Find these points using the first-derivative test. Use a variation chart as in Example 1.

1. $f(x) = x^3 - 27x$

2. $f(x) = x^3 - 6x^2 + 1$

3. $f(x) = -x^3 + 6x^2 - 9x + 1$

4. $f(x) = -6x^3 - \frac{3}{2}x^2 + 3x - 3$

5. $f(x) = \frac{1}{3}x^3 - x^2 + 1$

6. $f(x) = \frac{4}{3}x^3 - x + 2$

7. $f(x) = -x^3 - 12x^2 - 2$

8. $f(x) = 2x^3 + 3x^2 - 3$

Each of the graphs of the functions in Exercises 9–16 has one relative extreme point. Plot this point and check the concavity there.

Using only this information, sketch the graph. [Recall that if $f(x) = ax^2 + bx + c$, then $f(x)$ has a relative minimum point when $a > 0$ and a relative maximum point when $a < 0$.]

9. $f(x) = 2x^2 - 8$

10. $f(x) = x^2$

11. $f(x) = \frac{1}{2}x^2 + x - 4$

12. $f(x) = -3x^2 + 12x + 2$

13. $f(x) = 1 + 6x - x^2$

14. $f(x) = \frac{1}{2}x^2 + \frac{1}{2}$

15. $f(x) = -x^2 - 8x - 10$

16. $f(x) = -x^2 + 2x - 5$

Each of the graphs of the functions in Exercises 17–24 has one relative maximum and one relative minimum point. Plot these two points and check the concavity there. Using only this information, sketch the graph.

17. $f(x) = x^3 + 6x^2 + 9x$

18. $f(x) = \frac{1}{9}x^3 - x^2$

19. $f(x) = x^3 - 12x$

20. $f(x) = -\frac{1}{3}x^3 + 9x - 2$

21. $f(x) = -\frac{1}{9}x^3 + x^2 + 9x$

22. $f(x) = 2x^3 - 15x^2 + 36x - 24$

23. $f(x) = -\frac{1}{3}x^3 + 2x^2 - 12$

24. $f(x) = \frac{1}{3}x^3 + 2x^2 - 5x + \frac{8}{3}$

Sketch the following curves, indicating all relative extreme points and inflection points.

25. $y = x^3 - 3x + 2$

26. $y = x^3 - 6x^2 + 9x + 3$

27. $y = 1 + 3x^2 - x^3$

28. $y = -x^3 + 12x - 4$

29. $y = \frac{1}{3}x^3 - x^2 - 3x + 5$

30. $y = x^4 + \frac{1}{3}x^3 - 2x^2 - x + 1$

  [*Hint:* $4x^3 + x^2 - 4x - 1 = (x^2 - 1)(4x + 1)$]

31. $y = 2x^3 - 3x^2 - 36x + 20$

32. $y = x^4 - \frac{4}{3}x^3$

33. Let a, b, c be fixed numbers with $a \neq 0$, and let $f(x) = ax^2 + bx + c$. Is it possible for the graph of $f(x)$ to have an inflection point? Explain your answer.

34. Let a, b, c, d be fixed numbers with $a \neq 0$, and let $f(x) = ax^3 + bx^2 + cx + d$. Is it possible for the graph of $f(x)$ to have more than one inflection point? Explain your answer.

The graph of each function in Exercises 35–40 has one relative extreme point. Find it (giving both x- and y-coordinates) and determine if it is a relative maximum or a relative minimum point. Do not include a sketch of the graph of the function.

35. $f(x) = \frac{1}{4}x^2 - 2x + 7$

36. $f(x) = 5 - 12x - 2x^2$

37. $g(x) = 3 + 4x - 2x^2$

38. $g(x) = x^2 + 10x + 10$

39. $f(x) = 5x^2 + x - 3$

40. $f(x) = 30x^2 - 1800x + 29{,}000$

In Exercises 41 and 42, determine which function is the derivative of the other.

41.

42.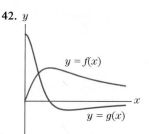

43. Consider the graph of $g(x)$ in Fig. 17.
  (a) If $g(x)$ is the first derivative of $f(x)$, describe $f(x)$ when $x = 2$?
  (b) If $g(x)$ is the second derivative of $f(x)$, describe $f(x)$ when $x = 2$?

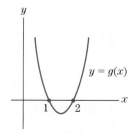

Figure 17

44. U.S. Population The population (in millions) of the United States (excluding Alaska and Hawaii) t years after 1800 is given by the function $f(t)$ in Fig. 18(a). The graphs of $f'(t)$ and $f''(t)$ are shown in Figs. 18(b) and 18(c).

(a)

(b)

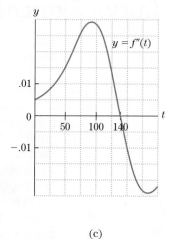

(c)

Figure 18 (a) Population (in millions) of the United States from 1800 to 1998.

(a) What was the population in 1925?

(b) Approximately when was the population 25 million?

(c) How fast was the population growing in 1950?

(d) When during the last 50 years was the population growing at the rate of 1.8 million people per year?

(e) In what year was the population growing at the greatest rate?

45. Index-Fund Fees When a mutual fund company charges a fee of 0.47% on its index funds, its assets in the fund are $41 billion. And when it charges a fee of 0.18%, its assets in the fund are $300 billion. (*Source: The Boston Globe.*)

(a) Let $x\%$ denote the fee that the company charges the index fund and $A(x)$ its assets in the fund. Express $A(x)$ as a linear function of x.

(b) In September 2004, Fidelity Mutual lowered its fees on various index funds from an average of 0.3% to 0.10%. Let $R(x)$ denote the revenue of the company from fees when its index-fund fee is $x\%$. Compare the revenue of the company before and after lowering the fees. [*Hint:* Revenue is $x\%$ of assets.]

(c) Find the fee that maximizes the revenue of the company and determine the maximum revenue.

46. Index-Fund Fees Suppose that the cost function in Exercise 45 is $C(x) = -2.5x + 1$, where $x\%$ is the index-fund fee. (The company has a fixed cost of $1 billion, and the cost decreases as a function of the index-fund fee.) Find the value of x that maximizes profit. How well did Fidelity Mutual do before and after lowering the index-fund fees?

TECHNOLOGY EXERCISES

47. Draw the graph of $f(x) = \frac{1}{6}x^3 - x^2 + 3x + 3$ in the window $[-2, 6]$ by $[-10, 20]$. It has an inflection point when $x = 2$, but no relative extreme points. Enlarge the window a few times to convince yourself that there are no relative extreme points anywhere. What does this tell you about $f'(x)$?

48. Draw the graph of $f(x) = \frac{1}{6}x^3 - \frac{5}{2}x^2 + 13x - 20$ in the window $[0, 10]$ by $[-20, 30]$. Algebraically determine the coordinates of the inflection point. Zoom in and zoom out to convince yourself that there are no relative extreme points anywhere.

49. Draw the graph of

$$f(x) = 2x + \frac{18}{x} - 10$$

in the window $[0, 16]$ by $[0, 16]$. In what ways is this graph like the graph of a parabola that opens upward? In what ways is it different?

50. Draw the graph of

$$f(x) = 3x + \frac{75}{x} - 25$$

in the window $[0, 25]$ by $[0, 50]$. Use the trace feature of the calculator or computer to estimate the coordinates of the relative minimum point. Then determine the coordinates algebraically. Explain both graphically (with the calculator or computer) and algebraically why this function has no inflection points.

Solutions to Check Your Understanding 2.3

1. Answer: (a) and (d). Curve (b) has the shape of a parabola, but it is not the graph of any function, since vertical lines cross it twice. Curve (c) has two relative extreme points, but the derivative of $f(x)$ is a linear function, which could not be zero for two different values of x.

2. Answer: (a), (c), (d). Curve (b) has three relative extreme points, but the derivative of $f(x)$ is a quadratic function, which could not be zero for three different values of x.

2.4   Curve Sketching (Conclusion)

In Section 2.3, we discussed the main techniques for curve sketching. Here, we add a few finishing touches and examine some slightly more complicated curves.

The more points we plot on a graph, the more accurate the graph becomes. This statement is true even for the quadratic and cubic curves in Section 2.3. Of course, the most important points on a curve are the relative extreme points and the inflection points. In addition, the x- and y-intercepts often have some meaning in an applied problem. The y-intercept is $(0, f(0))$. To find the x-intercepts on the graph of $f(x)$, we must find those values of x for which $f(x) = 0$. Since this can be a difficult (or impossible) problem, we shall find x-intercepts only when they are easy to find or when a problem specifically requires us to find them.

When $f(x)$ is a quadratic function, as in Example 1, we can compute the x-intercepts (if they exist) either by factoring the expression for $f(x)$ or by using the quadratic formula, which we now recall from Section 0.4:

The solutions of $ax^2 + bx + c = 0$ are given by the Quadratic Formula

$$x = \frac{-b \pm \sqrt{b^2 - 4ac}}{2a}$$

The sign $\pm$ tells us to form two expressions, one with $+$ and one with $-$. The equation has two distinct roots if $b^2 - 4ac > 0$; one double root if $b^2 - 4ac = 0$; and no (real) roots if $b^2 - 4ac < 0$.

EXAMPLE 1   **Applying the Second-Derivative Test**   Sketch the graph of $f(x) = \frac{1}{2}x^2 - 4x + 7$.

SOLUTION   We start by finding the first and second derivatives:

$$f'(x) = x - 4$$
$$f''(x) = 1.$$

Since $f'(x) = 0$ only when $x = 4$, and since $f''(4)$ is positive, $f(x)$ must have a relative minimum point at $x = 4$ (the second-derivative test). The relative minimum point is $(4, f(4)) = (4, -1)$.

The y-intercept is $(0, f(0)) = (0, 7)$. To find the x-intercepts, we set $f(x) = 0$ and solve for x:

$$\frac{1}{2}x^2 - 4x + 7 = 0.$$

The expression for $f(x)$ is not easily factored, so we use the quadratic formula to solve the equation:

$$x = \frac{-(-4) \pm \sqrt{(-4)^2 - 4(\frac{1}{2})(7)}}{2(\frac{1}{2})} = 4 \pm \sqrt{2}.$$

The x-intercepts are $(4 - \sqrt{2}, 0)$ and $(4 + \sqrt{2}, 0)$. To plot these points, we use the approximation $\sqrt{2} \approx 1.4$. (See Fig. 1.)   **» Now Try Exercise 3**

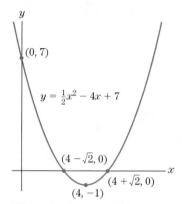

$(0, 7)$

$y = \frac{1}{2}x^2 - 4x + 7$

$(4 - \sqrt{2}, 0)$

$(4 + \sqrt{2}, 0)$

$(4, -1)$

Figure 1

The next example is interesting. The function that we consider has no critical points.

EXAMPLE 2   **A Function with No Critical Points**   Sketch the graph of $f(x) = \frac{1}{6}x^3 - \frac{3}{2}x^2 + 5x + 1$.

SOLUTION   Again, we start by finding the first and second derivatives:

$$f'(x) = \frac{1}{2}x^2 - 3x + 5$$
$$f''(x) = x - 3$$

Searching for critical points, we set $f'(x) = 0$ and try to solve for x:

$$\frac{1}{2}x^2 - 3x + 5 = 0. \tag{1}$$

If we apply the quadratic formula with $a = \frac{1}{2}$, $b = -3$, and $c = 5$, we see that $b^2 - 4ac$ is negative, and so there is no solution to (1). In other words, $f'(x)$ is never zero. Thus, the graph cannot have relative extreme points. If we evaluate $f'(x)$ at some x—say, $x = 0$—we see that the first derivative is positive, and so $f(x)$ is increasing there. Since the graph of $f(x)$ is a smooth curve with no relative extreme points and no breaks, $f(x)$ must be increasing for all x. (If a function were increasing at $x = a$ and decreasing at $x = b$, it would have a relative extreme point between a and b.)

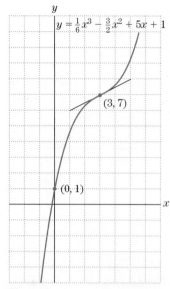

Figure 2

Now let us check the concavity.

	$f''(x) = x - 3$	Graph of $f(x)$
$x < 3$	Negative	Concave down
$x = 3$	Zero	Concavity reverses
$x > 3$	Positive	Concave up

Since $f''(x)$ changes sign at $x = 3$, we have an inflection point at $(3, 7)$. The y-intercept is $(0, f(0)) = (0, 1)$. We omit the x-intercept because it is difficult to solve the cubic equation $\frac{1}{6}x^3 - \frac{3}{2}x^2 + 5x + 1 = 0$.

The quality of our sketch of the curve will be improved if we first sketch the tangent line at the inflection point. To do this, we need to know the slope of the graph at $(3, 7)$:

$$f'(3) = \frac{1}{2}(3)^2 - 3(3) + 5 = \frac{1}{2}.$$

We draw a line through $(3, 7)$ with slope $\frac{1}{2}$ and then complete the sketch as shown in Fig. 2.

» Now Try Exercise 9

EXAMPLE 3 **Using the First-Derivative Test** Sketch the graph of $f(x) = (x - 2)^4 - 1$.

SOLUTION Let's compute:

$$f'(x) = 4(x - 2)^3 \quad \text{General power rule}$$
$$f''(x) = 12(x - 2)^2 \quad \text{General power rule}$$

Clearly, $f'(x) = 0$ only if $x = 2$. So the curve has a horizontal tangent at $(2, f(2)) = (2, -1)$. Since $f''(2) = 0$, the second-derivative test is inconclusive. We shall apply the first-derivative test. Note that

$$f'(x) = 4(x - 2)^3 \begin{cases} \text{negative} & \text{if } x < 2 \\ \text{positive} & \text{if } x > 2, \end{cases}$$

since the cube of a negative number is negative and the cube of a positive number is positive. Therefore, as x goes from left to right in the vicinity of 2, the first derivative changes sign and goes from negative to positive. By the first-derivative test, the point $(2, -1)$ is a relative minimum.

The y-intercept is $(0, f(0)) = (0, 15)$. To find the x-intercepts, we set $f(x) = 0$ and solve for x:

$$(x - 2)^4 - 1 = 0$$
$$(x - 2)^4 = 1$$
$$x - 2 = 1 \quad \text{or} \quad x - 2 = -1$$
$$x = 3 \quad \text{or} \quad x = 1.$$

The sketch in Fig. 3 incorporates all the information we have generated.

» Now Try Exercise 21

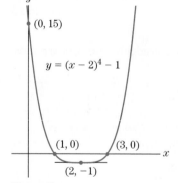

Figure 3

A Graph with Asymptotes

Graphs similar to the one in the next example will arise in several applications later in this chapter.

EXAMPLE 4 **Asymptotes** Sketch the graph of $f(x) = x + (1/x)$, for $x > 0$.

SOLUTION Think of $f(x)$ as $f(x) = x + x^{-1}$ and find the first and second derivatives:

$$f'(x) = 1 - x^{-2} = 1 - \frac{1}{x^2}$$

$$f''(x) = -(-2)x^{-3} = \frac{2}{x^3}.$$

We set $f'(x) = 0$ and solve for x:

$$1 - \frac{1}{x^2} = 0$$

$$1 = \frac{1}{x^2}$$

$$x^2 = 1$$

$$x = 1.$$

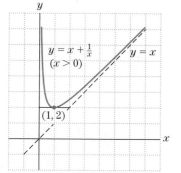

Figure 4

(We exclude the case $x = -1$ because we are considering only positive values of x.) The graph has a horizontal tangent at $(1, f(1)) = (1, 2)$. Since $f''(1) = 2 > 0$, the graph is concave up at $x = 1$ and, by the second-derivative test, $(1, 2)$ is a relative minimum point. In fact, $f''(x) = (2/x^3) > 0$ for all positive x, and therefore, the graph is concave up at all points.

Before sketching the graph, notice that as x approaches zero [a point at which $f(x)$ is not defined], $f(x)$ gets very large. Thus, $f(x)$ has the y-axis as an asymptote. For large values of x, $\frac{1}{x}$ gets close to 0, so $x + \frac{1}{x}$ approaches x for large values of x. Hence, as x increases, the graph of $f(x)$ has the line $y = x$ as an asymptote. (See Fig. 4.)

>> *Now Try Exercise 27*

Summary of Curve-Sketching Techniques

1. Compute $f'(x)$ and $f''(x)$.
2. Find all relative extreme points.
   (a) Find the critical values and critical points: Set $f'(x) = 0$ and solve for x. Suppose that $x = a$ is a solution (a critical value). Substitute $x = a$ into $f(x)$ to find $f(a)$, plot the critical point $(a, f(a))$, and draw a small horizontal tangent line through the point. Compute $f''(a)$.

   The Second-Derivative Test
   (i)   If $f''(a) > 0$, the curve has a relative minimum at $x = a$. Draw a small concave-up arc with $(a, f(a))$ as its lowest point.
   (ii) If $f''(a) < 0$, the curve has a relative maximum at $x = a$. Draw a small concave-down arc with $(a, f(a))$ as its peak.

   The First-Derivative Test
   (iii) If $f''(a) = 0$, examine $f'(x)$ to the left and right of $x = a$ to determine if the function changes from increasing to decreasing, or vice versa. If a relative extreme point is indicated, draw an appropriate arc as in parts (i) and (ii).

   (b) Repeat the preceding steps for each solution to $f'(x) = 0$.
3. Find all the inflection points of $f(x)$.
   (a) Set $f''(x) = 0$ and solve for x. Suppose that $x = b$ is a solution. Compute $f(b)$ and plot the point $(b, f(b))$.
   (b) Test the concavity of $f(x)$ to the left and right of b. If the concavity changes at $x = b$, then $(b, f(b))$ is an inflection point.
4. Consider other properties of the function and complete the sketch.
   (a) If $f(x)$ is defined at $x = 0$, the y-intercept is $(0, f(0))$.
   (b) Does the partial sketch suggest that there are x-intercepts? If so, they are found by setting $f(x) = 0$ and solving for x. (Solve only in easy cases or when a problem essentially requires you to calculate the x-intercepts.)
   (c) Observe where $f(x)$ is defined. Sometimes, the function is given only for restricted values of x. Sometimes, the formula for $f(x)$ is meaningless for certain values of x.

(d) Look for possible asymptotes.

    (i) Examine the formula for $f(x)$. If some terms become insignificant as x gets large, and if the rest of the formula gives the equation of a straight line, then that straight line is an asymptote.

    (ii) Suppose that there is some point a such that $f(x)$ is defined for x near a, but not at a (for example, $1/x$ at $x = 0$). If $f(x)$ gets arbitrarily large (in the positive or negative sense) as x approaches a, the vertical line $x = a$ is an asymptote for the graph.

(e) Complete the sketch.

Check Your Understanding 2.4

Solutions can be found following the section exercises.

Determine whether each of the following functions has an asymptote as x gets large. If so, give the equation of the straight line that is the asymptote.

1. $f(x) = \dfrac{3}{x} - 2x + 1$ **2.** $f(x) = \sqrt{x} + x$ **3.** $f(x) = \dfrac{1}{2x}$

EXERCISES 2.4

Find the x-intercepts of the given function.

1. $y = x^2 - 3x + 1$

2. $y = x^2 + 5x + 5$

3. $y = 2x^2 + 5x + 2$

4. $y = 4 - 2x - x^2$

5. $y = 4x - 4x^2 - 1$

6. $y = 3x^2 + 10x + 3$

7. Show that the function $f(x) = \frac{1}{3}x^3 - 2x^2 + 5x$ has no relative extreme points.

8. Show that the function $f(x) = -x^3 + 2x^2 - 6x + 3$ is always decreasing.

Sketch the graphs of the following functions.

9. $f(x) = x^3 - 6x^2 + 12x - 6$

10. $f(x) = -x^3$

11. $f(x) = x^3 + 3x + 1$

12. $f(x) = x^3 + 2x^2 + 4x$

13. $f(x) = 5 - 13x + 6x^2 - x^3$

14. $f(x) = 2x^3 + x - 2$

15. $f(x) = \frac{4}{3}x^3 - 2x^2 + x$

16. $f(x) = -3x^3 - 6x^2 - 9x - 6$

17. $f(x) = 1 - 3x + 3x^2 - x^3$

18. $f(x) = \frac{1}{3}x^3 - 2x^2$

19. $f(x) = x^4 - 6x^2$

20. $f(x) = 3x^4 - 6x^2 + 3$

21. $f(x) = (x - 3)^4$

22. $f(x) = (x + 2)^4 - 1$

Sketch the graphs of the following functions for $x > 0$.

23. $y = \dfrac{1}{x} + \dfrac{1}{4}x$

24. $y = \dfrac{2}{x}$

25. $y = \dfrac{9}{x} + x + 1$

26. $y = \dfrac{12}{x} + 3x + 1$

27. $y = \dfrac{2}{x} + \dfrac{x}{2} + 2$

28. $y = \dfrac{1}{x^2} + \dfrac{x}{4} - \dfrac{5}{4}$ [*Hint:* $(1, 0)$ is an x-intercept.]

29. $y = 6\sqrt{x} - x$

30. $y = \dfrac{1}{\sqrt{x}} + \dfrac{x}{2}$

In Exercises 31 and 32, determine which function is the derivative of the other.

31.

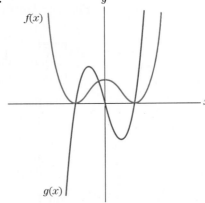

32.

33. Find the quadratic function $f(x) = ax^2 + bx + c$ that goes through $(2, 0)$ and has a local maximum at $(0, 1)$.

34. Find the quadratic function $f(x) = ax^2 + bx + c$ that goes through $(0, 1)$ and has a local minimum at $(1, -1)$.

35. If $f'(a) = 0$ and $f'(x)$ is increasing at $x = a$, explain why $f(x)$ must have a local minimum at $x = a$. [*Hint:* Use the first-derivative test.]

36. If $f'(a) = 0$ and $f'(x)$ is decreasing at $x = a$, explain why $f(x)$ must have a local maximum at $x = a$.

37. Changes in Body Weight In a medical experiment, the body weight of a baby rat in the control group after t days was $f(t) = 4.96 + .48t + .17t^2 - .0048t^3$ grams. (*Source: Growth, Development and Aging.*)

(a) Graph $f(t)$ in the window [0, 20] by [−12, 50].

(b) Approximately how much did the rat weigh after 7 days?

(c) Approximately when did the rat's weight reach 27 grams?

(d) Approximately how fast was the rat gaining weight after 4 days?

(e) Approximately when was the rat gaining weight at the rate of 2 grams per day?

(f) Approximately when was the rat gaining weight at the fastest rate?

38. Height of Tropical Grass The canopy height (in meters) of the tropical bunch-grass elephant millet t days after mowing (for $t \geq 32$) is $f(t) = -3.14 + .142t - .0016t^2 + .0000079t^3 - .0000000133t^4$. (*Source: Crop Science.*)

(a) Graph $f(t)$ in the window [32, 250] by [−1.2, 4.5].

(b) How tall was the canopy after 100 days?

(c) When was the canopy 2 meters high?

(d) How fast was the canopy growing after 80 days?

(e) When was the canopy growing at the rate of .02 meters per day?

(f) Approximately when was the canopy growing slowest?

(g) Approximately when was the canopy growing fastest?

Solutions to Check Your Understanding 2.4

From the solution of Example 4, we observe that functions with asymptotes as x gets large have the form $f(x) = g(x) + mx + b$, where $g(x)$ approaches zero as x gets large. The function $g(x)$ often looks like c/x or $c/(ax + d)$. The asymptote will be the straight line $y = mx + b$.

1. Here, $g(x)$ is $3/x$ and the asymptote is $y = -2x + 1$.

2. This function has no asymptote as x gets large. Of course, it can be written as $g(x) + mx + b$, where $m = 1$ and $b = 0$. However, $g(x) = \sqrt{x}$ does not approach 0 as x gets large.

3. Here, $g(x)$ is $\dfrac{1}{2x}$ and the asymptote is $y = 0$. That is, the function has the x-axis as an asymptote.

2.5 Optimization Problems

One of the most important applications of the derivative concept is to "optimization" problems, in which some quantity must be maximized or minimized. Examples of such problems abound in many areas of life. An airline must decide how many daily flights to schedule between two cities to maximize its profits. A doctor wants to find the minimum amount of a drug that will produce a desired response in one of her patients. A manufacturer needs to determine how often to replace certain equipment to minimize maintenance and replacement costs.

Our purpose in this section is to illustrate how calculus can be used to solve optimization problems. In each example, we will find or construct a function that provides a mathematical model for the problem. Then, by sketching the graph of this function, we will be able to determine the answer to the original optimization problem by locating the highest or lowest point on the graph. The y-coordinate of this point will be the maximum value or minimum value of the function.

The first two examples are very simple because the functions to be studied are given explicitly.

EXAMPLE 1   **Finding Extreme Points** Find the minimum value of the function $f(x) = 2x^3 - 15x^2 + 24x + 19$ for $x \geq 0$.

SOLUTION   Using the curve-sketching techniques from Section 2.3, we obtain the graph in Fig. 1. As part of that process, we compute the derivatives:

$$f'(x) = 6x^2 - 30x + 24$$
$$f''(x) = 12x - 30.$$

The x-coordinate of the minimum point satisfies the equation

$$f'(x) = 0$$
$$6x^2 - 30x + 24 = 0$$
$$6(x - 4)(x - 1) = 0$$
$$x = 1, 4 \quad \text{Critical values}$$

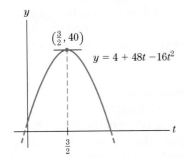

Figure 1

The corresponding critical points on the curve are

$$(1, f(1)) = (1, 30)$$
$$(4, f(4)) = (4, 3).$$

Applying the second-derivative test, we see that

$$f''(1) = 12 \cdot 1 - 30 = -18 < 0 \quad \text{Maximum}$$
$$f''(4) = 12 \cdot 4 - 30 = 18 > 0 \quad \text{Minimum}$$

That is, the point $(1, 30)$ is a relative maximum and the point $(4, 3)$ is a relative minimum. At the left end point, $x = 0$, $f(0) = 19$, which is clearly larger than the relative minimum value 3. Thus, the lowest point on the graph is $(4, 3)$. The minimum *value* of the function $f(x)$ is the y-coordinate of this point, 3. **» Now Try Exercise 1**

EXAMPLE 2 **Maximum Height Reached by a Ball** Suppose that a ball is thrown straight up into the air and its height after t seconds is $4 + 48t - 16t^2$ feet. Determine how long it will take for the ball to reach its maximum height and determine the maximum height.

SOLUTION Consider the function $f(t) = 4 + 48t - 16t^2$. For each value of t, $f(t)$ is the height of the ball at time t. We want to find the value of t for which $f(t)$ is the greatest. To do this, we use the techniques of Section 2.3 to sketch the graph of $f(t)$. Note that we may neglect the portions of the graph corresponding to points for which either $t < 0$ or $f(t) < 0$. [A negative value of $f(t)$ would correspond to the ball being underneath the ground.] The t-coordinate giving the maximum height is the solution of the equation:

$$f'(t) = 48 - 32t = 0$$
$$t = \frac{3}{2}.$$

Figure 2

Since

$$f''(t) = -32$$
$$f''\left(\frac{3}{2}\right) = -32 < 0,$$

we see from the second-derivative test that $t = \frac{3}{2}$ is the location of a relative maximum. So $f(t)$ is greatest when $t = \frac{3}{2}$. At this value of t, the ball attains a height of 40 feet. Thus, the ball reaches its maximum height of 40 feet in 1.5 seconds. A sketch of the height of the ball over time is in Fig. 2. **» Now Try Exercise 3**

EXAMPLE 3 **Maximizing an Area** You want to plant a rectangular garden along one side of a house, with a picket fence on the other three sides of the garden. Find the dimensions of the largest garden that can be enclosed using 40 feet of fencing.

SOLUTION Let us think about this problem before embarking on the solution. With 40 feet of fencing, you can enclose a rectangular garden along your house in many different ways. Here are three illustrations: In Fig. 3(a), the enclosed area is $10 \times 15 = 150$ square feet; in Fig. 3(b), it is $16 \times 12 = 192$ square feet; and in Fig. 3(c), it is $32 \times 4 = 128$ square

(a) Area = 10 × 15 = 150 sq. ft.

(b) Area = 16 × 12 = 192 sq. ft.

(c) Area = 32 × 4 = 128 sq. ft.

Figure 3 Rectangular gardens enclosed with 40 feet of fencing. Note that the house itself encloses one side of the garden.

feet. Clearly, the enclosed area varies with your choice of dimensions. Which dimensions of that total 40 feet yield the largest area? We next show how we can use calculus and optimization techniques to solve this problem.

Since we do not know the dimensions, the first step is to make a simple diagram that represents the general case and assign letters to the quantities that may vary. Let us denote the dimensions of the rectangular garden by w and x (Fig. 4). The phrase "largest garden" indicates that we must maximize the area, A, of the garden. In terms of the variables w and x,

$$A = wx. \qquad (1)$$

Area $= w \cdot x$ sq. ft.

Figure 4 General case.   Fencing required $= 2x + w$ ft.

The fencing on three sides must total 40 running feet; that is,

$$2x + w = 40. \qquad (2)$$

We now solve equation (2) for w in terms of x:

$$w = 40 - 2x. \qquad (3)$$

Substituting this expression for w into equation (1), we have

$$A = (40 - 2x)x = 40x - 2x^2. \qquad (4)$$

We now have a formula for the area A that depends on just one variable. From the statement of the problem, the value of $2x$ can be at most 40, so the domain of the function consists of x in the interval $(0, 20)$. Thus, the function that we wish to maximize is

$$A(x) = 40x - 2x^2 \quad 0 \le x \le 20.$$

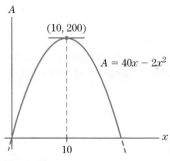

$A = 40x - 2x^2$

Figure 5

Its graph is a parabola facing downward (Fig. 5). To find the maximum point, we compute

$$A'(x) = 40 - 4x = 0$$
$$x = 10.$$

Since this value is in the domain of our function, we conclude that the absolute maximum of $A(x)$ occurs at $x = 10$. Alternatively, since $A''(x) = -4 < 0$, the concavity of the curve is always downward, and so the local maximum at $x = 10$ must be an absolute maximum. [The maximum area is $A(10) = 200$ square feet, but this fact is not needed for the problem.] From equation (3) we find that, when $x = 10$,

$$w = 40 - 2(10) = 20.$$

In conclusion, the rectangular garden with the largest area that can be enclosed using 40 feet of fencing has dimensions $x = 10$ feet and $w = 20$ feet.   **»› Now Try Exercise 11**

The equation $A = wx$ in Example 3 is called an **objective equation**. It expresses the quantity to be optimized (the area of the garden) in terms of the variables w and x. Equation (2) is called a **constraint equation** because it places a limit or constraint on the way x and w may vary.

| EXAMPLE 4 | **Minimizing Cost** The manager of a department store wants to build a 600-square-foot rectangular enclosure on the store's parking lot to display some equipment. Three sides of the enclosure will be built of redwood fencing at a cost of \$14 per running foot. The fourth side will be built of cement blocks, at a cost of \$28 per running foot. Find the dimensions of the enclosure that will minimize the total cost of the building materials. |

SOLUTION Let x be the length of the side built out of cement blocks, and let y be the length of an adjacent side, as shown in Fig. 6. The phrase "minimize the total cost" tells us that the objective equation should be a formula giving the total cost of the building materials:

$$[\text{cost of redwood}] = [\text{length of redwood fencing}] \times [\text{cost per foot}]$$
$$= (x + 2y) \cdot 14 = 14x + 28y$$

$$[\text{cost of cement blocks}] = [\text{length of cement wall}] \times [\text{cost per foot}]$$
$$= x \cdot 28.$$

Figure 6 Rectangular enclosure.

If C denotes the total cost of the materials, then the objective equation is:

$$C = (14x + 28y) + 28x$$
$$C = 42x + 28y. \quad \text{Combine like terms.} \qquad (5)$$

Since the area of the enclosure must be 600 square feet, the constraint equation is

$$xy = 600. \qquad (6)$$

We simplify the objective equation by solving equation (6) for one of the variables—say, y—and substituting into (5). Since $y = 600/x$,

$$C = 42x + 28\left(\frac{600}{x}\right) = 42x + \frac{16,800}{x}.$$

We now have C as a function of the single variable x. From the context, we must have $x > 0$, since a length must be positive. However, to any positive value for x, there is a corresponding value for C. So the domain of C consists of all $x > 0$. We may now sketch the graph of C. (See Fig. 7.) (A similar curve was sketched in Example 4 of Section 2.4.) The x-coordinate of the minimum point is a solution of

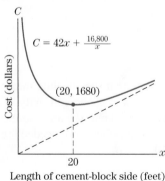

Length of cement-block side (feet)

Figure 7

$$C'(x) = 42 - \frac{16,800}{x^2} = 0$$

$$42 = \frac{16,800}{x^2}$$

$$42x^2 = 16,800 \qquad \text{Multiply by } x^2.$$
$$x^2 = 400 \qquad \text{Divide by 42.}$$
$$x = 20. \qquad \text{Take the positive square root.}$$
$$\text{(Ignore the negative root, since } x > 0.)$$

The corresponding value of C is

$$C(20) = 42(20) + \frac{16,800}{20} = \$1680.$$

That is, the minimum total cost of $1680 occurs where $x = 20$. From equation (6) we find that the corresponding value of y is $\frac{600}{20} = 30$. To minimize the total cost of building the 600-square-foot rectangular enclosure, the manager must take the dimensions to be $x = 20$ feet and $y = 30$ feet. **» Now Try Exercise 15**

EXAMPLE 5 **Maximizing Volume** Postal regulations state that packages must have length plus girth of no more than 84 inches. Find the dimensions of the cylindrical package of greatest volume that is mailable by parcel post.

SOLUTION Let l be the length of the package, and let r be the radius of the circular end. (See Fig. 8.) The phrase "greatest volume" tells us that the objective equation should express the volume of the package in terms of the dimensions l and r. Let V denote the volume. Then,

$$V = [\text{area of base}] \cdot [\text{length}]$$

$$V = \pi r^2 l \, (\text{objective equation}). \qquad (7)$$

The girth equals the circumference of the end, that is, $2\pi r$. Since we want the package to be as large as possible, we must use the entire 84 inches allowable:

$$\text{length} + \text{girth} = 84$$

$$l + 2\pi r = 84 \quad (\text{constraint equation}). \qquad (8)$$

We now solve equation (8) for one of the variables—say $l = 84 - 2\pi r$. Substituting this expression into (7), we obtain

$$V = \pi r^2 (84 - 2\pi r) = 84\pi r^2 - 2\pi^2 r^3. \qquad (9)$$

Let $f(r) = 84\pi r^2 - 2\pi^2 r^3$. Then, for each value of r, $f(r)$ is the volume of the parcel with end radius r that meets the postal regulations. We want to find that value of r for which $f(r)$ is as large as possible.

Using curve-sketching techniques, we obtain the graph of $f(r)$ in Fig. 9. The domain excludes values of r that are negative and values of r for which the volume $f(r)$ is negative. Points corresponding to values of r not in the domain are shown with a dashed curve. We see that the volume is greatest when $r = 28/\pi$.

From equation (8), we find that the corresponding value of l is

$$l = 84 - 2\pi r = 84 - 2\pi \left(\frac{28}{\pi} \right) = 84 - 56 = 28.$$

The girth when $r = 28/\pi$ is

$$2\pi r = 2\pi \left(\frac{28}{\pi} \right) = 56.$$

Thus, the dimensions of the cylindrical package of greatest volume are $l = 28$ inches and $r = 28/\pi$ inches. **» Now Try Exercise 23**

Figure 8 Cylindrical mailing package.

$V = 84\pi r^2 - 2\pi^2 r^3$

$\left(\frac{28}{\pi}, \frac{(28)^3}{\pi} \right)$

$\frac{28}{\pi}$

Figure 9

Suggestions for Solving an Optimization Problem

1. Draw a picture, if possible.
2. Decide what quantity, Q, is to be maximized or minimized.
3. Assign letters to other quantities that may vary.
4. Determine the "objective equation" that expresses Q as a function of the variables assigned in step 3.
5. Find the "constraint equation" that relates the variables to each other and to any constants that are given in the problem.
6. Use the constraint equation to simplify the objective equation in such a way that Q becomes a function of only one variable. Determine the domain of this function.
7. Sketch the graph of the function obtained in step 6 and use this graph to solve the optimization problem. Alternatively, you can use the second-derivative test.

NOTE ► Optimization problems often involve geometric formulas. The most common formulas are shown in Fig. 10 and on the inside back cover of the book. ◄◄

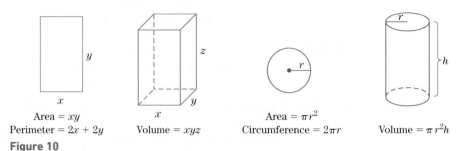

Area = xy
Perimeter = $2x + 2y$

Volume = xyz

Area = πr^2
Circumference = $2\pi r$

Volume = $\pi r^2 h$

Figure 10

Check Your Understanding 2.5

Solutions can be found following the section exercises.

1. **Volume** A canvas wind shelter for the beach has a back, two square sides, and a top (Fig. 11). If 96 square feet of canvas is to be used, find the dimensions of the shelter for which the space inside the shelter (the volume) is maximized.

2. In Check Your Understanding 1, what are the objective equation and the constraint equation?

Figure 11 Wind shelter.

EXERCISES 2.5

1. For what x does the function $g(x) = 10 + 40x - x^2$ have its maximum value?

2. Find the maximum value of the function $f(x) = 12x - x^2$, and give the value of x where this maximum occurs.

3. Find the minimum value of $f(t) = t^3 - 6t^2 + 40$, $t \geq 0$, and give the value of t where this minimum occurs.

4. For what t does the function $f(t) = t^2 - 24t$ have its minimum value?

5. **Optimization with Constraint** Find the maximum of $Q = xy$ if $x + y = 2$.

6. **Optimization with Constraint** Find two positive numbers x and y that maximize $Q = x^2 y$ if $x + y = 2$.

7. **Optimization with Constraint** Find the minimum of $Q = x^2 + y^2$ if $x + y = 6$.

8. In Exercise 7, can there be a maximum for $Q = x^2 + y^2$ if $x + y = 6$? Justify your answer.

9. **Minimizing a Sum** Find the positive values of x and y that minimize $S = x + y$ if $xy = 36$, and find this minimum value.

10. **Maximizing a Product** Find the positive values of x, y, and z that maximize $Q = xyz$, if $x + y = 1$ and $y + z = 2$. What is this maximum value?

11. **Area** There are $320 available to fence in a rectangular garden. The fencing for the side of the garden facing the road costs $6 per foot and the fencing for the other three sides costs $2 per foot. [See Fig. 12(a).] Consider the problem of finding the dimensions of the largest possible garden.
    (a) Determine the objective and constraint equations.
    (b) Express the quantity to be maximized as a function of x.
    (c) Find the optimal values of x and y.

12. **Volume** Figure 12(b) shows an open rectangular box with a square base. Consider the problem of finding the values of x and h for which the volume is 32 cubic feet and the total

surface area of the box is minimal. (The surface area is the sum of the areas of the five faces of the box.)
    (a) Determine the objective and constraint equations.
    (b) Express the quantity to be minimized as a function of x.
    (c) Find the optimal values of x and h.

(a)   (b)

Figure 12

13. **Volume** Postal requirements specify that parcels must have length plus girth of at most 84 inches. Consider the problem of finding the dimensions of the square-ended rectangular package of greatest volume that is mailable.
    (a) Draw a square-ended rectangular box. Label each edge of the square end with the letter x and label the remaining dimension of the box with the letter h.
    (b) Express the length plus the girth in terms of x and h.
    (c) Determine the objective and constraint equations.
    (d) Express the quantity to be maximized as a function of x.
    (e) Find the optimal values of x and h.

14. **Perimeter** Consider the problem of finding the dimensions of the rectangular garden of area 100 square meters for which the amount of fencing needed to surround the garden is as small as possible.
    (a) Draw a picture of a rectangle and select appropriate letters for the dimensions.
    (b) Determine the objective and constraint equations.
    (c) Find the optimal values for the dimensions.

15. **Cost** A rectangular garden of area 75 square feet is to be surrounded on three sides by a brick wall costing $10 per foot and on one side by a fence costing $5 per foot. Find the dimensions of the garden that minimize the cost of materials.

16. **Cost** A closed rectangular box with a square base and a volume of 12 cubic feet is to be constructed from two different types of materials. The top is made of a metal costing $2 per square foot and the remainder of wood costing $1 per square foot. Find the dimensions of the box for which the cost of materials is minimized.

17. **Surface Area** Find the dimensions of the closed rectangular box with square base and volume 8000 cubic centimeters that can be constructed with the least amount of material.

18. **Volume** A canvas wind shelter for the beach has a back, two square sides, and a top. Find the dimensions for which the volume will be 250 cubic feet and that requires the least possible amount of canvas.

19. **Area** A farmer has $1500 available to build an E-shaped fence along a straight river so as to create two identical rectangular pastures. (See Fig. 13.) The materials for the side parallel to the river cost $6 per foot, and the materials for the three sections perpendicular to the river cost $5 per foot. Find the dimensions for which the total area is as large as possible.

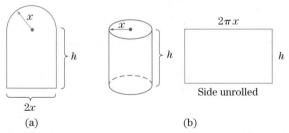

Figure 13 Rectangular pastures along a river.

20. **Area** Find the dimensions of the rectangular garden of greatest area that can be fenced off (all four sides) with 300 meters of fencing.

21. **Maximizing a Product** Find two positive numbers, x and y, whose sum is 100 and whose product is as large as possible.

22. **Minimizing a Sum** Find two positive numbers, x and y, whose product is 100 and whose sum is as small as possible.

23. **Area** Figure 14(a) shows a Norman window, which consists of a rectangle capped by a semicircular region. Find the value of x such that the perimeter of the window will be 14 feet and the area of the window will be as large as possible.

(a)   (b)

Figure 14

24. **Surface Area** A large soup can is to be designed so that the can will hold 16π cubic inches (about 28 ounces) of soup. [See Fig. 14(b).] Find the values of x and h for which the amount of metal needed is as small as possible.

25. In Example 3 we can solve the constraint equation (2) for x instead of w to get $x = 20 - \frac{1}{2}w$. Substituting this for x in (1), we get

$$A = xw = \left(20 - \frac{1}{2}w\right)w.$$

Sketch the graph of the equation

$$A = 20w - \frac{1}{2}w^2,$$

and show that the maximum occurs when $w = 20$ and $x = 10$.

26. **Cost** A ship uses $5x^2$ dollars of fuel per hour when traveling at a speed of x miles per hour. The other expenses of operating the ship amount to $2000 per hour. What speed minimizes the cost of a 500-mile trip? [*Hint:* Express cost in terms of speed and time. The constraint equation is *distance* = *speed* $\times$ *time*.]

27. **Cost** A cable is to be installed from one corner, C, of a rectangular factory to a machine, M, on the floor. The cable will run along one edge of the floor from C to a point, P, at a cost of $6 per meter, and then from P to M in a straight line buried under the floor at a cost of $10 per meter (see Fig. 15). Let x denote the distance from C to P. If M is 24 meters from the nearest point, A, on the edge of the floor on which P lies, and A is 20 meters from C, find the value of x that minimizes the cost of installing the cable and determine the minimum cost.

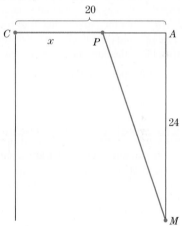

Figure 15

28. **Area** A rectangular page is to contain 50 square inches of print. The page has to have a 1-inch margin on top and at the bottom and a $\frac{1}{2}$-inch margin on each side (see Fig. 16). Find the dimensions of the page that minimize the amount of paper used.

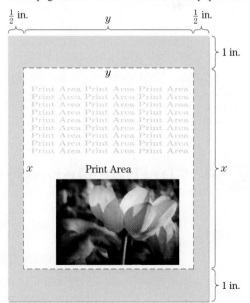

Figure 16

29. Distance Find the point on the graph of $y = \sqrt{x}$ that is closest to the point $(2, 0)$. See Fig. 17. [*Hint:* $\sqrt{(x-2)^2 + y^2}$ has its smallest value when $(x-2)^2 + y^2$ does. Therefore, just minimize the second expression.]

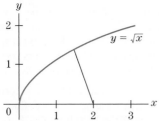

Figure 17 Shortest distance from a point to a curve.

30. Distance Developers from two cities, A and B, want to connect their cities to a major highway and plan to build rest stops and gas stations at the highway entrance. To minimize the cost of road construction, the developers must find the location for the highway entrance that minimizes the total distance, $d_1 + d_2$, from the cities A and B to the highway. Let x be as in Fig. 18. Find the value of x that solves the developers' problem, and find the minimum total distance.

31. Distance Find the point on the line $y = -2x + 5$ that is closest to the origin.

Figure 18

TECHNOLOGY EXERCISES

32. Inscribed Rectangle of Maximum Area Find the value of x for which the rectangle inscribed in the semicircle of radius 3 in Fig. 19 has the greatest area. [*Note:* The equation of the semicircle is $y = \sqrt{9 - x^2}$.]

Figure 19

Solutions to Check Your Understanding 2.5

1. Since the sides of the wind shelter are square, we may let x represent the length of each side of the square. The remaining dimension of the wind shelter can be denoted by the letter h. (See Fig. 20.)

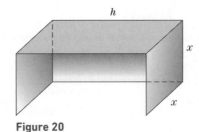

Figure 20

The volume of the shelter is x^2h, and this is to be maximized. We use the information that 96 feet of canvas are used; that is, $2x^2 + 2xh = 96$. (*Note:* The roof and the back each have an area xh, and each end has an area x^2.)

We now solve this equation for h:

$$2x^2 + 2xh = 96$$
$$2xh = 96 - 2x^2$$
$$h = \frac{96}{2x} - \frac{2x^2}{2x} = \frac{48}{x} - x.$$

The volume, V, is

$$x^2h = x^2\left(\frac{48}{x} - x\right) = 48x - x^3.$$

By sketching the graph of $V = 48x - x^3$, we see that V has a maximum value when $x = 4$. Then, $h = \frac{48}{4} - 4 = 12 - 4 = 8$. So each end of the shelter should be a 4-foot by 4-foot square, and the top should be 8 feet long.

2. The objective equation is $V = x^2h$, since it expresses the volume (the quantity to be maximized) in terms of the variables. The constraint equation is $2x^2 + 2xh = 96$, for it relates the variables to each other; that is, it can be used to express one of the variables in terms of the other.

2.6 Further Optimization Problems

In this section, we apply the optimization techniques developed in Section 2.5 to some additional practical situations.

Inventory Control

When a firm regularly orders and stores supplies for later use or resale, it must decide on the size of each order. If it orders enough supplies to last an entire year, the business will incur heavy *carrying costs*. Such costs include insurance, storage costs, and cost of capital that is tied up in inventory. To reduce these carrying costs, the firm could order small quantities of the supplies at frequent intervals. However, such a policy increases the *ordering costs*. These might consist of minimum freight charges, the clerical costs of preparing the orders, and the costs of receiving and checking the orders when they arrive. Clearly, the firm must find an inventory-ordering policy that lies between these two extremes.

To appreciate the inventory-control problem, let us first consider a simple problem that does not involve optimization.

EXAMPLE 1   **An Inventory Problem** A supermarket manager anticipates that 1200 cases of frozen orange juice will be sold at a steady rate during the next year. The manager plans on ordering these cases by placing equally spaced orders of the same size throughout the year. Given that it costs $8 to carry one case of orange juice in inventory for 1 year, find the carrying cost if

(a) The manager places only one order during the year.

(b) The manager places two orders during the year.

(c) The manager places four orders during the year.

Carrying costs should be computed on the average inventory during the order–reorder period.

SOLUTION   (a) This plan calls for ordering 1200 cases all at once. The inventory of frozen orange juice in this case is shown in Fig. 1(a). It starts at 1200 and decreases steadily to 0 during this one order–reorder period. At any given time during the year, the inventory is between 1200 and 0. So there are $\frac{1200}{2}$, or 600 cases on average in inventory. Since carrying costs are computed on the average inventory, the carrying cost in this case is $C = 600 \times 8 = \$4800$.

(b) This plan calls for ordering 1200 cases by placing two equally spaced orders of the same size. Thus the size of each order is equal to $\frac{1200}{2}$ or 600 cases. The inventory of frozen orange juice in this case is shown in Fig. 1(b). At the beginning of each order–reorder period, the inventory starts at 600 and decreases steadily to 0 at the end of the order–reorder period. At any given time during the order–reorder period, the inventory is between 600 and 0. So there are $\frac{600}{2}$, or 300 cases on average in inventory. Since the carrying costs are computed on the average inventory, the carrying cost in this case is $C = 300 \times 8 = \$2400$.

(c) This plan calls for ordering 1200 cases by placing four equally spaced orders of the same size. The size of each order is $\frac{1200}{4} = 300$ cases. The inventory in this case is shown in Fig. 1(c). The average inventory during an order–reorder period is $\frac{300}{2}$, or 150 cases, and so the carrying cost in this case is $C = 150 \times 8 = \$1200$.

» Now Try Exercise 1

By increasing the number of orders in Example 1, the manager was able to reduce the carrying cost from $4800 down to $2400 down to $1200. Clearly, the manager can reduce the carrying cost even further by ordering more frequently. In reality, things are not so simple. There is a cost attached to each order, and by increasing the number of orders, the manager will increase the annual inventory cost, where

[inventory cost] = [ordering cost] + [carrying cost].

The inventory-control problem consists of finding the *economic order quantity*, commonly referred to in business as the EOQ, that minimizes the inventory cost. We now show how we can use calculus to solve this problem. (*Source: Financial Management and Policy*, 6th ed.)

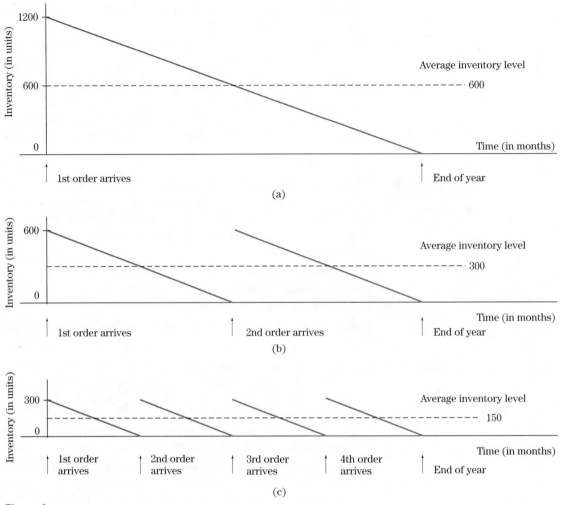

Figure 1

EXAMPLE 2 **Inventory Control** Suppose that the manager in Example 1 wants to establish an optimal inventory policy for frozen orange juice. Again, it is estimated that a total of 1200 cases will be sold at a steady rate during the next year. The manager plans to place several orders of the same size spaced equally throughout the year. Use the following data to determine the economic order quantity; that is, the order size that minimizes the total ordering and carrying cost.

1. The ordering cost for each delivery is $75.
2. It costs $8 to carry one case of orange juice in inventory for 1 year.
   (Carrying costs should be computed on the average inventory during the period.)

SOLUTION Since we are not given the size of each order or the number of orders that will be placed throughout the year, we let x be the order quantity and r the number of orders placed during the year. As in Example 1, the number of cases of orange juice in inventory declines steadily from x cases (each time a new order is filled) to 0 cases at the end of each order–reorder period. Figure 2 shows that the average number of cases in storage during the year is $x/2$. Since the carrying cost for one case is $8 per year, the cost for $\frac{x}{2}$ cases is $8 \cdot \left(\frac{x}{2}\right)$ dollars. Now,

$$[\text{inventory cost}] = [\text{ordering cost}] + [\text{carrying cost}]$$

$$= 75r + 8 \cdot \frac{x}{2}$$

$$= 75r + 4x.$$

Figure 2 Average inventory level.

1st order arrives 2nd order arrives 3rd order arrives 4th order arrives

Figure 3 shows a slanted asymptote. For a discussion of this topic, see Example 4, Sec. 2.4, p. 160.

If C denotes the inventory cost, then the objective equation is

$$C = 75r + 4x.$$

Since there are r orders of x cases each, the total number of cases ordered during the year is $r \cdot x$. Therefore, the constraint equation is

$$r \cdot x = 1200.$$

The constraint equation says that $r = \frac{1200}{x}$. Substitution into the objective equation yields

$$C = \frac{90{,}000}{x} + 4x.$$

Figure 3 is the graph of C as a function of x, for $x > 0$. The minimum point of the graph occurs where the first derivative equals zero. We find this point using the calculation

$$C'(x) = -\frac{90{,}000}{x^2} + 4 = 0$$

$$4 = \frac{90{,}000}{x^2}$$

$$4x^2 = 90{,}000$$

$$x^2 = 22{,}500$$

$$x = 150 \quad \text{(ignore the negative root since } x > 0\text{)}.$$

The total cost is at a minimum when $x = 150$. Therefore, the optimum inventory policy is to order 150 cases at a time and to place $\frac{1200}{150} = 8$ orders during the year.

» Now Try Exercise 3

$C = \dfrac{90{,}000}{x} + 4x$

Economic order quantity

150

Order quantity

Figure 3 Cost function for inventory problem.

EXAMPLE 3 **Inventory Control** What should the inventory policy of Example 2 be if sales of frozen orange juice increase fourfold (i.e., 4800 cases are sold each year), but all other conditions are the same?

SOLUTION The only change in our previous solution is in the constraint equation, which now becomes

$$r \cdot x = 4800.$$

The objective equation is, as before,

$$C = 75r + 4x.$$

Since $r = \frac{4800}{x}$,

$$C = 75 \cdot \frac{4800}{x} + 4x = \frac{360{,}000}{x} + 4x.$$

Now,

$$C' = -\frac{360{,}000}{x^2} + 4.$$

Setting $C' = 0$ yields

$$\frac{360{,}000}{x^2} = 4$$

$$90{,}000 = x^2$$

$$x = 300.$$

(Again, we ignore the negative root, since $x > 0$.)

Therefore, the economic order quantity is 300 cases.　　**≫ Now Try Exercise 5**

Notice that although the sales increased by a factor of 4, the economic order quantity increased by only a factor of 2 ($= \sqrt{4}$). In general, a store's inventory of an item should be proportional to the square root of the expected sales. (See Exercise 9 for a derivation of this result.) Many stores tend to keep their average inventories at a fixed percentage of sales. For example, each order may contain enough goods to last for 4 or 5 weeks. This policy is likely to create excessive inventories of high-volume items and uncomfortably low inventories of slower-moving items.

Manufacturers have an inventory-control problem similar to that of retailers. They have the carrying costs of storing finished products and the start-up costs of setting up each production run. The size of the production run that minimizes the sum of these two costs is called the **economic lot size**. See Exercises 6 and 7.

When solving optimization problems, we look for the maximum or minimum point on a graph. This point occurs either at a relative extreme point or at an endpoint of the domain of definition. In all our optimization problems so far, the maximum or minimum points were at relative extreme points. In the next example, the optimum point is an endpoint. This occurs often in applied problems.

EXAMPLE 4　　**Endpoint Extremum** A rancher has 204 meters of fencing from which to build two corrals: one square and the other rectangular with length that is twice the width. Find the dimensions that result in the greatest combined area.

SOLUTION　　Let x be the width of the rectangular corral and h be the length of each side of the square corral. (See Fig. 4.) Let A be the combined area. Then,

$$A = [\text{area of square}] + [\text{area of rectangle}] = h^2 + 2x^2.$$

The constraint equation is

$$204 = [\text{perimeter of square}] + [\text{perimeter of rectangle}] = 4h + 6x.$$

Since the perimeter of the rectangle cannot exceed 204, we must have $0 \le 6x \le 204$, or $0 \le x \le 34$. Solving the constraint equation for h and substituting into the objective equation leads to the function graphed in Fig. 5. The graph reveals that the area

Figure 4 Two corrals.

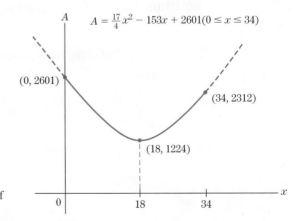

$$A = \tfrac{17}{4}x^2 - 153x + 2601 \quad (0 \le x \le 34)$$

$(0, 2601)$

$(34, 2312)$

$(18, 1224)$

Figure 5 Combined area of the corrals.

is minimized when $x = 18$. However, the problem asks for the *maximum* possible area. From Fig. 5, we see that this occurs at the endpoint where $x = 0$. Therefore, the rancher should build only the square corral, with $h = 204/4 = 51$ meters. In this example, the objective function has an endpoint extremum: The maximum value occurs at the endpoint where $x = 0$.   **» Now Try Exercise 15**

Check Your Understanding 2.6

Solutions can be found following the section exercises.

1. In the inventory problem of Example 2, suppose that the ordering cost is the same, but that it costs $9 to carry one case of orange juice in inventory for 1 year, and carrying costs are computed on the *maximum* inventory during the order–reorder period. What is the new economic order quantity?

2. In the inventory problem of Example 2, suppose that the sales of frozen orange juice increase ninefold; that is, 10,800 cases are sold each year. What is the new economic order quantity?

EXERCISES 2.6

1. **Inventory Problem** Figure 6 shows the inventory levels of dried Rainier cherries at a natural food store in Seattle and the order–reorder periods over 1 year. Refer to the figure to answer the following questions.
   (a) What is the average amount of cherries in inventory during one order–reorder period?
   (b) What is the maximum amount of cherries in inventory during one order–reorder period?
   (c) How many orders were placed during the year?
   (d) How many pounds of dried cherries were sold during 1 year?

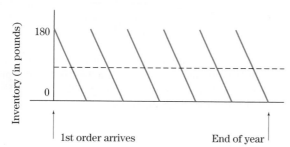

Figure 6

2. Refer to Fig. 6. Suppose that
   (i) the ordering cost for each delivery of dried cherries is $50, and
   (ii) it costs $7 to carry 1 pound of dried cherries in inventory for 1 year.
   (a) What is the inventory cost (carrying cost plus ordering cost) if carrying costs are computed on the average inventory during the order–reorder period?
   (b) What is the inventory cost if carrying costs are computed on the maximum inventory during the order–reorder period?

3. **Inventory Control** A pharmacist wants to establish an optimal inventory policy for a new antibiotic that requires refrigeration in storage. The pharmacist expects to sell 800 packages of this antibiotic at a steady rate during the next year. She plans to place several orders of the same size spaced equally throughout the year. The ordering cost for each delivery is $16, and carrying costs, based on the average number of packages in inventory, amount to $4 per year for one package.
   (a) Let x be the order quantity and r the number of orders placed during the year. Find the inventory cost (ordering cost plus carrying cost) in terms of x and r.
   (b) Find the constraint function.
   (c) Determine the economic order quantity that minimizes the inventory cost, and then find the minimum inventory cost.

4. **Inventory Control** A furniture store expects to sell 640 sofas at a steady rate next year. The manager of the store plans to order these sofas from the manufacturer by placing several orders of the same size spaced equally throughout the year. The ordering cost for each delivery is $160, and carrying costs, based on the average number of sofas in inventory, amount to $32 per year for one sofa.
   (a) Let x be the order quantity and r the number of orders placed during the year. Find the inventory cost in terms of x and r.
   (b) Find the constraint function.
   (c) Determine the economic order quantity that minimizes the inventory cost, and then find the minimum inventory cost.

5. **Inventory Control** A California distributor of sporting equipment expects to sell 10,000 cases of tennis balls during the coming year at a steady rate. Yearly carrying costs (to be computed on the average number of cases in stock during the year) are $10 per case, and the cost of placing an order with the manufacturer is $80.
   (a) Find the inventory cost incurred if the distributor orders 500 cases at a time during the year.
   (b) Determine the economic order quantity, that is, the order quantity that minimizes the inventory cost.

6. **Economic Lot Size** The Great American Tire Co. expects to sell 600,000 tires of a particular size and grade during the next year. Sales tend to be roughly the same from month to month. Setting up each production run costs the company $15,000. Carrying costs, based on the average number of tires in storage, amount to $5 per year for one tire.
   (a) Determine the costs incurred if there are 10 production runs during the year.
   (b) Find the economic lot size (that is, the production run size that minimizes the overall cost of producing the tires).

7. **Economic Lot Size** Foggy Optics, Inc., makes laboratory microscopes. Setting up each production run costs $2500. Insurance costs, based on the average number of microscopes in the warehouse, amount to $20 per microscope per year. Storage costs, based on the maximum number of microscopes in the warehouse, amount to $15 per microscope per year.

If the company expects to sell 1600 microscopes at a fairly uniform rate throughout the year, determine the number of production runs that will minimize the company's overall expenses.

8. **Inventory Control** A bookstore is attempting to determine the most economical order quantity for a popular book. The store sells 8000 copies of this book per year. The store figures that it costs $40 to process each new order for books. The carrying cost (due primarily to interest payments) is $2 per book, to be figured on the maximum inventory during an order–reorder period. How many times a year should orders be placed?

9. **Inventory Control** A store manager wants to establish an optimal inventory policy for an item. Sales are expected to be at a steady rate and should total Q items sold during the year. Each time an order is placed a cost of h dollars is incurred. Carrying costs for the year will be s dollars per item, to be figured on the average number of items in storage during the year. Show that the total inventory cost is minimized when each order calls for $\sqrt{2hQ/s}$ items.

10. Refer to the inventory problem of Example 2. If the distributor offers a discount of $1 per case for orders of 600 or more cases, should the manager change the quantity ordered?

11. **Area** Starting with a 100-foot-long stone wall, a farmer would like to construct a rectangular enclosure by adding 400 feet of fencing, as shown in Fig. 7(a). Find the values of x and w that result in the greatest possible area.

(a)   (b)

Figure 7 Rectangular enclosures.

12. Rework Exercise 11 for the case where only 200 feet of fencing is added to the stone wall.

13. **Length** A rectangular corral of 54 square meters is to be fenced off and then divided by a fence into two sections, as shown in Fig. 7(b). Find the dimensions of the corral so that the amount of fencing required is minimized.

14. Refer to Exercise 13. If the cost of the fencing for the boundary is $5 per meter and the dividing fence costs $2 per meter, find the dimensions of the corral that minimize the cost of the fencing.

15. **Revenue** Shakespear's Pizza sells 1000 large vegi pizzas per week for $18 a pizza. When the owner offers a $5 discount, the weekly sales increase to 1500.
   (a) Assume a linear relation between the weekly sales $A(x)$ and the discount x. Find $A(x)$.
   (b) Find the value of x that maximizes the weekly revenue. [*Hint:* Revenue = $A(x) \cdot$ (Price).]
   (c) Answer parts (a) and (b) if the price of one pizza is $9 and all other data are unchanged.

16. **Surface Area** Design an open rectangular box with square ends, having volume 36 cubic inches, that minimizes the amount of material required for construction.

17. **Cost** A storage shed is to be built in the shape of a box with a square base. It is to have a volume of 150 cubic feet. The

concrete for the base costs $4 per square foot, the material for the roof costs $2 per square foot, and the material for the sides costs $2.50 per square foot. Find the dimensions of the most economical shed.

18. **Cost** A supermarket is to be designed as a rectangular building with a floor area of 12,000 square feet. The front of the building will be mostly glass and will cost $70 per running foot for materials. The other three walls will be constructed of brick and cement block, at a cost of $50 per running foot. Ignore all other costs (labor, cost of foundation and roof, and the like) and find the dimensions of the base of the building that will minimize the cost of the materials for the four walls of the building.

19. **Volume** A certain airline requires that rectangular packages carried on an airplane by passengers be such that the sum of the three dimensions (i.e., length, width, and height), is at most 120 centimeters. Find the dimensions of the square-ended rectangular package of greatest volume that meets this requirement.

20. **Area** An athletic field [Fig. 8] consists of a rectangular region with a semicircular region at each end. The perimeter will be used for a 440-yard track. Find the value of x for which the area of the rectangular region is as large as possible.

Figure 8

21. **Volume** An open rectangular box is to be constructed by cutting square corners out of a 16- by 16-inch piece of cardboard and folding up the flaps. [See Fig. 9.] Find the value of x for which the volume of the box will be as large as possible.

Figure 9

22. **Volume** A closed rectangular box is to be constructed with a base that is twice as long as it is wide. If the total surface area must be 27 square feet, find the dimensions of the box that will maximize the volume.

23. **Amount of Oxygen in a Lake** Let $f(t)$ be the amount of oxygen (in suitable units) in a lake t days after sewage is dumped into the lake, and suppose that $f(t)$ is given approximately by

$$f(t) = 1 - \frac{10}{t+10} + \frac{100}{(t+10)^2}.$$

At what time is the oxygen content increasing the fastest?

24. **Daily Output of a Factory** The daily output of a coal mine after t hours of operation is approximately $40t + t^2 - \frac{1}{15}t^3$ tons, $0 \le t \le 12$. Find the maximum rate of output (in tons of coal per hour).

25. **Area** Consider a parabolic arch whose shape may be represented by the graph of $y = 9 - x^2$, where the base of the arch lies on the x-axis from $x = -3$ to $x = 3$. Find the dimensions of the rectangular window of maximum area that can be constructed inside the arch.

26. **Advertising and Sales** Advertising for a certain product is terminated, and t weeks later, the weekly sales are $f(t)$ cases, where

$$f(t) = 1000(t + 8)^{-1} - 4000(t + 8)^{-2}.$$

At what time is the weekly sales amount falling the fastest?

27. **Surface Area** An open rectangular box of volume 400 cubic inches has a square base and a partition down the middle. See Fig. 10. Find the dimensions of the box for which the amount of material needed to construct the box is as small as possible.

Figure 10 Open rectangular box with dividing partition.

28. If $f(x)$ is defined on the interval $0 \le x \le 5$ and $f'(x)$ is negative for all x, for what value of x will $f(x)$ have its greatest value? Explain why.

TECHNOLOGY EXERCISES

29. **Volume** A pizza box is formed from a 20-cm by 40-cm rectangular piece of cardboard by cutting out six squares of equal size, three from each long side of the rectangle, and then folding the cardboard in the obvious manner to create a box. See Fig. 11. Let x be the length of each side of the six squares. For what value of x will the box have greatest volume?

Figure 11

30. **Consumption of Coffee in the U.S.** Coffee consumption in the United States is greater on a per capita basis than anywhere else in the world. However, due to price fluctuations of coffee beans and worries over the health effects of caffeine, coffee consumption has varied considerably over the years. According to data published in *The Wall Street Journal*, the number of cups $f(x)$ consumed daily per adult in year x (with 1955 corresponding to $x = 0$) is given by the mathematical model

$$f(x) = 2.77 + 0.0848x - 0.00832x^2 + 0.000144x^3.$$

(a) Graph $y = f(x)$ to show daily coffee consumption from 1955 through 1994.

(b) Use $f'(x)$ to determine the year in which coffee consumption was least during this period. What was the daily coffee consumption at that time?

(c) Use $f'(x)$ to determine the year in which coffee consumption was greatest during this period. What was the daily coffee consumption at that time?

(d) Use $f''(x)$ to determine the year in which coffee consumption was decreasing at the greatest rate.

Solutions to Check Your Understanding 2.6

1. Let x denote the order quantity and r the number of orders placed during the year. It is clear from Fig. 2 that the maximum inventory at any time during the order–reorder period is x cases. This is the number of cases in inventory at the moment an order arrives. Since the carrying cost for one case is $9, the carrying cost for x cases is $9 \cdot x$ dollars. This is the new carrying cost. The ordering cost is the same as in Example 2. So the objective function becomes

   $$C = [\text{ordering cost}] + [\text{carrying cost}] = 75r + 9x.$$

   The constraint equation is also the same as in Example 2, because the total number of cases ordered during the year has not changed. Thus, $r \cdot x = 1200$.

   From this point, the solution proceeds as in Example 2. We have

   $$r = \frac{1200}{x}$$

   $$C = 75 \cdot \frac{1200}{x} + 9x = \frac{90{,}000}{x} + 9x$$

   $$C'(x) = -\frac{90{,}000}{x^2} + 9.$$

   Solving $C'(x) = 0$, we find

   $$-\frac{90{,}000}{x^2} + 9 = 0$$

   $$9x^2 = 90{,}000$$

   $$x^2 = 10{,}000$$

   $$x = \sqrt{10{,}000} = 100 \quad \begin{array}{l}\text{(ignore the}\\\text{negative root}\\\text{since } x > 0).\end{array}$$

   The total cost is at a minimum when $x = 100$. The optimum inventory policy is to order 100 cases at a time and to place $\frac{1200}{100}$ or 12 orders during the year.

2. This problem can be solved in the same manner that Example 3 was solved. However, the comment made at the end of Example 3 indicates that the economic order quantity should increase by a factor of 3, since $3 = \sqrt{9}$. Therefore, the economic order quantity is $3 \cdot 150 = 450$ cases.

2.7 Applications of Derivatives to Business and Economics

Figure 1 A cost function.

All our applications will center on what economists call the *theory of the firm*. In other words, we study the activity of a business (or possibly a whole industry) and restrict our analysis to a time period during which background conditions (such as supplies of raw materials, wage rates, and taxes) are fairly constant. We then show how derivatives can help the management of such a firm make vital production decisions.

Management, whether or not it knows calculus, utilizes many functions of the sort we have been considering. Examples of such functions are

$C(x) =$ cost of producing x units of the product,

$R(x) =$ revenue generated by selling x units of the product,

$P(x) = R(x) - C(x) =$ the profit (or loss) generated by producing and selling x units of the product.

Note that the functions $C(x)$, $R(x)$, and $P(x)$ are often defined only for nonnegative integers, that is, for $x = 0, 1, 2, 3, \ldots$. The reason is that it does not make sense to speak about the cost of producing -1 cars or the revenue generated by selling 3.62 refrigerators. Thus, each function may give rise to a set of discrete points on a graph, as in Fig. 1(a). In studying these functions, however, economists usually draw a smooth curve through the points and assume that $C(x)$ is actually defined for all positive x. Of course, we must often interpret answers to problems in light of the fact that x is, in most cases, a nonnegative integer.

Cost Functions

If we assume that a cost function, $C(x)$, has a smooth graph as in Fig. 1(b), we can use the tools of calculus to study it. A typical cost function is analyzed in Example 1.

EXAMPLE 1   **Marginal Cost Analysis** Suppose that the cost function for a manufacturer is given by $C(x) = (10^{-6})x^3 - .003x^2 + 5x + 1000$ dollars.
(a) Describe the behavior of the marginal cost.
(b) Sketch the graph of $C(x)$.

SOLUTION   The first two derivatives of $C(x)$ are given by

$$C'(x) = (3 \cdot 10^{-6})x^2 - .006x + 5.$$
$$C''(x) = (6 \cdot 10^{-6})x - .006$$

Let us sketch the marginal cost $C'(x)$ first. From the behavior of $C'(x)$, we will be able to graph $C(x)$. The marginal cost function $C'(x) = (3 \cdot 10^{-6})x^2 - .006x + 5$ has as its graph a parabola that opens upward. Since $C''(x) = .000006(x - 1000)$, we see that the parabola has a horizontal tangent at $x = 1000$. So the minimum value of $C'(x)$ occurs at $x = 1000$. The corresponding y-coordinate is

$$(3 \cdot 10^{-6})(1000)^2 - .006 \cdot (1000) + 5 = 3 - 6 + 5 = 2.$$

The graph of $C'(x)$ is shown in Fig. 2. Consequently, at first, the marginal cost decreases. It reaches a minimum of 2 dollars at unit production level of 1000 units and increases thereafter. This answers part (a). Let us now graph $C(x)$. Since the graph shown in Fig. 2 is the graph of the derivative of $C(x)$, we see that $C'(x)$ is never zero, so there are no relative extreme points. Since $C'(x)$ is always positive, $C(x)$ is always increasing (as any cost curve should). Moreover, since $C'(x)$ decreases for x less than 1000 and increases for x greater than 1000, we see that $C(x)$ is concave down for x less than 1000, is concave up for x greater than 1000, and has an inflection point at $x = 1000$. The graph of $C(x)$ is drawn in Fig. 3. Note that the inflection point of $C(x)$ occurs at the value of x for which marginal cost is a minimum.

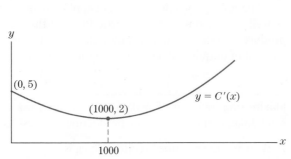

Figure 2 A marginal cost function.

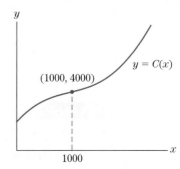

Figure 3 A cost function.

» Now Try Exercise 1

Actually, most marginal cost functions have the same general shape as the marginal cost curve of Example 1. For when x is small, production of additional units is subject to economies of production, which lowers unit costs. Thus, for x small, marginal cost decreases. However, increased production eventually leads to overtime; use of less efficient, older plants; and competition for scarce raw materials. As a result, the cost of additional units will increase for very large x. So we see that $C'(x)$ initially decreases and then increases.

Revenue Functions

In general, a business is concerned not only with its costs, but also with its revenues. Recall that, if $R(x)$ is the revenue received from the sale of x units of some commodity, then the derivative $R'(x)$ is called the **marginal revenue**. Economists use this to measure the rate of increase in revenue per unit increase in sales.

> If x units of a product are sold at a price, p, per unit, the total revenue, $R(x)$, is given by
>
> $$R(x) = x \cdot p.$$

If a firm is small and is in competition with many other companies, its sales have little effect on the market price. Then, since the price is constant as far as the one firm is concerned, the marginal revenue $R'(x)$ equals the price, p [that is, $R'(x)$ is the amount that the firm receives from the sale of one additional unit]. In this case, the revenue function will have a graph as in Fig. 4.

Figure 4 A revenue curve.

An interesting problem arises when a single firm is the only supplier of a certain product or service, that is, when the firm has a monopoly. Consumers will buy large amounts of the commodity if the price per unit is low and less if the price is raised. For each quantity x, let $f(x)$ be the highest price per unit that can be set to sell all x units to customers. Since selling greater quantities requires a lowering of the price, $f(x)$ will be a decreasing function. Figure 5 shows a typical demand curve that relates the quantity demanded, x, to the price, $p = f(x)$.

The **demand equation** $p = f(x)$ determines the total revenue function. If the firm wants to sell x units, the highest price it can set is $f(x)$ dollars per unit, and so the total revenue from the sale of x units is

> $$R(x) = x \cdot p = x \cdot f(x). \tag{1}$$

Figure 5 A demand curve.

The concept of a demand curve applies to an entire industry (with many producers) as well as to a single monopolistic firm. In this case, many producers offer the same product for sale. If x denotes the total output of the industry, $f(x)$ is the market price per unit of output and $x \cdot f(x)$ is the total revenue earned from the sale of the x units.

EXAMPLE 2

Maximizing Revenue The demand equation for a certain product is $p = 6 - \frac{1}{2}x$ thousand dollars. Find the level of production that results in maximum revenue.

SOLUTION

In this case, the revenue function, $R(x)$, is

$$R(x) = x \cdot p = x\left(6 - \frac{1}{2}x\right) = 6x - \frac{1}{2}x^2$$

thousand dollars. The marginal revenue is given by

$$R'(x) = 6 - x.$$

The graph of $R(x)$ is a parabola that opens downward. (See Fig. 6.) It has a horizontal tangent precisely at those x for which $R'(x) = 0$—that is, for those x at which marginal revenue is 0. The only such x is $x = 6$. The corresponding value of revenue is

$$R(6) = 6 \cdot 6 - \frac{1}{2}(6)^2 = \$18,000.$$

Thus, the rate of production resulting in maximum revenue is $x = 6$, which results in total revenue of \$18,000.　　　　　　　　　　　　　　　　　　**» Now Try Exercise 3**

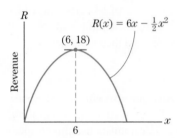

Figure 6 Maximizing revenue.

EXAMPLE 3

Setting Up a Demand Equation The WMA Bus Lines offers sightseeing tours of Washington, D.C. One tour, priced at \$7 per person, had an average demand of about 1000 customers per week. When the price was lowered to \$6, the weekly demand jumped to about 1200 customers. Assuming that the demand equation is linear, find the tour price that should be charged per person to maximize the total revenue each week.

SOLUTION

First, we must find the demand equation. Let x be the number of customers per week and let p be the price of a tour ticket. Then $(x_1, p_1) = (1000, 7)$ and $(x_2, p_2) = (1200, 6)$ are on the demand curve. (See Fig. 7.) Using the point–slope formula for the line through these two points, we have

$$p - 7 = \frac{7 - 6}{1000 - 1200} \cdot (x - 1000)$$

$$= -\frac{1}{200}(x - 1000)$$

$$= -\frac{1}{200}x + 5,$$

so

$$p = 12 - \frac{1}{200}x. \tag{2}$$

Figure 7 A demand curve.

From equation (1), we obtain the revenue function:

$$R(x) = x \cdot p = x\left(12 - \frac{1}{200}x\right) = 12x - \frac{1}{200}x^2.$$

The marginal revenue function is

$$R'(x) = 12 - \frac{1}{100}x = -\frac{1}{100}(x - 1200).$$

Using $R(x)$ and $R'(x)$, we can sketch the graph of $R(x)$. (See Fig. 8.) The maximum revenue occurs when the marginal revenue is zero, that is, when $x = 1200$. The price corresponding to this number of customers is found from demand equation (2):

$$p = 12 - \frac{1}{200}(1200) = 6 \text{ dollars}.$$

Thus, the price of $6 is most likely to bring the greatest revenue per week.

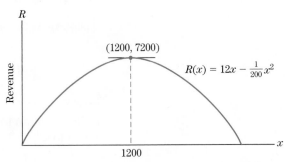

Figure 8 Maximizing revenue.

>> *Now Try Exercise 11*

Profit Functions

Once we know the cost function $C(x)$ and the revenue function $R(x)$, we can compute the profit function $P(x)$ from

$$P(x) = R(x) - C(x).$$

EXAMPLE 4   **Maximizing Profits** Suppose that the demand equation for a monopolist is $p = 100 - .01x$ and the cost function is $C(x) = 50x + 10,000$. (See Fig. 9.) Find the value of x that maximizes the profit, and determine the corresponding price and total profit for this level of production.

Figure 9 A demand curve.

SOLUTION   The total revenue function is

$$R(x) = x \cdot p = x(100 - .01x) = 100x - .01x^2.$$

Hence, the profit function is

$$P(x) = R(x) - C(x)$$
$$= 100x - .01x^2 - (50x + 10,000)$$
$$= -.01x^2 + 50x - 10,000.$$

The graph of this function is a parabola that opens downward. (See Fig. 10.) Its highest point will be where the curve has zero slope, that is, where the marginal profit $P'(x)$ is zero. Now,

$$P'(x) = -.02x + 50 = -.02(x - 2500).$$

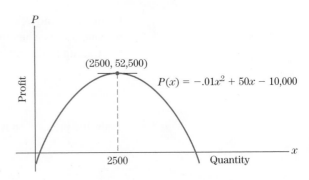

Figure 10 Maximizing profit.

So $P'(x) = 0$ when $x = 2500$. The profit for this level of production is

$$P(2500) = -.01(2500)^2 + 50(2500) - 10,000 = \$52,500.$$

Finally, we return to the demand equation to find the highest price that can be charged per unit to sell all 2500 units:

$$p = 100 - .01(2500) = 100 - 25 = \$75.$$

Thus, to maximize the profit, produce 2500 units and sell them at \$75 per unit. The profit will be \$52,500.   **≫ Now Try Exercise 17**

EXAMPLE 5   Rework Example 4 under the condition that the government has imposed an excise tax of \$10 per unit.

SOLUTION   For each unit sold, the manufacturer will have to pay \$10 to the government. In other words, $10x$ dollars are added to the cost of producing and selling x units. The cost function is now

$$C(x) = (50x + 10,000) + 10x = 60x + 10,000.$$

The demand equation is unchanged by this tax, so the revenue is still

$$R(x) = 100x - .01x^2.$$

Proceeding as before, we have

$$\begin{aligned}
P(x) &= R(x) - C(x) \\
&= 100x - .01x^2 - (60x + 10,000) \\
&= -.01x^2 + 40x - 10,000. \\
P'(x) &= -.02x + 40 = -.02(x - 2000).
\end{aligned}$$

The graph of $P(x)$ is still a parabola that opens downward, and the highest point is where $P'(x) = 0$, that is, where $x = 2000$. (See Fig. 11.) The corresponding profit is

$$P(2000) = -.01(2000)^2 + 40(2000) - 10,000 = \$30,000.$$

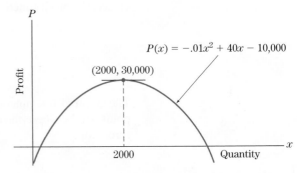

Figure 11 Profit after an excise tax.

From the demand equation, $p = 100 - .01x$, we find the price that corresponds to $x = 2000$:

$$p = 100 - .01(2000) = 80 \text{ dollars.}$$

To maximize profit, produce 2000 units and sell them at $80 per unit. The profit will be $30,000.   **≪**

Notice in Example 5 that the optimal price is raised from $75 to $80. If the monopolist wishes to maximize profits, he or she should pass only half the $10 tax on to the customer. The monopolist cannot avoid the fact that profits will be substantially lowered by the imposition of the tax. This is one reason why industries lobby against taxation.

Setting Production Levels

Suppose that a firm has cost function $C(x)$ and revenue function $R(x)$. In a free-enterprise economy, the firm will set production x in such a way as to maximize the profit function

$$P(x) = R(x) - C(x).$$

We have seen that if $P(x)$ has a maximum at $x = a$, then $P'(a) = 0$. In other words, since $P'(x) = R'(x) - C'(x)$,

$$R'(a) - C'(a) = 0$$
$$R'(a) = C'(a).$$

Thus, profit is maximized at a production level for which marginal revenue equals marginal cost. (See Fig. 12.)

Figure 12

Check Your Understanding 2.7

Solutions can be found following the section exercises.

1. Rework Example 4 by finding the production level at which marginal revenue equals marginal cost.

2. Rework Example 4 under the condition that the fixed cost is increased from $10,000 to $15,000.

3. On a certain route, a regional airline carries 8000 passengers per month, each paying $50. The airline wants to increase the fare. However, the market research department estimates that for each $1 increase in fare, the airline will lose 100 passengers. Determine the price that maximizes the airline's revenue.

EXERCISES 2.7

1. **Minimizing Marginal Cost** Given the cost function $C(x) = x^3 - 6x^2 + 13x + 15$, find the minimum marginal cost.

2. **Minimizing Marginal Cost** If a total cost function is $C(x) = .0001x^3 - .06x^2 + 12x + 100$, is the marginal cost increasing, decreasing, or not changing at $x = 100$? Find the minimum marginal cost.

3. **Maximizing Revenue Cost** The revenue function for a one-product firm is

$$R(x) = 200 - \frac{1600}{x + 8} - x.$$

Find the value of x that results in maximum revenue.

4. **Maximizing Revenue** The revenue function for a particular product is $R(x) = x(4 - .0001x)$. Find the largest possible revenue.

5. **Cost and Profit** A one-product firm estimates that its daily total cost function (in suitable units) is $C(x) = x^3 - 6x^2 + 13x + 15$ and its total revenue function is $R(x) = 28x$. Find the value of x that maximizes the daily profit.

6. **Maximizing Profit** A small tie shop sells ties for $3.50 each. The daily cost function is estimated to be $C(x)$ dollars, where x is the number of ties sold on a typical day and $C(x) = .0006x^3 - .03x^2 + 2x + 20$. Find the value of x that will maximize the store's daily profit.

7. **Demand and Revenue** The demand equation for a certain commodity is

$$p = \frac{1}{12}x^2 - 10x + 300,$$

$0 \le x \le 60$. Find the value of x and the corresponding price p that maximize the revenue.

8. **Maximizing Revenue** The demand equation for a product is $p = 2 - .001x$. Find the value of x and the corresponding price, p, that maximize the revenue.

9. **Profit** Some years ago, it was estimated that the demand for steel approximately satisfied the equation $p = 256 - 50x$, and the total cost of producing x units of steel was $C(x) = 182 + 56x$. (The quantity x was measured in millions of tons and the price and total cost were measured in millions of dollars.) Determine the level of production and the corresponding price that maximize the profits.

10. **Maximizing Area** Consider a rectangle in the xy-plane, with corners at $(0, 0)$, $(a, 0)$, $(0, b)$, and (a, b). If (a, b) lies on the graph of the equation $y = 30 - x$, find a and b such that the area of the rectangle is maximized. What economic interpretations can be given to your answer if the equation $y = 30 - x$ represents a demand curve and y is the price corresponding to the demand x?

11. **Demand, Revenue, and Profit** Until recently hamburgers at the city sports arena cost $4 each. The food concessionaire sold an average of 10,000 hamburgers on a game night. When the price was raised to $4.40, hamburger sales dropped off to an average of 8000 per night.
    (a) Assuming a linear demand curve, find the price of a hamburger that will maximize the nightly hamburger revenue.
    (b) If the concessionaire has fixed costs of $1000 per night and the variable cost is $.60 per hamburger, find the price of a hamburger that will maximize the nightly hamburger profit.

12. **Demand and Revenue** The average ticket price for a concert at the opera house was $50. The average attendance was 4000. When the ticket price was raised to $52, attendance declined to an average of 3800 persons per performance. What should the ticket price be to maximize the revenue for the opera house? (Assume a linear demand curve.)

13. **Demand and Revenue** An artist is planning to sell signed prints of her latest work. If 50 prints are offered for sale, she can charge $400 each. However, if she makes more than 50 prints, she must lower the price of all the prints by $5 for each print in excess of the 50. How many prints should the artist make to maximize her revenue?

14. **Demand and Revenue** A swimming club offers memberships at the rate of $200, provided that a minimum of 100 people join. For each member in excess of 100, the membership fee will be reduced $1 per person (for each member). At most, 160 memberships will be sold. How many memberships should the club try to sell to maximize its revenue?

15. **Profit** In the planning of a sidewalk café, it is estimated that for 12 tables, the daily profit will be $10 per table. Because of overcrowding, for each additional table the profit per table (for every table in the café) will be reduced by $.50. How many tables should be provided to maximize the profit from the café?

16. **Demand and Revenue** A certain toll road averages 36,000 cars per day when charging $1 per car. A survey concludes that increasing the toll will result in 300 fewer cars for each cent of increase. What toll should be charged to maximize the revenue?

17. **Price Setting** The monthly demand equation for an electric utility company is estimated to be

$$p = 60 - (10^{-5})x,$$

where p is measured in dollars and x is measured in thousands of kilowatt-hours. The utility has fixed costs of 7 million dollars per month and variable costs of $30 per 1000 kilowatt-hours of electricity generated, so the cost function is

$$C(x) = 7 \cdot 10^6 + 30x.$$

(a) Find the value of x and the corresponding price for 1000 kilowatt-hours that maximize the utility's profit.
(b) Suppose that rising fuel costs increase the utility's variable costs from $30 to $40, so its new cost function is

$$C_1(x) = 7 \cdot 10^6 + 40x.$$

Should the utility pass all this increase of $10 per thousand kilowatt-hours on to consumers? Explain your answer.

18. **Taxes, Profit, and Revenue** The demand equation for a company is $p = 200 - 3x$, and the cost function is

$$C(x) = 75 + 80x - x^2, \qquad 0 \le x \le 40.$$

(a) Determine the value of x and the corresponding price that maximize the profit.
(b) If the government imposes a tax on the company of $4 per unit quantity produced, determine the new price that maximizes the profit.
(c) The government imposes a tax of T dollars per unit quantity produced (where $0 \le T \le 120$), so the new cost function is

$$C(x) = 75 + (80 + T)x - x^2, \quad 0 \le x \le 40.$$

Determine the new value of x that maximizes the company's profit as a function of T. Assuming that the company cuts back production to this level, express the tax revenues received by the government as a function of T. Finally, determine the value of T that will maximize the tax revenue received by the government.

19. Interest Rate A savings and loan association estimates that the amount of money on deposit will be 1 million times the percentage rate of interest. For instance, a 4% interest rate will generate $4 million in deposits. If the savings and loan association can loan all the money it takes in at 10% interest, what interest rate on deposits generates the greatest profit?

20. Analyzing Profit Let $P(x)$ be the annual profit for a certain product, where x is the amount of money spent on advertising. (See Fig. 13.)

(a) Interpret $P(0)$

(b) Describe how the marginal profit changes as the amount of money spent on advertising increases.

(c) Explain the economic significance of the inflection point.

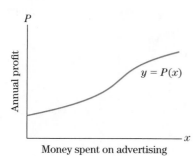

Figure 13 Profit as a function of advertising.

21. Revenue The revenue for a manufacturer is $R(x)$ thousand dollars, where x is the number of units of goods produced (and sold) and R and R' are the functions given in Figs. 14(a) and 14(b).

(a)

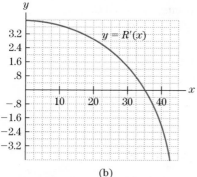

(b)

Figure 14 Revenue function and its first derivative.

(a) What is the revenue from producing 40 units of goods?

(b) What is the marginal revenue when 17.5 units of goods are produced?

(c) At what level of production is the revenue $45,000?

(d) At what level(s) of production is the marginal revenue $800?

(e) At what level of production is the revenue greatest?

22. Cost and Marginal Cost The cost function for a manufacturer is $C(x)$ dollars, where x is the number of units of goods produced and C, C', and C'' are the functions given in Fig. 15.

(a)

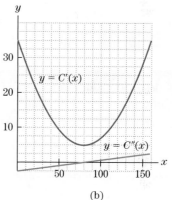

(b)

Figure 15 Cost function and its derivatives.

(a) What is the cost of manufacturing 60 units of goods?

(b) What is the marginal cost when 40 units of goods are manufactured?

(c) At what level of production is the cost $1200?

(d) At what level(s) of production is the marginal cost $22.50?

(e) At what level of production does the marginal cost have the least value? What is the marginal cost at this level of production?

Solutions to Check Your Understanding 2.7

1. The revenue function is $R(x) = 100x - .01x^2$, so the marginal revenue function is $R'(x) = 100 - .02x$. The cost function is $C(x) = 50x + 10,000$, so the marginal cost function is $C'(x) = 50$. Let us now equate the two marginal functions and solve for x:

$$R'(x) = C'(x)$$
$$100 - .02x = 50$$
$$-.02x = -50$$
$$x = \frac{-50}{-.02} = \frac{5000}{2} = 2500.$$

Of course, we obtain the same level of production as before.

2. If the fixed cost is increased from \$10,000 to \$15,000, the new cost function will be $C(x) = 50x + 15,000$, but the marginal cost function will still be $C'(x) = 50$. Therefore, the solution will be the same: 2500 units should be produced and sold at \$75 per unit. (Increases in fixed costs should not necessarily be passed on to the consumer if the objective is to maximize the profit.)

3. Let x denote the number of passengers per month and p the price per ticket. We obtain the number of passengers lost due to a fare increase by multiplying the number of dollars of fare increase, $p - 50$, by the number of passengers lost for each dollar of fare increase. So

$$x = 8000 - (p - 50)100 = -100p + 13,000.$$

Solving for p, we get the demand equation

$$p = -\frac{1}{100}x + 130.$$

From equation (1), the revenue function is

$$R(x) = x \cdot p = x\left(-\frac{1}{100}x + 130\right).$$

The graph is a parabola that opens downward, with x-intercepts at $x = 0$ and $x = 13,000$. (See Fig. 16.) Its maximum is located at the midpoint of the x-intercepts, or $x = 6500$. The price corresponding to this number of passengers is $p = -\frac{1}{100}(6500) + 130 = \$65$. Thus the price of \$65 per ticket will bring the highest revenue to the airline company per month.

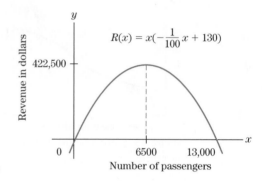

Figure 16

CHAPTER 2 Summary

KEY TERMS AND CONCEPTS	EXAMPLES

2.1 Describing Graphs of Functions

To describe the graph of a function, we can use the following six categories:

1. Intervals in which the function is increasing (respectively decreasing), relative maximum points, relative minimum points
2. Maximum value, minimum value
3. Intervals in which the function is concave up (respectively, concave down), inflection points
4. x-intercepts, y-intercept
5. Undefined points
6. Asymptotes

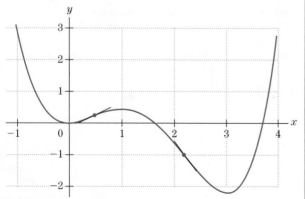

Figure 1

Describe the graph in Fig. 1, using the six categories listed previously.

1. Decreasing for $x < 0$ and $1 < x < 3$.
   Increasing for $0 < x < 1$ and $x > 3$.
   Relative minimum points: $(0, 0)$ and $(3, -2.2)$.
   Relative maximum point: $(1, .4)$.
2. Absolute minimum point: $(3, -2.2)$.
   No absolute maximum.
3. Concave up for $x < .5$ and $x > 2.2$.
   Concave down on $.5 < x < 2.2$.
   Inflection points: $(.5, .2)$ and $(2.2, -1)$.
4. x-intercepts: $(0, 1.6, 3.7)$.
   y-intercept: $(0, 0)$.
5. No undefined points.
6. No asymptotes.

KEY TERMS AND CONCEPTS	**EXAMPLES**

2.2 The First- and Second-Derivative Rules

To decide whether a function is increasing or decreasing at a point $x = c$, we can use the **first-derivative rule:** If $f'(c) > 0$, then $f(x)$ is increasing at $x = c$. If $f'(c) < 0$, then $f(x)$ is decreasing at $x = c$.

To decide whether a function is concave up or down at a point $x = c$, we can use the **second-derivative rule:** If $f''(c) > 0$, then $f(x)$ is concave up at $x = c$. If $f''(c) < 0$, then $f(x)$ is concave down at $x = c$.

The graph of $f'(x) = 1 - x$ is shown in the Fig. 2. We are also told that $f(1) = 2$.

(a) $f(x)$ is increasing where $f'(x) > 0$; that is, $x < 1$.

   $f(x)$ is decreasing where $f'(x) < 0$; that is, $x > 1$.

(b) $f''(x)$ is the derivative of $f'(x)$. So $f''(x) = -1$ for all x. Hence f is concave down for all x.

(c) The equation $f'(1) = 0$ tells us that the graph of $f(x)$ has a horizontal tangent line at the point $(1, 2)$. Since the concavity is negative at this point, we have a maximum at the point $(1, 2)$. We can use this information to plot the graph of $f(x)$ in Fig. 3.

Figure 2

Figure 3

2.3 The First- and Second-Derivative Tests and Curve Sketching

The **first-derivative test** requires studying the sign of f' and may be used to decide whether a critical value is a maximum, minimum, or neither. The **second-derivative test** is based on testing the second derivative at the critical value $x = a$. The test may be inconclusive if $f''(a) = 0$.

An **inflection point** is a point on the graph of f where f is continuous and changes concavity. To find an inflection point, look for values $x = a$ where $f''(a) = 0$ (or $f''(x)$ does not exist) and f'' changes sign at $x = a$.

(a) To find the critical values of $f(x) = x^3 + x^2 + 1$, set $f'(x) = 0$.

$$f'(x) = 3x^2 + 2x,$$
$$f'(x) = 0;\quad 3x^2 + 2x = 0$$
$$x(3x + 2) = 0, \text{ so } x = 0 \text{ or } x = -\frac{2}{3}.$$

(b) To decide whether we have a maximum or a minimum at a critical value, we will use the second-derivative test: $f''(x) = 6x + 2$.

Since $f''(0) = 2 > 0$, the concavity is positive. We have a minimum point at $x = 0$.

Since $f''\left(-\frac{2}{3}\right) = -2 < 0$, the concavity is negative. We have a maximum point at $x = -2/3$.

(c) To find the inflection point, set $f''(x) = 0$: $6x + 2 = 0$, $x = -\frac{1}{3}$.

If $x < -\frac{1}{3}$, $f''(x) < 0$, and if $x > -\frac{1}{3}$, $f''(x) > 0$. Inflection point at $x = -\frac{1}{3}$.

2.4 Curve Sketching (Conclusion)

To sketch the graph of $f(x)$, organize your solution by including steps from the following list.

- Compute $f'(x)$ and $f''(x)$.
- Find the critical values and critical points.
- Find all relative extreme points.
- Determine where f is increasing and where it is decreasing.
- Find all the inflection points of $f(x)$.

Sketch the graph of $y = x + 1 + \frac{1}{x-1}$, $x > 1$.
We have

$$f'(x) = 1 - \frac{1}{(x-1)^2} = 1 - (x-1)^{-2}$$

$$f''(x) = 2(x-1)^{-3} = \frac{2}{(x-1)^3}.$$

| KEY TERMS AND CONCEPTS | EXAMPLES |

KEY TERMS AND CONCEPTS

- Determine the concavity of the graph.
- If $f(x)$ is defined at $x = 0$, find the y-intercept $(0, f(0))$.
- Try to determine the x-intercepts by setting $f(x) = 0$ and solving for x.
- Look for possible asymptotes.
- Complete the sketch.

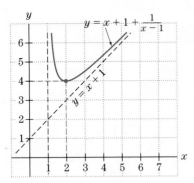

Figure 4

EXAMPLES

Critical values:

$$f'(x) = 0$$

$$1 - \frac{1}{(x-1)^2} = 0$$

$$1 = (x-1)^2$$

$$1 = x^2 - 2x + 1$$

$$x^2 - 2x = 0$$

$$x(x-2) = 0, \quad x = 0 \text{ or } x = 2.$$

We take $x = 2$, since $x > 1$. Since $f''(2) = 2 > 0$, relative minimum point at $x = 2$, $f(2) = 4$. Determine the sign of $f'(x)$ using a table of variation:

Thus, $f(x)$ is decreasing in $1 < x < 2$ and increasing in $x > 2$. The concavity is determined by the sign of $f''(x)$, which is always positive for $x > 1$. Thus, the graph is concave up for $x > 1$. As x gets large, $1/(x-1)$ gets very small and the graph of $f(x)$ is very close to the line $y = x + 1$. Thus, $y = x + 1$ is a slanted asymptote. Also, $x = 1$ is a vertical asymptote. (See Fig. 4.)

2.5 Optimization Problems

Setting up a typical optimization problem with constraint requires determining an **objective equation** and a **constraint equation**. We solve the problem by finding the maximum or minimum of the objective equation, subject to the constraint equation.

Figure 5

Figure 6

Suppose we want to find the maximum volume of an open-topped box with a square base, given 3 square feet of material.

Start by drawing the box and labeling its sides by x, and y (Fig. 5). The volume of the box is

$$V = x^2y. \tag{1}$$

This is the objective equation to be maximized. The surface area of the open box determines the amount of material needed to make it. The surface area is $x^2 + 4xy$. This surface area is not to exceed 3 square feet. This determines the constraint equation

$$x^2 + 4xy = 3. \tag{2}$$

We solve (2) for one of the variables

$$3 - x^2 = 4xy \text{ or } y = \frac{3 - x^2}{4x},$$

then substitute the value into (1) and find the derivative:

$$V = x^2\left(\frac{3 - x^2}{4x}\right) = -\frac{1}{4}x^3 + \frac{3}{4}x$$

$$V' = -\frac{3}{4}x^2 + \frac{3}{4} = \frac{3}{4}(1 - x^2).$$

Thus, $V' = 0$ when $x = \pm 1$. We take $x = 1$. Since $V''(1) < 0$, we have a maximum value at $x = 1$ (see Fig. 6). The maximum volume is $V(1) = -\frac{1}{4} + \frac{3}{4} = \frac{1}{2}$ ft^3.

KEY TERMS AND CONCEPTS	EXAMPLES
2.6 Further Optimization Problems	Suppose a department store manager wants to establish an optimal inventory policy for an item. Sales are expected to be at a steady rate and should total 10,000 items sold during the year. Each time an order is placed, a cost of $200 is incurred. Carrying costs for the year will be $4 per item, to be figured on the average number of items in storage during the year. Derive the objective and constraint equations for this inventory control problem. Let x be the order quantity and n the number of orders placed during the year. Since sales are at a steady rate, the average number of items in storage is $x/2$. The objective equation to be minimized is the total inventory cost:
Some optimization problems require specific skills to set up the objective and constraint equations. The inventory control problem is one such example.	

$$C = \text{ordering cost} + \text{carrying cost}$$

$$= 200n + 4\left(\frac{x}{2}\right) = 200n + 2x.$$

Since there are n orders of x items each, the total number of items ordered during the year is $n \cdot x$. Thus, the constraint equation is $n \cdot x = 10{,}000$.

2.7 Applications of Derivatives to Business and Economics	Suppose that the demand equation for a certain product is $p = 100 - .2x$ (p is the price, and x is the number of items sold at that price), and the cost function is $C(x) = 4x + 3000$. To find the value of x that maximizes the profit, recall that profit is equal to total revenue minus cost,
Derivatives are essential tools in solving certain decision-making problems in economics, especially those that involve maximizing or minimizing quantities. For example, we might be interested in maximizing a profit function or minimizing a cost function. Under certain assumptions on these functions, we can use the derivative to locate their maximum and minimum values.	

$$P(x) = R(x) - C(x),$$

and the total revenue function is the number of items sold times the price:

$$R(x) = x \cdot p = x(100 - .2x) = -.2x^2 + 100x.$$

So

$$P(x) = R(x) - C(x) = -.2x^2 + 100x - (4x + 3000)$$
$$= -.2x^2 + 96x - 3000.$$

The graph of this function is a parabola that opens downward. Its highest point will be where $P'(x)$ is zero:

$$P'(x) = -.4x + 96.$$

So $P'(x) = 0$ when $x = \frac{96}{.4} = 240$. So producing a quantity of 240 will maximize profit.

CHAPTER 2   Fundamental Concept Check Exercises

1. State as many terms used to describe graphs of functions as you can recall.

2. What is the difference between having a relative maximum at $x = 2$ and having an absolute maximum at $x = 2$?

3. Give three characterizations of what it means for the graph of $f(x)$ to be concave up at $x = 2$. Concave down.

4. What does it mean to say that the graph of $f(x)$ has an inflection point at $x = 2$?

5. What is the difference between an x-intercept and a zero of a function?

6. How do you determine the y-intercept of a function?

7. What is an asymptote? Give an example.

8. State the first-derivative rule. The second-derivative rule.

9. Give two connections between the graphs of $f(x)$ and $f'(x)$.

10. Outline a method for locating the relative extreme points of a function.

11. Outline a method for locating the inflection points of a function.

12. Outline a procedure for sketching the graph of a function.

13. What is an objective equation?

14. What is a constraint equation?

15. Outline the procedure for solving an optimization problem.

16. How are the cost, revenue, and profit functions related?

CHAPTER 2   Review Exercises

1. Figure 1 contains the graph of $f'(x)$, the derivative of $f(x)$. Use the graph to answer the following questions about the graph of $f(x)$.
   (a) For what values of x is the graph of $f(x)$ increasing? Decreasing?
   (b) For what values of x is the graph of $f(x)$ concave up? Concave down?

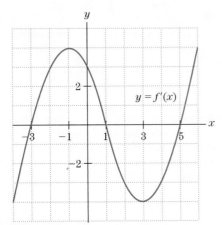

Figure 1

2. Figure 2 shows the graph of the function $f(x)$ and its tangent line at $x = 3$. Find $f(3)$, $f'(3)$, and $f''(3)$.

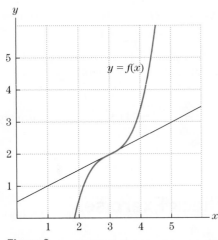

Figure 2

In Exercises 3–6, draw the graph of a function $f(x)$ for which the function and its first derivative have the stated property for all x.

3. $f(x)$ and $f'(x)$ increasing

4. $f(x)$ and $f'(x)$ decreasing

5. $f(x)$ increasing and $f'(x)$ decreasing

6. $f(x)$ decreasing and $f'(x)$ increasing

Exercises 7–12 refer to the graph in Fig. 3. List the labeled values of x at which the derivative has the stated property.

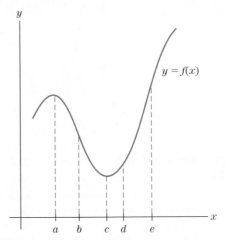

Figure 3

7. $f'(x)$ is positive.

8. $f'(x)$ is negative.

9. $f''(x)$ is positive.

10. $f''(x)$ is negative.

11. $f'(x)$ is maximized.

12. $f'(x)$ is minimized.

Properties of various functions are described next. In each case, draw some conclusion about the graph of the function.

13. $f(1) = 2, f'(1) > 0$

14. $g(1) = 5, g'(1) = -1$

15. $h(3) = 4, h''(3) = 1$

16. $F'(2) = -1, F''(2) < 0$

17. $G(10) = 2, G'(10) = 0, G''(10) > 0$

18. $f(4) = -2, f'(4) > 0, f''(4) = -1$

19. $g(5) = -1, g'(5) = -2, g''(5) = 0$

20. $H(0) = 0, H'(0) = 0, H''(0) = 1$

21. In Figs. 4(a) and 4(b), the t-axis represents time in hours.
    (a) When is $f(t) = 1$?
    (b) Find $f(5)$.
    (c) When is $f(t)$ changing at the rate of $-.08$ unit per hour?
    (d) How fast is $f(t)$ changing after 8 hours?

22. **U.S. Electric Energy** United States electrical energy production (in trillions of kilowatt-hours) in year t (with 1900 corresponding to $t = 0$) is given by $f(t)$, where f and its derivatives are graphed in Figs. 5(a) and 5(b).
    (a) How much electrical energy was produced in 1950?
    (b) How fast was energy production rising in 1950?
    (c) When did energy production reach 3000 trillion kilowatt-hours?
    (d) When was the level of energy production rising at the rate of 10 trillion kilowatt-hours per year?
    (e) When was energy production growing at the greatest rate? What was the level of production at that time?

(a)

(b)

Figure 4

(a)

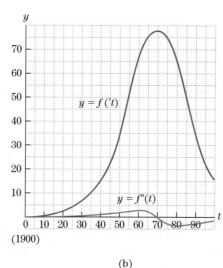

(b)

Figure 5 U.S. electrical energy production.

Sketch the following parabolas. Include their x- and y-intercepts.

23. $y = 3 - x^2$

24. $y = 7 + 6x - x^2$

25. $y = x^2 + 3x - 10$

26. $y = 4 + 3x - x^2$

27. $y = -2x^2 + 10x - 10$

28. $y = x^2 - 9x + 19$

29. $y = x^2 + 3x + 2$

30. $y = -x^2 + 8x - 13$

31. $y = -x^2 + 20x - 90$

32. $y = 2x^2 + x - 1$

Sketch the following curves.

33. $y = 2x^3 + 3x^2 + 1$

34. $y = x^3 - \frac{3}{2}x^2 - 6x$

35. $y = x^3 - 3x^2 + 3x - 2$

36. $y = 100 + 36x - 6x^2 - x^3$

37. $y = \frac{11}{3} + 3x - x^2 - \frac{1}{3}x^3$

38. $y = x^3 - 3x^2 - 9x + 7$

39. $y = -\frac{1}{3}x^3 - 2x^2 - 5x$

40. $y = x^3 - 6x^2 - 15x + 50$

41. $y = x^4 - 2x^2$

42. $y = x^4 - 4x^3$

43. $y = \dfrac{x}{5} + \dfrac{20}{x} + 3 \ (x > 0)$

44. $y = \dfrac{1}{2x} + 2x + 1 \ (x > 0)$

45. Let $f(x) = (x^2 + 2)^{3/2}$. Show that the graph of $f(x)$ has a possible relative extreme point at $x = 0$.

46. Show that the function $f(x) = (2x^2 + 3)^{3/2}$ is decreasing for $x < 0$ and increasing for $x > 0$.

47. Let $f(x)$ be a function whose *derivative* is

$$f'(x) = \frac{1}{1 + x^2}.$$

Note that $f'(x)$ is always positive. Show that the graph of $f(x)$ has an inflection point at $x = 0$.

48. Let $f(x)$ be a function whose *derivative* is

$$f'(x) = \sqrt{5x^2 + 1}.$$

Show that the graph of $f(x)$ has an inflection point at $x = 0$.

49. Position Velocity and Acceleration A car is traveling on a straight road and $s(t)$ is the distance traveled after t hours. Match each set of information about $s(t)$ and its derivatives with the corresponding description of the car's motion.

Information

   A. $s(t)$ is a constant function.

   B. $s'(t)$ is a positive constant function.

   C. $s'(t)$ is positive at $t = a$.

   D. $s'(t)$ is negative at $t = a$.

   E. $s'(t)$ and $s''(t)$ are positive at $t = a$.

   F. $s'(t)$ is positive and $s''(t)$ is negative at $t = a$.

Descriptions

   a. The car is moving forward and speeding up at time a.

   b. The car is backing up at time a.

c. The car is standing still.
d. The car is moving forward, but slowing down, at time a.
e. The car is moving forward at a steady rate.
f. The car is moving forward at time a.

50. The water level in a reservoir varies during the year. Let $h(t)$ be the depth (in feet) of the water at time t days, where $t = 0$ at the beginning of the year. Match each set of information about $h(t)$ and its derivatives with the corresponding description of the reservoir's activity.

Information
A. $h(t)$ has the value 50 for $1 \le t \le 2$.
B. $h'(t)$ has the value .5 for $1 \le t \le 2$.
C. $h'(t)$ is positive at $t = a$.
D. $h'(t)$ is negative at $t = a$.
E. $h'(t)$ and $h''(t)$ are positive at $t = a$.
F. $h'(t)$ is positive and $h''(t)$ is negative at $t = a$.

Descriptions
a. The water level is rising at an increasing rate at time a.
b. The water level is receding at time a.
c. The water level is constant at 50 feet on January 2.
d. At time a the water level is rising, but the rate of increase is slowing down.
e. On January 2 the water rose steadily at a rate of .5 foot per day.
f. The water level is rising at time a.

51. **Population near New York City** Let $f(x)$ be the number of people living within x miles of the center of New York City.
(a) What does $f(10 + h) - f(10)$ represent?
(b) Explain why $f'(10)$ cannot be negative.

52. For what x does the function $f(x) = \frac{1}{4}x^2 - x + 2, 0 \le x \le 8$, have its maximum value?

53. Find the maximum value of the function $f(x) = 2 - 6x - x^2$, $0 \le x \le 5$, and give the value of x where this maximum occurs.

54. Find the minimum value of the function $g(t) = t^2 - 6t + 9$, $1 \le t \le 6$.

55. **Surface Area** An open rectangular box is to be 4 feet long and have a volume of 200 cubic feet. Find the dimensions for which the amount of material needed to construct the box is as small as possible.

56. **Volume** A closed rectangular box with a square base is to be constructed using two different types of wood. The top is made of wood costing $3 per square foot and the remainder is made of wood costing $1 per square foot. If $48 is available to spend, find the dimensions of the box of greatest volume that can be constructed.

57. **Volume** A long rectangular sheet of metal 30 inches wide is to be made into a gutter by turning up strips vertically along the two sides (Fig. 6). How many inches should be turned up on each side to maximize the amount of water that the gutter can carry?

30 in

Fold along this line

Fold along this line

l

x $30 - 2x$ x

Figure 6

58. **Maximizing the Total Yield** A small orchard yields 25 bushels of fruit per tree when planted with 40 trees. Because of overcrowding, the yield per tree (for each tree in the orchard) is reduced by $\frac{1}{2}$ bushel for each additional tree that is planted. How many trees should be planted to maximize the total yield of the orchard?

59. **Inventory Control** A publishing company sells 400,000 copies of a certain book each year. Ordering the entire amount printed at the beginning of the year ties up valuable storage space and capital. However, printing the copies in several partial runs throughout the year results in added costs for setting up each printing run, which costs $1000. The carrying costs, figured on the average number of books in storage, are $.50 per book. Find the most economical lot size, that is, the production run size that minimizes the total setting up and carrying costs.

60. **Profit** If the demand equation for a monopolist is $p = 150 - .02x$ and the cost function is $C(x) = 10x + 300$, find the value of x that maximizes the profit.

61. **Minimizing Time** Jane wants to drive her tractor from point A on one side of her 5-mile-wide field to a point, B, on the opposite side of the field, as shown in Fig. 7. Jane could drive her tractor directly across the field to point C and then drive 15 miles down a road to B, or she could drive to some point, P, between B and C and then drive to B. If Jane drives her tractor 8 miles per hour across the field and 17 miles per hour on the road, find the point, P, where she should cut across the field to minimize the time it takes to reach B. [*Hint:* time $= \frac{\text{distance}}{\text{rate}}$.]

Figure 7

62. **Maximizing Revenue** A travel agency offers a boat tour of several Caribbean islands for 3 days and 2 nights. For a group of 12 people, the cost per person is $800. For each additional person above the 12-person minimum, the cost per person is reduced by $20 for each person in the group. The maximum tour group size is 25. What tour group size produces the greatest revenue for the travel agency?

Techniques of Differentiation

3.1   The Product and Quotient Rules

3.3   Implicit Differentiation and Related Rates

3.2   The Chain Rule and the General Power Rule

We have seen that the derivative is useful in many applications. However, our ability to differentiate functions is somewhat limited. For example, we cannot yet readily differentiate the functions

$$f(x) = (x^2 - 1)^4 (x^2 + 1)^5, \qquad g(x) = \frac{x^3}{(x^2 + 1)^4}.$$

In this chapter, we develop differentiation techniques that apply to functions like those just given. Two new rules are the *product rule* and the *quotient rule*. In Section 3.2, we extend the general power rule into a powerful formula called the *chain rule*.

3.1   The Product and Quotient Rules

We observed in our discussion of the sum rule for derivatives that the derivative of the sum of two differentiable functions is the sum of the derivatives. Unfortunately, however, the derivative of the product $f(x)g(x)$ is *not* the product of the derivatives. Rather, the derivative of a product is determined from the following rule:

FOR REVIEW

The notation $\dfrac{d}{dx}$ means take the derivative with respect to x. See Sec. 1.3, p. 77.

Product Rule

$$\frac{d}{dx}[f(x)g(x)] = f(x)g'(x) + g(x)f'(x)$$

For simplicity, we also write: $(fg)' = fg' + gf'$.

The derivative of the product of two functions is the first function times the derivative of the second plus the second function times the derivative of the first. At the end of this section, we show why this statement is true.

EXAMPLE 1   **Verifying the Product Rule** Show that the product rule works for the case $f(x) = x^2$, $g(x) = x^3$.

SOLUTION   Since $x^2 \cdot x^3 = x^5$, then

$$\frac{d}{dx}(x^2 \cdot x^3) = \frac{d}{dx}(x^5) = 5x^4. \quad \text{Power rule}$$

We also have

$$\frac{d}{dx}(x^2 \cdot x^3) = x^2 \cdot \frac{d}{dx}(x^3) + x^3 \cdot \frac{d}{dx}(x^2) \quad \text{Product rule with } f(x) = x^2, g(x) = x^3$$

$$= x^2(3x^2) + x^3(2x)$$

$$= 3x^4 + 2x^4 = 5x^4.$$

Thus, the product rule gives the correct answer.   «

EXAMPLE 2   **Finding the Derivative of a Product** Differentiate $y = (2x^3 - 5x)(3x + 1)$.

SOLUTION   Let $f(x) = 2x^3 - 5x$ and $g(x) = 3x + 1$. Then,

$$\frac{d}{dx}[(2x^3 - 5x)(3x + 1)] = (2x^3 - 5x) \cdot \frac{d}{dx}(3x + 1) + (3x + 1) \cdot \frac{d}{dx}(2x^3 - 5x)$$

$$= (2x^3 - 5x)(3) + (3x + 1)(6x^2 - 5)$$

$$= 6x^3 - 15x + 18x^3 - 15x + 6x^2 - 5$$

$$= 24x^3 + 6x^2 - 30x - 5. \qquad \text{» Now Try Exercise 3}$$

EXAMPLE 3   **The Product Rule** Apply the product rule to $y = g(x) \cdot g(x)$.

SOLUTION

$$\frac{d}{dx}[g(x) \cdot g(x)] = g(x) \cdot g'(x) + g(x) \cdot g'(x)$$

$$= 2g(x)g'(x)$$

This answer is the same as that given by the general power rule:

$$\frac{d}{dx}[g(x) \cdot g(x)] = \frac{d}{dx}[g(x)]^2 = 2g(x)g'(x). \qquad «$$

EXAMPLE 4   **Differentiating and Simplifying** Find

$$\frac{dy}{dx}, \quad \text{where } y = (x^2 - 1)^4(x^2 + 1)^5.$$

SOLUTION   Let $f(x) = (x^2 - 1)^4$, $g(x) = (x^2 + 1)^5$, and use the product rule. The general power rule is needed to compute $f'(x)$ and $g'(x)$:

$$\frac{dy}{dx} = (x^2 - 1)^4 \cdot \frac{d}{dx}(x^2 + 1)^5 + (x^2 + 1)^5 \cdot \frac{d}{dx}(x^2 - 1)^4$$

$$= (x^2 - 1)^4 \cdot 5(x^2 + 1)^4(2x) + (x^2 + 1)^5 \cdot 4(x^2 - 1)^3(2x). \qquad (1)$$

FOR REVIEW

The notation $\dfrac{dy}{dx}\Big|_{x=2}$

means take the derivative with respect to x, first, then evaluate the derivative at $x = 2$.

This form of $\dfrac{dy}{dx}$ is suitable for some purposes. For example, if we need to compute $\dfrac{dy}{dx}\Big|_{x=2}$, it is easier just to substitute 2 for x than to simplify and then substitute. However, it is often helpful to simplify the formula for $\dfrac{dy}{dx}$, such as in the case when we need to find x so that $\dfrac{dy}{dx} = 0$.

To simplify the answer in equation (1), we shall write $\dfrac{dy}{dx}$ as a single product, rather than as a sum of two products. The first step is to identify the common factors:

$$\frac{dy}{dx} = (x^2 - 1)^4 \cdot 5(x^2 + 1)^4(2x) + (x^2 + 1)^5 \cdot 4(x^2 - 1)^3(2x).$$

Both terms contain $2x$ and powers of $x^2 - 1$ and $x^2 + 1$. The most we can factor out of each term is $2x(x^2 - 1)^3(x^2 + 1)^4$. We obtain

$$\frac{dy}{dx} = 2x(x^2 - 1)^3(x^2 + 1)^4[5(x^2 - 1) + 4(x^2 + 1)].$$

Simplifying the right-most factor in this product, we have

$$\frac{dy}{dx} = 2x(x^2 - 1)^3(x^2 + 1)^4[9x^2 - 1]. \tag{2}$$

(See Fig. 3 and Exercise 34 for a related application.)   **≫ Now Try Exercise 7**

The answers to the exercises for this section appear in two forms, similar to those in (1) and (2). The unsimplified answers will allow you to check if you have mastered the differentiation rules. In each case, you should make an effort to transform your original answer into the simplified version. Examples 4 and 6 show how to do this.

The Quotient Rule

Another useful formula for differentiating functions is the quotient rule.

Quotient Rule

$$\frac{d}{dx}\left[\frac{f(x)}{g(x)}\right] = \frac{g(x)f'(x) - f(x)g'(x)}{[g(x)]^2}$$

We must be careful to remember the order of the terms in this formula because of the minus sign in the numerator.

EXAMPLE 5   **Using the Quotient Rule** Differentiate

$$y = \frac{x}{2x + 3}.$$

SOLUTION   Let $f(x) = x$ and $g(x) = 2x + 3$.

$$\frac{d}{dx}\left(\frac{x}{2x + 3}\right) = \frac{(2x + 3) \cdot \frac{d}{dx}(x) - (x) \cdot \frac{d}{dx}(2x + 3)}{(2x + 3)^2}$$

$$= \frac{(2x + 3) \cdot 1 - x \cdot 2}{(2x + 3)^2} = \frac{3}{(2x + 3)^2}$$

≫ Now Try Exercise 11

EXAMPLE 6   **Simplifying after the Quotient Rule** Find

$$\frac{dy}{dx}, \quad \text{where } y = \frac{x^3}{(x^2 + 1)^4}.$$

SOLUTION Let $f(x) = x^3$ and $g(x) = (x^2 + 1)^4$.

$$\frac{d}{dx}\left(\frac{x^3}{(x^2+1)^4}\right) = \frac{(x^2+1)^4 \cdot \frac{d}{dx}(x^3) - (x^3) \cdot \frac{d}{dx}(x^2+1)^4}{[(x^2+1)^4]^2}$$

$$= \frac{(x^2+1)^4 \cdot 3x^2 - x^3 \cdot 4(x^2+1)^3(2x)}{(x^2+1)^8}$$

If a simplified form of $\dfrac{dy}{dx}$ is desired, we can divide the numerator and the denominator by the common factor $(x^2 + 1)^3$:

$$\frac{dy}{dx} = \frac{(x^2+1) \cdot 3x^2 - x^3 \cdot 4(2x)}{(x^2+1)^5}$$

$$= \frac{3x^4 + 3x^2 - 8x^4}{(x^2+1)^5}$$

$$= \frac{3x^2 - 5x^4}{(x^2+1)^5} = \frac{x^2(3 - 5x^2)}{(x^2+1)^5}. \qquad \text{» Now Try Exercise 15}$$

CAUTION ▶ The derivative in Example 6 is a rational function. To simplify it, you can cancel common factors of the entire numerator and denominator, but terms in fractions should never be canceled. Here is another illustration of these rules. «

EXAMPLE 7 **Simplifying after the Quotient Rule** Differentiate

$$y = \frac{x}{x + (x+1)^3}.$$

SOLUTION Let $f(x) = x$ and $g(x) = x + (x+1)^3$. Then, $f'(x) = 1$, $g'(x) = 1 + 3(x+1)^2$, and so,

$$\frac{d}{dx}\left(\frac{x}{x + (x+1)^3}\right) = \frac{(x + (x+1)^3) - x \cdot (1 + 3(x+1)^2)}{(x + (x+1)^3)^2}.$$

Although there are several common terms in the denominator and numerator, none of them can be canceled because none of them are common factors of the entire numerator and denominator. You can continue simplifying as follows:

$$\frac{dy}{dx} = \frac{\cancel{x} + (x+1)^3 - \cancel{x} - 3x(x+1)^2}{(x + (x+1)^3)^2}$$

$$= \frac{(x+1)^3 - 3x(x+1)^2}{(x + (x+1)^3)^2}$$

$$= \frac{(x+1)^2[(x+1) - 3x]}{(x + (x+1)^3)^2} = \frac{(x+1)^2(-2x + 1)}{(x + (x+1)^3)^2}.$$

» Now Try Exercise 19

In the following example, we use several rules to compute the derivative.

EXAMPLE 8 **Using Multiple Differentiation Rules** Differentiate

$$f(x) = \sqrt{\frac{x^2 + 7}{x + 1}}.$$

SOLUTION   Start by writing f using powers, and apply the general power rule to compute $f'(x)$:

$$f(x) = \left(\frac{x^2 + 7}{x + 1}\right)^{1/2}$$

$$f'(x) = \frac{1}{2}\left(\frac{x^2 + 7}{x + 1}\right)^{\frac{1}{2}-1} \frac{d}{dx}\left[\frac{x^2 + 7}{x + 1}\right] = \frac{1}{2}\left(\frac{x^2 + 7}{x + 1}\right)^{-\frac{1}{2}} \frac{d}{dx}\left[\frac{x^2 + 7}{x + 1}\right].$$

Now use the quotient rule to compute the last derivative:

$$\frac{d}{dx}\left[\frac{x^2 + 7}{x + 1}\right] = \frac{(x + 1)\frac{d}{dx}(x^2 + 7) - (x^2 + 7)\frac{d}{dx}(x + 1)}{(x + 1)^2}$$

$$= \frac{(x + 1)(2x) - (x^2 + 7)(1)}{(x + 1)^2}$$

$$= \frac{x^2 + 2x - 7}{(x + 1)^2}.$$

Putting this into the expression for $f'(x)$, we find

$$f'(x) = \frac{1}{2}\left(\frac{x^2 + 7}{x + 1}\right)^{-\frac{1}{2}} \frac{(x^2 + 2x - 7)}{(x + 1)^2}.$$

To simplify the expression further, we use the fact that

$$\left(\frac{a}{b}\right)^{-c} = \frac{a^{-c}}{b^{-c}} = \frac{b^c}{a^c} = \left(\frac{b}{a}\right)^c,$$

and so:

$$\left(\frac{x^2 + 7}{x + 1}\right)^{-\frac{1}{2}} = \left(\frac{x + 1}{x^2 + 7}\right)^{\frac{1}{2}} = \frac{(x + 1)^{\frac{1}{2}}}{(x^2 + 7)^{\frac{1}{2}}}.$$

Hence:

$$f'(x) = \frac{1}{2}\frac{(x + 1)^{\frac{1}{2}}}{(x^2 + 7)^{\frac{1}{2}}}\frac{(x^2 + 2x - 7)}{(x + 1)^2}$$

$$= \frac{1}{2}\frac{x^2 + 2x - 7}{(x^2 + 7)^{\frac{1}{2}}(x + 1)^{\frac{3}{2}}}. \qquad \text{Cancel powers of } (x + 1): \frac{(x + 1)^{\frac{1}{2}}}{(x + 1)^2} = \frac{1}{(x + 1)^{\frac{3}{2}}}$$

» Now Try Exercise 25

EXAMPLE 9   **Minimizing the Average Cost** Suppose that the total cost of manufacturing x units of a certain product is given by the function $C(x)$. Then, the **average cost per unit**, AC, is defined by

$$AC = \frac{C(x)}{x}.$$

Recall that the *marginal cost*, MC, is defined by

$$MC = C'(x).$$

Show that, at the level of production where the average cost is at a minimum, the average cost equals the marginal cost.

SOLUTION   In practice, the marginal cost and average cost curves will have the general shapes shown in Fig. 1. So the minimum point on the average cost curve will occur when $\frac{d}{dx}(AC) = 0$. To compute the derivative, we need the quotient rule,

$$\frac{d}{dx}(AC) = \frac{d}{dx}\left(\frac{C(x)}{x}\right) = \frac{x \cdot C'(x) - C(x)}{x^2}.$$

Figure 1 Marginal cost and average cost functions.

Setting the derivative equal to zero and multiplying by x^2, we obtain

$$0 = x \cdot C'(x) - C(x)$$
$$C(x) = x \cdot C'(x)$$
$$\frac{C(x)}{x} = C'(x)$$
$$AC = MC.$$

Thus, when the output x is chosen so that the average cost is minimized, the average cost equals the marginal cost.   ≪

Appendix: Verification of the Product and Quotient Rules

Verification of the Product Rule   From our discussion of limits, we compute the derivative of $f(x)g(x)$ at $x = a$ as the limit

$$\frac{d}{dx}[f(x)g(x)]\bigg|_{x=a} = \lim_{h \to 0} \frac{f(a+h)g(a+h) - f(a)g(a)}{h}.$$

Let us add and subtract the quantity $f(a)g(a + h)$ in the numerator. After factoring and applying Limit Theorem III, Section 1.4, we obtain

$$\lim_{h \to 0} \frac{[f(a+h)g(a+h) - f(a)g(a+h)] + [f(a)g(a+h) - f(a)g(a)]}{h}$$
$$= \lim_{h \to 0} g(a+h) \cdot \frac{f(a+h) - f(a)}{h} + \lim_{h \to 0} f(a) \cdot \frac{g(a+h) - g(a)}{h}.$$

This expression may be rewritten by Limit Theorem V, Section 1.4, as

$$\lim_{h \to 0} g(a+h) \cdot \lim_{h \to 0} \frac{f(a+h) - f(a)}{h} + \lim_{h \to 0} f(a) \cdot \lim_{h \to 0} \frac{g(a+h) - g(a)}{h}.$$

Note, however, that since $g(x)$ is differentiable at $x = a$, it is continuous there, so $\lim_{h \to 0} g(a + h) = g(a)$. Therefore, the preceding expression equals

$$g(a)f'(a) + f(a)g'(a).$$

That is, we have proved that

$$\frac{d}{dx}[f(x)g(x)]\bigg|_{x=a} = g(a)f'(a) + f(a)g'(a),$$

which is the product rule. An alternative verification of the product rule, not involving limit arguments, is outlined in Exercise 66.

Verification of the Quotient Rule   From the general power rule, we know that

$$\frac{d}{dx}\left[\frac{1}{g(x)}\right] = \frac{d}{dx}[g(x)]^{-1} = (-1)[g(x)]^{-2} \cdot g'(x).$$

We can now derive the quotient rule from the product rule:

$$\frac{d}{dx}\left[\frac{f(x)}{g(x)}\right] = \frac{d}{dx}\left[\frac{1}{g(x)} \cdot f(x)\right]$$
$$= \frac{1}{g(x)} \cdot f'(x) + f(x) \cdot \frac{d}{dx}\left[\frac{1}{g(x)}\right]$$
$$= \frac{g(x)f'(x)}{[g(x)]^2} + f(x) \cdot (-1)[g(x)]^{-2} \cdot g'(x)$$
$$= \frac{g(x)f'(x) - f(x)g'(x)}{[g(x)]^2}.$$

INCORPORATING TECHNOLOGY

A quotient of continuous functions typically has a vertical asymptote at each zero of the denominator. Graphing calculators do not do well when displaying graphs with vertical asymptotes. For example, the graph of

$$y = \frac{1}{x - 1}$$

has a vertical asymptote at $x = 1$, which is shown on the calculator screen as a spike. The problem is that the standard graphing mode on most calculators (**Connected** mode on TI calculators) connects consecutive dots along the graph. There is no point over $x = 1$ since the function is not defined there. When the calculator connects the dots on either side of the asymptote, a spike results. To display a graph more accurately with an asymptote, most calculators have a **Dot** mode in which consecutive dots are displayed but are not connected with line segments. On the TI-84, the **Dot** mode is selected from the **Mode** screen. The **Dot** mode setting is retained until you change it.

Check Your Understanding 3.1

Solutions can be found following the section exercises.

1. Consider the function $y = (\sqrt{x} + 1)x$.
   (a) Differentiate y by the product rule.
   (b) First, multiply out the expression for y then differentiate.

2. Differentiate $y = \dfrac{5}{x^4 - x^3 + 1}$.

EXERCISES 3.1

Differentiate the functions in Exercises 1–28.

1. $y = (x + 1)(x^3 + 5x + 2)$

2. $y = (-x^3 + 2)\left(\dfrac{x}{2} - 1\right)$

3. $y = (2x^4 - x + 1)(-x^5 + 1)$

4. $y = (x^2 + x + 1)^3(x - 1)^4$

5. $y = x(x^2 + 1)^4$

6. $y = x\sqrt{x}$

7. $y = (x^2 + 3)(x^2 - 3)^{10}$

8. $y = [(-2x^3 + x)(6x - 3)]^4$

9. $y = (5x + 1)(x^2 - 1) + \dfrac{2x + 1}{3}$

10. $y = x^7(3x^4 + 12x - 1)^2$

11. $y = \dfrac{x - 1}{x + 1}$

12. $y = \dfrac{1}{x^2 + x + 7}$

13. $y = \dfrac{x^2 - 1}{x^2 + 1}$

14. $y = \dfrac{x}{x + \frac{1}{x}}$

15. $y = \dfrac{x + 3}{(2x + 1)^2}$

16. $y = x^4 + 4\sqrt[4]{x}$

17. $y = \dfrac{1}{\pi} + \dfrac{2}{x^2 + 1}$

18. $y = \dfrac{ax + b}{cx + d}$

19. $y = \dfrac{x^2}{(x^2 + 1)^2}$

20. $y = \dfrac{(x + 1)^3}{(x - 5)^2}$

21. $y = [(3x^2 + 2x + 2)(x - 2)]^2$

22. $y = \dfrac{1}{\sqrt{x} - 2}$

23. $y = \dfrac{1}{\sqrt{x} + 1}$

24. $y = \dfrac{3}{\sqrt[3]{x} + 1}$

25. $y = \left(\dfrac{x + 11}{x - 3}\right)^3$

26. $y = \sqrt{\dfrac{x + 3}{x^2 + 1}}$

27. $y = \sqrt{x + 2}(2x + 1)^2$

28. $y = \dfrac{\sqrt{3x - 1}}{x}$

29. Find the equation of the tangent line to the curve $y = (x - 2)^5(x + 1)^2$ at the point $(3, 16)$.

30. Find the equation of the tangent line to the curve $y = (x + 1)/(x - 1)$ at the point $(2, 3)$.

31. Find all x-coordinates of points (x, y) on the curve $y = (x - 2)^5/(x - 4)^3$ where the tangent line is horizontal.

32. Find the inflection points on the graph of $y = \dfrac{1}{x^2 + 1}$. (See Fig. 2.)

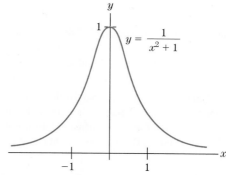

$y = \dfrac{1}{x^2 + 1}$

Figure 2

33. Find all x such that $\dfrac{dy}{dx} = 0$, where

$$y = (x^2 - 4)^3(2x^2 + 5)^5.$$

34. The graph of $y = (x^2 - 1)^4(x^2 + 1)^5$ is shown in Fig. 3. Find the coordinates of the local maxima and minima. [*Hint:* Use Example 4.]

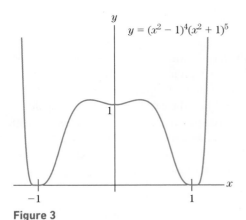

$$y = (x^2 - 1)^4(x^2 + 1)^5$$

Figure 3

35. Find the point(s) on the graph of $y = (x^2 + 3x - 1)/x$ where the slope is 5.

36. Find the point(s) on the graph of $y = (2x^4 + 1)(x - 5)$ where the slope is 1.

Find $\frac{d^2y}{dx^2}$.

37. $y = (x^2 + 1)^4$

38. $y = \sqrt{x^2 + 1}$

39. $y = x\sqrt{x + 1}$

40. $y = \dfrac{2}{2 + x^2}$

In Exercises 41–44, a function $h(x)$ is defined in terms of a differentiable $f(x)$. Find an expression for $h'(x)$.

41. $h(x) = xf(x)$

42. $h(x) = (x^2 + 2x - 1)f(x)$

43. $h(x) = \dfrac{f(x)}{x^2 + 1}$

44. $h(x) = \left(\dfrac{f(x)}{x}\right)^2$

45. Volume An open rectangular box is 3 feet long and has a surface area of 16 square feet. Find the dimensions of the box for which the volume is as large as possible.

46. Volume A closed rectangular box is to be constructed with one side 1 meter long. The material for the top costs $20 per square meter, and the material for the sides and bottom costs $10 per square meter. Find the dimensions of the box with the largest possible volume that can be built at a cost of $240 for materials.

47. Average Cost A sugar refinery can produce x tons of sugar per week at a weekly cost of $.1x^2 + 5x + 2250$ dollars. Find the level of production for which the average cost is at a minimum and show that the average cost equals the marginal cost at that level of production.

48. Average Cost A cigar manufacturer produces x cases of cigars per day at a daily cost of $50x(x + 200)/(x + 100)$ dollars. Show that his cost increases and his average cost decreases as the output x increases.

49. Average Revenue Let $R(x)$ be the revenue received from the sale of x units of a product. The *average revenue per unit* is defined by $AR = R(x)/x$. Show that at the level of production where the average revenue is maximized, the average revenue equals the marginal revenue.

50. Average Velocity Let $s(t)$ be the number of miles a car travels in t hours. Then, the average velocity during the first t hours is $\bar{v}(t) = s(t)/t$ miles per hour. If the average velocity is maximized at time t_0, show that at this time the average velocity $\bar{v}(t_0)$ equals the instantaneous velocity $s'(t_0)$. [*Hint:* Compute the derivative of $\bar{v}(t)$.]

51. Rate of Change The width of a rectangle is increasing at a rate of 3 inches per second and its length is increasing at the rate of 4 inches per second. At what rate is the area of the rectangle increasing when its width is 5 inches and its length is 6 inches? [*Hint:* Let $W(t)$ and $L(t)$ be the widths and lengths, respectively, at time t.]

52. Cost–Benefit of Emission Control A manufacturer plans to decrease the amount of sulfur dioxide escaping from its smokestacks. The estimated cost–benefit function is

$$f(x) = \dfrac{3x}{105 - x}, \qquad 0 \le x \le 100,$$

where $f(x)$ is the cost in millions of dollars for eliminating $x\%$ of the total sulfur dioxide. (See Fig. 4.) Find the value of x at which the rate of increase of the cost–benefit function is $1.4 million per unit. (Each unit is a 1 percentage point increase in pollutant removed.)

Figure 4 A cost-benefit function.

In Exercises 53 and 54, use the fact that at the beginning of 1998, the population of the United States was 268,924,000 people and growing at the rate of 1,856,000 people per year.

53. Gas Consumption. At the beginning of 1998, the annual per capita consumption of gasoline in the United States was 52.3 gallons and growing at the rate of .2 gallons per year. At what rate was the total annual consumption of gasoline in the United States increasing at that time? (*Hint:* [total annual consumption] = [population] · [annual per capita consumption].)

54. Consumption of Ice Cream in the U.S. At the beginning of 1998, the annual consumption of ice cream in the United States was 12,582,000 pints and growing at the rate of 212 million pints per year. At what rate was the annual per capita consumption of ice cream increasing at that time? (*Hint:* [annual per capita consumption] = $\frac{\text{[annual consumption]}}{\text{[population size]}}$.)

55. Figure 5 shows the graph of $y = \dfrac{10x}{1 + .25x^2}$ for $x \geq 0$.
Find the coordinates of the maximum point.

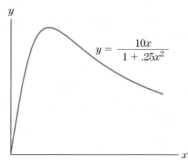

Figure 5

56. Figure 6 shows the graph of

$$y = \frac{1}{2} + \frac{x^2 - 2x + 1}{x^2 - 2x + 2}$$

for $0 \leq x \leq 2$. Find the coordinates of the minimum point.

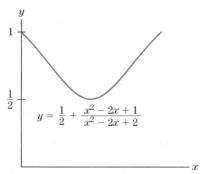

Figure 6

If $f(x)$ and $g(x)$ are differentiable functions such that $f(2) = f'(2) = 3$, $g(2) = 3$, and $g'(2) = \frac{1}{3}$, compute the following derivatives:

57. $\dfrac{d}{dx}[f(x)g(x)]\Big|_{x=2}$

58. $\dfrac{d}{dx}\left[\dfrac{f(x)}{g(x)}\right]\Big|_{x=2}$

59. $\dfrac{d}{dx}[(f(x))^2]\Big|_{x=2}$

60. $\dfrac{d}{dx}[(g(x))^2]\Big|_{x=2}$

61. $\dfrac{d}{dx}[xf(x)]\Big|_{x=2}$

62. $\dfrac{d}{dx}[x(g(x) - f(x))]\Big|_{x=2}$

63. Let $f(x) = 1/x$ and $g(x) = x^3$.
   (a) Show that the product rule yields the correct derivative of $(1/x)x^3 = x^2$.
   (b) Compute the product $f'(x)g'(x)$, and note that it is *not* the derivative of $f(x)g(x)$.

64. The derivative of $(x^3 - 4x)/x$ is obviously $2x$ for $x \neq 0$, because $(x^3 - 4x)/x = x^2 - 4$ for $x \neq 0$. Verify that the quotient rule gives the same derivative.

65. Let $f(x)$, $g(x)$, and $h(x)$ be differentiable functions. Find a formula for the derivative of $f(x)g(x)h(x)$. (*Hint:* First, differentiate $[f(x)g(x)] \cdot h(x)$.)

66. Alternative Verification of the Product Rule Apply the special case of the general power rule

$$\frac{d}{dx}[h(x)]^2 = 2h(x)h'(x)$$

and the identity

$$fg = \tfrac{1}{4}[(f + g)^2 - (f - g)^2]$$

to prove the product rule.

67. Body Mass Index The body mass index, or BMI, is a ratio of a person's weight divided by the square of his or her height. Let $b(t)$ denote the BMI; then,

$$b(t) = \frac{w(t)}{[h(t)]^2},$$

where t is the age of the person, $w(t)$ the weight in kilograms, and $h(t)$ the height in meters. Find an expression for $b'(t)$.

68. Body Mass Index The BMI is usually used as a guideline to determine whether a person is overweight or underweight. For example, according to the Centers for Disease Control, a 12-year-old boy is at risk of being overweight if his BMI is between 21 and 24 and is considered overweight if his BMI is above 24. A 13-year-old boy is at risk of being overweight if his BMI is between 22 and 25 and is overweight if his BMI is above 25.
   (a) At age 12, a boy's weight was 50 kilograms and his height was 1.55 meters. Find his BMI and determine whether he is overweight or at risk of being overweight.
   (b) Suppose that the boy's weight was increasing at the rate of 7 kilograms per year and his height was increasing at the rate of 5 centimeters per year. Find the rate of change of his BMI at age 12 or $b'(12)$.
   (c) Use the values $b(12)$ and $b'(12)$ to estimate $b(13)$, the BMI of the child at age 13. Is the child at risk of being overweight when he turns 13? [*Hint:* Use formula (2), Section 1.8, to approximate $b(13)$.]

TECHNOLOGY EXERCISES

69. Area of the Pupil The relationship between the area of the pupil of the eye and the intensity of light was analyzed by B. H. Crawford. Crawford concluded that the area of the pupil is

$$f(x) = \frac{160x^{-0.4} + 94.8}{4x^{-0.4} + 15.8} \quad (0 \leq x \leq 37)$$

square millimeters when x units of light are entering the eye per unit time. (*Source: Proceedings of the Royal Society.*)
   (a) Graph $f(x)$ and $f'(x)$ in the window $[0, 6]$ by $[-5, 20]$.
   (b) How large is the pupil when 3 units of light are entering the eye per unit time?
   (c) For what light intensity is the pupil size 11 square millimeters?
   (d) When 3 units of light are entering the eye per unit time, what is the rate of change of pupil size with respect to a unit change in light intensity?

Solutions to Check Your Understanding 3.1

1. (a) Apply the product rule to $y = (\sqrt{x} + 1)x$ with

$$f(x) = \sqrt{x} + 1 = x^{1/2} + 1$$
$$g(x) = x.$$
$$\frac{dy}{dx} = (x^{1/2} + 1) \cdot 1 + x \cdot \tfrac{1}{2} x^{-1/2}$$
$$= x^{1/2} + 1 + \tfrac{1}{2} x^{1/2}$$
$$= \tfrac{3}{2} \sqrt{x} + 1.$$

(b) $y = (\sqrt{x} + 1)x = (x^{1/2} + 1)x = x^{3/2} + x$
$$\frac{dy}{dx} = \tfrac{3}{2} x^{1/2} + 1.$$

Comparing parts (a) and (b), we note that it is helpful to simplify the function before differentiating.

2. We may apply the quotient rule to $y = \dfrac{5}{x^4 - x^3 + 1}$.

$$\frac{dy}{dx} = \frac{(x^4 - x^3 + 1) \cdot 0 - 5 \cdot (4x^3 - 3x^2)}{(x^4 - x^3 + 1)^2}$$
$$= \frac{-5x^2(4x - 3)}{(x^4 - x^3 + 1)^2}.$$

However, it is slightly faster to use the general power rule, since $y = 5(x^4 - x^3 + 1)^{-1}$. Thus,

$$\frac{dy}{dx} = -5(x^4 - x^3 + 1)^{-2}(4x^3 - 3x^2)$$
$$= -5x^2(x^4 - x^3 + 1)^{-2}(4x - 3).$$

3.2 The Chain Rule

In this section, we show that the general power rule is a special case of a powerful differentiation technique called the *chain rule*. Applications of the chain rule appear throughout the text.

Composition of Functions

A useful way of combining functions $f(x)$ and $g(x)$ is to replace each occurrence of the variable x in $f(x)$ by the function $g(x)$. The resulting function is called the **composition** (or **composite**) of $f(x)$ and $g(x)$ and is denoted by $f(g(x))$. (See Section 0.3 for additional information on function composition.)

EXAMPLE 1   **Composition of Functions** Let

$$f(x) = \frac{x - 1}{x + 1}, \quad g(x) = x^3.$$

What is $f(g(x))$?

SOLUTION   Replace each occurrence of x in $f(x)$ by $g(x)$ to obtain

$$f(g(x)) = \frac{g(x) - 1}{g(x) + 1} = \frac{x^3 - 1}{x^3 + 1}.$$ **» Now Try Exercise 1**

Given a composite function $f(g(x))$, we may think of $f(x)$ as the "outside" function that acts on the values of the "inside" function $g(x)$. This point of view often helps us to recognize a composite function.

EXAMPLE 2   **Decomposing Composite Functions** Write the following functions as composites of simpler functions.
  (a) $h(x) = (x^5 + 9x + 3)^8$ **(b)** $k(x) = \sqrt{4x^2 + 1}$

SOLUTION   **(a)** $h(x) = f(g(x))$, where the outside function is the power function, $(\dots)^8$, that is, $f(x) = x^8$. Inside this power function is $g(x) = x^5 + 9x + 3$.

  (b) $k(x) = f(g(x))$, where the outside function is the square root function, $f(x) = \sqrt{x}$, and the inside function is $g(x) = 4x^2 + 1$.   **» Now Try Exercise 5**

Chain Rule

A function of the form $[g(x)]^r$ is a composite $f(g(x))$, where the outside function is $f(x) = x^r$. We have already given a rule for differentiating this function:

$$\frac{d}{dx}[g(x)]^r = r[g(x)]^{r-1}g'(x).$$

The *chain rule* has the same form, except that the outside function $f(x)$ can be *any* differentiable function.

> **The Chain Rule**   To differentiate $f(g(x))$, differentiate first the outside function $f(x)$ and substitute $g(x)$ for x in the result. Then, multiply by the derivative of the inside function $g(x)$. Symbolically,
>
> $$\frac{d}{dx}f(g(x)) = f'(g(x))g'(x).$$

EXAMPLE 3   **The General Power Rule as a Chain Rule** Use the chain rule to compute the derivative of $f(g(x))$, where $f(x) = x^8$ and $g(x) = x^5 + 9x + 3$.

SOLUTION
$$f'(x) = 8x^7, \qquad g'(x) = 5x^4 + 9$$
$$f'(g(x)) = 8(x^5 + 9x + 3)^7$$

By the chain rule,

$$\frac{d}{dx}f(g(x)) = f'(g(x))g'(x) = 8(x^5 + 9x + 3)^7(5x^4 + 9).$$

Since, in this example, the outside function is a power function, the calculations are the same as when the general power rule is used to compute the derivative of $y = (x^5 + 9x + 3)^8$. However, the organization of the solution here emphasizes the notation of the chain rule.   **» Now Try Exercise 13**

EXAMPLE 4   **Derivative of a Function That Is Not Specified** If $h(x) = f(\sqrt{x})$, where f is a differentiable function that is not specified, find $h'(x)$ in terms of f'.

SOLUTION Let $g(x) = \sqrt{x}$; then, h is the composition of $f(x)$ and $g(x)$: $h(x) = f(g(x))$. We have $g'(x) = \dfrac{1}{2\sqrt{x}}$ and, by the chain rule,

$$h'(x) = \frac{d}{dx}(f(g(x)) = f'(g(x))g'(x)$$

$$= f'(\sqrt{x}) \cdot \frac{1}{2\sqrt{x}} = \frac{f'(\sqrt{x})}{2\sqrt{x}}. \qquad \text{» Now Try Exercise 21}$$

There is another way to write the chain rule. Given the function $y = f(g(x))$, set $u = g(x)$ so that $y = f(u)$. Then, y may be regarded either as a function of u or, indirectly through u, as a function of x. With this notation, we have $\dfrac{du}{dx} = g'(x)$ and $\dfrac{dy}{du} = f'(u) = f'(g(x))$. Thus, the chain rule says that

> **Theorem**
>
> $$\frac{dy}{dx} = \frac{dy}{du}\frac{du}{dx}. \tag{1}$$

Although the derivative symbols in equation (1) are not really fractions, the apparent cancellation of the "du" symbols provides a mnemonic device for remembering this form of the chain rule.

Here is an illustration that makes (1) a plausible formula. Suppose that y, u, and x are three varying quantities, with y varying three times as fast as u and u varying two times as fast as x. It seems reasonable that y should vary six times as fast as x. That is, $\dfrac{dy}{dx} = \dfrac{dy}{du}\dfrac{du}{dx} = 3 \cdot 2 = 6$.

EXAMPLE 5

The Chain Rule Find $\dfrac{dy}{dx}$ if $y = u^5 - 2u^3 + 8$ and $u = x^2 + 1$.

SOLUTION Since y is not given directly as a function of x, we cannot compute $\dfrac{dy}{dx}$ by differentiating y directly with respect to x. We can, however, differentiate, with respect to u, the relation $y = u^5 - 2u^3 + 8$, and get

$$\frac{dy}{du} = 5u^4 - 6u^2.$$

Similarly, we can differentiate with respect to x the relation $u = x^2 + 1$ and get

$$\frac{du}{dx} = 2x.$$

Applying the chain rule, as prescribed by equation (1), we obtain

$$\frac{dy}{dx} = \frac{dy}{du}\frac{du}{dx} = (5u^4 - 6u^2) \cdot (2x).$$

It is usually desirable to express $\dfrac{dy}{dx}$ as a function of x alone, so we substitute $x^2 + 1$ for u to obtain

$$\frac{dy}{dx} = [5(x^2 + 1)^4 - 6(x^2 + 1)^2] \cdot 2x.$$

>> **Now Try Exercise 37**

You may have noticed another entirely mechanical way to work Example 5: First, substitute $u = x^2 + 1$ into the original formula for y and obtain $y = (x^2 + 1)^5 - 2(x^2 + 1)^3 + 8$. Then, $\dfrac{dy}{dx}$ is easily calculated by the sum rule, the general power rule, and the constant-multiple rule. The solution in Example 5, however, lays the foundation for applications of the chain rule in Section 3.3 and elsewhere.

In many situations involving composition of functions, the basic variable is time, t. It may happen that x is a function of t—say, $x = g(t)$—and some other variable, such as R, is a function of x—say, $R = f(x)$. Then, $R = f(g(t))$, and the chain rule says that

$$\frac{dR}{dt} = \frac{dR}{dx}\frac{dx}{dt}.$$

EXAMPLE 6

Marginal Revenue and Time Rate of Change A store sells ties for \$12 apiece. Let x be the number of ties sold in one day, and let R be the revenue received from the sale of x ties, so $R = 12x$. If daily sales are rising at the rate of four ties per day, how fast is the revenue rising?

SOLUTION Clearly, revenue is rising at the rate of $48 per day, because each of the additional four ties brings in $12. This intuitive conclusion also follows from the chain rule,

$$\frac{dR}{dt} = \frac{dR}{dx} \cdot \frac{dx}{dt}$$

$$\begin{bmatrix} \text{rate of change} \\ \text{of revenue with} \\ \text{respect to time} \end{bmatrix} = \begin{bmatrix} \text{rate of change} \\ \text{of revenue with} \\ \text{respect to sales} \end{bmatrix} \cdot \begin{bmatrix} \text{rate of change} \\ \text{of sales with} \\ \text{respect to time} \end{bmatrix}$$

$$\begin{bmatrix} \$48 \text{ increase} \\ \text{per day} \end{bmatrix} = \begin{bmatrix} \$12 \text{ increase} \\ \text{per add'l tie} \end{bmatrix} \cdot \begin{bmatrix} \text{four additional} \\ \text{ties per day} \end{bmatrix}.$$

Notice that $\frac{dR}{dx}$ is actually the marginal revenue, studied earlier. This example shows that the time rate of change of revenue, $\frac{dR}{dt}$, is the marginal revenue multiplied by the time rate of change of sales. **≫ Now Try Exercise 49**

EXAMPLE 7 **Time Rate of Change of Revenue** The demand equation for a certain brand of graphing calculator is $p = 86 - .002x$, where p is the price of one calculator (in dollars) and x is the number of calculators produced and sold. Determine the (time) rate of change of the total revenue if the company increases production by 200 calculators per day, when the production level is at 6000 calculators.

SOLUTION The total revenue received from the production and sale of x calculators is $R(x) = x \cdot p = x(86 - .002x) = -.002x^2 + 86x$. We are told that, when $x = 6000$, the production will increase at the rate of 200 calculators per day; that is, $\frac{dx}{dt} = 200$ when $x = 6000$. We are asked to find the time rate of change of revenue, $\frac{dR}{dt}$, when $x = 6000$. Since R is not given directly as a function of t, to differentiate R with respect to t, we appeal to the chain rule. We have

$$\frac{dR}{dt} = \frac{dR}{dx} \cdot \frac{dx}{dt}.$$

(As in the previous example, the time rate of change of revenue, $\frac{dR}{dt}$, is the marginal revenue, $\frac{dR}{dx}$, multiplied by the time rate of change of sales, $\frac{dx}{dt}$.) Now,

$$\frac{dR}{dx} = \frac{d}{dx} \overbrace{(-.002x^2 + 86x)}^{R(x)} = -.004x + 86.$$

So,

$$\frac{dR}{dt} = \overbrace{(-.004x + 86)}^{dR/dx} \cdot \frac{dx}{dt}.$$

Substituting $x = 6000$ and $\frac{dx}{dt} = 200$, we find

$$\frac{dR}{dt} = [-.004(6000) + 86](200) = \$12{,}400 \text{ per day}.$$

Thus, if at production level $x = 6000$ the company increases production by 200 calculators per day, its total revenue will increase at the rate of $12,400 per day.

≫ Now Try Exercise 53

Appendix: Verification of the Chain Rule

Suppose that $f(x)$ and $g(x)$ are differentiable, and let $x = a$ be a number in the domain of $f(g(x))$. Since every differentiable function is continuous, we have

$$\lim_{h \to 0} g(a + h) = g(a),$$

which implies that

$$\lim_{h \to 0} [g(a + h) - g(a)] = 0. \tag{2}$$

Now $g(a)$ is a number in the domain of f, and the limit definition of the derivative gives us

$$f'(g(a)) = \lim_{k \to 0} \frac{f(g(a) + k) - f(g(a))}{k}. \tag{3}$$

Let $k = g(a + h) - g(a)$. By equation (2), k approaches zero as h approaches zero. Also, $g(a + h) = g(a) + k$. Therefore, (3) may be rewritten in the form

$$f'(g(a)) = \lim_{h \to 0} \frac{f(g(a + h)) - f(g(a))}{g(a + h) - g(a)}. \tag{4}$$

[Strictly speaking, we must assume that the denominator in (4) is never zero. We may avoid this assumption by a somewhat different and more technical argument that we omit.] Finally, we show that the function $f(g(x))$ has a derivative at $x = a$. We use the limit definition of the derivative, Limit Theorem V, and equation (4):

$$
\begin{aligned}
\frac{d}{dx} f(g(x)) \bigg|_{x=a} &= \lim_{h \to 0} \frac{f(g(a + h)) - f(g(a))}{h} \\
&= \lim_{h \to 0} \left[\frac{f(g(a + h)) - f(g(a))}{g(a + h) - g(a)} \cdot \frac{g(a + h) - g(a)}{h} \right] \\
&= \lim_{h \to 0} \frac{f(g(a + h)) - f(g(a))}{g(a + h) - g(a)} \cdot \lim_{h \to 0} \frac{g(a + h) - g(a)}{h} \\
&= f'(g(a)) \cdot g'(a).
\end{aligned}
$$

Check Your Understanding 3.2

Solutions can be found following the section exercises.

Consider the function $h(x) = (2x^3 - 5)^5 + (2x^3 - 5)^4$.

1. Write $h(x)$ as a composite function, $f(g(x))$.

2. Compute $f'(x)$ and $f'(g(x))$.

3. Use the chain rule to differentiate $h(x)$.

EXERCISES 3.2

Compute $f(g(x))$, where $f(x)$ and $g(x)$ are the following:

1. $f(x) = \dfrac{x}{x + 1}, g(x) = x^3$

2. $f(x) = x - 1, g(x) = \dfrac{1}{x + 1}$

3. $f(x) = x(x^2 + 1), g(x) = \sqrt{x}$

4. $f(x) = \dfrac{x + 1}{x - 3}, g(x) = x + 3$

Each of the following functions may be viewed as a composite function $h(x) = f(g(x))$. Find $f(x)$ and $g(x)$.

5. $h(x) = (x^3 + 8x - 2)^5$

6. $h(x) = (9x^2 + 2x - 5)^7$

7. $h(x) = \sqrt{4 - x^2}$

8. $h(x) = (5x^2 + 1)^{-1/2}$

9. $h(x) = \dfrac{1}{x^3 - 5x^2 + 1}$

10. $h(x) = (4x - 3)^3 + \dfrac{1}{4x - 3}$

Differentiate the functions in Exercises 11–20 using one or more of the differentiation rules discussed thus far.

11. $y = (x^2 + 5)^{15}$

12. $y = (x^4 + x^2)^{10}$

13. $y = 6x^2(x - 1)^3$

14. $y = 5x^3(2 - x)^4$

15. $y = 2(x^3 - 1)(3x^2 + 1)^4$

16. $y = 2(2x - 1)^{5/4}(2x + 1)^{3/4}$

Given $f(1) = 1$, $f'(1) = 5$, $g(1) = 3$, $g'(1) = 4$, $f'(3) = 2$, and $g'(3) = 6$, compute the following derivatives:

17. $\dfrac{d}{dx}[f(g(x))]\bigg|_{x=1}$

18. $\dfrac{d}{dx}[g(f(x))]\bigg|_{x=1}$

19. $\dfrac{d}{dx}[f(f(x))]\bigg|_{x=1}$

20. $\dfrac{d}{dx}[g(g(x))]\bigg|_{x=1}$

In Exercises 21–26, a function $h(x)$ is defined in terms of a differentiable $f(x)$. Find an expression for $h'(x)$.

21. $h(x) = f(x^2)$

22. $h(x) = 2f(2x + 1)$

23. $h(x) = -f(-x)$

24. $h(x) = f(f(x))$

25. $h(x) = \dfrac{f(x^2)}{x}$

26. $h(x) = \sqrt{f(x^2)}$

27. Sketch the graph of $y = 4x/(x + 1)^2$, $x > -1$.

28. Sketch the graph of $y = 2/(1 + x^2)$.

Compute $\dfrac{d}{dx} f(g(x))$, where $f(x)$ and $g(x)$ are the following:

29. $f(x) = x^5$, $g(x) = 6x - 1$

30. $f(x) = \sqrt{x}$, $g(x) = x^2 + 1$

31. $f(x) = \dfrac{1}{x}$, $g(x) = 1 - x^2$

32. $f(x) = \dfrac{1}{1 + \sqrt{x}}$, $g(x) = \dfrac{1}{x}$

33. $f(x) = x^4 - x^2$, $g(x) = x^2 - 4$

34. $f(x) = \dfrac{4}{x} + x^2$, $g(x) = 1 - x^4$

35. $f(x) = (x^3 + 1)^2$, $g(x) = x^2 + 5$

36. $f(x) = x(x - 2)^4$, $g(x) = x^3$

Compute $\dfrac{dy}{dx}$ using the chain rule in formula (1). State your answer in terms of x only.

37. $y = u^{3/2}$, $u = 4x + 1$

38. $y = \sqrt{u + 1}$, $u = 2x^2$

39. $y = \dfrac{u}{2} + \dfrac{2}{u}$, $u = x - x^2$

40. $y = \dfrac{u^2 + 2u}{u + 1}$, $u = x(x + 1)$

Compute $\dfrac{dy}{dt}\bigg|_{t = t_0}$.

41. $y = x^2 - 3x$, $x = t^2 + 3$, $t_0 = 0$

42. $y = (x^2 - 2x + 4)^2$, $x = \dfrac{1}{t + 1}$, $t_0 = 1$

43. $y = \dfrac{x + 1}{x - 1}$, $x = \dfrac{t^2}{4}$, $t_0 = 3$

44. $y = \sqrt{x + 1}$, $x = \sqrt{t + 1}$, $t_0 = 0$

45. Find the equation of the line tangent to the graph of $y = 2x(x - 4)^6$ at the point $(5, 10)$.

46. Find the equation of the line tangent to the graph of $y = \dfrac{x}{\sqrt{2 - x^2}}$ at the point $(1, 1)$.

47. Find the x-coordinates of all points on the curve $y = (-x^2 + 4x - 3)^3$ with a horizontal tangent line.

48. The function $f(x) = \sqrt{x^2 - 6x + 10}$ has one relative minimum point for $x \geq 0$. Find it.

49. The length, x, of the edge of a cube is increasing.

   (a) Write the chain rule for $\dfrac{dV}{dt}$, the time rate of change of the volume of the cube.

(b) For what value of x is $\dfrac{dV}{dt}$ equal to 12 times the rate of increase of x?

50. Allometric Equation Many relations in biology are expressed by power functions, known as *allometric equations*, of the form $y = kx^a$, where k and a are constants. For example, the weight of a male hognose snake is approximately $446x^3$ grams, where x is its length in meters. If a snake has length .4 meters and is growing at the rate of .2 meters per year, at what rate is the snake gaining weight? (*Source: Museum of Natural History.*)

51. Suppose that P, y, and t are variables, where P is a function of y and y is a function of t.

   (a) Write the derivative symbols for the following quantities: (i) the rate of change of y with respect to t; (ii) the rate of change of P with respect to y; (iii) the rate of change of P with respect to t. Select your answers from the following:

$$\frac{dP}{dy}, \quad \frac{dy}{dP}, \quad \frac{dy}{dt}, \quad \frac{dP}{dt}, \quad \frac{dt}{dP}, \quad \text{and} \quad \frac{dt}{dy}.$$

   (b) Write the chain rule for $\dfrac{dP}{dt}$.

52. Suppose that Q, x, and y are variables, where Q is a function of x and x is a function of y. (Read this carefully.)

   (a) Write the derivative symbols for the following quantities: (i) the rate of change of x with respect to y; (ii) the rate of change of Q with respect to y; (iii) the rate of change of Q with respect to x. Select your answers from the following:

$$\frac{dy}{dx}, \quad \frac{dx}{dy}, \quad \frac{dQ}{dx}, \quad \frac{dx}{dQ}, \quad \frac{dQ}{dy}, \quad \text{and} \quad \frac{dy}{dQ}.$$

   (b) Write the chain rule for $\dfrac{dQ}{dy}$.

53. Marginal Profit and Time Rate of Change When a company produces and sells x thousand units per week, its total weekly profit is P thousand dollars, where

$$P = \frac{200x}{100 + x^2}.$$

The production level at t weeks from the present is $x = 4 + 2t$.

   (a) Find the marginal profit, $\dfrac{dP}{dx}$.

   (b) Find the time rate of change of profit, $\dfrac{dP}{dt}$.

   (c) How fast (with respect to time) are profits changing when $t = 8$?

54. Marginal Cost and Time Rate of Change The cost of manufacturing x cases of cereal is C dollars, where $C = 3x + 4\sqrt{x} + 2$. Weekly production at t weeks from the present is estimated to be $x = 6200 + 100t$ cases.

   (a) Find the marginal cost, $\dfrac{dC}{dx}$.

   (b) Find the time rate of change of cost, $\dfrac{dC}{dt}$.

   (c) How fast (with respect to time) are costs rising when $t = 2$?

55. A Model for Carbon Monoxide Levels Ecologists estimate that, when the population of a certain city is x thousand persons, the average level L of carbon monoxide in the air above the city will be L ppm (parts per million), where $L = 10 + .4x + .0001x^2$. The population of the city is estimated to be $x = 752 + 23t + .5t^2$ thousand persons t years from the present.

(a) Find the rate of change of carbon monoxide with respect to the population of the city.

(b) Find the time rate of change of the population when $t = 2$.

(c) How fast (with respect to time) is the carbon monoxide level changing at time $t = 2$?

56. **Profit** A manufacturer of microcomputers estimates that t months from now it will sell x thousand units of its main line of microcomputers per month, where $x = .05t^2 + 2t + 5$. Because of economies of scale, the profit P from manufacturing and selling x thousand units is estimated to be $P = .001x^2 + .1x - .25$ million dollars. Calculate the rate at which the profit will be increasing 5 months from now.

57. If $f(x)$ and $g(x)$ are differentiable functions, find $g(x)$ if you know that

$$\frac{d}{dx} f(g(x)) = 3x^2 \cdot f'(x^3 + 1).$$

58. If $f(x)$ and $g(x)$ are differentiable functions, find $g(x)$ if you know that $f'(x) = 1/x$ and

$$\frac{d}{dx} f(g(x)) = \frac{2x + 5}{x^2 + 5x - 4}.$$

59. If $f(x)$ and $g(x)$ are differentiable functions, such that $f(1) = 2$, $f'(1) = 3$, $f'(5) = 4$, $g(1) = 5$, $g'(1) = 6$, $g'(2) = 7$, and $g'(5) = 8$, find $\dfrac{d}{dx} f(g(x)) \Big|_{x=1}$.

60. Consider the functions of Exercise 59. Find $\dfrac{d}{dx} g(f(x)) \Big|_{x=1}$.

61. **Effect of Stocks on Total Assets of a Company** After a computer software company went public, the price of one share of its stock fluctuated according to the graph in Fig. 1(a). The total worth of the company depended on the value of one share and was estimated to be

$$W(x) = 10 \frac{12 + 8x}{3 + x},$$

where x is the value of one share (in dollars) and $W(x)$ is the total value of the company in millions of dollars. [See Fig. 1(b).]

(a) Find the total value of the company when $t = 1.5$ and when $t = 3.5$.

(b) Find $\dfrac{dx}{dt} \Big|_{t=1.5}$ and $\dfrac{dx}{dt} \Big|_{t=3.5}$. Give an interpretation for these values.

Company went public. (in months)
(a)

(b)

Figure 1

62. Refer to Exercise 61. Use the chain rule to find $\dfrac{dW}{dt} \Big|_{t=1.5}$ and $\dfrac{dW}{dt} \Big|_{t=3.5}$. Give an interpretation for these values.

63. Refer to Exercise 61.

(a) Find $\dfrac{dx}{dt} \Big|_{t=2.5}$ and $\dfrac{dx}{dt} \Big|_{t=4}$. Give an interpretation for these values.

(b) Use the chain rule to find $\dfrac{dW}{dt} \Big|_{t=2.5}$ and $\dfrac{dW}{dt} \Big|_{t=4}$. Give an interpretation for these values.

64. Refer to Exercise 61.

(a) What was the maximum value of the company during the first 6 months since it went public, and when was that maximum value attained?

(b) Assuming that the value of one share will continue to increase at the rate that it did during the period following the 6th month, what is the limit of the total value of the company as t increases?

65. In an expression of the form $f(g(x))$, $f(x)$ is called the *outer function* and $g(x)$ is called the *inner function*. Give a written description of the chain rule using the words *inner* and *outer*.

Solutions to Check Your Understanding 3.2

1. Let $f(x) = x^5 + x^4$ and $g(x) = 2x^3 - 5$.

2. $f'(x) = 5x^4 + 4x^3$, $f'(g(x)) = 5(2x^3 - 5)^4 + 4(2x^3 - 5)^3$.

3. We have $g'(x) = 6x^2$. Then, from the chain rule and the result of the previous problem 2, we have

$$h'(x) = f'(g(x))g'(x)$$
$$= [5(2x^3 - 5)^4 + 4(2x^3 - 5)^3](6x^2).$$

3.3 Implicit Differentiation and Related Rates

This section presents two different applications of the chain rule. In each case, we will need to differentiate one or more composite functions where the "inside" functions are not known explicitly.

Implicit Differentiation

In some applications, the variables are related by an equation rather than a function. In these cases, we can still determine the rate of change of one variable with respect to the other by the technique of implicit differentiation. As an illustration, consider the equation

$$x^2 + y^2 = 4. \tag{1}$$

The graph of this equation is the circle in Fig. 1. This graph obviously is not the graph of a function since, for instance, there are two points on the graph whose x-coordinate is 1. (Functions must satisfy the vertical line test. See Section 0.1.)

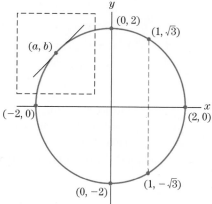

Figure 1 Graph of $x^2 + y^2 = 4$.

FOR REVIEW

The notation $\left.\dfrac{dy}{dx}\right|_{\substack{x=a \\ y=b}}$ means evaluate the derivative $\dfrac{dy}{dx}$ at $x = a$ and $y = b$.

We denote the slope of the curve at the point $(1, \sqrt{3})$ by

$$\left.\frac{dy}{dx}\right|_{\substack{x=1 \\ y=\sqrt{3}}}$$

In general, the slope at the point (a, b) is denoted by

$$\left.\frac{dy}{dx}\right|_{\substack{x=a \\ y=b}}$$

In a small vicinity of the point (a, b), the curve looks like the graph of a function. [That is, on this part of the curve, $y = g(x)$ for some function $g(x)$.] We say that this function is defined *implicitly* by the equation. (Of course, the tangent line at the point (a, b) must not be vertical. In this section, we assume that the given equations implicitly determine differentiable functions.) We obtain a formula for $\dfrac{dy}{dx}$ by differentiating both sides of the equation with respect to x while treating y as a function of x.

EXAMPLE 1 **Finding the Slope Using Implicit Differentiation** Consider the graph of the equation $x^2 + y^2 = 4$.

(a) Use implicit differentiation to compute $\dfrac{dy}{dx}$.

(b) Find the slope of the graph at the points $(1, \sqrt{3})$ and $(1, -\sqrt{3})$.

SOLUTION **(a)** The first term x^2 has derivative $2x$, as usual. We think of the second term y^2 as having the form $[g(x)]^2$. To differentiate, we use the chain rule (specifically, the general power rule):

$$\frac{d}{dx}[g(x)]^2 = 2[g(x)]g'(x)$$

or, equivalently,

$$\frac{d}{dx}y^2 = 2y\frac{dy}{dx}.$$

On the right side of the original equation, the derivative of the constant function 4 is zero. Thus, implicit differentiation of $x^2 + y^2 = 4$ yields

$$2x + 2y\frac{dy}{dx} = 0.$$

Solving for $\frac{dy}{dx}$, we have

$$2y\frac{dy}{dx} = -2x.$$

If $y \neq 0$, then

$$\frac{dy}{dx} = \frac{-2x}{2y} = -\frac{x}{y}.$$

Notice that this slope formula involves y as well as x, reflecting the fact that the slope of the circle at a point depends on the y-coordinate of the point as well as the x-coordinate.

(b) At the point $(1, \sqrt{3})$, the slope is

$$\left.\frac{dy}{dx}\right|_{\substack{x=1\\y=\sqrt{3}}} = \left.-\frac{x}{y}\right|_{\substack{x=1\\y=\sqrt{3}}} = -\frac{1}{\sqrt{3}}.$$

At the point $(1, -\sqrt{3})$, the slope is

$$\left.\frac{dy}{dx}\right|_{\substack{x=1\\y=-\sqrt{3}}} = \left.-\frac{x}{y}\right|_{\substack{x=1\\y=-\sqrt{3}}} = -\frac{1}{-\sqrt{3}} = \frac{1}{\sqrt{3}}.$$

(See Fig. 2.) The formula for $\frac{dy}{dx}$ gives the slope at every point on the graph of $x^2 + y^2 = 4$ except $(-2, 0)$ and $(2, 0)$. At these two points the tangent line is vertical and the slope of the curve is undefined. **» Now Try Exercise 3**

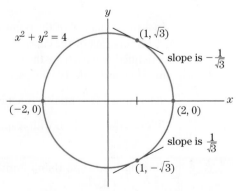

Figure 2 Slope of the tangent line.

The difficult step in Example 1(a) was to differentiate y^2 correctly. The derivative of y^2 *with respect to y* would be $2y$, by the ordinary power rule. But the derivative of y^2

with respect to x must be computed by the general power rule. To understand this distinction, think of y as a function of x, say, $y = f(x)$. Then,

$$\frac{dy}{dx} = f'(x)$$

$$y^2 = [f(x)]^2$$

$$\frac{d}{dx}y^2 = \frac{d}{dx}([f(x)]^2) = 2f(x)f'(x),$$

by the general power rule. Substituting $y = f(x)$ and $f'(x) = \dfrac{dy}{dx}$, we obtain

$$\frac{d}{dx}y^2 = 2y\frac{dy}{dx}.$$

Similarly, if a is a constant, then,

$$\frac{d}{dx}(ay) = a\frac{dy}{dx}$$

$$\frac{d}{dx}(ay^2) = 2ay\frac{dy}{dx}.$$

Also, if r is a constant, the general power rule yields

Theorem

$$\frac{d}{dx}y^r = ry^{r-1}\frac{dy}{dx}. \tag{2}$$

This rule is used to compute slope formulas in the next two examples.

EXAMPLE 2

Implicit Differentiation Use implicit differentiation to calculate $\dfrac{dy}{dx}$ for the equation $x^2y^6 = 1$.

SOLUTION

Differentiate each side of the equation $x^2y^6 = 1$ with respect to x. On the left side of the equation, use the product rule and treat y as a function of x:

$$x^2\frac{d}{dx}(y^6) + y^6\frac{d}{dx}(x^2) = \frac{d}{dx}(1)$$

$$x^2 \cdot 6y^5\frac{dy}{dx} + y^6 \cdot 2x = 0.$$

Solve for $\dfrac{dy}{dx}$:

$$6x^2y^5\frac{dy}{dx} = -2xy^6 \qquad \text{Move the term not involving } \frac{dy}{dx} \text{ to the right side.}$$

$$\frac{dy}{dx} = \frac{-2xy^6}{6x^2y^5} = -\frac{y}{3x}. \qquad \text{Divide by the factor that multiplies } \frac{dy}{dx}.$$

» Now Try Exercise 9

EXAMPLE 3

Implicit Differentiation Use implicit differentiation to calculate $\dfrac{dy}{dx}$ when y is related to x by the equation $x^2y + xy^3 - 3x = 5$.

SOLUTION   Differentiate the equation term by term, taking care to differentiate $x^2 y$ and xy^3 by the product rule:

$$x^2 \frac{d}{dx}(y) + y \frac{d}{dx}(x^2) + x \frac{d}{dx}(y^3) + y^3 \frac{d}{dx}(x) - 3 = 0 \qquad \text{Differentiate both sides with respect to } x$$

$$x^2 \frac{dy}{dx} + y \cdot 2x + x \cdot 3y^2 \frac{dy}{dx} + y^3 \cdot 1 - 3 = 0 \qquad \text{Note: } \frac{d}{dx}(y) = \frac{dy}{dx} \text{ and }$$

$$\frac{d}{dx}(y^3) = 3y^2 \frac{dy}{dx}$$

Solve for $\dfrac{dy}{dx}$ in terms of x and y following these steps:

Step 1   Keep all terms involving $\dfrac{dy}{dx}$ on the left side of the equation and move the other terms to the right side.

$$x^2 \frac{dy}{dx} + 3xy^2 \frac{dy}{dx} = 3 - y^3 - 2xy.$$

Step 2   Factor out $\dfrac{dy}{dx}$ on the left side of the equation.

$$(x^2 + 3xy^2) \frac{dy}{dx} = 3 - y^3 - 2xy.$$

Step 3   Divide both sides of the equation by the factor that multiplies $\dfrac{dy}{dx}$.

$$\frac{dy}{dx} = \frac{3 - y^3 - 2xy}{x^2 + 3xy^2}.$$

» Now Try Exercise 17

NOTE   When a power of y is differentiated with respect to x, the result must include the factor $\dfrac{dy}{dx}$. When a power of x is differentiated, there is no factor $\dfrac{dy}{dx}$. «

Here is the general procedure for implicit differentiation:

Finding $\dfrac{dy}{dx}$ by Implicit Differentiation   Differentiate each term of the equation *with respect to x*, treating y as a function of x.

Move all terms involving $\dfrac{dy}{dx}$ to the left side of the equation and move the other terms to the right side.

Factor out $\dfrac{dy}{dx}$ on the left side of the equation.

Divide both sides of the equation by the factor that multiplies $\dfrac{dy}{dx}$.

Equations that implicitly define functions arise frequently in economic models. The economic background for the equation in the next example is discussed in Section 7.1.

EXAMPLE 4   **Isoquants and Marginal Rate of Substitution**   Suppose that x and y represent the amounts of two basic inputs into a production process, and the equation

$$60x^{3/4}y^{1/4} = 3240$$

describes all input amounts (x, y) for which the output of the process is 3240 units. (The graph of this equation is called a **production isoquant** or **constant product curve**. See Fig. 3.) Use implicit differentiation to calculate the slope of the graph at the point on the curve where $x = 81$, $y = 16$.

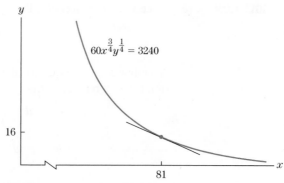

Figure 3 A production isoquant.

SOLUTION   We use the product rule and treat y as a function of x:

$$60x^{3/4}\frac{d}{dx}(y^{1/4}) + y^{1/4}\frac{d}{dx}(60x^{3/4}) = \frac{d}{dx}(3240)$$

$$60x^{3/4}\cdot\left(\frac{1}{4}\right)y^{-3/4}\frac{dy}{dx} + y^{1/4}\cdot 60\left(\frac{3}{4}\right)x^{-1/4} = 0$$

$$15x^{3/4}y^{-3/4}\frac{dy}{dx} = -45x^{-1/4}y^{1/4}$$ Move the term not involving $\frac{dy}{dx}$ to the right.

$$\frac{dy}{dx} = \frac{-45x^{-1/4}y^{1/4}}{15x^{3/4}y^{-3/4}} = \frac{-3y}{x}.$$ Divide by the term that multiplies $\frac{dy}{dx}$. Simplify.

When $x = 81$ and $y = 16$, we have

$$\left.\frac{dy}{dx}\right|_{\substack{x=81\\y=16}} = \frac{-3(16)}{81} = -\frac{16}{27}.$$

The number $-\frac{16}{27}$ is the slope of the production isoquant at the point $(81, 16)$. If the first input (corresponding to x) is increased by 1 unit, the second input (corresponding to y) must decrease by approximately $\frac{16}{27}$ unit to keep the production output unchanged [that is, to keep (x, y) on the curve]. In economic terminology, the absolute value of $\frac{dy}{dx}$ is called the **marginal rate of substitution** of the first input for the second input.

» Now Try Exercise 27

Related Rates

In implicit differentiation, we differentiate an equation involving x and y, with y treated as a function of x. However, in some applications where x and y are related by an equation, both variables are functions of a third variable, t (which may represent time). Often, the formulas for x and y as functions of t are not known. When we differentiate such an equation with respect to t, we derive a relationship between the rates of change $\frac{dy}{dt}$ and $\frac{dx}{dt}$. We say that these derivatives are **related rates**. The equation relating the rates may be used to find one of the rates when the other is known.

EXAMPLE 5   **A Point Moving along a Graph** Suppose that x and y are both differentiable functions of t and are related by the equation

$$x^2 + 5y^2 = 36. \tag{3}$$

(a) Differentiate each term in the equation with respect to t, and solve the resulting equation for $\frac{dy}{dt}$.

(b) Calculate $\frac{dy}{dt}$ at a time when $x = 4$, $y = 2$, and $\frac{dx}{dt} = 5$.

SOLUTION   **(a)** Since x is a function of t, the general power rule gives

$$\frac{d}{dt}(x^2) = 2x\frac{dx}{dt}.$$

A similar formula holds for the derivative of y^2. Differentiating each term in equation (3) with respect to t, we obtain

$$\frac{d}{dt}(x^2) + \frac{d}{dt}(5y^2) = \frac{d}{dt}(36)$$

$$2x\frac{dx}{dt} + 5\cdot 2y\frac{dy}{dt} = 0$$

$$10y\frac{dy}{dt} = -2x\frac{dx}{dt}$$

$$\frac{dy}{dt} = -\frac{x}{5y}\frac{dx}{dt}.$$

(b) When $x = 4$, $y = 2$, and $\dfrac{dx}{dt} = 5$,

$$\frac{dy}{dt} = -\frac{4}{5(2)}\cdot(5) = -2.$$

>> *Now Try Exercise 37*

Figure 4 Graph of $x^2 + 5y^2 = 36$.

There is a helpful graphic interpretation of the calculations in Example 5. Imagine a point that is moving clockwise along the graph of the equation $x^2 + 5y^2 = 36$, which is an ellipse. (See Fig. 4.) Suppose that, when the point is at $(4, 2)$, the x-coordinate of the point is changing at the rate of 5 units per minute, so $\dfrac{dx}{dt} = 5$. In Example 5(b), we found that $\dfrac{dy}{dt} = -2$. This means that the y-coordinate of the point is decreasing at the rate of 2 units per minute when the moving point reaches $(4, 2)$.

EXAMPLE 6   **Relating Weekly Sales to Price** Suppose that x thousand units of a commodity can be sold weekly when the price is p dollars per unit and that x and p satisfy the demand equation

$$p + 2x + xp = 38.$$

(See Fig. 5.) How fast are weekly sales changing at a time when $x = 4$, $p = 6$, and the price is falling at the rate of \$.40 per week?

Figure 5 A demand curve.

SOLUTION Assume that p and x are differentiable functions of t, and differentiate the demand equation with respect to t:

$$\frac{d}{dt}(p) + \frac{d}{dt}(2x) + \frac{d}{dt}(xp) = \frac{d}{dt}(38)$$

$$\frac{dp}{dt} + 2\frac{dx}{dt} + x\frac{dp}{dt} + p\frac{dx}{dt} = 0. \tag{4}$$

We want to know $\frac{dx}{dt}$ at a time when $x = 4$, $p = 6$, and $\frac{dp}{dt} = -.40$. (The derivative $\frac{dp}{dt}$ is negative because the price is decreasing.) We could solve equation (4) for $\frac{dx}{dt}$ and then substitute the given values, but since we do not need a general formula for $\frac{dx}{dt}$, it is easier to substitute first and then solve:

$$-.40 + 2\frac{dx}{dt} + 4(-.40) + 6\frac{dx}{dt} = 0$$

$$8\frac{dx}{dt} = 2$$

$$\frac{dx}{dt} = .25.$$

Thus, sales are rising at the rate of .25 thousand units (or 250 units) per week.

» Now Try Exercise 39

Suggestions for Solving Related-Rates Problems

1. Draw a picture, if possible.
2. Assign letters to quantities that vary, and identify one variable—say, t—on which the other variables depend.
3. Find an equation that relates the variables to each other.
4. Differentiate the equation with respect to the independent variable t. Use the chain rule whenever appropriate.
5. Substitute all specified values for the variables and their derivatives.
6. Solve for the derivative that gives the unknown rate.

INCORPORATING

TECHNOLOGY

The graph of an equation in x and y can be easily obtained when y can be expressed as one or more functions of x. For instance, the graph of $x^2 + y^2 = 4$ can be plotted by simultaneously graphing $y = \sqrt{4 - x^2}$ and $y = -\sqrt{4 - x^2}$.

Check Your Understanding 3.3

Solutions can be found following the section exercises.

Suppose that x and y are related by the equation $3y^2 - 3x^2 + y = 1$.

1. Use implicit differentiation to find a formula for the slope of the graph of the equation.

2. Suppose that x and y in the preceding equation are both functions of t. Differentiate both sides of the equation with respect to t, and find a formula for $\frac{dy}{dt}$ in terms of x, y, and $\frac{dx}{dt}$.

EXERCISES 3.3

In Exercises 1–18, suppose that x and y are related by the given equation and use implicit differentiation to determine $\frac{dy}{dx}$.

1. $x^2 - y^2 = 1$

2. $x^3 + y^3 - 6 = 0$

3. $y^5 - 3x^2 = x$

4. $x^4 + (y + 3)^4 = x^2$

5. $y^4 - x^4 = y^2 - x^2$

6. $x^3 + y^3 = x^2 + y^2$

7. $2x^3 + y = 2y^3 + x$

8. $x^4 + 4y = x - 4y^3$

9. $xy = 5$

10. $xy^3 = 2$

11. $x(y + 2)^5 = 8$

12. $x^2y^3 = 6$

13. $x^3y^2 - 4x^2 = 1$

14. $(x + 1)^2(y - 1)^2 = 1$

15. $x^3 + y^3 = x^3y^3$

16. $x^2 + 4xy + 4y = 1$

17. $x^2y + y^2x = 3$

18. $x^3y + xy^3 = 4$

Use implicit differentiation of the equations in Exercises 19–24 to determine the slope of the graph at the given point.

19. $4y^3 - x^2 = -5;\ x = 3,\ y = 1$

20. $y^2 = x^3 + 1;\ x = 2,\ y = -3$

21. $xy^3 = 2;\ x = -\frac{1}{4},\ y = -2$

22. $\sqrt{x} + \sqrt{y} = 7;\ x = 9,\ y = 16$

23. $xy + y^3 = 14;\ x = 3,\ y = 2$

24. $y^2 = 3xy - 5;\ x = 2,\ y = 1$

25. Find the equation of the tangent line to the graph of $x^2y^4 = 1$ at the point $(4, \frac{1}{2})$ and at the point $(4, -\frac{1}{2})$.

26. Find the equation of the tangent line to the graph of $x^4y^2 = 144$ at the point $(2, 3)$ and at the point $(2, -3)$.

27. **Slope of the Lemniscate** The graph of
$x^4 + 2x^2y^2 + y^4 = 4x^2 - 4y^2$ is the *lemniscate* in Fig. 6.

(a) Find $\dfrac{dy}{dx}$ by implicit differentiation.

(b) Find the slope of the tangent line to the lemniscate at $(\sqrt{6}/2, \sqrt{2}/2)$.

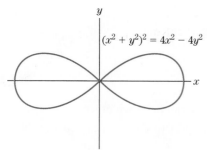

Figure 6 A lemniscate.

28. The graph of $x^4 + 2x^2y^2 + y^4 = 9x^2 - 9y^2$ is a lemniscate similar to that in Fig. 6.

(a) Find $\dfrac{dy}{dx}$ by implicit differentiation.

(b) Find the slope of the tangent line to the lemniscate at $(\sqrt{5}, -1)$.

29. **Marginal Rate of Substitution** Suppose that x and y represent the amounts of two basic inputs for a production process and that the equation

$$30x^{1/3}y^{2/3} = 1080$$

describes all input amounts where the output of the process is 1080 units.

(a) Find $\dfrac{dy}{dx}$.

(b) What is the marginal rate of substitution of x for y when $x = 16$ and $y = 54$? (See Example 4.)

30. **Demand Equation** Suppose that x and y represent the amounts of two basic inputs for a production process and

$$10x^{1/2}y^{1/2} = 600.$$

Find $\dfrac{dy}{dx}$ when $x = 50$ and $y = 72$.

In Exercises 31–36, suppose that x and y are both differentiable functions of t and are related by the given equation. Use implicit differentiation with respect to t to determine $\dfrac{dy}{dt}$ in terms of x, y, and $\dfrac{dx}{dt}$.

31. $x^4 + y^4 = 1$

32. $y^4 - x^2 = 1$

33. $3xy - 3x^2 = 4$

34. $y^2 = 8 + xy$

35. $x^2 + 2xy = y^3$

36. $x^2y^2 = 2y^3 + 1$

37. **Point on a Curve** A point is moving along the graph of $x^2 - 4y^2 = 9$. When the point is at $(5, -2)$, its x-coordinate is increasing at the rate of 3 units per second. How fast is the y-coordinate changing at that moment?

38. **Point on a Curve** A point is moving along the graph of $x^3y^2 = 200$. When the point is at $(2, 5)$, its x-coordinate is changing at the rate of -4 units per minute. How fast is the y-coordinate changing at that moment?

39. **Demand Equation** Suppose that the price p (in dollars) and the weekly sales x (in thousands of units) of a certain commodity satisfy the demand equation

$$2p^3 + x^2 = 4500.$$

Determine the rate at which sales are changing at a time when $x = 50$, $p = 10$, and the price is falling at the rate of $.50 per week.

40. **Demand Equation** Suppose that the price p (in dollars) and the weekly demand, x (in thousands of units) of a commodity satisfy the demand equation

$$6p + x + xp = 94.$$

How fast is the demand changing at a time when $x = 4$, $p = 9$, and the price is rising at the rate of $2 per week?

41. **Advertising Affects Revenue** The monthly advertising revenue, A, and the monthly circulation, x, of a magazine are related approximately by the equation

$$A = 6\sqrt{x^2 - 400}, \qquad x \geq 20,$$

where A is given in thousands of dollars and x is measured in thousands of copies sold. At what rate is the advertising revenue changing if the current circulation is $x = 25$ thousand copies and the circulation is growing at the rate of 2 thousand copies per month?

$$\left[\textit{Hint:} \text{ Use the chain rule } \frac{dA}{dt} = \frac{dA}{dx}\frac{dx}{dt}. \right]$$

42. **Rate of Change of Price** Suppose that in Boston the wholesale price, p, of oranges (in dollars per crate) and the daily supply, x (in thousands of crates), are related by the equation $px + 7x + 8p = 328$. If there are 4 thousand crates available today at a price of $25 per crate, and if the supply is changing at the rate of $-.3$ thousand crates per day, at what rate is the price changing?

43. **Related Rates** Figure 7 shows a 10-foot ladder leaning against a wall.

(a) Use the Pythagorean theorem to find an equation relating x and y.

(b) If the foot of the ladder is being pulled along the ground at the rate of 3 feet per second, how fast is the top end of the ladder sliding down the wall at the time when the foot of the ladder is 8 feet from the wall? That is, what is $\dfrac{dy}{dt}$ at the time when $\dfrac{dx}{dt} = 3$ and $x = 8$?

Figure 7

44. **Related Rates** An airplane flying 390 feet per second at an altitude of 5000 feet flew directly over an observer. Figure 8 shows the relationship of the airplane to the observer at a later time.
    (a) Find an equation relating x and y.
    (b) Find the value of x when y is 13,000.
    (c) How fast is the distance from the observer to the airplane changing at the time when the airplane is 13,000 feet from the observer? That is, what is $\dfrac{dy}{dt}$ at the time when $\dfrac{dx}{dt} = 390$ and $y = 13{,}000$?

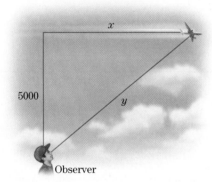

Figure 8

45. **Related Rates** A baseball diamond is a 90-foot by 90-foot square. (See Fig. 9.) A player runs from first to second base at the speed of 22 feet per second. How fast is the player's distance from third base changing when he is halfway between first and second base? [*Hint:* If x is the distance from the player to second base and y is his distance from third base, then $x^2 + 90^2 = y^2$.]

Figure 9 A baseball diamond.

46. **Related Rates** A motorcyclist is driving over a ramp as shown in Fig. 10 at the speed of 80 miles per hour. How fast is she rising? [*Hint:* Use similar triangles to relate x and h, and then, compute dh/dt.]

Figure 10

Solutions to Check Your Understanding 3.3

1. $\dfrac{d}{dx}(3y^2) - \dfrac{d}{dx}(3x^2) + \dfrac{d}{dx}(y) = \dfrac{d}{dx}(1)$

$$6y\frac{dy}{dx} - 6x + \frac{dy}{dx} = 0$$

$$(6y + 1)\frac{dy}{dx} = 6x$$

$$\frac{dy}{dx} = \frac{6x}{6y + 1}.$$

2. First, here is the solution without reference to the previous problem

$$\frac{d}{dt}(3y^2) - \frac{d}{dt}(3x^2) + \frac{d}{dt}(y) = \frac{d}{dt}(1)$$

$$6y\frac{dy}{dt} - 6x\frac{dx}{dt} + \frac{dy}{dt} = 0$$

$$(6y + 1)\frac{dy}{dt} = 6x\frac{dx}{dt}$$

$$\frac{dy}{dt} = \frac{6x}{6y + 1}\frac{dx}{dt}.$$

Alternatively, we can use the chain rule and the formula for $\dfrac{dy}{dx}$ from the previous problem:

$$\frac{dy}{dt} = \frac{dy}{dx}\frac{dx}{dt} = \frac{6x}{6y + 1}\frac{dx}{dt}.$$

CHAPTER 3   Summary

KEY TERMS AND CONCEPTS	EXAMPLES

3.1 The Product and Quotient Rules

The **product rule**:

$$(fg)' = fg' + gf'.$$

Other notation:

$$\frac{d}{dx}[f(x)g(x)] = f(x)\frac{d}{dx}[g(x)] + g(x)\frac{d}{dx}[f(x)].$$

The **quotient rule**:

$$\left(\frac{f}{g}\right)' = \frac{gf' - fg'}{g^2}.$$

$\dfrac{d}{dx}[(x^2 + 1)^2(3x + 7)]$

$= (x^2 + 1)^2\dfrac{d}{dx}[(3x + 7)] + (3x + 7)\dfrac{d}{dx}[(x^2 + 1)^2]$ Product rule

$= (x^2 + 1)^2(3) + (3x + 7)2(x^2 + 1)\dfrac{d}{dx}[(x^2 + 1)]$ General power rule

$= 3(x^2 + 1)^2 + (3x + 7)2(x^2 + 1)2x$

$= (x^2 + 1)[3(x^2 + 1) + 4x(3x + 7)]$ $(x^2 + 1)$ is a common factor.

$= (x^2 + 1)(15x^2 + 28x + 3)$

$\dfrac{d}{dx}\left[\dfrac{x^3 - 1}{x^2 + 7}\right]$

$= \dfrac{(x^2 + 7)\frac{d}{dx}(x^3 - 1) - (x^3 - 1)\frac{d}{dx}(x^2 + 7)}{(x^2 + 7)^2}$ Quotient rule

$= \dfrac{(x^2 + 7)(3x^2) - (x^3 - 1)(2x)}{(x^2 + 7)^2}$

$= \dfrac{3x^4 + 21x^2 - 2x^4 + 2x}{(x^2 + 7)^2}$

$= \dfrac{x^4 + 21x^2 + 2x}{(x^2 + 7)^2}.$

3.2 The Chain Rule and the General Power Rule

The **chain rule**:

$$[f(g(x))]' = f'(g(x))g'(x).$$

In words, the derivative of $f(g(x))$ is the derivative of the outer function, f, evaluated at the inner function, g, times the derivative of the inner function, g.

Another notation for the chain rule: If y is a function of u and u is a function of x, then,

$$\frac{dy}{dx} = \frac{dy}{du}\frac{du}{dx}.$$

To find $\frac{dy}{dx}$ if $y = 3u^2 - 2u$ and $u = 6x + 1$, observe that

$$\frac{dy}{du} = \frac{d}{du}(3u^2 - 2u) = 6u - 2 \text{ and } \frac{du}{dx} = \frac{d}{dx}(6x + 1) = 6.$$

Applying the chain rule, we find

$$\frac{dy}{dx} = \frac{dy}{du}\frac{du}{dx} = (6u - 2)(6) = 36u - 12.$$

In terms of x, we can use $u = 6x + 1$ and get

$$\frac{dy}{dx} = 36(6x + 1) - 12 = 216x + 24.$$

3.3 Implicit Differentiation and Related Rates

Useful facts to remember when using implicit differentiation:

$$\frac{d}{dx}[y^n] = ny^{n-1}\frac{dy}{dx}.$$

We also have

$$\frac{d}{dx}[ky] = k\frac{dy}{dx}$$

$$\frac{d}{dx}[y^2] = 2y\frac{dy}{dx}.$$

To find $\frac{dy}{dx}$ if $x^2y^3 = (1 + 2y)^2$, take the derivative and use the product rule on the left and the general power rule on the right:

$$\frac{d}{dx}[x^2y^3] = \frac{d}{dx}[(1 + 2y)^2]$$

$$x^2\frac{d}{dx}[y^3] + y^3(2x) = 2(1 + 2y)\frac{d}{dx}[(1 + 2y)]$$

$$x^2(3y^2)\frac{dy}{dx} + y^3(2x) = 2(1 + 2y)2\frac{dy}{dx}.$$

Collecting all terms with $\frac{dy}{dx}$ on the left and all others on the right, we find

$$3x^2y^2\frac{dy}{dx} - 4(1 + 2y)\frac{dy}{dx} = -2xy^3.$$

KEY TERMS AND CONCEPTS	EXAMPLES

Factoring $\frac{dy}{dx}$ and then solving for $\frac{dy}{dx}$, we find:

$$[3x^2y^2 - 4(1 + 2y)]\frac{dy}{dx} = -2xy^3$$

$$\frac{dy}{dx} = \frac{-2xy^3}{3x^2y^2 - 4(1 + 2y)}$$

$$\frac{dy}{dx} = \frac{-2xy^3}{3x^2y^2 - 8y - 4}.$$

CHAPTER 3   Fundamental Concept Check Exercises

1. State the product rule and the quotient rule.

2. State the chain rule. Give an example.

3. What is the relationship between the chain rule and the general power rule?

4. What does it mean for a function to be defined implicitly by an equation?

5. State the formula for $\frac{d}{dx}y^r$, where y is defined implicitly as a function of x.

6. Outline the procedure for solving a related-rates problem.

CHAPTER 3   Review Exercises

Differentiate the following functions.

1. $y = (4x - 1)(3x + 1)^4$

2. $y = 2(5 - x)^3(6x - 1)$

3. $y = x(x^5 - 1)^3$

4. $y = (2x + 1)^{5/2}(4x - 1)^{3/2}$

5. $y = 5(\sqrt{x} - 1)^4(\sqrt{x} - 2)^2$

6. $y = \dfrac{\sqrt{x}}{\sqrt{x} + 4}$

7. $y = 3(x^2 - 1)^3(x^2 + 1)^5$

8. $y = \dfrac{1}{(x^2 + 5x + 1)^6}$

9. $y = \dfrac{x^2 - 6x}{x - 2}$

10. $y = \dfrac{2x}{2 - 3x}$

11. $y = \left(\dfrac{3 - x^2}{x^3}\right)^2$

12. $y = \dfrac{x^3 + x}{x^2 - x}$

13. Let $f(x) = (3x + 1)^4(3 - x)^5$. Find all x such that $f'(x) = 0$.

14. Let $f(x) = (x^2 + 1)/(x^2 + 5)$. Find all x such that $f'(x) = 0$.

15. Find the equation of the line tangent to the graph of $y = (x^3 - 1)(x^2 + 1)^4$ at the point where $x = -1$.

16. Find the equation of the line tangent to the graph of $y = \dfrac{x - 3}{\sqrt{4 + x^2}}$ at the point where $x = 0$.

17. Minimizing Area A botanical display is to be constructed as a rectangular region with a river as one side and a sidewalk 2 meters wide along the inside edges of the other three sides. (See Fig. 1.) The area for the plants must be 800 square meters. Find the outside dimensions of the region that minimizes the area of the sidewalk (and hence, minimizes the amount of concrete needed for the sidewalk).

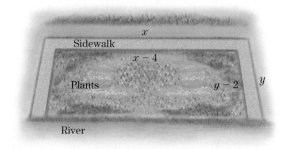

Figure 1 A botanical display.

18. Repeat Exercise 17, with the sidewalk on the inside of all four sides. In this case, the 800-square-meter planted region has dimensions $x - 4$ meters by $y - 4$ meters.

19. Cost Function A store estimates that its cost when selling x lamps per day is C dollars, where $C = 40x + 30$ (the marginal cost per lamp is \$40). If daily sales are rising at the rate of three lamps per day, how fast are the costs rising? Explain your answer using the chain rule.

20. Rate of Change of Taxes A company pays y dollars in taxes when its annual profit is P dollars. If y is some (differentiable) function of P and P is some function of time t, give a chain rule formula for the time rate of change of taxes dy/dt.

In Exercises 21–23, find a formula for $\dfrac{d}{dx}f(g(x))$, where $f(x)$ is a function such that $f'(x) = 1/(x^2 + 1)$.

21. $g(x) = x^3$ **22.** $g(x) = \dfrac{1}{x}$ **23.** $g(x) = x^2 + 1$

In Exercises 24–26, find a formula for $\dfrac{d}{dx}f(g(x))$, where $f(x)$ is a function such that $f'(x) = x\sqrt{1 - x^2}$.

24. $g(x) = x^2$ **25.** $g(x) = \sqrt{x}$ **26.** $g(x) = x^{3/2}$

In Exercises 27–29, find $\dfrac{dy}{dx}$, where y is a function of u such that $\dfrac{dy}{du} = \dfrac{u}{u^2 + 1}$. State the answer in terms of x only.

27. $u = x^{3/2}$ **28.** $u = x^2 + 1$ **29.** $u = \dfrac{5}{x}$

In Exercises 30–32, find $\dfrac{dy}{dx}$, where y is a function of u such that $\dfrac{dy}{du} = \dfrac{u}{\sqrt{1 + u^4}}$.

30. $u = x^2$ **31.** $u = \sqrt{x}$ **32.** $u = \dfrac{2}{x}$

Exercises 33–38 refer to the graphs of the functions $f(x)$ and $g(x)$ in Fig. 2. Determine $h(1)$ and $h'(1)$.

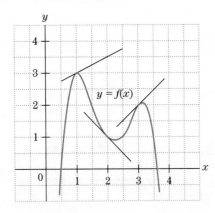

33. $h(x) = 2f(x) - 3g(x)$ **34.** $h(x) = f(x) \cdot g(x)$

35. $h(x) = \dfrac{f(x)}{g(x)}$ **36.** $h(x) = [f(x)]^2$

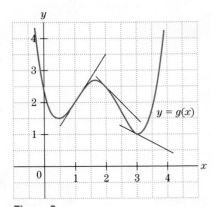

Figure 2

37. $h(x) = f(g(x))$ **38.** $h(x) = g(f(x))$

39. Revenue Function The revenue, R, that a company receives is a function of the weekly sales, x. Also, the sales level, x, is a function of the weekly advertising expenditures, A, and A, in turn, is a varying function of time.

(a) Write the derivative symbols for the following quantities: (i) rate of change of revenue, R, with respect to advertising expenditures; (ii) time rate of change of advertising expenditures; (iii) marginal revenue; (iv) rate of change of sales with respect to advertising expenditures. Select your answers from the following:

$$\frac{dR}{dx}, \quad \frac{dR}{dt}, \quad \frac{dA}{dt}, \quad \frac{dA}{dR}, \quad \frac{dA}{dx}, \quad \frac{dx}{dA}, \quad \text{and} \quad \frac{dR}{dA}.$$

(b) Write a type of chain rule that expresses the time rate of change of revenue, $\dfrac{dR}{dt}$, in terms of three of the derivatives described in part (a).

40. Amount of Drug Usage The amount, A, of anesthetics that a certain hospital uses each week is a function of the number, S, of surgical operations performed each week. Also, S, in turn, is a function of the population, P, of the area served by the hospital, while P is a function of time, t.

(a) Write the derivative symbols for the following quantities: (i) population growth rate; (ii) rate of change of anesthetic use with respect to population size; (iii) rate of change of surgical operations with respect to population size; (iv) rate of change of anesthetic use with respect to number of surgical operations. Select your answers from the following:

$$\frac{dS}{dP}, \quad \frac{dS}{dt}, \quad \frac{dP}{dS}, \quad \frac{dP}{dt}, \quad \frac{dA}{dS}, \quad \frac{dA}{dP}, \quad \text{and} \quad \frac{dS}{dA}.$$

(b) Write a type of chain rule that expresses the time rate of change of anesthetic usage, $\dfrac{dA}{dt}$, in terms of three of the derivatives described in part (a).

41. The graph of $x^{2/3} + y^{2/3} = 8$ is the *astroid* in Fig. 3.

(a) Find $\dfrac{dy}{dx}$ by implicit differentiation.

(b) Find the slope of the tangent line at $(8, -8)$.

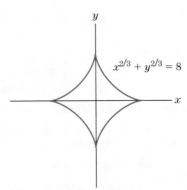

Figure 3 Astroid.

42. Slope of the Folium of Descartes The graph of $x^3 + y^3 = 9xy$ is the folium of Descartes shown in Fig. 4.

(a) Find $\dfrac{dy}{dx}$ by implicit differentiation.

(b) Find the slope of the curve at $(2, 4)$.

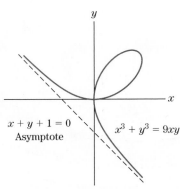

Figure 4 Folium of Descartes.

In Exercises 43–46, x and y are related by the given equation. Use implicit differentiation to calculate the value of $\dfrac{dy}{dx}$ for the given values of x and y.

43. $x^2 y^2 = 9$; $x = 1$, $y = 3$

44. $xy^4 = 48$; $x = 3$, $y = 2$

45. $x^2 - xy^3 = 20$; $x = 5$, $y = 1$

46. $xy^2 - x^3 = 10$; $x = 2$, $y = 3$

47. Cost Analysis and Production A factory's weekly production costs y and its weekly production quantity x are related by the equation $y^2 - 5x^3 = 4$, where y is in thousands of dollars and x is in thousands of units of output.

  (a) Use implicit differentiation to find a formula for $\dfrac{dy}{dx}$, the marginal cost of production.

  (b) Find the marginal cost of production when $x = 4$ and $y = 18$.

  (c) Suppose that the factory begins to vary its weekly production level. Assuming that x and y are differentiable functions of time t, use the method of related rates to find a formula for $\dfrac{dy}{dt}$, the time rate of change of production costs.

  (d) Compute $\dfrac{dy}{dt}$ when $x = 4$, $y = 18$, and the production level is rising at the rate of .3 thousand units per week (that is, when $\dfrac{dx}{dt} = .3$).

48. Use of Books at a Library A town library estimates that, when the population is x thousand persons, approximately y thousand books will be checked out of the library during 1 year, where x and y are related by the equation $y^3 - 8000x^2 = 0$.

  (a) Use implicit differentiation to find a formula for $\dfrac{dy}{dx}$, the rate of change of library circulation with respect to population size.

  (b) Find the value of $\dfrac{dy}{dx}$ when $x = 27$ thousand persons and $y = 180$ thousand books per year.

  (c) Assume that x and y are both differentiable functions of time t, and use the method of related rates to find a formula for $\dfrac{dy}{dt}$, the time rate of change of library circulation.

  (d) Compute $\dfrac{dy}{dt}$ when $x = 27$, $y = 180$, and the population is rising at the rate of 1.8 thousand persons per year $\left(\dfrac{dx}{dt} = 1.8 \right)$. Either use part (c) or use part (b) and the chain rule.

49. Demand Equation Suppose that the price p and quantity x of a certain commodity satisfy the demand equation $6p + 5x + xp = 50$ and that p and x are functions of time, t. Determine the rate at which the quantity x is changing when $x = 4$, $p = 3$, and $\dfrac{dp}{dt} = -2$.

50. Volume of an Oil Spill An offshore oil well is leaking oil onto the ocean surface, forming a circular oil slick about .005 meter thick. If the radius of the slick is r meters, the volume of oil spilled is $V = .005\pi r^2$ cubic meters. If the oil is leaking at a constant rate of 20 cubic meters per hour so that $\dfrac{dV}{dt} = 20$, find the rate at which the radius of the oil slick is increasing at a time when the radius is 50 meters. $\left[\text{Hint: Find a relation between } \dfrac{dV}{dt} \text{ and } \dfrac{dr}{dt}. \right]$

51. Weight and Surface Area Animal physiologists have determined experimentally that the weight W (in kilograms) and the surface area S (in square meters) of a typical horse are related by the empirical equation $S = 0.1W^{2/3}$. How fast is the surface area of a horse increasing at a time when the horse weighs 350 kg and is gaining weight at the rate of 200 kg per year? [Hint: Use the chain rule.]

52. Sales and Advertising Suppose that a kitchen appliance company's monthly sales and advertising expenses are approximately related by the equation $xy - 6x + 20y = 0$, where x is thousands of dollars spent on advertising and y is thousands of dishwashers sold. Currently, the company is spending 10 thousand dollars on advertising and is selling 2 thousand dishwashers each month. If the company plans to increase monthly advertising expenditures at the rate of \$1.5 thousand per month, how fast will sales rise? Use implicit differentiation to answer the question.

When an investment grows steadily at 15% per year, the rate of growth of the investment at any time is proportional to the value of the investment at that time. When a bacteria culture grows in a laboratory dish, the rate of growth of the culture at any moment is proportional to the total number of bacteria in the dish at that moment. These situations are examples of what is called **exponential growth**. A pile of radioactive uranium ^{235}U decays at a rate that at each moment is proportional to the amount of ^{235}U present. This decay of uranium (and of radioactive elements in general) is called **exponential decay**. Both exponential growth and exponential decay can be described and studied in terms of exponential functions and the natural logarithm function. The properties of these functions are investigated in this chapter. Subsequently, we shall explore a wide range of applications in fields such as business, biology, archaeology, public health, and psychology.

4.1   Exponential Functions

Laws of Exponents

FOR REVIEW

To practice working with exponents, see Sec. 0.5, pp. 32–35.

Throughout this section, b will denote a positive number. The function

$$f(x) = b^x$$

is called an **exponential function**, because the variable x is in the exponent. The number b is called the **base** of the exponential function. In Section 0.5 we reviewed the definition of b^x for various values of b and x (although we used the letter r there instead of x). For instance, if $f(x)$ is the exponential function with base 2,

$$f(x) = 2^x,$$

then, $\quad f(0) = 2^0 = 1, \quad f(1) = 2^1 = 2, \quad f(4) = 2^4 = 2 \cdot 2 \cdot 2 \cdot 2 = 16,$

and $\quad f(-1) = 2^{-1} = \frac{1}{2}, \quad f\left(\frac{1}{2}\right) = 2^{1/2} = \sqrt{2}, \quad f\left(\frac{3}{5}\right) = \left(2^{1/5}\right)^3 = \left(\sqrt[5]{2}\right)^3.$

Actually, in Section 0.5, we defined b^x only for rational (that is, integer or fractional) values of x. For other values of x (such as $\sqrt{3}$ or π), it is possible to define b^x by first approximating x with rational numbers and then applying a limiting process. We shall omit the details and simply assume henceforth that b^x can be defined for all numbers x in such a way that the usual laws of exponents remain valid.

Let us state the laws of exponents for reference.

Laws of Exponents

(i) $b^x \cdot b^y = b^{x+y}$ $\qquad\qquad$ **Product rule.**

(ii) $b^{-x} = \dfrac{1}{b^x}$ $\qquad\qquad$ **Changing sign of exponents.**

(iii) $\dfrac{b^x}{b^y} = b^x \cdot b^{-y} = b^{x-y}$ $\qquad$ **Quotient rule.**

(iv) $(b^y)^x = b^{xy}$ $\qquad\qquad$ **Power of a power.**

(v) $a^x b^x = (ab)^x$ $\qquad\qquad$ **Power of a product.**

(vi) $\dfrac{a^x}{b^x} = \left(\dfrac{a}{b}\right)^x$ $\qquad\qquad$ **Power of a quotient.**

Property (iv) may be used to change the appearance of an exponential function. For instance, the function $f(x) = 8^x$ may also be written as $f(x) = (2^3)^x = 2^{3x}$, and $g(x) = \left(\frac{1}{9}\right)^x$ may be written as $g(x) = (1/3^2)^x = (3^{-2})^x = 3^{-2x}$.

EXAMPLE 1   **Using the Laws of Exponents** Use properties of exponents to write the following expressions in the form 2^{kx} for a suitable constant k.

(a) $4^{5x/2}$ $\quad$ **(b)** $(2^{4x} \cdot 2^{-x})^{1/2}$ $\quad$ **(c)** $8^{x/3} \cdot 16^{3x/4}$ $\quad$ **(d)** $\dfrac{10^x}{5^x}$

SOLUTION   **(a)** First, express the base 4 as a power of 2, then use Property (iv):

$$4^{5x/2} = (2^2)^{5x/2} = 2^{2(5x/2)} = 2^{5x}.$$

(b) First, use Property (i) to simplify the quantity inside the parentheses, then use Property (iv):

$$(2^{4x} \cdot 2^{-x})^{1/2} = (2^{4x-x})^{1/2} = (2^{3x})^{1/2} = 2^{(3/2)x}.$$

(c) First, express the bases 8 and 16 as powers of 2, then use (iv) and (i):

$$8^{x/3} \cdot 16^{3x/4} = (2^3)^{x/3} \cdot (2^4)^{3x/4} = 2^x \cdot 2^{3x} = 2^{4x}.$$

(d) Use Property (v) to change the numerator 10^x, then cancel the common term 5^x:

$$\frac{10^x}{5^x} = \frac{(2 \cdot 5)^x}{5^x} = \frac{2^x \cdot 5^x}{5^x} = 2^x.$$

An alternative method is to use Property (vi):

$$\frac{10^x}{5^x} = \left(\frac{10}{5}\right)^x = 2^x.$$

» Now Try Exercise 5

Exponential Functions

Let us now study the graph of the exponential function $y = b^x$ for various values of b. We begin with the special case $b = 2$.

We have tabulated the values of 2^x for $x = 0, \pm 1, \pm 2$, and ± 3 and plotted these values in Fig. 1. Other intermediate values of 2^x for $x = \pm .1, \pm .2, \pm .3, \ldots$, may be obtained from a table or from a graphing calculator. [See Fig. 2(a).] By passing a smooth curve through these points, we obtain the graph of $y = 2^x$, Fig. 2(b).

x	2^x
-3	.125
-2	.25
-1	.5
0	1.0
1	2.0
2	4.0
3	8.0

Figure 1 Values of 2^x.

(a)

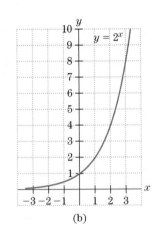

(b)

Figure 2 Graph of $y = 2^x$.

In the same manner, we have sketched the graph of $y = 3^x$ (Fig. 3). The graphs of $y = 2^x$ and $y = 3^x$ have the same basic shape. Also note that they both pass through the point $(0, 1)$ (because $2^0 = 1$, $3^0 = 1$).

Figure 3

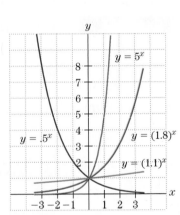

Figure 4

In Fig. 4, we have sketched the graphs of several more exponential functions. Notice that the graph of $y = 5^x$ has a large slope at $x = 0$, since the graph at $x = 0$ is quite steep; however, the graph of $y = (1.1)^x$ is nearly horizontal at $x = 0$, and hence, the slope is close to zero.

There is an important property of the function 3^x that is readily apparent from its graph. Since the graph is always increasing, the function 3^x never assumes the same y-value twice. That is, the only way 3^r can equal 3^s is to have $r = s$. This fact is true for any function of the form $y = b^x$ $(b \neq 0, 1)$ and is useful when you are solving certain equations involving exponentials.

EXAMPLE 2 **Solving Equations with Exponentials** Let $f(x) = 3^{5x}$. Determine all x for which $f(x) = 27$.

SOLUTION Since $27 = 3^3$, we must determine all x for which

$$3^{5x} = 3^3.$$

Equating exponents, we have

$$5x = 3$$

$$x = \frac{3}{5}.$$

>> *Now Try Exercise 17*

In general, for $b > 1$, the equation $b^r = b^s$ implies that $r = s$. The reason is that the graph of $y = b^x$ has the same basic shape as $y = 2^x$ and $y = 3^x$. Similarly, when $0 < b < 1$, the equation $b^r = b^s$ implies that $r = s$, because the graph of $y = b^x$ resembles the graph of $y = .5^x$ and is always decreasing.

There is no need at this point to become familiar with the graphs of the function b^x. We have shown a few graphs merely to make the reader more comfortable with the concept of an exponential function. The main purpose of this section has been to review properties of exponents in a context that is appropriate for our future work.

INCORPORATING TECHNOLOGY

Figure 5

Solving Equations and Intersection of Graphs To determine the value of x at which $2^{2-4x} = 8$, we can proceed as follows. First, enter $Y_1 = 2^{\wedge}(2 - 4X)$ and $Y_2 = 8$. Next, press 2nd [CALC] 5 for **Intersect**. Hit ENTER twice to accept Y_1 as the first function and Y_2 as the second function. Next, enter a guess by either hitting ENTER to accept the value given or entering your own value for X. Finally, hit ENTER to have the calculator find the intersection point of the two graphs. From Fig. 5, we see that $2^{2-4x} = 8$ at $x = -0.25$.

Check Your Understanding 4.1

Solutions can be found following the section exercises.

1. Can a function such as $f(x) = 5^{3x}$ be written in the form $f(x) = b^x$? If so, what is b?

2. Solve the equation $7 \cdot 2^{6-3x} = 28$.

EXERCISES 4.1

Write each expression in Exercises 1–14 in the form 2^{kx} or 3^{kx}, for a suitable constant k.

1. 4^x, $(\sqrt{3})^x$, $(\frac{1}{9})^x$

2. 27^x, $(\sqrt[3]{2})^x$, $(\frac{1}{8})^x$

3. $8^{2x/3}$, $9^{3x/2}$, $16^{-3x/4}$

4. $9^{-x/2}$, $8^{4x/3}$, $27^{-2x/3}$

5. $(\frac{1}{4})^{2x}$, $(\frac{1}{8})^{-3x}$, $(\frac{1}{81})^{x/2}$

6. $(\frac{1}{9})^{2x}$, $(\frac{1}{27})^{x/3}$, $(\frac{1}{16})^{-x/2}$

7. $6^x \cdot 3^{-x}$, $\dfrac{15^x}{5^x}$, $\dfrac{12^x}{2^{2x}}$

8. $7^{-x} \cdot 14^x$, $\dfrac{2^x}{6^x}$, $\dfrac{3^{2x}}{18^x}$

9. $\dfrac{3^{4x}}{3^{2x}}$, $\dfrac{2^{5x+1}}{2 \cdot 2^{-x}}$, $\dfrac{9^{-x}}{27^{-x/3}}$

10. $\dfrac{2^x}{6^x}$, $\dfrac{3^{-5x}}{3^{-2x}}$, $\dfrac{16^x}{8^{-x}}$

11. $2^{3x} \cdot 2^{-5x/2}$, $3^{2x} \cdot (\frac{1}{3})^{2x/3}$

12. $2^{5x/4} \cdot (\frac{1}{2})^x$, $3^{-2x} \cdot 3^{5x/2}$

13. $(2^{-3x} \cdot 2^{-2x})^{2/5}$, $(9^{1/2} \cdot 9^4)^{x/9}$

14. $(3^{-x} \cdot 3^{x/5})^5$, $(16^{1/4} \cdot 16^{-3/4})^{3x}$

15. Find a number b such that the function $f(x) = 3^{-2x}$ can be written in the form $f(x) = b^x$.

16. Find b so that $8^{-x/3} = b^x$ for all x.

Solve the following equations for x.

17. $5^{2x} = 5^2$

18. $10^{-x} = 10^2$

19. $(2.5)^{2x+1} = (2.5)^5$

20. $(3.2)^{x-3} = (3.2)^5$

21. $10^{1-x} = 100$

22. $2^{4-x} = 8$

23. $3(2.7)^{5x} = 8.1$

24. $4(2.7)^{2x-1} = 10.8$

25. $(2^{x+1} \cdot 2^{-3})^2 = 2$

26. $(3^{2x} \cdot 3^2)^4 = 3$

27. $2^{3x} = 4 \cdot 2^{5x}$

28. $3^{5x} \cdot 3^x - 3 = 0$

29. $(1 + x)2^{-x} - 5 \cdot 2^{-x} = 0$

30. $(2 - 3x)5^x + 4 \cdot 5^x = 0$

31. $2^x - \dfrac{8}{2^{2x}} = 0$

32. $2^x - \dfrac{1}{2^x} = 0$

[*Hint:* In Exercises 33–36, let $X = 2^x$ or $X = 3^x$.]

33. $2^{2x} - 6 \cdot 2^x + 8 = 0$ **34.** $2^{2x+2} - 17 \cdot 2^x + 4 = 0$

35. $3^{2x} - 12 \cdot 3^x + 27 = 0$ **36.** $2^{2x} - 4 \cdot 2^x - 32 = 0$

The expressions in Exercises 37–42 may be factored as shown. Find the missing factors.

37. $2^{3+h} = 2^3(\quad)$ **38.** $5^{2+h} = 25(\quad)$

39. $2^{x+h} - 2^x = 2^x(\quad)$ **40.** $5^{x+h} + 5^x = 5^x(\quad)$

41. $3^{x/2} + 3^{-x/2} = 3^{-x/2}(\quad)$ **42.** $5^{7x/2} - 5^{x/2} = \sqrt{5^x}(\quad)$

TECHNOLOGY EXERCISES

43. Graph the function $f(x) = 2^x$ in the window $[-1, 2]$ by $[-1, 4]$, and estimate the slope of the graph at $x = 0$.

44. Graph the function $f(x) = 3^x$ in the window $[-1, 2]$ by $[-1, 8]$, and estimate the slope of the graph at $x = 0$.

45. By trial and error, find a number of the form $b = 2.\square$ (just one decimal place) with the property that the slope of the graph of $y = b^x$ at $x = 0$ is as close to 1 as possible.

Solutions to Check Your Understanding 4.1

1. If $5^{3x} = b^x$, then, when $x = 1$, $5^{3(1)} = b^1$, which says that $b = 125$. This value of b certainly works, because

$$5^{3x} = (5^3)^x = 125^x.$$

2. Divide both sides of the equation by 7, obtaining

$$2^{6-3x} = 4.$$

Now 4 can be written as 2^2. So we have

$$2^{6-3x} = 2^2.$$

Equate exponents to obtain

$$6 - 3x = 2$$
$$4 = 3x$$
$$x = \tfrac{4}{3}.$$

4.2 The Exponential Function e^x

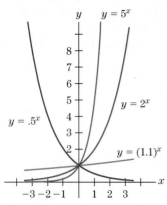

Figure 1 Several exponential functions.

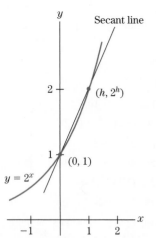

Figure 2 A secant line to the graph of $y = 2^x$.

We begin by examining the graphs of the exponential functions shown in Fig. 1. They all pass through $(0, 1)$, but with different slopes there. Notice that the graph of $y = 5^x$ is quite steep at $x = 0$, while the graph of $y = (1.1)^x$ is nearly horizontal at $x = 0$. It turns out that, at $x = 0$, the graph of $y = 2^x$ has a slope of approximately .693, while the graph of 3^x has a slope of approximately 1.1.

Evidently, there is a particular value of the base b, between 2 and 3, where the graph of $y = b^x$ has slope *exactly* 1 at $x = 0$. We take this fact for granted and denote this special value of b by the letter e, and we call

$$f(x) = e^x$$

the exponential function. The number e is an important constant of nature that has been calculated to thousands of decimal places. To 10 significant digits, we have $e = 2.718281828$. For our purposes, it is usually sufficient to think of e as "approximately 2.7."

Our goal in this section is to find a formula for the derivative of $y = e^x$. It turns out that the calculations for e^x and 2^x are very similar. Since many people are more comfortable working with 2^x rather than e^x, we shall first analyze the graph of $y = 2^x$. Then we shall draw the appropriate conclusions about the graph of $y = e^x$.

Before computing the slope of $y = 2^x$ at an arbitrary x, let us consider the special case $x = 0$. Denote the slope at $x = 0$ by m. We shall use the secant-line approximation of the derivative to approximate m. We proceed by constructing the secant line in Fig. 2. The slope of the secant line through $(0, 1)$ and $(h, 2^h)$ is $\dfrac{2^h - 1}{h}$. As h approaches zero, the slope of the secant line approaches the slope of $y = 2^x$ at $x = 0$. That is,

$$m = \lim_{h \to 0} \frac{2^h - 1}{h}.$$

Assuming that this limit exists, we can estimate the value of m by taking h smaller and smaller. Table 1 shows the values of the expressions for $h = .1, .01, \ldots, .0000001$. From the table it is reasonable to conclude that $m \approx .693$.

Table 1 $Y_1 = (2^{\wedge}X - 1)/X$

NORMAL FLOAT AUTO REAL RADIAN MP				
X	Y_1			
.1	.71773			
.01	.69556			
.001	.69339			
1E⁻4	.69317			
1E⁻5	.69315			
1E⁻6	.69315			
1E⁻7	.69315			
X=				

Since m equals the slope of $y = 2^x$ at $x = 0$, we have

$$m = \lim_{h \to 0} \frac{2^h - 1}{h} = \frac{d}{dx}(2^x)\Big|_{x=0} \approx .693. \tag{1}$$

Now that we have estimated the slope of $y = 2^x$ at $x = 0$, let us compute the slope for an arbitrary value of x. We construct a secant line through $(x, 2^x)$ and a nearby point $(x + h, 2^{x+h})$ on the graph. The slope of the secant line is

$$\frac{2^{x+h} - 2^x}{h}.$$

To obtain the slope of the tangent line, we take the limit of the slope of the secant line as h approaches 0. We get

$$\lim_{h \to 0} \frac{2^{x+h} - 2^x}{h} = \lim_{h \to 0} \frac{2^x 2^h - 2^x}{h} \qquad \text{Product rule for exponents.}$$

$$= \lim_{h \to 0} \frac{2^x(2^h - 1)}{h} \qquad \text{Factor } 2^x.$$

$$= 2^x \lim_{h \to 0} \frac{2^h - 1}{h} \qquad \begin{array}{l}2^x \text{ does not affect } h, \text{ so you can} \\ \text{move it outside the limit.}\end{array}$$

$$= m2^x, \text{ where } m \approx .693. \qquad \text{Use Equation (1).}$$

Since the slope of the tangent line is the derivative of 2^x, we have

Derivative of $y = 2^x$

$$\frac{d}{dx}(2^x) = m2^x, \quad \text{where } m = \frac{d}{dx}(2^x)\Big|_{x=0} \approx .693. \tag{2}$$

EXAMPLE 1 **Derivative of an Exponential** Calculate

(a) $\dfrac{d}{dx}(2^x)\Big|_{x=3}$ and (b) $\dfrac{d}{dx}(2^x)\Big|_{x=-1}$.

SOLUTION Using (2), we have

(a) $\dfrac{d}{dx}(2^x)\Big|_{x=3} = m \cdot 2^3 = 8m \approx 8(.693) = 5.544.$

(b) $\dfrac{d}{dx}(2^x)\Big|_{x=-1} = m \cdot 2^{-1} = .5m \approx .5(.693) = .3465.$ **» Now Try Exercise 3**

The calculations just carried out for $y = 2^x$ can be carried out for $y = b^x$, where b is any positive number. Equation (2) will read exactly the same except that 2 will be replaced by b. Thus, we have the following formula for the derivative of the function $f(x) = b^x$.

FOR REVIEW

Exponential functions are not power functions. Contrast this derivative with The Power Rule in Sec. 1.3, p. 73.

Derivative of $y = b^x$

$$\frac{d}{dx}(b^x) = mb^x, \quad \text{where } m = \frac{d}{dx}(b^x)\Big|_{x=0} \tag{3}$$

Our calculations showed that if $b = 2$ then $m \approx .693$. If $b = 3$, it turns out that $m \approx 1.1$. (See Exercise 1.) The derivative formula in (3) is simple when $m = 1$, that is, when the graph of $y = b^x$ has slope 1 at $x = 0$. As we said earlier, this special value of b is denoted by the letter e. Thus, the number e has the property that

$$\frac{d}{dx}(e^x)\Big|_{x=0} = 1 \qquad (4)$$

and

> **Derivative of** $y = e^x$
>
> $$\frac{d}{dx}(e^x) = 1 \cdot e^x = e^x. \qquad (5)$$

The geometric interpretation of equation (4) is that the curve $y = e^x$ has slope 1 at $x = 0$. The geometric interpretation of (5) is that the slope of the curve $y = e^x$ at an arbitrary value of x is exactly equal to the value of the function e^x at that point. (See Fig. 3.)

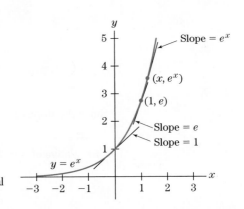

Figure 3 Fundamental properties of e^x.

The function e^x is the same type of function as 2^x and 3^x except that differentiating e^x is much easier. In fact, the next section shows that 2^x can be written as e^{kx} for a suitable constant k. The same is true for 3^x. For this reason, functions of the form e^{kx} are used in almost all applications that require an exponential type of function to describe a physical, economic, or biological phenomenon.

EXAMPLE 2   **Applying the Laws of Exponents with** e^x Write the given expressions in the form Ae^{kx} for some constants A and k.

(a) $\dfrac{(3e^{2x})^2 \cdot e^x}{e^{-2x}}$ (b) $\sqrt{\dfrac{e^x}{e^{7x}}}$

SOLUTION   (a) $\dfrac{(3e^{2x})^2 \cdot e^x}{e^{-2x}} = 3^2(e^{2x})^2 \cdot e^x \cdot e^{-(-2x)}$ Power of product. Change sign of exponent in denominator.

$= 9 \cdot e^{(2x)(2)} \cdot e^{2x+x}$ Power of a power and product rule.

$= 9e^{4x}e^{3x} = 9e^{7x}$ Product rule.

Here, $A = 9$ and $k = 7$.

(b) $\sqrt{\dfrac{e^x}{e^{7x}}} = (e^x \cdot e^{-7x})^{\frac{1}{2}}$ Change radical to power. Change sign of exponent in denominator.

$= (e^{x-7x})^{\frac{1}{2}}$ Product rule.

$= (e^{-6x})^{\frac{1}{2}} = e^{-6x \cdot \frac{1}{2}} = e^{-3x}$ Power of a power.

Here, $A = 1$ and $k = -3$.

▶▶ Now Try Exercise 9

| **EXAMPLE 3** | **Tangent Line to an Exponential Graph** Find the tangent line to the graph of $f(x) = e^x$ when $x = 1$. |

SOLUTION   When $x = 1$, $y = f(1) = e$. So, the point $(1, e)$ is on the tangent line. Since $\frac{d}{dx}[e^x] = e^x$, the slope of the tangent line when $x = 1$ is

$$m = \frac{d}{dx}[e^x]\Big|_{x=1} = e^x|_{x=1} = e^1 = e.$$

Therefore, the equation of the tangent line in point–slope form is $y - e = e(x - 1)$ or $y = ex$ (see Fig. 3.)   **» Now Try Exercise 19**

| **EXAMPLE 4** | **Derivatives Involving e^x** Differentiate |

(a) $(1 + x^2)e^x$ and   **(b)** $\dfrac{1 + e^x}{2x}$.

SOLUTION   **(a)** $\dfrac{d}{dx}[(1 + x^2)e^x] = (1 + x^2)\dfrac{d}{dx}[e^x] + e^x\dfrac{d}{dx}(1 + x^2)$ Product rule.

$$= (1 + x^2)e^x + e^x(2x)$$

$$= e^x(x^2 + 2x + 1) = e^x(1 + x)^2.$$

(b) $\dfrac{d}{dx}\left[\dfrac{1 + e^x}{2x}\right] = \dfrac{(2x)\frac{d}{dx}[e^x + 1] - (e^x + 1)\frac{d}{dx}[2x]}{(2x)^2}$ Quotient rule.

$$= \frac{2xe^x - (e^x + 1)(2)}{4x^2} = \frac{2xe^x - 2e^x - 2}{4x^2}$$

$$= \frac{2(xe^x - e^x - 1)}{4x^2} = \frac{xe^x - e^x - 1}{2x^2}.$$ **» Now Try Exercise 27**

In the following section, we will compute the derivative of $y = e^{g(x)}$ for any differentiable function g and, in particular, $y = e^{kx}$. However, as we now illustrate, by relating e^{kx} to e^x, we can also differentiate $y = e^{kx}$ by using the generalized power rule for differentiation.

| **EXAMPLE 5** | **Differentiating an Exponential** Find the derivative of $y = e^{-x}$. |

SOLUTION   Let $f(x) = e^x$. Then,

$$e^{-x} = (e^x)^{-1} = (f(x))^{-1};$$

and so,

$$\frac{d}{dx}[e^{-x}] = \frac{d}{dx}[(f(x))^{-1}]$$

$$= (-1)(f(x))^{-2}\frac{d}{dx}[f(x)]$$ General power rule.

$$= (-1)(e^x)^{-2}\frac{d}{dx}[e^x]$$ Plug in $f(x) = e^x$.

$$= (-1)e^{-2x}e^x$$ Use Equation (5).

$$= -e^{-x}.$$ **» Now Try Exercise 45**

EXAMPLE 6 **Graph of an Exponential** Sketch the graph of $y = e^{-x}$.

SOLUTION Note that, since $e > 1$, we have $0 < 1/e < 1$. In fact, $1/e \approx .37$. Now,

$$y = e^{-x} = \frac{1}{e^x} = \left(\frac{1}{e}\right)^x ;$$

and so, $f(x) = e^{-x}$ is an exponential function with base $b = \dfrac{1}{e}$, which is less than 1.

Hence, the graph of $y = e^{-x}$ is always decreasing and looks like the graph of the exponential function $y = .5^x$. (See Fig. 1.) With the help of a calculator, we form a table of values of $y = e^{-x}$ for selected values of x, as shown in Table 1.

Using these values, we sketch the graph of $y = e^{-x}$ in Fig. 4.

Table 1

x	e^{-x}
-3	20.09
-2	7.39
-1	2.72
0	1
1	.37
2	.14
3	.05

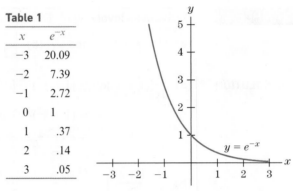

Figure 4

>> *Now Try Exercise 51*

We end the section by listing important properties of exponential functions that we have encountered so far.

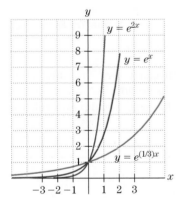

Figure 5

Properties of the Functions $f(x) = e^{kx}$ Figure 5 shows the graphs of several functions of this type when k is a positive number. These curves $y = e^{kx}$, k positive, have several properties in common:

1. $(0, 1)$ is on the graph.
2. The graph lies strictly above the x-axis (e^{kx} is never zero).
3. The x-axis is an asymptote as x becomes large negatively.
4. The graph is always increasing and concave up.

When k is negative, the graph of $y = e^{kx}$ is decreasing. (See Fig. 6.) Note the following properties of the curves $y = e^{kx}$, k negative:

1. $(0, 1)$ is on the graph.
2. The graph lies strictly above the x-axis.
3. The x-axis is an asymptote as x becomes large positively.
4. The graph is always decreasing and concave up.

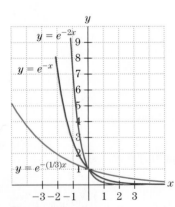

Figure 6

The Functions $f(x) = b^x$ If b is a positive number, the function $f(x) = b^x$ may be written in the form $f(x) = e^{kx}$ for some k. For example, take $b = 2$. From Fig. 3, it is clear that there is some value of x such that $e^x = 2$. Call this value k, so $e^k = 2$. Then,

$$2^x = (e^k)^x = e^{kx}$$

for all x. In general, if b is any positive number, there is a value of x—say, $x = k$—such that $e^k = b$. In this case, $b^x = (e^k)^x = e^{kx}$. Thus, all the curves $y = b^x$ discussed in Section 4.1 can be written in the form $y = e^{kx}$. This is one reason why we have focused on exponential functions with base e, instead of studying $y = 2^x$, $y = 3^x$, and so on.

INCORPORATING TECHNOLOGY

NORMAL FLOAT AUTO REAL RADIAN MP

$e^{(2/3)}$

　　　　1.947734041

$e^{(-2)}/3$

　　　　.0451117611

$\frac{d}{dx}(e^x)|_{x=0}$

　　　　1.000000167

Figure 7

All the techniques we have introduced in previous sections for analyzing functions with the TI-83/84 apply to exponential functions as well. To input an exponential function, press [2nd] [e^x]. For example, in Fig. 7, we evaluate $e^{2/3}$, $e^{-2}/3$, and the derivative of e^x at $x = 0$. We emphasize the importance of using parentheses when evaluating e^x.

Check Your Understanding 4.2

Solutions can be found following the section exercises.

1. Solve the following equation for x:

$$e^{6x} = e^3.$$

2. Differentiate $y = (x + e^x)^4$

EXERCISES 4.2

1. Show that

$$\frac{d}{dx}(3^x)\Big|_{x=0} \approx 1.1$$

by calculating the slope $\dfrac{3^h - 1}{h}$ of the secant line passing through the points $(0, 1)$ and $(h, 3^h)$. Take $h = .1, .01,$ and $.001$.

2. Show that

$$\frac{d}{dx}(2.7^x)\Big|_{x=0} \approx .99$$

by calculating the slope $\dfrac{2.7^h - 1}{h}$ of the secant line passing through the points $(0, 1)$ and $(h, 2.7^h)$. Take $h = .1, .01,$ and $.001$.

In Exercises 3–6, compute the given derivatives with the help of formulas (1)–(4).

3. (a) $\dfrac{d}{dx}(2^x)\Big|_{x=1}$　　(b) $\dfrac{d}{dx}(2^x)\Big|_{x=-2}$

4. (a) $\dfrac{d}{dx}(2^x)\Big|_{x=1/2}$　　(b) $\dfrac{d}{dx}(2^x)\Big|_{x=2}$

5. (a) $\dfrac{d}{dx}(e^x)\Big|_{x=1}$　　(b) $\dfrac{d}{dx}(e^x)\Big|_{x=-1}$

6. (a) $\dfrac{d}{dx}(e^x)\Big|_{x=e}$　　(b) $\dfrac{d}{dx}(e^x)\Big|_{x=1/e}$

Write each expression in the form e^{kx} for a suitable constant k.

7. $(e^2)^x$, $\left(\dfrac{1}{e}\right)^x$

8. $(e^3)^{x/5}$, $\left(\dfrac{1}{e^2}\right)^x$

9. $\left(\dfrac{1}{e^3}\right)^{2x}$, $e^{1-x} \cdot e^{3x-1}$

10. $\left(\dfrac{e^5}{e^3}\right)^x$, $e^{4x+2} \cdot e^{x-2}$

11. $(e^{4x} \cdot e^{6x})^{3/5}$, $\dfrac{1}{e^{-2x}}$

12. $\sqrt{e^{-x} \cdot e^{7x}}$, $\dfrac{e^{-3x}}{e^{-4x}}$

Solve each equation for x.

13. $e^{5x} = e^{20}$

14. $e^{1-x} = e^2$

15. $e^{x^2-2x} = e^8$

16. $e^{-x} = 1$

17. $e^x(x^2 - 1) = 0$

18. $4e^x(x^2 + 1) = 0$

19. Find an equation of the tangent line to the graph of $f(x) = e^x$, where $x = -1$. (Use $1/e = .37$.)

20. Find the point on the graph of $f(x) = e^x$, where the tangent line is parallel to $y = x$.

21. Use the first and second derivative rules from Section 2.2 to show that the graph of $y = e^x$ has no relative extreme points and is always concave up.

22. Estimate the slope of e^x at $x = 0$ by calculating the slope $\frac{e^h - 1}{h}$ of the secant line passing through the points $(0, 1)$ and (h, e^h). Take $h = .01, .001,$ and $.0001$.

23. Suppose that $A = (a, b)$ is a point on the graph of e^x. What is the slope of the graph of e^x at the point A?

24. Find the slope–point form of the equation of the tangent line to the graph of e^x at the point (a, e^a).

Differentiate the following functions.

25. $y = 3e^x - 7x$

26. $y = \dfrac{2x + 4 - 5e^x}{4}$

27. $y = xe^x$

28. $y = (x^2 + x + 1)e^x$

29. $y = 8e^x(1 + 2e^x)^2$

30. $y = (1 + e^x)(1 - e^x)$

31. $y = \dfrac{e^x}{x + 1}$

32. $y = \dfrac{x + 1}{e^x}$

33. $y = \dfrac{e^x - 1}{e^x + 1}$

34. $y = \sqrt{e^x + 1}$

35. The graph of $y = x - e^x$ has one extreme point. Find its coordinates and decide whether it is a maximum or a minimum. (Use the second derivative test.)

36. Find the extreme points on the graph of $y = x^2 e^x$, and decide which one is a maximum and which one is a minimum.

37. Find the point on the graph of $y = (1 + x^2)e^x$ where the tangent line is horizontal.

38. Show that the tangent line to the graph of $y = e^x$ at the point (a, e^a) is perpendicular to the tangent line to the graph of $y = e^{-x}$ at the point (a, e^{-a}).

39. Find the slope of the tangent line to the curve $y = xe^x$ at $(0, 0)$.

40. Find the slope of the tangent line to the curve $y = xe^x$ at $(1, e)$.

41. Find the equation of the tangent line to the curve $y = \dfrac{e^x}{1 + 2e^x}$ at $(0, \frac{1}{3})$.

42. Find the equation of the tangent line to the curve $y = \dfrac{e^x}{x + e^x}$ at $(0, 1)$.

Find the first and second derivatives.

43. $f(x) = e^x(1 + x)^2$ 44. $f(x) = \dfrac{e^x}{x}$

45. Compute the following derivatives.

   (a) $\dfrac{d}{dx}(5e^x)$

   (b) $\dfrac{d}{dx}(e^x)^{10}$ [*Hint:* The general power rule.]

   (c) Use the fact that $e^{2+x} = e^2 \cdot e^x$ to find $\dfrac{d}{dx}(e^{x+2})$.

46. (a) Use the fact that $e^{4x} = (e^x)^4$ to find $\dfrac{d}{dx}(e^{4x})$. Simplify the derivative as much as possible.

   (b) Take an approach similar to the one in (a) and show that, if k is a constant, $\dfrac{d}{dx}(e^{kx}) = ke^{kx}$.

Sketch the graphs of the following functions.

47. $y = e^{2x}$ 48. $y = 1 - e^x$

49. $y = e^{x/2}$ 50. $y = e^{x-1}$

51. $y = -e^{-x} + 1$ 52. $y = 2e^{-x}$

TECHNOLOGY EXERCISES

53. Find the equation of the tangent line to the graph of $y = e^x$ at $x = 0$. Then, graph the function and the tangent line together to confirm that your answer is correct.

54. (a) Graph $y = e^x$.
   (b) Zoom in on the region near $x = 0$ until the curve appears as a straight line and estimate the slope of the line. This number is an estimate of $\dfrac{d}{dx}e^x$ at $x = 0$. Compare your answer with the actual slope, 1.
   (c) Repeat parts (a) and (b) for $y = 2^x$. Observe that the slope at $x = 0$ is not 1.

55. Set $Y_1 = e^x$ and use your calculator's derivative command to specify Y_2 as the derivative of Y_1. Graph the two functions simultaneously in the window $[-1, 3]$ by $[-3, 20]$ and observe that the graphs overlap.

56. Calculate values of $\dfrac{10^x - 1}{x}$ for small values of x, and use them to estimate $\dfrac{d}{dx}(10^x)\Big|_{x=0}$. What is the formula for $\dfrac{d}{dx}(10^x)$?

Solutions to Check Your Understanding 4.2

1. This problem is similar to Check Your Understanding Problem 2 of Section 4.1. The idea is to equate exponents and then solve for x:

$$e^{6x} = e^3$$
$$6x = 3$$
$$x = \tfrac{1}{2}.$$

2. $\dfrac{d}{dx}(x + e^x)^4 = 4(x + e^x)^3 \dfrac{d}{dx}(x + e^x)$ Generalized power rule.

   $= 4(x + e^x)^3(1 + e^x)$ $\dfrac{d}{dx}(x) = 1, \dfrac{d}{dx}(e^x) = e^x$

4.3 Differentiation of Exponential Functions

We have shown that $\dfrac{d}{dx}(e^x) = e^x$. Using this fact and the chain rule, we can differentiate functions of the form $y = e^{g(x)}$, where $g(x)$ is any differentiable function. The reason is that $e^{g(x)}$ is the composite of two functions. Indeed, if $f(x) = e^x$, then

$$e^{g(x)} = f(g(x)).$$

FOR REVIEW

The Chain Rule, Sec. 3.2, p. 203.

Thus, by the chain rule, we have

$$\dfrac{d}{dx}(e^{g(x)}) = f'(g(x))g'(x)$$
$$= f(g(x))g'(x) \qquad \text{Since } f'(x) = f(x)$$
$$= e^{g(x)}g'(x).$$

So, we have the following result:

Chain Rule for Exponential Functions   Let $g(x)$ be any differentiable function. Then,

$$\frac{d}{dx}(e^{g(x)}) = e^{g(x)}g'(x). \tag{1}$$

If we write $u = g(x)$, then equation (1) can be written in the form

$$\frac{d}{dx}(e^u) = e^u\frac{du}{dx}. \tag{1a}$$

In words, formula (1) says that the derivative of $e^{g(x)}$ is $e^{g(x)}$ times the derivative of $g(x)$. So to find the derivative of $e^{g(x)}$, the only computation required is the derivative of $g(x)$. When applying formula (1), it is sometimes more convenient to use the derivative notation $\frac{d}{dx}(g(x))$ instead of $g'(x)$.

EXAMPLE 1   **Derivative of Functions of the Form $e^{g(x)}$** Differentiate

(a) $y = e^{5x}$ **(b)** $y = e^{x^2-1}$ **(c)** $y = e^{x-1/x}$

SOLUTION   Keep in mind that the derivative of $e^{g(x)}$ is $e^{g(x)}$ times $g'(x)$. So, in each part, identify $g(x)$, find $g'(x)$, and form the product $e^{g(x)}g'(x)$.

(a) Here $g(x) = 5x, g'(x) = 5$, so

$$\frac{d}{dx}(e^{5x}) = e^{5x}\frac{d}{dx}(5x) = e^{5x} \cdot 5 = 5e^{5x}.$$

(b) Here $g(x) = x^2 - 1, g'(x) = 2x$, so

$$\frac{d}{dx}(e^{x^2-1}) = e^{x^2-1}\frac{d}{dx}(x^2 - 1) = e^{x^2-1} \cdot (2x) = 2xe^{x^2-1}.$$

(c) Here $g(x) = x - \dfrac{1}{x}, g'(x) = 1 + \dfrac{1}{x^2}$, so

$$\frac{d}{dx}(e^{x-1/x}) = e^{x-1/x} \cdot \left(1 + \frac{1}{x^2}\right) = \left(1 + \frac{1}{x^2}\right)e^{x-1/x}.$$

>> *Now Try Exercise 7*

We can use various differentiation rules in conjunction with the chain rule to compute more challenging derivatives. As an illustration, we derive the following useful formula:

Let C and k be arbitrary constants, then

$$\frac{d}{dx}(Ce^{kx}) = kCe^{kx} \tag{2}$$

The proof of (2) is straightforward.

$$\frac{d}{dx}(Ce^{kx}) = C\frac{d}{dx}(e^{kx}) \qquad \text{Constant multiple rule for derivatives.}$$

$$= Ce^{kx}\frac{d}{dx}(kx) \qquad \text{Formula (1).}$$

$$= kCe^{kx} \qquad \text{Derivative of } kx \text{ is } k.$$

EXAMPLE 2   **Quotient Rule with Exponential Functions** Differentiate $y = \frac{3e^{2x}}{1 + x^2}$.

SOLUTION

$$\frac{d}{dx}\left[\frac{3e^{2x}}{1+x^2}\right] = \frac{(1+x^2)\frac{d}{dx}(3e^{2x}) - 3e^{2x}\frac{d}{dx}(1+x^2)}{(1+x^2)^2}$$

Quotient rule with $f(x) = 3e^{2x}$; $g(x) = 1 + x^2$.

$$= \frac{(1+x^2)(3)(2)e^{2x} - 3e^{2x}(2x)}{(1+x^2)^2}$$

Formula (2) with $C = 3$ and $k = 2$.

$$= \frac{e^{2x}(6(1+x^2) - 6x)}{(1+x^2)^2}$$

Factor e^{2x} in the numerator.

$$= \frac{e^{2x}(6x^2 - 6x + 6)}{(1+x^2)^2} = 6e^{2x}\frac{x^2 - x + 1}{(1+x^2)^2}$$

Factor 6 in the numerator.

>> *Now Try Exercise 15*

The following application illustrates the important role of the exponential function in modeling applications in finance and analyzing investment portfolios. Applications of this type will be developed in greater detail in Chapter 5.

EXAMPLE 3   **Combined Returns from Two Investments** Joan invested $25,000 in two separate stocks A and B. She invested $15,000 in A and $10,000 in B. The value of her account, in thousands of dollars, t years from its inception, is given by

$$f(t) = 15e^{0.02t} + 10e^{-0.06t}.$$

(As you will learn in Chapter 5, the expression $e^{0.02t}$ indicates a return of 2% compounded continuously, while $e^{-0.06t}$ indicates a negative return of 6%.)
(a) What is the value of the account after 4 years? After 12 years?
(b) At what rate is the account changing after 4 years? After 12 years?

SOLUTION   **(a)** We evaluate $f(t)$ when $t = 4$ and when $t = 12$:

$$f(4) = 15e^{0.02(4)} + 10e^{-0.06(4)} = 15e^{0.08} + 10e^{-0.24} \approx 24.1156 \quad \text{thousand dollars}$$

$$f(12) = 15e^{0.02(12)} + 10e^{-0.06(12)} = 15e^{0.24} + 10e^{-0.72} \approx 23.9363 \quad \text{thousand dollars.}$$

Thus, the combined value of the stocks is about $24,115.60 after 4 years and about $23,936.60 after 12 years.
(b) The rate of change of the account is given by $f'(t)$. With the help of (2), we find

$$f'(t) = 15(.02)e^{0.02t} + 10(-.06)e^{-0.06t} = .3e^{0.02t} - .6e^{-0.06t}$$

So the rate of change of the account after 4 years is

$$f'(4) = .3e^{0.02(4)} - .6e^{-0.06(4)} = .3e^{0.08} - .6e^{-0.24} \approx -.14699 \text{ thousand dollars per year.}$$

Thus the account is losing money at the rate of almost $147 per year. After 12 years,

$$f'(12) = .3e^{0.02(12)} - .6e^{-0.06(12)} = .3e^{0.24} - .6e^{-0.72} \approx .08932 \text{ thousand dollars per year.}$$

Thus, after 12 years, the account is making money at the rate of about $89 per year. The graph of $f(t)$ is shown in Fig. 1. It shows that the account lost money for about 9 years but then slowly started increasing.

To find the exact value of t where the graph in Fig. 1 has a minimum, we have to solve the equation $f'(t) = 0$ or $.3e^{0.02t} - .6e^{-0.06t} = 0$. The solution requires knowledge of the inverse of the exponential function. This will be the topic of our next section.

>> *Now Try Exercise 33*

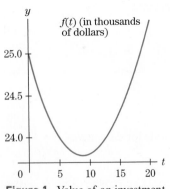

Figure 1 Value of an investment portfolio.

Check Your Understanding 4.3

Solutions can be found following the section exercises.

1. Differentiate te^{-t^2}

2. Differentiate $[e^{-3x}(1 + e^{6x})]^{12}$.

EXERCISES 4.3

Differentiate the following functions.

1. $f(x) = e^{2x+3}$

2. $f(x) = e^{-3x-2}$

3. $f(x) = e^{4x^2-x}$

4. $f(x) = e^{(1+x)^3}$

5. $f(x) = e^{e^x}$

6. $f(x) = e^{\frac{1}{x}}$

7. $f(x) = e^{\sqrt{x}}$

8. $f(x) = e^{\sqrt{x^2+1}}$

9. $f(x) = -7e^{\frac{x}{7}}$

10. $f(x) = 10e^{\frac{-x-2}{5}}$

11. $f(t) = 4e^{0.05t} - 23e^{0.01t}$

12. $f(t) = 2e^{t/2} - .4e^{0.001\,t}$

13. $f(t) = (t^2 + 2e^t)e^{t-1}$

14. $f(t) = (t^3 - 3t)e^{1+t}$

15. $f(x) = \left(x + \dfrac{1}{x}\right)e^{2x}$

16. $f(x) = e^{e^{e^x}}$

17. $f(x) = \dfrac{e^x + e^{-x}}{e^x - e^{-x}}$

18. $f(x) = \dfrac{e^x - e^{-x}}{e^x + e^{-x}}$

19. $f(x) = \sqrt{e^x + 1}$

20. $f(x) = e^{ex}$

In Exercises 21–26, simplify the function before differentiating.

21. $f(x) = (e^{3x})^5$

22. $f(x) = e^x e^{2x} e^{3x}$

23. $f(x) = \dfrac{1}{\sqrt{e^x}}$

24. $f(t) = e^{3t}(e^{2t} - e^{4t})$

25. $f(x) = \dfrac{e^x + 5e^{2x}}{e^x}$

26. $f(t) = \sqrt{e^{3x}}$

In Exercises 27–32, find the values of x at which the function has a possible relative maximum or minimum point. (Recall that e^x is positive for all x.) Use the second derivative to determine the nature of the function at these points.

27. $f(x) = (1 + x)e^{-3x}$

28. $f(x) = (1 - x)e^{2x}$

29. $f(x) = \dfrac{3 - 4x}{e^{2x}}$

30. $f(x) = \dfrac{4x - 1}{e^{x/2}}$

31. $f(x) = (5x - 2)e^{1-2x}$

32. $f(x) = (2x - 5)e^{3x-1}$

33. An Investment Portfolio The value of an investment portfolio consisting of two stocks is given by $f(t) = 3e^{0.06t} + 2e^{0.02t}$, where t is the number of years since the inception of the portfolio, and $f(t)$ is in thousands of dollars.

(a) What is the initial dollar amount invested?

(b) What is the value of the portfolio after 5 years?

(c) At what rate is the investment appreciating after 5 years?

34. Depreciation of Assets The value of a computer t years after purchase is $v(t) = 2000e^{-0.35t}$ dollars. At what rate is the computer's value falling after 3-years?

35. The Most Expensive Artwork to Date The highest price ever paid for an artwork at auction was for Pablo Picasso's 1955 painting *Les femmes d'Alger*, which fetched $179.4 million in a Christie's auction in 2015. The painting was last sold in 1997 for $31.9 million. If the painting keeps on appreciating at its current rate, then a model for its value is given by $f(t) = 31.87e^{0.096t}$, where $f(t)$ is in millions of dollars and t is the number of years since 1997.

(a) Verify that this function gives an approximation of the value of the painting in 2015.

(b) At what rate is the painting appreciating in 2015?

(c) Estimate the value of the painting in 2020 and the rate at which it is appreciating.

36. Appreciation of Assets A painting purchased in 2015 for $100,000 is estimated to be worth $v(t) = 100,000e^{t/5}$ dollars after t years. At what rate will the painting be appreciating in 2020?

37. Velocity and Acceleration The velocity of a parachutist during free fall is

$$f(t) = 60(1 - e^{-0.17t})$$

meters per second. Answer the following questions by reading the graph in Fig. 2. (Recall that acceleration is the derivative of velocity.)

(a) What is the velocity when $t = 8$ seconds?

(b) What is the acceleration when $t = 0$?

(c) When is the parachutist's velocity 30 m/sec?

(d) When is the acceleration 5 m/sec²?

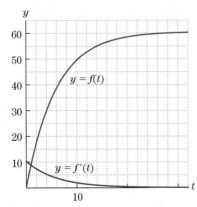

Figure 2

38. **Velocity and Acceleration** Suppose that the velocity of a parachutist is

$$v(t) = 65(1 - e^{-0.16t})$$

meters per second. The graph of $v(t)$ is similar to that in Fig. 2. Calculate the parachutist's velocity and acceleration when $t = 9$ seconds.

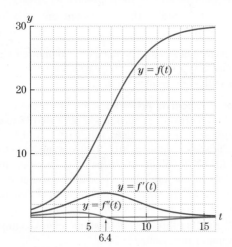

Figure 3

39. **Height of a Plant** The height of a certain plant, in inches, after t weeks is

$$f(t) = \frac{1}{.05 + e^{-0.4t}}.$$

The graph of $f(t)$ resembles the graph in Fig. 3. Calculate the rate of growth of the plant after 7 weeks.

40. **Height of a Plant** The length of a certain weed, in centimeters, after t weeks is

$$f(t) = \frac{6}{.2 + 5e^{-0.5t}}.$$

Answer the following questions by reading the graph in Fig. 3.
(a) How fast is the weed growing after 10 weeks?
(b) When is the weed 10 centimeters long?
(c) When is the weed growing at the rate of 2 cm/week?
(d) What is the maximum rate of growth?

41. **Gompertz Growth Curve** Let a and b be positive numbers. A curve whose equation is

$$y = e^{-ae^{-bx}}$$

is called a **Gompertz growth curve**. These curves are used in biology to describe certain types of population growth. Compute the derivative of $y = e^{-2e^{-0.01x}}$.

42. Find $\dfrac{dy}{dx}$ if $y = e^{-(1/10)e^{-x/2}}$.

TECHNOLOGY EXERCISES

43. **Size of a Tumor** In a study, a cancerous tumor was found to have a volume of

$$f(t) = 1.825^3(1 - 1.6e^{-0.4196t})^3$$

milliliters after t weeks, with $t > 1$. (*Source: Growth, Development and Aging.*)
(a) Sketch the graphs of $f(t)$ and $f'(t)$ for $1 \le t \le 15$. What do you notice about the tumor's volume?
(b) How large is the tumor after 5 weeks?
(c) When will the tumor have a volume of 5 milliliters?
(d) How fast is the tumor growing after 5 weeks?
(e) When is the tumor growing at the fastest rate?
(f) What is the fastest rate of growth of the tumor?

44. **Height of a Plant** Let $f(t)$ be the function from Exercise 39 that gives the height (inches) of a plant at time t (weeks).
(a) When is the plant 11 inches tall?
(b) When is the plant growing at the rate of 1 inch per week?
(c) What is the fastest rate of growth of the plant, and when does this occur?

Solutions to Check Your Understanding 4.3

1. $\dfrac{d}{dt}(te^{-t^2}) = t \cdot \dfrac{d}{dt}e^{-t^2} + e^{-t^2} \cdot \dfrac{d}{dt}(t)$ Product rule for differentiation.

     $= t \cdot (-2t)e^{-t^2} + e^{-t^2} \cdot (1)$ Formula (1).

     $= -2t^2e^{-t^2} + e^{-t^2}$

     $= e^{-t^2}(1 - 2t^2)$ Factor e^{-t^2}.

2. We must use the general power rule. However, this is most easily done if we first use the laws of exponents to simplify the function inside the brackets:

$$e^{-3x}(1 + e^{6x}) = e^{-3x} + e^{-3x} \cdot e^{6x} = e^{-3x} + e^{3x}.$$

Now,

$$\frac{d}{dx}[e^{-3x} + e^{3x}]^{12} = 12 \cdot [e^{-3x} + e^{3x}]^{11} \cdot (-3e^{-3x} + 3e^{3x})$$

$$= 36 \cdot [e^{-3x} + e^{3x}]^{11} \cdot (-e^{-3x} + e^{3x}).$$

4.4 The Natural Logarithm Function

As a preparation for the definition of the natural logarithm, we shall make a geometric digression. In Fig. 1, we have plotted several pairs of points. Observe how they are related to the line $y = x$.

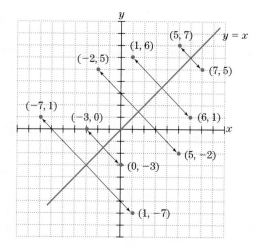

Figure 1 Reflections of points through the line $y = x$.

The points $(5, 7)$ and $(7, 5)$, for example, are the same distance from the line $y = x$. If we were to plot the point $(5, 7)$ with wet ink and then fold the page along the line $y = x$, the ink blot would produce a second blot at the point $(7, 5)$. If we think of the line $y = x$ as a mirror, then $(7, 5)$ is the mirror image of $(5, 7)$. We say that $(7, 5)$ is the *reflection* of $(5, 7)$ through the line $y = x$. Similarly, $(5, 7)$ is the reflection of $(7, 5)$ through the line $y = x$.

Now let us consider all points lying on the graph of the exponential function $y = e^x$. [See Fig. 2(a).] If we reflect each such point through the line $y = x$, we obtain a new graph. [See Fig. 2(b).] For each positive x, there is exactly one value of y such that (x, y) is on the new graph. We call this value of y the **natural logarithm of x**, denoted $\ln x$. Thus, the reflection of the graph of $y = e^x$ through the line $y = x$ is the graph of the natural logarithm function $y = \ln x$.

From this, we derive the following definition of the natural logarithm.

(a)

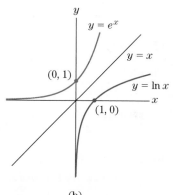

(b)

Figure 2 Obtaining the graph of $y = \ln x$ as a reflection of $y = e^x$.

DEFINITION   Natural Logarithm

For $x > 0$, $y = \ln x$ if, and only if, $x = e^y$

Note that $y = \ln x$ is defined for $x > 0$ only, because $x = e^y$ and e^y is always positive.

We may deduce some properties of the natural logarithm function from an inspection of its graph and using the fact that (a, b) is on the graph of $y = \ln x$ if, and only if, (b, a) is on the graph of $y = e^x$.

Properties of the natural logarithm function

Property	Reason
(1) $\ln 1 = 0$; equivalently, $(1, 0)$ on graph of $y = \ln x$	Because $(0, 1)$ on graph of $y = e^x$; equivalently, $e^0 = 1$
(2) $\ln e = 1$; equivalently, $(e, 1)$ on graph of $y = \ln x$	Because $(1, e)$ on graph of $y = e^x$; equivalently, $e^1 = e$
(3) $\ln e^x = x$; equivalently, (e^x, x) on graph of $y = \ln x$	Because (x, e^x) on graph of $y = e^x$
(4) $e^{\ln x} = x$; equivalently, $(\ln x, x)$ on graph of $y = e^x$	Because $(x, \ln x)$ on graph of $y = \ln x$

The identities (3) and (4) express the fact that the natural logarithm function $\ln x$ is the **inverse** of the exponential function e^x for $x > 0$. For instance, if we take a number x and compute e^x, we can, by (3), undo the effect of the exponential by taking the natural logarithm; that is, the natural logarithm of e^x is the original number x. Similarly, if we take a positive number x and compute $\ln x$, we can, by (4), undo the effect of the logarithm by raising e to the $\ln x$ power; that is $e^{\ln x}$ equals the original number x.

EXAMPLE 1   **Using Properties of Exponential and Logarithm Functions**   Simplify

(a) $e^{\ln 4 + \ln 5}$ (b) $e^{\ln 4 - \ln 3}$ (c) $e^{\ln 3 + 2\ln 4}$ (d) $\ln\left(\frac{1}{e^2}\right)$

SOLUTION

(a) $e^{\ln 4 + \ln 5} = e^{\ln 4} \cdot e^{\ln 5}$ Product rule for exponents.

$\qquad\qquad = (4) \cdot (5)$ Property (4): $e^{\ln x} = x$.

$\qquad\qquad = 20$

(b) $e^{\ln 4 - \ln 3} = \dfrac{e^{\ln 4}}{e^{\ln 3}}$ Quotient rule for exponents.

$\qquad\qquad = \dfrac{4}{3}$ Property (4): $e^{\ln x} = x$.

(c) $e^{\ln 3 + 2\ln 4} = e^{\ln 3} \cdot e^{2\ln 4}$ Product rule for exponents.

$\qquad\qquad = 3 \cdot e^{(\ln 4)(2)}$ Property (4): $e^{\ln x} = x$.

$\qquad\qquad = 3 \cdot (e^{\ln 4})^2$ Power of a power: $e^{ab} = (e^a)^b$.

$\qquad\qquad = (3)(4)^2 = 48$

(d) $\ln\left(\dfrac{1}{e^2}\right) = \ln(e^{-2})$ Changing signs of exponents.

$\qquad\qquad = -2$ Property (3): $\ln e^x = x$.   **》Now Try Exercise 17**

The inverse function relationships between e^x and $\ln x$ (Properties (3) and (4)) may be used to solve equations, as the next examples show.

EXAMPLE 2   **Solving Equations with Exponentials**   Solve the equation $5e^{x-3} = 4$ for x.

SOLUTION

$\qquad 5e^{x-3} = 4$ Given equation.

$\qquad e^{x-3} = .8.$ Divide by 5.

$\qquad \ln(e^{x-3}) = \ln .8$ Take ln of each side.

$\qquad x - 3 = \ln .8$ Use (3).

$\qquad x = 3 + \ln .8 \approx 2.777$ Add 3 to each side.

》Now Try Exercise 25

EXAMPLE 3   **Solving Equations with Logarithms**   Solve the equation $2 \ln x + 7 = 0$ for x.

SOLUTION

$\qquad 2 \ln x = -7$ Subtract 7.

$\qquad \ln x = -3.5$ Divide by 2.

$\qquad e^{\ln x} = e^{-3.5}$ Take exponential of each side.

$\qquad x = e^{-3.5}$ By Property (4).   **》Now Try Exercise 29**

Other Exponential and Logarithm Functions

In our discussion of the exponential function, we mentioned that all exponential functions of the form $y = b^x$, where b is a fixed positive number, can be expressed in terms of *the* exponential function $y = e^x$. Now, we can be quite explicit: Since $b = e^{\ln b}$, we see that

$$b^x = (e^{\ln b})^x = e^{(\ln b)x}.$$

Hence, we have shown that

$$b^x = e^{kx}, \quad \text{where } k = \ln b.$$

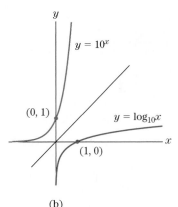

Figure 3 Graphs of $y = \log_2 x$ and $y = \log_{10} x$ as reflections of $y = 2^x$ and $y = 10^x$.

The natural logarithm function is sometimes called the **logarithm to the base e,** for it is the inverse of the exponential function $y = e^x$. If we reflect the graph of the function $y = 2^x$ through the line $y = x$, we obtain the graph of a function called the **logarithm to the base 2**, denoted by $\log_2 x$. Similarly, if we reflect the graph of $y = 10^x$ through the line $y = x$, we obtain the graph of a function called the **logarithm to the base 10**, denoted by $\log_{10} x$. (See Fig. 3.)

Logarithms to the base 10 are sometimes called **common logarithms**. Common logarithms are usually introduced into algebra courses for the purpose of simplifying certain arithmetic calculations. However, with the advent of the modern computer and the widespread availability of graphing calculators, the need for common logarithms has diminished considerably. It can be shown that

$$\log_{10} x = \frac{1}{\ln 10} \cdot \ln x,$$

so $\log_{10} x$ is simply a constant multiple of $\ln x$. However, we shall not need this fact.

EXAMPLE 4   **An Investment Portfolio** In Example 3 of the previous section, we considered an investment portfolio whose value in thousands of dollars is given by $f(t) = 15e^{0.02t} + 10e^{-0.06t}$, where t is the number of years since the inception of the portfolio. The graph of $f(t)$ (Fig. 1 of the previous section) has a minimum value. Find the value of t where this minimum occurs.

SOLUTION   We have to set $f'(t) = 0$ and solve for t. Recall, from Example 3 of the previous section, $f'(t) = .3e^{0.02t} - .6e^{-0.06t}$, so

$$.3e^{0.02t} - .6e^{-0.06t} = 0$$

$$.3e^{0.02t} = .6e^{-0.06t} \qquad \text{Add } .6e^{-0.06t} \text{ to each side.}$$

$$.3e^{0.02t}e^{0.06t} = .6e^{-0.06t}e^{0.06t} \qquad \text{Multiply by } e^{0.06t}.$$

$$.3e^{0.02t+0.06t} = .6e^{-0.06t+0.06t} = .6e^0 \qquad \text{Product rule for exponents.}$$

$$.3e^{0.08t} = .6 \qquad e^0 = 1$$

$$e^{0.08t} = 2 \qquad \text{Divide by } .3.$$

$$.08t = \ln 2 \qquad \text{Take ln of each side.}$$

$$t = \frac{\ln 2}{.08} \approx 8.66$$

Thus, the minimum occurs when $t \approx 8.66$ or in almost 9 years.   **≫ Now Try Exercise 47**

The natural logarithm function is used in calculus because differentiation and integration formulas are simpler than for $\log_{10} x$ or $\log_2 x$, and so on. (Recall that we prefer the function e^x over the functions 10^x and 2^x for the same reason.) Also, $\ln x$ arises "naturally" in the process of solving certain differential equations that describe various growth processes.

Check Your Understanding 4.4

Solutions can be found following the section exercises.

1. Find $\ln e$.

2. Solve $e^{-3x} = 2$ using the natural logarithm function.

EXERCISES 4.4

1. Find $\ln(\sqrt{e}\,)$.

2. Find $\ln\left(\dfrac{1}{e^2}\right)$.

3. If $e^x = 5$, write x in terms of the natural logarithm.

4. If $e^{-x} = 3.2$, write x in terms of the natural logarithm.

5. If $\ln x = -1$, write x using the exponential function.

6. If $\ln x = 4.5$, write x using the exponential function.

Simplify the following expressions.

7. $\ln e^{-3}$

8. $e^{\ln 4.1}$

9. $e^{e^{\ln 1}}$

10. $\ln\left(e^{-2\ln e}\right)$

11. $\ln(\ln e)$

12. $e^{4\ln 1}$

13. $e^{2\ln x}$

14. $e^{x\ln 2}$

15. $e^{-2\ln 7}$

16. $\ln\left(e^{-2}e^4\right)$

17. $e^{\ln x + \ln 2}$

18. $e^{\ln 3 - 2\ln x}$

Solve the following equations for x.

19. $e^{2x} = 5$

20. $e^{1-3x} = 4$

21. $\ln(4 - x) = \frac{1}{2}$

22. $\ln 3x = 2$

23. $\ln x^2 = 9$

24. $e^{x^2} = 25$

25. $6e^{-0.00012x} = 3$

26. $4 - \ln x = 0$

27. $\ln 3x = \ln 5$

28. $\ln(x^2 - 5) = 0$

29. $\ln(\ln 3x) = 0$

30. $2\ln x = 7$

31. $2e^{x/3} - 9 = 0$

32. $e^{\sqrt{x}} = \sqrt{e^x}$

33. $5\ln 2x = 8$

34. $750e^{-0.4x} = 375$

35. $(e^2)^x \cdot e^{\ln 1} = 4$

36. $e^{5x} \cdot e^{\ln 5} = 2$

37. $4e^x \cdot e^{-2x} = 6$

38. $(e^x)^2 \cdot e^{2-3x} = 4$

39. The graph of $f(x) = -5x + e^x$ is shown in Fig. 4. Find the coordinates of the minimum point.

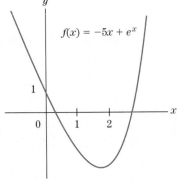

Figure 4

40. Use the second derivative to show that the graph in Fig. 4 is always concave up.

41. (a) Find the first coordinates of the points on the graph in Fig. 4 where the tangent line has slope 3.

    (b) Are there any points on the graph where the tangent line has slope -7? Explain.

42. The graph of $f(x) = -1 + (x - 1)^2 e^x$ is shown in Fig. 5. Find the coordinates of the relative maximum and minimum points.

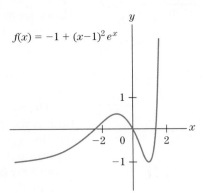

Figure 5

43. (a) Find the point on the graph of $y = e^{-x}$ where the tangent line has slope -2.

    (b) Plot the graphs of $y = e^{-x}$ and the tangent line in part (a).

44. Find the x-intercepts of $y = (x - 1)^2 \ln(x + 1)$, $x > -1$.

In Exercises 45–46, find the coordinates of each relative extreme point of the given function, and determine if the point is a relative maximum point or a relative minimum point.

45. $f(x) = e^{-x} + 3x$

46. $f(x) = 5x - 2e^x$

Solve for t.

47. $e^{0.05t} - 4e^{-0.06t} = 0$

48. $4e^{0.01t} - 3e^{0.04t} = 0$

49. **Concentration of a Drug in the Blood** When a drug or vitamin is administered intramuscularly (into a muscle), the concentration in the blood at time t after injection can be approximated by a function of the form $f(t) = c(e^{-k_1 t} - e^{-k_2 t})$. The graph of $f(t) = 5(e^{-0.01t} - e^{-0.51t})$, for $t \geq 0$, is shown in Fig. 6. Find the value of t at which this function reaches its maximum value.

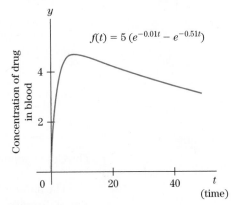

Figure 6

50. **Wind Velocity** Under certain geographic conditions, the wind velocity v at a height x centimeters above the ground is given by $v = K \ln(x/x_0)$, where K is a positive constant (depending on the air density, average wind velocity, and the like) and x_0 is a roughness parameter (depending on the roughness of the vegetation on the ground). Suppose that $x_0 = .7$ centimeter (a value that applies to lawn grass 3 centimeters high) and $K = 300$ centimeters per second. (*Source: Dynamic Ecology.*)

(a) At what height above the ground is the wind velocity zero?

(b) At what height is the wind velocity 1200 centimeters per second?

51. Find k such that $2^x = e^{kx}$ for all x.

52. Find k such that $2^{-x/5} = e^{kx}$ for all x.

TECHNOLOGY EXERCISES

53. Graph the function $y = \ln(e^x)$, and use trace to convince yourself that it is the same as the function $y = x$. What do you observe about the graph of $y = e^{\ln x}$?

54. Graph $y = e^{2x}$ and $y = 5$ together, and determine the x-coordinate of their point of intersection (to four decimal places). Express this number in terms of a logarithm.

55. Graph $y = \ln 5x$ and $y = 2$ together and determine the x-coordinate of their point of intersection (to four decimal places). Express this number in terms of a power of e.

Solutions to Check Your Understanding 4.4

1. Answer: 1. The number ln *e* is that exponent to which *e* must be raised to obtain *e*. Alternatively,

$$\ln e = \ln e^1$$
$$= 1 \qquad \text{By Property (3).}$$

2.

$\ln e^{-3x} = \ln 2$	Take ln of each side.
$-3x = \ln 2$	By property (4).
$x = -\dfrac{\ln 2}{3}.$	Divide by -3.

4.5 The Derivative of ln *x*

Let us now compute the derivative of $y = \ln x$ for $x > 0$. Start with the equation

$$e^{\ln x} = x.$$

We will differentiate each side of this equation and then equate our results. Differentiate the left side, and get

$$\frac{d}{dx}(e^{\ln x}) = e^{\ln x} \cdot \frac{d}{dx}(\ln x) \qquad \text{Use chain rule.}$$

$$= x \cdot \frac{d}{dx}(\ln x). \qquad \text{Use } e^{\ln x} = x.$$

Differentiate the right side, and get

$$\frac{d}{dx}(x) = 1.$$

Equating the two derivatives, we get

$$x \cdot \frac{d}{dx}(\ln x) = 1.$$

Dividing each side by $x > 0$, we obtain

Derivative of $y = \ln x$

$$\frac{d}{dx}(\ln x) = \frac{1}{x}, \quad x > 0. \tag{1}$$

The function $y = \ln x$, $x > 0$, is always increasing, since its derivative $y' = \dfrac{1}{x}$ is always positive for $x > 0$. Let us compute the second derivative:

$$\frac{d^2}{dx^2}(\ln x) = \frac{d}{dx}\left(\frac{1}{x}\right) = -\frac{1}{x^2}.$$

The second derivative of $y = \ln x$ is negative for all $x > 0$, and so the graph of $y = \ln x$ is always concave down for $x > 0$. (See Fig. 1.)

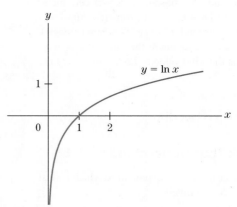

Figure 1 Graph of $y = \ln x$, $x > 0$.

EXAMPLE 1	**Derivatives Involving ln x** Differentiate

(a) $y = (\ln x)^5$ (b) $y = x \ln x$

SOLUTION   (a) By the general power rule,

$$\frac{d}{dx}(\ln x)^5 = 5(\ln x)^4 \cdot \frac{d}{dx}(\ln x) = 5(\ln x)^4 \cdot \frac{1}{x} = \frac{5(\ln x)^4}{x}.$$

(b) By the product rule,

$$\frac{d}{dx}(x \ln x) = x \cdot \frac{d}{dx}(\ln x) + (\ln x) \cdot 1 = x \cdot \frac{1}{x} + \ln x = 1 + \ln x.$$

≫ Now Try Exercise 9

Let $g(x)$ be any differentiable function. For any value of x for which $g(x)$ is positive, the function $y = \ln(g(x))$ is defined. For such a value of x, the derivative is given by the chain rule as

---

Chain Rule for Log Function

$$\frac{d}{dx}[\ln g(x)] = \frac{1}{g(x)} \cdot \frac{d}{dx} g(x) = \frac{g'(x)}{g(x)}. \tag{2}$$

If $u = g(x)$, equation (4) can be written in the form

$$\frac{d}{dx}[\ln u] = \frac{1}{u} \frac{du}{dx}. \tag{2a}$$

EXAMPLE 2	**Derivatives of Functions of the Form $y = \ln[g(x)]$** Differentiate

(a) $y = \ln(2x + 1)$ (b) $y = \ln(4x^2 - 2x + 9)$ (c) $y = \ln(xe^x)$

SOLUTION   In each case we use (2).

(a) Here, $g(x) = 2x + 1$, $g'(x) = 2$, and so,

$$\frac{d}{dx}[\ln(2x + 1)] = \frac{1}{2x + 1} \frac{d}{dx}(2x + 1) = \frac{2}{2x + 1}.$$

(b) Here, $g(x) = 4x^2 - 2x + 9$, $g'(x) = 8x - 2$, and so,

$$\frac{d}{dx}[\ln(4x^2 - 2x + 9)] = \frac{1}{4x^2 - 2x + 9} \frac{d}{dx}(4x^2 - 2x + 9) = \frac{8x - 2}{4x^2 - 2x + 9}.$$

(c) To compute the derivative of $g(x) = xe^x$, we use the product rule: $g'(x) = xe^x + e^x = e^x(x + 1)$. So,

$$\frac{d}{dx}(xe^x) = \frac{1}{xe^x} \frac{d}{dx}(xe^x) = \frac{e^x(x + 1)}{xe^x} = \frac{x + 1}{x}.$$ **≫ Now Try Exercise 19**

EXAMPLE 3   **Analyzing a Function Involving ln x** The function $f(x) = (\ln x)/x$ has a relative extreme point for some $x > 0$. Find the point and determine whether it is a relative maximum or a relative minimum point.

SOLUTION   By the quotient rule,

$$f'(x) = \frac{x \cdot \dfrac{1}{x} - (\ln x) \cdot 1}{x^2} = \frac{1 - \ln x}{x^2}$$

$$f''(x) = \frac{x^2 \cdot \left(-\dfrac{1}{x}\right) - (1 - \ln x)(2x)}{x^4}$$ Quotient rule.

$$= \frac{-x - (2x - 2x \ln x)}{x^4}$$ Because $x^2 \cdot \left(-\dfrac{1}{x}\right) = -x$

$$= \frac{-3x + 2x \ln x}{x^4} = \frac{x(-3 + 2\ln x)}{x^4}$$ Further simplify, then factor x.

$$= \frac{-3 + 2\ln x}{x^3}$$ Cancel x from numerator and denominator.

If we set $f'(x) = 0$, then,

$$\frac{1 - \ln x}{x^2} = 0$$

$$1 - \ln x = 0$$ Multiply by $x^2 \neq 0$.

$$\ln x = 1$$ Add $\ln x$ to each side.

$$e^{\ln x} = e^1 = e$$ Take exponential of each side.

$$x = e.$$ $e^{\ln x} = x$.

Therefore, the only possible relative extreme point is at $x = e$. When $x = e$, $f(e) = (\ln e)/e = 1/e$. Furthermore,

$$f''(e) = \frac{2\ln e - 3}{e^3} = -\frac{1}{e^3} < 0,$$

which implies that the graph of $f(x)$ is concave down at $x = e$. Therefore, $(e, 1/e)$ is a relative maximum point of the graph of $f(x)$. (See Fig. 2.)

Figure 2   **≫ Now Try Exercise 23**

The next example introduces a function that will be needed later when we study integration.

EXAMPLE 4

Derivative of ln|x| The function $y = \ln|x|$ is defined for all nonzero values of x. Its graph is sketched in Fig. 3. Compute the derivative of $y = \ln|x|$.

SOLUTION

If x is positive, $|x| = x$, so,

$$\frac{d}{dx}\ln|x| = \frac{d}{dx}\ln x = \frac{1}{x}.$$

If x is negative, $|x| = -x$ and, by the chain rule,

$$\frac{d}{dx}\ln|x| = \frac{d}{dx}\ln(-x) = \frac{1}{-x} \cdot \frac{d}{dx}(-x) = \frac{1}{-x} \cdot (-1) = \frac{1}{x}.$$

Figure 3 Graph of $y = \ln|x|$.

» Now Try Exercise 28

Therefore, we have established the following useful fact:

$$\frac{d}{dx}\ln|x| = \frac{1}{x}, \qquad x \neq 0.$$

Check Your Understanding 4.5

Solutions can be found following the section exercises.

Differentiate

1. $f(x) = \dfrac{1}{\ln(x^4 + 5)}$.

2. $f(x) = \ln(\ln x)$.

EXERCISES 4.5

Differentiate the following functions.

1. $y = 3\ln x + \ln 2$

2. $y = \dfrac{\ln x}{\ln 3}$

3. $y = \dfrac{x^2 \ln x}{2}$

4. $y = 3\dfrac{\ln x}{x}$

5. $y = e^x \ln x$

6. $y = e^{1 + \ln x}$

7. $y = \dfrac{\ln x}{\sqrt{x}}$

8. $y = \dfrac{1}{2 + 3\ln x}$

9. $y = \ln x^2$

10. $y = \ln \sqrt{x}$

11. $y = \ln\left(\dfrac{1}{x}\right)$

12. $y = \ln\left(\dfrac{1}{x^2}\right)$

13. $y = \ln(3x^4 - x^2)$

14. $y = \ln(e^x + e^{-x})$

15. $y = \dfrac{1}{\ln x}$

16. $y = \ln x \ln 2x$

17. $y = \dfrac{\ln x}{\ln 2x}$

18. $y = (\ln x)^2$

19. $y = (x^3 + 1)\ln(x^3 + 1)$

20. $y = \dfrac{\ln(x^2 + 1)}{x^2 + 1}$

Find the second derivatives.

21. $\dfrac{d^2}{dt^2}(t^2 \ln t)$

22. $\dfrac{d^2}{dt^2}\ln(\ln t)$

23. The graph of $f(x) = (\ln x)/\sqrt{x}$ is shown in Fig. 4. Find the coordinates of the maximum point.

Figure 4

24. The graph of $f(x) = x/(\ln x + x)$ is shown in Fig. 5. Find the coordinates of the minimum point.

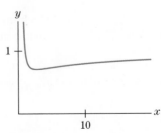

Figure 5

25. Write the equation of the tangent line to the graph of $y = \ln(x^2 + e)$ at $x = 0$.

26. The function $f(x) = (\ln x + 1)/x$ has a relative extreme point for $x > 0$. Find the coordinates of the point. Is it a relative maximum point?

27. Determine the domain of definition of the given function.
    (a) $f(t) = \ln(\ln t)$ (b) $f(t) = \ln(\ln(\ln t))$

28. Find the equations of the tangent lines to the graph of $y = \ln |x|$ at $x = 1$ and $x = -1$.

29. Find the coordinates of the relative extreme point of $y = x^2 \ln x$, $x > 0$. Then, use the second derivative test to decide if the point is a relative maximum point or a relative minimum point.

30. Repeat the previous exercise with $y = \sqrt{x} \ln x$.

31. The graphs of $y = x + \ln x$ and $y = \ln 2x$ are shown in Fig. 6.
    (a) Show that both functions are increasing for $x > 0$.
    (b) Find the point of intersection of the graphs.

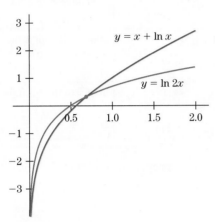

Figure 6

32. Repeat Exercise 31 with the functions $y = x + \ln x$ and $y = \ln 5x$. (See Fig. 7).

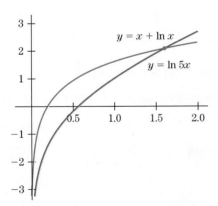

Figure 7

33. The graph of the function $y = x^2 - \ln x$ is shown in Fig. 8. Find the coordinates of its minimum point.

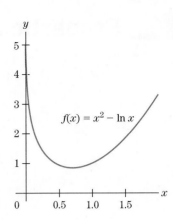

Figure 8

34. The function $y = 2x^2 - \ln 4x \, (x > 0)$ has one minimum point. Find its first coordinate.

35. **A Demand Equation** If the demand equation for a certain commodity is $p = 45/(\ln x)$, determine the marginal revenue function for this commodity, and compute the marginal revenue when $x = 20$.

36. **Total Revenue** Suppose that the total revenue function for a manufacturer is $R(x) = 300 \ln(x + 1)$, so the sale of x units of a product brings in about $R(x)$ dollars. Suppose also that the total cost of producing x units is $C(x)$ dollars, where $C(x) = 2x$. Find the value of x at which the profit function $R(x) - C(x)$ will be maximized. Show that the profit function has a relative maximum and not a relative minimum point at this value of x.

37. **An Area Problem** Find the maximum area of a rectangle in the first quadrant with one corner at the origin, an opposite corner on the graph of $y = -\ln x$, and two sides on the coordinate axes.

TECHNOLOGY EXERCISES

38. **Analysis of the Effectiveness of an Insect Repellent** Human hands covered with cotton fabrics impregnated with the insect repellent DEPA were inserted for 5 minutes into a test chamber containing 200 female mosquitoes. The function $f(x) = 26.48 - 14.09 \ln x$ gives the number of mosquito bites received when the concentration was x percent. [*Note:* The answers to parts (b)–(e) can be obtained either algebraically or from the graphs. You might consider trying both methods.] (*Source: Journal of Medical Entomology.*)
    (a) Graph $f(x)$ and $f'(x)$ for $0 < x \leq 6$.
    (b) How many bites were received when the concentration was 3.25%?
    (c) What concentration resulted in 15 bites?
    (d) At what rate is the number of bites changing with respect to concentration of DEPA when $x = 2.75$?
    (e) For what concentration does the rate of change of bites with respect to concentration equal -10 bites per percentage increase in concentration?

Solutions to Check Your Understanding 4.5

1. Here, $f(x) = [\ln(x^4 + 5)]^{-1}$. By the chain rule,

$$f'(x) = (-1) \cdot [\ln(x^4 + 5)]^{-2} \cdot \frac{d}{dx} \ln(x^4 + 5)$$

$$= -[\ln(x^4 + 5)]^{-2} \cdot \frac{4x^3}{x^4 + 5}.$$

2. $f'(x) = \frac{d}{dx} \ln(\ln x) = \frac{1}{\ln x} \cdot \frac{d}{dx} \ln x = \frac{1}{\ln x} \cdot \frac{1}{x} = \frac{1}{x \ln x}$

4.6   Properties of the Natural Logarithm Function

The natural logarithm function $\ln x$ possesses many of the familiar properties of logarithms to base 10 (or common logarithms) that are encountered in algebra.

Properties of natural logarithm functions		
Property: For $x, y > 0$, and b any number	**Examples**	**In words**
(1) ln of a Product: $\ln(x \cdot y) = \ln x + \ln y$	$\ln(6) = \ln(2 \cdot 3) = \ln 2 + \ln 3$ $\ln(27) = \ln(3 \cdot 9) = \ln 3 + \ln 9$	ln of a product is the sum of ln's
(2) ln of an Inverse: $\ln\left(\dfrac{1}{x}\right) = -\ln x$	$\ln(\frac{1}{2}) = -\ln 2$ $\ln(\frac{1}{\sqrt{2}}) = -\ln(\sqrt{2})$	ln of $1/x$ is minus ln of x, or ln of an inverse is minus the ln
(3) ln of a Quotient: $\ln\left(\dfrac{x}{y}\right) = \ln x - \ln y$	$\ln(\frac{4}{3}) = \ln 4 - \ln 3$ $\ln\left(\dfrac{x}{5}\right) = \ln x - \ln 5$	ln of a quotient is the difference of ln's
(4) ln of a Power: $\ln(x^b) = b \ln x$	$\ln(2^3) = 3 \ln 2$ $\ln(x^2) = 2 \ln x$ $\ln \sqrt{x} = \ln(x^{1/2}) = \frac{1}{2} \ln x$	ln of x to the b is b times ln x

Verification of these properties follows from properties of e^x and the inverse function relationships between $\ln x$ and e^x.

Verification of (1):

$$\ln(x \cdot y) = \ln(e^{\ln x} e^{\ln y}) \quad \text{Inverse property: } x = e^{\ln x}, \, y = e^{\ln y}.$$
$$= \ln(e^{[\ln x + \ln y]}) \quad \text{Product of exponentials: } e^a e^b = e^{a+b}.$$
$$= \ln x + \ln y \quad \text{Inverse property: } \ln e^a = a, \text{ where } a = \ln x + \ln y. \quad \text{«}$$

Verification of (2):

$$\ln\left(\frac{1}{x}\right) = \ln\left(\frac{1}{e^{\ln x}}\right) \quad \text{Inverse property: } x = e^{\ln x}.$$
$$= \ln(e^{[-\ln x]}) \quad \text{Changing sign of exponents: } \frac{1}{e^b} = e^{-b}.$$
$$= -\ln x \quad \text{Inverse property: } \ln e^a = a, \text{ where } a = -\ln x. \quad \text{«}$$

Verification of (3):   By (1) and (2), we have

$$\ln\left(\frac{x}{y}\right) = \ln\left(x \cdot \frac{1}{y}\right) = \ln x + \ln\left(\frac{1}{y}\right) = \ln x - \ln y. \quad \text{«}$$

Verification of (4):

$$\ln(x^b) = \ln\big((e^{\ln x})^b\big) \quad \text{Inverse property: } x = e^{\ln x}.$$
$$= \ln\big(e^{[b \ln x]}\big) \quad \text{Power of a power: } (e^a)^b = e^{ab}.$$
$$= b \ln x \quad \text{Inverse property: } \ln e^a = a, \text{ where } a = b \ln x. \quad \text{«}$$

These properties of the natural logarithm should be learned thoroughly. You will find them useful in many calculations involving $\ln x$ and the exponential function.

EXAMPLE 1   **Simplifying Expressions with Logarithms**   Write $\ln 5 + 2 \ln 3$ as a single logarithm.

SOLUTION

$$\ln 5 + 2 \ln 3 = \ln 5 + \ln 3^2 \quad \text{ln of a power.}$$
$$= \ln 5 + \ln 9$$
$$= \ln(5 \cdot 9) \quad \text{ln of a product.}$$
$$= \ln 45 \qquad \qquad \text{» Now Try Exercise 1}$$

EXAMPLE 2   **Simplifying Expressions with Logarithms** Write $\frac{1}{2}\ln(4t) - \ln(t^2 + 1)$ as a single logarithm.

SOLUTION

$$\frac{1}{2}\ln(4t) - \ln(t^2 + 1) = \ln[(4t)^{1/2}] - \ln(t^2 + 1) \quad \text{ln of a power.}$$
$$= \ln(2\sqrt{t}) - \ln(t^2 + 1)$$
$$= \ln\left(\frac{2\sqrt{t}}{t^2 + 1}\right) \quad \text{ln of a quotient.}$$

» Now Try Exercise 5

EXAMPLE 3   **Using Properties of the Logarithms** Simplify $\ln x + \ln 3 + \ln y - \ln 5$.

SOLUTION

$$(\ln x + \ln 3) + \ln y - \ln 5 = \ln 3x + \ln y - \ln 5 \quad \text{ln of a product.}$$
$$= \ln 3xy - \ln 5 \quad \text{ln of a product.}$$
$$= \ln\left(\frac{3xy}{5}\right) \quad \text{ln of a quotient.}$$

» Now Try Exercise 9

EXAMPLE 4   **Differentiating after Simplifying** Differentiate $f(x) = \ln[x(x + 1)(x + 2)]$.

SOLUTION   First, rewrite $f(x)$ using (1):

$$f(x) = \ln[x(x + 1)(x + 2)] = \ln x + \ln(x + 1) + \ln(x + 2).$$

Then $f'(x)$ is easily calculated:

$$f'(x) = \frac{1}{x} + \frac{1}{x + 1} + \frac{1}{x + 2}. \qquad \text{**» Now Try Exercise 33**}$$

Logarithmic Differentiation

The natural logarithm function can be used to simplify the task of differentiating products. Suppose, for example, that we wish to differentiate the function

$$g(x) = x(x + 1)(x + 2).$$

As we showed in Example 4,

$$\frac{d}{dx}\ln g(x) = \frac{1}{x} + \frac{1}{x + 1} + \frac{1}{x + 2}.$$

However,

$$\frac{d}{dx}\ln g(x) = \frac{g'(x)}{g(x)}.$$

Therefore, equating the two expressions for $\frac{d}{dx}\ln g(x)$, we have

$$\frac{g'(x)}{g(x)} = \frac{1}{x} + \frac{1}{x + 1} + \frac{1}{x + 2}.$$

Finally, we solve for $g'(x)$:

$$g'(x) = g(x) \cdot \left(\frac{1}{x} + \frac{1}{x + 1} + \frac{1}{x + 2}\right)$$
$$= x(x + 1)(x + 2)\left(\frac{1}{x} + \frac{1}{x + 1} + \frac{1}{x + 2}\right).$$

In a similar way, we differentiate the product of any number of factors by first taking natural logarithms, then differentiating, and finally solving for the desired derivative. This procedure is called **logarithmic differentiation.**

EXAMPLE 5 **Logarithmic Differentiation** Differentiate the function $g(x) = (x^2 + 1)(x^3 - 3)(2x + 5)$ using logarithmic differentiation.

SOLUTION **Step 1** Take the natural logarithm of both sides of the given equation:

$$\ln g(x) = \ln[(x^2 + 1)(x^3 - 3)(2x + 5)]$$
$$= \ln(x^2 + 1) + \ln(x^3 - 3) + \ln(2x + 5). \quad \text{ln of a product.}$$

Step 2 Differentiate:

$$\frac{d}{dx}(\ln g(x)) = \frac{g'(x)}{g(x)} = \frac{2x}{x^2 + 1} + \frac{3x^2}{x^3 - 3} + \frac{2}{2x + 5}.$$

Step 3 Solve for $g'(x)$:

$$g'(x) = g(x)\left(\frac{2x}{x^2 + 1} + \frac{3x^2}{x^3 - 3} + \frac{2}{2x + 5}\right)$$
$$= (x^2 + 1)(x^3 - 3)(2x + 5)\left(\frac{2x}{x^2 + 1} + \frac{3x^2}{x^3 - 3} + \frac{2}{2x + 5}\right).$$

» Now Try Exercise 43

Let us now use logarithmic differentiation to finally establish the power rule:

$$\frac{d}{dx}(x^r) = rx^{r-1}.$$

Verification of the Power Rule for $x > 0$ Let $f(x) = x^r$. Then,

$$\ln f(x) = \ln x^r = r \ln x.$$

Differentiation of this equation yields

$$\frac{f'(x)}{f(x)} = r \cdot \frac{1}{x}$$

$$f'(x) = r \cdot \frac{1}{x} \cdot f(x) = r \cdot \frac{1}{x} \cdot x^r = rx^{r-1}.$$

«

Check Your Understanding 4.6

Solutions can be found following the section exercises.

1. Differentiate $f(x) = \ln\left[\dfrac{e^x\sqrt{x}}{(x + 1)^6}\right]$.

2. Use logarithmic differentiation to differentiate

$$f(x) = (x + 1)^7(x + 2)^8(x + 3)^9.$$

EXERCISES 4.6

Simplify the following expressions.

1. $\ln 5 + \ln x$

2. $\ln x^5 - \ln x^3$

3. $\frac{1}{2}\ln 9$

4. $3\ln\frac{1}{2} + \ln 16$

5. $\ln 4 + \ln 6 - \ln 12$

6. $\ln 2 - \ln x + \ln 3$

7. $e^{2\ln x}$

8. $\frac{3}{2}\ln 4 - 5\ln 2$

9. $5\ln x - \frac{1}{2}\ln y + 3\ln z$

10. $e^{\ln x^2 + 3\ln y}$

11. $\ln x - \ln x^2 + \ln x^4$

12. $\frac{1}{2}\ln xy + \frac{3}{2}\ln\frac{x}{y}$

13. Which is larger, $2\ln 5$ or $3\ln 3$? Explain.

14. Which is larger, $\frac{1}{2}\ln 16$ or $\frac{1}{3}\ln 27$? Explain.

Evaluate the given expressions. Use $\ln 2 = .69$ and $\ln 3 = 1.1$.

15. (a) $\ln 4$ (b) $\ln 6$ (c) $\ln 54$

16. (a) $\ln 12$ (b) $\ln 16$ (c) $\ln(9 \cdot 2^4)$

17. (a) $\ln\frac{1}{6}$ (b) $\ln\frac{2}{9}$ (c) $\ln\frac{1}{\sqrt{2}}$

18. (a) $\ln 100 - 2\ln 5$ (b) $\ln 10 + \ln\frac{1}{5}$ (c) $\ln\sqrt{108}$

19. Which of the following is the same as $4\ln 2x$?

(a) $\ln 8x$ (b) $8\ln x$

(c) $\ln 8 + \ln x$ (d) $\ln 16x^4$

20. Which of the following is the same as $\ln(9x) - \ln(3x)$?

(a) $\ln 6x$ (b) $\ln(9x)/\ln(3x)$

(c) $6 \cdot \ln(x)$ (d) $\ln 3$

21. Which of the following is the same as $\dfrac{\ln 8x^2}{\ln 2x}$?

(a) $\ln 4x$ (b) $4x$

(c) $\ln 8x^2 - \ln 2x$ (d) none of these

22. Which of the following is the same as $\ln 9x^2$?

(a) $2 \cdot \ln 9x$ (b) $3x \cdot \ln 3x$

(c) $2 \cdot \ln 3x$ (d) none of these

Solve the given equation for x.

23. $\ln x - \ln x^2 + \ln 3 = 0$ **24.** $\ln \sqrt{x} - 2\ln 3 = 0$

25. $\ln x^4 - 2\ln x = 1$ **26.** $\ln x^2 - \ln 2x + 1 = 0$

27. $(\ln x)^2 - 1 = 0$ **28.** $3\ln x - \ln 3x = 0$

29. $\ln \sqrt{x} = \sqrt{\ln x}$

30. $2(\ln x)^2 + \ln x - 1 = 0$

31. $\ln(x + 1) - \ln(x - 2) = 1$

32. $\ln[(x - 3)(x + 2)] - \ln(x + 2)^2 - \ln 7 = 0$

Differentiate.

33. $y = \ln[(x + 5)(2x - 1)(4 - x)]$

34. $y = \ln[(x + 1)(2x + 1)(3x + 1)]$

35. $y = \ln[(1 + x)^2(2 + x)^3(3 + x)^4]$

36. $y = \ln[e^{2x}(x^3 + 1)(x^4 + 5x)]$

37. $y = \ln\left[\sqrt{xe^{x^2+1}}\right]$

38. $y = \ln \dfrac{x + 1}{x - 1}$

39. $y = \ln \dfrac{(x + 1)^4}{e^{x-1}}$

40. $y = \ln \dfrac{(x + 1)^4(x^3 + 2)}{x - 1}$

41. $y = \ln(3x + 1)\ln(5x + 1)$

42. $y = (\ln 4x)(\ln 2x)$

Use logarithmic differentiation to differentiate the following functions.

43. $f(x) = (x + 1)^4(4x - 1)^2$ **44.** $f(x) = e^x(3x - 4)^8$

45. $f(x) = \dfrac{(x + 1)(2x + 1)(3x + 1)}{\sqrt{4x + 1}}$

46. $f(x) = \dfrac{(x - 2)^3(x - 3)^4}{(x + 4)^5}$ **47.** $f(x) = 2^x$

48. $f(x) = \sqrt[x]{3}$ **49.** $f(x) = x^x$ **50.** $f(x) = \sqrt[x]{x}$

51. Allometric Equation Substantial empirical data show that, if x and y measure the sizes of two organs of a particular animal, then x and y are related by an *allometric equation* of the form

$$\ln y - k\ln x = \ln c,$$

where k and c are positive constants that depend only on the type of parts or organs that are measured and are constant among animals belonging to the same species. Solve this equation for y in terms of x, k, and c. (*Source: Introduction to Mathematics for Life Scientists*)

52. Epidemics Model In the study of epidemics, we find the equation

$$\ln(1 - y) - \ln y = C - rt,$$

where y is the fraction of the population that has a specific disease at time t. Solve the equation for y in terms of t and the constants C and r.

53. Determine the values of h and k for which the graph of $y = he^{kx}$ passes through the points $(1, 6)$ and $(4, 48)$.

54. Find values of k and r for which the graph of $y = kx^r$ passes through the points $(2, 3)$ and $(4, 15)$.

Solutions to Check Your Understanding 4.6

1. Use the properties of the natural logarithm to express $f(x)$ as a sum of simple functions before differentiating:

$$f(x) = \ln\left[\frac{e^x\sqrt{x}}{(x + 1)^6}\right]$$

$$f(x) = \ln e^x + \ln \sqrt{x} - \ln(x + 1)^6$$

$$= x + \tfrac{1}{2}\ln x - 6\ln(x + 1)$$

$$f'(x) = 1 + \frac{1}{2x} - \frac{6}{x + 1}.$$

2. $f(x) = (x + 1)^7(x + 2)^8(x + 3)^9$

$$\ln f(x) = 7\ln(x + 1) + 8\ln(x + 2) + 9\ln(x + 3).$$

Now we differentiate both sides of the equation:

$$\frac{f'(x)}{f(x)} = \frac{7}{x + 1} + \frac{8}{x + 2} + \frac{9}{x + 3}$$

$$f'(x) = f(x)\left(\frac{7}{x + 1} + \frac{8}{x + 2} + \frac{9}{x + 3}\right)$$

$$= (x + 1)^7(x + 2)^8(x + 3)^9 \cdot$$

$$\left(\frac{7}{x + 1} + \frac{8}{x + 2} + \frac{9}{x + 3}\right).$$

| CHAPTER 4 | Summary |

| **KEY TERMS AND CONCEPTS** | **EXAMPLES** |

4.1 Exponential Functions

The laws of exponents are used to simplify expressions and solve equations involving exponential functions.

(i) $b^x \cdot b^y = b^{x+y}$ **Product rule.**

(ii) $b^{-x} = \dfrac{1}{b^x}$ **Changing sign of exponents.**

(iii) $\dfrac{b^x}{b^y} = b^{x-y}$ **Quotient rule.**

(iv) $(b^y)^x = b^{xy}$ **Power of a power.**

(v) $a^x b^x = (ab)^x$ **Power of a product.**

(vi) $\dfrac{a^x}{b^x} = \left(\dfrac{a}{b}\right)^x$ **Power of a quotient.**

$$\frac{4^x}{64^{2x}} = \frac{4^x}{(4^3)^{2x}} \qquad (64 = 4^3)$$

$$= \frac{4^x}{4^{6x}} \qquad \text{Power of a power.}$$

$$= 4^{x-6x} \qquad \text{Quotient rule.}$$

$$= 4^{-5x} = \frac{1}{4^{5x}}$$

4.2 The Exponential Function e^x

The function $y = e^x$ is equal to its own derivative:

$$\frac{d}{dx}[e^x] = e^x.$$

For exponential functions in base b, we have

$$\frac{d}{dx}[b^x] = mb^x,$$

where m is the slope of the graph of $y = b^x$ at $x = 0$.

In base 2, we have

$$\frac{d}{dx}[2^x] = m2^x, \text{ where } m \approx .69.$$

1. $\dfrac{d}{dx}[e^x + 2x] = \dfrac{d}{dx}[e^x] + \dfrac{d}{dx}[2x] = e^x + 2$

2. $\dfrac{d}{dx}[x^4 e^x] = x^4 \dfrac{d}{dx}[e^x] + e^x \dfrac{d}{dx}[x^4]$ Product rule.

$\qquad = x^4 e^x + e^x(4x^3) = e^x x^3(x + 4)$

3. $\dfrac{d}{dx}[2^x e^x] = 2^x \dfrac{d}{dx}[e^x] + e^x \dfrac{d}{dx}[2^x]$ Product rule.

$\qquad = 2^x e^x + e^x(m2^x) = e^x 2^x(1 + m),$

where $m \approx .69$.

4.3 Differentiation of Exponential Functions

The following are differentiation formulas for exponential functions:

$$\frac{d}{dx}[e^{g(x)}] = e^{g(x)}\frac{d}{dx}[g(x)]$$

$$\frac{d}{dx}[Ce^{kx}] = Cke^{kx}$$

1. $\dfrac{d}{dx}\left[e^{\frac{3}{x}}\right] = e^{\frac{3}{x}}\dfrac{d}{dx}\left[\dfrac{3}{x}\right] = e^{\frac{3}{x}}\left(-\dfrac{3}{x^2}\right) = -\dfrac{3}{x^2}e^{\frac{3}{x}}.$

2. $\dfrac{d}{dx}\left[\dfrac{e^{4x+2}}{e^{2x}+1}\right]$

$\qquad = \dfrac{(e^{2x}+1)\frac{d}{dx}[e^{4x+2}] - e^{4x+2}\frac{d}{dx}[e^{2x}+1]}{(e^{2x}+1)^2}$ Quotient rule.

$\qquad = \dfrac{(e^{2x}+1)4e^{4x+2} - e^{4x+2}2e^{2x}}{(e^{2x}+1)^2}$

$\qquad = \dfrac{2e^{4x+2}(2(e^{2x}+1) - e^{2x})}{(e^{2x}+1)^2}$

$\qquad = \dfrac{2e^{4x+2}(e^{2x}+2)}{(e^{2x}+1)^2}.$

KEY TERMS AND CONCEPTS	EXAMPLES

4.4 The Natural Logarithm Function

The natural logarithm function $y = \ln x$ is the inverse function of $y = e^x$. That is,

$$\ln(e^x) = x \quad \text{and} \quad e^{\ln x} = x.$$

The following are properties of the natural logarithm:

- $\ln x$ is defined for $x > 0$;
- $\ln 1 = 0$;
- $\ln x$ tends to $-\infty$ as x approaches 0 from the right;
- $\ln x$ tends to ∞ as x increases;
- The graph of $\ln x$ is the reflection of the graph of $y = e^x$ through the line $y = x$.

1. Simplify the expression $e^{4\ln(2x)}$.

$$e^{4\ln(2x)} = (e^{\ln(2x)})^4 = (2x)^4 = 2^4 x^4 = 16\, x^4.$$

2. Solve for x: $e^{3x+1} = 8$.

$$e^{3x+1} = 8$$
$$\ln(e^{3x+1}) = \ln 8 \qquad \text{Take } \ln \text{ of each side.}$$
$$3x + 1 = \ln 8 \qquad \text{Inverse property: } \ln e^a = a.$$
$$x = \frac{\ln 8 - 1}{3} \approx \frac{2.08 - 1}{3} = .3598.$$

4.5 The Derivative of ln x

$$\frac{d}{dx}[\ln x] = \frac{1}{x}, \quad x > 0$$
$$\frac{d}{dx}[\ln g(x)] = \frac{g'(x)}{g(x)},$$
$$\frac{d}{dx}[\ln |x|] = \frac{1}{x}, \quad x \neq 0.$$

1. $\dfrac{d}{dx}[(6x + 1)\ln x]$

$$= (6x + 1)\frac{d}{dx}[\ln x] + \ln x \frac{d}{dx}[(6x + 1)] \quad \text{Product Rule.}$$
$$= (6x + 1)\frac{1}{x} + \ln x(6)$$
$$= 6 \ln x + \frac{6x + 1}{x}.$$

2. $\dfrac{d}{dx}[\ln(2x^2 - 1)] = \dfrac{(2x^2 - 1)'}{(2x^2 - 1)} = \dfrac{4x}{2x^2 - 1}.$

4.6 Properties of the Natural Logarithm Function

The following are properties of the natural logarithm function:

- $\ln(xy) = \ln x + \ln y$
- $\ln\left(\dfrac{1}{x}\right) = -\ln x$
- $\ln\left(\dfrac{x}{y}\right) = \ln x - \ln y$
- $\ln(x^b) = b \ln x$

1. Simplify the expressions

$$e^{\ln 3 + \ln 4} = e^{\ln(3 \cdot 4)} = e^{\ln(12)} = 12$$
$$2\, e^{\ln 3 - \ln 4} = 2\, e^{\ln \frac{3}{4}} = 2\left(\frac{3}{4}\right) = \frac{3}{2}.$$

2. Differentiate $y = \ln \dfrac{(x + 1)(2x^2 - 7)}{x^2 + 1}$.

First, note that

$$\ln \frac{(x + 1)(2x^2 - 7)}{x^2 + 1} = \ln(x + 1) + \ln(2x^2 - 7) - \ln(x^2 + 1).$$

So,

$$\frac{d}{dx}\left[\ln \frac{(x + 1)(2x^2 - 7)}{x^2 + 1}\right]$$
$$= \frac{d}{dx}[\ln(x + 1)] + \frac{d}{dx}[\ln(2x^2 - 7)] - \frac{d}{dx}[\ln(x^2 + 1)]$$
$$= \frac{1}{x + 1} + \frac{4x}{2x^2 - 7} - \frac{2x}{x^2 + 1}.$$

CHAPTER 4   Fundamental Concept Check Exercise

1. State as many laws of exponents as you can recall.

2. What is e?

3. Write the differential equation satisfied by $y = Ce^{kt}$.

4. State the properties that graphs of the form $y = e^{kx}$ have in common when k is positive and when k is negative.

5. What are the coordinates of the reflection of the point (a, b) across the line $y = x$?

6. What is a logarithm?

7. What is the x-intercept of the graph of the natural logarithm function?

8. State the main features of the graph of $y = \ln x$.

9. State the two key equations giving the relationships between e^x and $\ln x$. [*Hint:* The right side of each equation is just x.]

10. What is the difference between a natural logarithm and a common logarithm?

11. Give the formula that converts a function of the form b^x to an exponential function with base e.

12. State the differentiation formula for each of the following functions
   (a)
   $$f(x) = e^{kx} \qquad \textbf{(b) } f(x) = e^{g(x)} \qquad \textbf{(c) } f(x) = \ln g(x)$$

13. State the four algebraic properties of the natural logarithm function.

14. Give an example of the use of logarithmic differentiation.

CHAPTER 4 Review Exercises

Calculate the following.

1. $27^{4/3}$
2. $4^{1.5}$
3. 5^{-2}
4. $16^{-0.25}$

5. $(2^{5/7})^{14/5}$
6. $8^{1/2} \cdot 2^{1/2}$
7. $\dfrac{9^{5/2}}{9^{3/2}}$
8. $4^{0.2} \cdot 4^{0.3}$

Simplify the following.

9. $\left(e^{x^2}\right)^3$
10. $e^{5x} \cdot e^{2x}$
11. $\dfrac{e^{3x}}{e^x}$

12. $2^x \cdot 3^x$
13. $(e^{8x} + 7e^{-2x})e^{3x}$
14. $\dfrac{e^{5x/2} - e^{3x}}{\sqrt{e^x}}$

Solve the following equations for x.

15. $e^{-3x} = e^{-12}$
16. $e^{x^2 - x} = e^2$
17. $(e^x \cdot e^2)^3 = e^{-9}$
18. $e^{-5x} \cdot e^4 = e$

Differentiate the following functions.

19. $y = 10e^{7x}$
20. $y = e^{\sqrt{x}}$

21. $y = xe^{x^2}$
22. $y = \dfrac{e^x + 1}{x - 1}$

23. $y = e^{e^x}$
24. $y = (\sqrt{x} + 1)e^{-2x}$

25. $y = \dfrac{x^2 - x + 5}{e^{3x} + 3}$
26. $y = x^e$

27. The graph of the functions $f(x) = e^{x^2} - 4x^2$ is shown in Fig. 1. Find the first coordinates of the relative extreme points.

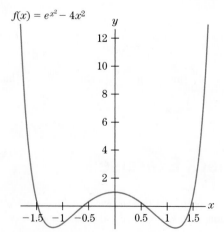

$f(x) = e^{x^2} - 4x^2$

Figure 1

28. Show that the function in Fig. 1 has a relative maximum at $x = 0$ by determining the concavity of the graph at $x = 0$.

Solve the following equations for t.

29. $4e^{0.03t} - 2e^{0.06t} = 0$
30. $e^t - 8e^{0.02t} = 0$

31. Solve the equation $4 \cdot 2^x = e^x$. [*Hint:* Express 2^x as an exponential in base e.]

32. Solve the equation $3^x = 2e^x$. [*Hint:* Express 3^x as an exponential in base e.]

33. Find the points on the graph of $y = e^x$ where the tangent line has slope 4.

34. Find the points on the graph of $y = e^x + e^{-2x}$ where the tangent line is horizontal.

35. Determine the intervals where the function $f(x) = \ln(x^2 + 1)$ is increasing and where it is decreasing.

36. Determine the intervals where the function $f(x) = x \ln x \ (x > 0)$ is increasing and where it is decreasing.

37. Find the equation of the tangent line to the graph of $y = \dfrac{e^x}{1 + e^x}$ at $(0, .5)$.

38. Show that the tangent lines to the graph of $y = \dfrac{e^x - e^{-x}}{e^x + e^{-x}}$ at $x = 1$ and $x = -1$ are parallel.

Simplify the following expressions.

39. $e^{(\ln 5)/2}$
40. $e^{\ln(x^2)}$
41. $\dfrac{\ln x^2}{\ln x^3}$

42. $e^{2 \ln 2}$
43. $e^{-5 \ln 1}$
44. $[e^{\ln x}]^2$

Solve the following equations for t.

45. $t^{\ln t} = e$
46. $\ln(\ln 3t) = 0$

47. $3e^{2t} = 15$
48. $3e^{t/2} - 12 = 0$

49. $2 \ln t = 5$
50. $2e^{-0.3t} = 1$

Differentiate the following functions.

51. $y = \ln(x^6 + 3x^4 + 1)$
52. $y = \dfrac{x}{\ln x}$

53. $y = \ln(5x - 7)$
54. $y = \ln(9x)$

55. $y = (\ln x)^2$
56. $y = (x \ln x)^3$

57. $y = \ln\left(\dfrac{xe^x}{\sqrt{1 + x}}\right)$
58. $y = \ln[e^{6x}(x^2 + 3)^5(x^3 + 1)^{-4}]$

59. $y = x \ln x - x$

60. $y = e^{2 \ln(x+1)}$

61. $y = \ln(\ln \sqrt{x})$

62. $y = \dfrac{1}{\ln x}$

63. $y = e^x \ln x$

64. $y = \ln(x^2 + e^x)$

65. $y = \ln \sqrt{\dfrac{x^2 + 1}{2x + 3}}$

66. $y = \ln|-2x + 1|$

67. $y = \ln\left(\dfrac{e^{x^2}}{x}\right)$

68. $y = \ln \sqrt[3]{x^3 + 3x - 2}$

69. $y = \ln(2^x)$

70. $y = \ln(3^{x+1}) - \ln 3$

71. $y = \ln|x - 1|$

72. $y = e^{2 \ln(2x+1)}$

73. $y = \ln\left(\dfrac{1}{e^{\sqrt{x}}}\right)$

74. $y = \ln(e^x + 3e^{-x})$

Use logarithmic differentiation to differentiate the following functions.

75. $f(x) = \sqrt[5]{\dfrac{x^5 + 1}{x^5 + 5x + 1}}$

76. $f(x) = 2^x$

77. $f(x) = x^{\sqrt{x}}$

78. $f(x) = b^x$, where $b > 0$

79. $f(x) = (x^2 + 5)^6(x^3 + 7)^8(x^4 + 9)^{10}$

80. $f(x) = x^{1+x}$

81. $f(x) = 10^x$

82. $f(x) = \sqrt{x^2 + 5}\,e^{x^2}$

83. $f(x) = \sqrt{\dfrac{xe^x}{x^3 + 3}}$

84. $f(x) = \dfrac{e^x \sqrt{x + 1}(x^2 + 2x + 3)^2}{4x^2}$

85. $f(x) = e^{x+1}(x^2 + 1)x$

86. $f(x) = e^x x^2 2^x$

87. The atmospheric pressure at an altitude of x kilometers is $f(x)$ g/cm^2 (grams per square centimeter), where $f(x) = 1035e^{-0.12x}$. Give approximate answers to the following questions using the graphs of $f(x)$ and $f'(x)$ shown in Fig. 2.

(a) What is the pressure at an altitude of 2 kilometers?

(b) At what altitude is the pressure 200 g/cm^2?

(c) At an altitude of 8 kilometers, at what rate is the atmospheric pressure changing (with respect to change in altitude)?

(d) At what altitude is the atmospheric pressure falling at the rate of 100 g/cm^2 per kilometer?

88. Health Expenditures The health expenditures (in billions of dollars) for a certain country from 1990 to 2010 are given approximately by $f(t) = 27e^{0.106t}$, with time in years measured from 1990. Give approximate answers to the following questions using the graphs of $f(t)$ and $f'(t)$ shown in Fig. 3.

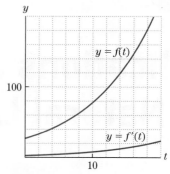

Figure 3

(a) How much money was spent in 2008?

(b) How fast were expenditures rising in 2002?

(c) When did expenditures reach $120 billion?

(d) When were expenditures rising at the rate of $20 billion per year?

Figure 2

Applications of the Exponential and Natural Logarithm Functions

In Chapter 4, we introduced the exponential function $y = e^x$ and the natural logarithm function $y = \ln x$, and we studied their most important properties. It is by no means clear that these functions have any substantial connection with the physical world. However, as this chapter will demonstrate, the exponential and natural logarithm functions are involved in the study of many physical problems, often in a very curious and unexpected way.

5.1 Exponential Growth and Decay

Exponential Growth

You walk into your kitchen one day and you notice that the overripe bananas that you left on the counter invited unwanted guests: fruit flies. To take advantage of this pesky situation, you decide to study the growth of the fruit flies colony. It didn't take you too long to make your first observation: The colony is increasing at a *rate that is proportional to its size*. That is, the more fruit flies, the faster their number grows.

To help us model this population growth, we introduce some notation. Let $P(t)$ denote the number of fruit flies in your kitchen, t days from the moment you first noticed them. A very important fact that we learned about derivatives tells us that

the rate of change of $P(t)$ is $P'(t)$.

FOR REVIEW

"The derivative is a rate of change." See Sec. 1.7, p. 107.

Translating into mathematical language our observation that the rate of change of $P(t)$ is proportional to $P(t)$, we obtain the equation

$$\overbrace{P'(t)}^{\text{rate of change}} = \overbrace{k\,P(t),}^{\text{proportional to } P(t)}$$

where k is a positive constant of proportionality. If we let $y = P(t)$, the equation becomes

$$y' = ky. \tag{1}$$

Equation (1) expresses a relationship between the function y and its derivative y'. Any equation expressing a relationship between a function and its derivatives is called a **differential equation**. Differential equations will be discussed in greater detail in Chapter 10.

    A **solution** of (1) is any *function* whose derivative is equal to k times itself. This is clearly a new type of equation unlike any algebraic equation that we have encountered earlier. To find a solution, we recall a useful derivative from Section 4.3.

FOR REVIEW

$\dfrac{d}{dx}(e^{g(x)}) = e^{g(x)}g'(x)$

See Sec. 4.3, p. 233.

> Let C and k be any constants, and let $y = Ce^{kt}$. Then, $y' = Cke^{kt}$.

(We use the independent variable t instead of x throughout this chapter. The reason is that, in most applications, the variable of our exponential function is time.) Note that

$$y' = Cke^{kt} = k \cdot \overbrace{Ce^{kt}}^{y} = ky.$$

Hence, $y = Ce^{kt}$ is a solution of (1). The converse is also true in the following sense.

> **Theorem 1   Exponential Function Solution of a Differential Equation** The function $y = Ce^{kt}$ satisfies the differential equation
>
> $$y' = ky.$$
>
> Conversely, if $y = f(t)$ satisfies the differential equation $y' = ky$, then, $y = Ce^{kt}$ for some constant C.

In biology, chemistry, and economics, if, at every instant, the rate of increase of a quantity is proportional to the quantity at that instant, as expressed by (1) (with $k > 0$), then the quantity is said to be *growing exponentially* or is *exhibiting exponential growth*. Theorem 1 justifies this terminology, since, in this case, the quantity is an exponential function. The proportionality constant k is also called the **growth constant**.

EXAMPLE 1

Solving a Differential Equation Determine all functions $y = f(t)$ such that $y' = .3\,y$.

SOLUTION

The equation $y' = .3y$ has the form $y' = ky$ with $k = .3$. Therefore, according to Theorem 1, any solution of the equation has the form

$$y = Ce^{0.3t},$$

where C is a constant.   **▶▶ Now Try Exercise 3**

Note that even if the constant k is known in Example 1 ($k = .3$), the equation $y' = .3y$ has *infinitely many solutions* of the form $y = Ce^{0.3t}$, one for each choice of the *arbitrary constant C*. For example, the functions $y = e^{0.3t}$, $y = 3e^{0.3t}$, and $y = 6e^{0.3t}$ are all solutions that correspond to the choices $C = 1, 3,$ and $6,$ respectively. (See Fig. 1.) The fact that a differential equation has infinitely many solutions allows us to select a *particular* solution that fits the situation under study. We illustrate this important fact by returning to our fruit flies problem.

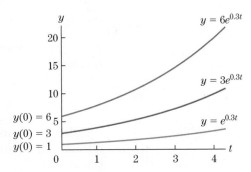

Figure 1 Some solutions of $y' = .3y$.

EXAMPLE 2   **Exponential Growth of a Fruit Flies Population**   Let $y = P(t)$ denote the number of fruit flies in the kitchen, t days since you first observed them. It is known that this species of flies exhibits exponential growth with growth constant $k = .3$. Suppose that the initial number of fruit flies is 6.
(a) Find $P(t)$.
(b) Estimate the number of fruit flies after 7 days.

SOLUTION   (a) Because $y = P(t)$ exhibits exponential growth with growth constant $k = .3$, it satisfies the differential equation

$$y' = .3y.$$

We are also told that the initial number of fruit flies is 6. That is, at time $t = 0$, there are 6 fruit flies. Hence, $P(0) = 6$. From Example 1, the differential equation $y' = .3y$ has infinitely many solutions of the form $y = Ce^{0.3t}$, but only one of these solutions will satisfy the condition $P(0) = 6$. Indeed, taking $t = 0$ in $P(t) = Ce^{0.3t}$, we get

$$6 = P(0) = Ce^{(0.3)(0)} = Ce^0 = C.$$

Thus, $C = 6$ and so $P(t) = 6e^{0.3t}$.
(b) After 7 days, we have

$$P(7) = 6e^{0.3(7)} = 6e^{2.1} \approx 48.997.$$

Thus after 7 days, there are approximately 49 fruit flies in the kitchen.

>> Now Try Exercise 19

The condition $P(0) = 6$ in Example 2 is called an **initial condition**. The initial condition describes the initial size of the population, which, in turn, can be used to determine a unique solution of the differential equation. Fig. 1 shows many solutions of the differential equation $y' = .3y$ but only one goes through the point $(0, 6)$ and so satisfies the initial condition $y(0) = 6$. For future reference, we state the following useful result.

Theorem 2   Solution of a Differential Equation with Initial Condition
The *unique* solution, $y = P(t)$, of the differential equation with initial condition,

$$y' = ky, \qquad y(0) = P_0,$$

is $y = P(t) = P_0 e^{kt}$.

EXAMPLE 3	**A Differential Equation with Initial Condition** Solve $y' = 3y$, $y(0) = 2$.

SOLUTION   Here, $k = 3$ and $P_0 = 2$. By Theorem 2, the (unique) solution is $y = 2e^{3t}$. Note the initial condition on the graph of the solution in Fig. 2.

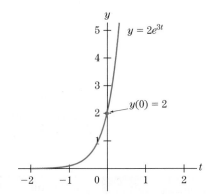

Figure 2 Unique solution of $y' = 3y$, $y(0) = 2$.

» Now Try Exercise 13

The following examples illustrate different ways to determine the constants C and k in exponential growth problems arising in real-world situations.

EXAMPLE 4	**Exponential Growth** A colony of fruit flies grows at a rate proportional to its size. At time $t = 0$, approximately 20 fruit flies are present. In 5 days there are 400 fruit flies. Determine a function that expresses the size of the colony as a function of time, measured in days.

SOLUTION   Let $P(t)$ be the number of fruit flies present at time t. By assumption, $P(t)$ satisfies a differential equation of the form $y' = ky$, so $P(t)$ has the form

$$P(t) = P_0 e^{kt},$$

where the constants P_0 and k must be determined. The values of P_0 and k can be obtained from the data that give the population size at two different times. We are told that

$$P(0) = 20, \qquad P(5) = 400.$$

The first condition immediately implies that $P_0 = 20$, so

$$P(t) = 20e^{kt}.$$

Using the second condition, we have

$$P(5) = 20e^{k(5)} = 400$$
$$e^{5k} = 20 \qquad \text{Divide by 20.}$$
$$5k = \ln 20 \qquad \text{Take ln of each side.}$$
$$k = \frac{\ln 20}{5} \approx .60.$$

So, using the values of P and k, we get

$$P(t) = 20e^{0.6t}.$$

This function is a mathematical model of the growth of the colony of flies. (See Fig. 3.)

FOR REVIEW

e^x and $\ln x$ are inverse of each other: $\ln e^x = x$ and $e^{\ln x} = x$. See Sec. 4.4.

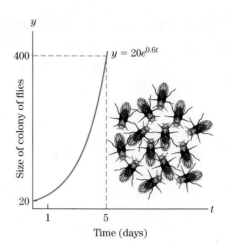

Figure 3 A model for a colony of fruit flies as a function of time $t \geq 0$.

y axis: Size of colony of flies — values 20, 400

$y = 20e^{0.6t}$

t axis: Time (days) — values 1, 5

>> *Now Try Exercise 20*

EXAMPLE 5   **Determining the Growth Constant** A colony of fruit flies is growing according to the exponential law $P(t) = P_0 e^{kt}$, and the size of the colony doubles in 9 days. Determine the growth constant k.

SOLUTION   We do not know the initial size of the population at $t = 0$. However, we are told that the colony doubles in 9 days. Mathematically this is represented by $P(9) = 2P(0)$; that is,

$$P_0 e^{k(9)} = 2P_0$$

$$e^{9k} = 2 \qquad \text{Divide by } P_0 \neq 0.$$

$$9k = \ln 2 \qquad \text{Take ln of each side.}$$

$$k = \frac{\ln 2}{9} \approx .077. \quad \text{Solve for } k.$$

>> *Now Try Exercise 21*

The initial size P_0 of the population was not given in Example 5. But we were able to determine the growth constant because we were told the amount of time required for the colony to double in size. Thus, the growth constant does not depend on the initial size of the population. This property is characteristic of exponential growth.

EXAMPLE 6   **Working with a Differential Equation** The initial size of the colony in Example 5 was 100.
(a) How large will the colony be after 41 days?
(b) How fast will the colony be growing at that time?
(c) At what time will the colony contain 800 fruit flies?
(d) How large is the colony when it is growing at the rate of 200 fruit flies per day?

SOLUTION   (a) From Example 5, we have $P(t) = P_0 e^{0.077t}$. Since $P(0) = 100$, we conclude that

$$P(t) = 100 e^{0.077t}.$$

Therefore, after 41 days, the size of the colony is

$$P(41) = 100 e^{0.077(41)} = 100 e^{3.157} \approx 2350 \text{ fruit flies.}$$

(b) Recall from Example 5 that $k = .077$. Since the function $P(t)$ satisfies the differential equation $y' = .077y$,

$$P'(t) = .077 P(t).$$

In particular, when $t = 41$,

$$P'(41) = .077 P(41) = (.077)(2350) \approx 181.$$

Therefore, after 41 days, the colony is growing at the rate of about 181 fruit flies per day.

(c)

$$100e^{0.077t} = 800 \qquad \text{Set } P(t) = 800.$$

$$e^{0.077t} = 8 \qquad \text{Divide by 100.}$$

$$.077t = \ln 8 \qquad \text{Take ln of each side.}$$

$$t = \frac{\ln 8}{.077} \approx 27 \text{ days.} \qquad \text{Solve for } t.$$

(d) When the colony is growing at the rate of 200 fruit flies per day, $P'(t) = 200$. As in (b), we use the differential equation $P'(t) = .077P(t)$ and set $P'(t) = 200$. Then,

$$200 = .077P(t),$$

$$P(t) = \frac{200}{.077} \approx 2597.$$

Therefore, there are 2597 fruit flies in the colony when it is growing at the rate of 200 fruit flies per day.   **» Now Try Exercise 23**

Exponential Decay

To solve the differential equation $y' = ky$ when the constant k is negative we can still appeal to Theorem 1 and obtain the solution $y = Ce^{kt}$, where k is negative and C is an arbitrary constant. In this case, we are dealing with a negative exponential growth, or **exponential decay**. An example of exponential decay is given by the disintegration of a radioactive element such as uranium 235. It is known that, at any instant, the rate at which a radioactive substance is decaying is proportional to the amount of the substance that has not yet disintegrated. If $P(t)$ is the quantity present at time t, then $P'(t)$ is the rate of decay. Since $P(t)$ is decreasing, $P'(t)$ must be negative. Thus, we may write $P'(t) = kP(t)$ for some negative constant k. To emphasize the fact that the constant is negative, k is often replaced by $-\lambda$, where λ is a positive constant (λ is the Greek lowercase letter lambda). Then, $P(t)$ satisfies the differential equation

$$P'(t) = -\lambda P(t).$$

By Theorem 1, the solution has the form

$$P(t) = P_0 e^{-\lambda t}$$

for some positive number P_0. We call such a function an **exponential decay function**. The constant λ is called the **decay constant**.

EXAMPLE 7   **Exponential Decay** The decay constant for the radioactive element **strontium** 90 is $\lambda = .0244$, where time is measured in years. How long will it take for a quantity P_0 of strontium 90 to decay to one-half its original mass?

SOLUTION   Since $\lambda = .0244$, we have

$$P(t) = P_0 e^{-0.0244t}.$$

Next, set $P(t)$ equal to $\frac{1}{2}P_0$ and solve for t:

$$P_0 e^{-0.0244t} = \tfrac{1}{2}P_0$$

$$e^{-0.0244t} = \tfrac{1}{2} = .5 \qquad \text{Divide by } P_0.$$

$$-.0244t = \ln .5 \qquad \text{Take ln of each side.}$$

$$t = \frac{\ln .5}{-.0244} \approx 28 \text{ years.} \qquad \text{Solve for } t.$$

» Now Try Exercise 35

The **half-life** of a radioactive element is the length of time required for a given quantity of that element to decay to one-half its original mass. Thus, strontium 90 has a half-life of about 28 years. It takes 28 years for it to decay to half its original mass and another 28 years for it to decay to $\frac{1}{4}$ its original mass, another 28 years to decay to $\frac{1}{8}$, and so forth. (See Fig. 4.) Notice from Example 7 that the half-life does not depend on the initial amount P_0.

Figure 4 Half-life of radioactive strontium 90.

One problem connected with above ground nuclear explosions is the radioactive debris, or fallout, that contaminates plants and grass, the food supply of animals. Strontium 90 is one of the most dangerous components of radioactive debris because it has a relatively long half-life. Also, it is chemically similar to calcium and is absorbed into the bone structure of animals (including humans) who eat contaminated food. Iodine 131 is also produced by nuclear explosions, but it presents less of a hazard because it has a half-life of 8 days. See Exercise 41.

EXAMPLE 8	**Half-Life and Decay Constant** Radioactive carbon 14 has a half-life of about 5730 years. Find its decay constant.

SOLUTION

If P_0 denotes the initial amount of carbon 14, the amount after t years will be

$$P(t) = P_0 e^{-\lambda t}.$$

After 5730 years, $P(t)$ will equal $\frac{1}{2}P_0$. That is,

$$P_0 e^{-\lambda(5730)} = P(5730) = \tfrac{1}{2}P_0 = .5P_0.$$

Solving for λ gives

$$P_0 e^{-\lambda(5730)} = .5P_0$$
$$e^{-5730\lambda} = .5 \qquad \text{Divide by } P_0.$$
$$-5730\lambda = \ln.5 \qquad \text{Take ln of each side.}$$
$$\lambda = \frac{\ln.5}{-5730} \approx .00012. \qquad \text{Solve for } \lambda.$$

>> Now Try Exercise 37

FOR REVIEW

- $\ln x < 0$ if $0 < x < 1$, so $\ln (.5) < 0$.
- $\ln \dfrac{1}{x} = -\ln x$ so $\ln (.5) = \ln \left(\tfrac{1}{2}\right) = -\ln 2$
- $\ln 2 \approx .69$.

See the graph of $y = \ln x$, p. 237.

Radiocarbon Dating

Knowledge about radioactive decay is valuable to archaeologists and anthropologists who want to estimate the age of objects belonging to ancient civilizations. Several different substances are useful for radioactive-dating techniques; the most common is radiocarbon, ^{14}C. Carbon 14 is produced in the upper atmosphere when cosmic rays react with atmospheric nitrogen. Because the ^{14}C eventually decays, the concentration of ^{14}C cannot rise above certain levels. An equilibrium is reached where ^{14}C is produced at the same rate as it decays. Scientists usually assume that the total amount of ^{14}C in the biosphere has remained constant over the past 50,000 years. Consequently, it is assumed that the *ratio* of ^{14}C to ordinary nonradioactive carbon 12, ^{12}C, has been constant during this same period. (The ratio is about one part ^{14}C to 10^{12} parts of ^{12}C.) Both ^{14}C and ^{12}C are in the atmosphere as constituents of carbon dioxide. All living vegetation and most forms of animal life contain ^{14}C and ^{12}C in the same proportion as the atmosphere because plants absorb carbon dioxide through photosynthesis. The ^{14}C and ^{12}C in plants are distributed through the food chain to almost all animal life.

When an organism dies, it stops replacing its carbon; therefore, the amount of ^{14}C begins to decrease through radioactive decay, but the ^{12}C in the dead organism remains constant. The ratio of ^{14}C to ^{12}C can be later measured to determine when the organism died.

EXAMPLE 9   **Carbon Dating**   A parchment fragment made from animal skin was discovered that had about 80% of the ^{14}C level found today in living matter. Estimate the age of the parchment.

SOLUTION   We assume that the original ^{14}C level in the parchment was the same as the level in living organisms today. Consequently, about eight-tenths of the original ^{14}C remains. From Example 8 we obtain the formula for the amount of ^{14}C present t years after the parchment was made from an animal skin:

$$P(t) = P_0 e^{-0.00012t},$$

where $P_0 = $ initial amount. We want to find t such that $P(t) = .8P_0$:

$$P_0 e^{-0.00012t} = .8P_0$$
$$e^{-0.00012t} = .8 \qquad \text{Divide by } P_0.$$
$$-.00012t = \ln .8 \qquad \text{Take ln of each side.}$$
$$t = \frac{\ln .8}{-.00012} \approx 1860 \text{ years old.} \qquad \text{Solve for } t.$$

» Now Try Exercise 47

Thus the parchment is about 1860 years old.

The Time Constant

Consider an exponential decay function $y = Ce^{-\lambda t}$. Figure 5 shows the tangent line to the decay curve when $t = 0$. The slope there is the initial rate of decay. If the decay process were to continue at this rate, the decay curve would follow the tangent line, and y would be zero at some time T. This time is called the **time constant** of the decay curve. It can be shown (see Exercise 52) that $T = 1/\lambda$ for the curve $y = Ce^{-\lambda t}$. Thus, $\lambda = 1/T$ and the decay curve can be written in the form

$$y = Ce^{-t/T}.$$

If we have experimental data that tend to lie along an exponential decay curve, the numerical constants for the curve may be obtained from Fig. 5. First, sketch the curve and estimate the y-intercept, C. Then, sketch an approximate tangent line, and from this, estimate the time constant, T. This procedure is sometimes used in biology and medicine.

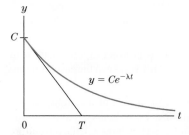

Figure 5 The time constant T in exponential decay: $T = 1/\lambda$.

Check Your Understanding 5.1

Solutions can be found following the section exercises.

1. (a) Solve the differential equation $P'(t) = -.6P(t)$, $P(0) = 50$.
  (b) Solve the differential equation $P'(t) = kP(t)$, $P(0) = 4000$, where k is some constant.
  (c) Interpret the meaning of $P(2) = 100P(0)$, where t is in hours.
  (d) Find the value of k in part (b) for which $P(2) = 100P(0)$.

2. Under ideal conditions a colony of *Escherichia coli* bacteria can grow by a factor of 100 every 2 hours. If 4000 bacteria are present initially, how long will it take before there are 1 million bacteria?

EXERCISES 5.1

In Exercises 1–10, determine the growth constant k, then find all solutions of the given differential equation.

1. $y' = y$

2. $y' = .4y$

3. $y' = 1.7y$

4. $y' = \dfrac{y}{4}$

5. $y' - \dfrac{y}{2} = 0$

6. $y' - 6y = 0$

7. $2y' - \dfrac{y}{2} = 0$

8. $y = 1.6y'$

9. $\dfrac{y}{3} = 4y'$

10. $5y' - 6y = 0$

In Exercises 11–18, solve the given differential equation with initial condition.

11. $y' = 3y, y(0) = 1$

12. $y' = 4y, y(0) = 0$

13. $y' = 2y, y(0) = 2$

14. $y' = y, y(0) = 4$

15. $y' - .6y = 0, y(0) = 5$

16. $y' - \dfrac{y}{7} = 0, y(0) = 6$

17. $6y' = y, y(0) = 12$

18. $5y = 3y', y(0) = 7$

19. Population with Exponential Growth Let $P(t)$ be the population (in millions) of a certain city t years after 2015, and suppose that $P(t)$ satisfies the differential equation

$$P'(t) = .01P(t), P(0) = 2.$$

(a) Find a formula for $P(t)$.

(b) What was the initial population, that is, the population in 2015?

(c) Estimate the population in 2019.

20. Growth of a Colony of Fruit Flies A colony of fruit flies exhibits exponential growth. Suppose that 500 fruit flies are present. Let $P(t)$ denote the number of fruit flies t days later, and let $k = .08$ denote the growth constant.

(a) Write a differential equation and initial condition that model the growth of this colony.

(b) Find a formula for $P(t)$.

(c) Estimate the size of the colony 5 days later.

21. Growth Constant for a Bacteria Culture A bacteria culture that exhibits exponential growth quadruples in size in 2 days.

(a) Find the growth constant if time is measured in days.

(b) If the initial size of the bacteria culture was 20,000, what is its size after just 12 hours?

22. Growth of a Bacteria Culture The initial size of a bacteria culture that grows exponentially was 10,000. After 1 day, there are 15,000 bacteria.

(a) Find the growth constant if time is measured in days.

(b) How long will it take for the culture to double in size?

23. Using the Differential Equation Let $P(t)$ be the population (in millions) of a certain city t years after 2015, and suppose that $P(t)$ satisfies the differential equation

$$P'(t) = .03P(t), P(0) = 4.$$

(a) Use the differential equation to determine how fast the population is growing when it reaches 5 million people.

(b) Use the differential equation to determine the population size when it is growing at the rate of 400,000 people per year.

(c) Find a formula for $P(t)$.

24. Growth of Bacteria Approximately 10,000 bacteria are placed in a culture. Let $P(t)$ be the number of bacteria present in the culture after t hours, and suppose that $P(t)$ satisfies the differential equation

$$P'(t) = .55P(t).$$

(a) What is $P(0)$?

(b) Find the formula for $P(t)$.

(c) How many bacteria are there after 5 hours?

(d) What is the growth constant?

(e) Use the differential equation to determine how fast the bacteria culture is growing when it reaches 100,000.

(f) What is the size of the bacteria culture when it is growing at a rate of 34,000 bacteria per hour?

25. Growth of Cells After t hours there are $P(t)$ cells present in a culture, where $P(t) = 5000e^{0.2t}$.

(a) How many cells were present initially?

(b) Give a differential equation satisfied by $P(t)$.

(c) When will the initial number of cells double?

(d) When will 20,000 cells be present?

26. Insect Population The size of a certain insect population is given by $P(t) = 300e^{0.01t}$, where t is measured in days.

(a) How many insects were present initially?

(b) Give a differential equation satisfied by $P(t)$.

(c) At what time will the initial population double?

(d) At what time will the population equal 1200?

27. Population Growth Determine the growth constant of a population that is growing at a rate proportional to its size, where the population doubles in size every 40 days and time is measured in days.

28. Time to Triple Determine the growth constant of a population that is growing at a rate proportional to its size, where the population triples in size every 10 years and time is measured in years.

29. Exponential Growth A population is growing exponentially with growth constant .05. In how many years will the current population triple?

30. Time to Double A population is growing exponentially with growth constant .04. In how many years will the current population double?

31. Exponential Growth The rate of growth of a certain cell culture is proportional to its size. In 10 hours a population of 1 million cells grew to 9 million. How large will the cell culture be after 15 hours?

32. World's Population The world's population was 5.51 billion on January 1, 1993, and 5.88 billion on January 1, 1998. Assume that, at any time, the population grows at a rate proportional to the population at that time. In what year will the world's population reach 7 billion?

33. Population of Mexico City At the beginning of 1990, 20.2 million people lived in the metropolitan area of Mexico City, and the population was growing exponentially. The 1995 population was 23 million. (Part of the growth is due to immigration.) If this trend continues, how large will the population be in the year 2010?

34. A Population Model The population (in millions) of a state t years after 2010 is given by the graph of the exponential function $y = P(t)$ with growth constant .025 in Fig. 6. [In parts (c) and (d) use the differential equation satisfied by $P(t)$.]

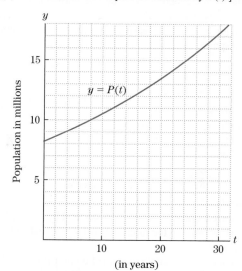

Figure 6

(a) What is the population in 2020?
(b) When is the population 10 million?
(c) How fast is the population growing in 2020?
(d) When is the population growing at the rate of 275,000 people per year?

35. Radioactive Decay A sample of 8 grams of radioactive material is placed in a vault. Let $P(t)$ be the amount remaining after t years, and let $P(t)$ satisfy the differential equation $P'(t) = -.021P(t)$.
(a) Find the formula for $P(t)$
(b) What is $P(0)$?
(c) What is the decay constant?
(d) How much of the material will remain after 10 years?
(e) Use the differential equation to determine how fast the sample is disintegrating when just 1 gram remains.
(f) What amount of radioactive material remains when it is disintegrating at the rate of .105 gram per year?
(g) The radioactive material has a half-life of 33 years. How much will remain after 33 years? 66 years? 99 years?

36. Radioactive Decay Radium 226 is used in cancer radiotherapy. Let $P(t)$ be the number of grams of radium 226 in a sample remaining after t years, and let $P(t)$ satisfy the differential equation

$$P'(t) = -.00043P(t), \qquad P(0) = 12.$$

(a) Find the formula for $P(t)$.
(b) What was the initial amount?
(c) What is the decay constant?
(d) Approximately how much of the radium will remain after 943 years?
(e) How fast is the sample disintegrating when just 1 gram remains? Use the differential equation.
(f) What is the weight of the sample when it is disintegrating at the rate of .004 gram per year?
(g) The radioactive material has a half-life of about 1612 years. How much will remain after 1612 years? 3224 years? 4836 years?

37. Decay of Penicillin in the Bloodstream A person is given an injection of 300 milligrams of penicillin at time $t = 0$. Let $f(t)$ be the amount (in milligrams) of penicillin present in the person's bloodstream t hours after the injection. Then, the amount of penicillin decays exponentially, and a typical formula is $f(t) = 300e^{-0.6t}$.
(a) Give the differential equation satisfied by $f(t)$
(b) How much will remain at time $t = 5$ hours?
(c) What is the biological half-life of the penicillin (that is, the time required for half of a given amount to decompose) in this case?

38. Radioactive Decay Ten grams of a radioactive substance with decay constant .04 is stored in a vault. Assume that time is measured in days, and let $P(t)$ be the amount remaining at time t.
(a) Give the formula for $P(t)$
(b) Give the differential equation satisfied by $P(t)$.
(c) How much will remain after 5 days?
(d) What is the half-life of this radioactive substance?

39. Radioactive Decay The decay constant for the radioactive element cesium 137 is .023 when time is measured in years. Find its half-life.

40. Drug Constant Radioactive cobalt 60 has a half-life of 5.3 years. Find its decay constant.

41. Iodine Level in Dairy Products If dairy cows eat hay containing too much iodine 131, their milk will be unfit to drink. Iodine 131 has half-life of 8 days. If the hay contains 10 times the maximum allowable level of iodine 131, how many days should the hay be stored before it is fed to dairy cows?

42. Half-Life Ten grams of a radioactive material disintegrates to 3 grams in 5 years. What is the half-life of the radioactive material?

43. Decay of Sulfate in the Bloodstream In an animal hospital, 8 units of a sulfate were injected into a dog. After 50 minutes, only 4 units remained in the dog. Let $f(t)$ be the amount of sulfate present after t minutes. At any time, the rate of change of $f(t)$ is proportional to the value of $f(t)$. Find the formula for $f(t)$.

44. Radioactive Decay Forty grams of a certain radioactive material disintegrates to 16 grams in 220 years. How much of this material is left after 300 years?

45. Radioactive Decay A sample of radioactive material decays over time (measured in hours) with decay constant .2. The graph of the exponential function $y = P(t)$ in Fig. 7 gives the number of grams remaining after t hours. [*Hint:* In parts (c) and (d) use the differential equation satisfied by $P(t)$.]

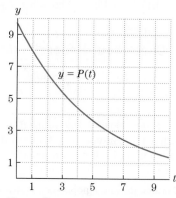

Figure 7

(a) How much was remaining after 1 hour?
(b) Approximate the half-life of the material.
(c) How fast was the sample decaying after 6 hours?
(d) When was the sample decaying at the rate of .4 grams per hour?

46. Rate of Decay A sample of radioactive material has decay constant .25, where time is measured in hours. How fast will the sample be disintegrating when the sample size is 8 grams? For what sample size will the sample size be decreasing at the rate of 2 grams per day?

47. Carbon Dating In 1947, a cave with beautiful prehistoric wall paintings was discovered in Lascaux, France. Some charcoal found in the cave contained 20% of the ^{14}C expected in living trees. How old are the Lascaux cave paintings? (Recall that the decay constant for ^{14}C is .00012.)

48. King Arthur's Round Table According to legend, in the fifth century King Arthur and his knights sat at a huge round table. A round table alleged to have belonged to King Arthur was found at Winchester Castle in England. In 1976, carbon dating revealed the amount of radiocarbon in the table to be 91% of the radiocarbon present in living wood. Could the table possibly have belonged to King Arthur? Why? (Recall that the decay constant for ^{14}C is .00012.)

49. Radioactive Decay A 4500-year-old wooden chest was found in the tomb of the twenty-fifth century B.C. Chaldean king Meskalumdug of Ur. What percentage of the original ^{14}C would you expect to find in the wooden chest?

50. Population of the Pacific Northwest In 1938, sandals woven from strands of tree bark were found in Fort Rock Creek Cave in Oregon. The bark contained 34% of the level of ^{14}C found in living bark. Approximately how old were the sandals? [*Note:* This discovery by University of Oregon anthropologist Luther Cressman forced scientists to double their estimate of how long ago people came to the Pacific Northwest.]

51. Time of the Fourth Ice Age Many scientists believe there have been four ice ages in the past 1 million years. Before the technique of carbon dating was known, geologists erroneously believed that the retreat of the Fourth Ice Age began about 25,000 years ago. In 1950, logs from ancient spruce trees were found under glacial debris near Two Creeks, Wisconsin. Geologists determined that these trees had been crushed by the advance of ice during the Fourth Ice Age. Wood from the spruce trees contained 27% of the level of ^{14}C found in living trees. Approximately how long ago did the Fourth Ice Age actually occur?

52. Time Constant Let T be the time constant of the curve $y = Ce^{-\lambda t}$ as defined in Fig. 5. Show that $T = 1/\lambda$. [*Hint:*

Express the slope of the tangent line in Fig. 5 in terms of C and T. Then, set this slope equal to the slope of the curve $y = Ce^{-\lambda t}$ at $t = 0$.]

53. Differential Equation and Decay The amount in grams of a certain radioactive material present after t years is given by the function $P(t)$. Match each of the following answers with its corresponding question.

Answers
a. Solve $P(t) = .5P(0)$ for t.
b. Solve $P(t) = .5$ for t.
c. $P(.5)$
d. $P'(.5)$
e. $P(0)$
f. Solve $P'(t) = -.5$ for t.
g. $y' = ky$
h. $P_0 e^{kt}, k < 0$

Questions
A. Give a differential equation satisfied by $P(t)$.
B. How fast will the radioactive material be disintegrating in $\frac{1}{2}$ year?
C. Give the general form of the function $P(t)$.
D. Find the half-life of the radioactive material.
E. How many grams of the material will remain after $\frac{1}{2}$ year?
F. When will the radioactive material be disintegrating at the rate of $\frac{1}{2}$ gram per year?
G. When will there be $\frac{1}{2}$ gram remaining?
H. How much radioactive material was present initially?

54. Time Constant and Half-life Consider an exponential decay function $P(t) = P_0 e^{-\lambda t}$, and let T denote its time constant. Show that, at $t = T$, the function $P(t)$ decays to about one-third of its initial size. Conclude that the time constant is always larger than the half-life.

55. An Initial Value Problem Suppose that the function $P(t)$ satisfies the differential equation

$$y'(t) = -.5y(t), \quad y(0) = 10.$$

(a) Find an equation of the tangent line to the graph of $y = P(t)$ at $t = 0$. [*Hint:* What are $P'(0)$ and $P(0)$?]
(b) Find $P(t)$.
(c) What is the time constant of the decay curve $y = P(t)$?

56. Time to Finish Consider the exponential decay function $y = P_0 e^{-\lambda t}$, with time constant T. We define the time to finish to be the time it takes for the function to decay to about 1% of its initial value P_0. Show that the time to finish is about four times the time constant T.

Solutions to Check Your Understanding 5.1

1. (a) Answer: $P(t) = 50e^{-0.6t}$. Differential equations of the type $y' = ky$ have as their solution $P(t) = Ce^{kt}$, where C is $P(0)$.

(b) Answer: $P(t) = 4000e^{kt}$. This problem is like part (a) except that the constant is not specified. Additional information is needed if we want to determine a specific value for k.

(c) After 2 hours, the initial population increased 100-fold.

(d) Answer: $P(t) = 4000e^{2.3t}$. From the solution to part (b), we know that $P(t) = 4000e^{kt}$. We are given that $P(2) = 100P(0) = 100(4000) = 400,000$. So,

$$P(2) = 4000e^{k(2)} = 400,000$$
$$e^{2k} = 100$$
$$2k = \ln 100$$
$$k = \frac{\ln 100}{2} \approx 2.3.$$

2. Let $P(t)$ be the number of bacteria present after t hours. We must first find an expression for $P(t)$ and then determine the value of t for which $P(t) = 1,000,000$. From the discussion at the beginning of the section, we know that $P'(t) = k \cdot P(t)$. Also, we are given that $P(2)$ (the population after 2 hours) is $100P(0)$ (100 times the initial population). From part (d) of the previous problem, we have an expression for $P(t)$:

$$P(t) = 4000e^{2.3t}.$$

Now we must solve $P(t) = 1,000,000$ for t:

$$4000e^{2.3t} = 1,000,000$$
$$e^{2.3t} = 250$$
$$2.3t = \ln 250$$
$$t = \frac{\ln 250}{2.3} \approx 2.4.$$

Therefore, after 2.4 hours, there will be 1,000,000 bacteria.

5.2 Compound Interest

Continuous Compounding

FOR REVIEW

Compound interest was introduced in Sec. 0.5, pp. 35–38.

The subject of compound interest was introduced in Section 0.5, where we derived a formula for the compound amount in a savings account if interest is compounded at stated intervals of time per year (compound periods). In an era of online banking, it is possible to compound interest every month, or every day, or every hour, or perhaps even more frequently. We say that interest is **compounded continuously** if the number of compound periods per year is increased indefinitely. As we will show momentarily, if interest is compounded continuously, your savings account will grow exponentially, much like the fruit flies in your kitchen, or the bacteria in a petri dish (Sec. 5.1).

Let P_0 denote your initial deposit in dollars (also called the principal amount), r the annual rate of interest, and $y = A(t)$ the compound amount or balance in your savings account at the end of time t, where t is measured in years. Even though it is difficult to describe $A(t)$, it is not difficult to describe its *rate of change* at any time t. Indeed, since the interest rate is r, if at time t you have $A(t)$ dollars in the account, then the account is growing at a rate of r times $A(t)$ dollars per year. Since the rate of change is $A'(t)$, we get

$$\overbrace{A'(t)}^{\text{[rate of change]}} = r \times A(t).$$

Thus, the balance in your savings account satisfies the differential equation and initial condition

$$A'(t) = rA(t), \qquad A(0) = P_0. \tag{1}$$

The solution of this equation follows from Theorem 2 of the previous section:

$$A(t) = P_0 e^{rt}. \tag{2}$$

This is the **continuous compound interest formula**. It gives the balance of your savings account or the compound amount after t years, when interest is compounded continuously.

The formula $A(t) = P_0 e^{rt}$ contains four variables. (Remember that the letter e represents a specific constant, $e = 2.718\ldots.$) In a typical problem, we are given values for three of these variables and must solve for the remaining variable.

EXAMPLE 1 **Continuous Compound Interest** One thousand dollars is invested at 5% interest compounded continuously.
(a) Give the formula for $A(t)$, the compound amount after t years.
(b) How much will be in the account after 6 years?
(c) After 6 years, at what rate will $A(t)$ be growing?
(d) How long is required for the initial investment to double?

SOLUTION
(a) $P_0 = 1000$ and $r = .05$. By the continuous compound formula (2), $A(t) = 1000e^{0.05t}$.

(b) $A(6) = 1000e^{0.05(6)} = 1000e^{0.3} \approx \$1349.86$.

(c) Rate of growth is different from interest rate. Interest rate is fixed at 5% and does not change with time. However, the rate of growth $A'(t)$ is always changing. Since $A(t) = 1000e^{0.05t}$, $A'(t) = (1000) \cdot (.05)e^{0.05t} = 50e^{0.05t}$. So, after 6 years,

$$A'(6) = 50e^{0.05(6)} = 50e^{0.3} \approx 67.49 \text{ dollars per year.}$$

After 6 years, the investment is growing at the rate of \$67.49 per year.

There is an easier way to answer part (c), given that we have already calculated $A(6)$. Since $A(t)$ satisfies the differential equation $A'(t) = rA(t)$,

$$A'(6) = .05A(6) = .05 \cdot 1349.86 \approx \$67.49 \text{ per year.}$$

(d) We must find t such that $A(t) = \$2000$. So we set $1000e^{0.05t} = 2000$ and solve for t.

$1000e^{0.05t} = 2000$	Given equation.
$e^{0.05t} = 2$	Divide by 1000.
$\ln e^{0.05t} = \ln 2$	Take ln of each side.
$.05t = \ln 2$	$\ln e^{0.05t} = .05t$.
$t = \dfrac{\ln 2}{.05} \approx 13.86 \text{ years}$	Solve for t.

≫ Now Try Exercise 1

NOTE ▶ The calculations in Example 1(d) would be essentially unchanged after the first step if the initial amount of the investment were changed from \$1000 to any arbitrary amount P. When this investment doubles, the compound amount will be $2P$. So, we set $2P = Pe^{0.05t}$ and solve for t as we did previously to conclude that, at 5% interest compounded continuously, any amount doubles in about 13.86 years. ◀

EXAMPLE 2 **Appreciation of a Painting** Pablo Picasso's *The Dream* was purchased in 1941 for a war-distressed price of \$7000. The painting was sold in 1997 for \$48.4 million, the second highest price ever paid for a Picasso painting at auction. What rate of interest compounded continuously did this investment earn?

SOLUTION Let P_0e^{rt} be the value (in millions) of the painting t years after 1941. Since the initial value is .007 million, $P_0 = .007$. Since the value after 56 years is 48.4 million dollars, $.007e^{r(56)} = 48.4$. Now solve for r:

$.007e^{r(56)} = 48.4$	Given equation.
$e^{r(56)} = \dfrac{48.4}{.007} \approx 6914.29$	Divide by .007.
$r(56) = \ln(6914.29)$	Take ln of each side.
$r = \dfrac{\ln(6914.29)}{56} \approx .158$	Solve for r.

Therefore, as an investment, the painting earned an interest rate of about 15.8%.

≫ Now Try Exercise 17

Ordinary Versus Continuous Compounding

How much is gained from continuous compounding as opposed to ordinary compounding? Our intuition tells us that if we compound interest frequently enough, then the compound amount should be close to the compound amount from continuous compounding. Let $A(t)$ denote the continuous compound amount that we derived in (2), and let B denote the compound amount if we use m compound periods per year. In Section 0.5, we derived the formula

$$B = P_0\left(1 + \frac{r}{m}\right)^{mt}. \tag{3}$$

For instance, suppose that $1000 is invested at 6% interest for 1 year, and that interest is compounded once per year. In formula (3), this corresponds to $P_0 = \$1000$, $r = .06$, $m = 1$, and $t = 1$ year. The amount at the end of one year is

$$B = P_0(1 + r)^1 = 1000(1 + .06) = \$1060.$$

If interest is compounded quarterly ($m = 4$),

$$B = P_0\left(1 + \frac{r}{4}\right)^{4t} = 1000\left(1 + \frac{.06}{4}\right)^4 \approx \$1061.36.$$

If interest is compounded monthly ($m = 12$),

$$B = P_0\left(1 + \frac{r}{12}\right)^{12t} = 1000\left(1 + \frac{.06}{12}\right)^{12} \approx \$1061.68.$$

If interest is compounded continuously, we use (2) and get

$$A = P_0 e^{rt} = 1000 e^{0.06(1)} \approx \$1061.84.$$

Table 1 contains these results along with the one for daily compounding for 1 year.

Table 5.1 Effect of Increased Compounding Periods

Frequency of Compounding	Annually	Quarterly	Monthly	Daily	Continuous
m	1	4	12	365	
Balance after 1 year ($)	1060.00	1061.36	1061.68	1061.83	1061.84

Comparing the results from Table 1, we note that continuous compounding yields only one cent more than the result of the daily compounding. Consequently, frequent compounding (such as every hour or every second) will produce, at most, 1 cent more in our case.

In many computations, it is simpler to use the formula for interest compounded continuously than the formula for ordinary compound interest. In these instances, it is commonplace to use interest compounded continuously as an approximation to ordinary compound interest.

Negative Interest Rates

In 2015, some European banks began "paying" negative interest on short-term deposits to encourage customers to invest their savings. Let $-r$ denote the negative interest rate, P_0 the principal amount, and $A(t)$ the compound amount after time t in years. To describe the account in this case, we modify equation (1) as follows:

$$A'(t) = -rA(t), \qquad A(0) = P_0, \tag{4}$$

where the minus sign indicates that $A'(t)$, the rate of change of the compound amount $A(t)$, is negative. Using Theorem 2, Sec. 5.1, we see that the solution is an exponentially decaying function

$$A(t) = P_0 e^{-rt}. \tag{5}$$

EXAMPLE 3   **Negative Interest Rate** In 2015, The Swiss National Bank used a negative interest rate of $-.75\%$ on savings deposits. A customer makes an initial deposit of 10,000 Swiss Francs (SFr).
(a) What is the formula for $A(t)$, the balance in SFr after t years?
(b) How much money is in the account after 2 years?

SOLUTION   **(a)** Use (5) with $r = 0.0075$ and $P_0 = 10,000$:

$$A(t) = 10{,}000e^{-0.0075t}.$$

(b) After 2 years, the balance is

$$A(2) = 10{,}000e^{-0.0075(2)} \approx 9851.12\ \text{SFr}.$$

Thus, the balance decreased by about 49 Swiss Francs in 2 years.   «

Present Value

If P dollars are invested today, the formula $A = Pe^{rt}$ gives the value of this investment after t years (assuming continuously compounded interest). We say that P is the **present value** of the amount A to be received in t years. If we solve for P in terms of A, we obtain

$$P = Ae^{-rt}. \tag{6}$$

The concept of the present value of money is an important theoretical tool in business and economics. Problems involving depreciation of equipment, for example, may be analyzed by calculus techniques when the present value of money is computed from (2), using continuously compounded interest.

EXAMPLE 4   **Present Value** Suppose you want to invest some amount of money now (i.e., the present value) so that you have $5000 in 2 years. Assume you can earn 12% compounded continuously. How much would you need to invest? Find the present value of $5000 to be received in 2 years if money can be invested at 12% compounded continuously.

SOLUTION   Use formula (6) with $A = 5000$, $r = .12$, and $t = 2$.

$$P = 5000e^{-(0.12)(2)} = 5000e^{-0.24}$$
$$\approx \$3933.14 \qquad \text{» Now Try Exercise 19}$$

A Limit Formula for *e*

We have defined continuous compounding as the limit of ordinary compounding if the number of compound periods per year increases indefinitely. So, if we let m tend to infinity in (3), the formula for ordinary compounding, we should get (2), the formula for continuous compounding. In other words,

$$\lim_{m \to \infty} P_0\left(1 + \frac{r}{m}\right)^{mt} = P_0e^{rt}.$$

In this formula, take $P_0 = 1$, $r = 1$, $t = 1$, and get

$$\lim_{m \to \infty} \left(1 + \frac{1}{m}\right)^{m} = e^1 = e.$$

This gives a limit formula for the number e that can be used to approximate the number e, and can be verified with the help of a calculator.

Check Your Understanding 5.2

Solutions can be found following the section exercises.

1. One thousand dollars is to be invested in a bank for 4 years. Would 8% interest compounded semiannually be better than $7\frac{3}{4}$% interest compounded continuously?

2. A building was bought for $150,000 and sold 10 years later for $400,000. What interest rate (compounded continuously) was earned on the investment?

EXERCISES 5.2

1. **Savings Account** Let $A(t) = 5000e^{0.04t}$ be the balance in a savings account after t years.
   (a) How much money was originally deposited?
   (b) What is the interest rate?
   (c) How much money will be in the account after 10 years?
   (d) What differential equation is satisfied by $y = A(t)$?
   (e) Use the results of parts (c) and (d) to determine how fast the balance is growing after 10 years.
   (f) How large will the balance be when it is growing at the rate of $280 per year?

2. **Savings Account** Let $A(t)$ be the balance in a savings account after t years, and suppose that $A(t)$ satisfies the differential equation

$$A'(t) = .045A(t), \qquad A(0) = 3000.$$

   (a) How much money was originally deposited in the account?
   (b) What interest rate is being earned?
   (c) Find the formula for $A(t)$.
   (d) What is the balance after 5 years?
   (e) Use part (d) and the differential equation to determine how fast the balance is growing after 5 years.
   (f) How large will the balance be when it is growing at the rate of $270 per year?

3. **Savings Account** Four thousand dollars is deposited in a savings account at 3.5% yearly interest compounded continuously.
   (a) What is the formula for $A(t)$, the balance after t years?
   (b) What differential equation is satisfied by $A(t)$, the balance after t years?
   (c) How much money will be in the account after 2 years?
   (d) When will the balance reach $5000?
   (e) How fast is the balance growing when it reaches $5000?

4. **Savings Account** Ten thousand dollars is deposited in a savings account at 4.6% yearly interest compounded continuously.
   (a) What differential equation is satisfied by $A(t)$, the balance after t years?
   (b) What is the formula for $A(t)$?
   (c) How much money will be in the account after 3 years?
   (d) When will the balance triple?
   (e) How fast is the balance growing when it triples?

5. **Investment Analysis** An investment earns 4.2% yearly interest compounded continuously. How fast is the investment growing when its value is $9000?

6. **Investment Analysis** An investment earns 5.1% yearly interest compounded continuously and is currently growing at the rate of $765 per year. What is the current value of the investment?

7. **Continuous Compound** One thousand dollars is deposited in a savings account at 6% yearly interest compounded continuously. How many years are required for the balance in the account to reach $2500?

8. **Continuous Compound** Ten thousand dollars is invested at 6.5% interest compounded continuously. When will the investment be worth $41,787?

9. **Technology Stock** One hundred shares of a technology stock were purchased on January 2, 1990, for $1200 and sold on January 2, 1998, for $12,500. What rate of interest compounded continuously did this investment earn?

10. **Appreciation of Art Work** Pablo Picasso's *Angel Fernandez de Soto* was acquired in 1946 for a postwar splurge of $22,220. The painting was sold in 1995 for $29.1 million. What yearly rate of interest compounded continuously did this investment earn?

11. **Investment Analysis** How many years are required for an investment to double in value if it is appreciating at the yearly rate of 4% compounded continuously?

12. **Doubling an Investment** What yearly interest rate (compounded continuously) is earned by an investment that doubles in 10 years?

13. **Tripling an Investment** If an investment triples in 15 years, what yearly interest rate (compounded continuously) does the investment earn?

14. **Real Estate Investment** If real estate in a certain city appreciates at the yearly rate of 15% compounded continuously, when will a building purchased in 2010 triple in value?

15. **Negative Interest Rates** Suppose that the bank in Example 3 increased its fees by charging a negative annual interest rate of $-.9$%. Find the balance after two years in a savings account if $P_0 = 10,000$ SFr.

16. **Negative Interest Rates** How is the account in Exercise 15 changing when the balance is 9,500 SFr?

17. **Real Estate Investment** A farm purchased in 2000 for $1 million was valued at $3 million in 2010. If the farm continues to appreciate at the same rate (with continuous compounding), when will it be worth $10 million?

18. **Real Estate Investment** A parcel of land bought in 1990 for $10,000 was worth $16,000 in 1995. If the land continues to appreciate at this rate, in what year will it be worth $45,000?

19. **Present Value** Find the present value of $1000 payable at the end of 3 years, if money may be invested at 8% with interest compounded continuously.

20. **Present Value** Find the present value of $2000 to be received in 10 years, if money may be invested at 8% with interest compounded continuously.

21. **Present Value** How much money must you invest now at 4.5% interest compounded continuously to have $10,000 at the end of 5 years?

22. **Present Value** If the present value of $1000 to be received in 5 years is $559.90, what rate of interest, compounded continuously, was used to compute this present value?

23. **Comparing Two Investments** Investment A is currently worth $70,200 and is growing at the rate of 13% per year compounded continuously. Investment B is currently worth $60,000 and is growing at the rate of 14% per year compounded continuously. After how many years will the two investments have the same value?

24. **Compound Interest** Ten thousand dollars is deposited in a money market fund paying 8% interest compounded continuously. How much interest will be earned during the second year of the investment?

25. **Differential Equation and Interest** A small amount of money is deposited in a savings account with interest compounded continuously. Let $A(t)$ be the balance in the account after t years. Match each of the following answers with its corresponding question.

Answers
a. Pe^{rt} b. $A(3)$ c. $A(0)$ d. $A'(3)$
e. Solve $A'(t) = 3$ for t.
f. Solve $A(t) = 3$ for t.
g. $y' = ry$
h. Solve $A(t) = 3A(0)$ for t.

Questions
A. How fast will the balance be growing in 3 years?
B. Give the general form of the function $A(t)$.
C. How long will it take for the initial deposit to triple?
D. Find the balance after 3 years.
E. When will the balance be 3 dollars?
F. When will the balance be growing at the rate of 3 dollars per year?
G. What was the principal amount?
H. Give a differential equation satisfied by $A(t)$.

26. **Growth of a Savings Account** The curve in Fig. 1 shows the growth of money in a savings account with interest compounded continuously.
(a) What is the balance after 20 years?
(b) At what rate is the money growing after 20 years?
(c) Use the answers to parts (a) and (b) to determine the interest rate.

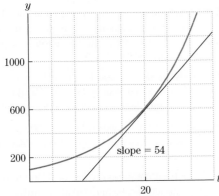

Figure 1 Growth of money in a savings account.

27. **Savings Account** The function $A(t)$ in Fig. 2(a) gives the balance in a savings account after t years with interest compounded continuously. Figure 2(b) shows the derivative of $A(t)$.
(a) What is the balance after 20 years?
(b) How fast is the balance increasing after 20 years?

 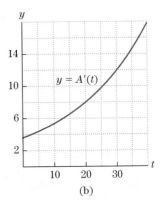

(a)   (b)

Figure 2

(c) Use the answers to parts (a) and (b) to determine the interest rate.
(d) When is the balance $300?
(e) When is the balance increasing at the rate of $12 per year?
(f) Why do the graphs of $A(t)$ and $A'(t)$ look the same?

28. When $1000 is invested at $r\%$ interest (compounded continuously) for 10 years, the balance is $f(r)$ dollars, where f is the function shown in Fig. 3.
(a) What will the balance be at 7% interest?
(b) For what interest rate will the balance be $3000?
(c) If the interest rate is 9%, what is the growth rate of the balance with respect to a unit increase in interest?

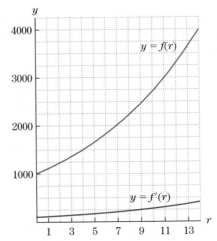

Figure 3 Effect of interest rate on balance.

TECHNOLOGY EXERCISES

29. Verify that $\lim_{m \to \infty} \left(1 + \dfrac{1}{m}\right)^m = e$ by taking m increasingly large and noticing that $\left(1 + \dfrac{1}{m}\right)^m$ approaches 2.718.

30. Verify that daily compounding is nearly the same as continuous compounding by graphing $y = 100[1 + (.05/360)]^{360x}$, together with $y = 100e^{0.05x}$ in the window $[0, 64]$ by $[250, 2500]$. The two graphs should appear the same on the screen. Approximately how far apart are they when $x = 32$? When $x = 64$?

31. Internal Rate of Return An investment of $2000 yields payments of $1200 in 3 years, $800 in 4 years, and $500 in 5 years. Thereafter, the investment is worthless. What constant rate of return r would the investment need to produce to yield the payments specified? The number r is called the *internal rate of return* on the investment. We can consider the investment as consisting of three parts, each part yielding one payment. The sum of the present values of the three parts must total $2000. This yields the equation

$$2000 = 1200e^{-3r} + 800e^{-4r} + 500e^{-5r}.$$

Solve this equation to find the value of r.

Solutions to Check Your Understanding 5.2

1. Let us compute the balance after 4 years for each type of interest.

8% *compounded semiannually*: Use the formula at the beginning of this section. Here $P = 1000$, $r = .08$, $m = 2$ (semiannually means there are two interest periods per year), and $t = 4$. Therefore,

$$A = 1000\left(1 + \frac{.08}{2}\right)^{2 \cdot 4} = 1000(1.04)^8 \approx \$1368.57.$$

$7\frac{3}{4}$% *compounded continuously*: Use the formula $A = Pe^{rt}$, where $P = 1000$, $r = .0775$, and $t = 4$. Then,

$$A = 1000e^{(0.0775) \cdot 4} = 1000e^{0.31} \approx \$1363.43.$$

Therefore, 8% compounded semiannually is better.

2. If the $150,000 had been compounded continuously for 10 years at interest rate r, the balance would be $150,000e^{r \cdot 10}$. The question is at what value of r will the balance be $400,000? We need to solve an equation for r.

$$150,000e^{r \cdot 10} = 400,000$$
$$e^{r \cdot 10} \approx 2.67$$
$$r \cdot 10 = \ln 2.67$$
$$r = \frac{\ln 2.67}{10} \approx .098$$

Therefore, the investment earned 9.8% interest per year.

5.3 Applications of the Natural Logarithm Function to Economics

In this section, we consider two applications of the natural logarithm to the field of economics. Our first application is concerned with relative rates of change and the second with elasticity of demand.

In 2015, the price of eggs began to climb due to a shortage of supply caused by the Avian flu, commonly known as bird flu. The average price of one dozen eggs rose to $2.80 and was increasing at the rate of $1.20 per year. At the same time, the price of a new compact car rose to $12,500 and was increasing at the rate of $1100 per year. As a consumer, you want to know which price is increasing more quickly. It is not meaningful to say that the car price is increasing faster simply because $1100 is larger than $1.20. We must take into account the vast difference between the actual cost of a car and the cost of one dozen eggs. A more meaningful basis of comparison of price increase is the **percentage rate of increase**, which compares the rate of change with the actual price. We can say that, in 2015, the price of one dozen eggs is increasing at the percentage rate

$$\frac{[\text{rate of increase}]}{[\text{actual price}]} = \frac{1.2}{2.8} \approx .43 = 43\% \text{ per year.}$$

At the same time, the price of a new compact car is increasing at the percentage rate

$$\frac{[\text{rate of increase}]}{[\text{actual price}]} = \frac{1100}{12,500} \approx .09 = 9\% \text{ per year.}$$

Thus, the price of one dozen eggs is increasing at a faster percentage rate than the price of a new compact car.

The concept of percentage rate of change is useful to economists. To state a general definition, let us recall that the rate of change of a function is given by its derivative. We can now introduce the following useful concept.

> **DEFINITION**   For a given function $f(t)$, the **relative rate of change** of $f(t)$ per unit change of t is defined to be
>
> $$\frac{[\text{rate of change}]}{[\text{actual value}]} = \frac{f'(t)}{f(t)}. \tag{1}$$
>
> The **percentage rate of change** is the relative rate of change of $f(t)$ expressed as a percentage.

The relative rate of change of $f(t)$ is also called the **logarithmic derivative** of $f(t)$, because of the derivative formula you saw in Section 4.5:

$$\frac{d}{dt}\ln[f(t)] = \frac{f'(t)}{f(t)} \tag{2}$$

EXAMPLE 1   **Log Derivative and Relative Rate of Change**   Find the logarithmic derivative and then compute the relative rate of change and percentage rate of change at the given value of t.
(a) $f(t) = t^3 + 2t^2 - 11$, $t = 1$.
(b) $f(t) = e^{\sqrt{t}}$, $t = 4$.

SOLUTION   (a) Differentiating $f(t)$, we find $f'(t) = 3t^2 + 4t$. So, from (2), the logarithmic derivative of f is

$$\frac{f'(t)}{f(t)} = \frac{3t^2 + 4t}{t^3 + 2t^2 - 11}.$$

When $t = 1$,

$$\frac{f'(1)}{f(1)} = \frac{7}{-8} = -\frac{7}{8} = -.875.$$

Thus, when $t = 1$, the relative rate of change of $f(t)$ with respect to t is $-.875$, which corresponds to a percentage rate of change of -87.5%. (A negative percentage corresponds to a decrease of 87.5%.)

(b) Because of the exponential in $f(t)$, we can simplify our computations by using (2). The logarithmic derivative is

$$\frac{d}{dt}\ln[f(t)] = \frac{d}{dt}\ln\left[e^{\sqrt{t}}\right]$$

$$= \frac{d}{dt}(\sqrt{t}) \qquad \text{Because } \ln(e^{\sqrt{t}}) = \sqrt{t}$$

$$= \frac{1}{2\sqrt{t}}.$$

When $t = 4$, the logarithmic derivative is equal to

$$\frac{1}{2\sqrt{4}} = \frac{1}{4} = .25.$$

Thus, when $t = 4$, the relative rate of change of $f(t)$ is $.25$, and so the percentage rate of change is 25%.   **» Now Try Exercise 3**

Economists often use percentage rates of change when discussing the growth of various economic quantities, such as national income or national debt, because such rates of change can be meaningfully compared.

EXAMPLE 2   **Gross Domestic Product** A certain school of economists modeled the nominal gross domestic product of the United States at time t (measured in years from January 1, 2005) by the formula

$$f(t) = 13.2 + .7t - .11t^2 + .01t^3,$$

where $f(t)$ is measured in trillions of dollars. (See Fig. 1.) What was the predicted percentage rate of growth (or decline) of the economy at $t = 3$ and $t = 9$?

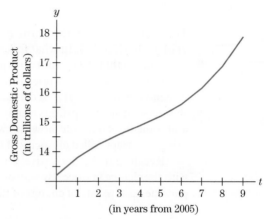

Figure 1

SOLUTION   Since

$$f'(t) = .7 - .22t + .03t^2,$$

we see that

$$\frac{f'(3)}{f(3)} = \frac{.7 - .22(3) + .03(9)}{13.2 + .7(3) - .11(9) + .01(27)} = \frac{.31}{14.58} \approx .021$$

$$\frac{f'(9)}{f(9)} = \frac{.7 - .22(9) + .03(81)}{13.2 + .7(9) - .11(81) + .01(729)} = \frac{1.15}{17.88} \approx .064.$$

So on January 1, 2008 ($t = 3$), the economy is predicted to grow at a relative rate of about 2.1% per year. On January 1, 2014 ($t = 9$), the economy is predicted to be still growing, but at a relative rate of about 6.4% per year.   **» Now Try Exercise 9**

EXAMPLE 3   **Constant Relative Rate of Change** If the function $f(t)$ has a constant relative rate of change k, show that $f(t) = Ce^{kt}$ for some constant C.

SOLUTION   We are given that

$$\frac{f'(t)}{f(t)} = k.$$

Hence, $f'(t) = kf(t)$. But this is just the differential equation satisfied by the exponential function (Theorem 1, Section 5.1). Therefore, we must have $f(t) = Ce^{kt}$ for some constant C.   **«**

Elasticity of Demand

In Section 2.7, we considered demand equations for companies and for entire industries. Recall that a demand equation expresses, for each quantity x to be produced, the market price that will generate a demand of exactly x. For instance, the demand equation

$$p = 150 - .01x$$

Figure 2

says that, to sell x units, the price must be set at $150 - .01x$ dollars. To be specific, to sell 6000 units, the price must be set at $150 - .01(6000) = \$90$ per unit.

The demand equation may be solved for x in terms of p to yield

$$x = 100(150 - p).$$

This last equation gives quantity in terms of price. If we let the letter q represent quantity, the equation becomes

$$q = 100(150 - p).$$

This equation is of the form $q = f(p)$, where, in this case, $f(p)$ is the function $f(p) = 100(150 - p)$. In what follows, it will be convenient to always write our demand functions so that the quantity q is expressed as a function $f(p)$ of the price p.

Usually, raising the price of a commodity lowers demand. Therefore, the typical demand function $q = f(p)$ is decreasing and has a negative slope everywhere. (See Fig. 2.) But does raising the price typically also lowers revenue? The answer is "sometimes yes, and sometimes no." But how can we predict the answer to this important question? By using a concept called elasticity of demand.

Recall that the derivative $f'(p)$ compares the change in the quantity demanded with the change in price. By way of contrast, the concept of elasticity is designed to compare the *relative* rate of change of the quantity demanded with the *relative* rate of change of price.

Let us be more explicit. Consider a particular demand function $q = f(p)$. From our interpretation of the logarithmic derivative in (1), we know that the relative rate of change of the quantity demanded with respect to p is

$$\frac{(d/dp)\,f(p)}{f(p)} = \frac{f'(p)}{f(p)}.$$

Similarly, the relative rate of change of price with respect to p is

$$\frac{(d/dp)p}{p} = \frac{1}{p}.$$

Hence, the ratio of the relative rate of change of the quantity demanded to the relative rate of change of price is

$$\frac{[\text{relative rate of change of quantity}]}{[\text{relative rate of change of price}]} = \frac{f'(p)/f(p)}{1/p} = \frac{pf'(p)}{f(p)}.$$

Since $f'(p)$ is always negative for a typical demand function, the quantity $pf'(p)/f(p)$ will be negative for all values of p. For convenience, economists prefer to work with positive numbers, and therefore, the *elasticity of demand* is taken to be this quantity multiplied by -1.

Elasticity of Demand The **elasticity of demand** $E(p)$ at price p for the demand function $q = f(p)$ is defined to be

$$E(p) = \frac{-pf'(p)}{f(p)}.$$

EXAMPLE 4   **Elasticity of Demand** The demand function for a certain metal is $q = 100 - 2p$, where p is the price per pound and q is the quantity demanded (in millions of pounds).

(a) What quantity can be sold at \$30 per pound?

(b) Determine the function $E(p)$.

(c) Determine and interpret the elasticity of demand at $p = 30$.

(d) Determine and interpret the elasticity of demand at $p = 20$.

SOLUTION

(a) In this case, $q = f(p)$, where $f(p) = 100 - 2p$. When $p = 30$, we have $q = f(30) = 100 - 2(30) = 40$. Therefore, 40 million pounds of the metal can be sold. We also say that the *demand* is 40 million pounds.

(b) $E(p) = \dfrac{-pf'(p)}{f(p)} = \dfrac{-p(-2)}{100 - 2p} = \dfrac{2p}{100 - 2p}$

(c) The elasticity of demand at price $p = 30$ is $E(30)$.

$$E(30) = \frac{2(30)}{100 - 2(30)} = \frac{60}{40} = \frac{3}{2}.$$

When the price is set at \$30 per pound, a small increase in price will result in a relative rate of decrease in quantity demanded of about $\frac{3}{2}$ times the relative rate of increase in price. For example, if the price is increased from \$30 by 1%, the quantity demanded will decrease by about 1.5%.

(d) When $p = 20$, we have

$$E(20) = \frac{2(20)}{100 - 2(20)} = \frac{40}{60} = \frac{2}{3}.$$

When the price is set at \$20 per pound, a small increase in price will result in a relative rate of decrease in quantity demanded of only $\frac{2}{3}$ of the relative rate of increase of price. For example, if the price is increased from \$20 by 1%, the quantity demanded will decrease by $\frac{2}{3}$ of 1%.   **» Now Try Exercise 21**

Elasticity, Price and Revenue

The significance of the concept of elasticity may perhaps be best appreciated by a study of how revenue, $R(p)$, responds to change in price. Let's start by expressing the revenue function as a function of price:

$$R(p) = f(p) \cdot p,$$

where $f(p)$ is the demand function. Differentiate $R(p)$ using the product rule and get:

$$
\begin{aligned}
R'(p) &= \frac{d}{dp}[f(p) \cdot p] = f(p) \cdot 1 + p \cdot f'(p) \\
&= f(p)\left[1 + \frac{pf'(p)}{f(p)}\right] && \text{Factor } f(p). \\
&= f(p)[1 - E(p)]. && \text{Because } E(p) = -\frac{pf'(p)}{f(p)}. \quad (3)
\end{aligned}
$$

This equation relates the rate of change of revenue to elasticity of demand. Note that if $E(p) = 1$ then $R'(p) = 0$. The cases $E(p) < 1$ and $E(p) > 1$ have interesting implications. Let us introduce a terminology used by economists.

> **DEFINITION   Elastic and Inelastic Demand**   We say that demand is **elastic** at price p_0 if $E(p_0) > 1$ and **inelastic** if $E(p_0) < 1$.

Now, suppose that demand is elastic at some price p_0. Then $E(p_0) > 1$ and $1 - E(p_0) < 0$. Since $f(p)$ is always positive, we see from (3) that $R'(p_0) < 0$. Therefore, by the first derivative rule, $R(p)$ is decreasing at p_0. So, an increase in price will result in a decrease in revenue, and a decrease in price will result in an increase in revenue. In a similar way, we can show that if demand is inelastic, then $R'(p)$ will be positive. In this case, an increase in price will result in an increase in revenue, and a decrease in price will result in a decrease in revenue. This can be summarized as follows.

Elasticity Rule When demand is elastic ($E(p) > 1$), the change in revenue is in the opposite direction of the change in price. And, when demand is inelastic ($E(p) < 1$), the change in revenue is in the same direction of the change in price.

As noted previously, when $E(p_0) = 1$, then, from (3), $R'(p_0) = 0$, and so p_0 is a critical value of R.

EXAMPLE 5 **Elasticity of Demand** Figure 3 shows the elasticity of demand for the metal in Example 4:

$$E(p) = \frac{2p}{100 - 2p}.$$

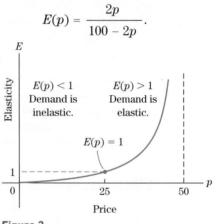

Figure 3

(a) For what values of p is demand elastic? Inelastic?
(b) Find and plot the revenue function for $0 < p < 50$.
(c) Verify the elasticity rule by analyzing how revenue responds to an increase in price when demand is elastic or, respectively, inelastic.

SOLUTION (a) In Example 4(b), we found the elasticity of demand to be

$$E(p) = \frac{2p}{100 - 2p}.$$

Let us solve $E(p) = 1$ for p.

$$\frac{2p}{100 - 2p} = 1 \qquad \text{Given equation.}$$

$$2p = 100 - 2p \quad \text{Multiply by } 100 - 2p.$$

$$4p = 100 \qquad \text{Add } 2p \text{ to both sides.}$$

$$p = 25 \qquad \text{Divide by 4.}$$

By definition, demand is elastic at price p if $E(p) > 1$ and inelastic if $E(p) < 1$. From Figure 4, we see that demand is elastic if $25 < p < 50$ and inelastic if $0 < p < 25$.

(b) Recall that

$$[\text{revenue}] = [\text{quantity}] \cdot [\text{price per unit}].$$

Using the formula for demand (in millions of pounds) from Example 4, we obtain the revenue function

$$R = (100 - 2p) \cdot p = p(100 - 2p) \quad \text{(in millions of dollars)}.$$

This is a parabola opening down with p-intercepts at $p = 0$ and $p = 50$. Its maximum is located at the midpoint of the p-intercepts, or $p = 25$. (See Fig. 4.)

Figure 4

(c) In part (a), we determined that demand is elastic for $25 < p < 50$. For p in this price range, Figure 4 shows that an increase in price results in a decrease in revenue, and a decrease in price results in an increase in revenue. Hence, we conclude that, when demand is elastic, the change of revenue is in the opposite direction of the change in price. Similarly, when demand is inelastic ($0 < p < 25$), Figure 4 shows that the change of revenue is in the same direction as the change in price.

» Now Try Exercise 23

Check Your Understanding 5.3

Solutions can be found following the section exercises.

The current toll for the use of a certain toll road is $2.50. A study conducted by the state highway department determined that, with a toll of p dollars, q cars will use the road each day, where $q = 60{,}000e^{-0.5p}$.

1. Compute the elasticity of demand at $p = 2.5$.

2. Is demand elastic or inelastic at $p = 2.5$?

3. If the state increases the toll slightly, will the revenue increase or decrease?

EXERCISES 5.3

Find the logarithmic derivative and then determine the percentage rate of change of the functions at the points indicated.

1. $f(t) = t^2$ at $t = 10$ and $t = 50$

2. $f(t) = t^{10}$ at $t = 10$ and $t = 50$

3. $f(x) = e^{0.3x}$ at $x = 10$ and $x = 20$

4. $f(x) = e^{-0.05x}$ at $x = 1$ and $x = 10$

5. $f(t) = e^{0.3t^2}$ at $t = 1$ and $t = 5$

6. $G(s) = e^{-0.05s^2}$ at $s = 1$ and $s = 10$

7. $f(p) = 1/(p + 2)$ at $p = 2$ and $p = 8$

8. $g(p) = 5/(2p + 3)$ at $p = 1$ and $p = 11$

9. **Percentage Rate of Growth** The annual sales S (in dollars) of a company may be approximated by the formula

$$S = 50{,}000\sqrt{e^{\sqrt{t}}},$$

where t is the number of years beyond some fixed reference date. Use a logarithmic derivative to determine the percentage rate of growth of sales at $t = 4$.

10. **Percentage Rate of Change** The price of wheat per bushel at time t (in months) is approximated by

$$f(t) = 4 + .001t + .01e^{-t}.$$

What is the percentage rate of change of $f(t)$ at $t = 0$? $t = 1$? $t = 2$?

11. **Price of Ground Beef** The wholesale price in dollars of one pound of ground beef is modeled by the function $f(t) = 3.08 + .57t - .1t^2 + .01t^3$, where t is measured in years from January 1, 2010.

(a) Estimate the price in 2011 and find the rate in dollars per year at which the price was rising in 2011.

(b) What is the percentage rate of increase of the price of one pound of beef in 2011?

(c) Answer parts (a) and (b) for the year 2016.

12. **Price of Pork** The wholesale price in dollars of one pound of pork is modeled by the function $f(t) = 1.4 + .26t - .1t^2 + .01t^3$, where t is measured in years from January 1, 2010.

(a) Estimate the price in 2012 and find the percentage rate of increase of the price in 2012?

(b) Answer part (a) for the year 2017.

For each demand function, find $E(p)$ and determine if demand is elastic or inelastic (or neither) at the indicated price.

13. $q = 700 - 5p$, $p = 80$

14. $q = 600e^{-0.2p}$, $p = 10$

15. $q = 400(116 - p^2)$, $p = 6$

16. $q = (77/p^2) + 3$, $p = 1$

17. $q = p^2e^{-(p+3)}$, $p = 4$

18. $q = 700/(p + 5)$, $p = 15$

19. Elasticity of Demand Currently, 1800 people ride a certain commuter train each day and pay $4 for a ticket. The number of people q willing to ride the train at price p is $q = 600(5 - \sqrt{p})$. The railroad would like to increase its revenue.
(a) Is demand elastic or inelastic at $p = 4$?
(b) Should the price of a ticket be raised or lowered?

20. Elasticity of Demand An electronic store can sell $q = 10,000/(p + 50) - 30$ cellular phones at a price p dollars per phone. The current price is $150.
(a) Is demand elastic or inelastic at $p = 150$?
(b) If the price is lowered slightly, will revenue increase or decrease?

21. Elasticity of Demand A movie theater has a seating capacity of 3000 people. The number of people attending a show at price p dollars per ticket is $q = (18,000/p) - 1500$. Currently, the price is $6 per ticket.
(a) Is demand elastic or inelastic at $p = 6$?
(b) If the price is lowered, will revenue increase or decrease?

22. Elasticity of Demand A subway charges 65 cents per person and has 10,000 riders each day. The demand function for the subway is $q = 2000\sqrt{90 - p}$.
(a) Is demand elastic or inelastic at $p = 65$?
(b) Should the price of a ride be raised or lowered to increase the amount of money taken in by the subway?

23. Elasticity of Demand A country that is the major supplier of a certain commodity wishes to improve its balance-of-trade position by lowering the price of the commodity. The demand function is $q = 1000/p^2$.
(a) Compute $E(p)$.
(b) Will the country succeed in raising its revenue?

24. Show that any demand function of the form $q = a/p^m$ has constant elasticity m.

Relative Rate of Change of Cost A cost function $C(x)$ gives the total cost of producing x units of a product. The elasticity of cost at quantity x, $E_c(x)$, is defined to be the ratio of the relative rate of change of cost (with respect to x) divided by the relative rate of change of quantity (with respect to x).

25. Show that $E_c(x) = x \cdot C'(x)/C(x)$.

26. Show that E_c is equal to the marginal cost divided by the average cost.

27. Let $C(x) = (1/10)x^2 + 5x + 300$. Show that $E_c(50) < 1$. (Hence, when 50 units are produced, a small relative increase in production results in an even smaller relative increase in total cost. Also, the average cost of producing 50 units is greater than the marginal cost at $x = 50$.)

28. Let $C(x) = 1000e^{0.02x}$. Determine and simplify the formula for $E_c(x)$. Show that $E_c(60) > 1$, and interpret this result.

TECHNOLOGY EXERCISES

29. Consider the demand function $q = 60,000e^{-0.5p}$ from Check Your Understanding 5.3.
(a) Determine the value of p for which the value of $E(p)$ is 1. For what values of p is demand inelastic?
(b) Graph the revenue function in the window $[0, 4]$ by $[-5000, 50,000]$, and determine where its maximum value occurs. For what values of p is the revenue an increasing function?

Solutions to Check Your Understanding 5.3

1. The demand function is $f(p) = 60,000e^{-0.5p}$.

$$f'(p) = -30,000e^{-0.5p}$$

$$E(p) = \frac{-pf'(p)}{f(p)} = \frac{-p(-30,000)e^{-0.5p}}{60,000e^{-0.5p}} = \frac{p}{2}$$

$$E(2.5) = \frac{2.5}{2} = 1.25$$

2. The demand is elastic, because $E(2.5) > 1$.

3. Since demand is elastic at $2.50, a slight change in price causes revenue to change in the *opposite* direction. Hence, revenue will decrease.

5.4 Further Exponential Models

Terminal Velocity After jumping out of an airplane, a skydiver falls at an increasing rate. However, the wind rushing past the skydiver's body creates an upward force that begins to counterbalance the downward force of gravity. This air friction finally becomes so great that the skydiver's velocity reaches a limiting speed called the **terminal velocity**. If we let $v(t)$ be the downward velocity of the skydiver after t seconds of free fall, a good mathematical model for $v(t)$ is given by

$$v(t) = M(1 - e^{-kt}), \tag{1}$$

where M is the terminal velocity and k is some positive constant. (See Fig. 1.) When t is close to zero, e^{-kt} is close to 1, and the velocity is small. As t increases, e^{-kt} becomes small, and so $v(t)$ approaches M.

FOR REVIEW

The line $y = M$ is a horizontal asymptote. See Sec. 2.1, p. 137.

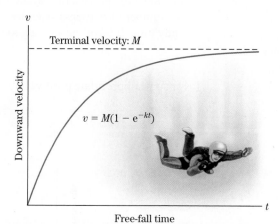

Figure 1 Downward velocity of a skydiver.

EXAMPLE 1   **Velocity of a Skydiver** Show that the velocity given in equation (1) satisfies the equations

$$v'(t) = k[M - v(t)], \quad v(0) = 0. \tag{2}$$

SOLUTION   From (1) we have $v(t) = M - Me^{-kt}$. Then,

$$v'(t) = Mke^{-kt}.$$

However,

$$k[M - v(t)] = k[M - (M - Me^{-kt})] = kMe^{-kt},$$

so the differential equation $v'(t) = k[M - v(t)]$ holds. Also,

$$v(0) = M - Me^0 = M - M = 0. \qquad \text{»» Now Try Exercise 3}$$

The differential equation (2) says that the rate of change in v is proportional to the difference between the terminal velocity M and the actual velocity v. It can be shown that the only solution of (2) is given by the formula in (1). We summarize this useful fact as follows.

Solution of a Differential Equation   The unique solution of the differential equation and initial condition

$$y'(t) = k(M - y(t)), \quad y(0) = 0 \quad \text{is}$$

$$y(t) = M(1 - e^{-kt})$$

The two equations (1) and (2) arise as mathematical models in a variety of situations. Some of these application are described next.

The Learning Curve   Psychologists have found that, in many learning situations, a person's rate of learning is rapid at first and then slows down. Finally, as the task is mastered, the person's level of performance reaches a level above which it is almost impossible to rise. For example, within reasonable limits, each person seems to have a certain maximum capacity for memorizing a list of nonsense syllables. Suppose that a subject can memorize M syllables in a row if given sufficient time—say, an hour—to study the list but cannot memorize $M + 1$ syllables in a row even if allowed several hours of study. By giving the subject different lists of syllables and varying lengths of time to study the lists, the psychologist can determine an empirical relationship between the number of nonsense syllables memorized accurately and the number of minutes of study time. It turns out that a good model for this situation is

$$f(t) = M(1 - e^{-kt})$$

for some appropriate positive constant k. (See Fig. 2.)

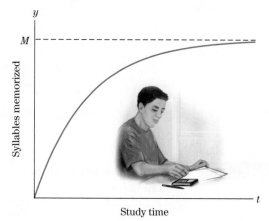

Figure 2 Learning curve, $f(t) = M(1 - e^{-kt})$.

The *slope* of this learning curve at time t is approximately the number of additional syllables that can be memorized if the subject is given 1 more minute of study time. Thus, the slope is a measure of the *rate of learning*. The differential equation satisfied by the function $y = f(t)$ is

$$y' = k(M - y), \qquad f(0) = 0.$$

This equation says that, if the subject is given a list of M nonsense syllables, the rate of memorization is proportional to the number of syllables remaining to be memorized.

Diffusion of Information by Mass Media   Sociologists have found that the differential equation (2) provides a good model for the way information is spread (or "diffused") through a population when the information is being publicized constantly by mass media, such as television or online. (*Source: Introduction to Mathematical Sociology.*) Given a fixed population P, let $f(t)$ be the number of people who have already heard a certain piece of information by time t. Then, $P - f(t)$ is the number who have not yet heard the information by time t. Also, $f'(t)$ is the rate of increase of the number of people who have heard the news (the "rate of diffusion" of the information). If the information is being publicized often by some mass media, it is likely that the number of *newly informed* people per unit time is proportional to the number of people who have not yet heard the news. Therefore,

$$f'(t) = k[P - f(t)]. \tag{3}$$

Assume that $f(0) = 0$ (that is, there was a time $t = 0$ when nobody had heard the news). Then, the solution box following Example 1 shows that

$$f(t) = P(1 - e^{-kt}). \tag{4}$$

(See Fig. 3.)

Figure 3 Diffusion of information by mass media.

EXAMPLE 2

Diffusion of Information The news of the resignation of a public official is broadcast frequently by internet news media and television stations. Also, one-half of the residents of a city have heard the news within 4 hours of its initial release. Use the exponential model (4) to estimate when 90% of the residents will have heard the news.

SOLUTION

We must find the value of k in (4). If P is the number of residents, the number who have heard the news in the first 4 hours is given by (4), with $t = 4$. By assumption, this number is half the population. So,

$$\tfrac{1}{2}P = P(1 - e^{-k\cdot 4})$$

$$.5 = 1 - e^{-4k} \qquad \text{Divide by } P.$$

$$e^{-4k} = 1 - .5 = .5 \qquad \text{Solve for } e^{-4k}.$$

$$\ln(e^{-4k}) = \ln(.5) \qquad \text{Take ln.}$$

$$-4k = \ln(.5)$$

$$k = -\frac{1}{4}\ln(.5) \qquad \text{Solve for } k.$$

$$\approx .173$$

Note that

$$-\frac{1}{4}\ln(.5) = -\frac{1}{4}\ln\left(\frac{1}{2}\right) = \frac{1}{4}\ln(2).$$

So the model for this particular situation is

$$f(t) = P(1 - e^{-kt}), \quad \text{where } k = \frac{\ln 2}{4}.$$

Now we want to find t such that $f(t) = .90P$. We solve for t:

$$.90P = P(1 - e^{-kt})$$

$$.90 = 1 - e^{-kt}$$

$$e^{-kt} = 1 - .90 = .10$$

$$-kt = \ln .10$$

$$t = \frac{\ln .10}{-k} = -4\frac{\ln .10}{\ln 2} \approx 13.29.$$

Therefore, 90% of the residents will hear the news in 13.3 hours of its initial release.

»» Now Try Exercise 7

Intravenous Infusion of Glucose

The human body both manufactures and uses glucose (blood sugar). Usually, there is a balance in these two processes, so the bloodstream has a certain equilibrium level of glucose. Suppose that a patient is given a single intravenous injection of glucose, and let $A(t)$ be the amount of glucose (in milligrams) above the equilibrium level. Then, the body will start using up the excess glucose at a rate proportional to the amount of excess glucose; that is,

$$A'(t) = -\lambda A(t), \tag{5}$$

where λ is a positive constant called the **velocity constant of elimination**. This constant depends on how fast an individual patient's metabolic processes eliminate the excess glucose from the blood. Equation (5) describes a simple exponential decay process.

Now suppose that, instead of a single shot, the patient receives a continuous intravenous infusion of glucose. A bottle of glucose solution is suspended above the patient, and a small tube carries the glucose down to a needle that runs into a vein. In this case, there are two influences on the amount of excess glucose in the blood: the glucose being added steadily from the bottle and the glucose being removed from the body by metabolic processes. Let r be the rate of infusion of glucose (often, from 10 to 100 milligrams per minute). If the body did not remove any glucose, the excess glucose would increase at a constant rate of r milligrams per minute; that is,

$$A'(t) = r. \tag{6}$$

Taking into account the two influences on $A'(t)$ described by (5) and (6), we can write

$$A'(t) = r - \lambda A(t). \tag{7}$$

Define M to be r/λ, and note that initially there is no excess glucose; then,

$$A'(t) = \lambda(M - A(t)), \qquad A(0) = 0.$$

As stated in the solution box following Example 1, a solution of this differential equation is given by

$$A(t) = M(1 - e^{-\lambda t}) = \frac{r}{\lambda}(1 - e^{-\lambda t}). \tag{8}$$

Note that M is the limiting value of the glucose level. Reasoning as in Example 1, we conclude that the amount of excess glucose rises until it reaches a stable level. (See Fig. 4.)

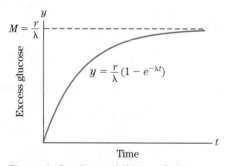

Figure 4 Continuous infusion of glucose.

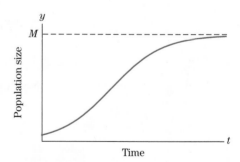

Figure 5 Logistic growth.

Logistic Growth

The model for simple exponential growth discussed in Section 5.1 is adequate for describing the growth of many types of populations, but obviously, a population cannot increase exponentially forever. The simple exponential growth model becomes inapplicable when the environment begins to inhibit the growth of the population. The logistic growth curve is an important exponential model that takes into account some of the effects of the environment on a population. (See Fig. 5.) For small values of t, the curve has the same basic shape as an exponential growth curve. Then, when the population begins to suffer from overcrowding or lack of food, the growth rate (the slope of the population curve) begins to slow down. Eventually, the growth rate tapers off to zero as the population reaches the maximum size that the environment will support. This latter part of the curve resembles the growth curves studied earlier in this section.

The equation for logistic growth has the general form

Logistic Growth

$$y = \frac{M}{1 + Be^{-Mkt}},$$ (9)

---

where B, M, and k are positive constants. We can show that y satisfies the differential equation

$$y' = ky(M - y).$$ (10)

The factor y reflects the fact that the growth rate (y') depends in part on the size y of the population. The factor $M - y$ reflects the fact that the growth rate also depends on how close y is to the maximum level M.

The logistic curve is often used to fit experimental data that lie along an S-shaped curve. Examples are given by the growth of a fish population in a lake and the growth of a fruit fly population in a laboratory container. Also, certain enzyme reactions in animals follow a logistic law. One of the earliest applications of the logistic curve occurred in about 1840, when the Belgian sociologist P. Verhulst fit a logistic curve to six U.S. census figures, 1790 to 1840, and predicted the U.S. population for 1940. His prediction missed by fewer than 1 million people (an error of about 1%).

EXAMPLE 3   **Logistic Growth** A lake is stocked with 100 fish. After 3 months, there are 250 fish. A study of the ecology of the lake predicts that the lake can support 1000 fish. Find a formula for the number $P(t)$ of fish in the lake t months after it has been stocked.

SOLUTION   The limiting population M is 1000. Therefore, we have

$$P(t) = \frac{1000}{1 + Be^{-1000kt}}.$$

At $t = 0$ there are 100 fish, so that

$$100 = P(0) = \frac{1000}{1 + Be^0} = \frac{1000}{1 + B}.$$

Thus, $1 + B = 10$, or $B = 9$. Finally, since $P(3) = 250$, we have

$$250 = \frac{1000}{1 + 9e^{-3000k}}$$

$$1 + 9e^{-3000k} = 4$$

$$e^{-3000k} = \frac{1}{3}$$

$$-3000k = \ln\frac{1}{3}$$

$$k \approx .00037.$$

Therefore,

$$P(t) = \frac{1000}{1 + 9e^{-0.37t}}.$$

≫ Now Try Exercise 9

An Epidemic Model

It will be instructive to actually "build" a mathematical model. Our example concerns the spread of a highly contagious disease. We begin by making several simplifying assumptions:

1. The population is a fixed number P, and each member of the population is susceptible to the disease.
2. The duration of the disease is long, so that no cures occur during the time period under study.
3. All infected individuals are contagious and circulate freely among the population.
4. During each time period (such as 1 day or 1 week), each infected person makes c contacts, and each contact with an uninfected person results in transmission of the disease.

Consider a short period of time from t to $t + h$. Each infected person makes $c \cdot h$ contacts. How many of these contacts are with uninfected persons? If $f(t)$ is the number of infected persons at time t, then $P - f(t)$ is the number of uninfected persons, and $[P - f(t)]/P$ is the fraction of the population that is uninfected. Thus, of the $c \cdot h$ contacts made,

$$\left[\frac{P - f(t)}{P}\right] \cdot c \cdot h$$

will be with uninfected persons. This is the number of new infections produced by one infected person during the time period of length h. The total number of *new* infections during this period is

$$f(t)\left[\frac{P - f(t)}{P}\right] ch.$$

But this number must equal $f(t + h) - f(t)$, where $f(t + h)$ is the total number of infected persons at time $t + h$. So,

$$f(t + h) - f(t) = f(t)\left[\frac{P - f(t)}{P}\right] ch.$$

Dividing by h, the length of the time period, we obtain the average number of new infections per unit time (during the small time period):

$$\frac{f(t + h) - f(t)}{h} = \frac{c}{P} f(t)[P - f(t)].$$

If we let h approach zero and let y stand for $f(t)$, the left-hand side approaches the rate of change in the number of infected persons, and we derive the following equation:

$$\frac{dy}{dt} = \frac{c}{P} y(P - y). \tag{11}$$

This is the same type of equation as that used in (10) for logistic growth, although the two situations leading to this model appear to be quite dissimilar.

Comparing (11) with (10), we see that the number of infected individuals at time t is described by a logistic curve with $M = P$ and $k = c/P$. Therefore, by (9), we can write

$$f(t) = \frac{P}{1 + Be^{-ct}}.$$

B and c can be determined from the characteristics of the epidemic. (See Example 4.)

The logistic curve has an inflection point at that value of t for which $f(t) = P/2$. The position of this inflection point has great significance for applications of the logistic curve. From inspecting a graph of the logistic curve, we see that the inflection point is the point at which the curve has greatest slope. In other words, the inflection point corresponds to the instant of fastest growth of the logistic curve. This means, for example, that, in the foregoing epidemic model, the disease is spreading with the greatest rapidity precisely when half the population is infected. Any attempt at disease control (through immunization, for example) must strive to reduce the incidence of the disease to as low a point as possible, but, in any case, at least below the inflection point at $P/2$, at which point the epidemic is spreading fastest.

EXAMPLE 4 **Spread of an Epidemic** The Department of Public Health monitors the spread of an epidemic of a particularly long-lasting strain of flu in a city of 500,000 people. At the beginning of the first week of monitoring, 200 cases had been reported; during the first week, 300 new cases are reported. Estimate the number of infected individuals after 6 weeks.

SOLUTION Here, $P = 500,000$. If $f(t)$ denotes the number of cases at the end of t weeks, then,

$$f(t) = \frac{P}{1 + Be^{-ct}} = \frac{500,000}{1 + Be^{-ct}}.$$

Moreover, $f(0) = 200$, so

$$200 = \frac{500,000}{1 + Be^0} = \frac{500,000}{1 + B}.$$

$$200(1 + B) = 500,000$$

$$1 + B = 2500$$

$$B = 2499.$$

Consequently, since $f(1) = 300 + 200 = 500$, we have

$$500 = f(1) = \frac{500,000}{1 + 2499e^{-c}},$$

$$500(1 + 2499e^{-c}) = 500,000$$

$$1 + 2499\,e^{-c} = 1000$$

$$2499\,e^{-c} = 999$$

$$e^{-c} = \frac{999}{2499} \approx .4.$$

so $-c \approx \ln(.4)$ or $c \approx .92$. Finally,

$$f(t) = \frac{500,000}{1 + 2499e^{-0.92t}}$$

and

$$f(6) = \frac{500,000}{1 + 2499e^{-0.92(6)}} \approx 45,000.$$

After 6 weeks, about 45,000 individuals are infected. **» Now Try Exercise 11**

This epidemic model is used by sociologists (who still call it an epidemic model) to describe the spread of a rumor. In economics, the model is used to describe the diffusion of knowledge about a product. An "infected person" represents an individual who possesses knowledge of the product. In both cases, it is assumed that the members of the population are themselves primarily responsible for the spread of the rumor or knowledge of the product. This situation is in contrast to the model described earlier, in which information was spread through a population by external sources, such as online news outlets, radio, and television.

There are several limitations to this epidemic model. Each of the four simplifying assumptions made at the outset is unrealistic in varying degrees. More complicated models can be constructed that rectify one or more of these defects, but they require more advanced mathematical tools.

Check Your Understanding 5.4

Solutions can be found following the section exercises.

1. A sociological study was made to examine the process by which doctors decide to adopt a new drug. The doctors were divided into two groups. The doctors in group A had little interaction with other doctors and so received most of their information through mass media. The doctors in group B had extensive interaction with other doctors and so received most of their information through word of mouth. For each group, let $f(t)$ be the number who have learned about the new drug after t months. Examine the appropriate differential equations to explain why the two graphs were of the types shown in Fig. 6. (*Source: Sociometry.*)

Group A   Group B

Figure 6 Results of a sociological study.

EXERCISES 5.4

1. Consider the function $f(x) = 5(1 - e^{-2x})$, $x \geq 0$.
   (a) Show that $f(x)$ is increasing and concave down for all $x \geq 0$.
   (b) Explain why $f(x)$ approaches 5 as x gets large.
   (c) Sketch the graph of $f(x)$, $x \geq 0$.

2. Consider the function $g(x) = 10 - 10e^{-0.1x}$, $x \geq 0$.
   (a) Show that $g(x)$ is increasing and concave down for $x \geq 0$.
   (b) Explain why $g(x)$ approaches 10 as x gets large.
   (c) Sketch the graph of $g(x)$, $x \geq 0$.

3. If $y = 2(1 - e^{-x})$, compute y' and show that $y' = 2 - y$.

4. If $y = 5(1 - e^{-2x})$, compute y' and show that $y' = 10 - 2y$.

5. If $f(x) = 3(1 - e^{-10x})$, show that $y = f(x)$ satisfies the differential equation

$$y' = 10(3 - y), \quad f(0) = 0.$$

6. **Ebbinghaus Model for Forgetting** A student learns a certain amount of material for some class. Let $f(t)$ denote the percentage of the material that the student can recall t weeks later. The psychologist Hermann Ebbinghaus found that this percentage of retention can be modeled by a function of the form

$$f(t) = (100 - a)e^{-\lambda t} + a,$$

where λ and a are positive constants and $0 < a < 100$. Sketch the graph of the function $f(t) = 85e^{-0.5t} + 15$, $t \geq 0$.

7. **Spread of News** When a grand jury indicted the mayor of a certain town for accepting bribes, the newspaper, online news outlets, radio, and television immediately began to publicize the news. Within an hour, one-quarter of the citizens heard about the indictment. Estimate when three-quarters of the town heard the news.

8. Examine formula (8) for the amount $A(t)$ of excess glucose in the bloodstream of a patient at time t. Describe what would happen if the rate r of infusion of glucose were doubled.

9. **Spread of News** A news item is spread by word of mouth to a potential audience of 10,000 people. After t days,

$$f(t) = \frac{10,000}{1 + 50e^{-0.4t}}$$

people will have heard the news. The graph of this function is shown in Fig. 7.
   (a) Approximately how many people will have heard the news after 7 days?
   (b) At approximately what rate will the news spread after 14 days?
   (c) Approximately when will 7000 people have heard the news?
   (d) Approximately when will the news spread at the rate of 600 people per day?
   (e) When will the news spread at the greatest rate?
   (f) Use equations (9) and (10) to determine the differential equation satisfied by $f(t)$.
   (g) At what rate will the news spread when half the potential audience has heard the news?

10. **Concentration of Glucose in the Bloodstream** Physiologists usually describe the continuous intravenous infusion of glucose in terms of the excess *concentration* of glucose, $C(t) = A(t)/V$, where V is the total volume of blood in the patient. In this case, the rate of increase in the concentration of glucose due to the continuous injection is r/V. Find a differential equation that gives a model for the rate of change of the excess concentration of glucose.

11. **Spread of News** A news item is broadcast by mass media to a potential audience of 50,000 people. After t days,

$$f(t) = 50,000(1 - e^{-0.3t})$$

people will have heard the news. The graph of this function is shown in Fig. 8.
   (a) How many people will have heard the news after 10 days?
   (b) At what rate is the news spreading initially?
   (c) When will 22,500 people have heard the news?

(a)

(b)

Figure 7

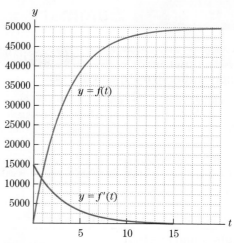

Figure 8

13. **Amount of a Drug in the Bloodstream** After a drug is taken orally, the amount of the drug in the bloodstream after t hours is $f(t) = 122(e^{-0.2t} - e^{-t})$ units.
    (a) Graph $f(t)$, $f'(t)$, and $f''(t)$ in the window $[0, 12]$ by $[-20, 75]$
    (b) How many units of the drug are in the bloodstream after 7 hours?
    (c) At what rate is the level of drug in the bloodstream increasing after 1 hour?
    (d) While the level is decreasing, when is the level of drug in the bloodstream 20 units?
    (e) What is the greatest level of drug in the bloodstream, and when is this level reached?
    (f) When is the level of drug in the bloodstream decreasing the fastest?

14. **Growth with Restriction** A model incorporating growth restrictions for the number of bacteria in a culture after t days is given by $f(t) = 5000(20 + te^{-0.04t})$.
    (a) Graph $f'(t)$ and $f''(t)$ in the window $[0, 100]$ by $[-700, 300]$.
    (b) How fast is the culture changing after 100 days?
    (c) Approximately when is the culture growing at the rate of 76.6 bacteria per day?
    (d) When is the size of the culture greatest?
    (e) When is the size of the culture decreasing the fastest?

(d) Approximately when will the news spread at the rate of 2500 people per day?
(e) Use equations (3) and (4) to determine the differential equation satisfied by $f(t)$
(f) At what rate will the news spread when half the potential audience has heard the news?

12. **Glucose Elimination** Describe an experiment that a doctor could perform to determine the velocity constant of elimination of glucose for a particular patient.

Solutions to Check Your Understanding 5.4

1. The difference between transmission of information by mass media (Group A) and by word of mouth (Group B) is that in Group B the rate of transmission depends not only on the number of people who have not yet received the information, but also on the number of people who know the information and therefore are capable of spreading it. Therefore, for Group A, $f'(t) = k[P - f(t)]$, and for Group B, $f'(t) = kf(t)[P - f(t)]$. Note that the spread of information by word of mouth follows the same pattern as the spread of an epidemic.

| CHAPTER 5 | Summary |

KEY TERMS AND CONCEPTS	EXAMPLES

5.1 Exponential Growth and Decay

- The solutions of the differential equation $y' = ky$ are all of the form $y = Ce^{kt}$, where C is an arbitrary constant.
- If $k > 0$, the solution is an exponential growth function, and k is called the growth constant.
- If $k < 0$, write $k = -\lambda$, then the solution, $y = Ce^{-\lambda t}$, is an exponential decay function, and the constant λ is called the decay constant.
- The unique solution of the differential equation $y' = ky$ with initial condition $y(0) = P_0$ is the function $y = P_0 e^{kt}$.

- The differential equation $y' = .02y$ has infinitely many solutions. They are all of the form $y = Ce^{0.02t}$, where C is an arbitrary constant.
- The unique solution of the differential equation with initial condition, $y' = 3y$, $y(0) = 413$, is $y(t) = 413e^{3t}$.
- Radioactive cobalt has a decay constant $\lambda = .13$. If 2 grams of cobalt is present, let $P(t)$ be the number of grams remaining after t years. Then $y = P(t)$ satisfies $y' = -.13y$, $y(0) = 2$. The formula for $P(t)$ is $P(t) = 2e^{-0.13t}$.

5.2 Compound Interest

When the interest rate is compounded continuously in a savings account, the account grows exponentially. If r is the annual interest rate, P_0 the initial deposit, and $y = A(t)$ the balance in the account after t years, then y satisfies the equation

$$y' = ry, \quad y(0) = P_0$$

and is given by

$$y = A(t) = P_0 e^{rt}.$$

Present Value If P dollars are invested today, then $A = Pe^{rt}$ gives the value of this investment after t years. We say that P is the present value of the amount A to be received in t years. We have $P = Ae^{-rt}$.

An amount of \$2000 dollars is deposited in an account that earns 6% annual interest rate, compounded continuously.

(a) Give a formula for $A(t)$, the compound amount in the account after t years.

(b) How long is required for the amount to reach \$3000?

To answer (a), we appeal to the solution of the differential equation $y' = .06y$, $y(0) = 2000$. Then, $A(t) = 2000e^{0.06t}$.

(b) To find t, we set $A(t) = 3000$ and solve for t:

$$2000e^{0.06t} = 3000$$

$$e^{0.06t} = \frac{3000}{2000} = \frac{3}{2}$$

$$\ln[e^{0.06t}] = \ln\left(\frac{3}{2}\right)$$

$$.06t = \ln\left(\frac{3}{2}\right)$$

$$t = \frac{\ln\left(\frac{3}{2}\right)}{.06} \approx 6.8 \text{ years.}$$

5.3 Applications of the Natural Logarithm Function to Economics

The **relative rate of change** of a function is

$$\frac{f'(t)}{f(t)}.$$

We note that this is also $\frac{d}{dt}[\ln(f(t))]$.

The **percentage rate of change** is the relative rate of change expressed as a percentage.

The value of an investment t years later is approximated by the formula

$$f(t) = 10{,}000e^{\frac{t}{t+1}}.$$

Determine the percentage rate of change of the investment when $t = 1$ and when $t = 5$.

The percentage rate of change is given by $\frac{f'(t)}{f(t)} = \frac{d}{dt}[\ln(f(t))]$. In our case,

$$\ln(f(t)) = \ln\left[10{,}000e^{\frac{t}{t+1}}\right]$$

$$= \ln(10{,}000) + \ln\left[e^{\frac{t}{t+1}}\right] = \ln(10{,}000) + \frac{t}{t+1}$$

$$\frac{d}{dt}[\ln(f(t))] = \frac{d}{dt}\left[\ln(10{,}000) + \frac{t}{t+1}\right] = 0 + \frac{d}{dt}\left[\frac{t}{t+1}\right],$$

because $\ln(10{,}000)$ is a constant, its derivative is 0. Now, by the quotient rule,

$$\frac{d}{dt}\left[\frac{t}{t+1}\right] = \frac{(t+1) - t}{(t+1)^2} = \frac{1}{(t+1)^2}.$$

KEY TERMS AND CONCEPTS	EXAMPLES
	So, $$\frac{f'(t)}{f(t)} = \frac{d}{dt}[\ln(f(t))] = \frac{1}{(t+1)^2}.$$ When $t = 1$, $$\frac{f'(1)}{f(1)} = \frac{1}{(1+1)^2} = \frac{1}{4} = .25 = 25\%.$$ When $t = 5$, $$\frac{f'(5)}{f(5)} = \frac{1}{(1+5)^2} = \frac{1}{36} \approx .03 = 3\%.$$

5.4 Further Exponential Models

Several applications are discussed in this section, where the exponential function plays a central role: the velocity of a skydiver; the learning curve; exponential growth applications with a limiting capacity, such as a population of fish in a lake with a maximum capacity; the spread of an epidemic in a limited environment; the diffusion of information by mass media. These applications are modeled by one of the two following differential equation: $y' = k(M - y)$ and $y' = ky(M - y)$.

In a model of diffusion of information by mass media, the number of people who have heard a certain piece of information by time t is denoted by $y = f(t)$, where $f(t)$ satisfies the differential equation $y' = .2(1 - y)$, $y(0) = 0$. The solution of this differential equation is given following Example 1 of Sec. 5.4. It is $f(t) = 1000(1 - e^{-0.2t})$.

CHAPTER 5 Fundamental Concept Check Exercises

1. What differential equation is key to solving exponential growth and decay problems? State a result about the solution to this differential equation.

2. What is a growth constant? A decay constant?

3. What is meant by the half-life of a radioactive element?

4. Explain how radiocarbon dating works.

5. State the formula for each of the following quantities:
   (a) The compound amount of P dollars in t years at interest rate r, compounded continuously
   (b) The present value of A dollars in n years at interest rate r, compounded continuously

6. What is the difference between a relative rate of change and a percentage rate of change?

7. Define the elasticity of demand, $E(p)$, for a demand function. How is $E(p)$ used?

8. Describe an application of the differential equation $y' = k(M - y)$.

9. Describe an application of the differential equation $y' = ky(M - y)$.

CHAPTER 5 Review Exercises

1. **Atmospheric Pressure** The atmospheric pressure $P(x)$ (measured in inches of mercury) at height x miles above sea level satisfies the differential equation $P'(x) = -.2P(x)$. Find the formula for $P(x)$ if the atmospheric pressure at sea level is 29.92.

2. **Population Model** The herring gull population in North America has been doubling every 13 years since 1900. Give a differential equation satisfied by $P(t)$, the population t years after 1900.

3. **Present Value** Find the present value of $10,000 payable at the end of 5 years if money can be invested at 12% with interest compounded continuously.

4. **Compound Interest** One thousand dollars is deposited in a savings account at 10% interest compounded continuously.

How many years are required for the balance in the account to reach $3000?

5. **Half-Life** The half-life of the radioactive element tritium is 12 years. Find its decay constant.

6. **Carbon Dating** A piece of charcoal found at Stonehenge contained 63% of the level of ^{14}C found in living trees. Approximately how old is the charcoal?

7. **Population Model** From January 1, 2010, to January 1, 2017, the population of a state grew from 17 million to 19.3 million.
   (a) Give the formula for the population t years after 2010.
   (b) If this growth continues, how large will the population be in 2020?
   (c) In what year will the population reach 25 million?

8. **Compound Interest** A stock portfolio increased in value from $100,000 to $117,000 in 2 years. What rate of interest, compounded continuously, did this investment earn?

9. **Comparing Investments** An investor initially invests $10,000 in a risky venture. Suppose that the investment earns 20% interest, compounded continuously, for 5 years and then 6% interest, compounded continuously, for 5 years thereafter.
   (a) How much does the $10,000 grow to after 10 years?
   (b) The investor has the alternative of an investment paying 14% interest compounded continuously. Which investment is superior over a 10-year period, and by how much?

10. **Bacteria Growth** Two different bacteria colonies are growing near a pool of stagnant water. The first colony initially has 1000 bacteria and doubles every 21 minutes. The second colony has 710,000 bacteria and doubles every 33 minutes. How much time will elapse before the first colony becomes as large as the second?

11. **Population Model** The population of a city t years after 1990 satisfies the differential equation $y' = .02y$. What is the growth constant? How fast will the population be growing when the population reaches 3 million people? At what level of population will the population be growing at the rate of 100,000 people per year?

12. **Bacteria Growth** A colony of bacteria is growing exponentially with growth constant .4, with time measured in hours. Determine the size of the colony when the colony is growing at the rate of 200,000 bacteria per hour. Determine the rate at which the colony will be growing when its size is 1 million.

13. **Population Model** The population of a certain country is growing exponentially. The total population (in millions) in t years is given by the function $P(t)$. Match each of the following answers with its corresponding question.

   Answers
   a. Solve $P(t) = 2$ for t.
   b. $P(2)$
   c. $P'(2)$
   d. Solve $P'(t) = 2$ for t.
   e. $y' = ky$
   f. Solve $P(t) = 2P(0)$ for t
   g. $P_0 e^{kt}$, $k > 0$
   h. $P(0)$

   Questions
   A. How fast will the population be growing in 2 years?
   B. Give the general form of the function $P(t)$.
   C. How long will it take for the current population to double?
   D. What will be the size of the population in 2 years?
   E. What is the initial size of the population?
   F. When will the size of the population be 2 million?
   G. When will the population be growing at the rate of 2 million people per year?
   H. Give a differential equation satisfied by $P(t)$.

14. **Radioactive Decay** You have 80 grams of a certain radioactive material, and the amount remaining after t years is given by the function $f(t)$ shown in Fig. 1.
   (a) How much will remain after 5 years?
   (b) When will 10 grams remain?
   (c) What is the half-life of this radioactive material?
   (d) At what rate will the radioactive material be disintegrating after 1 year?
   (e) After how many years will the radioactive material be disintegrating at the rate of about 5 grams per year?

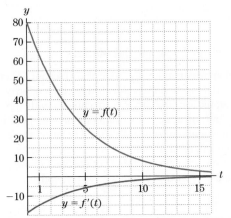

Figure 1

15. **Compound Interest** A few years after money is deposited into a bank, the compound amount is $1000, and it is growing at the rate of $60 per year. What interest rate (compounded continuously) is the money earning?

16. **Compound Interest** The current balance in a savings account is $1230, and the interest rate is 4.5%. At what rate is the compound amount currently growing?

17. Find the percentage rate of change of the function $f(t) = 50e^{0.2t^2}$ at $t = 10$.

18. Find $E(p)$ for the demand function $q = 4000 - 40p^2$, and determine if demand is elastic or inelastic at $p = 5$.

19. **Elasticity of Demand** For a certain demand function, $E(8) = 1.5$. If the price is increased to $8.16, estimate the percentage decrease in the quantity demanded. Will the revenue increase or decrease?

20. Find the percentage rate of change of the function $f(p) = \dfrac{1}{3p + 1}$ at $p = 1$.

21. **Elasticity of Demand** A company can sell $q = 1000p^2 e^{-0.02(p+5)}$ calculators at a price of p dollars per calculator. The current price is $200. If the price is decreased, will the revenue increase or decrease?

22. **Elasticity of Demand** Consider a demand function of the form $q = ae^{-bp}$, where a and b are positive numbers. Find $E(p)$, and show that the elasticity equals 1 when $p = 1/b$.

23. Refer to Check Your Understanding 5.4. Out of 100 doctors in Group A, none knew about the drug at time $t = 0$, but 66 of them were familiar with the drug after 13 months. Find the formula for $f(t)$.

24. **Height of a Weed** The growth of the yellow nutsedge weed is described by a logistic growth formula $f(t)$ of type (9) in Section 5.4. A typical weed has length 8 centimeters after 9 days and length 48 centimeters after 25 days and reaches length 55 centimeters at maturity. Find the formula for $f(t)$.

25. **Temperature of a Rod** When a rod of molten steel with a temperature of 1800°F is placed in a large vat of water at temperature 60°F, the temperature of the rod after t seconds is

$$f(t) = 60(1 + 29e^{-0.15t})°F.$$

The graph of this function is shown in Fig. 2.

(a) What is the temperature of the rod after 11 seconds?

(b) At what rate is the temperature of the rod changing after 6 seconds?

(c) Approximately when is the temperature of the rod 200°F?

(d) Approximately when is the rod cooling at the rate of 200°F per second?

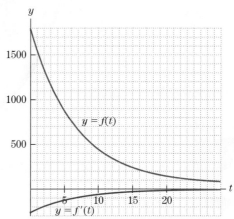

Figure 2

26. **Bacteria Growth** A certain bacteria culture grows at a rate proportional to its size. If 10,000 bacteria grow at the rate of 500 bacteria per day, how fast is the culture growing when it reaches 15,000 bacteria?

The Definite Integral

There are two fundamental problems of calculus: (1) to find the slope of a curve at a point and (2) to find the area of a region under a curve. These problems are quite simple when the curve is a straight line, as in Fig. 1. Both the slope of the line and the area of the shaded trapezoid can be calculated by geometric principles. When the graph consists of several line segments, as in Fig. 2, the slope of each line segment can be computed separately, and the area of the region can be found by adding the areas of the regions under each line segment.

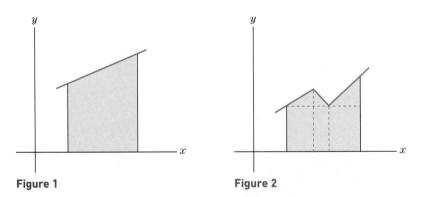

Figure 1   Figure 2

Calculus is needed when the curves are not straight lines. We have seen that the slope problem is resolved with the derivative of a function. In this chapter, we describe how the area problem is connected with the notion of the "integral" or the "antiderivative" of a function, which is, roughly speaking, the reverse of the derivative of a function. Both the slope problem and the area problem were studied by the ancient Greeks and solved in special cases. But it was not until the development of calculus in the seventeenth century that the intimate connection between the two problems was discovered. In this chapter, we will discuss this connection as stated in the fundamental theorem of calculus.

6.1 Antidifferentiation

We have developed several techniques for calculating the derivative $F'(x)$ of a function $F(x)$. As we saw in the Introduction, in many applications, however, it is necessary to proceed in reverse. We are given the derivative $F'(x)$ and must determine the function $F(x)$. The process of determining $F(x)$ from $F'(x)$ is called **antidifferentiation**. The next example gives a typical application involving antidifferentiation.

EXAMPLE 1   **Determining the Height of a Rocket from Its Velocity** A rocket is fired vertically into the air. Its velocity at t seconds after liftoff is $v(t) = 6t + .5$ meters per second. Before launch, the top of the rocket is 8 meters above the launch pad. Find the height of the rocket (measured from the top of the rocket to the launch pad) at time t.

SOLUTION (PART 1)   If $s(t)$ denotes the height of the rocket at time t, then $s'(t)$ is the rate at which the height is changing; that is, $s'(t) = v(t)$. Thus, although we do not yet have a formula for $s(t)$, we do know that

$$s'(t) = v(t) = 6t + .5.$$

Thus, the problem of determining a formula for $s(t)$ has been reduced to a problem of antidifferentiation: Find a function whose derivative is $v(t)$. We shall solve this particular problem after developing some techniques for solving antidifferentiation problems in general.   «

> **DEFINITION Antiderivative**   Suppose that $f(x)$ is a given function and $F(x)$ is a function having $f(x)$ as its derivative, that is, $F'(x) = f(x)$. We call $F(x)$ an **antiderivative** of $f(x)$.

EXAMPLE 2   **A Power Function** Find an antiderivative of $f(x) = x^2$.

SOLUTION   The derivative of x^3 is $3x^2$, which is almost the same as x^2 except for a factor of 3. To make this factor 1 instead of 3, consider the function $\frac{1}{3}x^3$. Then

$$\frac{d}{dx}\left(\frac{1}{3}x^3\right) = \frac{1}{3}\left(\frac{d}{dx}x^3\right) \qquad \text{Constant multiple rule for differentiation.}$$

$$= \frac{1}{3} \cdot 3x^2 = x^2 \quad \text{Differentiate } x^3 \text{ then simplify.}$$

So, $\frac{1}{3}x^3$ is an antiderivative of x^2. Is this the only antiderivative of x^2? Consider the function $\frac{1}{3}x^3 + 5$. Since

$$\frac{d}{dx}\left(\frac{1}{3}x^3 + 5\right) = \frac{d}{dx}\left(\frac{1}{3}x^3\right) + \frac{d}{dx}(5) \quad \text{Sum rule for differentiation.}$$

$$= x^2 + 0 \qquad\qquad \text{Derivative of a constant is 0.}$$

$$= x^2.$$

Since the derivative of $\frac{1}{3}x^3 + 5$ is x^2, we conclude that $\frac{1}{3}x^3 + 5$ is also an antiderivative of x^2. More generally, if C is *any* constant, then $F(x) = \frac{1}{3}x^3 + C$ is an antiderivative of x^2, because

$$F'(x) = \frac{d}{dx}\left(\frac{1}{3}x^3 + C\right)$$

$$= \frac{d}{dx}\left(\frac{1}{3}x^3\right) + \frac{d}{dx}(C) \quad \text{Sum rule for differentiation.}$$

$$= x^2 + 0 \qquad\qquad \text{Derivative of a constant is 0.}$$

$$= x^2.$$

» Now Try Exercise 1

EXAMPLE 3   **An Exponential Function** Find an antiderivative of $f(x) = e^{-2x}$.

SOLUTION   Recall that the derivative of e^{rx} is just a constant times e^{rx}. For an antiderivative of e^{-2x}, try a function of the form ke^{-2x}, where k is some constant to be determined. Then,

$$\frac{d}{dx}(ke^{-2x}) = k \cdot \left(\frac{d}{dx}e^{-2x}\right) \quad \text{Constant multiple rule for differentiation.}$$

$$= -2ke^{-2x} \qquad \text{Derivative of } e^{-2x} \text{ is } -2e^{-2x}.$$

We want this derivative to be e^{-2x}, so choose k to make $-2k = 1$; that is, choose $k = -\frac{1}{2}$. Then,

$$\frac{d}{dx}\left(-\frac{1}{2}e^{-2x}\right) = \left(-\frac{1}{2}\right)(-2e^{-2x}) = 1 \cdot e^{-2x} = e^{-2x}.$$

Thus, $-\frac{1}{2}e^{-2x}$ is an antiderivative of e^{-2x}. Also, for any constant C, the function $F(x) = -\frac{1}{2}e^{-2x} + C$ is an antiderivative of e^{-2x}, because

$$\frac{d}{dx}\left(-\frac{1}{2}e^{-2x} + C\right) = e^{-2x} + 0 = e^{-2x}.$$

>> Now Try Exercise 3

Examples 2 and 3 illustrate the fact that if $F(x)$ is an antiderivative of $f(x)$, so is $F(x) + C$, where C is any constant. The next theorem says that *all* antiderivatives of $f(x)$ can be produced in this way.

Theorem I   Suppose that $f(x)$ is a continuous function on an interval I. If $F_1(x)$ and $F_2(x)$ are two antiderivatives of $f(x)$, then $F_1(x)$ and $F_2(x)$ differ by a constant on I. In other words, there is a constant C such that

$$F_2(x) = F_1(x) + C \quad \text{for all } x \text{ in } I.$$

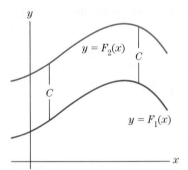

Figure 1 Two antiderivatives of the same function.

Geometrically, we obtain the graph of any antiderivative $F_2(x)$ by shifting the graph of $F_1(x)$ vertically. (See Fig. 1.)

Our verification of Theorem I is based on the following fact, which is important in its own right.

Theorem II   If $F'(x) = 0$ for all x in an interval I, there is a constant C such that $F(x) = C$ for all x in I.

It is easy to see why Theorem II is reasonable. (A formal proof of the theorem requires an important theoretical result called the *mean value theorem*.) If $F'(x) = 0$ for all x, the curve $y = F(x)$ has slope equal to zero at every point. Thus, the tangent line to $y = F(x)$ at any point is horizontal, which implies that the graph of $y = F(x)$ is a horizontal line. (Try to draw the graph of a function with a horizontal tangent everywhere. There is no choice but to keep your pencil moving on a constant, horizontal line!) If the horizontal line is $y = C$, then $F(x) = C$ for all x.

Verification of Theorem I   If $F_1(x)$ and $F_2(x)$ are two antiderivatives of $f(x)$, the function $F(x) = F_2(x) - F_1(x)$ has the derivative

$$F'(x) = F_2'(x) - F_1'(x) = f(x) - f(x) = 0.$$

So, by Theorem II, we know that $F(x) = C$ for some constant C. In other words, $F_2(x) - F_1(x) = C$, so

$$F_2(x) = F_1(x) + C,$$

which is Theorem I.   **«**

Using Theorem I, we can find *all* antiderivatives of a given function once we know one antiderivative. For instance, since one antiderivative of x^2 is $\frac{1}{3}x^3$ (Example 2), all antiderivatives of x^2 have the form $\frac{1}{3}x^3 + C$, where C is a constant.

> **DEFINITION** **Indefinite Integral** Suppose that $f(x)$ is a function whose antiderivatives are $F(x) + C$. The standard way to express this fact is to write
>
> $$\int f(x)dx = F(x) + C. \tag{1}$$
>
> The symbol $\int$ is called an **integral sign**. The entire notation $\int f(x)dx$ is called an **indefinite integral** and stands for antidifferentiation of the function $f(x)$.

We always record the independent variable by prefacing it by the letter d. For example, if the independent variable is t rather than x, we would write $\int f(t)dt$ for the antiderivative. In the integral $\int f(x)dx$, the function $f(x)$ is called the **integrand**.

EXAMPLE 4 **Integrals of Powers and Exponentials** Determine

(a) $\displaystyle \int x^r \, dx$, r a constant $\neq -1$ (b) $\displaystyle \int e^{kx} \, dx$, k a constant $\neq 0$

SOLUTION (a) By the constant-multiple and power rules for differentiation,

$$\frac{d}{dx}\left(\frac{1}{r+1}x^{r+1}\right) = \frac{1}{r+1} \cdot \frac{d}{dx}x^{r+1} = \frac{1}{r+1} \cdot (r+1)x^r = x^r.$$

Thus, $x^{r+1}/(r+1)$ is an antiderivative of x^r. Letting C represent any constant, we have

> **Power Rule**
>
> $$\int x^r \, dx = \frac{1}{r+1}x^{r+1} + C, \qquad r \neq -1. \tag{2}$$

(b) An antiderivative of e^{kx} is e^{kx}/k, since

$$\frac{d}{dx}\left(\frac{1}{k}e^{kx}\right) = \frac{1}{k} \cdot \frac{d}{dx}e^{kx} = \frac{1}{k}(ke^{kx}) = e^{kx}.$$

Hence,

> **Exponential Rule**
>
> $$\int e^{kx} \, dx = \frac{1}{k}e^{kx} + C, \qquad k \neq 0. \tag{3}$$

«

EXAMPLE 5 **Using the Power Rule** Determine

(a) $\displaystyle \int \sqrt{x} \, dx$ (b) $\displaystyle \int \frac{1}{x^2} \, dx$

SOLUTION Both integrals follow by applying the power rule (2).

(a) Take $r = \frac{1}{2}$ in (2); then

$$\int x^{\frac{1}{2}} \, dx = \frac{1}{\frac{1}{2}+1}x^{\frac{1}{2}+1} + C = \frac{1}{\frac{3}{2}}x^{\frac{3}{2}} + C = \frac{2}{3}x^{\frac{3}{2}} + C.$$

(b) Take $r = -2$ in (2), then,

$$\int x^{-2}\, dx = \frac{1}{-2+1} x^{-2+1} + C = -x^{-1} + C = -\frac{1}{x} + C.$$

» **Now Try Exercise 7**

There is a subtle point to be made concerning the integral of a discontinuous function. Consider the integral of $f(x) = \frac{1}{x^2}$ from Example 5(b). Since $f(x)$ is not defined at $x = 0$ and is continuous everywhere else, Theorem I tells us only that the antiderivatives of $f(x)$ are $(-1/x) + C$ on any interval that does not contain 0. So to express the antiderivatives of $f(x)$, we exclude 0 and write

$$\int \frac{1}{x^2}\, dx = \begin{cases} -\frac{1}{x} + C_1 & \text{if } x > 0, \\ -\frac{1}{x} + C_2 & \text{if } x < 0. \end{cases}$$

Note that, since we have two separate intervals ($x < 0$ and $x > 0$), we need two separate constants of integration, C_1 and C_2. For notational convenience, however, we will continue to use only one constant of integration, as we did in Example 5(b).

Formula (2) does not give an antiderivative of x^{-1} because $1/(r + 1)$ is undefined for $r = -1$. However, we know that for $x \neq 0$ the derivative of $\ln|x|$ is $1/x$. Hence, $\ln|x|$ is an antiderivative of $1/x$, and we have

---

Log Rule

$$\int \frac{1}{x}\, dx = \ln|x| + C, \qquad x \neq 0. \tag{4}$$

The log rule supplies the missing case ($r = -1$) of formula (2). Because $1/x$ is not defined at $x = 0$, as we explained following Example 5, we should distinguish two intervals for the validity of the log rule and interpret it to mean

$$\int \frac{1}{x}\, dx = \begin{cases} \ln x + C_1 & \text{if } x > 0, \\ \ln(-x) + C_2 & \text{if } x < 0. \end{cases}$$

We obtain each of formulas (2), (3), and (4) by "reversing" a familiar differentiation rule. In a similar fashion, we may use the sum rule and constant-multiple rule for derivatives to obtain corresponding rules for antiderivatives:

Basic Properties of Integrals

Sum rule. $$\int [f(x) + g(x)]\, dx = \int f(x)\, dx + \int g(x)\, dx. \tag{5}$$

Constant-multiple rule. $$\int kf(x)\, dx = k \int f(x)\, dx, \quad k \text{ a constant} \tag{6}$$

In words, the sum rule says that a sum of functions may be antidifferentiated term by term, and the constant-multiple rule says that a constant multiple may be moved through the integral sign.

EXAMPLE 6   **Using Basic Properties**

Compute $\int \left(x^{-3} + 7e^{5x} + \frac{4}{x} \right) dx.$

SOLUTION   Using the preceding rules, we have

$$\int\left(x^{-3} + 7e^{5x} + \frac{4}{x}\right)dx = \int x^{-3}\,dx + \int 7e^{5x}\,dx + \int \frac{4}{x}\,dx \quad \text{Sum rule.}$$

$$= \int x^{-3}\,dx + 7\int e^{5x}\,dx + 4\int \frac{1}{x}\,dx \quad \begin{array}{l}\text{Constant-multiple}\\ \text{rule.}\end{array}$$

$$= \frac{1}{-2}x^{-2} + 7\left(\frac{1}{5}e^{5x}\right) + 4\ln|x| + C \quad \begin{array}{l}\text{Power, exponential}\\ \text{and Log rules.}\end{array}$$

$$= -\frac{1}{2}x^{-2} + \frac{7}{5}e^{5x} + 4\ln|x| + C.$$

>> *Now Try Exercise 17*

CAUTION   Do not forget to add the arbitrary constant C when evaluating integrals. «

After some practice, most of the intermediate steps shown in the solution of Example 6 can be omitted.

A function has infinitely many different antiderivatives, corresponding to various choices of the constant C. In applications, it is often necessary to satisfy an additional condition, which then determines a specific value of C.

EXAMPLE 7   **Solving a Differential Equation**   Find the function $f(x)$ for which $f'(x) = x^2 - 2$ and $f(1) = \frac{4}{3}$. [Equivalently, in this problem, you are asked to solve the differential equation $y' = x^2 - 2$, $y(1) = \frac{4}{3}$.]

SOLUTION   The unknown function $f(x)$ is an antiderivative of $x^2 - 2$. One antiderivative of $x^2 - 2$ is $\frac{1}{3}x^3 - 2x$. Therefore, by Theorem I,

$$f(x) = \tfrac{1}{3}x^3 - 2x + C, \quad C \text{ a constant.}$$

Figure 2 shows the graphs of $f(x)$ for several choices of C. We want the function whose graph passes through $(1, \frac{4}{3})$. To find the value of C that makes $f(1) = \frac{4}{3}$, we set

$$\tfrac{4}{3} = f(1) = \tfrac{1}{3}(1)^3 - 2(1) + C = -\tfrac{5}{3} + C$$

and find $C = \frac{4}{3} + \frac{5}{3} = 3$. Therefore, $f(x) = \frac{1}{3}x^3 - 2x + 3$.

Figure 2   Several antiderivatives of $x^2 - 2$.   >> *Now Try Exercise 43*

Having introduced the basics of antidifferentiation, let us now solve the problem in Example 1.

SOLUTION OF EXAMPLE 1
(continued)

The position function $s(t)$ is an antiderivative of the velocity function $v(t)$. So,

$$s(t) = \int v(t)dt = \int (6t + .5)dt = 3t^2 + .5t + C,$$

where C is a constant. When $t = 0$, the rocket's height is 8 meters. That is, $s(0) = 8$, and so,

$$8 = s(0) = 3(0)^2 + .5(0) + C = C.$$

Thus, $C = 8$ and

$$s(t) = 3t^2 + .5t + 8. \qquad \text{\textgg\textit{ Now Try Exercise 55}}$$

EXAMPLE 8   **Cost Function**   A company's marginal cost function is $C'(x) = .015x^2 - 2x + 80$, dollars, where x denotes the number of units produced in 1 day. The company has fixed costs of \$1000 per day.
(a) Find the cost of producing x units per day.
(b) If the current production level is $x = 30$, determine the amount that costs will rise if the production level is raised to $x = 60$ units.

SOLUTION   (a) Let $C(x)$ be the cost of producing x units in 1 day. The derivative $C'(x)$ is the marginal cost. In other words, $C(x)$ is an antiderivative of the marginal cost function. Thus,

$$C(x) = \int (.015x^2 - 2x + 80)dx = .005x^3 - x^2 + 80x + C.$$

The \$1000 fixed costs are the costs incurred when 0 units are produced. That is, $C(0) = 1000$. So,

$$1000 = C(0) = .005(0)^3 - (0)^2 + 80(0) + C.$$

Therefore, $C = 1000$, and

$$C(x) = .005x^3 - x^2 + 80x + 1000.$$

(b) The cost when $x = 30$ is $C(30)$, and the cost when $x = 60$ is $C(60)$. So, the *increase* in cost when production is raised from $x = 30$ to $x = 60$ is $C(60) - C(30)$. We compute

$$C(60) = .005(60)^3 - (60)^2 + 80(60) + 1000 = 3280$$
$$C(30) = .005(30)^3 - (30)^2 + 80(30) + 1000 = 2635.$$

Thus, the increase in cost is $3280 - 2635 = \$645.$ **«**

Antidifferentiation or integration is a process that involves reversing a derivative formula and often involves guessing and adjusting your guess. To verify an integral formula, we can think of it as the reverse of a derivative formula and then use differentiation rules to check it. This useful process is illustrated by the following two examples.

EXAMPLE 9   **Verifying an Integral Formula**   Verify the integral formula

$$\int xe^{-x^2}\, dx = -\frac{1}{2}e^{-x^2} + C.$$

SOLUTION   The formula says that the antiderivative of xe^{-x^2} is $-\frac{1}{2}e^{-x^2} + C$. So to verify it, it is enough to show that the derivative of $-\frac{1}{2}e^{-x^2} + C$ is xe^{-x^2}. Indeed,

$$\frac{d}{dx}\left[-\frac{1}{2}e^{-x^2} + C\right] = -\frac{1}{2}\frac{d}{dx}[e^{-x^2}] + \frac{d}{dx}[C] \qquad \text{Constant-multiple and sum rules.}$$

$$= -\frac{1}{2}e^{-x^2}\frac{d}{dx}[-x^2] + 0 \qquad \text{Chain rule; derivative of a constant is 0.}$$

$$= -\frac{1}{2}e^{-x^2}(-2x) = xe^{-x^2}.$$

\textgg\textit{ Now Try Exercise 49}

EXAMPLE 10   **Verifying and Completing an Integral Formula** Find the value of k that makes the following antidifferentiation formula true:

$$\int (1 - 2x)^3 \, dx = k(1 - 2x)^4 + C.$$

SOLUTION   The given formula is equivalent to saying that the derivative of $k(1 - 2x)^4 + C$ is $(1 - 2x)^3$. Differentiating, we obtain

$$\frac{d}{dx}(k(1 - 2x)^4 + C) = \frac{d}{dx}(k(1 - 2x)^4) + \frac{d}{dx}(C)$$

$$= k\frac{d}{dx}(1 - 2x)^4 + 0$$

$$= k(4)(1 - 2x)^3\frac{d}{dx}(1 - 2x) \qquad \text{Generalized power rule.}$$

$$= k(4)(1 - 2x)^3(-2) = -8k(1 - 2x)^3.$$

This is supposed to equal $(1 - 2x)^3$. Therefore, $-8k(1 - 2x)^3 = (1 - 2x)^3$, which implies $-8k = 1$, so $k = -\frac{1}{8}$. Setting $k = -\frac{1}{8}$, we obtain the integral formula

$$\int (1 - 2x)^3 \, dx = -\frac{1}{8}(1 - 2x)^4 + C.$$

If you wish, you can verify this formula by taking the derivative of both sides, as we did in Example 9.   **» Now Try Exercise 29**

INCORPORATING
TECHNOLOGY

Graphing an Antiderivative and Solving a Differential Equation To graph the solution of the differential equation in Example 7, proceed as follows. First, press $\boxed{Y=}$; then set $Y_1 = \textbf{fnInt}(X^2 - 2, X, 1, X) + 4/3$. To display **fnInt**, press $\boxed{\text{MATH}}$ $\boxed{9}$ in classic mode. Upon pressing $\boxed{\text{GRAPH}}$, you will notice that graphing Y_1 proceeds very slowly. You can speed it by setting xRes (in the $\boxed{\text{WINDOW}}$ screen) to a higher value. Figure 3 shows the graph of the solution to Example 7.

Figure 3

Check Your Understanding 6.1

Solutions can be found following the section exercises.

1. Determine the following:

(a) $\int t^{7/2} \, dt$

(b) $\int \left(\frac{x^3}{3} + \frac{3}{x^3} + \frac{3}{x} \right) dx$

2. Find a function $f(t)$ that satisfies $f'(t) = 3t + 5$ and $f(0) = 5$.

EXERCISES 6.1

Find all antiderivatives of each following function:

1. $f(x) = x$

2. $f(x) = 9x^8$

3. $f(x) = e^{3x}$

4. $f(x) = e^{-3x}$

5. $f(x) = 3$

6. $f(x) = -4x$

Determine the following:

7. $\int 4x^3 \, dx$

8. $\int \frac{x}{3} \, dx$

9. $\int 7 \, dx$

10. $\int k^2 \, dx$ (k a constant)

11. $\int \frac{x}{c} \, dx$ (c a constant $\neq 0$)

12. $\int x \cdot x^2 \, dx$

13. $\int \left(\frac{2}{x} + \frac{x}{2} \right) dx$

14. $\displaystyle\int \frac{1}{7x}\,dx$

15. $\displaystyle\int x\sqrt{x}\,dx$

16. $\displaystyle\int \left(\frac{2}{\sqrt{x}} + 2\sqrt{x}\right)dx$

17. $\displaystyle\int \left(x - 2x^2 + \frac{1}{3x}\right)dx$

18. $\displaystyle\int \left(\frac{7}{2x^3} - \sqrt[3]{x}\right)dx$

19. $\displaystyle\int 3e^{-2x}\,dx$

20. $\displaystyle\int e^{-x}\,dx$

21. $\displaystyle\int e\,dx$

22. $\displaystyle\int \frac{7}{2e^{2x}}\,dx$

23. $\displaystyle\int -2(e^{2x} + 1)dx$

24. $\displaystyle\int \left(-3e^{-x} + 2x - \frac{e^{0.5x}}{2}\right)dx$

In Exercises 25–36, find the value of k that makes the antidifferentiation formula true. [*Note:* You can check your answer without looking in the answer section. How?]

25. $\displaystyle\int 5e^{-2t}\,dt = ke^{-2t} + C$ **26.** $\displaystyle\int 3e^{t/10}\,dt = ke^{t/10} + C$

27. $\displaystyle\int 2e^{4x-1}\,dx = ke^{4x-1} + C$

28. $\displaystyle\int \frac{4}{e^{3x+1}}\,dx = \frac{k}{e^{3x+1}} + C$

29. $\displaystyle\int (5x - 7)^{-2}\,dx = k(5x - 7)^{-1} + C$

30. $\displaystyle\int \sqrt{x + 1}\,dx = k(x + 1)^{3/2} + C$

31. $\displaystyle\int (4 - x)^{-1}\,dx = k \ln |4 - x| + C$

32. $\displaystyle\int \frac{7}{(8 - x)^4}\,dx = \frac{k}{(8 - x)^3} + C$

33. $\displaystyle\int (3x + 2)^4\,dx = k(3x + 2)^5 + C$

34. $\displaystyle\int (2x - 1)^3\,dx = k(2x - 1)^4 + C$

35. $\displaystyle\int \frac{3}{2 + x}\,dx = k \ln |2 + x| + C$

36. $\displaystyle\int \frac{5}{2 - 3x}\,dx = k \ln |2 - 3x| + C$

Find all functions $f(t)$ that satisfy the given condition.

37. $f'(t) = t^{3/2}$ **38.** $f'(t) = \dfrac{4}{6 + t}$

39. $f'(t) = 0$ **40.** $f'(t) = t^2 - 5t - 7$

Find all functions $f(x)$ that satisfy the given conditions.

41. $f'(x) = .5e^{-0.2x}$, $f(0) = 0$ **42.** $f'(x) = 2x - e^{-x}$, $f(0) = 1$

43. $f'(x) = x$, $f(0) = 3$ **44.** $f'(x) = 8x^{1/3}$, $f(1) = 4$

45. $f'(x) = \sqrt{x} + 1$, $f(4) = 0$ **46.** $f'(x) = x^2 + \sqrt{x}$, $f(1) = 3$

47. Figure 4 shows the graphs of several functions $f(x)$ for which $f'(x) = \frac{2}{x}$. Find the expression for the function $f(x)$ whose graph passes through (1, 2).

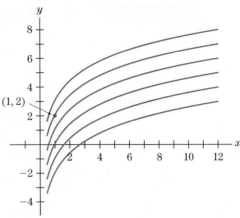

Figure 4

48. Figure 5 shows the graphs of several functions $f(x)$ for which $f'(x) = \frac{1}{3}$. Find the expression for the function $f(x)$ whose graph passes through (6, 3).

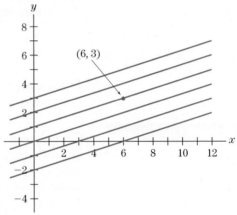

Figure 5

49. Which of the following is $\int \ln x\,dx$?

(a) $\dfrac{1}{x} + C$ (b) $x \cdot \ln x - x + C$

(c) $\dfrac{1}{2} \cdot (\ln x)^2 + C$

50. Which of the following is $\int x\sqrt{x + 1}\,dx$?

(a) $\frac{2}{5}(x + 1)^{5/2} - \frac{2}{3}(x + 1)^{3/2} + C$

(b) $\frac{1}{2}x^2 \cdot \frac{2}{3}(x + 1)^{3/2} + C$

51. Figure 6 contains the graph of a function $F(x)$. On the same coordinate system, draw the graph of the function $G(x)$ having the properties $G(0) = 0$ and $G'(x) = F'(x)$ for each x.

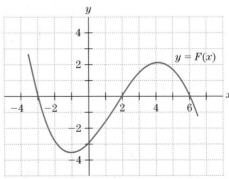

Figure 6

52. Figure 7 contains an antiderivative of the function $f(x)$. Draw the graph of another antiderivative of $f(x)$.

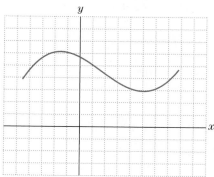

Figure 7

53. The function $g(x)$ in Fig. 8 resulted from shifting the graph of $f(x)$ up 3 units. If $f'(5) = \frac{1}{4}$, what is $g'(5)$?

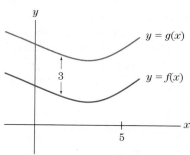

Figure 8

54. The function $g(x)$ in Fig. 9 resulted from shifting the graph of $f(x)$ up 2 units. What is the derivative of $h(x) = g(x) - f(x)$?

Figure 9

55. Height of a Ball A ball is thrown upward from a height of 256 feet above the ground, with an initial velocity of 96 feet per second. From physics it is known that the velocity at time t is $v(t) = 96 - 32t$ feet per second.
   (a) Find $s(t)$, the function giving the height above the ground of the ball at time t.
   (b) How long will the ball take to reach the ground?
   (c) How high will the ball go?

56. Free Fall A rock is dropped from the top of a 400-foot cliff. Its velocity at time t seconds is $v(t) = -32t$ feet per second.
   (a) Find $s(t)$, the height of the rock above the ground at time t.
   (b) How long will the rock take to reach the ground?
   (c) What will be its velocity when it hits the ground?

57. Rate of Production Let $P(t)$ be the total output of a factory assembly line after t hours of work. If the rate of production at time t is $P'(t) = 60 + 2t - \frac{1}{4}t^2$ units per hour, find the formula for $P(t)$.

58. Rate of Production After t hours of operation, a coal mine is producing coal at the rate of $C'(t) = 40 + 2t - \frac{1}{5}t^2$ tons of coal per hour. Find a formula for the total output of the coal mine after t hours of operation.

59. Heat Diffusion A package of frozen strawberries is taken from a freezer at $-5°$C into a room at $20°$C. At time t, the average temperature of the strawberries is increasing at the rate of $T'(t) = 10e^{-0.4t}$ degrees Celsius per hour. Find the temperature of the strawberries at time t.

60. Epidemic A flu epidemic hits a town. Let $P(t)$ be the number of persons sick with the flu at time t, where time is measured in days from the beginning of the epidemic and $P(0) = 100$. After t days, if the flu is spreading at the rate of $P'(t) = 120t - 3t^2$ people per day, find the formula for $P(t)$.

61. Profit A small tie shop finds that at a sales level of x ties per day, its marginal profit is $MP(x)$ dollars per tie, where $MP(x) = 1.30 + .06x - .0018x^2$. Also, the shop will lose \$95 per day at a sales level of $x = 0$. Find the profit from operating the shop at a sales level of x ties per day.

62. Cost A soap manufacturer estimates that its marginal cost of producing soap powder is $C'(x) = .2x + 100$ dollars per ton at a production level of x tons per day. Fixed costs are \$200 per day. Find the cost of producing x tons of soap powder per day.

63. U.S. Consumption of Iron Ore The United States has been consuming iron ore at the rate of $R(t)$ million metric tons per year at time t, where $t = 0$ corresponds to 1980 and $R(t) = 94e^{0.016t}$. Find a formula for the total U.S. consumption of iron ore from 1980 until time t.

64. U.S. Natural Gas Production Since 1987, the rate of production of natural gas in the United States has been approximately $R(t)$ quadrillion British thermal units per year at time t, with $t = 0$ corresponding to 1987 and $R(t) = 17.04e^{0.016t}$. Find a formula for the total U.S. production of natural gas from 1987 until time t.

65. Cost Drilling of an oil well has a fixed cost of \$10,000 and a marginal cost of $C'(x) = 1000 + 50x$ dollars per foot, where x is the depth in feet. Find the expression for $C(x)$, the total cost of drilling x feet. [*Note:* $C(0) = 10,000$.]

TECHNOLOGY EXERCISES

In Exercises 66 and 67, find an antiderivative of $f(x)$, call it $F(x)$, and compare the graphs of $F(x)$ and $f(x)$ in the given window to check that the expression for $F(x)$ is reasonable. [That is, determine whether the two graphs are consistent. When $F(x)$ has a relative extreme point, $f(x)$ should be zero; when $F(x)$ is increasing, $f(x)$ should be positive, and so on.]

66. $f(x) = 2x - e^{-0.02x}$, $[-10, 10]$ by $[-20, 100]$

67. $f(x) = e^{2x} + e^{-x} + \frac{1}{2}x^2$, $[-2.4, 1.7]$ by $[-10, 10]$

68. Plot the graph of the solution of the differential equation $y' = e^{-x^2}$, $y(0) = 0$. Observe that the graph approaches the value $\sqrt{\pi}/2 \approx .9$ as x increases.

Solutions to Check Your Understanding 6.1

1. (a) $\int t^{7/2}\,dt = \frac{1}{\frac{9}{2}}t^{9/2} + C = \frac{2}{9}t^{9/2} + C$

(b) $\int\left(\frac{x^3}{3} + \frac{3}{x^3} + \frac{3}{x}\right)dx$

$= \int\left(\frac{1}{3}\cdot x^3 + 3x^{-3} + 3\cdot\frac{1}{x}\right)dx$

$= \frac{1}{3}\left(\frac{1}{4}x^4\right) + 3\left(-\frac{1}{2}x^{-2}\right) + 3\ln|x| + C$

$= \frac{1}{12}x^4 - \frac{3}{2}x^{-2} + 3\ln|x| + C$

2. The unknown function $f(t)$ is an antiderivative of $3t + 5$. Since all antiderivatives of $3t + 5$ are given by

$$\int (3t + 5)\,dt = \frac{3}{2}t^2 + 5t + C,$$

we conclude that $f(t) = \frac{3}{2}t^2 + 5t + C$ for some constant C. To find the value of C that makes $f(0) = 5$, we set

$$5 = f(0) = \frac{3}{2}(0)^2 + 5(0) + C = C.$$

So $C = 5$ and hence $f(t) = \frac{3}{2}t^2 + 5t + 5$.

6.2 The Definite Integral and Net Change of a Function

In Example 1 of the previous section, you were asked to find $s(t)$, the height at time t of a rocket, given its velocity $v(t)$ and its initial height $s(0)$. From $v(t)$ you determined $s(t)$ up to an arbitrary constant C, then you determined C by using $s(0)$. In this section, we are interested in a related problem, where $v(t)$ is given for t in an interval $[a, b]$, but the initial height may not be known, and we are asked to compute the *net change* in the rocket's height $s(t)$, as t varies from a to b. That is, given $v(t)$, for t in $[a, b]$, without knowing the initial height, can you compute the net change in height $s(b) - s(a)$? As we will see in this section, any antiderivative of $v(t)$, not just $s(t)$, can be used to compute the net change, $s(b) - s(a)$. We start with an important definition.

> **DEFINITION The Definite Integral** Suppose that f is a continuous function on an interval $[a, b]$ with an antiderivative F; that is, $F'(x) = f(x)$. The **definite integral of f from a to b** is
>
> $$\int_a^b f(x)\,dx = F(b) - F(a). \tag{1}$$
>
> The numbers a and b are called the **limits of integration**, with a the **lower limit** and b the **upper limit**. The number $F(b) - F(a)$ is the **net change** of the function F, as x varies from a to b, and is abbreviated by the symbol $F(x)\big|_a^b$.

Note 1: While the indefinite integral of a function f is itself a function (or, more precisely, a family of functions), the definite integral is a *number* that does not depend on x. The symbol dx indicates that the independent variable of integration is x. Because the answer in the definite integral does not depend on x, we could have used any other variable of integration without affecting the value of the integral in (1). So, for example,

$$\int_a^b f(x)\,dx = \int_a^b f(t)\,dt = F(t)\Big|_a^b = F(b) - F(a).$$

EXAMPLE 1 **Computing a Definite Integral**

Evaluate $\int_1^2 x\,dx$.

SOLUTION An antiderivative of $f(x) = x$ is $F(x) = \frac{1}{2}x^2 + C$, so from (1)

$$\int_1^2 x\,dx = F(2) - F(1) = \left(\frac{1}{2}(2)^2 + C\right) - \left(\frac{1}{2}(1)^2 + C\right) = 2 - \frac{1}{2} = \frac{3}{2}.$$

>> *Now Try Exercise 1*

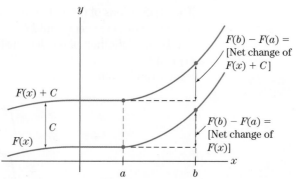

Figure 1 Net changes of $F(x)$ and its vertical shift $F(x) + C$.

Note 2: In Example 1, when computing $F(2) - F(1)$, the C in $F(1)$ cancels the C in $F(2)$. Thus, the value of $F(2) - F(1)$ does not depend on the choice of C. In general, if instead of using $F(x)$ you use $F(x) + C$ as an antiderivative of $f(x)$, then (1) becomes

$$\int_a^b f(x)\,dx = (F(b) + C) - (F(a) + C) = F(b) - F(a) + C - C = F(b) - F(a).$$

Thus, the value of the definite integral in (1) does not depend on the constant C (Fig. 1). For this reason, we may simplify our computation by taking $C = 0$ when evaluating the definite integral.

EXAMPLE 2

Definite Integral of an Exponential

Evaluate $\displaystyle\int_{-1}^{1} e^{-t}\,dt$.

SOLUTION

To find an antiderivative of $f(t) = e^{-t}$, we use the exponential rule of the previous section, with $k = -1$, set the arbitrary constant $C = 0$, and get $F(t) = -e^{-t}$. Thus,

$$\int_{-1}^{1} e^{-t}\,dt = -e^{-t}\Big|_{-1}^{1} = -e^{-1} - (-e^{-(-1)}) = e - e^{-1} = e - \frac{1}{e} \approx 2.35.$$

» Now Try Exercise 9

Properties of the Definite Integral

Let $f(x)$ and $g(x)$ be functions and a, b, and k be any constants. Then,

Integral of sums. $\displaystyle\int_a^b [f(x) + g(x)]dx = \int_a^b f(x)dx + \int_a^b g(x)dx$ (2)

Integral of differences. $\displaystyle\int_a^b [f(x) - g(x)]dx = \int_a^b f(x)dx - \int_a^b g(x)dx$ (3)

Integral of constant multiple. $\displaystyle\int_a^b kf(x)dx = k\int_a^b f(x)dx.$ (4)

Proof of (2)   Let $F(x)$ and $G(x)$ be antiderivatives of $f(x)$ and $g(x)$, respectively. Then, $F(x) + G(x)$ is an antiderivative of $f(x) + g(x)$. By (1),

$$\int_a^b [f(x) + g(x)]dx = [F(x) + G(x)]\Big|_a^b$$

$$= [F(b) + G(b)] - [F(a) + G(a)]$$

$$= [F(b) - F(a)] + [G(b) - G(a)]$$

$$= \int_a^b f(x)dx + \int_a^b g(x)dx. \qquad \text{«}$$

The verifications of (3) and (4) are similar and use the facts that $F(x) - G(x)$ is an antiderivative of $f(x) - g(x)$ and $kF(x)$ is an antiderivative of $kf(x)$.

In our definition of the definite integral, it is implied that $a < b$, but the right side of (1) can be evaluated also in the cases $a > b$ or $a = b$. For future reference, we record these cases as follows:

$$\int_a^b f(x)dx = -\int_b^a f(x)dx \tag{5}$$

$$\int_a^a f(x)dx = 0 \tag{6}$$

Property (5) follows because $F(b) - F(a) = -(F(a) - F(b))$, and (6) follows because $F(a) - F(a) = 0$.

EXAMPLE 3   **Using Properties of the Integral**

Evaluate $\displaystyle\int_1^3 (6x - 18x^3)dx - 2\int_1^3 (x - x^3)dx$.

SOLUTION   Integrals over the *same* interval can be combined by the use of (2). We have

$$\int_1^3 (6x - 18x^3)dx - 2\int_1^3 (x - x^3)dx$$

$$= \int_1^3 (6x - 18x^3)dx + \int_1^3 (-2x + 2x^3)dx \quad \text{By (4)}$$

$$= \int_1^3 ((6x - 18x^3) + (-2x + 2x^3))dx \quad \text{By (2)}$$

$$= \int_1^3 (4x - 16x^3)dx.$$

An antiderivative of $f(x) = 4x - 16x^3$ is $F(x) = \frac{4}{2}x^2 - \frac{16}{4}x^4 = 2x^2 - 4x^4$. Thus, by (1),

$$\int_1^3 (4x - 16x^3)dx = F(3) - F(1)$$

$$= (2(3^2) - 4(3^4)) - (2(1)^2 - 4(1)^4)$$

$$= (18 - 324) - (2 - 4) = -306 - (-2) = -304.$$

» Now Try Exercise 19

Two definite integrals of the same function over adjacent intervals can be combined as follows:

Integrals over adjacent intervals. $\displaystyle\int_a^c f(x)dx + \int_c^b f(x)dx = \int_a^b f(x)dx. \tag{7}$

In terms of net change, (7) says that the net change from a to b is equal to the net change from a to c plus the net change from c to b.

Proof of (7)

$$\int_a^c f(x)dx + \int_c^b f(x)dx = (F(c) - F(a)) + (F(b) - F(c)) \quad \text{By (1)}$$

$$= F(b) - F(a) = \int_a^b f(x)dx. \quad \text{By (1)} \quad \text{«}$$

EXAMPLE 4   **Using Properties of the Integral**

Given that $\int_0^6 f(x)dx = 8$ and $\int_5^6 f(x)dx = 7$, find $\int_0^5 f(x)dx$.

SOLUTION   By (7), we have

$$\int_0^6 f(x)dx = \int_0^5 f(x)dx + \int_5^6 f(x)dx.$$

Using the given values of the integrals, we find

$$8 = \int_0^5 f(x)dx + 7,$$

and so, $\int_0^5 f(x)dx = 1$

» Now Try Exercise 15

Property (7) is particularly useful when you are evaluating the integral of functions that are defined piecewise, as illustrated by the following example.

EXAMPLE 5   **Integral of a Piecewise-Defined Function**

Evaluate the integral $\int_0^2 f(x)dx$, where $f(x)$ is shown in Fig. 2.

Figure 2 A piecewise-defined function.

SOLUTION   We use (7) and write

$$\int_0^2 f(x)dx = \int_0^1 f(x)dx + \int_1^2 f(x)dx.$$

On the interval $[0, 1]$, the function $f(x)$ is equal to x^2, while on the interval $[1, 2]$, it is equal to $2 - x$. We have,

$$\int_0^1 f(x)dx = \int_0^1 x^2\,dx = \frac{1}{3}x^3\Big|_0^1 = \frac{1}{3}(1)^3 - \frac{1}{3}(0)^3 = \frac{1}{3},$$

and

$$\int_1^2 f(x)dx = \int_1^2 (2 - x)\,dx = 2x - \frac{1}{2}x^2\Big|_1^2 = \left(2(2) - \frac{1}{2}(2^2)\right) - \left(2(1) - \frac{1}{2}(1^2)\right) = \frac{1}{2}.$$

Putting the two integrals together, we find

$$\int_0^2 f(x)\,dx = \int_0^1 f(x)\,dx + \int_1^2 f(x)\,dx = \frac{1}{3} + \frac{1}{2} = \frac{5}{6}.$$

» Now Try Exercise 27

The Definite Integral as a Net Change of a Function

In the definition of the definite integral (1), since $f = F'$, we can rewrite (1) as

$$\int_a^b F'(x)\,dx = F(b) - F(a). \tag{8}$$

Recall that the derivative F' is the rate of change of F. Thus, (8) states that *the integral of the rate of change of F is its net change as x varies from a to b*. The remaining examples of this section illustrate this interesting connection between the integral of a rate of change and the net change of a function.

Suppose that an object is moving in a straight line. Denote by $s(t)$ its position at time t, measured from some reference point, and by $v(t)$ its velocity. Recall from Section 1.8, if t varies in an interval $[a, b]$, the net change in position of the object, or its *displacement*, is $s(b) - s(a)$.

FOR REVIEW

The velocity, $v(t)$, is the rate of change of position, $s(t)$. So $v(t) = s'(t)$. See p. 115.

EXAMPLE 6 **Net Change in Position** The velocity at time t seconds of an object moving in a straight line is $v(t) = 4t - 1$ meters per second. Find the displacement of the object during the time interval $1 \le t \le 3$.

SOLUTION Let $s(t)$ denote the position function. We are asked to compute the net change in position $s(3) - s(1)$. Using (8) and the fact that $s'(t) = v(t) = 4t - 1$, we find

$$s(3) - s(1) = \int_1^3 s'(t)\,dt = \int_1^3 (4t - 1)\,dt$$

$$= (2t^2 - t)\Big|_1^3 = (2(3^2) - 3) - (2(1^2) - 1) = 14.$$

Thus, the object moved 14 meters to the right (because 14 is positive) as t changed from 1 to 3. **≫ Now Try Exercise 31**

EXAMPLE 7 **Marginal Revenue Analysis** A company's marginal revenue function is $R'(x) = .03x^2 - 2x + 25$ dollars per unit, where x denotes the number of units produced in 1 day. Determine the net change in revenue if the production level is raised from $x = 20$ to $x = 25$ units.

SOLUTION Let $R(x)$ denote the revenue function. We are asked to compute the net change in revenue as x varies from 20 to 25. This net change is $R(25) - R(20) = \int_{20}^{25} R'(x)\,dx$. Since an antiderivative of $R'(x) = .03x^2 - 2x + 25$ is $.01x^3 - x^2 + 25x$, we have

$$\int_{20}^{25} R'(x)\,dx = \int_{20}^{25} (.03x^2 - 2x + 25)\,dx$$

$$= .01x^3 - x^2 + 25x\,\Big|_{20}^{25}$$

$$= [.01(25^3) - 25^2 + 25(25)] - [.01(20^3) - 20^2 + 25(20)]$$

$$= 156.25 - 180 = -23.75.$$

Thus, the revenue will decrease by \$23.75, if the company increases production from 20 to 25 units per day. **≫ Now Try Exercise 35**

EXAMPLE 8 **Net Increase in Federal Health Expenditures** Based on data from the Centers for Medicare and Medicaid Services, during the late 1990s and early 2000s, the federal health expenditures grew at an exponential rate with a growth constant of about .12. Let $R(t)$ denote the rate (in billions of dollars per year) of health expenditures at time t, where t is the number of years since the beginning of 2000. Then, a reasonable model for $R(t)$ is given by

$$R(t) = 380e^{0.12t}.$$

Find the total amount of federal health expenditures from the year 2000 ($t = 0$) to the year 2010 ($t = 10$).

SOLUTION   Let $T(t)$ denote the federal health expenditures from time 0 (2000) until time t. So, $T'(t) = R(t)$. We wish to calculate $T(10)$, the health expenditures from 2000 to 2010. Because $T(0)$ represents the expenditures from time 0 to time 0, we have $T(0) = 0$, and so $T(10) = T(10) - T(0)$. This shows that $T(10)$ is the net change in T as t varies from 0 to 10. This net change is given by

$$T(10) - T(0) = \int_0^{10} T'(t)\, dt = \int_0^{10} R(t)\, dt$$

$$= \int_0^{10} 380 e^{0.12t}\, dt = \frac{380}{.12} e^{0.12t} \Big|_0^{10}$$

$$= \frac{380}{.12} e^{0.12(10)} - \frac{380}{.12} e^{0.12(0)} = \frac{380}{.12} (e^{0.12(10)} - 1).$$

Thus, $T(10) = 3166.67(e^{0.12(10)} - 1) \approx 7347$ billion dollars. Hence, the federal health expenditures from 2000 to 2010 exceeded 7 trillion dollars.

INCORPORATING TECHNOLOGY

Computing a Definite Integral The definite integral in Example 7 is evaluated in Fig. 3. To do this evaluation, select MATH 9. Complete the integral, as shown in Fig. 3. Then press ENTER.

Figure 3

Check Your Understanding 6.2

Solutions can be found following the section exercises.

1. Evaluate $\displaystyle\int_0^1 \frac{e^{2x} - 1}{e^x}\, dx$.

2. If $f'(t) = 1 - t$, find $f(2) - f(0)$.

EXERCISES 6.2

In Exercises 1–14, evaluate the given integral.

1. $\displaystyle\int_0^1 \left(2x - \frac{3}{4}\right) dx$

2. $\displaystyle\int_{-1}^2 \left(\frac{x^2}{3} - \frac{2}{9}x\right) dx$

3. $\displaystyle\int_1^4 (3\sqrt{t} + 4t)\, dt$

4. $\displaystyle\int_1^9 \frac{1}{\sqrt{x}}\, dx$

5. $\displaystyle\int_1^2 -\frac{3}{x^2}\, dx$

6. $\displaystyle\int_1^8 (-x + \sqrt[3]{x})\, dx$

7. $\displaystyle\int_1^2 \frac{5 - 2x^3}{x^6}\, dx$

8. $\displaystyle\int_1^4 \frac{x^2 - \sqrt{x}}{x}\, dx$

9. $\displaystyle\int_{-1}^0 (3e^{3t} + t)\, dt$

10. $\displaystyle\int_{-2}^2 \frac{2}{e^{2t}}\, dt$

11. $\displaystyle\int_1^2 \frac{2}{x}\, dx$

12. $\displaystyle\int_{-2}^{-1} \frac{1 + x}{x}\, dx$

13. $\displaystyle\int_0^1 \frac{e^x + e^{0.5x}}{e^{2x}}\, dx$

14. $\displaystyle\int_0^{\ln 2} \frac{e^x + e^{-x}}{2}\, dx$

15. Given $\displaystyle\int_0^1 f(x)dx = 3.5$ and $\displaystyle\int_1^4 f(x)dx = 5$, find $\displaystyle\int_0^4 f(x)dx$.

16. Given $\displaystyle\int_{-1}^1 f(x)dx = 0$ and $\displaystyle\int_{-1}^{10} f(x)dx = 4$, find $\displaystyle\int_1^{10} f(x)dx$.

17. Given $\displaystyle\int_1^3 f(x)dx = 3$ and $\displaystyle\int_1^3 g(x)dx = -1$, find

$$\int_1^3 (2f(x) - 3g(x))dx.$$

18. Given $\displaystyle\int_{-0.5}^3 f(x)dx = 0$ and $\displaystyle\int_{-0.5}^3 (2g(x) + f(x))dx = -4$,

find $\displaystyle\int_{-0.5}^3 g(x)dx$.

In Exercises 19–22, combine the integrals into one integral, then evaluate the integral.

19. $2\int_1^2\left(3x + \frac{1}{2}x^2 - x^3\right)dx + 3\int_1^2(x^2 - 2x + 7)dx$

20. $\int_0^1(4x - 2)dx + 3\int_0^1(x - 1)dx$

21. $\int_{-1}^0(x^3 + x^2)dx + \int_0^1(x^3 + x^2)dx$

22. $\int_0^1(7x + 4)dx + \int_1^2(7x + 5)dx$

In Exercises 23–26, use formula (8) to help you answer the question.

23. Given $f'(x) = -2x + 3$, compute $f(3) - f(1)$.

24. Given $f'(x) = 73$, compute $f(4) - f(2)$.

25. Given $f'(t) = -.5t + e^{-2t}$, compute $f(1) - f(-1)$.

26. Given $f'(t) = -12t - \frac{1}{e^t}$, compute $f(3) - f(0)$.

27. Refer to Fig. 4 and evaluate $\int_0^2 f(x)dx$.

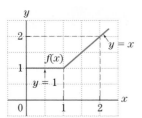

Figure 4

28. Refer to Fig. 5 and evaluate $\int_0^3 f(x)dx$.

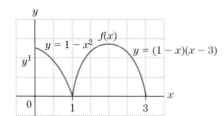

Figure 5

29. Refer to Fig. 6 and evaluate $\int_{-1}^1 f(t)dt$.

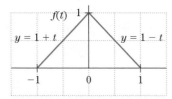

Figure 6

30. Refer to Fig. 7 and evaluate $\int_{-1}^2 f(t)dt$.

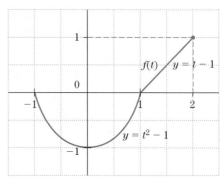

Figure 7

31. Net Change in Position A rock is dropped from the top of a 400-foot cliff. Its velocity at time t seconds is $v(t) = -32t$ feet per second. Find the displacement of the rock during the time interval $2 \le t \le 4$.

32. Net Change in Position The velocity at time t seconds of a ball thrown up into the air is $v(t) = -32t + 75$ feet per second.
(a) Find the displacement of the ball during the time interval $0 \le t \le 3$.
(b) Given that the initial position of the ball is $s(0) = 6$ feet, use (a) to determine its position at time $t = 3$.

33. Net Change in Position The velocity at time t seconds of a ball thrown up into the air is $v(t) = -32t + 75$ feet per second.
(a) Compute the displacement of the ball during the time interval $1 \le t \le 3$.
(b) Is the position of the ball at time $t = 3$ higher than its position at time $t = 1$? Justify your answer.
(c) Repeat part (a) using the time interval $1 \le t \le 5$.

34. Velocity of a Skydiver The velocity of a skydiver at time t seconds is $v(t) = 45 - 45e^{-0.2t}$ meters per second. Find the distance traveled by the skydiver the first 9 seconds.

35. Net Change in Cost A company's marginal cost function is $.1x^2 - x + 12$ dollars, where x denotes the number of units produced in 1 day.
(a) Determine the increase in cost if the production level is raised from $x = 1$ to $x = 3$ units.
(b) If $C(1) = 15$, determine $C(3)$ using your answer in (a).

36. Cost Increase A company's marginal cost function is given by $C'(x) = 32 + \frac{x}{20}$, where x denotes the number of items produced in 1 day and $C(x)$ is in thousands of dollars. Determine the increase in cost if the company goes from a production level of 15 to 20 items per day.

37. Net Increase of an Investment An investment grew at an exponential rate $R(t) = 700e^{0.07t} + 1000$, where t is in years and $R(t)$ is in dollars per year. Approximate the net increase in value of the investment after the first 10 years (as t varies from 0 to 10).

38. Depreciation of Real Estate A property with an appraised value of \$200,000 in 2015 is depreciating at the rate $R(t) = -8e^{-0.04t}$, where t is in years since 2015 and $R(t)$ is in thousands of dollars per year. Estimate the loss in value of the property between 2015 and 2021 (as t varies from 0 to 6).

39. Population Model with Emigration The rate of change of a population with emigration is given by $P'(t) = \frac{7}{300}e^{t/25} - \frac{1}{80}e^{t/16}$, where $P(t)$ is the population in millions, t years after the year 2000.
(a) Estimate the change in population as t varies from 2000 to 2010.
(b) Estimate the change in population as t varies from 2010 to 2040. Compare and explain your answers in (a) and (b).

40. Paying Down a Mortgage You took a $200,000 home mortgage at an annual interest rate of 3%. Suppose that the loan is amortized over a period of 30 years, and let $P(t)$ denote the amount of money (in thousands of dollars) that you owe on the loan after t years. A reasonable estimate of the rate of change of P is given by $P'(t) = -4.1107e^{0.03t}$.

(a) Approximate the net change in P after 20 years.

(b) What is the amount of money owed on the loan after 20 years?

(c) Verify that the loan is paid off in 30 years by computing the net change in P after 30 years.

41. Mortgage Using the data from the previous exercise, find $P(t)$.
[*Hint:* $P(0) = 200$.]

42. Radioactive Decay A sample of radioactive material with decay constant .1 is decaying at a rate $R(t) = -e^{-0.1t}$ grams per year. How many grams of this material decayed after the first 10 years?

43. Saline Solution A saline solution is being flushed with fresh water in such a way that salt is eliminated at the rate $r(t) = -(t + \frac{1}{2})$ grams per minute. Find the amount of salt that is eliminated during the first 2 minutes.

44. Level of Water in a Tank A conical-shaped tank is being drained. The height of the water level in the tank is decreasing at the rate $h'(t) = -\frac{t}{2}$ inches per minute. Find the decrease in the depth of the water in the tank during the time interval $2 \le t \le 4$.

Solution to Check Your Understanding 6.2

1. Start by simplifying

$$\frac{e^{2x} - 1}{e^x} = \frac{e^{2x}}{e^x} - \frac{1}{e^x} = e^x - e^{-x}.$$

An antiderivative of e^x is e^x, and an antiderivative of e^{-x} is $-e^{-x}$. Hence, an antiderivative of $e^x - e^{-x}$ is $e^x - (-e^{-x}) = e^x + e^{-x}$, and so,

$$\int_0^1 \frac{e^{2x} - 1}{e^x} dx = \int_0^1 (e^x - e^{-x}) dx$$

$$= (e^x + e^{-x}) \Big|_0^1$$

$$= (e + e^{-1}) - (1 + 1) = e + 1/e - 2 \approx 1.09.$$

2. According to formula (8),

$$f(2) - f(0) = \int_0^2 f'(t)dt = \int_0^2 (1 - t)dt$$

$$= (t - \tfrac{1}{2}t^2) \Big|_0^2 = \left(2 - \tfrac{1}{2}(2)^2\right) - 0 = 2 - 2 = 0.$$

6.3 The Definite Integral and Area under a Graph

This section and the next reveal the important connection between definite integrals and areas of regions under curves. We start by defining one type of region that we will be considering.

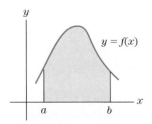

Figure 1 Area under a graph.

> **Area under a Graph**   If $f(x)$ is a continuous nonnegative function on the interval $a \le x \le b$, we refer to the area of the region shown in Fig. 1 as the *area under the graph of $f(x)$ from a to b*, or the *area bounded by the graph of $f(x)$, the x-axis, and the (vertical) lines $x = a$ and $x = b$*.

In this section we solve the area problem, which consists of finding the area of the region under the graph of a continuous function $f(x)$ from a to b as illustrated in Fig. 1.

Many areas of this type are easy to compute with geometric formulas. In Fig. 2(a), the shaded rectangular region is under the graph of the constant function $f(x) = 4$ from $x = 0$ to $x = 3$. Its area is $3 \times 4 = 12$.

In Fig. 2(b), the shaded region under the graph of the function $g(x) = -x + 4$ from $x = 2$ to $x = 3$ is a trapezoid that consists of a right triangle on top of a square. Its total area is $\frac{1}{2} + 1 = \frac{3}{2}$.

In Fig. 2(c), we shaded a region under the graph of a "ramp function." Its area is also the sum of the areas of a triangle plus a rectangle and is equal to $2 + 4 = 6$.

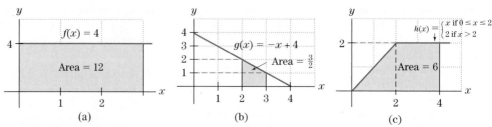

Figure 2 Areas under graphs.

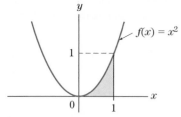

Figure 3 Area under the parabola.

In all three examples in Fig. 2, the top boundary of the region consists of line segments. The areas in these cases can be computed with simple geometric constructions. The computation of the area such as graphed in Fig. 1 is not a trivial matter when the top boundary of the region is curved. Consider, for example, the region under the graph of the parabola $f(x) = x^2$ from $x = 0$ to $x = 1$ (Fig. 3). It is not hard to see that this area is less than 1/2. But what is the exact value of the area? Obviously, the answer cannot be derived from simple geometric formulas. We will show how to solve this area problem using important techniques based on approximations with rectangles. These same techniques will also be used to establish the following fundamental result in calculus, which provides the solution to the area problem.

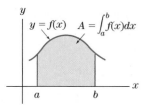

Figure 4 Area under the graph is a definite integral.

> **Theorem I: Area under a Graph** If $f(x)$ is a continuous nonnegative function on the interval $a \leq x \leq b$, then the area under the graph of $f(x)$, above the x-axis, from $x = a$ to $x = b$ is equal to the definite integral of f from a to b:
>
> $$\begin{bmatrix} \text{Area under the graph of } f \\ \text{from } x = a \text{ to } x = b \end{bmatrix} = \int_a^b f(x)dx = F(b) - F(a),$$
>
> where F is an antiderivative of f (Fig. 4).

An explanation of why this theorem is true is given later in this section. For now, let's verify the theorem for the regions in Fig. 2.

EXAMPLE 1 **Verifying Theorem I** Use Theorem I to compute the areas of the three shaded regions in Fig. 2.

SOLUTION (a) The shaded region in Fig. 2(a) is under the graph of the constant function $f(x) = 4$ from $x = 0$ to $x = 3$. According to Theorem I, its area is equal to the definite integral of the function $f(x) = 4$ from 0 to 3. So,

$$[\text{Area}] = \int_0^3 4\,dx = 4x \Big|_0^3 = (4)(3) - (4)(0) = 12.$$

Thus, Theorem I implies that the area of the region is equal to 12, which agrees with our earlier findings.

(b) The shaded region in Fig. 2(b) is under the graph of the linear function $g(x) = -x + 4$ from 2 to 3. By Theorem I, its area is equal to the definite integral of the function $g(x) = -x + 4$ from 2 to 3. An antiderivative of $-x + 4$ is $-\frac{1}{2}x^2 + 4x$, and so,

$$[\text{Area}] = \int_2^3 (-x + 4)dx = \left(-\frac{1}{2}x^2 + 4x\right)\Big|_2^3 = \left(-\frac{9}{2} + 12\right) - \left(-\frac{4}{2} + 8\right)$$

$$= 4 - \frac{5}{2} = \frac{3}{2}.$$

Thus, Theorem I implies that the area of the shaded region in Fig. 2(b) is $\frac{3}{2}$, as we found earlier.

(c) The shaded region in Fig. 2(c) is under the graph of the ramp function $h(x)$ from $x = 0$ to $x = 4$. According to Theorem I, the area is equal to the definite integral of $h(x)$ from 0 to 4. Since $h(x)$ is defined piecewise, we write its integral as the sum of two integrals over two adjacent intervals as follows:

$$\int_0^4 h(x)dx = \int_0^2 h(x)dx + \int_2^4 h(x)dx \quad \text{By (7), Sec. 6.2.}$$

Since $h(x) = x$ on $[0, 2]$ and $h(x) = 2$ on $[2, 4]$, we get

$$\int_0^4 h(x)dx = \int_0^2 x\,dx + \int_2^4 2\,dx$$

$$= \frac{1}{2}x^2 \Big|_0^2 + 2x \Big|_2^4$$

$$= 2 + (8 - 4) = 6.$$

Thus, Theorem I implies that the area of the shaded region in Fig. 2(c) is 6, which agrees with our earlier finding.   **>> Now Try Exercise 3**

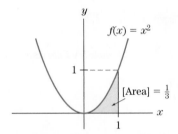

Figure 5 Area under the parabola.

The three cases that we just considered illustrate how Theorem I can be applied to compute areas under graphs. The importance of Theorem I, however, is in computing more challenging areas, such as the one under the parabola $f(x) = x^2$ from 0 to 1. In this case, Theorem I tells us that the area is the definite integral of x^2 over the interval $[0, 1]$:

$$\int_0^1 x^2\,dx = \frac{1}{3}x^3 \Big|_0^1 = \frac{1}{3}.$$

Thus, the area of the shaded region in Fig. 5 is $\frac{1}{3}$. This simple application of Theorem I to evaluate an area for which we don't have a geometric formula demonstrates the power of this theorem. We give two more applications.

EXAMPLE 2   **Area under the Graph of a Function** Use Theorem I to compute the area of the shaded region under the graph of $f(x) = 3x^2 + e^x$, from $x = -1$ to $x = 1$ (Fig. 6).

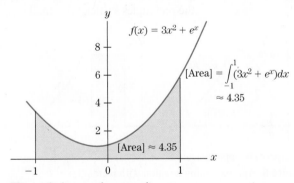

Figure 6 Area under a graph.

SOLUTION   From Fig. 6, $f(x)$ is a nonnegative function. By Theorem I, the area under the graph, from -1 to 1, is

$$\int_{-1}^1 (3x^2 + e^x)dx = (x^3 + e^x)\Big|_{-1}^1 = (1 + e^1) - (-1 + e^{-1})$$

$$= 2 + e - 1/e \approx 4.35.$$

Thus, the area is approximately 4.35.   **>> Now Try Exercise 7**

EXAMPLE 3 **Area under a Cubic** Use Theorem I to compute the area of the shaded region under the graph of $f(x) = x^3 + 2x + 1$, from $x = 0$ to $x = 1$ (Fig. 7).

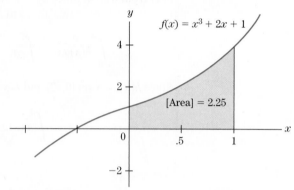

Figure 7 Area under a cubic.

SOLUTION By approximating, it appears from Fig. 7 that the area is slightly larger than 2. By Theorem I, the exact value of this area is the definite integral of $f(x) = x^3 + 2x + 1$ from $x = 0$ to $x = 1$. Since one antiderivative of f is $\frac{1}{4}x^4 + x^2 + x$, we get

$$\int_0^1 (x^3 + 2x + 1)\,dx = \left(\frac{1}{4}x^4 + x^2 + x\right)\Big|_0^1 = \frac{1}{4} + 1 + 1 = 2.25.$$

Thus, the area is 2.25.

» Now Try Exercise 21

We now turn our attention to the proof of Theorem I.

Riemann Sums

Consider the shaded region under the graph of $y = f(x)$ in Fig. 8. We will describe a process to *estimate* the area of this region to any desired degree of accuracy by using rectangles whose total area is approximately the same as the area to be computed. The area of each rectangle, of course, is easy to compute. Let n denote the number of rectangles used in each approximation. Figure 8 shows rectangular approximations to the area with $n = 2, 4$ and 10. When the rectangles are thin, the mismatch between the rectangles and the region under the graph is quite small. In general, we can make a rectangular approximation as close as desired to the exact area simply by making the width of the rectangles sufficiently small.

   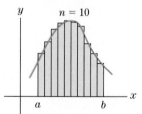

Area to be approximated

Figure 8 Approximating a region with rectangles.

Given a continuous nonnegative function $f(x)$ on the interval $a \le x \le b$, divide the x-axis interval into n equal subintervals, where n represents some positive integer. Such a subdivision is called a **partition** of the interval from a to b. Since the entire interval is of width $b - a$, the width of each of the n subintervals is $(b - a)/n$. For brevity, denote this width by Δx. That is,

$$\Delta x = \frac{b - a}{n} \quad \text{(width of one subinterval).}$$

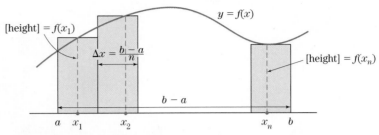

Figure 9 Rectangles with heights $f(x_1), \ldots, f(x_n)$.

In each subinterval, select a point. (Any point in the subinterval will do.) Let x_1 be the point selected from the first subinterval, x_2 the point from the second subinterval, and so on. These points are used to form rectangles that approximate the region under the graph of $f(x)$. Construct the first rectangle with height $f(x_1)$ and the first subinterval as base, as in Fig. 9. The top of the rectangle touches the graph directly above x_1. Notice that

$$[\text{Area of first rectangle}] = [\text{height}][\text{width}] = f(x_1)\,\Delta x.$$

The second rectangle rests on the second subinterval and has height $f(x_2)$. Thus,

$$[\text{Area of second rectangle}] = [\text{height}][\text{width}] = f(x_2)\,\Delta x.$$

Continuing in this way, we construct n rectangles with a combined area of

Riemann sum.	$f(x_1)\,\Delta x + f(x_2)\,\Delta x + \cdots + f(x_n)\,\Delta x.$	(1)

A sum as in (1) is called a **Riemann sum**. Riemann sums are named after the nineteenth-century German mathematician G. B. Riemann (pronounced "Reemahn"), who used them extensively in his work on calculus. The concept of a Riemann sum has several uses: to approximate areas under curves, to construct mathematical models in applied problems, and to give a formal definition of area. The value in (1) is easier to calculate when written as

$$[f(x_1) + f(x_2) + \cdots + f(x_n)]\,\Delta x. \tag{2}$$

This calculation requires only one multiplication. The Riemann sums provide an approximation to the area under the graph of $f(x)$ when $f(x)$ is nonnegative and continuous. In fact, as the number of subintervals increases indefinitely; that is, as $\Delta x \to 0$, it can be shown that the Riemann sums in (1) approach a limiting value, the area under the graph. We record this fact for future reference:

$$\begin{bmatrix} \text{Area under the graph} \\ \text{of } f \text{ from } a \text{ to } b \end{bmatrix} = \lim_{\Delta x \to 0} [f(x_1) + f(x_2) + \cdots + f(x_n)]\,\Delta x. \tag{3}$$

As a first application of the use of Riemann sums, let us consider a surveying problem.

EXAMPLE 4   **Approximating the Area of a Waterfront Lot** To estimate the area of a 100-foot-wide waterfront lot, a surveyor measured the distance from the road to the waterline at 20-foot intervals, starting 10 feet from one corner of the lot. Use the data to construct a Riemann sum approximation to the area of the lot. See Fig. 10.

SOLUTION   Treat the road as the x-axis, and consider the waterline along the property as the graph of a function $f(x)$ over the interval from 0 to 100. The five "vertical" distances give $f(x_1), \ldots, f(x_5)$, where $x_1 = 10, \ldots, x_5 = 90$. Since there are five points $x_1, \ldots, x_5$

Figure 10 Survey of a waterfront property.

spread across the interval $0 \le x \le 100$, we partition the interval into five subintervals, with $\Delta x = \frac{100}{5} = 20$. Fortunately, each subinterval contains one x_i. (In fact, each x_i is the midpoint of a subinterval.) Thus, the area of the lot is approximated by the Riemann sum

$$f(x_1)\,\Delta x + \cdots + f(x_5)\,\Delta x = [f(x_1) + \cdots + f(x_5)]\,\Delta x$$
$$= [39 + 46 + 44 + 40 + 41.5]\cdot(20)$$
$$= (210.5)\cdot(20) = 4210 \text{ square feet.}$$

For a better estimate of the area, the surveyor will have to make more measurements from the street to the waterline. Note that we are able to approximate the area without ever knowing an algebraic expression for the function $f(x)$. **» Now Try Exercise 45**

We now return to the region under the parabola in Fig. 5 and use Riemann sums to approximate its area.

EXAMPLE 5   **A Riemann Sum Approximation of an Area** Use a Riemann sum with $n = 4$ to estimate the area under the graph of $f(x) = x^2$ from 0 to 1. Select the right endpoints of the subintervals as $x_1,\ x_2,\ x_3,\ x_4$.

SOLUTION   Here, $\Delta x = (1 - 0)/4 = .25$. The right endpoint of the first subinterval is $0 + \Delta x = .25$. We obtain subsequent right endpoints by successively adding .25, as follows:

The corresponding Riemann sum is

$$f(x_1)\,\Delta x + f(x_2)\,\Delta x + f(x_3)\,\Delta x + f(x_4)\,\Delta x$$
$$= [f(x_1) + f(x_2) + f(x_3) + f(x_4)]\Delta x$$
$$= [(.25)^2 + (.5)^2 + (.75)^2 + (1)^2](.25)$$
$$= [.0625 + .25 + .5625 + 1](.25)$$
$$= 1.875\cdot(.25) = .46875.$$

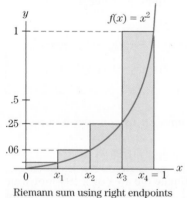

Riemann sum using right endpoints

Figure 11

(The rectangles used for this Riemann sum are shown in Fig. 11.) The right endpoints here give an area estimate that is obviously greater than the exact area. Midpoints would work better. But if the rectangles are sufficiently narrow, even a Riemann sum using right endpoints will be close to the exact area. **» Now Try Exercise 31**

EXAMPLE 6   **Improving a Riemann Sum Approximation** Use a Riemann sum with $n = 10$ to estimate the area under the graph of $f(x) = x^2$ from 0 to 1. Select the right endpoints of the subintervals as $x_1,\ x_2, \ldots, x_{10}$.

SOLUTION Here, $\Delta x = (1-0)/10 = .1$. The right endpoints of the first and last subintervals are $x_1 = .1$ and $x_{10} = 1$. Starting from $x_1 = .1$, we obtain subsequent right endpoints by successively adding .1. The corresponding Riemann sum is

$$f(x_1)\,\Delta x + f(x_2)\,\Delta x + f(x_3)\,\Delta x + \cdots + f(x_{10})\,\Delta x$$
$$= [f(x_1) + f(x_2) + f(x_3) + \cdots + f(x_{10})]\Delta x$$
$$= [(.1)^2 + (.2)^2 + (.3)^2 + \cdots + (1)^2](.1)$$
$$= [.01 + .04 + .09 + \cdots + 1](.1)$$
$$= 3.85 \cdot (.1) = .385.$$

The rectangles used for this Riemann sum are shown in Fig. 12. As in the previous example, the right endpoints here give an area estimate that is greater than the exact area (which we know is equal to $\frac{1}{3} \approx .33$).

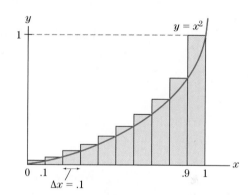

Figure 12 Improving the Riemann sum approximation.

You can show that, by increasing indefinitely the number of subintervals, the Riemann sum approximation in Example 6 approaches $\frac{1}{3}$, which is the exact value of the area (see Exercises 47 and 48).

Even though our discussion centered around nonnegative continuous functions, the definition of a Riemann sum applies as well to continuous functions with both positive and negative values.

Our next step is to argue that the limit of the Riemann sums is equal to the definite integral of f. The result is known as the **fundamental theorem of calculus** and has many important applications.

> **Theorem II: Fundamental Theorem of Calculus** Let f be a continuous function on the interval $[a, b]$, with an antiderivative F. Then, the Riemann sums (1) approach the definite integral of f on $[a, b]$, as the number of subintervals increases indefinitely. That is,
>
> $$\lim_{\Delta x \to 0} [f(x_1) + f(x_2) + \cdots + f(x_n)]\,\Delta x = \int_a^b f(x)\,dx = F(b) - F(a). \qquad (4)$$

Note that Theorem I follows from the fundamental theorem of calculus: On the one hand, from (3), the area under the graph is equal to the limit of the Riemann sums. On the other hand, from (4), the limit of the Riemann sums is equal to the definite integral.

Justification of Fundamental Theorem of Calculus We next explain why the Riemann sums in (4) approach $F(b) - F(a)$ and thus complete the verification of Theorem II.

Partition $a \le x \le b$ into n subintervals of width $\Delta x = \frac{b-a}{n}$ each. To simplify the presentation, instead of selecting arbitrary points from the subintervals, as stated in (4), let $x_1, x_2, \ldots, x_n$ denote the left endpoints of the subintervals. A similar proof works for any choice of points $x_1, x_2, \ldots, x_n$ from the subintervals. So, in our proof, $x_1 = a$, $x_i = x_{i-1} + \Delta x$, and $x_n + \Delta x = b$. Since $f(x) = F'(x)$, we see that (4) is equivalent to

$$\lim_{\Delta x \to 0} [F'(x_1)\Delta x + F'(x_2)\Delta x + \cdots + F'(x_n)\Delta x] = F(b) - F(a). \tag{5}$$

From the definition of the derivative, we have

$$F'(x) = \lim_{h \to 0} \frac{F(x + h) - F(x)}{h}.$$

So, for small $h = \Delta x$,

$$F'(x) \approx \frac{F(x + \Delta x) - F(x)}{\Delta x} \quad \text{or} \quad F'(x)\Delta x \approx F(x + \Delta x) - F(x).$$

Using this approximation, we estimate the terms on the left of (5) as follows:

$$F'(x_1)\Delta x \approx F(x_2) - F(x_1),$$
$$F'(x_2)\Delta x \approx F(x_3) - F(x_2),$$
$$F'(x_3)\Delta x \approx F(x_4) - F(x_3),$$
$$\vdots$$
$$F'(x_n)\Delta x \approx F(x_n + \Delta x) - F(x_n).$$

When we add, the intermediate terms on the right cancel, and we get

$$F'(x_1)\Delta x + F'(x_2)\Delta x + F'(x_3)\Delta x + \cdots + F'(x_n)\Delta x$$
$$\approx (F(x_2) - F(x_1)) + (F(x_3) - F(x_2))$$
$$+ (F(x_4) - F(x_3)) + \cdots + (F(x_n + \Delta x) - F(x_n))$$
$$= -F(x_1) + F(x_n + \Delta x)$$
$$= F(b) - F(a),$$

because $x_1 = a$ and $x_n + \Delta x = b$. Thus,

$$F'(x_1)\Delta x + F'(x_2)\Delta x + F'(x_3)\Delta x + \cdots + F'(x_n)\Delta x \approx F(b) - F(a).$$

Since the approximation improves as $\Delta x \to 0$, we see that the limit (5) and, thus, (4) must hold. ≪

INCORPORATING
TECHNOLOGY

Figure 13

Riemann Sums When the points selected for a Riemann sum are all midpoints, all left endpoints, or all right endpoints, $[f(x_1) + f(x_2) + \cdots + f(x_n)]$ is the sum of the sequence of values $f(x_1), f(x_2), \ldots, f(x_n)$, where successive numbers $x_1, x_2, \ldots, x_n$ each differ by the value Δx. In such a case, this sum of a sequence can be evaluated on a graphing calculator. Figure 13 shows the computation of the Riemann sum asked for in Example 6. To do this, first set $Y_1 = X^2$. Return to the home screen and press [2nd] [LIST], and move the cursor right to MATH. Press [5] to display **sum(** and press [2nd] [LIST]; then, move the cursor right to OPS. Press [5] to display **seq(**. Now complete the expression as in Fig. 13. The numbers 0, 1, and .1 are determined as explained in Example 6.

Check Your Understanding 6.3
Solutions can be found following the section exercises.

1. Repeat Example 6 using midpoints of the subintervals.

2. Repeat Example 6 using left endpoints of the subintervals.

EXERCISES 6.3

In Exercises 1–6, compute the area of the shaded region in two different ways: (a) by using simple geometric formulas; (b) by applying Theorem I.

1.

2.

3.

4.

5.

6.

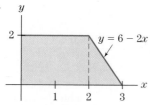

In Exercises 7–12, set-up the definite integral that gives the area of the shaded region. Do not evaluate the integral.

7.

8.

9.

10.

11.

12.

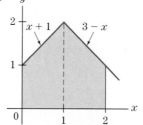

13. Use Theorem I to compute the shaded area in Exercise 7.

14. Use Theorem I to compute the shaded area in Exercise 8.

15. Use Theorem I to compute the shaded area in Exercise 11.

In Exercises 16–18, draw the region whose area is given by the definite integral.

16. $\displaystyle\int_2^4 x^2\,dx$ **17.** $\displaystyle\int_0^4 (8-2x)dx$ **18.** $\displaystyle\int_0^4 \sqrt{x}\,dx$

Find the area under each of the given curves.

19. $y = 4x$; $x = 2$ to $x = 3$

20. $y = 3x^2$; $x = -1$ to $x = 1$

21. $y = 3x^2 + x + 2e^{x/2}$; $x = 0$ to $x = 1$

22. $y = \sqrt{x}$; $x = 0$ to $x = 4$

23. $y = (x - 3)^4$; $x = 1$ to $x = 4$

24. $y = e^{3x}$; $x = -\frac{1}{3}$ to $x = 0$

25. Find the real number $b > 0$ so that the area under the graph of $y = x^3$ from 0 to b is equal to 4.

26. Find the real number $b > 0$ so that the area under the graph of $y = x^2$ from 0 to b is equal to the area under the graph of $y = x^3$ from 0 to b.

Determine Δx and the midpoints of the subintervals formed by partitioning the given interval into n subintervals.

27. $0 \le x \le 2$; $n = 4$ **28.** $0 \le x \le 3$; $n = 6$

29. $1 \le x \le 4$; $n = 5$ **30.** $3 \le x \le 5$; $n = 5$

In Exercises 31–36, use a Riemann sum to approximate the area under the graph of $f(x)$ on the given interval, with selected points as specified.

31. $f(x) = x^2$; $1 \le x \le 3$, $n = 4$, midpoints of subintervals

32. $f(x) = x^2$; $-2 \le x \le 2$, $n = 4$, midpoints of subintervals

33. $f(x) = x^3$; $1 \le x \le 3$, $n = 5$, left endpoints

34. $f(x) = x^3$; $0 \le x \le 1$, $n = 5$, right endpoints

35. $f(x) = e^{-x}$; $2 \le x \le 3$, $n = 5$, right endpoints

36. $f(x) = \ln x$; $2 \le x \le 4$, $n = 5$, left endpoints

In Exercises 37–40, use a Riemann sum to approximate the area under the graph of $f(x)$ in Fig. 14 on the given interval, with selected points as specified. Draw the approximating rectangles.

37. $0 \le x \le 8$, $n = 4$, midpoints of subintervals

38. $3 \le x \le 7$, $n = 4$, left endpoints

39. $4 \le x \le 9$, $n = 5$, right endpoints

40. $1 \le x \le 7$, $n = 3$, midpoints of subintervals

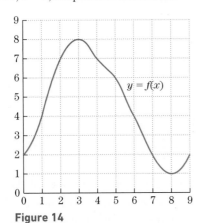

Figure 14

41. Use a Riemann sum with $n = 4$ and left endpoints to estimate the area under the graph of $f(x) = 4 - x$ on the interval $1 \le x \le 4$. Then repeat with $n = 4$ and midpoints. Compare the answers with the exact answer, 4.5, which can be computed from the formula for the area of a triangle.

42. Use a Riemann sum with $n = 4$ and right endpoints to estimate the area under the graph of $f(x) = 2x - 4$ on the interval $2 \le x \le 3$. Then, repeat with $n = 4$ and midpoints. Compare the answers with the exact answer, 1, which can be computed from the formula for the area of a triangle.

43. The graph of the function $f(x) = \sqrt{1 - x^2}$ on the interval $-1 \le x \le 1$ is a semicircle. The area under the graph is $\frac{1}{2}\pi(1)^2 = \pi/2 = 1.57080$, to five decimal places. Use a Riemann sum with $n = 5$ and midpoints to estimate the area under the graph. See Fig. 15. Carry out the calculations to five decimal places and compute the error (the difference between the estimate and 1.57080).

Figure 15   **Figure 16**

44. Use a Riemann sum with $n = 5$ and midpoints to estimate the area under the graph of $f(x) = \sqrt{1 - x^2}$ on the interval $0 \le x \le 1$. The graph is a quarter circle, and the area under the graph is .78540, to five decimal places. See Fig. 16. Carry out the calculations to five decimal places and compute the error.

45. Estimate the area (in square feet) of the residential lot in Fig. 17.

Figure 17

46. A farmer wants to divide the lot in Fig. 18 into two lots of equal area by erecting a fence that extends from the road to the river as shown. Determine the location of the fence.

Figure 18

In Exercises 47 and 48, we show that, as the number of subintervals increases indefinitely, the Riemann sum approximation of the area under the graph of $f(x) = x^2$ from 0 to 1 approaches the value $\frac{1}{3}$, which is the exact value of the area.

47. Verify the given formula for $n = 1, 2, 3, 4$:

$$1^2 + 2^2 + \cdots + n^2 = \frac{n(n + 1)(2n + 1)}{6}.$$

48. Partition the interval $[0, 1]$ into n equal subintervals of length $\Delta x = 1/n$ each, and let $x_1, x_2, \ldots, x_n$ denote the right endpoints of the subintervals. Let

$$S_n = [f(x_1) + f(x_2) + \cdots + f(x_n)]\Delta x$$

denote the Riemann sum that estimates the area under the graph of $f(x) = x^2$ on the interval $0 \le x \le 1$.

(a) Show that $S_n = \frac{1}{n^3}(1^2 + 2^2 + \cdots + n^2)$.

(b) Using the previous exercise, conclude that

$$S_n = \frac{n(n + 1)(2n + 1)}{6n^3}.$$

(c) As n increases indefinitely, S_n approaches the area under the curve. Show that this area is 1/3.

TECHNOLOGY EXERCISES

49. The area under the graph of the function e^{-x^2} plays an important role in probability. Compute this area from -1 to 1.

50. Compute the area under the graph of $y = \frac{1}{1 + x^2}$ from 0 to 5.

Evaluate a Riemann sum to approximate the area under the graph of $f(x)$ on the given interval, with points selected as specified.

51. $f(x) = x\sqrt{1 + x^2}$; $1 \le x \le 3$, $n = 20$, midpoints of subintervals

52. $f(x) = \sqrt{1 - x^2}$; $-1 \le x \le 1$, $n = 20$, left endpoints of subintervals

Solutions to Check Your Understanding 6.3

1. As in Example 6, $n = 4$ and $\Delta x = .25$. The first midpoint is $x_1 = \frac{.25}{2} = .125$. We find subsequent midpoints by successively adding $\Delta x = .25$. So, $x_2 = .125 + .25 = .375$, $x_3 = .375 + .25 = .625$, $x_4 = .625 + .25 = .875$. The Riemann sum is

$$(f(x_1) + f(x_2) + f(x_3) + f(x_4))\Delta x$$
$$= ((.125)^2 + (.375)^2 + (.625)^2 + (.875)^2)(.25)$$
$$= 0.328125.$$

2. We have $n = 4$ and $\Delta x = .25$. The first left endpoint is $x_1 = 0$. We find subsequent midpoints by successively adding .25. So, $x_2 = 0 + .25 = .25$, $x_3 = .25 + .25 = .5$, $x_4 = .5 + .25 = .75$. The Riemann sum is

$$(f(x_1) + f(x_2) + f(x_3) + f(x_4))\Delta x$$
$$= (0^2 + (.25)^2 + (.5)^2 + (.75)^2)(.25)$$
$$= 0.21875.$$

6.4 Areas in the *xy*-Plane

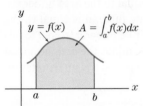

Figure 1 Region above the *x*-axis.

In Section 6.3, we saw that the area under the graph of a continuous nonnegative function $f(x)$ from a to b is the definite integral of f from a to b (Fig. 1). This gives us a concrete visual interpretation of the definite integral as an area. In case $f(x)$ is negative at some points in the interval, we may also give a geometric interpretation of the definite integral.

Consider first the case where $f(x) \le 0$ for all $a \le x \le b$, as shown in Fig. 2. The area A under the *x*-axis, bounded by the graph of $f(x)$, from a to b, is the same as the area A above the *x*-axis, bounded by the graph of $-f(x)$ from a to b. Since $-f(x)$ is nonnegative, we have

$$A = \int_a^b -f(x)dx = -\int_a^b f(x)dx, \tag{1}$$

where the second equality follows from property (4), Section 6.2, with $k = -1$. So, if $f(x) \le 0$, then,

$$\int_a^b f(x)dx = -A.$$

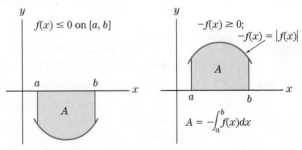

Figure 2

For this reason, the definite integral of f from a to b is referred to as a *signed area*, since it takes on negative values if the area is under the *x*-axis. We note that when $f(x) \le 0$, then, $-f(x) = |f(x)|$ and so, from (1), the area bounded by the curve and the *x*-axis is

$$A = \int_a^b |f(x)| \, dx, \tag{2}$$

It will follow from our discussion that formula (2) is also valid for an arbitrary continuous function $f(x)$. For an arbitrary continuous function $f(x)$, divide the interval $[a, b]$

into subintervals on which f does not change signs. For example, in Fig. 3, the interval $[a, b]$ is divided into four subintervals so that $f(x) \leq 0$ on $[a, c]$ and $[d, e]$; and $f(x) \geq 0$ on $[c, d]$ and $[e, b]$. By property (7), Section 6.2, we have

$$\int_a^b f(x)dx = \int_a^c f(x)dx + \int_c^d f(x)dx + \int_d^e f(x)dx + \int_e^b f(x)dx. \tag{3}$$

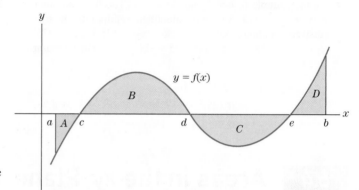

Figure 3 Regions above and below the x-axis.

On an interval where f is nonnegative, the definite integral is equal to the area under the curve, while on the interval where f is nonpositive, the definite integral is equal to the negative of the area under the curve. Thus, the value of the definite integral from a to b is equal to the area bounded by the graph that is above the x-axis, minus the area bounded by the graph that is below the x-axis. This gives us the following geometric interpretation of the definite integral.

Suppose that $f(x)$ is continuous on the interval $a \leq x \leq b$. Then,

$$\int_a^b f(x)dx$$

is equal to the area above the x-axis bounded by the graph of $y = f(x)$ from $x = a$ to $x = b$, minus the area below the x-axis.

Referring to Fig. 3, we have

$$\int_a^b f(x)dx = [\text{area of } B \text{ and } D] - [\text{area of } A \text{ and } C].$$

EXAMPLE 1

Area Bounded by a Curve and the x-Axis Find the shaded area in Fig. 4 bounded by the x-axis and the graph of $f(x) = x^2 - x$ from 0 to 2.

SOLUTION

Because the function takes on negative values in the interval $0 \leq x \leq 2$, we cannot evaluate the area by simply computing the definite integral of f from 0 to 2. We must distinguish the interval on which $f \geq 0$ from the one on which $f \leq 0$. Following the notation in Fig. 4, we have

$$A = -\int_0^1 f(x)dx \quad \text{and} \quad B = \int_1^2 f(x)dx.$$

So,

$$A = -\int_0^1 (x^2 - x)dx = -\left[\frac{1}{3}x^3 - \frac{1}{2}x^2\right]\Big|_0^1 = -\left[\frac{1}{3} - \frac{1}{2}\right] = \frac{1}{6}$$

$$B = \int_1^2 (x^2 - x)dx = \left[\frac{1}{3}x^3 - \frac{1}{2}x^2\right]\Big|_1^2 = \left(\frac{8}{3} - \frac{4}{2}\right) - \left(\frac{1}{3} - \frac{1}{2}\right) = \frac{4}{6} + \frac{1}{6} = \frac{5}{6}.$$

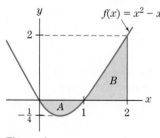

Figure 4

Thus, the desired area is $A + B = \frac{1}{6} + \frac{5}{6} = 1$. **» Now Try Exercise 7**

| EXAMPLE 2 | **Area Bounded by a Curve and the *x*-Axis** Find the shaded area in Fig. 5 bounded by the *x*-axis and the graph of $f(x) = 1 - e^{2x}$ from -1 to 1. |

SOLUTION   Following the notation in Fig. 5, we have

$$A = \int_{-1}^{0} f(x)dx \quad \text{and} \quad B = -\int_{0}^{1} f(x)dx.$$

Using the fact that an antiderivative of e^{2x} is $\frac{1}{2}e^{2x}$, we obtain

$$A = \int_{-1}^{0} (1 - e^{2x})dx = \left[x - \frac{1}{2}e^{2x} \right]\Big|_{-1}^{0}$$

$$= \left[0 - \frac{1}{2} \right] - \left[-1 - \frac{1}{2}e^{-2} \right] = \frac{1}{2}e^{-2} + \frac{1}{2} \approx .568$$

$$B = -\int_{0}^{1} (1 - e^{2x})dx = -\left(x - \frac{1}{2}e^{2x} \right)\Big|_{0}^{1}$$

$$= -\left(1 - \frac{1}{2}e^{2} \right) + \left(0 - \frac{1}{2} \right) = \frac{1}{2}e^{2} - \frac{3}{2} \approx 2.195.$$

Thus, the desired area is $A + B \approx .568 + 2.195 = 2.763$ **▷▷ Now Try Exercise 11**

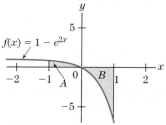

Figure 5

Area between Two Curves

The regions that we just considered were bounded by the *x*-axis and the graph of a function. These are particular cases of a more general type of regions that are bounded both above and below by graphs of functions.

Referring to Fig. 6(c), we would like to find a simple expression for the area of the shaded region under the graph of $y = f(x)$ and above the graph of $y = g(x)$ from $x = a$ to $x = b$. It is the region under the graph of $y = f(x)$ (Fig. 6(a)) with the region under the graph of $y = g(x)$ (Fig. 6(b)) taken away. Therefore,

[area of shaded region] = [area under $f(x)$] − [area under $g(x)$]

$$= \int_{a}^{b} f(x)dx - \int_{a}^{b} g(x)dx$$

$$= \int_{a}^{b} [f(x) - g(x)]dx \quad \text{By Property (3), Sec. 6.2.}$$

  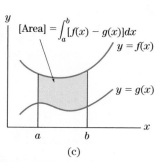

(a)   (b)   (c)

Figure 6

> **Area between Two Curves**   If $y = f(x)$ lies above $y = g(x)$ from $x = a$ to $x = b$, the area of the region between $f(x)$ and $g(x)$ from $x = a$ to $x = b$ is
>
> $$\int_{a}^{b} [f(x) - g(x)]dx.$$

| EXAMPLE 3 | **Area between Two Curves** Find the area of the region between $y = 2x^2 - 4x + 6$ and $y = -x^2 + 2x + 1$ from $x = 1$ to $x = 2$. |

SOLUTION Upon sketching the two graphs (Fig. 7), we see that $f(x) = 2x^2 - 4x + 6$ lies above $g(x) = -x^2 + 2x + 1$ for $1 \le x \le 2$. Therefore, our formula gives the area of the shaded region as

$$\int_1^2 [(2x^2 - 4x + 6) - (-x^2 + 2x + 1)]dx = \int_1^2 (3x^2 - 6x + 5)dx$$

$$= (x^3 - 3x^2 + 5x) \Big|_1^2 = 6 - 3 = 3.$$

>> *Now Try Exercise 13*

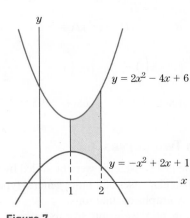

Figure 7   Figure 8

| EXAMPLE 4 | **Area between Two Curves** Find the area of the region between $y = x^2$ and $y = (x - 2)^2 = x^2 - 4x + 4$ from $x = 0$ to $x = 3$. |

SOLUTION Upon sketching the graphs (Fig. 8), we see that the two graphs cross; by setting $x^2 = x^2 - 4x + 4$, we find that they cross when $x = 1$. Thus, one graph does not always lie above the other from $x = 0$ to $x = 3$, so that we cannot directly apply our rule for finding the area between two curves. However, the difficulty can be surmounted if we break the region into two parts, namely, the area from $x = 0$ to $x = 1$ and the area from $x = 1$ to $x = 3$. From $x = 0$ to $x = 1$, $y = x^2 - 4x + 4$ is on top; and from $x = 1$ to $x = 3$, $y = x^2$ is on top. Consequently,

$$[\text{Area from } x = 0 \text{ to } x = 1] = \int_0^1 [(x^2 - 4x + 4) - (x^2)]dx$$

$$= \int_0^1 (-4x + 4)dx$$

$$= (-2x^2 + 4x) \Big|_0^1 = 2 - 0 = 2, \text{ and}$$

$$[\text{Area from } x = 1 \text{ to } x = 3] = \int_1^3 [(x^2) - (x^2 - 4x + 4)]dx$$

$$= \int_1^3 (4x - 4)dx$$

$$= (2x^2 - 4x) \Big|_1^3 = 6 - (-2) = 8.$$

Thus, the total area is $2 + 8 = 10$.

>> *Now Try Exercise 15*

In our derivation of the formula for the area between two curves, we examined functions that are nonnegative. However, the statement of the rule does not contain this stipulation, and rightfully so. Consider the case where $f(x)$ and $g(x)$ are not always positive. Let us determine the area of the shaded region in Fig. 9(a). Select some constant c such that the graphs of the functions $f(x) + c$ and $g(x) + c$ lie completely above the *x*-axis [Fig. 9(b)]. The region between them will have the same area as the original region. Using the rule as applied to nonnegative functions, we have

$$[\text{Area of the region}] = \int_a^b [(f(x) + c) - (g(x) + c)]dx$$

$$= \int_a^b [f(x) - g(x)]dx.$$

Therefore, we see that our rule is valid for any functions $f(x)$ and $g(x)$ as long as the graph of $f(x)$ lies above the graph of $g(x)$ for all x from $x = a$ to $x = b$.

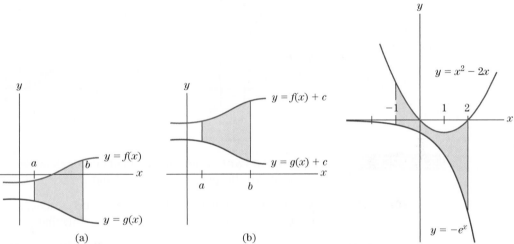

Figure 9 Shifting a region vertically does not change its area.   **Figure 10**

EXAMPLE 5 **Area between Two Curves** Set up the integral that gives the area between the curves $y = x^2 - 2x$ and $y = -e^x$ from $x = -1$ to $x = 2$.

SOLUTION Since $y = x^2 - 2x$ lies above $y = -e^x$ (Fig. 10), the rule for finding the area between two curves can be applied directly. The area between the curves is

$$\int_{-1}^{2} (x^2 - 2x + e^x)dx.$$

» Now Try Exercise 17

Sometimes, we are asked to find the area between two curves without being given the values of a and b. In these cases, there is a region that is completely enclosed by the two curves. As the next examples illustrate, we must first find the points of intersection of the two curves in order to obtain the values of a and b. In such problems, careful curve sketching is especially important.

EXAMPLE 6 **Area between Two Curves and Points of Intersection** Set up the integral that gives the area bounded by the curves $y = x^2 + 2x + 3$ and $y = 2x + 4$.

SOLUTION The two curves are sketched in Fig. 11, and the region bounded by them is shaded. In order to find the points of intersection, we set $x^2 + 2x + 3 = 2x + 4$ and solve for x. We obtain $x^2 = 1$, or $x = -1$ and $x = 1$. When $x = -1$, $2x + 4 = 2(-1) + 4 = 2$. When $x = 1$, $2x + 4 = 2(1) + 4 = 6$. Thus, the curves intersect at the points $(1, 6)$ and $(-1, 2)$.

Since $y = 2x + 4$ lies above $y = x^2 + 2x + 3$ from $x = -1$ to $x = 1$, the area between the curves is given by

$$\int_{-1}^{1} [(2x + 4) - (x^2 + 2x + 3)]dx = \int_{-1}^{1} (1 - x^2)dx.$$

≫ Now Try Exercise 19

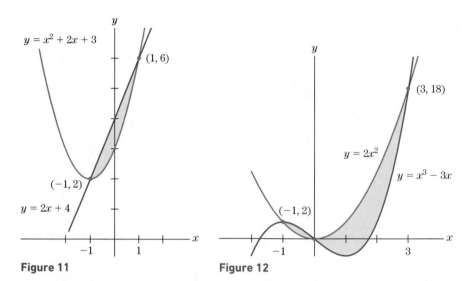

Figure 11 Figure 12

EXAMPLE 7 **Area between Two Curves and Points of Intersection** Set up the integral that gives the area bounded by the two curves $y = 2x^2$ and $y = x^3 - 3x$.

SOLUTION First, we make a rough sketch of the two curves, as in Fig. 12. The curves intersect where $x^3 - 3x = 2x^2$, or $x^3 - 2x^2 - 3x = 0$. Note that

$$x^3 - 2x^2 - 3x = x(x^2 - 2x - 3) = x(x - 3)(x + 1).$$

So the solutions to $x^3 - 2x^2 - 3x = 0$ are $x = 0, 3, -1$, and the curves intersect at $(-1, 2)$, $(0, 0)$, and $(3, 18)$. From $x = -1$ to $x = 0$, the curve $y = x^3 - 3x$ lies above $y = 2x^2$. But from $x = 0$ to $x = 3$, the reverse is true. Thus, the area between the curves is given by

$$\int_{-1}^{0} (x^3 - 3x - 2x^2)dx + \int_{0}^{3} (2x^2 - x^3 + 3x)dx.$$

≫ Now Try Exercise 29

Areas in Applications

So far, we have at least two interpretations of the definite integral. To recall these, suppose that f is continuous and nonnegative on $[a, b]$, and let F be an antiderivative of f, that is, $F' = f$. Then

$$\begin{bmatrix} \text{Area under the} \\ \text{graph of } f \text{ from} \\ a \text{ to } b \end{bmatrix} = \int_{a}^{b} f(x)dx = \int_{a}^{b} F'(x)dx = F(b) - F(a) = \begin{bmatrix} \text{Net change of} \\ F \text{ on } [a, b] \end{bmatrix} \quad (4)$$

Thus, *the area under the graph of the rate of change function f is equal to the net change in the function F as x varies from a to b*. This provides another geometric way to visualize the net change of the function (Fig. 13).

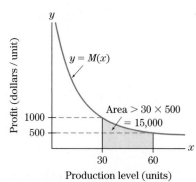

Figure 13 Area under the rate of change equals the net change.

EXAMPLE 8   **Marginal Profit** A company is trying to decide whether to increase its daily production of a certain commodity from 30 to 60 units per day. The company will not increase the production level unless there is an increase of at least $15,000 in profit. Based on the marginal profit curve in Fig. 14, should the company raise its production level from 30 to 60?

SOLUTION   Let $P(x)$ and $M(x)$ denote the profit and marginal profit at production level x, respectively. Since $P'(x) = M(x)$, it follows that

$$\begin{bmatrix} \text{Area under the} \\ \text{graph of } M(x) \\ \text{from 30 to 60} \end{bmatrix} = \int_{30}^{60} M(x)dx = \int_{30}^{60} P'(x)dx$$

$$= P(60) - P(30) = \begin{bmatrix} \text{Net change of} \\ P \text{ as } x \text{ varies} \\ \text{from 30 to 60.} \end{bmatrix}$$

It is clear from Fig. 14 that the area under the graph of $M(x)$ from 30 to 60 is greater than the shaded rectangular area $30 \times 500 = 15,000$. Thus, the net change in profit, which equals the area under the curve, will exceed $15,000, and so the company should increase its production level to 60 units.

Figure 14 Marginal profit curve.   **≫ Now Try Exercise 33**

We now consider an example where the rate of change takes on negative values.

EXAMPLE 9   **Displacement versus Distance Traveled** A rocket is fired vertically into the air. Its velocity at t seconds after liftoff is $v(t) = -32t + 160$ feet per second.

(a) Find the displacement of the rocket as t varies in the interval $0 \le t \le 8$. Interpret this displacement using area under the graph of $v(t)$.

(b) Find the total distance traveled by the rocket during the interval of time $0 \le t \le 8$. Interpret this distance as an area.

FOR REVIEW

See the discussion preceding Example 6, Sec. 6.2, p. 306.

SOLUTION   (a) Let $s(t)$ denote the position of the rocket at time t, measured from its initial position. The required displacement is $s(8) - s(0)$, which is the net change in position over the interval $0 \le t \le 8$. We have

$$s(8) - s(0) = \int_0^8 v(t)dt = \int_0^8 (-32t + 160)dt$$

$$= [-16t^2 + 160t \Big|_0^8 = (-16)(8^2) + (160)(8) = 256 \text{ feet}.$$

Thus, the rocket is 256 feet above its initial position. The value of the definite integral over the interval $[0, 8]$ is equal to the area A above the t-axis bounded by the graph of $v(t)$ *minus* the area B below the t-axis (Fig. 15).

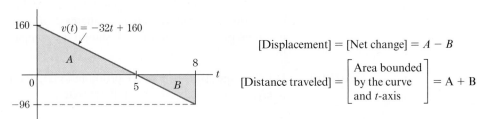

$$[\text{Displacement}] = [\text{Net change}] = A - B$$

$$[\text{Distance traveled}] = \begin{bmatrix} \text{Area bounded} \\ \text{by the curve} \\ \text{and } t\text{-axis} \end{bmatrix} = A + B$$

Figure 15 Displacement versus distance traveled for $0 \le t \le 8$.

(b) Note that $v(t) \le 0$ for $5 \le t \le 8$. So the rocket changes direction at time $t = 5$, and hence, to compute the total distance traveled, we must compute the displacements on two separate intervals, $[0, 5]$ and $[5, 8]$. On $[0, 5]$ the displacement is

$$\int_0^5 (-32t + 160)dt = [-16t^2 + 160t \Big|_0^5 = (-16)(5^2) + (160)(5) = 400 \text{ feet}.$$

Thus, the rocket moved 400 feet upward in the interval $0 \le t \le 5$. On $[5, 8]$ the displacement is

$$\int_5^8 (-32t + 160)dt = [-16t^2 + 160t \Big|_5^8$$

$$= [(-16)(8^2) + (160)(8)] - [(-16)(5^2) + (160)(5)] = -144 \text{ feet}.$$

Thus, the rocket moved 144 feet downward in the interval of time $5 \le t \le 8$. The total distance traveled is $400 + 144 = 544$ feet. In terms of area, the total distance traveled is equal to the area A above the t-axis bounded by the graph of $v(t)$ from 0 to 5 *plus* the area B below the t-axis bounded by the graph of $v(t)$ from 5 to 8 (Fig. 15).

≫ Now Try Exercise 45

From Example 9, we see that the displacement and total distance traveled are the same as long as $v(t)$ is nonnegative in $[a, b]$. If $v(t)$ takes on negative values in $[a, b]$, then the displacement and total distance traveled are not necessarily equal. The displacement is always given as $\int_a^b v(t)dt$ and is represented by the difference of the area above the t-axis and below the velocity curve minus the area below the t-axis and above the velocity curve. The total distance traveled is the total area bounded by the velocity curve and the t-axis. In terms of definite integrals, we have

$$[\text{Displacement}] = \int_a^b v(t)dt; \qquad [\text{Distance traveled}] = \int_a^b |v(t)|\,dt.$$

Our final example provides an interesting interpretation of an area between two curves.

EXAMPLE 10   **Launch of an Online Transportation Network** In 2010, after an online transportation network company launched its mobile app, its client base was expected to grow exponentially. A fairly good model for the rate of growth (in thousands of clients per year) from 2010 to 2015 was given by

$$R_1(t) = 65e^{0.07t}.$$

In 2015, due to growing competition, this rate of growth was adjusted down and a new model for the projected growth rate (in thousands of clients per year) was given by

$$R_2(t) = 79.4e^{0.03t}.$$

In both models, $t = 0$ corresponds to 2010. Estimate the total number of clients that will be lost between 2015 and 2020 due to competition.

SOLUTION   If the number of clients continued to grow as it did prior to 2015, then the net change in clients between 2015 and 2020 would have been

$$\int_5^{10} R_1(t)\, dt.$$

However, taking into account the slower rate of growth since 2015, the projected net change in clients between 2015 and 2020 will be

$$\int_5^{10} R_2(t)\, dt.$$

Both integrals of $R_1(t)$ and $R_2(t)$ may be interpreted as the areas under the curves $y = R_1(t)$ and $y = R_2(t)$, respectively, from $t = 5$ to $t = 10$. (See Fig. 16.) By superposing the two curves, we see that the area between them from $t = 5$ to $t = 10$ represents the total number of clients that was lost by using $R_2(t)$ instead of $R_1(t)$. (See Fig. 17.) The area between the two curves equals

$$\int_5^{10} [R_1(t) - R_2(t)]\, dt = \int_5^{10} [65e^{0.07t} - 79.4e^{0.03t}]\, dt$$

$$= \left(\frac{65}{0.07} e^{0.07t} - \frac{79.4}{0.03} e^{0.03t} \right)\Big|_5^{10}$$

$$\approx (928.571e^{0.7} - 2646.67e^{0.3}) - (928.571e^{0.35} - 2646.67e^{0.15})$$

$$\approx 54.5684.$$

Thus, the company is projected to have about 54,568 fewer clients if we use the model predicted by $R_2(t)$ instead of $R_1(t)$.   **» Now Try Exercise 41**

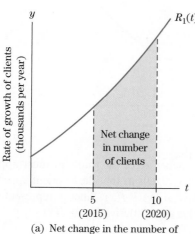

(a) Net change in the number of clients based on $R_1(t)$ model

Figure 16

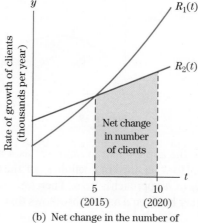

(b) Net change in the number of clients based on $R_2(t)$ model

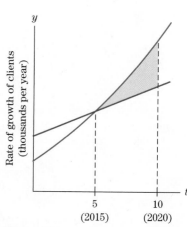

Figure 17

Areas in the Theory of Calculus: The Area Function

Question: Suppose that f is a continuous function on an interval $[a, b]$. Does f always have an antiderivative, and, if it does, how can we find it?

When f has an antiderivative F, we used F to compute the area under the graph of f, as described by Theorem 1, Section 6.3. As we now show, the concept of area can be used to define an antiderivative of f, and this will prove that *every continuous function has an antiderivative*.

Let f be a continuous function on an interval $[a, b]$. For any x in $[a, b]$, define $A(x)$ as the limit of the Riemann sums of f on the interval $[a, x]$, as in (3), Section 6.3. When f is nonnegative, this limit is equal to the area under the graph of f from a to x. For this reason, the function $A(x)$ is called the **area function** (Fig. 18). We now argue that $A'(x) = f(x)$; in other words, $A(x)$ is an antiderivative of f.

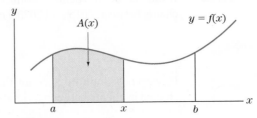

Figure 18 The area function.

> **Theorem I: The Area Function**   Let $f(x)$ be a continuous function for $a \le x \le b$. Let $A(x)$ be the area function. Then, $A(x)$ is an antiderivative of $f(x)$.

Justification of Theorem I   Suppose for simplicity that f is nonnegative. If h is a small positive number, then $A(x + h) - A(x)$ is the area of the darker shaded region in Fig. 19. This shaded region is approximately a rectangle of width h, height $f(x)$, and area $h \cdot f(x)$. Thus,

$$A(x + h) - A(x) \approx h \cdot f(x),$$

where the approximation becomes better as h approaches zero. Dividing by h, we have

$$\frac{A(x + h) - A(x)}{h} \approx f(x).$$

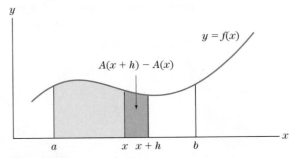

Figure 19

Since the approximation improves as h approaches zero, the quotient must approach $f(x)$. However, the limit definition of the derivative tells us that the quotient approaches $A'(x)$ as h approaches zero. Therefore, we have $A'(x) = f(x)$. Since x represented any number between a and b, this shows that $A(x)$ is an antiderivative of $f(x)$.   «

A Second Proof of the Fundamental Theorem of Calculus   We close this section by showing how the area function can be used to prove Theorem I of the previous section.

Indeed, if F is any antiderivative of f, since the area function $A(x)$ is also an antiderivative of $f(x)$, we have

$$A(x) = F(x) + C$$

for some constant C. Notice that $A(a)$ is 0 and $A(b)$ equals the area of the region under the graph of $f(x)$ for $a \leq x \leq b$. Therefore, if f is nonnegative,

$$
\begin{aligned}
[\text{Area under the graph of } f \text{ from } a \text{ to } b] = A(b) &= A(b) - A(a) \\
&= [F(b) + C] - [F(a) + C] \\
&= F(b) - F(a),
\end{aligned}
$$

which proves Theorem I of Section 6.3.   «

INCORPORATING TECHNOLOGY

Area between Two Curves We illustrate how to shade the area between two curves using the functions from Example 3. First, set $Y_1 = -X^2 + 2X + 1$ and $Y_2 = 2X^2 - 4X + 6$. Now press 2nd [DRAW] 7 to display **Shade(**. Complete the expression to read **Shade(Y_1, Y_2, 1, 2)** and press ENTER. This will shade the area above Y_1 and below Y_2 with $1 \leq X \leq 2$. The result is shown in Fig. 20(a). Note that, in the **Shade** routine, it is necessary to list the lower function first and the upper function second.

We evaluate the definite integral from Example 3 to find the area between the curves. Select MATH 9 and complete the expression as shown in Fig. 20(b), and press ENTER. The result is shown in Fig. 20(b).

(a)

(b)

Figure 20

Check Your Understanding 6.4

Solutions can be found following the section exercises.

1. Find the area between the curves $y = x + 3$ and $y = \frac{1}{2}x^2 + x - 7$ from $x = -2$ to $x = 1$.

2. A company plans to increase its production from 10 to 15 units per day. The present marginal cost function is $MC_1(x) = x^2 - 20x + 108$. By redesigning the production process and purchasing new equipment, the company can change the marginal cost function to $MC_2(x) = \frac{1}{2}x^2 - 12x + 75$. Determine the area between the graphs of the two marginal cost curves from $x = 10$ to $x = 15$. Interpret this area in economic terms.

EXERCISES 6.4

1. Write a definite integral or sum of definite integrals that gives the area of the shaded portions in Fig. 21.

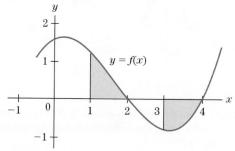

Figure 21

2. Write a definite integral or sum of definite integrals that gives the area of the shaded portions in Fig. 22.

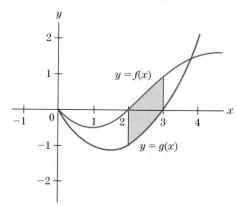

Figure 22

3. Shade the portion of Fig. 23 whose area is given by the integral

$$\int_0^2 [f(x) - g(x)]\,dx + \int_2^4 [h(x) - g(x)]\,dx.$$

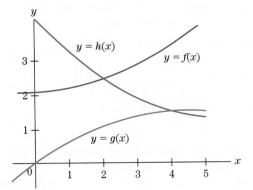

Figure 23

4. Shade the portion of Fig. 24 whose area is given by the integral

$$\int_0^1 [f(x) - g(x)] + \int_1^2 [g(x) - f(x)]\,dx.$$

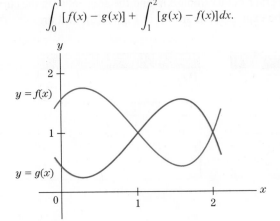

Figure 24

5. Let $f(x)$ be the function pictured in Fig. 25. Determine whether $\int_0^7 f(x)\,dx$ is positive, negative, or zero.

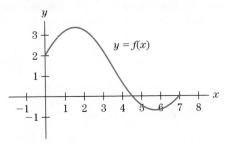

Figure 25

6. Let $g(x)$ be the function pictured in Fig. 26. Determine whether $\int_0^7 g(x)\,dx$ is positive, negative, or zero.

Figure 26

Find the area of the region between the curve and the x-axis.

7. $f(x) = 1 - x^2$, from -2 to 2

8. $f(x) = x(x^2 - 1)$, from -1 to 1.

9. $f(x) = x^2 - 2x - 3$, from 0 to 2.

10. $f(x) = x^2 + 6x + 5$ from 0 to 1.

11. $f(x) = e^x - 3$ from 0 to $\ln 3$

12. $f(x) = e^{-x} + 2$ from -1 to 2

Find the area of the region between the curves.

13. $y = 2x^2$ and $y = 8$ (a horizontal line) from $x = -2$ to $x = 2$

14. $y = x^2 + 1$ and $y = -x^2 - 1$ from $x = -1$ to $x = 1$

15. $y = x^2 - 6x + 12$ and $y = 1$ from $x = 0$ to $x = 4$

16. $y = x(2 - x)$ and $y = 2$ from $x = 0$ to $x = 2$

17. $y = e^x$ and $y = \frac{1}{x^2}$ from $x = 1$ to $x = 2$

18. $y = e^{2x}$ and $y = 1 - x$ from $x = 0$ to $x = 1$

Find the area of the region bounded by the curves.

19. $y = x^2$ and $y = x$

20. $y = 4x(1 - x)$ and $y = \frac{3}{4}$

21. $y = -x^2 + 6x - 5$ and $y = 2x - 5$

22. $y = x^2 - 1$ and $y = 3$

23. $y = x(x^2 - 1)$ and the x-axis

24. $y = x^3$ and $y = 2x^2$

25. $y = 8x^2$ and $y = \sqrt{x}$

26. $y = \frac{4}{x}$ and $y = 5 - x$

27. Find the area of the region between $y = x^2 - 3x$ and the x-axis
    (a) from $x = 0$ to $x = 3$,
    (b) from $x = 0$ to $x = 4$,
    (c) from $x = -2$ to $x = 3$.

28. Find the area of the region between $y = x^2$ and $y = 1/x^2$
    (a) from $x = 1$ to $x = 4$,
    (b) from $x = \frac{1}{2}$ to $x = 4$.

29. Find the area in Fig. 27 of the region bounded by $y = 1/x^2$, $y = x$, and $y = 8x$, for $x \geq 0$.

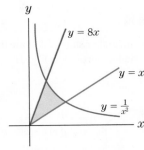

Figure 27

30. Find the area of the region bounded by $y = 1/x$, $y = 4x$, and $y = x/2$, for $x \geq 0$. (The region resembles the shaded region in Exercise 29.)

31. **Height of a Helicopter** A helicopter is rising straight up in the air. Its velocity at time t is $v(t) = 2t + 1$ feet per second.
    (a) How high does the helicopter rise during the first 5 seconds?
    (b) Represent the answer to part (a) as an area.

32. **Assembly Line Production** After t hours of operation, an assembly line is producing lawn mowers at the rate of $r(t) = 21 - \frac{4}{5}t$ mowers per hour.
    (a) How many mowers are produced during the time from $t = 2$ to $t = 5$ hours?
    (b) Represent the answer to part (a) as an area.

33. **Cost** Suppose that the marginal cost function of a handbag manufacturer is $C'(x) = \frac{3}{32}x^2 - x + 200$ dollars per unit at production level x (where x is measured in units of 100 handbags).
    (a) Find the total cost of producing 6 additional units if 2 units are currently being produced.
    (b) Describe the answer to part (a) as an area. (Give a written description rather than a sketch.)

34. **Profit** Suppose that the marginal profit function for a company is $P'(x) = 100 + 50x - 3x^2$ at production level x.
    (a) Find the extra profit earned from the sale of 3 additional units if 5 units are currently being produced.
    (b) Describe the answer to part (a) as an area. (Do not make a sketch.)

35. **Marginal Profit** Let $M(x)$ be a company's marginal profit at production level x. Give an economic interpretation of the number $\int_{44}^{48} M(x)\,dx$.

36. **Marginal Profit** Let $M(x)$ be a company's marginal cost at production level x. Give an economic interpretation of the number $\int_{0}^{100} M(x)\,dx$. (*Note:* At any production level, the total cost equals the fixed cost plus the total variable cost.)

37. **Heat Diffusion** Some food is placed in a freezer. After t hours, the temperature of the food is dropping at the rate of $r(t)$ degrees Fahrenheit per hour, where $r(t) = 12 + 4/(t + 3)^2$.
    (a) Compute the area under the graph of $y = r(t)$ over the interval $0 \le t \le 2$.
    (b) What does the area in part (a) represent?

38. **Velocity** Suppose that the velocity of a car at time t is $v(t) = 40 + 8/(t + 1)^2$ kilometers per hour.
    (a) Compute the area under the velocity curve from $t = 1$ to $t = 9$.
    (b) What does the area in part (a) represent?

39. **Deforestation and Fuel Wood** Deforestation is one of the major problems facing sub-Saharan Africa. Although the clearing of land for farming has been the major cause, the steadily increasing demand for fuel wood has also become a significant factor. Figure 28 summarizes projections of the World Bank. The rate of fuel wood consumption (in millions of cubic meters per year) in the Sudan t years after 1980 is given approximately by the function $c(t) = 76.2e^{0.03t}$. Determine the amount of fuel wood consumed from 1980 to 2000.

Rate of fuel wood consumption
$y = c(t)$
Rate of new tree growth

Figure 28

40. Refer to Exercise 39. The rate of new tree growth (in millions of cubic meters per year) in the Sudan t years after 1980 is given approximately by the function $g(t) = 50 - 6.03e^{0.09t}$. Set up the definite integral giving the amount of depletion of the forests due to the excess of fuel wood consumption over new growth from 1980 to 2000.

41. After an advertising campaign, a company's marginal profit was adjusted up from $M_1(x) = 2x^2 - 3x + 11$, before advertising, to $M_2(x) = 2x^2 - 2.4x + 8$, after advertising. Here x denotes the number of units produced, and $M_1(x)$ and $M_2(x)$ are measured in thousands of dollars per unit. Determine the net change in profit that resulted from increasing the production level from $x = 5$ to $x = 10$ due to the advertising campaign.

42. **Profit and Area** The marginal profit for a certain company is $MP_1(x) = -x^2 + 14x - 24$. The company expects the daily production level to rise from $x = 6$ to $x = 8$ units. The management is considering a plan that would have the effect of changing the marginal profit to $M_2(x) = -x^2 + 12x - 20$. Should the company adopt the plan? Determine the area between the graphs of the two marginal profit functions from $x = 6$ to $x = 8$. Interpret this area in economic terms.

43. **Velocity and Distance** Two rockets are fired simultaneously straight up into the air. Their velocities (in meters per second) are $v_1(t)$ and $v_2(t)$, and $v_1(t) \ge v_2(t)$ for $t \ge 0$. Let A denote the area of the region between the graphs of $y = v_1(t)$ and $y = v_2(t)$ for $0 \le t \le 10$. What physical interpretation may be given to the value of A?

44. **Distance Traveled** Cars A and B start at the same place and travel in the same direction, with velocities after t hours given by the functions $v_A(t)$ and $v_B(t)$ in Fig. 29.
    (a) What does the area between the two curves from $t = 0$ to $t = 1$ represent?
    (b) At what time will the distance between the cars be greatest?

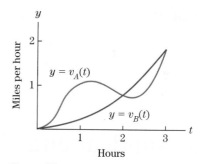

$y = v_A(t)$
$y = v_B(t)$
Miles per hour
Hours

Figure 29

45. **Displacement versus Distance Traveled** The velocity of an object moving along a line is given by $v(t) = 2t^2 - 3t + 1$ feet per second.
    (a) Find the displacement of the object as t varies in the interval $0 \le t \le 3$.
    (b) Find the total distance traveled by the object during the interval of time $0 \le t \le 3$.

46. **Displacement versus Distance Traveled** The velocity of an object moving along a line is given by $v(t) = t^2 + t - 2$ feet per second.
    (a) Find the displacement of the object as t varies in the interval $0 \le t \le 3$. Interpret this displacement using area under the graph of $v(t)$.
    (b) Find the total distance traveled by the object during the interval of time $0 \le t \le 3$. Interpret this distance as an area.

In Exercises 47–50, use a graphing utility to find the intersection points of the curves, and then use the utility to find the area of the region bounded by the curves.

47. $y = e^x$, $y = 4x + 1$

48. $y = 5 - (x - 2)^2$, $y = e^x$

49. $y = \sqrt{x + 1}$, $y = (x - 1)^2$

50. $y = 1/x$, $y = 3 - x$

Solutions to Check Your Understanding 6.4

1. First, graph the two curves, as shown in Fig. 30. The curve $y = x + 3$ lies on top. So, the area between the curves is

$$\int_{-2}^{1} \left[(x + 3) - \left(\tfrac{1}{2}x^2 + x - 7 \right) \right] dx$$

$$= \int_{-2}^{1} \left(-\tfrac{1}{2}x^2 + 10 \right) dx$$

$$= \left(-\tfrac{1}{6}x^3 + 10x \right) \Big|_{-2}^{1} = 28.5.$$

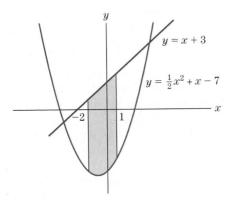

Figure 30

2. Graphing the two marginal cost functions yields the results shown in Fig. 31. So, the area between the curves equals

$$\int_{10}^{15} [MC_1(x) - MC_2(x)] dx$$

$$= \int_{10}^{15} [(x^2 - 20x + 108) - (\tfrac{1}{2}x^2 - 12x + 75)] dx$$

$$= \int_{10}^{15} \left[\tfrac{1}{2}x^2 - 8x + 33 \right] dx$$

$$= \left(\tfrac{1}{6}x^3 - 4x^2 + 33x \right) \Big|_{10}^{15} = 60\tfrac{5}{6}.$$

This amount represents the cost savings on the increased production (from 10 to 15), provided the new production process is used.

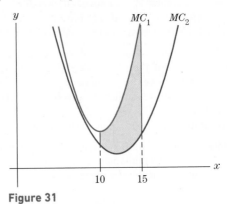

Figure 31

6.5   Applications of the Definite Integral

The applications in this section have two features in common. First, each example contains a quantity that is computed by evaluating a definite integral. Second, the formula for the definite integral is derived from Riemann sums.

In each application, we show that a certain quantity, call it Q, may be approximated if we divide an interval into equal subintervals and form an appropriate sum. We then observe that this sum is a Riemann sum for some function $f(x)$. [Sometimes $f(x)$ is not given beforehand.] The Riemann sums approach Q as the number of subintervals becomes large. Since the Riemann sums also approach a definite integral, we conclude that the value of Q is given by the definite integral.

Once we have expressed Q as a definite integral, we may calculate its value using the fundamental theorem of calculus, which reduces the calculation to antidifferentiation. It is not necessary to calculate the value of any Riemann sum. Rather, we use the Riemann sums only as a device to express Q as a definite integral. The key step is to recognize a given sum as a Riemann sum and to determine the corresponding definite integral. Example 1 illustrates this step.

EXAMPLE 1   **Using an Integral to Approximate a Sum** Suppose that the interval $1 \leq x \leq 2$ is divided into 50 subintervals, each of length Δx. Let $x_1, x_2, \ldots, x_{50}$ denote points selected from these subintervals. Find an approximate value for the sum

$$(8x_1^7 + 6x_1)\Delta x + (8x_2^7 + 6x_2)\Delta x + \cdots + (8x_{50}^7 + 6x_{50})\Delta x.$$

SOLUTION   The sum is a Riemann sum for the function $f(x) = 8x^7 + 6x$ on the interval $1 \leq x \leq 2$. Therefore, by Theorem II, Section 6.3, an approximation to the sum is given by the integral

$$\int_1^2 (8x^7 + 6x)dx.$$

We may evaluate this integral using the fundamental theorem of calculus:

$$\int_1^2 (8x^7 + 6x)\,dx = (x^8 + 3x^2)\Big|_1^2$$
$$= [2^8 + 3(2^2)] - [1^8 + 3(1^2)]$$
$$= 268 - 4 = 264.$$

Therefore, the sum is approximately 264.

> **FOR REVIEW**
>
> In the notation x_2^7, the 7 is an exponent, and the 2 is a subscript. The variable x_2 stands for some value of x in the interval $1 \leq x \leq 2$.

The Average Value of a Function

Let $f(x)$ be a continuous function on the interval $a \leq x \leq b$. The definite integral may be used to define the *average value* of $f(x)$ on this interval. To calculate the average of a collection of numbers $y_1, y_2, \ldots, y_n$, we add the numbers and divide by n to obtain

$$\frac{y_1 + y_2 + \cdots + y_n}{n}.$$

To determine the average value of $f(x)$, we proceed similarly. Choose n values of x, say $x_1, x_2, \ldots, x_n$, and calculate the corresponding function values $f(x_1), f(x_2), \ldots, f(x_n)$. The average of these values is

$$\frac{f(x_1) + f(x_2) + \cdots + f(x_n)}{n}. \tag{1}$$

Our goal now is to obtain a reasonable definition of the average of all the values of $f(x)$ on the interval $a \leq x \leq b$. If the points $x_1, x_2, \ldots, x_n$ are spread "evenly" throughout the interval, the average (1) should be a good approximation to our intuitive concept of the average value of $f(x)$. In fact, as n becomes large, the average (1) should approximate the average value of $f(x)$ to any arbitrary degree of accuracy. To guarantee that the points $x_1, x_2, \ldots, x_n$ are "evenly" spread out from a to b, we divide the interval from $x = a$ to $x = b$ into n subintervals of equal length $\Delta x = (b - a)/n$. Then, we choose x_1 from the first subinterval, x_2 from the second, and so forth. The average (1) that corresponds to these points may be arranged in the form of a Riemann sum as follows:

$$\frac{f(x_1) + f(x_2) + \cdots + f(x_n)}{n}$$
$$= f(x_1) \cdot \frac{1}{n} + f(x_2) \cdot \frac{1}{n} + \cdots + f(x_n) \cdot \frac{1}{n}$$
$$= \frac{1}{b - a}\left[f(x_1) \cdot \frac{b - a}{n} + f(x_2) \cdot \frac{b - a}{n} + \cdots + f(x_n) \cdot \frac{b - a}{n} \right]$$
$$= \frac{1}{b - a}[f(x_1)\Delta x + f(x_2)\Delta x + \cdots + f(x_n)\Delta x].$$

The sum inside the brackets is a Riemann sum for the definite integral of $f(x)$. Thus, we see that, for a large number of points x_i, the average in (1) approaches the quantity

$$\frac{1}{b-a}\int_a^b f(x)dx.$$

This argument motivates the following definition:

DEFINITION The **average value** of a continuous function $f(x)$ over the interval $a \leq x \leq b$ is defined as the quantity

$$\frac{1}{b-a}\int_a^b f(x)dx. \tag{2}$$

EXAMPLE 2 **Average Value** Compute the average value of $f(x) = \sqrt{x}$ over the interval $0 \leq x \leq 9$.

SOLUTION Using (2) with $a = 0$ and $b = 9$, we find the average value of $f(x) = \sqrt{x}$ over the interval $0 \leq x \leq 9$ is equal to

$$\frac{1}{9-0}\int_0^9 \sqrt{x}\, dx.$$

Since $\sqrt{x} = x^{1/2}$, an antiderivative of $f(x) = \sqrt{x}$ is $F(x) = \frac{2}{3}x^{3/2}$. Therefore,

$$\frac{1}{9}\int_0^9 \sqrt{x}\, dx = \frac{1}{9}\left(\frac{2}{3}x^{3/2}\right)\bigg|_0^9 = \frac{1}{9}\left(\frac{2}{3}\cdot 9^{3/2} - 0\right) = \frac{1}{9}\left(\frac{2}{3}\cdot 27\right) = 2,$$

so that the average value of $f(x) = \sqrt{x}$ over the interval $0 \leq x \leq 9$ is 2. The area of the shaded region is the same as the area of the rectangle pictured in Fig. 1.

>> Now Try Exercise 1

FOR REVIEW

If r is the average value, then from (2):

$$r(b-a) = \int_a^b f(x)dx$$
$$= \text{Area under curve.}$$

Here $r(b - a)$ is the area of the rectangle with length $(b - a)$ and height r.

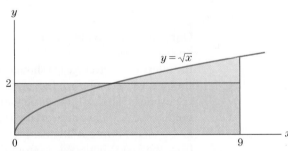

Figure 1 Average value of a function.

EXAMPLE 3 **Average World Population** On January 1, 1998, the world population was 5.9 billion and growing at the rate of 1.34% annually. Assuming that this growth rate continues, the population t years from then will be given by the exponential growth law

$$P(t) = 5.9e^{0.0134t}.$$

Determine the average world population during the next 30 years. (This average is important in long-range planning for agricultural production and the allocation of goods and services.)

SOLUTION   The average value of the population $P(t)$ from $t = 0$ to $t = 30$ is

$$\frac{1}{30 - 0} \int_0^{30} P(t)\,dt = \frac{1}{30} \int_0^{30} 5.9\, e^{0.0134t}\, dt$$

$$= \frac{1}{30}\left(\frac{5.9}{.0134} e^{0.0134t}\right)\Bigg|_0^{30} = \frac{5.9}{.402}(e^{0.402} - 1)$$

$$\approx 7.26 \text{ billion.}$$

>> Now Try Exercise 9

Consumers' Surplus

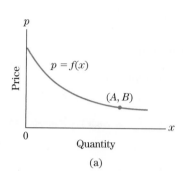

(a)

Using a demand curve from economics, we can derive a formula showing the amount that consumers benefit from an open system that has no price discrimination. Figure 2(a) is a *demand curve* for a commodity. It is determined by complex economic factors and gives a relationship between the quantity sold and the unit price of a commodity. Specifically, it says that, to sell x units, the price must be set at $f(x)$ dollars per unit. Since, for most commodities, selling larger quantities requires a lowering of the price, demand curves are usually decreasing. Interactions between supply and demand determine the amount of a quantity available. Let A designate the amount of the commodity currently available and $B = f(A)$ the current selling price.

Divide the interval from 0 to A into n subintervals, each of length $\Delta x = (A - 0)/n$, and take x_i to be the right-hand endpoint of the ith interval. Consider the first subinterval, from 0 to x_1. [See Fig. 2(b).] Suppose that only x_1 units had been available. Then the price per unit could have been set at $f(x_1)$ dollars and these x_1 units sold. Of course, at this price we could not have sold any more units. However, those people who paid $f(x_1)$ dollars had a great demand for the commodity. It was extremely valuable to them, and there was no advantage in substituting another commodity at that price. They were actually paying what the commodity was worth to them. In theory, then, the first x_1 units of the commodity could be sold to these people at $f(x_1)$ dollars per unit, yielding

$$[\text{price per unit}] \cdot [\text{number of units}] = f(x_1) \cdot (x_1) = f(x_1) \cdot \Delta x \text{ dollars.}$$

After the first x_1 units are sold, suppose that more units become available, so that now a total of x_2 units have been produced. With the price set at $f(x_2)$, the remaining $x_2 - x_1 = \Delta x$ units can be sold, yielding $f(x_2) \cdot \Delta x$ dollars. Here, again, the second group of buyers would have paid as much for the commodity as it was worth to them. Continue this process of price discrimination, and the amount of money paid by consumers would be

$$f(x_1)\Delta x + f(x_2)\Delta x + \cdots + f(x_n)\Delta x.$$

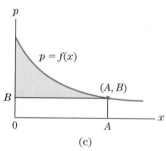

(c)

Figure 2 Consumers' surplus.

Taking n large, we see that this Riemann sum approaches $\int_0^A f(x)dx$. Since $f(x)$ is positive, this integral equals the area under the graph of $f(x)$ from $x = 0$ to $x = A$.

Of course, in an open system, everyone pays the same price, B, so the total amount paid is $[\text{price per unit}] \cdot [\text{number of units}] = BA$. Since BA is the area of the rectangle under the graph of the line $p = B$ from $x = 0$ to $x = A$, the amount of money saved by the consumers is the area of the shaded region in Fig. 2(c). That is, the area between the curves $p = f(x)$ and $p = B$ gives a numerical value to one benefit of a modern efficient economy.

> DEFINITION   The **consumers' surplus** for a commodity having demand curve $p = f(x)$ is
>
> $$\int_0^A [f(x) - B]\,dx,$$
>
> where the quantity demanded is A and the price is $B = f(A)$.

EXAMPLE 4   **Consumers' Surplus** Find the consumers' surplus for the demand curve $p = 50 - .06x^2$ at the sales level $x = 20$.

SOLUTION   Since 20 units are sold, the price must be

$$B = 50 - .06(20)^2 = 50 - 24 = 26.$$

Therefore, the consumers' surplus is

$$\int_0^{20} [(50 - .06x^2) - 26]\,dx = \int_0^{20} (24 - .06x^2)\,dx$$

$$= (24x - .02x^3)\Big|_0^{20}$$

$$= 24(20) - .02(20)^3$$

$$= 480 - 160 = 320.$$

That is, the consumers' surplus is $320.   **» Now Try Exercise 11**

Future Value of an Income Stream

The next example shows how the definite integral can be used to approximate the sum of a large number of terms.

EXAMPLE 5   **Future Value of an Income Stream** Suppose that money is deposited daily into a savings account so that $1000 is deposited each year. The account pays 6% interest compounded continuously. Approximate the amount of money in the account at the end of 5 years.

SOLUTION   Divide the time interval from 0 to 5 years into daily subintervals. Each subinterval is then of duration $\Delta t = \frac{1}{365}$ year. Let $t_1, t_2, \ldots, t_n$ be points chosen from these subintervals. Since we deposit money at an annual rate of $1000, the amount deposited during one subinterval is $1000\,\Delta t$ dollars. If this amount is deposited at time t_i, the $1000\,\Delta t$ dollars will earn interest for the remaining $5 - t_i$ years. The total amount resulting from this one deposit at time t_i is then

$$1000\,\Delta t e^{0.06(5 - t_i)}.$$

Add the effects of the deposits at times $t_1, t_2, \ldots, t_n$ to arrive at the total balance in the account:

$$A = 1000\,e^{0.06(5 - t_1)}\Delta t + 1000\,e^{0.06(5 - t_2)}\Delta t + \cdots + 1000\,e^{0.06(5 - t_n)}\Delta t.$$

This is a Riemann sum for the function $f(t) = 1000e^{0.06(5 - t)}$ on the interval $0 \le t \le 5$. Since Δt is very small when compared with the interval, the total amount in the account, A, is approximately

$$\int_0^5 1000\,e^{0.06(5 - t)}\,dt = \frac{1000}{-.06}\,e^{0.06(5 - t)}\Big|_0^5 = \frac{1000}{-.06}(1 - e^{0.3}) \approx 5831.$$

That is, the approximate balance in the account at the end of 5 years is $5831.

» Now Try Exercise 21

NOTE ▶   We computed the antiderivative by observing that since the derivative of $e^{0.06(5 - t)}$ is $e^{0.06(5 - t)}(-.06)$, we must divide the integrand by $-.06$ to obtain an antiderivative. **«**

In Example 5, money was deposited daily into the account. If the money had been deposited several times a day, the definite integral would have given an even better approximation to the balance. Actually, the more frequently the money is deposited, the better the approximation. Economists consider a hypothetical situation where money is deposited steadily throughout the year. This flow of money is called a **continuous income stream**, and the balance in the account is given exactly by the definite integral.

> **DEFINITION**   The **future value of a continuous income stream** of K dollars per year for N years at interest rate r compounded continuously is
>
> $$\int_0^N Ke^{r(N-t)}\, dt.$$

Solids of Revolution   When the region of Fig. 3(a) is revolved about the x-axis, it sweeps out a solid. [See Fig. 3(b).] Riemann sums can be used to derive a formula for the volume of this **solid of revolution**. Let us break the x-axis between a and b into a large number n of equal subintervals, each of length $\Delta x = (b - a)/n$. Using each subinterval as a base, we can divide the region into strips. (See Fig. 4.)

Let x_i be a point in the ith subinterval. Then, the volume swept out when we revolve the ith strip is approximately the same as the volume of the cylinder swept out when we revolve the rectangle of height $g(x_i)$ and base Δx around the x-axis. (See Fig. 5.) The volume of the cylinder is

$$[\text{Area of circular side}] \cdot [\text{width}] = \pi[g(x_i)]^2 \cdot \Delta x.$$

The total volume swept out by all the strips is approximated by the total volume swept out by the rectangles, which is

$$[\text{Volume}] \approx \pi[g(x_1)]^2 \Delta x + \pi[g(x_2)]^2 \Delta x + \cdots + \pi[g(x_n)]^2 \Delta x.$$

Figure 3

Figure 4

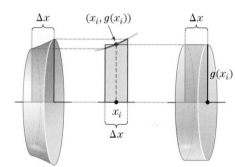

Figure 5

As n gets larger and larger, this approximation becomes arbitrarily close to the true volume. The expression on the right is a Riemann sum for the definite integral of $f(x) = \pi[g(x)]^2$. Therefore, the volume of the solid equals the value of the definite integral.

> The volume of the **solid of revolution** obtained from revolving the region below the graph of $y = g(x)$ from $x = a$ to $x = b$ about the x-axis is
>
> $$\int_a^b \pi[g(x)]^2\, dx.$$

EXAMPLE 6   **Volume of a Solid** Find the volume of the solid of revolution we obtain by revolving the region of Fig. 6 about the x-axis.

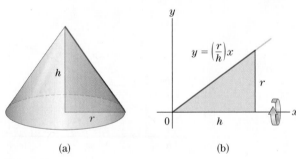

Figure 6

SOLUTION   Here, $g(x) = e^{kx}$, and

$$[\text{volume}] = \int_0^1 \pi(e^{kx})^2\, dx = \int_0^1 \pi e^{2kx}\, dx = \frac{\pi}{2k}e^{2kx}\bigg|_0^1 = \frac{\pi}{2k}(e^{2k} - 1).$$

»» Now Try Exercise 31

EXAMPLE 7   **Volume of a Cone**   Find the volume of a right circular cone of radius r and height h.

SOLUTION   The cone in Fig. 7(a) is the solid of revolution swept out when the shaded region in Fig. 7(b) is revolved about the x-axis. With the formula developed previously, the volume of the cone is

$$\int_0^h \pi\left(\frac{r}{h}x\right)^2 dx = \frac{\pi r^2}{h^2}\int_0^h x^2\, dx = \frac{\pi r^2}{h^2}\frac{x^3}{3}\bigg|_0^h = \frac{\pi r^2 h}{3}.$$

(a)

(b)

Figure 7

»» Now Try Exercise 35

Check Your Understanding 6.5

Solutions can be found following the section exercises.

1. A rock dropped from a bridge has a velocity of $-32t$ feet per second after t seconds. Find the average velocity of the rock during the first 3 seconds.

2. An investment yields a continuous income stream of $300 per year. This amount compounded continuously for 10 years at 10% interest. What is the (future) value of this income stream at the end of 10 years?

EXERCISES 6.5

Determine the average value of $f(x)$ over the interval from $x = a$ to $x = b$, where

1. $f(x) = x^2$; $a = 0$, $b = 3$.

2. $f(x) = 1 - x$; $a = -1$, $b = 1$.

3. $f(x) = 100e^{-0.5x}$; $a = 0$, $b = 4$.

4. $f(x) = 2$; $a = 0$, $b = 1$.

5. $f(x) = 1/x$; $a = 1/3$, $b = 3$.

6. $f(x) = \frac{1}{\sqrt{x}}$; $a = 1$, $b = 9$.

7. **Average Temperature** During a certain 12-hour period, the temperature at time t (measured in hours from the start of the period) was $T(t) = 47 + 4t - \frac{1}{3}t^2$ degrees. What was the average temperature during that period?

8. **Average Population** Assuming that a country's population is now 3 million and is growing exponentially with growth constant .02, what will be the average population during the next 50 years?

9. **Average Amount of Radium** One hundred grams of radioactive radium having a half-life of 1690 years is placed in a concrete vault. What will be the average amount of radium in the vault during the next 1000 years?

10. **Average Amount of Money** One hundred dollars is deposited in the bank at 5% interest compounded continuously. What will be the average value of the money in the account during the next 20 years?

Consumers' Surplus Find the consumers' surplus for each of the following demand curves at the given sales level, x.

11. $p = 3 - \dfrac{x}{10}; x = 20$

12. $p = \dfrac{x^2}{200} - x + 50; x = 20$

13. $p = \dfrac{500}{x + 10} - 3; x = 40$

14. $p = \sqrt{16 - .02x}; x = 350$

Producers' Surplus Figure 8 shows a supply curve for a commodity. It gives the relationship between the selling price of the commodity and the quantity that producers will manufacture. At a higher selling price, a greater quantity will be produced. Therefore, the curve is increasing. If (A, B) is a point on the curve, then, to stimulate the production of A units of the commodity, the price per unit must be B dollars. Of course, some producers will be willing to produce the commodity even with a lower selling price. Since everyone receives the same price in an open efficient economy, most producers are receiving more than their minimal required price. The excess is called the **producers' surplus**. Using an argument analogous to that of the *consumers' surplus*, we can show that the total producers' surplus when the price is B is the area of the shaded region in Fig. 8. Find the producers' surplus for each of the following supply curves at the given sales level, x.

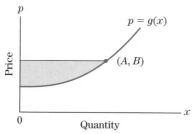

Figure 8 Producers' surplus.

15. $p = .01x + 3; x = 200$ 16. $p = \dfrac{x^2}{9} + 1; x = 3$

17. $p = \dfrac{x}{2} + 7; x = 10$ 18. $p = 1 + \frac{1}{2}\sqrt{x}; x = 36$

Consumers' and Producers' Surpluses For a particular commodity, the quantity produced and the unit price are given by the coordinates of the point where the supply and demand curves intersect. For each pair of supply and demand curves, determine the point of intersection (A, B) and the consumers' and producers' surplus. (See Fig. 9.)

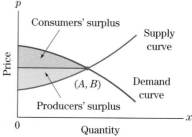

Figure 9

19. Demand curve: $p = 12 - (x/50)$; supply curve: $p = (x/20) + 5$.

20. Demand curve: $p = \sqrt{25 - .1x}$; supply curve: $p = \sqrt{.1x + 9} - 2$.

21. **Future Value** Suppose that money is deposited daily in a savings account at an annual rate of $1000. If the account pays 5% interest compounded continuously, estimate the balance in the account at the end of 3 years.

22. **Future Value** Suppose that money is deposited daily in a savings account at an annual rate of $2000. If the account pays 6% interest compounded continuously, approximately how much will be in the account at the end of 2 years?

23. **Future Value** Suppose that money is deposited steadily in a savings account so that $16,000 is deposited each year. Determine the balance at the end of 4 years if the account pays 8% interest compounded continuously.

24. **Future Value** Suppose that money is deposited steadily in a savings account so that $14,000 is deposited each year. Determine the balance at the end of 6 years if the account pays 4.5% interest compounded continuously.

25. **Future Value** An investment pays 10% interest compounded continuously. If money is invested steadily so that $5000 is deposited each year, how much time is required until the value of the investment reaches $140,000?

26. **Rate of a Savings Account** A savings account pays 4.25% interest compounded continuously. At what rate per year must money be deposited steadily in the account to accumulate a balance of $100,000 after 10 years?

Volume of Solids of Revolution Find the volume of the solid of revolution generated by revolving about the x-axis the region under each of the following curves.

27. $y = x + 1$ from $x = 0$ to $x = 2$.

28. $y = -x^2 + 1$ from $x = 0$ to $x = 1$.

29. $y = \sqrt{4 - x^2}$ from $x = -2$ to $x = 2$ (generates a sphere of radius 2)

30. $y = \sqrt{r^2 - x^2}$ from $x = -r$ to $x = r$ (generates a sphere of radius r)

31. $y = x^2$ from $x = 1$ to $x = 2$

32. $y = kx$ from $x = 0$ to $x = h$ (generates a cone)

33. $y = \sqrt{x}$ from $x = 0$ to $x = 4$ (The solid generated is called a *paraboloid*.)

34. $y = 2x - x^2$ from $x = 0$ to $x = 2$

35. $y = 2x + 1$ from $x = 0$ to $x = 1$ (The solid generated is called a *truncated cone*.)

36. $y = e^{-x}$ from $x = 0$ to $x = 1$

For the Riemann sums on an interval $[a, b]$, in Exercises 37–40, determine n, b, and $f(x)$.

37. $[(8.25)^3 + (8.75)^3 + (9.25)^3 + (9.75)^3](.5)$; $a = 8$

38. $\left[\dfrac{3}{1} + \dfrac{3}{1.5} + \dfrac{3}{2} + \dfrac{3}{2.5} + \dfrac{3}{3} + \dfrac{3}{3.5}\right](.5)$; $a = 1$

39. $[(5 + e^5) + (6 + e^6) + (7 + e^7)](1)$; $a = 4$

40. $[3(.3)^2 + 3(.9)^2 + 3(1.5)^2 + 3(2.1)^2 + 3(2.7)^2](.6)$; $a = 0$

41. Suppose that the interval $0 \le x \le 3$ is divided into 100 subintervals of width $\Delta x = .03$. Let $x_1, x_2, \ldots, x_{100}$ be points in these subintervals. Suppose that in a particular application we need to estimate the sum

$$(3 - x_1)^2 \Delta x + (3 - x_2)^2 \Delta x + \cdots + (3 - x_{100})^2 \Delta x.$$

Show that this sum is close to 9.

42. Suppose that the interval $0 \le x \le 1$ is divided into 100 subintervals of width $\Delta x = .01$. Show that the following sum is close to 5/4.

$$[2(.01) + (.01)^3]\Delta x + [2(.02) + (.02)^3]\Delta x$$
$$+ \cdots + [2(1.0) + (1.0)^3]\Delta x.$$

TECHNOLOGY EXERCISES

The following exercises ask for an unknown quantity x. After setting up the appropriate formula involving a definite integral, use the fundamental theorem to evaluate the definite integral as an expression in x. Because the resulting equation will be too complicated to solve algebraically, you must use a graphing utility to obtain the solution. (*Note:* If the quantity x is an interest rate paid by a savings account, it will most likely be between 0 and .10.)

43. A single deposit of $1000 is to be made into a savings account, and the interest (compounded continuously) is allowed to accumulate for 3 years. Therefore, the amount at the end of t years is $1000e^{rt}$.
   (a) Find an expression (involving r) that gives the average value of the money in the account during the 3-year time period $0 \le t \le 3$.
   (b) Find the interest rate r at which the average amount in the account during the 3-year period is $1070.60.

44. A single deposit of $100 is made into a savings account paying 4% interest compounded continuously. How long must the money be held in the account so that the average amount of money during that time period will be $122.96?

45. Money is deposited steadily so that $1000 is deposited each year into a savings account.
   (a) Find the expression (involving r) that gives the (future) balance in the account at the end of 6 years.
   (b) Find the interest rate that will result in a balance of $6997.18 after 6 years.

46. Money is deposited steadily so that $3000 is deposited each year into a savings account. After 10 years the balance is $36,887. What interest rate, with interest compounded continuously, did the money earn?

Solutions to Check Your Understanding 6.5

1. By definition, the average value of the function $v(t) = -32t$ for $t = 0$ to $t = 3$ is

$$\frac{1}{3-0}\int_0^3 -32t\, dt = \frac{1}{3}(-16t^2)\Big|_0^3 = \frac{1}{3}(-16 \cdot 3^2)$$
$$= -48 \text{ feet per second.}$$

Note: There is another way to approach this problem:

$$[\text{Average velocity}] = \frac{[\text{distance traveled}]}{[\text{time elapsed}]}.$$

As we discussed in Section 6.4, distance traveled equals the area under the velocity curve. Therefore,

$$[\text{Average velocity}] = \frac{\int_0^3 -32t\, dt}{3}.$$

2. According to the formula developed in the text, the future value of the income stream after 10 years is equal to

$$\int_0^{10} 300e^{0.1(10-t)}\, dt = -\frac{300}{.1}e^{0.1(10-t)}\Big|_0^{10}$$
$$= -3000e^0 - (-3000e^1)$$
$$= 3000e - 3000$$
$$\approx \$5154.85.$$

CHAPTER 6 Summary

KEY TERMS AND CONCEPTS	EXAMPLES
6.1 Antidifferentiation	
Given a function $f(x)$, an *antiderivative* of $f(x)$ is a function $F(x)$ such that $F'(x) = f(x)$.	Find all antiderivatives of $f(x) = \dfrac{3}{5\sqrt{x}} - \dfrac{8}{x^4}$.
We represent the family of antiderivatives by $F(x) + C$, where $F(x)$ is any antiderivative of $f(x)$ and C is a constant.	The antiderivatives are given by the indefinite integral $$\int \left[\frac{3}{5\sqrt{x}} - \frac{8}{x^4}\right] dx.$$

KEY TERMS AND CONCEPTS	EXAMPLES

The family of antiderivatives of $f(x)$ is also denoted by the indefinite integral symbol

$$\int f(x)dx.$$

$$\int x^r dx = \frac{1}{r+1}x^{r+1} + C, r \neq -1.$$

$$\int \frac{1}{x}dx = \ln|x| + C, x \neq 0$$

$$\int e^{kx}\,dx = \frac{1}{k}e^{kx} + C, k \neq 0$$

$$\int (f(x) + g(x))dx = \int f(x)dx + \int g(x)dx$$

$$\int kf(x)dx = k\int f(x)dx.$$

In evaluating the integral, we use properties from Section 6.1. Let's start by writing the integral using powers of x:

$$\int \left[\frac{3}{5}\frac{1}{\sqrt{x}} - \frac{8}{x^4} \right] dx = \int \left[\frac{3}{5}x^{-1/2} - 8x^{-4} \right] dx$$

$$= \int \left[\frac{3}{5}x^{-1/2} \right] dx + \int \left[(-8)x^{-4} \right] dx$$

$$\text{(by (5), Sec. 6.1)}$$

$$= \frac{3}{5}\int x^{-1/2}\,dx - 8\int x^{-4}\,dx$$

$$\text{(by (6), Sec. 6.1)}$$

$$= \frac{3}{5}\left[\frac{1}{-\frac{1}{2}+1}x^{-\frac{1}{2}+1} \right] - 8\left[\frac{1}{-4+1}x^{-4+1} \right] + C$$

$$\text{(by (2), Sec. 6.1)}$$

$$= \frac{3}{5}\left[\frac{1}{\frac{1}{2}}x^{\frac{1}{2}} \right] - 8\left[\frac{1}{-3}x^{-3} \right] + C$$

$$= \frac{6}{5}\sqrt{x} + \frac{8}{3x^3} + C.$$

6.2 The Definite Integral and Net Change of a Function

Suppose that $f(x)$ is a continuous function on $a \leq x \leq b$. The *definite integral* of $f(x)$ from a to b is the number

$$\int_a^b f(x)dx = F(b) - F(a), \qquad (1)$$

where F is any antiderivative of f. The number $F(b) - F(a)$ is abbreviated by $F(x)\big|_a^b$.

$$\int_a^b (f(x) + g(x))dx = \int_a^b f(x)dx + \int_a^b g(x)dx$$

$$\int_a^b kf(x)dx = k\int_a^b f(x)dx$$

$$\int_a^b f(x)dx = -\int_b^a f(x)dx$$

$$\int_a^a f(x)dx = 0$$

$$\int_a^c f(x)dx + \int_c^b f(x)dx = \int_a^b f(x)dx.$$

1. Evaluate the definite integral $\int_{\frac{1}{2}}^1 (7x + \frac{1}{x})dx$.

$$\int_{\frac{1}{2}}^1 \left(7x + \frac{1}{x} \right) dx = \left[\frac{7}{2}x^2 + \ln|x| \right]\Big|_{\frac{1}{2}}^1 \quad \text{(By (2), (4), Sec. 6.2)}$$

$$= \left[\frac{7}{2}1^2 + \ln 1 \right] - \left[\frac{7}{2}\left(\frac{1}{2}\right)^2 + \ln\left(\frac{1}{2}\right) \right]$$

$$= \left[\frac{7}{2} + 0 \right] - \left[\frac{7}{2}\frac{1}{4} - \ln 2 \right]$$

$$\text{(Because } \ln 1 = 0 \text{ and } \ln(\tfrac{1}{2}) = -\ln 2)$$

$$= \frac{7}{2} - \frac{7}{8} + \ln 2 = \frac{21}{8} + \ln 2 \approx 3.32$$

2. A company's marginal revenue function is given by $R'(x) = 4.03e^{0.05x} + 2$, where x is the number of items produced in 1 day and $R'(x)$ is measured in thousands of dollars per item. Determine the net increase in revenue if the company decides to increase production from $x = 3$ to $x = 6$ items per day.

The net change in revenue is

$$R(6) - R(3) = \int_3^6 R'(x)dx.$$

We have

$$\int_3^6 R'(x)dx = \int_3^6 (4.03e^{0.05x} + 2)dx$$

$$= \left(\frac{4.03}{.05}e^{0.05x} + 2x \right)\Big|_3^6$$

$$= \left(\frac{4.03}{.05}e^{0.05(6)} + 2(6) \right) - \left(\frac{4.03}{.05}e^{0.05(3)} + 2(3) \right) \approx 21.155.$$

Thus, the company's revenue will increase by \$21,155 if production is increased from 3 to 6 items per day.

KEY TERMS AND CONCEPTS	EXAMPLES

6.3 The Definite Integral and Area under a Graph

Suppose that $f(x)$ is a continuous nonnegative function on $a \leq x \leq b$. The area of the region in Fig. 1 is given by

$$A = \int_a^b f(x)dx$$

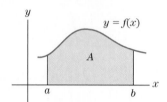

Figure 1

To approximate the area of the region under the graph of f, we can use Riemann sums.

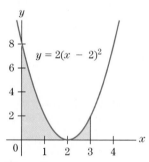

Figure 2

1. Find the area under the graph of the parabola $f(x) = 2(x - 2)^2$, above the x-axis, from $x = 0$ to $x = 3$ (Fig. 2).

   Since $f(x) \geq 0$ on the interval $0 \leq x \leq 3$, the area is equal to $\int_0^3 f(x)dx$. To evaluate the integral, expand the integrand first:

   $$\int_0^3 2(x - 2)^2 dx = 2\int_0^3 (x^2 - 4x + 4)dx$$

   $$= 2\left(\frac{1}{3}x^3 - 2x^2 + 4x\right)\Big|_0^3$$

   $$= 2\left(\frac{1}{3}3^3 - 2(3^2) + 4(3)\right) - 0$$

   $$= 2(9 - 18 + 12) = 6.$$

   Thus, the area is equal to 6.

2. Use a Riemann sum with $n = 4$ and midpoints to estimate the integral $\int_0^{1.2} \frac{dx}{1 + x}$.

   The integral is the area under the graph of $f(x) = \frac{1}{1 + x}$, above the x-axis, from 0 to 1.2. We approximate the area, and thus the integral, using Riemann sums.

   Step 1 Divide the interval $[0, 1.2]$ into $n = 4$ subintervals. The length of each subinterval is $\Delta x = \frac{1.2 - 0}{4} = .3$.

   Step 2 Select the midpoints of the subintervals:

   $$x_1 = \frac{.3}{2} = .15,$$

   $$x_2 = x_1 + .3 = .15 + .3 = .45,$$

   $$x_3 = x_2 + .3 = .45 + .3 = .75,$$

   $$x_4 = x_3 + .3 = .75 + .3 = 1.05.$$

   Step 3 Evaluate $f(x)$ at the midpoints

   $$f(x_1) = f(.15) = \frac{1}{1 + .15} \approx .870$$

   $$f(x_2) = f(.45) = \frac{1}{1 + .45} \approx .690$$

   $$f(x_3) = f(.75) = \frac{1}{1 + .75} \approx .571$$

   $$f(x_4) = f(1.05) = \frac{1}{1 + 1.05} \approx .488$$

   These are the heights of the rectangles in the Riemann sum.

KEY TERMS AND CONCEPTS	EXAMPLES

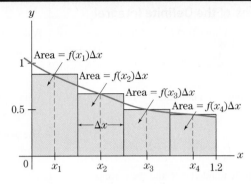

Step 4 The area of each rectangle in the Riemann sum is

$$[\text{height} \times \text{width}] = f(x_i)\Delta x.$$

The Riemann sum is the sum of the areas of all four rectangles:

$$f(x_1)\Delta x + f(x_2)\Delta x + f(x_3)\Delta x + f(x_4)\Delta x$$
$$= [f(x_1) + f(x_2) + f(x_3) + f(x_4)]\Delta x$$
$$\approx [.870 + .690 + .571 + .488](.3) = .7857.$$

6.4 Areas in the *xy*-Plane

Given an arbitrary continuous function $f(x)$ for $a \leq x \leq b$, the area bounded by the graph of f and the *x*-axis from a to b is given by

$$A = \int_a^b |f(x)|\, dx.$$

When f is nonnegative, the absolute value of $f(x)$ is equal to $f(x)$; and when $f(x)$ is negative, the absolute value of $f(x)$ is equal to $-f(x)$. Thus, in computing the area, we must distinguish the intervals on which $f(x) \geq 0$ and those on which $f(x) \leq 0$.

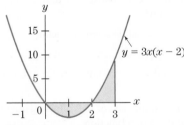

Figure 1

Suppose that $f(x) \geq g(x)$ for all $a \leq x \leq b$; equivalently, the graph of $f(x)$ is above the graph of $g(x)$ on the interval $a \leq x \leq b$. Then, the area of the region between the two graphs is given by

$$A = \int_a^b [f(x) - g(x)]dx.$$

$y = x^2 + x + 9$

$y = 3x^2 - 2x$

Figure 2

1. Find the area A bounded by the graph of $f(x) = 3x(x - 2)$, and the *x*-axis, from $x = 0$ to $x = 3$. (See Fig. 1.)

As shown in the figure, $f(x) \leq 0$ if $0 \leq x \leq 2$ and $f(x) \geq 0$ if $2 \leq x \leq 3$. Thus, in computing the area A, we must evaluate two separate integrals:

$$A = -\int_0^2 f(x)dx + \int_2^3 f(x)dx.$$

One antiderivative of $f(x) = 3x^2 - 6x$ is $x^3 - 3x^2$. So,

$$A = -(x^3 - 3x^2)\Big|_0^2 + (x^3 - 3x^2)\Big|_2^3$$
$$= -(2^3 - 3(2^2)) - 0 + (3^3 - 3(3^2)) - (2^3 - 3(2^2)) = 8.$$

2. Set up the integral that gives the area bounded by the curve $y = 3x^2 - 2x$ and $y = x^2 + x + 9$. (See Fig. 2.)

Make a rough sketch of the two curves. The curves intersect where $3x^2 - 2x = x^2 + x + 9$ or $2x^2 - 3x - 9 = 0$.

Note that $2x^2 - 3x - 9 = (x - 3)(2x + 3)$.

So, the solutions to $2x^2 - 3x - 9 = 0$ are $x = -3/2$ and $x = 3$, which are the first coordinates of the points of intersection. From the figure, we observe that the graph of $y = x^2 + x + 9$ is above the graph of $y = 3x^2 - 2x$ over the interval $-3/2 \leq x \leq 3$. Thus, the area between the two curves is

$$A = \int_{-3/2}^3 [\text{upper function} - \text{lower function}]dx$$

$$= \int_{-3/2}^3 [(x^2 + x + 9) - (3x^2 - 2x)]dx$$

$$= \int_{-3/2}^3 [-2x^2 + 3x + 9]dx.$$

(The value of this integral is $\frac{243}{8} = 30.375$.)

KEY TERMS AND CONCEPTS	EXAMPLES

6.5 Applications of the Definite Integral

The *average* of a continuous function $f(x)$ on $a \leq x \leq b$ is

$$\frac{1}{b-a}\int_a^b f(x)dx.$$

The *consumers' surplus* for a commodity having demand curve $p = f(x)$ is

$$\int_0^A [f(x) - B]dx,$$

where the quantity demanded is A and the price is $B = f(A)$.

The *future value of a continuous income stream* of K dollars per year for N years at interest rate r compounded continuously is

$$\int_0^N Ke^{r(N-t)}dt.$$

The volume of the *solid of revolution* obtained from revolving the region below the graph of $y = g(x)$ from $x = a$ to $x = b$ about the x-axis is

$$\int_a^b \pi[g(x)]^2\,dx.$$

A demand curve for a certain commodity is given by

$$p = \frac{x^3}{200} - 8x + 150.$$

What is the average price p if the number of items varies from $x = 10$ to $x = 20$?

The average is given by

$$\frac{1}{20-10}\int_{10}^{20}\left[\frac{x^3}{200} - 8x + 150\right]dx$$

$$= \frac{1}{10}\left[\frac{x^4}{(4)(200)} - \frac{8}{2}x^2 + 150x\right]\Big|_{10}^{20}$$

$$= \frac{1}{10}\left[\frac{x^4}{800} - 4x^2 + 150x\right]\Big|_{10}^{20}$$

$$= \frac{1}{10}\left[\left[\frac{20^4}{800} - (4)(20)^2 + (150)(20)\right]\right.$$

$$\left. -\left[\frac{10^4}{800} - (4)(10)^2 + (150)(10)\right]\right] = 48.75.$$

So, if the number of items varies between 10 and 20, the average price will be \$48.75.

CHAPTER 6   Fundamental Concept Check Exercises

1. What does it mean to antidifferentiate a function?

2. State the formula for $\int h(x)dx$ for each of the following functions.

   (a) $h(x)x^r, r \neq -1$ (b) $h(x) = e^{kx}$

   (c) $h(x) = \dfrac{1}{x}$ (d) $h(x) = f(x) + g(x)$

   (e) $h(x) = kf(x)$

3. In the formula $\Delta x = \dfrac{b-a}{n}$, what do a, b, n, and Δx denote?

4. What is a Riemann sum?

5. Give an interpretation of the area under a rate of change function. Give a concrete example.

6. What is a definite integral?

7. What is the difference between a definite integral and an indefinite integral?

8. State the fundamental theorem of calculus.

9. How is $F(x)\big|_a^b$ calculated, and what is it called?

10. Outline a procedure for finding the area of a region bounded by two curves.

11. State the formula for each of the following quantities:
   (a) average value of a function
   (b) consumers' surplus
   (c) future value of an income stream
   (d) volume of a solid of revolution

CHAPTER 6   Review Exercises

Calculate the following integrals.

1. $\displaystyle\int 3^2\,dx$

2. $\displaystyle\int (x^2 - 3x + 2)dx$

3. $\displaystyle\int \sqrt{x+1}\,dx$

4. $\displaystyle\int \frac{2}{x+4}\,dx$

5. $2\displaystyle\int (x^3 + 3x^2 - 1)dx$

6. $\displaystyle\int \sqrt[5]{x+3}\,dx$

7. $\displaystyle\int e^{-x/2}\,dx$

8. $\displaystyle\int \frac{5}{\sqrt{x-7}}\,dx$

9. $\displaystyle\int (3x^4 - 4x^3)dx$

10. $\displaystyle\int (2x+3)^7\,dx$

11. $\displaystyle\int \sqrt{4-x}\,dx$

12. $\displaystyle\int \left(\frac{5}{x} - \frac{x}{5}\right)dx$

13. $\int_{-1}^{1} (x + 1)^2 \, dx$

14. $\int_{0}^{1/8} \sqrt[5]{x} \, dx$

15. $\int_{-1}^{2} \sqrt{2x + 4} \, dx$

16. $2\int_{0}^{1} \left(\frac{2}{x + 1} - \frac{1}{x + 4} \right) dx$

17. $\int_{1}^{2} \frac{4}{x^5} \, dx$

18. $\frac{2}{3}\int_{0}^{8} \sqrt{x + 1} \, dx$

19. $\int_{1}^{4} \frac{1}{x^2} \, dx$

20. $\int_{3}^{6} e^{2-(x/3)} \, dx$

21. $\int_{0}^{5} (5 + 3x)^{-1} \, dx$

22. $\int_{-2}^{2} \frac{3}{2e^{3x}} \, dx$

23. $\int_{0}^{\ln 2} (e^x - e^{-x}) \, dx$

24. $\int_{\ln 2}^{\ln 3} (e^x + e^{-x}) \, dx$

25. $\int_{0}^{\ln 3} \frac{e^x + e^{-x}}{e^{2x}} \, dx$

26. $\int_{0}^{1} \frac{3 + e^{2x}}{e^x} \, dx$

27. Find the area under the curve $y = (3x - 2)^{-3}$ from $x = 1$ to $x = 2$.

28. Find the area under the curve $y = 1 + \sqrt{x}$ from $x = 1$ to $x = 9$.

In Exercises 29–36, find the area of the shaded region.

29.

30.

31.

32.

33.

34.

35.

36.

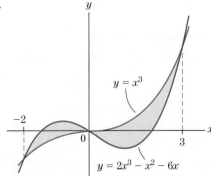

37. Find the area of the region bounded by the curves $y = x^3 - 3x + 1$ and $y = x + 1$.

38. Find the area of the region between the curves $y = 2x^2 + x$ and $y = x^2 + 2$ from $x = 0$ to $x = 2$.

39. Find the function $f(x)$ for which $f'(x) = (x - 5)^2$, $f(8) = 2$.

40. Find the function $f(x)$ for which $f'(x) = e^{-5x}$, $f(0) = 1$.

41. Describe all solutions of the following differential equations, where y represents a function of t.
   (a) $y' = 4t$
   (b) $y' = 4y$
   (c) $y' = e^{4t}$

42. Let k be a constant, and let $y = f(t)$ be a function such that $y' = kty$. Show that $y = Ce^{kt^2/2}$, for some constant C. [*Hint:* Use the product rule to evaluate $\dfrac{d}{dt}[f(t)e^{-kt^2/2}]$, and then apply Theorem II of Section 6.1.]

43. An airplane tire plant finds that its marginal cost of producing tires is $.04x + \$150$ at a production level of x tires per day. If fixed costs are $\$500$ per day, find the cost of producing x tires per day.

44. If the marginal revenue function for a company is $400 - 3x^2$, find the additional revenue received from doubling production if 10 units are currently being produced.

45. A drug is injected into a patient at the rate of $f(t)$ cubic centimeters per minute at time t. What does the area under the graph of $y = f(t)$ from $t = 0$ to $t = 4$ represent?

46. A rock thrown straight up into the air has a velocity of $v(t) = -9.8t + 20$ meters per second after t seconds.
   (a) Determine the distance the rock travels during the first 2 seconds.
   (b) Represent the answer to part (a) as an area.

47. Use a Riemann sum with $n = 4$ and left endpoints to estimate the area under the graph in Fig. 1 for $0 \le x \le 2$.

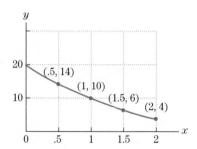

Figure 1

48. Redo Exercise 47 using right endpoints.

49. Use a Riemann sum with $n = 2$ and midpoints to estimate the area under the graph of

$$f(x) = \frac{1}{x + 2}$$

on the interval $0 \le x \le 2$. Then, use a definite integral to find the exact value of the area to five decimal places.

50. Use a Riemann sum with $n = 5$ and midpoints to estimate the area under the graph of $f(x) = e^{2x}$ on the interval $0 \le x \le 1$. Then, use a definite integral to find the exact value of the area to five decimal places.

51. Find the consumers' surplus for the demand curve $p = \sqrt{25 - .04x}$ at the sales level $x = 400$.

52. Three thousand dollars is deposited in the bank at 4% interest compounded continuously. What will be the average value of the money in the account during the next 10 years?

53. Find the average value of $f(x) = 1/x^3$ from $x = \frac{1}{3}$ to $x = \frac{1}{2}$.

54. Suppose that the interval $0 \le x \le 1$ is divided into 100 subintervals with a width of $\Delta x = .01$. Show that the sum

$$[3e^{-0.01}]\, \Delta x + [3e^{-0.02}]\, \Delta x + [3e^{-0.03}]\, \Delta x + \cdots + [3e^{-1}]\, \Delta x$$

is close to $3(1 - e^{-1})$.

55. In Fig. 2, three regions are labeled with their areas. Determine $\int_a^c f(x)dx$ and determine $\int_a^d f(x)dx$.

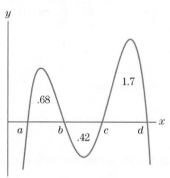

Figure 2

56. Find the volume of the solid of revolution generated by revolving about the x-axis the region under the curve $y = 1 - x^2$ from $x = 0$ to $x = 1$.

57. A store has an inventory of Q units of a certain product at time $t = 0$. The store sells the product at the steady rate of Q/A units per week and exhausts the inventory in A weeks.
   (a) Find a formula $f(t)$ for the amount of product in inventory at time t.
   (b) Find the average inventory level during the period $0 \le t \le A$.

58. A retail store sells a certain product at the rate of $g(t)$ units per week at time t, where $g(t) = rt$. At time $t = 0$, the store has Q units of the product in inventory.
   (a) Find a formula $f(t)$ for the amount of product in inventory at time t.
   (b) Determine the value of r in part (a) such that the inventory is exhausted in A weeks.
   (c) Using $f(t)$, with r as in part (b), find the average inventory level during the period $0 \le t \le A$.

59. Let x be any positive number, and define $g(x)$ to be the number determined by the definite integral

$$g(x) = \int_0^x \frac{1}{1 + t^2}\, dt.$$

   (a) Give a geometric interpretation of the number $g(3)$.
   (b) Find the derivative $g'(x)$.

60. For each number x satisfying $-1 \le x \le 1$, define $h(x)$ by

$$h(x) = \int_{-1}^x \sqrt{1 - t^2}\, dt.$$

(a) Give a geometric interpretation of the values $h(0)$ and $h(1)$.

(b) Find the derivative $h'(x)$.

61. Suppose that the interval $0 \le t \le 3$ is divided into 1000 sub-intervals of width Δt. Let $t_1, t_2, \ldots, t_{1000}$ denote the right endpoints of these subintervals. If we need to estimate the sum

$$5000e^{-0.1t_1}\Delta t + 5000e^{-0.1t_2}\Delta t + \cdots + 5000e^{-0.1t_{1000}}\Delta t,$$

show that this sum is close to 13,000. [*Note:* A sum such as this would arise if we wanted to compute the present value of a continuous stream of income of $5000 per year for 3 years, with interest compounded continuously at 10%.]

62. What number does

$$[e^0 + e^{1/n} + e^{2/n} + e^{3/n} + \cdots + e^{(n-1)/n}] \cdot \frac{1}{n}$$

approach as n gets very large?

63. What number does the sum

$$\left[1^3 + \left(1 + \frac{1}{n}\right)^3 + \left(1 + \frac{2}{n}\right)^3 + \left(1 + \frac{3}{n}\right)^3 \right.$$
$$\left. + \cdots + \left(1 + \frac{n-1}{n}\right)^3\right] \cdot \frac{1}{n}$$

approach as n gets very large?

64. In Fig. 3, the rectangle has the same area as the region under the graph of $f(x)$. What is the average value of $f(x)$ on the interval $2 \le x \le 6$?

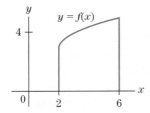

Figure 3

65. *True or false*: If $3 \le f(x) \le 4$ whenever $0 \le x \le 5$, then $3 \le \frac{1}{5}\int_0^5 f(x)dx \le 4$.

66. Suppose that water is flowing into a tank at a rate of $r(t)$ gallons per hour, where the rate depends on the time t according to the formula

$$r(t) = 20 - 4t, \qquad 0 \le t \le 5.$$

(a) Consider a brief period of time, say, from t_1 to t_2. The length of this time period is $\Delta t = t_2 - t_1$. During this period the rate of flow does not change much and is approximately $20 - 4t_1$ (the rate at the beginning of the brief time interval). Approximately how much water flows into the tank during the time from t_1 to t_2 ?

(b) Explain why the total amount of water added to the tank during the time interval from $t = 0$ to $t = 5$ is given by $\int_0^5 r(t)dt$.

67. The annual world rate of water use t years after 1960, for $t \le 35$, was approximately $860e^{0.04t}$ cubic kilometers per year. How much water was used between 1960 and 1995?

68. If money is deposited steadily in a savings account at the rate of $4500 per year, determine the balance at the end of 1 year if the account pays 9% interest compounded continuously.

69. Find a function $f(x)$ whose graph goes through the point $(1, 1)$ and whose slope at any point $(x, f(x))$ is $3x^2 - 2x + 1$.

70. For what value of a is the shaded area in Fig. 4 equal to 1?

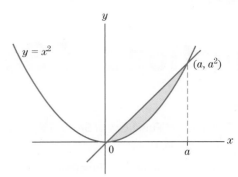

Figure 4

71. Show that for any positive number b we have

$$\int_0^{b^2} \sqrt{x}\, dx + \int_0^b x^2\, dx = b^3.$$

72. Generalize the result of Exercise 71 as follows: Let n be a positive integer. Show that for any positive number b we have

$$\int_0^{b^n} \sqrt[n]{x}\, dx + \int_0^b x^n\, dx = b^{n+1}.$$

73. Show that

$$\int_0^1 (\sqrt{x} - x^2)dx = \frac{1}{3}.$$

74. Generalize the result of Exercise 73 as follows: Let n be a positive integer. Show that

$$\int_0^1 (\sqrt[n]{x} - x^n)\, dx = \frac{n-1}{n+1}.$$

Functions of Several Variables

Until now, most of our applications of calculus have involved functions of one variable. In real life, however, a quantity of interest often depends on more than one variable. For instance, the sales level of a product may depend not only on its price, but also on the prices of competing products, the amount spent on advertising, and perhaps the time of year. The total cost of manufacturing the product depends on the cost of raw materials, labor, plant maintenance, and so on.

This chapter introduces the basic ideas of calculus for functions of more than one variable. Section 7.1 presents two examples that will be used throughout the chapter. Derivatives are treated in Section 7.2 and then used in Sections 7.3 and 7.4 to solve optimization problems more general than those in Chapter 2. The final two sections are devoted to least-squares problems and a brief introduction to the integration of functions of two variables.

7.1 Examples of Functions of Several Variables

A function $f(x, y)$ of the two variables x and y is a rule that assigns a number to each pair of values for the variables; for instance,

$$f(x, y) = e^x(x^2 + 2y).$$

An example of a function of three variables is

$$f(x, y, z) = 5xy^2z.$$

| EXAMPLE 1 | **A Function with Two Variables** A store sells butter at $4.50 per pound and margarine at $3.40 per pound. The revenue from the sale of x pounds of butter and y pounds of margarine is given by the function |

$$f(x, y) = 4.50x + 3.40y.$$

Determine and interpret $f(200, 300)$.

SOLUTION $f(200, 300) = 4.50(200) + 3.40(300) = 900 + 1020 = 1920$. The revenue from the sale of 200 pounds of butter and 300 pounds of margarine is $1920.   **>> Now Try Exercise 1**

A function $f(x, y)$ of two variables may be graphed in a manner analogous to that for functions of one variable. It is necessary to use a three-dimensional coordinate system, where each point is identified by three coordinates (x, y, z). For each choice of x, y, the graph of $f(x, y)$ includes the point $(x, y, f(x, y))$. This graph is usually a surface in three-dimensional space, with equation $z = f(x, y)$. (See Fig. 1.) Three graphs of specific functions are shown in Fig. 2.

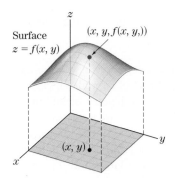

Figure 1 Graph of $f(x, y)$.

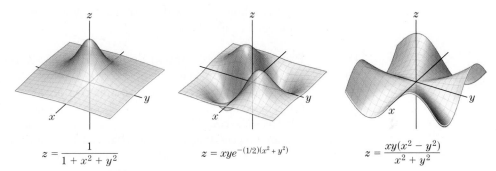

$$z = \frac{1}{1 + x^2 + y^2} \qquad z = xye^{-(1/2)(x^2 + y^2)} \qquad z = \frac{xy(x^2 - y^2)}{x^2 + y^2}$$

Figure 2

Application to Architectural Design

When designing a building, we would like to know, at least approximately, how much heat the building loses per day. The heat loss affects many aspects of the design, such as the size of the heating plant and the size and location of duct work. A building loses heat through its sides, roof, and floor. How much heat is lost will generally differ for each face of the building and will depend on such factors as insulation, materials used in construction, exposure (north, south, east, or west), and climate. It is possible to estimate how much heat is lost per square foot of each face. Using these data, we can construct a heat-loss function as in the following example.

| EXAMPLE 2 | **Heat-Loss Function** A rectangular industrial building of dimensions x, y, and z is shown in Fig. 3. In Table 1, we give the amount of heat lost per day by each side of the building, measured in suitable units of heat per square foot. Let $f(x, y, z)$ be the total daily heat loss for such a building. |

(a) Find a formula for $f(x, y, z)$.
(b) Find the total daily heat loss if the building has length 100 feet, width 70 feet, and height 50 feet.

Figure 3 Heat loss from an industrial building.

Table 1

	Roof	East side	West side	North side	South side	Floor
Daily heat loss (units per sq ft)	10	8	6	10	5	1
Area (sq ft)	xy	yz	yz	xz	xz	xy

SOLUTION   (a) The total heat loss is the sum of the amount of heat loss through each face of the building. The heat loss through the roof is

[heat loss per square foot of roof] · [area of roof in square feet] $= 10xy$.

Similarly, the heat loss through the east side is $8yz$. Continuing in this way, we see that the total daily heat loss is

$$f(x, y, z) = 10xy + 8yz + 6yz + 10xz + 5xz + 1 \cdot xy.$$

We collect terms to obtain

$$f(x, y, z) = 11xy + 14yz + 15xz.$$

(b) The amount of heat loss when $x = 100$, $y = 70$, and $z = 50$ is given by $f(100, 70, 50)$, which equals

$$f(100, 70, 50) = 11(100)(70) + 14(70)(50) + 15(100)(50)$$
$$= 77{,}000 + 49{,}000 + 75{,}000 = 201{,}000 \text{ units.}$$

>> *Now Try Exercise 7*

In Section 7.3, we will determine the dimensions x, y, z that minimize the heat loss for a building of specific volume.

Production Functions in Economics   The costs of a manufacturing process can generally be classified as one of two types: cost of labor and cost of capital. The meaning of the cost of labor is clear. By the cost of capital, we mean the cost of buildings, tools, machines, and similar items used in the production process. A manufacturer usually has some control over the relative portions of labor and capital utilized in its production process. It can completely automate production so that labor is at a minimum or utilize mostly labor and little capital. Suppose that x units of labor and y units of capital are used. (Economists normally use L and K, respectively, for labor and capital. However, for simplicity, we use x and y.) Let $f(x, y)$ denote the number of units of finished product that are manufactured. Economists have found that $f(x, y)$ is often a function of the form

$$f(x, y) = Cx^A y^{1-A},$$

where A and C are constants, $0 < A < 1$. Such a function is called a **Cobb–Douglas production function.**

EXAMPLE 3   **Production in a Firm** Suppose that, during a certain time period, the number of units of goods produced when x units of labor and y units of capital are used is $f(x, y) = 60x^{3/4}y^{1/4}$.

(a) How many units of goods will be produced by 81 units of labor and 16 units of capital?

(b) Show that, whenever the amounts of labor and capital being used are doubled, so is the production. (Economists say that the production function has "constant returns to scale.")

SOLUTION   (a) $f(81, 16) = 60(81)^{3/4} \cdot (16)^{1/4} = 60 \cdot 27 \cdot 2 = 3240$. There will be 3240 units of goods produced.

(b) Utilization of a units of labor and b units of capital results in the production of $f(a, b) = 60a^{3/4}b^{1/4}$ units of goods. Utilizing $2a$ and $2b$ units of labor and capital, respectively, results in $f(2a, 2b)$ units produced.

$$
\begin{aligned}
f(2a, 2b) &= 60(2a)^{3/4}(2b)^{1/4} & &\text{Set } x = 2a \text{ and } y = 2b.\\
&= 60 \cdot 2^{3/4} \cdot a^{3/4} \cdot 2^{1/4} \cdot b^{1/4} & &\text{Recall } (ab)^c = a^c b^c.\\
&= 60 \cdot 2^{(3/4+1/4)} \cdot a^{3/4}b^{1/4} & &\text{To multiply two powers in the}\\
&= 2^1 \cdot 60a^{3/4}b^{1/4} & &\text{same base, add the exponents.}\\
&= 2f(a, b).
\end{aligned}
$$

>> *Now Try Exercise 9*

Level Curves

It is possible graphically to depict a function $f(x, y)$ of two variables using a family of curves called **level curves**. Let c be any number. Then, the graph of the equation $f(x, y) = c$ is a curve in the xy-plane called the **level curve of height** c. This curve describes all points of height c on the graph of the function $f(x, y)$. As c varies, we have a family of level curves indicating the sets of points on which $f(x, y)$ assumes various values c. In Fig. 4, we have drawn the graph and various level curves for the function $f(x, y) = x^2 + y^2$.

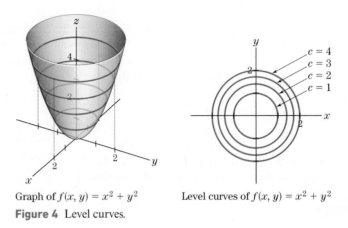

Graph of $f(x, y) = x^2 + y^2$ Level curves of $f(x, y) = x^2 + y^2$

Figure 4 Level curves.

Level curves often have interesting physical interpretations. For example, surveyors draw *topographic maps* that use level curves to represent points having equal altitude. Here $f(x, y) = $ the altitude at point (x, y). Figure 5(a) shows the graph of $f(x, y)$ for a typical hilly region. Figure 5(b) shows the level curves corresponding to various altitudes. Note that when the level curves are closer together the surface is steeper.

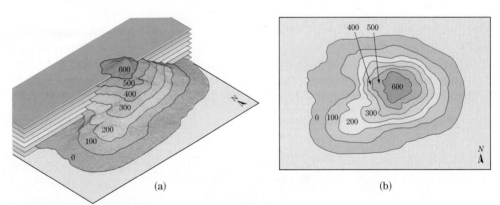

(a)   (b)

Figure 5 Topographic level curves show altitudes.

EXAMPLE 4   **Level Curves** Determine the level curve at height 600 for the production function $f(x, y) = 60x^{3/4}y^{1/4}$ of Example 3.

SOLUTION   The level curve is the graph of $f(x, y) = 600$, or

$$60x^{3/4}y^{1/4} = 600$$

$$y^{1/4} = \frac{10}{x^{3/4}}$$

$$y = \frac{10{,}000}{x^3}.$$

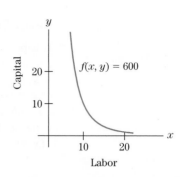

Figure 6 A level curve of a production function.

Of course, since x and y represent quantities of labor and capital, they must both be positive. We have sketched the graph of the level curve in Fig. 6. The points on the curve are precisely those combinations of capital and labor that yield 600 units of production. Economists call this curve an *isoquant*.

≫ Now Try Exercise 15

Check Your Understanding 7.1

Solutions can be found following the section exercises.

1. Let $f(x, y, z) = x^2 + y/(x - z) - 4$. Compute $f(3, 5, 2)$.

2. In a certain country, the daily demand for coffee is given by $f(p_1, p_2) = 16p_1/p_2$ thousand pounds, where p_1 and p_2 are the respective prices of tea and coffee in dollars per pound. Compute and interpret $f(3, 4)$.

EXERCISES 7.1

1. Let $f(x, y) = x^2 - 3xy - y^2$. Compute $f(5, 0)$, $f(5, -2)$, and $f(a, b)$.

2. Let $g(x, y) = \sqrt{x^2 + 2y^2}$. Compute $g(1, 1)$, $g(0, -1)$, and $g(a, b)$.

3. Let $g(x, y, z) = x/(y - z)$. Compute $g(2, 3, 4)$ and $g(7, 46, 44)$.

4. Let $f(x, y, z) = x^2 e^{\sqrt{y^2 + z^2}}$. Compute $f(1, -1, 1)$ and $f(2, 3, -4)$.

5. Let $f(x, y) = xy$. Show that $f(2 + h, 3) - f(2, 3) = 3h$.

6. Let $f(x, y) = xy$. Show that $f(2, 3 + k) - f(2, 3) = 2k$.

7. **Cost** Find a formula $C(x, y, z)$ that gives the cost of materials for the closed rectangular box in Fig. 7(a), with dimensions in feet. Assume that the material for the top and bottom costs $3 per square foot and the material for the sides costs $5 per square foot.

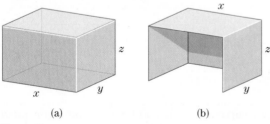

(a)   (b)

Figure 7

8. **Cost** Find a formula $C(x, y, z)$ that gives the cost of material for the rectangular enclosure in Fig. 7(b), with dimensions in feet. Assume that the material for the top costs $3 per square foot and the material for the back and two sides costs $5 per square foot.

9. Consider the Cobb–Douglas production function $f(x, y) = 20x^{1/3}y^{2/3}$. Compute $f(8, 1)$, $f(1, 27)$, and $f(8, 27)$. Show that, for any positive constant k, $f(8k, 27k) = kf(8, 27)$.

10. Let $f(x, y) = 10x^{2/5}y^{3/5}$. Show that $f(3a, 3b) = 3f(a, b)$.

11. **Present Value** The present value of A dollars to be paid t years in the future (assuming a 5% continuous interest rate) is $P(A, t) = Ae^{-0.05t}$. Find and interpret $P(100, 13.8)$.

12. Refer to Example 3. If labor costs $100 per unit and capital costs $200 per unit, express as a function of two variables, $C(x, y)$, the cost of utilizing x units of labor and y units of capital.

13. **Tax and Homeowner Exemption** The value of residential property for tax purposes is usually much lower than its actual market value. If v is the market value, the *assessed value* for real estate taxes might be only 40% of v. Suppose that the property tax, T, in a community is given by the function

$$T = f(r, v, x) = \frac{r}{100}(.40v - x),$$

where v is the market value of a property (in dollars), x is a *homeowner's exemption* (a number of dollars depending on the type of property), and r is the tax rate (stated in dollars per hundred dollars).

(a) Determine the real estate tax on a property valued at $200,000 with a homeowner's exemption of $5000, assuming a tax rate of $2.50 per hundred dollars of net assessed value.

(b) Determine the tax due if the tax rate increases by 20% to $3.00 per hundred dollars of net assessed value. Assume the same property value and homeowner's exemption. Does the tax due also increase by 20%?

14. **Tax and Homeowner Exemption** Let $f(r, v, x)$ be the real estate tax function of Exercise 13.

(a) Determine the real estate tax on a property valued at $100,000 with a homeowner's exemption of $5000, assuming a tax rate of $2.20 per hundred dollars of net assessed value.

(b) Determine the real estate tax when the market value rises 20% to $120,000. Assume the same homeowner's exemption and a tax rate of $2.20 per hundred dollars of net assessed value. Does the tax due also increase by 20%?

Draw the level curves of heights 0, 1, and 2 for the functions in Exercises 15 and 16.

15. $f(x, y) = 2x - y$ **16.** $f(x, y) = -x^2 + 2y$

17. Draw the level curve of the function $f(x, y) = x - y$ containing the point $(0, 0)$.

18. Draw the level curve of the function $f(x, y) = xy$ containing the point $\left(\frac{1}{2}, 4\right)$.

19. Find a function $f(x, y)$ that has the line $y = 3x - 4$ as a level curve.

20. Find a function $f(x, y)$ that has the curve $y = 2/x^2$ as a level curve.

21. Suppose that a topographic map is viewed as the graph of a certain function $f(x, y)$. What are the level curves?

22. Isocost Lines A certain production process uses units of labor and capital. If the quantities of these commodities are x and y, respectively, the total cost is $100x + 200y$ dollars. Draw the level curves of height 600, 800, and 1000 for this function. Explain the significance of these curves. (Economists frequently refer to these lines as *budget lines* or *isocost lines*.)

Match the graphs of the functions in Exercises 23–26 to the systems of level curves shown in Figs. 8(a)–(d).

23.

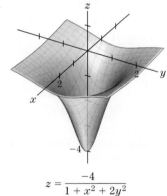

$$z = \frac{-4}{1 + x^2 + 2y^2}$$

24.

$$z = x^2 - x^4 - y^2$$

25.

$$z = \frac{15x^2y^2e^{-x^2 - y^2}}{x^2 + y^2}$$

26.

$$z = e^{-x^2} + e^{-4y^2}$$

(a)

(b)

(c)

(d)

Figure 8

Solutions to Check Your Understanding 7.1

1. Substitute 3 for x, 5 for y, and 2 for z:

$$f(3, 5, 2) = 3^2 + \frac{5}{3-2} - 4 = 10.$$

2. To compute $f(3, 4)$, substitute 3 for p_1 and 4 for p_2 into $f(p_1, p_2) = 16p_1/p_2$. Thus,

$$f(3, 4) = 16 \cdot \tfrac{3}{4} = 12.$$

Therefore, if the price of tea is \$3 per pound and the price of coffee is \$4 per pound, 12,000 pounds of coffee will be sold each day. (Notice that as the price of coffee increases the demand decreases.)

7.2 Partial Derivatives

In Chapter 1, we introduced the notion of a derivative to measure the rate at which a function $f(x)$ is changing with respect to changes in the variable x. Let us now study the analog of the derivative for functions of two (or more) variables.

Let $f(x, y)$ be a function of the two variables x and y. Since we want to know how $f(x, y)$ changes with respect to the changes in both the variable x and the variable y, we shall define two derivatives of $f(x, y)$ (to be called partial derivatives), one with respect to each variable.

» The ∂ is a rounded d called the partial derivative symbol.

> **DEFINITION** The **partial derivative of $f(x, y)$ with respect to x**, written $\dfrac{\partial f}{\partial x}$, is the derivative of $f(x, y)$, where y is treated as a constant and $f(x, y)$ is considered a function of x alone. The **partial derivative of $f(x, y)$ with respect to y**, written $\dfrac{\partial f}{\partial y}$, is the derivative of $f(x, y)$, where x is treated as a constant.

EXAMPLE 1

Computing Partial Derivatives Let $f(x, y) = 5x^3y^2$. Compute (a) $\dfrac{\partial f}{\partial x}$ and (b) $\dfrac{\partial f}{\partial y}$.

SOLUTION   (a) To compute $\dfrac{\partial f}{\partial x}$, we think of $f(x, y)$ written as

— Treat as constant.

$$f(x, y) = [5y^2]\, x^3,$$

where the brackets emphasize that $5y^2$ is to be treated as a constant. Therefore, when differentiating with respect to x, $f(x, y)$ is just a constant times x^3. Recall that if k is any constant, then

$$\frac{d}{dx}(kx^3) = 3 \cdot k \cdot x^2.$$

Thus,

— Treat as constant.

$$\frac{\partial f}{\partial x} = 3 \cdot [5y^2] \cdot x^2 = 15x^2y^2.$$

After some practice, it is unnecessary to place the y^2 in front of the x^3 before differentiating.

(b) Now, to compute $\dfrac{\partial f}{\partial y}$, we think of

— Treat as constant.

$$f(x, y) = [5x^3]y^2.$$

When we are differentiating with respect to y, $f(x, y)$ is simply a constant (that is, $5x^3$) times y^2. Hence,

$$\frac{\partial f}{\partial y} = 2 \cdot [5x^3] \cdot y = 10x^3y.$$

FOR REVIEW

See Sec. 1.7, Example 1, to review computing derivatives with constants.

» **Now Try Exercise 1**

EXAMPLE 2

Computing Partial Derivatives Let $f(x, y) = 3x^2 + 2xy + 5y$. Compute (a) $\dfrac{\partial f}{\partial x}$ and (b) $\dfrac{\partial f}{\partial y}$.

SOLUTION   (a) To compute $\dfrac{\partial f}{\partial x}$, we think of

Treat as constant.

$$f(x, y) = 3x^2 + [2y]x + [5y].$$

Now, we differentiate $f(x, y)$ as if it were a quadratic polynomial in x:

$$\frac{\partial f}{\partial x} = 6x + [2y] + 0 = 6x + 2y.$$

Note that we treat $5y$ as a constant when differentiating with respect to x, so the partial derivative of $5y$ with respect to x is zero.

(b) To compute $\dfrac{\partial f}{\partial y}$, we think of

Treat as constant.

$$f(x, y) = [3x^2] + [2x]y + 5y.$$

Then,

$$\frac{\partial f}{\partial y} = 0 + [2x] + 5 = 2x + 5.$$

Note that we treat $3x^2$ as a constant when differentiating with respect to y, so the partial derivative of $3x^2$ with respect to y is zero.   **≫ Now Try Exercise 3**

EXAMPLE 3

Differentiation Rules and Partial Derivatives Compute $\dfrac{\partial f}{\partial x}$ and $\dfrac{\partial f}{\partial y}$ for each of the following.

(a) $f(x, y) = (4x + 3y - 5)^8$ (b) $f(x, y) = e^{xy^2}$ (c) $f(x, y) = y/(x + 3y)$

SOLUTION   (a) To compute $\dfrac{\partial f}{\partial x}$, we think of

Treat as constant.

$$f(x, y) = (4x + [3y - 5])^8.$$

By the general power rule,

Treat as constant.

$$\frac{\partial f}{\partial x} = 8 \cdot (4x + [3y - 5])^7 \cdot \frac{d}{dx}(4x + [3y - 5])$$

$$= 8 \cdot (4x + [3y - 5])^7 \cdot (4 + 0)$$

$$= 32(4x + [3y - 5])^7.$$

To compute $\dfrac{\partial f}{\partial y}$, we think of

Treat as constant.

$$f(x, y) = ([4x] + 3y - 5)^8.$$

Then, by the general power rule,

$$\frac{\partial f}{\partial y} = 8 \cdot ([4x] + 3y - 5)^7 \cdot 3 = 24(4x + 3y - 5)^7.$$

(b) To compute $\dfrac{\partial f}{\partial x}$, we observe that

Treat as constant.

$$f(x, y) = e^{x[y^2]},$$

so that

$$\frac{\partial f}{\partial x} = [y^2]\, e^{x[y^2]} = y^2 e^{xy^2}. \qquad\qquad \frac{d}{dx}(e^{ax}) = ae^{ax}$$

To compute $\frac{\partial f}{\partial y}$, we think of

— Treat as constant.

$$f(x, y) = e^{[x]y^2}.$$

Thus,

$$\frac{\partial f}{\partial y} = e^{[x]y^2} \cdot 2[x]y = 2xye^{xy^2}. \qquad\qquad \frac{d}{dy}[e^{f(y)}] = e^{f(y)}f'(y)$$

(c) To compute $\frac{\partial f}{\partial x}$, we use the general power rule to differentiate $[y](x + [3y])^{-1}$ with respect to x:

$$\frac{\partial f}{\partial x} = (-1)\cdot[y](x + [3y])^{-2}\cdot 1 = -\frac{y}{(x + 3y)^2}.$$

To compute $\frac{\partial f}{\partial y}$, we use the quotient rule to differentiate

$$f(x, y) = \frac{y}{[x] + 3y}$$

with respect to y. We find that

— Treat as constant.

$$\frac{\partial f}{\partial y} = \frac{([x] + 3y)\cdot 1 - y\cdot 3}{([x] + 3y)^2} = \frac{x}{(x + 3y)^2}. \qquad \text{\textbf{\textgg} \textit{Now Try Exercise 9}}$$

The use of brackets to highlight constants is helpful initially when we compute partial derivatives. From now on, we shall merely form a mental picture of those terms to be treated as constants and dispense with brackets.

It is useful to note the following differentiation rule that is illustrated by our solution of part (b) of Example 3:

If $$f(x, y) = e^{g(x, y)}$$

then

$$\frac{\partial f}{\partial x} = e^{g(x, y)}\frac{\partial g}{\partial x} \quad \text{and} \quad \frac{\partial f}{\partial y} = e^{g(x, y)}\frac{\partial g}{\partial y}. \qquad (1)$$

A partial derivative of a function of several variables is also a function of several variables and hence can be evaluated at specific values of the variables. We write

$$\frac{\partial f}{\partial x}(a, b)$$

for $\frac{\partial f}{\partial x}$ evaluated at $x = a$, $y = b$. Similarly,

$$\frac{\partial f}{\partial y}(a, b)$$

denotes the function $\frac{\partial f}{\partial y}$ evaluated at $x = a$, $y = b$.

EXAMPLE 4 **Evaluating Partial Derivatives** Let $f(x, y) = 3x^2 + 2xy + 5y$.

(a) Calculate $\frac{\partial f}{\partial x}(1, 4)$. **(b)** Evaluate $\frac{\partial f}{\partial y}$ at $(x, y) = (1, 4)$.

SOLUTION   (a) $\dfrac{\partial f}{\partial x} = 6x + 2y,$ $\dfrac{\partial f}{\partial x}(1, 4) = 6 \cdot 1 + 2 \cdot 4 = 14$

(b) $\dfrac{\partial f}{\partial y} = 2x + 5,$ $\dfrac{\partial f}{\partial y}(1, 4) = 2 \cdot 1 + 5 = 7$

》 Now Try Exercise 19

Geometric Interpretation of Partial Derivatives

Consider the three-dimensional surface $z = f(x, y)$ in Fig. 1. If y is held constant at b and x is allowed to vary, the equation

$$z = f(x, b)$$

constant

describes a curve on the surface. [The curve is formed by cutting the surface $z = f(x, y)$ with a vertical plane parallel to the xz-plane.] The value of $\dfrac{\partial f}{\partial x}(a, b)$ is the slope of the tangent line to the curve at the point where $x = a$ and $y = b$.

Likewise, if x is held constant at a and y is allowed to vary, the equation

$$z = f(a, y)$$

constant

describes the curve on the surface $z = f(x, y)$ shown in Fig. 2. The value of the partial derivative $\dfrac{\partial f}{\partial y}(a, b)$ is the slope of this curve at the point where $x = a$ and $y = b$.

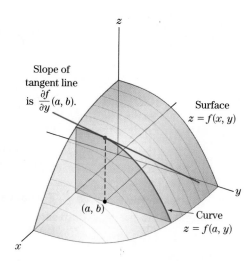

Figure 1 $\dfrac{\partial f}{\partial x}$ gives the slope of a curve formed by holding y constant at b.

Figure 2 $\dfrac{\partial f}{\partial y}$ gives the slope of a curve formed by holding x constant at a.

Partial Derivatives and Rates of Change

Since $\dfrac{\partial f}{\partial x}$ is simply the ordinary derivative with y held constant, $\dfrac{\partial f}{\partial x}$ gives the rate of change of $f(x, y)$ with respect to x for y held constant. In other words, keeping y constant and increasing x by one unit produces a change in $f(x, y)$ that is approximately given by $\dfrac{\partial f}{\partial x}$. An analogous interpretation holds for $\dfrac{\partial f}{\partial y}$.

EXAMPLE 5   **Interpreting Partial Derivatives** Interpret the partial derivatives of $f(x, y) = 3x^2 + 2xy + 5y$ calculated in Example 4.

SOLUTION   We showed in Example 4 that

$$\frac{\partial f}{\partial x}(1, 4) = 14, \qquad \frac{\partial f}{\partial y}(1, 4) = 7.$$

The fact that

$$\frac{\partial f}{\partial x}(1, 4) = 14$$

means that if y is kept constant at 4 and x is allowed to vary near 1, then $f(x, y)$ changes at a rate 14 times the change in x. That is, if x increases by one small unit, $f(x, y)$ increases by approximately 14 units. If x increases by h units (where h is small), $f(x, y)$ increases by approximately $14 \cdot h$ units. That is,

$$f(1 + h, 4) - f(1, 4) \approx 14 \cdot h.$$

Similarly, the fact that

$$\frac{\partial f}{\partial y}(1, 4) = 7$$

means that, if we keep x constant at 1 and let y vary near 4, then $f(x, y)$ changes at a rate equal to seven times the change in y. So, for a small value of k, we have

$$f(1, 4 + k) - f(1, 4) \approx 7 \cdot k. \qquad \text{**» Now Try Exercise 21**}$$

We can generalize the interpretations of $\dfrac{\partial f}{\partial x}$ and $\dfrac{\partial f}{\partial y}$ given in Example 5 to yield the following general fact:

Let $f(x, y)$ be a function of two variables. Then, if h and k are small, we have

$$f(a + h, b) - f(a, b) \approx \frac{\partial f}{\partial x}(a, b) \cdot h,$$

$$f(a, b + k) - f(a, b) \approx \frac{\partial f}{\partial y}(a, b) \cdot k.$$

Partial derivatives can be computed for functions of any number of variables. When taking the partial derivative with respect to one variable, we treat the other variables as constant.

EXAMPLE 6   **Partial Derivatives**   Let $f(x, y, z) = x^2yz - 3z$.

(a) Compute $\dfrac{\partial f}{\partial x}, \dfrac{\partial f}{\partial y},$ and $\dfrac{\partial f}{\partial z}$.   (b) Calculate $\dfrac{\partial f}{\partial z}(2, 3, 1)$.

SOLUTION   (a) $\dfrac{\partial f}{\partial x} = 2xyz, \dfrac{\partial f}{\partial y} = x^2z, \dfrac{\partial f}{\partial z} = x^2y - 3$

(b) $\dfrac{\partial f}{\partial z}(2, 3, 1) = 2^2 \cdot 3 - 3 = 12 - 3 = 9$

» Now Try Exercise 15

EXAMPLE 7   **Heat-Loss Function**   Let $f(x, y, z)$ be the heat-loss function computed in Example 2 of Section 7.1. That is, $f(x, y, z) = 11xy + 14yz + 15xz$. Calculate and interpret $\dfrac{\partial f}{\partial x}(10, 7, 5)$.

SOLUTION   We have

$$\frac{\partial f}{\partial x} = 11y + 15z$$

$$\frac{\partial f}{\partial x}(10, 7, 5) = 11 \cdot 7 + 15 \cdot 5 = 77 + 75 = 152.$$

The quantity $\dfrac{\partial f}{\partial x}$ is commonly referred to as the **marginal heat loss with respect to change in x**. Specifically, if x is changed from 10 by h units (where h is small) and the values of y and z remain fixed at 7 and 5, the amount of heat loss will change by approximately $152 \cdot h$ units.

>> *Now Try Exercise 31*

EXAMPLE 8 **Marginal Productivity of Capital** Consider the production function $f(x, y) = 60x^{3/4}y^{1/4}$, which gives the number of units of goods produced when x units of labor and y units of capital are used.

(a) Find $\dfrac{\partial f}{\partial x}$ and $\dfrac{\partial f}{\partial y}$.

(b) Evaluate $\dfrac{\partial f}{\partial x}$ and $\dfrac{\partial f}{\partial y}$ at $x = 81$, $y = 16$.

(c) Interpret the numbers computed in part (b).

SOLUTION

(a) $\dfrac{\partial f}{\partial x} = 60 \cdot \dfrac{3}{4}x^{-1/4}y^{1/4} = 45x^{-1/4}y^{1/4} = 45\dfrac{y^{1/4}}{x^{1/4}}$

$\dfrac{\partial f}{\partial y} = 60 \cdot \dfrac{1}{4}x^{3/4}y^{-3/4} = 15x^{3/4}y^{-3/4} = 15\dfrac{x^{3/4}}{y^{3/4}}$

(b) $\dfrac{\partial f}{\partial x}(81, 16) = 45 \cdot \dfrac{16^{1/4}}{81^{1/4}} = 45 \cdot \dfrac{2}{3} = 30$

$\dfrac{\partial f}{\partial y}(81, 16) = 15 \cdot \dfrac{81^{3/4}}{16^{3/4}} = 15 \cdot \dfrac{27}{8} = \dfrac{405}{8} = 50\dfrac{5}{8}$

(c) The quantities $\dfrac{\partial f}{\partial x}$ and $\dfrac{\partial f}{\partial y}$ are referred to as the *marginal productivity of labor* and the *marginal productivity of capital*. If the amount of capital is held fixed at $y = 16$ and the amount of labor increases by 1 unit from 81, the quantity of goods produced will increase by approximately 30 units. Similarly, an increase in capital of 1 unit (with labor fixed at 81) results in an increase in production of approximately $50\frac{5}{8}$ units of goods.

>> *Now Try Exercise 25*

Partial Derivatives of Higher Order

Just as we formed second derivatives and derivatives of higher order in the case of one variable, we can form second partial derivatives and derivatives of higher order of a function $f(x, y)$ of two variables. Since $\dfrac{\partial f}{\partial x}$ is a function of x and y, we can differentiate it with respect to x or y. The partial derivative of $\dfrac{\partial f}{\partial x}$ with respect to x is denoted by $\dfrac{\partial^2 f}{\partial x^2}$. The partial derivative of $\dfrac{\partial f}{\partial x}$ with respect to y is denoted by $\dfrac{\partial^2 f}{\partial y\, \partial x}$. Similarly, the partial derivative of the function $\dfrac{\partial f}{\partial y}$ with respect to x is denoted by $\dfrac{\partial^2 f}{\partial x\, \partial y}$, and the partial derivative of $\dfrac{\partial f}{\partial y}$ with respect to y is denoted by $\dfrac{\partial^2 f}{\partial y^2}$. Almost all functions $f(x, y)$ encountered in applications [and all functions $f(x, y)$ in this text] have the property that

$$\frac{\partial^2 f}{\partial y\, \partial x} = \frac{\partial^2 f}{\partial x\, \partial y}. \tag{2}$$

When computing $\dfrac{\partial^2 f}{\partial y\, \partial x}$ and $\dfrac{\partial^2 f}{\partial x\, \partial y}$, note that verifying the last equation is a check that you have done the differentiation correctly.

EXAMPLE 9

Partial Derivatives of Higher Order Let $f(x, y) = e^{x^2y}$. Find

(a) $\dfrac{\partial f}{\partial x}$ (b) $\dfrac{\partial f}{\partial y}$ (c) $\dfrac{\partial^2 f}{\partial y \, \partial x}$; and (d) $\dfrac{\partial^2 f}{\partial x \, \partial y}$

SOLUTION

(a) $\dfrac{\partial f}{\partial x} = \dfrac{\partial}{\partial x}(e^{x^2y})$

$= e^{x^2y}\dfrac{\partial}{\partial x}(x^2y)$ Use (1).

$= 2xye^{x^2y}$ Treat y as a constant.

(b) $\dfrac{\partial f}{\partial y} = \dfrac{\partial}{\partial y}(e^{x^2y})$

$= e^{x^2y}\dfrac{\partial}{\partial y}(x^2y)$ Use (1).

$= x^2e^{x^2y}$ Treat x^2 as a constant.

(c) $\dfrac{\partial^2 f}{\partial y \, \partial x} = \dfrac{\partial}{\partial y}\left(\dfrac{\partial f}{\partial x}\right)$

$= \dfrac{\partial}{\partial y}(2xye^{x^2y})$ From (a).

$= 2xy \cdot \dfrac{\partial}{\partial y}(e^{x^2y}) + e^{x^2y} \cdot \dfrac{\partial}{\partial y}(2xy)$ Product rule.

$= 2xy \cdot x^2e^{x^2y} + e^{x^2y} \cdot 2x$ Treat x as a constant.

$= 2x^3ye^{x^2y} + 2xe^{x^2y} = 2xe^{x^2y}(x^2y + 1)$ Simplify, then factor.

(d) $\dfrac{\partial^2 f}{\partial x \, \partial y} = \dfrac{\partial}{\partial x}\left(\dfrac{\partial f}{\partial y}\right)$

$= \dfrac{\partial}{\partial x}(x^2e^{x^2y})$ From (b).

$= x^2 \cdot \dfrac{\partial}{\partial x}(e^{x^2y}) + e^{x^2y} \cdot \dfrac{\partial}{\partial x}(x^2)$ Product rule.

$= x^2 \cdot 2xye^{x^2y} + e^{x^2y} \cdot 2x$ Treat y as a constant.

$= 2x^3ye^{x^2y} + 2xe^{x^2y} = 2xe^{x^2y}(x^2y + 1)$ Simplify, then factor.

Comparing (c) and (d), we see that the mixed partial derivatives are equal, as asserted by (2). **≫ Now Try Exercise 23**

INCORPORATING TECHNOLOGY

Evaluating Partial Derivatives The function from Example 4 and its first partial derivatives are specified in Fig. 3(a) and evaluated in Fig. 3(b). Recall that the expression $1 \to X$ is entered with $\boxed{1}$ $\boxed{\text{STO}\triangleright}$ $\boxed{\text{X,T, }\theta, n}$ and indicates that we are setting $X = 1$. The expression $4 \to Y$ has a similar meaning, but the variable Y is entered by means of $\boxed{\text{ALPHA}}$ $\boxed{\text{Y}}$. We can also evaluate other partial derivatives. For example, we can find the partial derivative $\dfrac{\partial^2 f}{\partial x \, \partial y}$ in this case by setting $Y_4 = \mathbf{nDeriv}(Y_3, X, X)$.

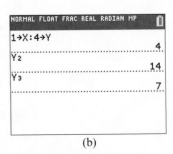

Figure 3   (a)   (b)

Check Your Understanding 7.2

Solutions can be found following the section exercises.

1. The number of TV sets an appliance store sells per week is given by a function of two variables, $f(x, y)$, where x is the price per TV set and y is the amount of money spent weekly on advertising. Suppose that the current price is $400 per set and that currently $2000 per week is being spent for advertising.

   (a) Would you expect $\dfrac{\partial f}{\partial x}(400, 2000)$ to be positive or negative? Explain.

   (b) Would you expect $\dfrac{\partial f}{\partial y}(400, 2000)$ to be positive or negative?

2. The monthly mortgage payment for a house is a function of two variables, $f(A, r)$, where A is the amount of the mortgage and the interest rate is $r\%$. For a 30-year mortgage, $f(92,000, 3.5) = 412.78$ and $\dfrac{\partial f}{\partial r}(92,000, 3.5) = 38.47$. What is the significance of the number 38.47?

EXERCISES 7.2

Find $\dfrac{\partial f}{\partial x}$ and $\dfrac{\partial f}{\partial y}$ for each of the following functions.

1. $f(x, y) = 5xy$

2. $f(x, y) = x^2 - y^2$

3. $f(x, y) = 2x^2 e^y$

4. $f(x, y) = xe^{xy}$

5. $f(x, y) = \dfrac{x}{y} + \dfrac{y}{x}$

6. $f(x, y) = \dfrac{1}{x + y}$

7. $f(x, y) = (2x - y + 5)^2$

8. $f(x, y) = \dfrac{e^x}{1 + e^y}$

9. $f(x, y) = xe^{x^2 y^2}$

10. $f(x, y) = \ln(xy)$

11. $f(x, y) = \dfrac{x - y}{x + y}$

12. $f(x, y) = \sqrt{x^2 + y^2}$

13. Let $f(L, K) = 3\sqrt{LK}$. Find $\dfrac{\partial f}{\partial L}$.

14. Let $f(p, q) = 1 - p(1 + q)$. Find $\dfrac{\partial f}{\partial q}$ and $\dfrac{\partial f}{\partial p}$.

15. Let $f(x, y, z) = (1 + x^2 y)/z$. Find $\dfrac{\partial f}{\partial x}, \dfrac{\partial f}{\partial y}$, and $\dfrac{\partial f}{\partial z}$.

16. Let $f(x, y, z) = ze^{x/y}$. Find $\dfrac{\partial f}{\partial x}, \dfrac{\partial f}{\partial y}$, and $\dfrac{\partial f}{\partial z}$.

17. Let $f(x, y, z) = xze^{yz}$. Find $\dfrac{\partial f}{\partial x}, \dfrac{\partial f}{\partial y}$, and $\dfrac{\partial f}{\partial z}$.

18. Let $f(x, y, z) = \dfrac{xy}{z}$. Find $\dfrac{\partial f}{\partial x}, \dfrac{\partial f}{\partial y}$, and $\dfrac{\partial f}{\partial z}$.

19. Let $f(x, y) = x^2 + 2xy + y^2 + 3x + 5y$. Find $\dfrac{\partial f}{\partial x}(2, -3)$ and $\dfrac{\partial f}{\partial y}(2, -3)$.

20. Let $f(x, y) = (x + y^2)^3$. Evaluate $\dfrac{\partial f}{\partial x}$ and $\dfrac{\partial f}{\partial y}$ at $(x, y) = (1, 2)$.

21. Let $f(x, y) = xy^2 + 5$. Evaluate $\dfrac{\partial f}{\partial y}$ at $(x, y) = (2, -1)$ and interpret your result.

22. Let $f(x, y) = \dfrac{x}{y - 6}$. Compute $\dfrac{\partial f}{\partial y}(2, 1)$ and interpret your result.

23. Let $f(x, y) = x^3 y + 2xy^2$. Find $\dfrac{\partial^2 f}{\partial x^2}, \dfrac{\partial^2 f}{\partial y^2}, \dfrac{\partial^2 f}{\partial x\,\partial y}$, and $\dfrac{\partial^2 f}{\partial y\,\partial x}$.

24. Let $f(x, y) = xe^y + x^4 y + y^3$. Find $\dfrac{\partial^2 f}{\partial x^2}, \dfrac{\partial^2 f}{\partial y^2}, \dfrac{\partial^2 f}{\partial x\,\partial y}$, and $\dfrac{\partial^2 f}{\partial y\,\partial x}$.

25. **Production** A farmer can produce $f(x, y) = 200\sqrt{6x^2 + y^2}$ units of produce by utilizing x units of labor and y units of capital. (The capital is used to rent or purchase land, materials, and equipment.)

    (a) Calculate the marginal productivities of labor and capital when $x = 10$ and $y = 5$.

    (b) Let h be a small number. Use the result of part (a) to determine the approximate effect on production of changing labor from 10 to $10 + h$ units while keeping capital fixed at 5 units.

    (c) Use part (b) to estimate the change in production when labor decreases from 10 to 9.5 units and capital stays fixed at 5 units.

26. **Productivity Labor and Capital** The productivity of a country is given by $f(x, y) = 300x^{2/3}y^{1/3}$, where x and y are the amount of labor and capital.

    (a) Compute the marginal productivities of labor and capital when $x = 125$ and $y = 64$.

    (b) Use part (a) to determine the approximate effect on productivity of increasing capital from 64 to 66 units, while keeping labor fixed at 125 units.

    (c) What would be the approximate effect of decreasing labor from 125 to 124 units while keeping capital fixed at 64 units?

27. **Modes of Transportation** In a certain suburban community, commuters have the choice of getting into the city by bus or train. The demand for these modes of transportation varies with their cost. Let $f(p_1, p_2)$ be the number of people who will take the bus when p_1 is the price of the bus ride and p_2 is the price of the train ride. For example, if $f(4.50, 6) = 7000$, then 7000 commuters will take the bus when the price of a bus ticket is $4.50 and the price of a train ticket is $6.00. Explain why $\dfrac{\partial f}{\partial p_1} < 0$ and $\dfrac{\partial f}{\partial p_2} > 0$.

28. Refer to Exercise 27. Let $g(p_1, p_2)$ be the number of people who will take the train when p_1 is the price of the bus ride and p_2 is the price of the train ride. Would you expect $\dfrac{\partial g}{\partial p_1}$ to be positive or negative? How about $\dfrac{\partial g}{\partial p_2}$?

29. Let p_1 be the average price of MP3 players, p_2 the average price of audio files, $f(p_1, p_2)$ the demand for MP3 players, and $g(p_1, p_2)$ the demand for audio files. Explain why $\dfrac{\partial f}{\partial p_2} < 0$ and $\dfrac{\partial g}{\partial p_1} < 0$.

30. The demand for a certain gas-guzzling car is given by $f(p_1, p_2)$, where p_1 is the price of the car and p_2 is the price of gasoline. Explain why $\dfrac{\partial f}{\partial p_1} < 0$ and $\dfrac{\partial f}{\partial p_2} < 0$.

31. The volume (V) of a certain amount of a gas is determined by the temperature (T) and the pressure (P) by the formula $V = .08(T/P)$. Calculate and interpret $\dfrac{\partial V}{\partial P}$ and $\dfrac{\partial V}{\partial T}$ when $P = 20$, $T = 300$.

32. Beer Consumption Using data collected from 1929 to 1941, Richard Stone determined that the yearly quantity Q of beer consumed in the United Kingdom was approximately given by the formula $Q = f(m, p, r, s)$, where

$$f(m, p, r, s) = (1.058)m^{0.136}p^{-0.727}r^{0.914}s^{0.816}$$

and m is the aggregate real income (personal income after direct taxes, adjusted for retail price changes), p is the average retail price of the commodity (in this case, beer), r is the average retail price level of all other consumer goods and services, and s is a measure of the strength of the beer. Determine which partial derivatives are positive and which are negative, and give interpretations. (For example, since $\dfrac{\partial f}{\partial r} > 0$, people buy more beer when the prices of other goods increase and the other factors remain constant.) (*Source: Journal of the Royal Statistical Society.*)

33. Richard Stone (see Exercise 32) determined that the yearly consumption of food in the United States was given by

$$f(m, p, r) = (2.186)m^{0.595}p^{-0.543}r^{0.922}.$$

Determine which partial derivatives are positive and which are negative, and give interpretations of these facts.

34. Distribution of Revenue For the production function $f(x, y) = 60x^{3/4}y^{1/4}$ considered in Example 8, think of $f(x, y)$ as the revenue when x units of labor and y units of capital are used. Under actual operating conditions, say, $x = a$ and $y = b$, $\dfrac{\partial f}{\partial x}(a, b)$ is referred to as the *wage per unit of labor* and $\dfrac{\partial f}{\partial y}(a, b)$ is referred to as the *wage per unit of capital*. Show that

$$f(a, b) = a \cdot \left[\frac{\partial f}{\partial x}(a, b)\right] + b \cdot \left[\frac{\partial f}{\partial y}(a, b)\right].$$

(This equation shows how the revenue is distributed between labor and capital.)

35. Compute $\dfrac{\partial^2 f}{\partial x^2}$, where $f(x, y) = 60x^{3/4}y^{1/4}$, a production function (where x is units of labor). Explain why $\dfrac{\partial^2 f}{\partial x^2}$ is always negative.

36. Compute $\dfrac{\partial^2 f}{\partial y^2}$, where $f(x, y) = 60x^{3/4}y^{1/4}$, a production function (where y is units of capital). Explain why $\dfrac{\partial^2 f}{\partial y^2}$ is always negative.

37. Let $f(x, y) = 3x^2 + 2xy + 5y$, as in Example 5. Show that

$$f(1 + h, 4) - f(1, 4) = 14h + 3h^2.$$

Thus, the error in approximating $f(1 + h, 4) - f(1, 4)$ by $14h$ is $3h^2$. (If $h = .01$, for instance, the error is only $.0003$.)

38. Body Surface Area Physicians, particularly pediatricians, sometimes need to know the body surface area of a patient. For instance, they use surface area to adjust the results of certain tests of kidney performance. Tables are available that give the approximate body surface area A in square meters of a person who weighs W kilograms and is H centimeters tall. The following empirical formula is also used:

$$A = .007W^{0.425}H^{0.725}.$$

Evaluate $\dfrac{\partial A}{\partial W}$ and $\dfrac{\partial A}{\partial H}$ when $W = 54$ and $H = 165$, and give a physical interpretation of your answers. You may use the approximations $(54)^{0.425} \approx 5.4$, $(54)^{-0.575} \approx .10$, $(165)^{0.725} \approx 40.5$, and $(165)^{-0.275} \approx .25$. (*Source: Mathematical Preparation for Laboratory Technicians.*)

Solutions to Check Your Understanding 7.2

1. (a) Negative. $\dfrac{\partial f}{\partial x}(400, 2000)$ is approximately the change in sales due to a \$1 increase in x (price). Since raising prices lowers sales, we would expect $\dfrac{\partial f}{\partial x}(400, 2000)$ to be negative.

(b) Positive. $\dfrac{\partial f}{\partial y}(400, 2000)$ is approximately the change in sales due to a \$1 increase in advertising. Since spending more money on advertising brings in more customers, we would expect sales to increase; that is, $\dfrac{\partial f}{\partial y}(400, 2000)$ is most likely positive.

2. If the interest rate is raised from 3.5% to 4.5%, the monthly payment will increase by about \$38.67. [An increase to 4% causes an increase in the monthly payment of about $\frac{1}{2} \cdot (38.47)$ or \$19.24, and so on.]

7.3 Maxima and Minima of Functions of Several Variables

In Chapter 2, we studied how to determine the maxima and minima of functions of a single variable. Let us extend that discussion to functions of several variables.

If $f(x, y)$ is a function of two variables, we say that $f(x, y)$ has a *relative maximum* when $x = a$, $y = b$ if $f(x, y)$ is at most equal to $f(a, b)$ whenever x is near a and y is near b. Geometrically, the graph of $f(x, y)$ has a peak at $(x, y) = (a, b)$. [See Fig. 1(a).] Similarly, we say that $f(x, y)$ has a *relative minimum* when $x = a$, $y = b$, if $f(x, y)$ is at least equal to $f(a, b)$ whenever x is near a and y is near b. Geometrically, the graph of $f(x, y)$ has a pit whose bottom occurs at $(x, y) = (a, b)$. [See Fig. 1(b).]

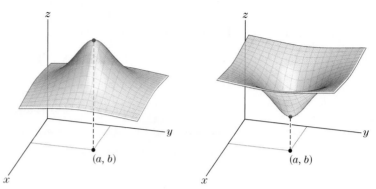

Figure 1 Maximum and minimum points.

First Derivative Test

Figure 2 Horizontal tangent lines at a relative minimum.

Suppose that the function $f(x, y)$ has a relative minimum at $(x, y) = (a, b)$, as in Fig. 2. When y is held constant at b, $f(x, y)$ is a function of x with a relative minimum at $x = a$. Therefore, the tangent line to the curve $z = f(x, b)$ is horizontal at $x = a$ and hence has slope 0. That is,

$$\frac{\partial f}{\partial x}(a, b) = 0.$$

Likewise, when x is held constant at a, then $f(x, y)$ is a function of y with a relative minimum at $y = b$. Therefore, its derivative with respect to y is zero at $y = b$. That is,

$$\frac{\partial f}{\partial y}(a, b) = 0.$$

Similar considerations apply when $f(x, y)$ has a relative maximum at $(x, y) = (a, b)$.

First-Derivative Test for Functions of Two Variables   If $f(x, y)$ has either a relative maximum or minimum at $(x, y) = (a, b)$, then

$$\frac{\partial f}{\partial x}(a, b) = 0 \quad \text{and} \quad \frac{\partial f}{\partial y}(a, b) = 0.$$

A relative maximum or minimum may or may not be an absolute maximum or minimum. However, to simplify matters in this text, the examples and exercises have been chosen so that, if an absolute extremum of $f(x, y)$ exists, it will occur at a point where $f(x, y)$ has a relative extremum.

EXAMPLE 1   **Locating a Minimum Point** The graph of $f(x, y) = 2x^2 - 2xy + 5y^2 - 6x + 5$ in Fig. 3 shows that $f(x, y)$ has a minimum point. Find its coordinates.

SOLUTION   We look for those values of x and y at which both partial derivatives are zero. The partial derivatives are

$$\frac{\partial f}{\partial x} = 4x - 2y - 6;$$

$$\frac{\partial f}{\partial y} = -2x + 10y.$$

Set the partial derivatives equal to zero and solve for x and y:

$$4x - 2y - 6 = 0 \quad \text{or} \quad 4x - 2y = 6,$$
$$-2x + 10y = 0 \quad \text{or} \quad 2x = 10y.$$

From the last equation, we have $x = 5y$. Substituting $x = 5y$ into $4x - 2y = 6$, we get

$$4(5y) - 2y = 6 \quad \text{or} \quad 18y = 6.$$

Hence, $y = \frac{1}{3}$ and, because $x = 5y$, we get $x = \frac{5}{3}$. Since $f(x, y)$ has a minimum, it must occur where $\dfrac{\partial f}{\partial x} = 0$ and $\dfrac{\partial f}{\partial y} = 0$. We have determined that the partial derivatives are zero only when $x = 5/3$ and $y = 1/3$. From Fig. 3 we know that $f(x, y)$ has a minimum, so it must be at $(x, y) = (5/3, 1/3)$.

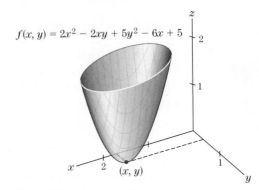

Figure 3

›› Now Try Exercise 1

EXAMPLE 2   **Profit Function of Two Variables** A firm markets a product in two countries and can charge different amounts in each country. Let x be the number of units to be sold in the first country and y the number of units to be sold in the second country. Due to the laws of demand, the firm must set the price at $97 - (x/10)$ dollars in the first country and $83 - (y/20)$ dollars in the second country to sell all the units. The cost of producing these units is $20{,}000 + 3(x + y)$. Find the values of x and y that maximize the profit.

SOLUTION   Let $f(x, y)$ be the profit derived from selling x units in the first country and y in the second. Then,

$$f(x, y) = [\text{revenue from first country}] + [\text{revenue from second country}] - [\text{cost}]$$

$$= \left(97 - \frac{x}{10} \right) x + \left(83 - \frac{y}{20} \right) y - [20{,}000 + 3(x + y)]$$

$$= 97x - \frac{x^2}{10} + 83y - \frac{y^2}{20} - 20{,}000 - 3x - 3y$$

$$= 94x - \frac{x^2}{10} + 80y - \frac{y^2}{20} - 20{,}000.$$

To find where $f(x, y)$ has its maximum value, we look for those values of x and y at which both partial derivatives are zero.

$$\frac{\partial f}{\partial x} = 94 - \frac{x}{5}$$

$$\frac{\partial f}{\partial y} = 80 - \frac{y}{10}$$

We set $\dfrac{\partial f}{\partial x} = 0$ and $\dfrac{\partial f}{\partial y} = 0$ to obtain

$$94 - \frac{x}{5} = 0 \quad \text{or} \quad x = 470,$$

$$80 - \frac{y}{10} = 0 \quad \text{or} \quad y = 800.$$

Therefore, the firm should adjust its prices to levels where it will sell 470 units in the first country and 800 units in the second country.   **>> Now Try Exercise 9**

EXAMPLE 3   **Heat Loss** Suppose that we want to design a rectangular building having a volume of 147,840 cubic feet. Assuming that the daily loss of heat is given by

$$w = 11xy + 14yz + 15xz,$$

where x, y, and z are, respectively, the length, width, and height of the building, find the dimensions of the building for which the daily heat loss is minimal.

SOLUTION   We must minimize the function

$$w = 11xy + 14yz + 15xz, \tag{1}$$

where x, y, z satisfy the constraint equation (refer to Section 2.5)

$$xyz = 147,840.$$

For simplicity, let us denote 147,840 by V. Then, $xyz = V$, so $z = V/xy$. We substitute this expression for z in the objective function (1) to obtain a heat-loss function $g(x, y)$ of two variables:

$$g(x, y) = 11xy + 14y\frac{V}{xy} + 15x\frac{V}{xy} = 11xy + \frac{14V}{x} + \frac{15V}{y}.$$

To minimize this function, we first compute the partial derivatives with respect to x and y; then we equate them to zero.

$$\frac{\partial g}{\partial x} = 11y - \frac{14V}{x^2} = 0$$

$$\frac{\partial g}{\partial y} = 11x - \frac{15V}{y^2} = 0$$

These two equations yield

$$y = \frac{14V}{11x^2}, \tag{2}$$

$$11xy^2 = 15V. \tag{3}$$

If we substitute the value of y from (2) in (3), we see that

$$11x\left(\frac{14V}{11x^2}\right)^2 = 15V$$

$$\frac{14^2 V^2}{11x^3} = 15V$$

$$x^3 = \frac{14^2 \cdot V^2}{11 \cdot 15 \cdot V} = \frac{14^2 \cdot V}{11 \cdot 15} \qquad \text{Isolate } x^3.$$

$$= \frac{14^2 \cdot 147,840}{11 \cdot 15} \qquad \text{Substitute 147,840 for } V.$$

$$= 175,616$$

$$x = 56. \qquad \text{Use a calculator to find the cube root.}$$

From equation (2) we find that

$$y = \frac{14 \cdot V}{11x^2} = \frac{14 \cdot 147{,}840}{11 \cdot 56^2} = 60.$$

Finally,

$$z = \frac{V}{xy} = \frac{147{,}840}{56 \cdot 60} = 44.$$

Thus, the building should be 56 feet long, 60 feet wide, and 44 feet high to minimize the heat loss.   **» Now Try Exercise 27**

Second-Derivative Test

When considering a function of two variables, we find points (x, y) at which $f(x, y)$ has a potential relative maximum or minimum by setting $\dfrac{\partial f}{\partial x}$ and $\dfrac{\partial f}{\partial y}$ equal to zero and solving for x and y. However, if we are given no additional information about $f(x, y)$, it may be difficult to determine whether we have found a maximum or a minimum (or neither). In the case of functions of one variable, we studied concavity and deduced the second-derivative test. There is an analog of the second-derivative test for functions of two variables, but it is much more complicated than the one-variable test. We state it without proof.

Second-Derivative Test for Functions of Two Variables   Suppose that $f(x, y)$ is a function and (a, b) is a point at which

$$\frac{\partial f}{\partial x}(a, b) = 0 \quad \text{and} \quad \frac{\partial f}{\partial y}(a, b) = 0,$$

and let

$$D(x, y) = \frac{\partial^2 f}{\partial x^2} \cdot \frac{\partial^2 f}{\partial y^2} - \left(\frac{\partial^2 f}{\partial x \, \partial y} \right)^2.$$

1. If

$$D(a, b) > 0 \quad \text{and} \quad \frac{\partial^2 f}{\partial x^2}(a, b) > 0,$$

then $f(x, y)$ has a relative minimum at (a, b).

2. If

$$D(a, b) > 0 \quad \text{and} \quad \frac{\partial^2 f}{\partial x^2}(a, b) < 0,$$

then $f(x, y)$ has a relative maximum at (a, b).

3. If

$$D(a, b) < 0,$$

then $f(x, y)$ has neither a relative maximum nor a relative minimum at (a, b).

4. If $D(a, b) = 0$, no conclusion can be drawn from this test.

The saddle-shaped graph in Fig. 4 illustrates a function $f(x, y)$ for which $D(a, b) < 0$. Both partial derivatives are zero at $(x, y) = (a, b)$, and yet the function has neither a relative maximum nor a relative minimum there. (Observe that the function has a relative maximum with respect to x when y is held constant and a relative minimum with respect to y when x is held constant.)

Figure 4

EXAMPLE 4

Application of the Second-Derivative Test Refer to Example 1, where we determined, with the help of Fig. 3, that at the point $(5/3, 1/3)$ the function $f(x, y) = 2x^2 - 2xy + 5y^2 - 6x + 5$ has a minimum. Verify this assertion using the second-derivative test.

SOLUTION

In Example 1, we showed that $\frac{\partial f}{\partial x}(5/3, 1/3) = 0$ and $\frac{\partial f}{\partial y}(5/3, 1/3) = 0$. Also, from Example 1,

$$\frac{\partial f}{\partial x} = 4x - 2y - 6 \quad \text{so} \quad \frac{\partial^2 f}{\partial x^2} = 4;$$

and

$$\frac{\partial f}{\partial y} = -2x + 10y \quad \text{so} \quad \frac{\partial^2 f}{\partial y^2} = 10.$$

Hence,

$$\frac{\partial^2 f}{\partial x \partial y} = \frac{\partial}{\partial x}\left(\frac{\partial f}{\partial y}\right) = \frac{\partial}{\partial x}(-2x + 10y) = -2.$$

We have

$$D(x, y) = \frac{\partial^2 f}{\partial x^2} \cdot \frac{\partial^2 f}{\partial y^2} - \left(\frac{\partial^2 f}{\partial x \partial y}\right)^2 = (4) \cdot (10) - (-2)^2 = 36 > 0.$$

Since $D(x, y) > 0$ for all (x, y), it follows that $D(5/3, 1/3) > 0$. By the second-derivative test, the function $f(x, y)$ has either a relative maximum or a relative minimum at $(5/3, 1/3)$. Now $\frac{\partial^2 f}{\partial x^2}(5/3, 1/3) = 4 > 0$ so, by case 1 of the test, the function has a relative minimum at $(5/3, 1/3)$, as we expected!   **》 Now Try Exercise 19**

In Example 4, all second partial derivatives and $D(x, y)$ were constant for all (x, y). This is not always the case, as illustrated by our next example.

EXAMPLE 5

Applying the Second-Derivative Test Let $f(x, y) = x^3 - y^2 - 12x + 6y + 5$. Find all possible relative maximum and minimum points of $f(x, y)$. Use the second-derivative test to determine the nature of each such point.

SOLUTION

Since

$$\frac{\partial f}{\partial x} = 3x^2 - 12, \qquad \frac{\partial f}{\partial y} = -2y + 6,$$

we find that $f(x, y)$ has a potential relative extreme point when

$$3x^2 - 12 = 0,$$
$$-2y + 6 = 0.$$

From the first equation, $3x^2 = 12$, $x^2 = 4$, and $x = \pm 2$. From the second equation, $y = 3$. Thus, $\frac{\partial f}{\partial x}$ and $\frac{\partial f}{\partial y}$ are both zero when $(x, y) = (2, 3)$ and when $(x, y) = (-2, 3)$. To apply the second-derivative test, compute

$$\frac{\partial^2 f}{\partial x^2} = 6x, \qquad \frac{\partial^2 f}{\partial y^2} = -2, \qquad \frac{\partial^2 f}{\partial x \partial y} = 0,$$

and

$$D(x, y) = \frac{\partial^2 f}{\partial x^2} \cdot \frac{\partial^2 f}{\partial y^2} - \left(\frac{\partial^2 f}{\partial x \partial y}\right)^2 = (6x)(-2) - 0^2 = -12x.$$

Since $D(2, 3) = -12(2) = -24$, which is negative, case 3 of the second-derivative test says that $f(x, y)$ has neither a relative maximum nor a relative minimum at $(2, 3)$. However,

$D(-2, 3) = -12(-2) = 24$. Since $D(-2, 3)$ is positive, the function $f(x, y)$ has either a relative maximum or a relative minimum at $(-2, 3)$. To determine which, we compute

$$\frac{\partial^2 f}{\partial x^2}(-2, 3) = 6(-2) = -12 < 0.$$

By case 2 of the second-derivative test, the function $f(x, y)$ has a relative maximum at $(-2, 3)$.

» Now Try Exercise 25

In this section, we have restricted ourselves to functions of two variables, but the case of three or more variables is handled in a similar fashion. For instance, here is the first-derivative test for a function of three variables.

If $f(x, y, z)$ has a relative maximum or minimum at $(x, y, z) = (a, b, c)$, then

$$\frac{\partial f}{\partial x}(a, b, c) = 0,$$

$$\frac{\partial f}{\partial y}(a, b, c) = 0,$$

$$\frac{\partial f}{\partial z}(a, b, c) = 0.$$

Check Your Understanding 7.3

Solutions can be found following the section exercises.

1. Find all points (x, y) where $f(x, y) = x^3 - 3xy + \frac{1}{2}y^2 + 8$ has a possible relative maximum or minimum.

2. Apply the second-derivative test to the function $g(x, y)$ of Example 3 to confirm that a relative minimum actually occurs when $x = 56$ and $y = 60$.

EXERCISES 7.3

Find all points (x, y) where $f(x, y)$ has a possible relative maximum or minimum.

1. $f(x, y) = x^2 - 3y^2 + 4x + 6y + 8$
2. $f(x, y) = \frac{1}{2}x^2 + y^2 - 3x + 2y - 5$
3. $f(x, y) = x^2 - 5xy + 6y^2 + 3x - 2y + 4$
4. $f(x, y) = -3x^2 + 7xy - 4y^2 + x + y$
5. $f(x, y) = 3x^2 + 8xy - 3y^2 - 2x + 4y - 1$
6. $f(x, y) = 4x^2 + 4xy - 3y^2 + 4y - 1$
7. $f(x, y) = x^3 + y^2 - 3x + 6y$
8. $f(x, y) = x^2 - y^3 + 5x + 12y + 1$
9. $f(x, y) = -8y^3 + 4xy + 9y^2 - 2y$
10. $f(x, y) = -8y^3 + 4xy + 4x^2 + 9y^2$
11. $f(x, y) = 2x^3 + 2x^2y - y^2 + y$
12. $f(x, y) = \frac{15}{4}x^2 + 6xy - 3y^2 + 3x + 6y$
13. $f(x, y) = \frac{1}{3}x^3 - 2y^3 - 5x + 6y - 5$
14. $f(x, y) = x^4 - 8xy + 2y^2 - 3$
15. $f(x, y) = x^3 + x^2y - y$
16. $f(x, y) = x^4 - 2xy - 7x^2 + y^2 + 3$

17. The function $f(x, y) = 2x + 3y + 9 - x^2 - xy - y^2$ has a maximum at some point (x, y). Find the values of x and y where this maximum occurs.

18. The function $f(x, y) = \frac{1}{2}x^2 + 2xy + 3y^2 - x + 2y$ has a minimum at some point (x, y). Find the values of x and y where this minimum occurs.

In Exercises 19–24, both first partial derivatives of the function $f(x, y)$ are zero at the given points. Use the second-derivative test to determine the nature of $f(x, y)$ at each of these points. If the second-derivative test is inconclusive, so state.

19. $f(x, y) = 3x^2 - 6xy + y^3 - 9y$; $(3, 3)$, $(-1, -1)$
20. $f(x, y) = 6xy^2 - 2x^3 - 3y^4$; $(0, 0)$, $(1, 1)$, $(1, -1)$
21. $f(x, y) = 2x^2 - x^4 - y^2$; $(-1, 0)$, $(0, 0)$, $(1, 0)$
22. $f(x, y) = x^4 - 4xy + y^4$; $(0, 0)$, $(1, 1)$, $(-1, -1)$
23. $f(x, y) = ye^x - 3x - y + 5$; $(0, 3)$
24. $f(x, y) = \dfrac{1}{x} + \dfrac{1}{y} + xy$; $(1, 1)$

Find all points (x, y) where $f(x, y)$ has a possible relative maximum or minimum. Then, use the second-derivative test to determine, if possible, the nature of $f(x, y)$ at each of these points. If the second-derivative test is inconclusive, so state.

25. $f(x, y) = -5x^2 + 4xy - 17y^2 - 6x + 6y + 2$
26. $f(x, y) = -2x^2 + 6xy - 17y^2 - 4x + 6y$
27. $f(x, y) = 3x^2 + 8xy - 3y^2 + 2x + 6y$
28. $f(x, y) = 8xy + 8y^2 - 2x + 2y - 1$
29. $f(x, y) = x^4 - x^2 - 2xy + y^2 + 1$
30. $f(x, y) = x^2 + 2xy + 10y^2$
31. $f(x, y) = 6xy - 3y^2 - 2x + 4y - 1$
32. $f(x, y) = 2xy + y^2 + 2x - 1$
33. $f(x, y) = -2x^2 + 2xy - 25y^2 - 2x + 8y - 1$
34. $f(x, y) = 3x^2 + 8xy - 3y^2 - 2x + 4y + 1$

35. $f(x, y) = x^4 - 12x^2 - 4xy - y^2 + 16$

36. $f(x, y) = \frac{17}{4}x^2 + 2xy + 5y^2 + 5x - 2y + 2$

37. $f(x, y) = x^2 - 2xy + 4y^2$

38. $f(x, y) = 2x^2 + 3xy + 5y^2$

39. $f(x, y) = -2x^2 + 2xy - y^2 + 4x - 6y + 5$

40. $f(x, y) = -x^2 - 8xy - y^2$

41. $f(x, y) = x^2 + 2xy + 5y^2 + 2x + 10y - 3$

42. $f(x, y) = x^2 - 2xy + 3y^2 + 4x - 16y + 22$

43. $f(x, y) = x^3 - y^2 - 3x + 4y$

44. $f(x, y) = x^3 - 2xy + 4y$

45. $f(x, y) = 2x^2 + y^3 - x - 12y + 7$

46. $f(x, y) = x^2 + 4xy + 2y^4$

47. Find the possible values of x, y, z at which
$$f(x, y, z) = 2x^2 + 3y^2 + z^2 - 2x - y - z$$
assumes its minimum value.

48. Find the possible values of x, y, z at which
$$f(x, y, z) = 5 + 8x - 4y + x^2 + y^2 + z^2$$
assumes its minimum value.

49. Maximizing Volume U.S. postal rules require that the length plus the girth of a package cannot exceed 84 inches. Find the dimensions of the rectangular package of greatest volume that can be mailed. [*Note:* From Fig. 5 we see that $84 = (\text{length}) + (\text{girth}) = l + (2x + 2y)$.]

Figure 5

50. Minimizing Surface Area Find the dimensions of the rectangular box of least surface area that has a volume of 1000 cubic inches.

51. Maximizing Profit A company manufactures and sells two products, I and II, that sell for $10 and $9 per unit, respectively. The cost of producing x units of product I and y units of product II is
$$400 + 2x + 3y + .01(3x^2 + xy + 3y^2).$$
Find the values of x and y that maximize the company's profits. [*Note:* Profit = (revenue) − (cost).]

52. Maximizing Profit A monopolist manufactures and sells two competing products, I and II, that cost $30 and $20 per unit, respectively, to produce. The revenue from marketing x units of product I and y units of product II is $98x + 112y - .04xy - .1x^2 - .2y^2$. Find the values of x and y that maximize the monopolist's profits.

53. Profit from Two Products A company manufactures and sells two products, I and II, that sell for p_1 and p_2 per unit, respectively. Let $C(x, y)$ be the cost of producing x units of product I and y units of product II. Show that if the company's profit is maximized when $x = a$, $y = b$, then
$$\frac{\partial C}{\partial x}(a, b) = p_1 \quad \text{and} \quad \frac{\partial C}{\partial y}(a, b) = p_2.$$

54. Revenue from Two Products A company manufactures and sells two competing products, I and II, that cost p_{I} and p_{II} per unit, respectively, to produce. Let $R(x, y)$ be the revenue from marketing x units of product I and y units of product II. Show that if the company's profit is maximized when $x = a$, $y = b$, then
$$\frac{\partial R}{\partial x}(a, b) = p_{\mathrm{I}} \quad \text{and} \quad \frac{\partial R}{\partial y}(a, b) = p_{\mathrm{II}}.$$

Solutions to Check Your Understanding 7.3

1. Compute the first partial derivatives of $f(x, y)$ and solve the system of equations that results from setting the partials equal to zero.
$$\frac{\partial f}{\partial x} = 3x^2 - 3y = 0$$
$$\frac{\partial f}{\partial y} = -3x + y = 0$$
Solve each equation for y in terms of x.
$$\begin{cases} y = x^2 \\ y = 3x \end{cases}$$
Equate expressions for y and solve for x.
$$x^2 = 3x$$
$$x^2 - 3x = 0$$
$$x(x - 3) = 0$$
$$x = 0 \quad \text{or} \quad x = 3.$$
When $x = 0$, $y = 0^2 = 0$. When $x = 3$, $y = 3^2 = 9$. Therefore, the possible relative maximum or minimum points are $(0, 0)$ and $(3, 9)$.

2. We have
$$g(x, y) = 11xy + \frac{14V}{x} + \frac{15V}{y},$$
$$\frac{\partial g}{\partial x} = 11y - \frac{14V}{x^2}, \quad \text{and} \quad \frac{\partial g}{\partial y} = 11x - \frac{15V}{y^2}.$$
Now,
$$\frac{\partial^2 g}{\partial x^2} = \frac{28V}{x^3}, \quad \frac{\partial^2 g}{\partial y^2} = \frac{30V}{y^3}, \quad \text{and} \quad \frac{\partial^2 g}{\partial x \, \partial y} = 11.$$
Therefore,
$$D(x, y) = \frac{28V}{x^3} \cdot \frac{30V}{y^3} - (11)^2,$$
$$D(56, 60) = \frac{28(147{,}840)}{(56)^3} \cdot \frac{30(147{,}840)}{(60)^3} - 121,$$
$$= 484 - 121 = 363 > 0,$$
and
$$\frac{\partial^2 g}{\partial x^2}(56, 60) = \frac{28(147{,}840)}{(56)^3} > 0.$$
It follows that $g(x, y)$ has a relative minimum at $x = 56$, $y = 60$.

7.4   Lagrange Multipliers and Constrained Optimization

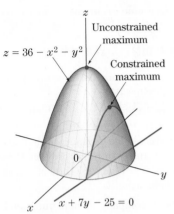

$z = 36 - x^2 - y^2$

Unconstrained maximum

Constrained maximum

$x + 7y - 25 = 0$

Figure 1 A constrained optimization problem.

We have seen a number of optimization problems in which we were required to minimize (or maximize) an objective function where the variables were subject to a constraint equation. For instance, in Example 4 of Section 2.5, we minimized the cost of a rectangular enclosure by minimizing the objective function $42x + 28y$, where x and y were subject to the constraint equation $600 - xy = 0$. In the preceding section (Example 3), we minimized the daily heat loss from a building by minimizing the objective function $11xy + 14yz + 15xz$, subject to the constraint equation $147{,}840 - xyz = 0$.

Figure 1 gives a graphical illustration of what happens when an objective function is maximized subject to a constraint. The graph of the objective function is the cone-shaped surface $z = 36 - x^2 - y^2$, and the colored curve on that surface consists of those points whose x- and y-coordinates satisfy the constraint equation $x + 7y - 25 = 0$. The constrained maximum is at the highest point on this curve. Of course, the surface itself has a higher "unconstrained maximum" at $(x, y, z) = (0, 0, 36)$, but these values of x and y do not satisfy the constraint equation.

In this section, we introduce a powerful technique for solving problems of this type. Let us begin with the following general problem, which involves two variables.

Problem   Let $f(x, y)$ and $g(x, y)$ be functions of two variables. Find values of x and y that maximize (or minimize) the objective function $f(x, y)$ and that also satisfy the constraint equation $g(x, y) = 0$.

Of course, if we can solve the equation $g(x, y) = 0$ for one variable in terms of the other and substitute the resulting expression in $f(x, y)$, we arrive at a function of a single variable that can be maximized (or minimized) by using the methods of Chapter 2. However, this technique can be unsatisfactory for two reasons. First, it may be difficult to solve the equation $g(x, y) = 0$ for x or for y. For example, if $g(x, y) = x^4 + 5x^3y + 7x^2y^3 + y^5 - 17 = 0$, it is difficult to write y as a function of x or x as a function of y. Second, even if $g(x, y) = 0$ can be solved for one variable in terms of the other, substitution of the result into $f(x, y)$ may yield a complicated function.

One clever idea for handling the preceding problem was discovered by the eighteenth-century mathematician Lagrange, and the technique that he pioneered today bears his name, the **method of Lagrange multipliers**. The basic idea of this method is to replace $f(x, y)$ by an auxiliary function of three variables $F(x, y, \lambda)$, defined as

$$F(x, y, \lambda) = f(x, y) + \lambda g(x, y).$$

The new variable λ (lambda) is called a **Lagrange multiplier** and always multiplies the constraint function $g(x, y)$. The following theorem is stated without proof.

> **Theorem**   Suppose that, subject to the constraint $g(x, y) = 0$, the function $f(x, y)$ has a relative maximum or minimum at $(x, y) = (a, b)$. Then, there is a value of λ—say, $\lambda = c$—such that the partial derivatives of $F(x, y, \lambda)$ all equal zero at $(x, y, \lambda) = (a, b, c)$.

The theorem implies that, if we locate all points (x, y, λ) where the partial derivatives of $F(x, y, \lambda)$ are all zero, among the corresponding points (x, y), we then will find all possible places where $f(x, y)$ may have a constrained relative maximum or minimum. Thus, the first step in the method of Lagrange multipliers is to set the partial derivatives of $F(x, y, \lambda)$ equal to zero and solve for x, y, and λ:

$$\frac{\partial F}{\partial x} = 0 \tag{L-1}$$

$$\frac{\partial F}{\partial y} = 0 \tag{L-2}$$

$$\frac{\partial F}{\partial \lambda} = 0. \tag{L-3}$$

From the definition of $F(x, y, \lambda)$, we see that $\dfrac{\partial F}{\partial \lambda} = g(x, y)$. Thus, the third equation (L-3) is just the original constraint equation $g(x, y) = 0$. So, when we find a point (x, y, λ) that satisfies (L-1), (L-2), and (L-3), the coordinates x and y will automatically satisfy the constraint equation.

The first example applies this method to the problem described in Fig. 1.

EXAMPLE 1   **Lagrange Multipliers**   Maximize $36 - x^2 - y^2$ subject to the constraint $x + 7y - 25 = 0$.

SOLUTION   Here, $f(x, y) = 36 - x^2 - y^2$, $g(x, y) = x + 7y - 25$, and

$$F(x, y, \lambda) = 36 - x^2 - y^2 + \lambda(x + 7y - 25).$$

Equations (L-1) to (L-3) read

$$\frac{\partial F}{\partial x} = -2x + \lambda = 0, \tag{1}$$

$$\frac{\partial F}{\partial y} = -2y + 7\lambda = 0, \tag{2}$$

$$\frac{\partial F}{\partial \lambda} = x + 7y - 25 = 0. \tag{3}$$

We solve the first two equations for λ:

$$\lambda = 2x$$

$$\lambda = \frac{2}{7}y. \tag{4}$$

If we equate these two expressions for λ, we obtain

$$2x = \frac{2}{7}y$$

$$x = \frac{1}{7}y. \tag{5}$$

Substituting this expression for x in equation (3), we have

$$\frac{1}{7}y + 7y - 25 = 0 \qquad \text{Substitute for } x = \tfrac{1}{7}y \text{ in (3)}.$$

$$\frac{50}{7}y = 25 \qquad \begin{array}{l} \text{Add 25 to both sides.} \\ \text{Combine the terms in } y. \end{array}$$

$$y = \frac{7}{2}. \qquad \text{Solve for } y.$$

With this value for y, equations (4) and (5) produce the values of x and λ:

$$x = \frac{1}{7}y = \frac{1}{7}\left(\frac{7}{2}\right) = \frac{1}{2}, \qquad \text{Substitute for } y \text{ in (5).}$$

$$\lambda = \frac{2}{7}y = \frac{2}{7}\left(\frac{7}{2}\right) = 1. \qquad \text{Substitute for } y \text{ in (4).}$$

Therefore, the partial derivatives of $F(x, y, \lambda)$ are zero when $x = \frac{1}{2}$, $y = \frac{7}{2}$, and $\lambda = 1$. So, the maximum value of $36 - x^2 - y^2$ subject to the constraint $x + 7y - 25 = 0$ occurs when $x = \frac{1}{2}$ and $y = \frac{7}{2}$. That maximum value is

$$36 - \left(\frac{1}{2}\right)^2 - \left(\frac{7}{2}\right)^2 = \frac{47}{2}.$$

>> *Now Try Exercise 1*

The preceding technique for solving three equations in the three variables x, y, and λ can usually be applied to solve Lagrange multiplier problems. Here is the basic procedure:

Using Lagrange Multipliers

1. Solve (L-1) and (L-2) for λ in terms of x and y; then, equate the resulting expressions for λ.
2. Solve the resulting equation for one of the variables.
3. Substitute the result of step 2 in the equation (L-3), and solve the resulting equation of one variable.
4. Use the one known variable and the equations of steps 1 and 2 to determine the other two variables.

In most applications, we know that an absolute (constrained) maximum or minimum exists. In the event that the method of Lagrange multipliers produces exactly one possible relative extreme value, we will assume that it is indeed the sought-after absolute extreme value. For instance, the statement of Example 1 is meant to imply that there is an absolute maximum value. Since we determined that there was just one possible relative extreme value, we concluded that it was the absolute maximum value.

EXAMPLE 2   **Lagrange Multipliers** Using Lagrange multipliers, minimize $42x + 28y$, subject to the constraint $600 - xy = 0$, where x and y are restricted to positive values. (This problem arose in Example 4 of Section 2.5, where $42x + 28y$ was the cost of building a 600-square-foot enclosure having dimensions x and y.)

SOLUTION   We have $f(x, y) = 42x + 28y$, $g(x, y) = 600 - xy$, and

$$F(x, y, \lambda) = 42x + 28y + \lambda(600 - xy).$$

The equations (L-1) to (L-3), in this case, are

$$\frac{\partial F}{\partial x} = 42 - \lambda y = 0,$$

$$\frac{\partial F}{\partial y} = 28 - \lambda x = 0,$$

$$\frac{\partial F}{\partial \lambda} = 600 - xy = 0.$$

From the first two equations, we see that

$$\lambda = \frac{42}{y} = \frac{28}{x}$$

$$42x = 28y \qquad \text{Cross-multiply.}$$

$$x = \frac{2}{3}y. \qquad \text{Solve for } x.$$

Substituting this expression for x in the third equation, we derive

$$600 - \left(\frac{2}{3}y\right)y = 0 \qquad \text{Substitute } x = \frac{2}{3}y.$$

$$y^2 = \frac{3}{2} \cdot 600 = 900 \qquad \text{Solve for } y.$$

$$y = \pm 30.$$

We discard the case $y = -30$ because we are interested only in positive values of x and y. Using $y = 30$, we find that

$$x = \frac{2}{3}(30) = 20$$

$$\lambda = \frac{28}{20} = \frac{7}{5}.$$

So the minimum value of $42x + 28y$ with x and y subject to the constraint occurs when $x = 20$, $y = 30$, and $\lambda = \frac{7}{5}$. That minimum value is

$$42 \cdot (20) + 28 \cdot (30) = 1680. \qquad \text{» Now Try Exercise 3}$$

EXAMPLE 3 **Maximizing Production** Suppose that x units of labor and y units of capital can produce $f(x, y) = 60x^{3/4}y^{1/4}$ units of a certain product. Also, suppose that each unit of labor costs \$100, whereas each unit of capital costs \$200. Assume that \$30,000 is available to spend on production. How many units of labor and how many units of capital should be utilized to maximize production?

SOLUTION The cost of x units of labor and y units of capital equals $100x + 200y$. Therefore, since we want to use all the available money (\$30,000), we must satisfy the constraint equation

$$100x + 200y = 30{,}000$$

or

$$g(x, y) = 30{,}000 - 100x - 200y = 0.$$

The objective function is $f(x, y) = 60x^{3/4}y^{1/4}$. In this case, we have

$$F(x, y, \lambda) = 60x^{3/4}y^{1/4} + \lambda(30{,}000 - 100x - 200y).$$

The equations (L-1) to (L-3) read

$$\frac{\partial F}{\partial x} = 45x^{-1/4}y^{1/4} - 100\lambda = 0,$$

$$\frac{\partial F}{\partial y} = 15x^{3/4}y^{-3/4} - 200\lambda = 0,$$

$$\frac{\partial F}{\partial \lambda} = 30{,}000 - 100x - 200y = 0. \qquad (6)$$

By solving the first two equations for λ, we see that

$$\lambda = \frac{45}{100}x^{-1/4}y^{1/4} = \frac{9}{20}x^{-1/4}y^{1/4},$$

$$\lambda = \frac{15}{200}x^{3/4}y^{-3/4} = \frac{3}{40}x^{3/4}y^{-3/4}.$$

Therefore, we must have

$$\frac{9}{20}x^{-1/4}y^{1/4} = \frac{3}{40}x^{3/4}y^{-3/4}.$$

To solve for y in terms of x, let us multiply both sides of this equation by $x^{1/4}y^{3/4}$:

$$\frac{9}{20}y = \frac{3}{40}x$$

or

$$y = \frac{1}{6}x.$$

Inserting this result in (6), we find that

$$100x + 200\left(\frac{1}{6}x\right) = 30{,}000$$

$$\frac{400x}{3} = 30{,}000$$

$$x = 225.$$

Hence,

$$y = \frac{225}{6} = 37.5.$$

So maximum production is achieved by the use of 225 units of labor and 37.5 units of capital. Note that we don't find the value of λ here. It's not necessary.

> **» Now Try Exercise 25**

In Example 3, it turns out that, at the optimum values of x and y,

$$\lambda = \frac{9}{20}x^{-1/4}y^{1/4} = \frac{9}{20}(225)^{-1/4}(37.5)^{1/4} \approx .2875,$$

$$\frac{\partial f}{\partial x} = 45x^{-1/4}y^{1/4} = 45(225)^{-1/4}(37.5)^{1/4}, \tag{7}$$

$$\frac{\partial f}{\partial y} = 15x^{3/4}y^{-3/4} = 15(225)^{3/4}(37.5)^{-3/4}. \tag{8}$$

It can be shown that the Lagrange multiplier λ can be interpreted as the **marginal productivity of money**. That is, if 1 additional dollar is available, approximately .2875 additional units of the product can be produced.

Recall that the partial derivatives $\dfrac{\partial f}{\partial x}$ and $\dfrac{\partial f}{\partial y}$ are called the **marginal productivity of labor** and **capital**, respectively. From equations (7) and (8) we have

$$\frac{[\text{marginal productivity of labor}]}{[\text{marginal productivity of capital}]} = \frac{45(225)^{-1/4}(37.5)^{1/4}}{15(225)^{3/4}(37.5)^{-3/4}}$$

$$= \frac{45}{15}(225)^{-1}(37.5)^{1}$$

$$= \frac{3(37.5)}{225} = \frac{37.5}{75} = \frac{1}{2}.$$

On the other hand,

$$\frac{[\text{cost per unit of labor}]}{[\text{cost per unit of capital}]} = \frac{100}{200} = \frac{1}{2}.$$

This result illustrates the following law of economics. *If labor and capital are at their optimal levels, the ratio of their marginal productivities equals the ratio of their unit costs.*

Lagrange Multipliers in Three Variables

The method of Lagrange multipliers generalizes to functions of any number of variables. For instance, we can maximize $f(x, y, z)$, subject to the constraint equation $g(x, y, z) = 0$, by considering the Lagrange function

$$F(x, y, z, \lambda) = f(x, y, z) + \lambda g(x, y, z).$$

The analogs of equations (L-1) to (L-3) are

$$\frac{\partial F}{\partial x} = 0, \quad \frac{\partial F}{\partial y} = 0, \quad \frac{\partial F}{\partial z} = 0, \quad \frac{\partial F}{\partial \lambda} = 0.$$

Let us now show how we can solve the heat-loss problem of Section 7.3 by using this method.

EXAMPLE 4

Lagrange Multipliers in Three Variables Use Lagrange multipliers to find the values of x, y, z that minimize the objective function

$$f(x, y, z) = 11xy + 14yz + 15xz,$$

subject to the constraint

$$xyz = 147{,}840.$$

SOLUTION The Lagrange function is

$$F(x, y, z, \lambda) = 11xy + 14yz + 15xz + \lambda(147{,}840 - xyz).$$

The conditions for a relative minimum are

$$\frac{\partial F}{\partial x} = 11y + 15z - \lambda yz = 0,$$

$$\frac{\partial F}{\partial y} = 11x + 14z - \lambda xz = 0,$$

$$\frac{\partial F}{\partial z} = 14y + 15x - \lambda xy = 0,$$

$$\frac{\partial F}{\partial \lambda} = 147{,}840 - xyz = 0. \tag{9}$$

From the first three equations, we have the system of equations

$$\left. \begin{array}{l} \lambda = \dfrac{11y + 15z}{yz} = \dfrac{11}{z} + \dfrac{15}{y} \\[2mm] \lambda = \dfrac{11x + 14z}{xz} = \dfrac{11}{z} + \dfrac{14}{x} \\[2mm] \lambda = \dfrac{14y + 15x}{xy} = \dfrac{14}{x} + \dfrac{15}{y} \end{array} \right\}. \tag{10}$$

Let us equate the first two expressions for λ:

$$\frac{11}{z} + \frac{15}{y} = \frac{11}{z} + \frac{14}{x}$$

$$\frac{15}{y} = \frac{14}{x}$$

$$x = \frac{14}{15}y.$$

Next, we equate the second and third expressions for λ in (10):

$$\frac{11}{z} + \frac{14}{x} = \frac{14}{x} + \frac{15}{y}$$

$$\frac{11}{z} = \frac{15}{y}$$

$$z = \frac{11}{15}y.$$

From (9),

$$xyz = 147{,}840$$

$$\frac{14}{15}y \cdot y \cdot \frac{11}{15}y = 147{,}840 \qquad \text{Substitute } x = \frac{14}{15}y \text{ and } z = \frac{11}{15}y.$$

$$y^3 = \frac{(147{,}840)(15)^2}{(14)(11)} = 216{,}000 \quad \text{Simplify, cross-multiply.}$$

$$y = 60.$$

From this, we find that

$$x = \frac{14}{15}(60) = 56 \quad \text{and} \quad z = \frac{11}{15}(60) = 44.$$

We conclude that the heat loss is minimized when $x = 56$, $y = 60$, and $z = 44$.

»Now Try Exercise 15

In the solution of Example 4, we found that, at the optimal values of x, y, and z,

$$\frac{14}{x} = \frac{15}{y} = \frac{11}{z}.$$

Referring to Example 2 of Section 7.1, we see that 14 is the combined heat loss through the east and west sides of the building, 15 is the heat loss through the north and south sides of the building, and 11 is the heat loss through the floor and roof. Thus, we have that, under optimal conditions,

$$\frac{[\text{heat loss through east and west sides}]}{[\text{distance between east and west sides}]} = \frac{[\text{heat loss through north and south sides}]}{[\text{distance between north and south sides}]}$$

$$= \frac{[\text{heat loss through floor and roof}]}{[\text{distance between floor and roof}]}.$$

This is a principle of optimal design: Minimal heat loss occurs when the distance between each pair of opposite sides is some fixed constant times the heat loss from the pair of sides.

The value of λ in Example 4 corresponding to the optimal values of x, y, and z is

$$\lambda = \frac{11}{z} + \frac{15}{y} = \frac{11}{44} + \frac{15}{60} = \frac{1}{2}.$$

We can show that the Lagrange multiplier λ is the marginal heat loss with respect to volume. That is, if a building of volume slightly more than 147,840 cubic feet is optimally designed, $\frac{1}{2}$ unit of additional heat will be lost for each additional cubic foot of volume.

Check Your Understanding 7.4

Solutions can be found following the section exercises.

1. Let $F(x, y, \lambda) = 2x + 3y + \lambda(90 - 6x^{1/3}y^{2/3})$. Find $\dfrac{\partial F}{\partial x}$.

2. Refer to Exercise 29 of Section 7.3. What is the function $F(x, y, l, \lambda)$ when the exercise is solved by means of the method of Lagrange multipliers?

EXERCISES 7.4

Solve the following exercises by the method of Lagrange multipliers.

1. Minimize $x^2 + 3y^2 + 10$, subject to the constraint $8 - x - y = 0$.

2. Maximize $x^2 - y^2$, subject to the constraint $2x + y - 3 = 0$.

3. Maximize $x^2 + xy - 3y^2$, subject to the constraint $2 - x - 2y = 0$.

4. Minimize $\frac{1}{2}x^2 - 3xy + y^2 + \frac{1}{2}$, subject to the constraint $3x - y - 1 = 0$.

5. Find the values of x, y that maximize

$$-2x^2 - 2xy - \tfrac{3}{2}y^2 + x + 2y,$$

subject to the constraint $x + y - \tfrac{5}{2} = 0$.

6. Find the values of x, y that minimize $x^2 + xy + y^2 - 2x - 5y$, subject to the constraint $1 - x + y = 0$.

7. Find the values of x and y that minimize $xy + y^2 - x - 1$ subject to the constraint $x - 2y = 0$.

8. Find the values of x and y that minimize $x^2 - 2xy + 2y^2$ subject to the constraint $2x - y + 5 = 0$.

9. Find the values of x and y that minimize $2x^2 + xy + y^2 - y$ subject to the constraint $x + y = 0$.

10. Find the values of x and y that minimize $2x^2 - 2xy + y^2 - 2x + 1$ subject to the constraint $x - y = 3$.

11. Find the values of x and y that minimize $18x^2 + 12xy + 4y^2 + 6x - 4y + 5$ subject to the constraint $3x + 2y - 1 = 0$.

12. Find the values of x and y that minimize $3x^2 - 2xy + x - 3y + 1$ subject to the constraint $x - 3y = 1$.

13. Find the values of x and y that minimize $f(x, y) = x - xy + 2y^2$ subject to the constraint $x - y + 1 = 0$.

14. Find the values of x and y that maximize xy subject to the constraint $x^2 - y = 3$.

15. Find the values of x, y and z that minimize $xy + xz - yz$ subject to the constraint $x + y + z = 1$.

16. Find the values of x, y, and z that minimize $xy + xz - 2yz$ subject to the constraint $x + y + z = 2$.

17. **Maximizing a Product** Find the two positive numbers whose product is 25 and whose sum is as small as possible.

18. **Maximizing Area** Four hundred eighty dollars are available to fence in a rectangular garden. The fencing for the north and south sides of the garden costs $10 per foot, and the fencing for the east and west sides costs $15 per foot. Find the dimensions of the largest possible garden.

19. **Maximizing Volume** Three hundred square inches of material are available to construct an open rectangular box with a square base. Find the dimensions of the box that maximize the volume.

20. **Minimizing Space in a Firm** The amount of space required by a particular firm is $f(x, y) = 1000\sqrt{6x^2 + y^2}$, where x and y are, respectively, the number of units of labor and capital utilized. Suppose that labor costs $480 per unit and capital costs $40 per unit and that the firm has $5000 to spend. Determine the amounts of labor and capital that should be utilized in order to minimize the amount of space required.

21. **Inscribed Rectangle with Maximum Area** Find the dimensions of the rectangle of maximum area that can be inscribed in the unit circle. [See Fig. 2(a).]

22. **Distance from a Point to a Parabola** Find the point on the parabola $y = x^2$ that has minimal distance from the point $(16, \tfrac{1}{2})$. [See Fig. 2(b).] [*Suggestion:* If d denotes the distance from (x, y) to $(16, \tfrac{1}{2})$, then $d^2 = (x - 16)^2 + (y - \tfrac{1}{2})^2$. If d^2 is minimized, then d will be minimized.]

(a)

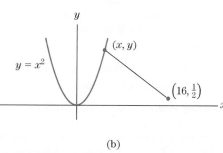

(b)

Figure 2

23. **Production Schedule and Production Possibilities Curve** Suppose that a firm makes two products, A and B, that use the same raw materials. Given a fixed amount of raw materials and a fixed amount of labor, the firm must decide how much of its resources should be allocated to the production of A and how much to B. If x units of A and y units of B are produced, suppose that x and y must satisfy

$$9x^2 + 4y^2 = 18{,}000.$$

The graph of this equation (for $x \geq 0$, $y \geq 0$) is called a *production possibilities curve* (Fig. 3). A point (x, y) on this curve represents a *production schedule* for the firm, committing it to produce x units of A and y units of B. The reason for the relationship between x and y involves the limitations on personnel and raw materials available to the firm. Suppose that each unit of A yields a $3 profit, whereas each unit of B yields a $4 profit. Then, the profit of the firm is

$$P(x, y) = 3x + 4y.$$

Find the production schedule that maximizes the profit function $P(x, y)$.

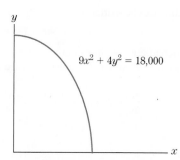

Figure 3 A production possibilities curve.

24. **Maximizing Profit** A firm makes x units of product A and y units of product B and has a production possibilities curve given by the equation $4x^2 + 25y^2 = 50{,}000$ for $x \geq 0$, $y \geq 0$. (See Exercise 23.) Suppose profits are $2 per unit for product A

A and $10 per unit for product B. Find the production schedule that maximizes the total profit.

25. **Optimal Amount of Labor** The production function for a firm is $f(x, y) = 64x^{3/4}y^{1/4}$, where x and y are the number of units of labor and capital utilized. Suppose that labor costs $96 per unit and capital costs $162 per unit and that the firm decides to produce 3456 units of goods.
    (a) Determine the amounts of labor and capital that should be utilized in order to minimize the cost. That is, find the values of x, y that minimize $96x + 162y$, subject to the constraint $3456 - 64x^{3/4}y^{1/4} = 0$.
    (b) Find the value of λ at the optimal level of production.
    (c) Show that, at the optimal level of production, we have

    $$\frac{[\text{marginal productivity of labor}]}{[\text{marginal productivity of capital}]} = \frac{[\text{unit price of labor}]}{[\text{unit price of capital}]}.$$

26. **Maximizing Profit** Consider the firm of Example 2, Section 7.3, that sells its goods in two countries. Suppose that the firm must set the same price in each country. That is, $97 - (x/10) = 83 - (y/20)$. Find the values of x and y that maximize profits under this new restriction.

27. **Maximizing a Product** Find the values of x, y, and z that maximize xyz subject to the constraint $36 - x - 6y - 3z = 0$.

28. Find the values of x, y, and z that maximize $xy + 3xz + 3yz$ subject to the constraint $9 - xyz = 0$.

29. Find the values of x, y, z that maximize
    $$3x + 5y + z - x^2 - y^2 - z^2,$$
    subject to the constraint $6 - x - y - z = 0$.

30. Find the values of x, y, z that minimize
    $$x^2 + y^2 + z^2 - 3x - 5y - z,$$
    subject to the constraint $20 - 2x - y - z = 0$.

31. **Minimizing Cost** The material for a closed rectangular box costs $2 per square foot for the top and $1 per square foot for the sides and bottom. Using Lagrange multipliers, find the dimensions for which the volume of the box is 12 cubic feet and the cost of the materials is minimized. [Refer to Fig. 4(a)] The cost will be $3xy + 2xz + 2yz$.

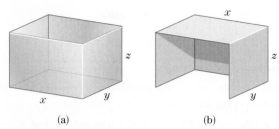

(a)   (b)

Figure 4

32. Use Lagrange multipliers to find the three positive numbers whose sum is 15 and whose product is as large as possible.

33. **Minimizing Surface Area** Find the dimensions of an open rectangular glass tank of volume 32 cubic feet for which the amount of material needed to construct the tank is minimized. [See Fig. 4(a).]

34. **Maximizing Volume** A shelter for use at the beach has a back, two sides, and a top made of canvas. [See Fig. 4(b).] Find the dimensions that maximize the volume and require 96 square feet of canvas.

35. **Production Function** Let $f(x, y)$ be any production function where x represents labor (costing a per unit) and y represents capital (costing b per unit). Assuming that c is available, show that, at the values of x, y that maximize production,

    $$\frac{\frac{\partial f}{\partial x}}{\frac{\partial f}{\partial y}} = \frac{a}{b}.$$

    Note: Let $F(x, y, \lambda) = f(x, y) + \lambda(c - ax - by)$. The result follows from (L-1) and (L-2).

36. **Production Function** By applying the result in Exercise 25 to the production function $f(x, y) = kx^\alpha y^\beta$, show that, for the values of x, y that maximize production, we have

    $$\frac{y}{x} = \frac{a\beta}{b\alpha}.$$

    (This tells us that the ratio of capital to labor does not depend on the amount of money available, nor on the level of production, but only on the numbers a, b, α, and β.)

Solutions to Check Your Understanding 7.4

1. The function can be written as
    $$F(x, y, \lambda) = 2x + 3y + \lambda \cdot 90 - \lambda \cdot 6x^{1/3}y^{2/3}.$$

    When differentiating with respect to x, treat both y and λ as constants (so $\lambda \cdot 90$ and $\lambda \cdot 6$ are also regarded as constants).

    $$\frac{\partial F}{\partial x} = 2 - \lambda \cdot 6 \cdot \frac{1}{3}x^{-2/3} \cdot y^{2/3}$$
    $$= 2 - 2\lambda x^{-2/3}y^{2/3}$$

(*Note:* It is not necessary to write out the multiplication by λ as we did. Most people just do this mentally and then differentiate.)

2. The quantity to be maximized is the volume xyl. The constraint is that length plus girth is 84. This translates to $84 = l + 2x + 2y$ or $84 - l - 2x - 2y = 0$. Therefore,
    $$F(x, y, l, \lambda) = xyl + \lambda(84 - l - 2x - 2y).$$

7.5 The Method of Least Squares

Today, people can compile graphs of literally thousands of different quantities: the purchasing value of the dollar as a function of time, the pressure of a fixed volume of air as a function of temperature, the average income of people as a function of their years of formal education, or the incidence of strokes as a function of blood pressure.

The observed points on such graphs tend to be irregularly distributed due to the complicated nature of the phenomena underlying them, as well as to errors made in observation. (For example, a given procedure for measuring average income may not count certain groups.)

In spite of the imperfect nature of the data, we are often faced with the problem of making assessments and predictions based on them. Roughly speaking, this problem amounts to filtering the sources of errors in the data and isolating the basic underlying trend. Frequently, on the basis of a suspicion or a working hypothesis, we may suspect that the underlying trend is linear; that is, the data should lie on a straight line. But which straight line? This is the problem that the *method of least squares* attempts to answer. To be more specific, let us consider the following problem:

Problem of Fitting a Straight Line to Data

Given observed data points (x_1, y_1), (x_2, y_2), $\ldots$, (x_N, y_N) on a graph, find the straight line that best fits these points.

To completely understand the statement of the problem being considered, we must define what it means for a line to "best" fit a set of points. If (x_i, y_i) is one of our observed points, we will measure how far it is from a given line $y = Ax + B$ by the vertical distance from the point to the line. Since the point on the line with x-coordinate x_i is $(x_i, Ax_i + B)$, this vertical distance is the distance between the y-coordinates $Ax_i + B$ and y_i. (See Fig. 1.) If $E_i = (Ax_i + B) - y_i$, either E_i or $-E_i$ is the vertical distance from (x_i, y_i) to the line. To avoid this ambiguity, we work with the square of this vertical distance:

$$E_i^2 = (Ax_i + B - y_i)^2.$$

The total error in approximating the data points $(x_1, y_1), \ldots, (x_N, y_N)$ by the line $y = Ax + B$ is usually measured by the sum E of the squares of the vertical distances from the points to the line,

$$E = E_1^2 + E_2^2 + \cdots + E_N^2.$$

E is called the **least-squares error** of the observed points with respect to the line. If all the observed points lie on the line $y = Ax + B$, all E_i are zero and the error E is zero. If a given observed point is far away from the line, the corresponding E_i^2 is large and hence makes a large contribution to the error E.

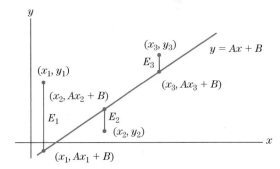

Figure 1 Fitting a line to data points.

In general, we cannot expect to find a line $y = Ax + B$ that fits the observed points so well that the error E is zero. Actually, this situation will occur only if the observed points lie on a straight line. However, we can rephrase our original problem as follows:

Problem   Given observed data points (x_1, y_1), (x_2, y_2), $\ldots$, (x_N, y_N), find a straight line $y = Ax + B$ for which the error E is as small as possible. This line is called the **least-squares line** or **regression line**.

It turns out that this problem is a minimization problem in the two variables A and B and can be solved with the methods of Section 7.3. Let us consider an example.

EXAMPLE 1 **Least-Squares Error** Find the straight line that minimizes the least-squares error for the points $(1, 4)$, $(2, 5)$, $(3, 8)$.

SOLUTION Let the straight line be $y = Ax + B$. When $x = 1, 2, 3$, the y-coordinate of the corresponding point of the line is $A + B$, $2A + B$, $3A + B$, respectively. Therefore, the squares of the vertical distances from the points $(1, 4)$, $(2, 5)$, $(3, 8)$ are, respectively,

$$E_1^2 = (A + B - 4)^2,$$
$$E_2^2 = (2A + B - 5)^2,$$
$$E_3^2 = (3A + B - 8)^2.$$

(See Fig. 2.) Thus, the least-squares error is

$$E = E_1^2 + E_2^2 + E_3^2 = (A + B - 4)^2 + (2A + B - 5)^2 + (3A + B - 8)^2.$$

This error obviously depends on the choice of A and B. Let $f(A, B)$ denote this least-squares error. We want to find values of A and B that minimize $f(A, B)$. To do so, we take partial derivatives with respect to A and B and set the partial derivatives equal to zero:

$$\frac{\partial f}{\partial A} = 2(A + B - 4) + 2(2A + B - 5) \cdot 2 + 2(3A + B - 8) \cdot 3$$
$$= 28A + 12B - 76 = 0,$$
$$\frac{\partial f}{\partial B} = 2(A + B - 4) + 2(2A + B - 5) + 2(3A + B - 8)$$
$$= 12A + 6B - 34 = 0.$$

To find A and B, we must solve the system of simultaneous linear equations

$$28A + 12B = 76$$
$$12A + 6B = 34.$$

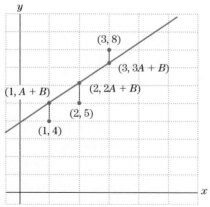

Figure 2

Multiplying the second equation by 2 and subtracting from the first equation, we have $4A = 8$, or $A = 2$. Therefore, $B = \frac{5}{3}$, and the straight line that minimizes the least-squares error is $y = 2x + \frac{5}{3}$.

➤➤ Now Try Exercise 1

For a general set of data points, the minimization process used in Example 1 can be generalized to any set of data points $(x_1, y_1), \ldots, (x_N, y_N)$ to yield the following algebraic formula for A and B:

$$A = \frac{N \cdot \Sigma xy - \Sigma x \cdot \Sigma y}{N \cdot \Sigma x^2 - (\Sigma x)^2},$$

$$B = \frac{\Sigma y - A \cdot \Sigma x}{N},$$

where

Σx = sum of the x-coordinates of the data points

Σy = sum of the y-coordinates of the data points

Σxy = sum of the products of the coordinates of the data points

Σx^2 = sum of the squares of the x-coordinates of the data points

N = number of data points.

That is,

$$\Sigma x = x_1 + x_2 + \cdots + x_N$$
$$\Sigma y = y_1 + y_2 + \cdots + y_N$$
$$\Sigma xy = x_1 \cdot y_1 + x_2 \cdot y_2 + \cdots + x_N \cdot y_N$$
$$\Sigma x^2 = x_1^2 + x_2^2 + \cdots + x_N^2.$$

EXAMPLE 2   **Car-Accident-Related Deaths in the U.S.** The following table gives the number in thousands of car-accident-related deaths in the U.S. for certain years.

Year	Number (in thousands)
1990	46.8
2000	43.4
2005	45.3
2007	43.9
2008	39.7
2009	35.9

(a) Use the preceding formulas to obtain the straight line that best fits these data.

(b) Use the straight line found in part (a) to estimate the number of car-accident-related deaths in 2012. (It is interesting that, while the number of drivers is obviously increasing with time, the number of car-accident-related deaths is actually decreasing, maybe because of improvements in car-manufacturing technologies and added safety measures.)

SOLUTION   **(a)** The data are plotted in Fig. 3, where x denotes the number of years since 1990. The sums are calculated in Table 1 and then used to determine the values of A and B.

Table 1 Car-Accident-Related Deaths in U.S.

x Years since 1990	y Number of deaths in thousands	xy	x^2
0	46.8	0	0
10	43.4	434	100
15	45.3	679.5	225
17	43.9	746.3	289
18	39.7	714.6	324
19	35.9	682.1	361
$\Sigma x = 79$	$\Sigma y = 255$	$\Sigma xy = 3256.5$	$\Sigma x^2 = 1299$

$$A = \frac{6 \cdot 3256.5 - 79 \cdot 255}{6 \cdot 1299 - 79^2} \approx -.39$$

$$B = \frac{255 + .39 \cdot 79}{6} \approx 47.64$$

Therefore, the equation of the least-squares line is $y = -.39x + 47.64$ (see Fig. 3).

Figure 3

(b) We use the straight line to estimate the number of car-accident-related deaths in 2012 by setting $x = 22$. Then, we get

$$y = (-.39)(22) + 47.64 = 39.06.$$

Therefore, we estimate the number of car-related accidental deaths to be 39.06 thousand in 2012.

>> *Now Try Exercise 9*

INCORPORATING TECHNOLOGY

Least-Squares Method To implement the least-squares method on your TI-83/84, select STAT 1 for the EDIT screen to obtain a table used for entering the data. If necessary, clear data from columns L_1 and/or L_2 by moving the cursor to the top of the column and pressing CLEAR ENTER. [See Fig. 4(a).]

After the x- and y-values are placed in lists on a graphing calculator, we use the statistical routine LinReg to calculate the coefficients of the least-squares line. Now press STAT ▷ for the CALC menu, and press 4 to place **LinReg(ax+b)** on the home screen. Press ENTER to obtain the slope and y-intercept of the least-squares line. [See Fig. 4(b).]

If desired, we can automatically assign the equation for the line to a function and graph it along with the original points. First, we assign the equation for the least-squares line to a function. Select Y=, move to the function, and press CLEAR to erase any current expression. Now, press VARS 5 to select the **Statistics** variables. Move your cursor over to the EQ menu, and press 1 for **RegEQ** (Regression Equation).

To graph this line, press GRAPH. To graph this line along with the original data points, we proceed as follows. From the Y=, and with only the least-squares line selected, press 2nd [STAT PLOT] ENTER to select **Plot1**, and press ENTER to turn **Plot1 ON**. Now, select the first plot from the six icons for the plot **Type**. This corresponds to a scatter plot. Finally, press GRAPH. [See Fig. 4(c).]

Figure 4   (a)   (b)   (c)

Check Your Understanding 7.5

Solutions can be found following the section exercises.

1. Let $E = (A + B + 2)^2 + (3A + B)^2 + (6A + B - 8)^2$. What is $\dfrac{\partial E}{\partial A}$?

2. Find the formula (of the type in Problem 1) that gives the least-squares error E for the points $(1, 10)$, $(5, 8)$, and $(7, 0)$.

EXERCISES 7.5

1. Find the least-squares error E for the least-squares line fit to the four points in Fig. 5.

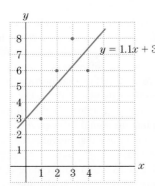

$y = 1.1x + 3$

Figure 5

2. Find the least-squares error E for the least-squares line fit to the five points in Fig. 6.

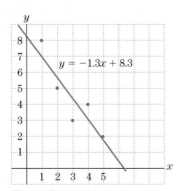

$y = -1.3x + 8.3$

Figure 6

3. Find the formula (of the type in Check Your Understanding Problem 1) that gives the least-squares error for the points $(2, 6)$, $(5, 10)$, and $(9, 15)$.

4. Find the formula (of the type in Check Your Understanding Problem 1) that gives the least-squares error for the points $(8, 4)$, $(9, 2)$, and $(10, 3)$.

In Exercises 5–8, use partial derivatives to obtain the formula for the best least-squares fit to the data points.

5. $(1, 2)$, $(2, 5)$, $(3, 11)$

6. $(1, 8)$, $(2, 4)$, $(4, 3)$

7. $(1, 9)$, $(2, 8)$, $(3, 6)$, $(4, 3)$

8. $(1, 5)$, $(2, 7)$, $(3, 6)$, $(4, 10)$

9. Complete Table 2 and find the values of A and B for the straight line that provides the best least-squares fit to the data.

Table 2

x	y	xy	x^2
1	7		
2	6		
3	4		
4	3		
$\sum x =$	$\sum y =$	$\sum xy =$	$\sum x^2 =$

10. Complete Table 3 and find the values of A and B for the straight line that provides the best least-squares fit to the data.

Table 3

x	y	xy	x^2
1	2		
2	3		
3	7		
4	9		
5	12		
$\sum x =$	$\sum y =$	$\sum xy =$	$\sum x^2 =$

In the remaining exercises, use one or more of the three methods discussed in this section (partial derivatives, formulas, or graphing utilities) to obtain the formula for the least-squares line.

11. Health Care Expenditures Table 4. *(Source: Center for Medicare and Medicaid Services.)*

Table 4 U.S. Per Capita Health Care Expenditures

Years (after 2000)	Dollars
9	8175
10	8428
12	8996
13	9255

(a) Find the least-squares line for these data.

(b) Use the least-squares line to predict the per capita health care expenditures for the year 2016.

(c) Use the least-squares line to predict when per capita health care expenditures will reach \$12,000.

12. Table 5 gives the number of students enrolled at the University of Illinois, at Urbana-Champaign (UIUC), for the fall semesters 2012–2015.

Table 5 Student Enrollment at UIUC

Fall semester	Number of students
2012	42,883
2013	43,398
2014	43,603
2015	44,087

(a) Find the least-squares line for these data.

(b) The university will build more student housing on campus once enrollment exceeds 46,000. Based on your model in part (a), in what year should the university build more student housing?

13. Table 6 gives the U.S. minimum wage in dollars for certain years.

Table 6 U.S. Federal Minimum Wage

Year	Wage
2000	$5.15
2005	$5.15
2010	$7.25
2016	$7.25

(a) Use the method of least squares to obtain the straight line that best fits these data. [*Hint:* First convert *Year* to *Years after 2000*.]

(b) Estimate the minimum wage for the year 2008.

(c) If the trend determined by the straight line in part (a) continues, when will the minimum wage reach $10?

14. Table 7 gives the number of visitors per year at Yosemite National Park.

(a) Find the least-squares line for these data.

(b) Estimate the number of visitors in 2017.

Table 7 Yosemite National Park Visitors

Year	Number of Visitors in millions
2010	3.901
2011	3.951
2012	3.853
2013	3.691
2014	3.882
2015	4.150

15. An ecologist wished to know whether certain species of aquatic insects have their ecological range limited by temperature. He collected the data in Table 8, relating the average daily temperature at different portions of a creek with the elevation (above sea level) of that portion of the creek. (The authors express their thanks to Dr. J. David Allen, formerly of the Department of Zoology at the University of Maryland, for providing the data for this exercise.)

(a) Find the straight line that provides the best least-squares fit to these data.

(b) Use the linear function to estimate the average daily temperature for this creek at altitude 3.2 kilometers.

Table 8 Relationship between Elevation and Temperature in a Creek

Elevation (kilometers)	Average Temperature (degrees Celsius)
2.7	11.2
2.8	10
3.0	8.5
3.5	7.5

Solutions to Check Your Understanding 7.5

1. $\dfrac{\partial E}{\partial A} = 2(A + B + 2) \cdot 1 + 2(3A + B) \cdot 3$

$\qquad + 2(6A + B - 8) \cdot 6$

$\quad = (2A + 2B + 4) + (18A + 6B)$

$\qquad + (72A + 12B - 96)$

$\quad = 92A + 20B - 92.$

(Notice that we used the general power rule when differentiating and so had to always multiply by the derivative of the quantity inside the parentheses. Also, you might be tempted to first square the terms in the expression for E and then differentiate. We recommend that you resist this temptation.)

2. $E = (A + B - 10)^2 + (5A + B - 8)^2 + (7A + B)^2$. In general, E is a sum of squares, one for each point being fitted. The point (a, b) gives rise to the term $(aA + B - b)^2$.

7.6 Double Integrals

Up to this point, our discussion of the calculus of several variables has been confined to the study of differentiation. Let us now take up the topic of the integration of functions of several variables. As has been the case throughout most of this chapter, we restrict our discussion to functions $f(x, y)$ of two variables.

We begin with some motivation. Before we define the concept of an integral for functions of several variables, we review the essential features of the integral in one variable.

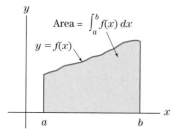

$$\text{Area} = \int_a^b f(x)\,dx$$

$y = f(x)$

Figure 1

Consider the definite integral $\int_a^b f(x)\,dx$. To write down this integral takes two pieces of information. The first is the function $f(x)$. The second is the interval over which the integration is to be performed. In this case, the interval is the portion of the x-axis from $x = a$ to $x = b$. The value of the definite integral is a number. In case the function $f(x)$ is nonnegative throughout the interval from $x = a$ to $x = b$, this number equals the area under the graph of $f(x)$ from $x = a$ to $x = b$. (See Fig. 1.) If $f(x)$ is negative for some values of x in the interval, the integral still equals the area bounded by the graph, but areas below the x-axis are counted as negative.

Let us generalize this to a function $f(x, y)$ of two variables. First, we must provide a two-dimensional analog of the interval from $x = a$ to $x = b$. This is easy. We take a two-dimensional region R of the plane, such as the region shown in Fig. 2. As our generalization of $f(x)$, we take a function $f(x, y)$ of two variables. Our generalization of the definite integral is denoted

$$\iint\limits_R f(x, y)\,dx\,dy$$

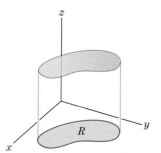

Figure 2 A region in the xy-plane.

and is called the **double integral of $f(x, y)$ over the region R**. The value of the double integral is a number defined as follows. For the sake of simplicity, let us begin by assuming that $f(x, y) \geq 0$ for all points (x, y) in the region R. [This is the analog of the assumption that $f(x) \geq 0$ for all x in the interval from $x = a$ to $x = b$.] This means that the graph of f lies above the region R in three-dimensional space. (See Fig. 3.) The portion of the graph over R determines a solid figure. (See Fig. 4.) This figure is called the *solid bounded by $f(x, y)$ over the region R*. We define the double integral $\iint\limits_R f(x, y)\,dx\,dy$ to be the volume of this solid. In case the graph of $f(x, y)$ lies partially above the region R and partially below, we define the double integral to be the volume of the solid above the region minus the volume of the solid below the region. That is, we count volumes below the xy-plane as negative.

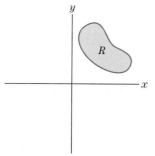

Figure 3 Graph of $f(x, y)$ above the region R.

Now that we have defined the notion of a double integral, we must learn how to calculate its value. To do so, let us introduce the notion of an iterated integral. Let $f(x, y)$ be a function of two variables, let $g(x)$ and $h(x)$ be two functions of x alone, and let a and b be numbers. Then, an **iterated integral** is an expression of the form

$$\int_a^b \left(\int_{g(x)}^{h(x)} f(x, y)\,dy \right) dx.$$

To explain the meaning of this collection of symbols, we proceed from the inside out. We evaluate the integral

$$\int_{g(x)}^{h(x)} f(x, y)\,dy$$

by considering $f(x, y)$ as a function of y alone. This is indicated by the dy in the inner integral. We treat x as a constant in this integration. So, we evaluate the integral by first finding an antiderivative $F(x, y)$ with respect to y. The integral above is then evaluated as

$$F(x, h(x)) - F(x, g(x)).$$

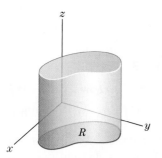

Figure 4 Solid bounded by $f(x, y)$ over R.

That is, we evaluate the antiderivative between the limits $y = g(x)$ and $y = h(x)$. This gives us a function of x alone. To complete the evaluation of the integral, we integrate this function from $x = a$ to $x = b$. The next two examples illustrate the procedure for evaluating iterated integrals.

EXAMPLE 1   **A Double Integral** Evaluate the iterated integral

$$\int_1^2 \left(\int_3^4 (y - x)\,dy \right) dx.$$

SOLUTION Here $g(x)$ and $h(x)$ are constant functions: $g(x) = 3$ and $h(x) = 4$. We evaluate the inner integral first. The variable in this integral is y, so we treat x as a constant:

$$\int_3^4 (y - x)\, dy = \left(\frac{1}{2}y^2 - xy\right)\Big|_3^4$$

$$= \left(\frac{1}{2} \cdot 16 - x \cdot 4\right) - \left(\frac{1}{2} \cdot 9 - x \cdot 3\right)$$

$$= 8 - 4x - \frac{9}{2} + 3x$$

$$= \frac{7}{2} - x.$$

Now, we carry out the integration with respect to x:

$$\int_1^2 \left(\frac{7}{2} - x\right) dx = \frac{7}{2}x - \frac{1}{2}x^2\Big|_1^2$$

$$= \left(\frac{7}{2} \cdot 2 - \frac{1}{2} \cdot 4\right) - \left(\frac{7}{2} - \frac{1}{2} \cdot 1\right)$$

$$= (7 - 2) - (3) = 2.$$

So, the value of the iterated integral is 2. **» Now Try Exercise 1**

EXAMPLE 2 **A Double Integral** Evaluate the iterated integral

$$\int_0^1 \left(\int_{\sqrt{x}}^{x+1} 2xy\, dy\right) dx.$$

SOLUTION We evaluate the inner integral first:

$$\int_{\sqrt{x}}^{x+1} 2xy\, dy = xy^2\Big|_{\sqrt{x}}^{x+1} = x(x + 1)^2 - x(\sqrt{x})^2$$

$$= x(x^2 + 2x + 1) - x \cdot x$$

$$= x^3 + 2x^2 + x - x^2$$

$$= x^3 + x^2 + x.$$

Now, we evaluate the outer integral:

$$\int_0^1 (x^3 + x^2 + x)\, dx = \frac{1}{4}x^4 + \frac{1}{3}x^3 + \frac{1}{2}x^2\Big|_0^1 = \frac{1}{4} + \frac{1}{3} + \frac{1}{2} = \frac{13}{12}.$$

So, the value of the iterated integral is $\frac{13}{12}$. **» Now Try Exercise 7**

Let us now return to the discussion of the double integral $\iint\limits_R f(x, y)\, dx\, dy$. When the region R has a special form, the double integral may be expressed as an iterated integral, as follows: Suppose that R is bounded by the graphs of $y = g(x)$ and $y = h(x)$, and by the vertical lines $x = a$ and $x = b$. (See Fig. 5.) In this case, we have the following fundamental result, which we cite without proof.

Let R be the region in the xy-plane bounded by the graphs of $y = g(x)$, $y = h(x)$, and the vertical lines $x = a$, $x = b$. Then,

$$\iint\limits_R f(x, y)\, dx\, dy = \int_a^b \left(\int_{g(x)}^{h(x)} f(x, y)\, dy\right) dx.$$

Since the value of the double integral gives the volume of the solid bounded by the graph of $f(x, y)$ over the region R, the preceding result may be used to calculate volumes, as the next two examples show.

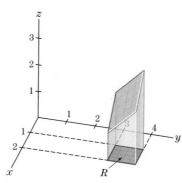

Figure 5

EXAMPLE 3   **Volume Using a Double Integral** Calculate the volume of the solid bounded above by the function $f(x, y) = y - x$ and lying over the rectangular region R: $1 \leq x \leq 2$, $3 \leq y \leq 4$. (See Fig. 6.)

SOLUTION   The desired volume is given by the double integral $\iint\limits_{R} f(y - x)\, dx\, dy$. By the result just cited, this double integral is equal to the iterated integral

$$\int_{1}^{2} \left(\int_{3}^{4} (y - x)\, dy \right) dx.$$

The value of this iterated integral was shown in Example 1 to be 2, so the volume of the solid shown in Fig. 6 is 2.   **>> Now Try Exercise 13**

Figure 6

EXAMPLE 4   **Double Integral over a Region** Calculate $\iint\limits_{R} 2xy\, dx\, dy$, where R is the region shown in Fig. 7.

SOLUTION   The region R is bounded below by $y = \sqrt{x}$, above by $y = x + 1$, on the left by $x = 0$, and on the right by $x = 1$. Therefore,

$$\iint\limits_{R} 2xy\, dx\, dy = \int_{0}^{1} \left(\int_{\sqrt{x}}^{x+1} 2xy\, dy \right) dx = \frac{13}{12} \quad \text{(by Example 2).}$$

>> Now Try Exercise 9

In our discussion, we have confined ourselves to iterated integrals in which the inner integral was with respect to y. In a completely analogous manner, we may treat iterated integrals in which the inner integral is with respect to x. Such iterated integrals may be used to evaluate double integrals over regions R bounded by curves of the form $x = g(y)$, $x = h(y)$ and horizontal lines $y = a$, $y = b$. The computations are analogous to those given in this section.

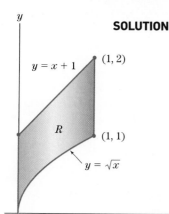

Figure 7

Check Your Understanding 7.6

Solutions can be found following the section exercises.

1. Calculate the iterated integral

$$\int_0^2 \left(\int_0^{x/2} e^{2y-x}\, dy \right) dx.$$

2. Calculate $\iint_R e_{2y-x}\, dx\, dy$, where R is the region in Fig. 8.

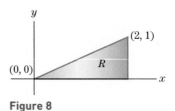

Figure 8

EXERCISES 7.6

Calculate the following iterated integrals.

1. $\displaystyle \int_0^1 \left(\int_0^1 e^{x+y}\, dy \right) dx$ 2. $\displaystyle \int_{-1}^1 \left(\int_{-1}^1 xy\, dx \right) dy$

3. $\displaystyle \int_{-2}^0 \left(\int_{-1}^1 xe^{xy}\, dy \right) dx$ 4. $\displaystyle \int_0^1 \left(\int_{-1}^1 \frac{1}{3}y^3 x\, dy \right) dx$

5. $\displaystyle \int_1^4 \left(\int_x^{x^2} xy\, dy \right) dx$ 6. $\displaystyle \int_0^3 \left(\int_x^{2x} y\, dy \right) dx$

7. $\displaystyle \int_{-1}^1 \left(\int_x^{2x} (x+y)\, dy \right) dx$ 8. $\displaystyle \int_0^1 \left(\int_0^x e^{x+y}\, dy \right) dx$

Let R be the rectangle consisting of all points (x, y) such that $0 \le x \le 2$, $2 \le y \le 3$. Calculate the following double integrals. Interpret each as a volume.

9. $\displaystyle \iint_R xy^2\, dx\, dy$ 10. $\displaystyle \iint_R (xy + y^2)\, dx\, dy$

11. $\displaystyle \iint_R e^{-x-y}\, dx\, dy$ 12. $\displaystyle \iint_R e^{y-x}\, dx\, dy$

Calculate the volumes over the following regions R bounded above by the graph of $f(x, y) = x^2 + y^2$.

13. R is the rectangle bounded by the lines $x = 1$, $x = 3$, $y = 0$, and $y = 1$.

14. R is the region bounded by the lines $x = 0$, $x = 1$ and the curves $y = 0$ and $y = \sqrt[3]{x}$.

Solutions to Check Your Understanding 7.6

1. $\displaystyle \int_0^2 \left(\int_0^{x/2} e^{2y-x}\, dy \right) dx = \int_0^2 \left(\frac{1}{2}e^{2y-x}\Big|_0^{x/2} \right) dx$

$$= \int_0^2 \left(\frac{1}{2}e^{2(x/2)-x} - \frac{1}{2}e^{2(0)-x} \right) dx$$

$$= \int_0^2 \left(\frac{1}{2} - \frac{1}{2}e^{-x} \right) dx$$

$$= \frac{1}{2}x + \frac{1}{2}e^{-x}\Big|_0^2$$

$$= \frac{1}{2}\cdot 2 + \frac{1}{2}e^{-2} - \left(\frac{1}{2}\cdot 0 + \frac{1}{2}e^{-0} \right)$$

$$= 1 + \frac{1}{2}e^{-2} - 0 - \frac{1}{2}$$

$$= \frac{1}{2} + \frac{1}{2}e^{-2}$$

2. The line passing through the points $(0, 0)$ and $(2, 1)$ has equation $y = x/2$. Hence, the region R is bounded below by $y = 0$, above by $y = x/2$, on the left by $x = 0$, and on the right by $x = 2$. Therefore,

$$\iint_R e^{2y-x}\, dx\, dy = \int_0^2 \left(\int_0^{x/2} e^{2y-x}\, dy \right) dx$$

$$= \frac{1}{2} + \frac{1}{2}e^{-2}$$

by Problem 1.

CHAPTER 7 Summary

KEY TERMS AND CONCEPTS	EXAMPLES

7.1 Examples of Functions of Several Variables

A function $f(x, y)$ of the two variables x and y is a rule that assigns a number to each pair of values for the variables.

Given a function of two variables $f(x, y)$, the graph of the equation $f(x, y) = c$ is a curve in the xy-plane called the **level curve of height** c.

Suppose that, during a certain time period, the number of units of goods produced with x units of labor and y units of capital is given by the Cobb–Douglas production function $f(x, y) = 40x^{1/2}y^{1/2}$.

(a) How many units of goods will be produced with 16 units of labor and 16 units of capital?

(b) Determine the isoquant or level curve at level 100 of the production function.

Solution

(a) $f(16, 16) = 40(16)^{1/2} \cdot (16)^{1/2} = 40 \cdot 4 \cdot 4 = 640$; 640 units of goods will be produced.

(b) The level curve is the graph of $f(x, y) = 100$, or

$$40x^{1/2}y^{1/2} = 100$$

$$y^{1/2} = \frac{100}{40x^{1/2}} = \frac{5}{2x^{1/2}}$$

$$y = \frac{25}{4x}.$$

Thus, the isoquant is the curve $y = \frac{25}{4x}$. Each point on the curve has coordinates $(x, \frac{25}{4x})$ and represents a combination of capital x and labor $\frac{25}{4x}$ that yields 100 units of production.

7.2 Partial Derivatives

The **partial derivative of $f(x, y)$ with respect to x**, written $\frac{\partial f}{\partial x}$, is the derivative of $f(x, y)$, where y is treated as a constant and $f(x, y)$ is considered as a function of x alone. The **partial derivative of $f(x, y)$ with respect to y**, written $\frac{\partial f}{\partial y}$, is the derivative of $f(x, y)$, where x is treated as a constant. Higher-order derivatives are defined similarly.

Let $f(x, y) = e^{x^2 + 7y}$.

(a) Compute $\frac{\partial f}{\partial x}, \frac{\partial f}{\partial y}$, and $\frac{\partial f}{\partial x}(1, -1)$.

(b) Compute $\frac{\partial^2 f}{\partial x^2}$.

Solution

(a) Thinking of y as a constant, we have

$$\frac{\partial f}{\partial x} = \frac{\partial}{\partial x}[e^{x^2 + 7y}] = e^{x^2 + 7y} \cdot \frac{\partial}{\partial x}[x^2 + 7y] = 2x\, e^{x^2 + 7y}.$$

Thinking of x as a constant, we have

$$\frac{\partial f}{\partial y} = \frac{\partial}{\partial y}[e^{x^2 + 7y}] = e^{x^2 + 7y} \cdot \frac{\partial}{\partial y}[x^2 + 7y] = 7e^{x^2 + 7y}.$$

Finally,

$$\frac{\partial f}{\partial x}(1, -1) = 2x\, e^{x^2 + 7y}\bigg|_{(1, -1)} = 2e^{1-7} = 2e^{-6}.$$

(b) Start with the formula for $\frac{\partial f}{\partial x}$, and think of y as a constant; then,

$$\frac{\partial^2 f}{\partial x^2} = \frac{\partial}{\partial x}\left(\frac{\partial f}{\partial x}\right) = \frac{\partial}{\partial x}[2x\, e^{x^2 + 7y}]$$

$$= e^{x^2 + 7y}(2) + 2x\frac{\partial}{\partial x}[e^{x^2 + 7y}] \quad \text{Product rule.}$$

$$= 2e^{x^2 + 7y} + 2xe^{x^2 + 7y}\frac{\partial}{\partial x}(x^2 + 7y)$$

$$= 2e^{x^2 + 7y} + 2xe^{x^2 + 7y}(2x)$$

$$= 2e^{x^2 + 7y}(1 + 2x^2).$$

KEY TERMS AND CONCEPTS	EXAMPLES

7.3 Maxima and Minima of Functions of Several Variables

First-Derivative Test for Functions of Two Variables If $f(x, y)$ has either a relative maximum or minimum at $(x, y) = (a, b)$, then

$$\frac{\partial f}{\partial x}(a, b) = 0 \quad \text{and} \quad \frac{\partial f}{\partial y}(a, b) = 0.$$

You can use this test to locate candidate points where the function has a relative extreme value. Once you have located a candidate point where the first derivatives are 0, you check whether this point is a maximum, a minimum, or neither by applying the second-derivative test for function of two variables. The outcomes of the second-derivative test depend on the signs of $\frac{\partial^2 f}{\partial x^2}(a, b)$ and

$$D(x, y) = \frac{\partial^2 f}{\partial x^2} \cdot \frac{\partial^2 f}{\partial y^2} - \left(\frac{\partial^2 f}{\partial x \, \partial y}\right)^2.$$

(See Section 7.3 for the full statement and the examples for an illustration.)

Let $f(x, y) = x^2 + y^2 - 4x - 6y + 10$.

(a) Find $\dfrac{\partial f}{\partial x}$ and $\dfrac{\partial f}{\partial y}$.

(b) Find the points (x, y) where the first derivatives are zero.

(c) Apply the second derivative test at the points in (b), and decide, if possible, the nature of $f(x, y)$ at each of these points.

Solution

(a) $\dfrac{\partial f}{\partial x} = \dfrac{\partial}{\partial x}\,[x^2 + y^2 - 4x - 6y + 10] = 2x - 4$

$\dfrac{\partial f}{\partial y} = 2y - 6$

(b) The partial derivatives are equal to 0 when $x = 2$ and $y = 3$, so the only point where both partial derivatives are 0 is $(2, 3)$.

(c) The nature of the function at the point $(2, 3)$ depends on the signs of $D(2, 3)$ and $\dfrac{\partial^2 f}{\partial x^2}(2, 3)$. We have

$$\frac{\partial^2 f}{\partial x^2} = 2, \frac{\partial^2 f}{\partial y^2} = 2, \quad \frac{\partial^2 f}{\partial x \, \partial y} = \frac{\partial}{\partial x}(2y - 6) = 0.$$

So,

$$D(x, y) = \frac{\partial^2 f}{\partial x^2} \cdot \frac{\partial^2 f}{\partial y^2} - \left(\frac{\partial^2 f}{\partial x \, \partial y}\right)^2 = (2)(2) - 0 = 4.$$

Since $D(2, 3) = 4 > 0$ and $\dfrac{\partial^2 f}{\partial x^2}(2, 3) = 2 > 0$, according to the second-derivative test, $f(x, y)$ has a relative minimum at $(2, 3)$.

7.4 Lagrange Multipliers and Constrained Optimization

To find a relative maximum or minimum of the function $f(x, y)$ subject to the constraint $g(x, y) = 0$, we can use the Lagrange multiplier method. We apply this method in steps as follows:

Step 1 Form the function

$$F(x, y, \lambda) = f(x, y) + \lambda g(x, y).$$

The number λ is called a **Lagrange multiplier**.

Step 2 Compute the partial derivatives $\dfrac{\partial F}{\partial x}$ and $\dfrac{\partial F}{\partial y}$.

Step 3 Solve the system of equations

$$\frac{\partial F}{\partial x} = 0; \frac{\partial F}{\partial y} = 0; g(x, y) = 0.$$

Step 4 If you found more than one point (x, y) in Step 3, evaluate f at all the points. The largest of these values is the maximum value for f, and the smallest is the minimum value of f.

Use Lagrange multipliers to find the minimum value of $f(x, y) = x^2 + y^2$ subject to the constraint $x + y = 4$.

Solution

Step 1 Write the constraint in the form $x + y - 4 = 0$. Then, $g(x, y) = x + y - 4$ and $F(x, y, \lambda) = x^2 + y^2 + \lambda(x + y - 4)$.

Step 2 $\dfrac{\partial F}{\partial x} = 2x + \lambda; \dfrac{\partial F}{\partial y} = 2y + \lambda$

Step 3 Solve

$$\begin{cases} 2x + \lambda = 0 & (1) \\ 2y + \lambda = 0 & (2) \\ x + y - 4 = 0 & (3) \end{cases}$$

Subtract (2) from (1) and get $2x - 2y = 0$ or $x = y$. Use $x = y$ in (3) and get $2y = 4$ or $y = 2$. Hence, $x = 2$.

Step 4 At the point $(2, 2)$, $f(x, y)$ takes on the value 8, which is the minimum value of $x^2 + y^2$ subject to $x + y = 4$

KEY TERMS AND CONCEPTS

7.5 The Method of Least Squares

Suppose that you have a set of N data points (x_1, y_1), (x_2, y_2), $\ldots$, (x_N, y_N), and you have reasons to believe that y is linearly related to x, or at least approximately. Then, you can look for a linear function $y = Ax + B$ that best fits the given data. This line is called the *least-squares line* or *regression line*. The coefficients A and B are computed as follows:

$$A = \frac{N \cdot \Sigma xy - \Sigma x \cdot \Sigma y}{N \cdot \Sigma x^2 - (\Sigma x)^2}$$

$$B = \frac{\Sigma y - A \cdot \Sigma x}{N},$$

where

Σx = sum of the x-coordinates of the data points

Σy = sum of the y-coordinates of the data points

Σxy = sum of the products of the coordinates of the data points

Σx^2 = sum of the squares of the x-coordinates of the data points

N = number of data points.

EXAMPLES

Table 1 shows the number of seniors who graduated from a high school in Jefferson City, Missouri, in the years 2007 to 2012.

(a) Find the line that best fits these data.

(b) Use the straight line that you found in part (a) to approximate the number of seniors who will graduate in 2013.

Table 1 Students' Data

Years (after 2000)	Graduating seniors
7	245
8	275
9	225
10	215
11	218
12	212

Solution

(a) Let x denote the number of years since 2000 and y the number of graduating seniors. Let $y = Ax + B$ denote the line of best fit through these points. The sums are calculated in Table 2 and then used to determine the values of A and B.

Table 2 Students' Data

x	y	xy	x^2
7	245	1715	49
8	275	2200	64
9	225	2025	81
10	215	2150	100
11	218	2398	121
12	212	2544	144
$\Sigma x = 57$	$\Sigma y = 1390$	$\Sigma xy = 13032$	$\Sigma x^2 = 559$

In this example, $N = 6$, since we have six data points. Applying the formulas for the coefficients, we find

$$A = \frac{N \cdot \Sigma xy - \Sigma x \cdot \Sigma y}{N \cdot \Sigma x^2 - (\Sigma x)^2}$$

$$= \frac{6 \cdot 13{,}032 - 57 \cdot 1390}{6 \cdot 559 - (57)^2}$$

$$= -\frac{346}{35} \approx -9.89$$

$$B = \frac{\Sigma y - A \cdot \Sigma x}{N}$$

$$= \frac{1390 + \frac{346}{35} \cdot 57}{6}$$

$$= \frac{34{,}186}{105} \approx 325.58.$$

Thus, the line of best fit is $y = -9.89x + 325.58$, which is a decreasing line.

KEY TERMS AND CONCEPTS	EXAMPLES
	(b) To approximate the number of graduating seniors in 2013, we set $x = 13$ in the formula for the line of best fit and get $$y = -9.89(13) + 325.58 \approx 197.01$$ Thus, in 2013, the number of graduating seniors will be approximately 197.

7.6 Double Integrals

For a function of two variables $f(x, y)$, we can define a double integral where (x, y) varies over a region R in the xy-plane. In the iterated double integral

$$\int_{-2}^{2}\left(\int_{0}^{3} f(x, y)dx \right) dy,$$

the inner variable x varies from 0 to 3, while the outer variable y varies from -2 to 2. The point (x, y) in this double integral varies over the rectangular region $0 \le x \le 3$, $-2 \le y \le 2$.

If, in the iterated integral, the symbol $dy\, dx$ appears, instead of $dx\, dy$, then you should integrate with respect to y first. This is the case with the iterated integral

$$\int_{-3}^{2}\left(\int_{1}^{5} f(x, y)dy \right) dx.$$

The inner integral, which you should compute first, is with respect to y.

Examples

Evaluate the iterated integral $\int_{0}^{1}\left(\int_{0}^{2} xy^2\, dx \right) dy$.

Solution

Step 1 Evaluate the inner integral in x, while treating y as a constant.

$$\int_{0}^{2} xy^2\, dx = \frac{y^2}{2}x^2\Big|_{0}^{2} = \frac{y^2}{2}(2^2 - 0) = 2y^2$$

Step 2 Evaluate the outer integral of the function $2y^2$ that we found in Step 1.

$$\int_{0}^{1} 2y^2\, dy = \frac{2}{3}y^3\Big|_{0}^{1} = \frac{2}{3}$$

Thus, the double integral is equal to $\frac{2}{3}$.

CHAPTER 7 Fundamental Concept Check Exercises

1. Give an example of a level curve of a function of two variables.

2. Explain how to find a first partial derivative of a function of two variables.

3. Explain how to find a second partial derivative of a function of two variables.

4. What expression involving a partial derivative gives an approximation to $f(a + h, b) - f(a, b)$?

5. Interpret $\dfrac{\partial f}{\partial y}(2, 3)$ as a rate of change.

6. Give an example of a Cobb–Douglas production function. What is the marginal productivity of labor? Of capital?

7. Explain how to find possible relative extreme points for a function of several variables.

8. State the second-derivative test for functions of two variables.

9. Outline how the method of Lagrange multipliers is used to solve an optimization problem.

10. What is the least-squares line approximation to a set of data points? How is the line determined?

11. Give a geometric interpretation for $\iint\limits_{R} f(x, y)\, dx\, dy$, where $f(x, y) \ge 0$.

12. Give a formula for evaluating a double integral in terms of an iterated integral.

CHAPTER 7 Review Exercises

1. Let $f(x, y) = x\sqrt{y}/(1 + x)$. Compute $f(2, 9)$, $f(5, 1)$, and $f(0, 0)$.

2. Let $f(x, y, z) = x^2 e^{y/z}$. Compute $f(-1, 0, 1)$, $f(1, 3, 3)$, and $f(5, -2, 2)$.

3. **A Savings Account** If A dollars are deposited in a bank at a 6% continuous interest rate, the amount in the account after t years is $f(A, t) = Ae^{0.06t}$. Find and interpret $f(10, 11.5)$.

4. Let $f(x, y, \lambda) = xy + \lambda(5 - x - y)$. Find $f(1, 2, 3)$.

5. Let $f(x, y) = 3x^2 + xy + 5y^2$. Find $\dfrac{\partial f}{\partial x}$ and $\dfrac{\partial f}{\partial y}$.

6. Let $f(x, y) = 3x - \frac{1}{2}y^4 + 1$. Find $\dfrac{\partial f}{\partial x}$ and $\dfrac{\partial f}{\partial y}$.

7. Let $f(x, y) = e^{x/y}$. Find $\dfrac{\partial f}{\partial x}$ and $\dfrac{\partial f}{\partial y}$.

8. Let $f(x, y) = x/(x - 2y)$. Find $\dfrac{\partial f}{\partial x}$ and $\dfrac{\partial f}{\partial y}$.

9. Let $f(x, y, z) = x^3 - yz^2$. Find $\dfrac{\partial f}{\partial x}, \dfrac{\partial f}{\partial y}$, and $\dfrac{\partial f}{\partial z}$.

10. Let $f(x, y, \lambda) = xy + \lambda(5 - x - y)$. Find $\dfrac{\partial f}{\partial x}, \dfrac{\partial f}{\partial y}$, and $\dfrac{\partial f}{\partial \lambda}$.

11. Let $f(x, y) = x^3y + 8$. Compute $\frac{\partial f}{\partial x}(1, 2)$ and $\frac{\partial f}{\partial y}(1, 2)$.

12. Let $f(x, y, z) = (x + y)z$. Evaluate $\frac{\partial f}{\partial y}$ at $(x, y, z) = (2, 3, 4)$.

13. Let $f(x, y) = x^5 - 2x^3y + \frac{1}{2}y^4$. Find $\frac{\partial^2 f}{\partial x^2}$, $\frac{\partial^2 f}{\partial y^2}$, $\frac{\partial^2 f}{\partial x\,\partial y}$, and $\frac{\partial^2 f}{\partial y\,\partial x}$.

14. Let $f(x, y) = 2x^3 + x^2y - y^2$. Compute $\frac{\partial^2 f}{\partial x^2}$, $\frac{\partial^2 f}{\partial y^2}$, and $\frac{\partial^2 f}{\partial x\,\partial y}$ at $(x, y) = (1, 2)$.

15. A dealer in a certain brand of electronic calculator finds that (within certain limits) the number of calculators she can sell per week is given by $f(p, t) = -p + 6t - .02pt$, where p is the price of the calculator and t is the number of dollars spent on advertising. Compute $\frac{\partial f}{\partial p}(25, 10{,}000)$ and $\frac{\partial f}{\partial t}(25, 10{,}000)$, and interpret these numbers.

16. The crime rate in a certain city can be approximated by a function $f(x, y, z)$, where x is the unemployment rate, y is the number of social services available, and z is the size of the police force. Explain why $\frac{\partial f}{\partial x} > 0$, $\frac{\partial f}{\partial y} < 0$, and $\frac{\partial f}{\partial z} < 0$.

In Exercises 17–20, find all points (x, y) where $f(x, y)$ has a possible relative maximum or minimum.

17. $f(x, y) = -x^2 + 2y^2 + 6x - 8y + 5$

18. $f(x, y) = x^2 + 3xy - y^2 - x - 8y + 4$

19. $f(x, y) = x^3 + 3x^2 + 3y^2 - 6y + 7$

20. $f(x, y) = \frac{1}{2}x^2 + 4xy + y^3 + 8y^2 + 3x + 2$

In Exercises 21–23, find all points (x, y) where $f(x, y)$ has a possible relative maximum or minimum. Then, use the second-derivative test to determine, if possible, the nature of $f(x, y)$ at each of these points. If the second-derivative test is inconclusive, so state.

21. $f(x, y) = x^2 + 3xy + 4y^2 - 13x - 30y + 12$

22. $f(x, y) = 7x^2 - 5xy + y^2 + x - y + 6$

23. $f(x, y) = x^3 + y^2 - 3x - 8y + 12$

24. Find the values of x, y, z at which

$$f(x, y, z) = x^2 + 4y^2 + 5z^2 - 6x + 8y + 3$$

assumes its minimum value.

Use the method of Lagrange multipliers to:

25. Maximize $3x^2 + 2xy - y^2$, subject to the constraint $5 - 2x - y = 0$.

26. Find the values of x, y that minimize $-x^2 - 3xy - \frac{1}{2}y^2 + y + 10$, subject to the constraint $10 - x - y = 0$.

27. Find the values of x, y, z that minimize

$$3x^2 + 2y^2 + z^2 + 4x + y + 3z, \text{ subject to the constraint } 4 - x - y - z = 0.$$

28. Find the dimensions of a rectangular box of volume 1000 cubic inches for which the sum of the dimensions is minimized.

29. A person wants to plant a rectangular garden along one side of a house and put a fence on the other three sides. (See Fig. 1.) Using the method of Lagrange multipliers, find the dimensions of the garden of greatest area that can be enclosed with 40 feet of fencing.

Figure 1 A garden.

30. The solution to Exercise 29 is $x = 10$, $y = 20$, $\lambda = 10$. If 1 additional foot of fencing becomes available, compute the new optimal dimensions and the new area. Show that the increase in area (compared with the area in Exercise 29) is approximately equal to 10 (the value of λ).

In Exercises 31–33, find the straight line that best fits the following data points, where "best" is meant in the sense of least squares.

31. $(1, 1), (2, 3), (3, 6)$

32. $(1, 1), (3, 4), (5, 7)$

33. $(0, 1), (1, -1), (2, -3), (3, -5)$

In Exercises 34 and 35, calculate the iterated integral.

34. $\int_0^1 \left(\int_0^4 (x\sqrt{y} + y)\, dy \right) dx$

35. $\int_0^5 \left(\int_1^4 (2xy^4 + 3)\, dy \right) dx$

In Exercises 36 and 37, let R be the rectangle consisting of all points (x, y), such that $0 \le x \le 4$, $1 \le y \le 3$, and calculate the double integral.

36. $\iint\limits_R (2x + 3y)\, dx\, dy$

37. $\iint\limits_R 5\, dx\, dy$

38. The present value of y dollars after x years at 15% continuous interest is $f(x, y) = ye^{-0.15x}$. Sketch some sample level curves. (Economists call this collection of level curves a **discount system**.)

The Trigonometric Functions

In this chapter, we expand the collection of functions to which we can apply calculus by introducing the trigonometric functions. As we shall see, these functions are **periodic**. That is, after a certain point, their graphs repeat themselves. This repetitive phenomenon is not displayed by any of the functions that we have considered until now. Yet, many natural phenomena are repetitive or cyclical, for example, the motion of the planets in our solar system, earthquake vibrations, and the natural rhythm of the heart. Thus, the functions introduced in this chapter add considerably to our capacity to describe physical processes.

8.1 Radian Measure of Angles

The ancient Babylonians introduced angle measurement in terms of degrees, minutes, and seconds, and these units are still generally used today for navigation and practical measurements. In calculus, however, it is more convenient to measure angles in terms of **radians**, for in this case the differentiation formulas for the trigonometric functions are easier to remember and use. Also, the radian is becoming more widely used today in scientific work because it is the unit of angle measurement in the international metric system (Système International d'Unités).

To define a radian, we consider a circle of radius 1 and measure angles in terms of distances around the circumference. The central angle determined by an arc of length 1 along the circumference is said to have a measure of 1 *radian*. (See Fig. 1.) Since the circumference of the circle of radius 1 has length 2π, there are 2π radians in one full revolution of the circle. Equivalently,

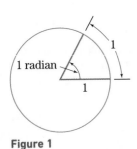

Figure 1

$$360° = 2\pi \text{ radians.} \tag{1}$$

The following important relations should be memorized (see Fig. 2):

$$90° = \frac{\pi}{2} \text{ radians} \qquad \text{(one quarter revolution)}$$

$$180° = \pi \text{ radians} \qquad \text{(one half revolution)}$$

$$270° = \frac{3\pi}{2} \text{ radians} \qquad \text{(three quarter revolutions)}$$

$$360° = 2\pi \text{ radians} \qquad \text{(one full revolution)}.$$

Figure 2

From formula (1), we see that

$$1° = \frac{2\pi}{360} \text{ radians} = \frac{\pi}{180} \text{ radians}.$$

If d is any number, then

$$d° = d \times \frac{\pi}{180} \text{ radians.} \qquad (2)$$

That is, to convert degrees to radians, multiply the number of degrees by $\pi/180$.

EXAMPLE 1 **Converting Degrees to Radians** Convert 45°, 60°, and 135° to radians.

SOLUTION

$$45° = \overset{1}{\cancel{45}} \times \frac{\pi}{\underset{4}{\cancel{180}}} \text{ radians} = \frac{\pi}{4} \text{ radians} \qquad \text{Use (2).}$$

$$60° = \overset{1}{\cancel{60}} \times \frac{\pi}{\underset{3}{\cancel{180}}} \text{ radians} = \frac{\pi}{3} \text{ radians}$$

$$135° = \overset{3}{\cancel{135}} \times \frac{\pi}{\underset{4}{\cancel{180}}} \text{ radians} = \frac{3\pi}{4} \text{ radians}$$

These three angles are shown in Fig. 3.

≫ Now Try Exercise 1

Figure 3

We usually omit the word *radian* when measuring angles because all our angle measurements will be in radians unless degrees are specifically indicated.

For our purposes, it is important to be able to speak of both negative and positive angles, so let us define what we mean by a negative angle. We shall usually consider angles that are in **standard position** on a coordinate system, with the vertex of the angle at $(0, 0)$ and one side, called the **initial side,** along the positive x-axis. We measure such an angle from the initial side to the **terminal side,** where a *counterclockwise angle is positive* and a *clockwise angle is negative*. Some examples are given in Fig. 4.

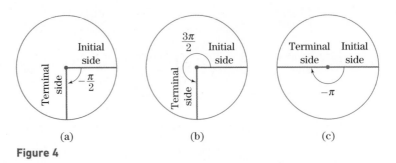

Figure 4

Notice in Figs. 4(a) and (b) how essentially the same picture can describe more than one angle.

By considering angles formed from more than one revolution (in the positive or negative direction), we can construct angles whose measure is of arbitrary size (that is, not necessarily between -2π and 2π). Three examples are illustrated in Fig. 5.

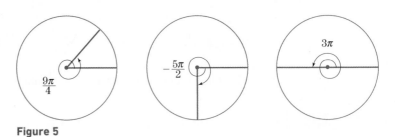

Figure 5

EXAMPLE 2

Measuring Angles in Radians
(a) What is the radian measure of the angle in Fig. 6?
(b) Construct an angle of $5\pi/2$ radians.

SOLUTION

(a) The angle described in Fig. 6 consists of one full revolution (2π radians) plus three quarter-revolutions [$3 \times (\pi/2)$ radians]. That is,

$$t = 2\pi + 3 \times \frac{\pi}{2} = 4 \times \frac{\pi}{2} + 3 \times \frac{\pi}{2} = \frac{7\pi}{2}.$$

(b) Think of $5\pi/2$ radians as $5 \times (\pi/2)$ radians, that is, five quarter-revolutions of the circle. This is one full revolution plus one quarter-revolution. An angle of $5\pi/2$ radians is shown in Fig. 7.

Figure 6   **Figure 7**   **» Now Try Exercise 5**

Check Your Understanding 8.1

Solutions can be found following the section exercises.

1. A right triangle has one angle of $\pi/3$ radians. What are the other angles?

2. How many radians are there in an angle of $-780°$? Draw the angle.

EXERCISES 8.1

Convert the following to radian measure.

1. $30°, 120°, 315°$

2. $18°, 72°, 150°$

3. $450°, -210°, -90°$

4. $990°, -270°, -540°$

Give the radian measure of each angle described.

5.

6.

7.

8.

9.

10.

11.

12.

Construct angles with the following radian measure.

13. $3\pi/2, 3\pi/4, 5\pi$

14. $\pi/3, 5\pi/2, 6\pi$

15. $-\pi/3, -3\pi/4, -7\pi/2$

16. $-\pi/4, -3\pi/2, -3\pi$

17. $\pi/6, -2\pi/3, -\pi$

18. $2\pi/3, -\pi/6, 7\pi/2$

Solutions to Check Your Understanding 8.1

1. The sum of the angles of a triangle is $180°$ or π radians. Since a right angle is $\pi/2$ radians and one angle is $\pi/3$ radians, the remaining angle is $\pi - (\pi/2 + \pi/3) = \pi/6$ radians.

2. $-780° = -780 \times (\pi/180)$ radians
$= -13\pi/3$ radians

Since $-13\pi/3 = -4\pi - \pi/3$, we draw the angle by first making two revolutions in the negative direction and then a rotation of $\pi/3$ in the negative direction. (See Fig. 8.)

Figure 8

8.2   The Sine and the Cosine

Given a number t, we consider an angle of t radians placed in standard position, as in Fig. 1, and we let P be a point on the terminal side of this angle. Denote the coordinates of P by (x, y), and let r be the length of the segment OP; that is, $r = \sqrt{x^2 + y^2}$. The **sine** and **cosine** of t, denoted by $\sin t$ and $\cos t$, respectively, are defined by the ratios

$$\sin t = \frac{y}{r}$$

$$\cos t = \frac{x}{r}.$$

(1)

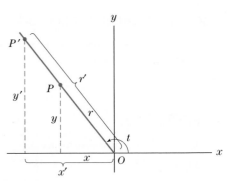

Figure 1 Diagram for the definitions of sine and cosine.

Figure 2

It does not matter which point on the ray through P we use to define $\sin t$ and $\cos t$. If $P' = (x', y')$ is another point on the same ray and if r' is the length of OP' (Fig. 2), then, by properties of similar triangles, we have

$$\frac{y'}{r'} = \frac{y}{r} = \sin t,$$

$$\frac{x'}{r'} = \frac{x}{r} = \cos t.$$

Three examples that illustrate the definition of $\sin t$ and $\cos t$ are shown in Fig. 3.

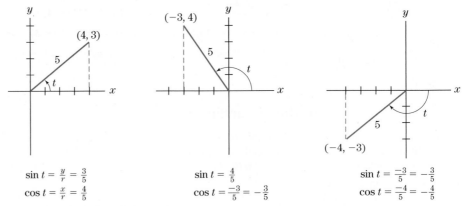

$$\sin t = \frac{y}{r} = \frac{3}{5} \qquad\qquad \sin t = \frac{4}{5} \qquad\qquad \sin t = \frac{-3}{5} = -\frac{3}{5}$$
$$\cos t = \frac{x}{r} = \frac{4}{5} \qquad\qquad \cos t = \frac{-3}{5} = -\frac{3}{5} \qquad\qquad \cos t = \frac{-4}{5} = -\frac{4}{5}$$

Figure 3 Calculation of $\sin t$ and $\cos t$.

When $0 < t < \pi/2$, the values of $\sin t$ and $\cos t$ may be expressed as ratios of the lengths of the sides of a right triangle. Indeed, if we are given the right triangle in Fig. 4, we have

$$\sin t = \frac{\text{opposite}}{\text{hypotenuse}}, \qquad \cos t = \frac{\text{adjacent}}{\text{hypotenuse}}. \tag{2}$$

A typical application of (2) appears in Example 1.

Figure 4

EXAMPLE 1

Solving a Right Triangle The hypotenuse of a right triangle is 4 units and one angle is .7 radian. Determine the length of the side opposite this angle.

SOLUTION   See Fig. 5.

$$\sin .7 = \frac{y}{4} \qquad\qquad \sin t = \frac{\text{opposite}}{\text{hypotenuse}}$$

$$y = 4\sin .7 \qquad\qquad \text{Multiply both sides by 4.}$$

$$\approx 4(.64422) = 2.57688. \qquad \text{Evaluate } \sin .7 \text{ using a calculator.}$$

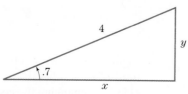

Figure 5   **≫ Now Try Exercise 15**

Figure 6

Another way to describe the sine and cosine functions is to choose the point P in Fig. 1 so that $r = 1$. That is, choose P on the unit circle. (See Fig. 6.) In this case,

$$\sin t = \frac{y}{1} = y,$$

$$\cos t = \frac{x}{1} = x.$$

So, the y-coordinate of P is $\sin t$, and the x-coordinate of P is $\cos t$. Thus, we have the following result:

Figure 7

> **Alternative Definition of Sine and Cosine Functions**   We can think of $\cos t$ and $\sin t$ as the x- and y-coordinates of the point P on the unit circle that is determined by an angle of t radians. (See Fig. 7.)

EXAMPLE 2

Values of the Cosine Function Find a value of t such that $0 < t < \pi/2$ and $\cos t = \cos(-\pi/3)$.

SOLUTION

Figure 8

On the unit circle, we locate the point P that is determined by an angle of $-\pi/3$ radians. The x-coordinate of P is $\cos(-\pi/3)$. There is another point Q on the unit circle with the same x-coordinate. (See Fig. 8.) Let t be the radian measure of the angle determined by Q. Then,

$$\cos t = \cos\left(-\frac{\pi}{3}\right)$$

because Q and P have the same x-coordinate. Also, $0 < t < \pi/2$. From the symmetry of the diagram, it is clear that $t = \pi/3$.   **≫ Now Try Exercise 27**

Properties of the Sine and Cosine Functions

Each number t determines a point $(\cos t, \sin t)$ on the unit circle $x^2 + y^2 = 1$, as in Fig. 7. Therefore, $(\cos t)^2 + (\sin t)^2 = 1$. It is convenient (and traditional) to write $\sin^2 t$ instead of $(\sin t)^2$ and $\cos^2 t$ instead of $(\cos t)^2$. Thus, we can write the last formula as follows:

$$\cos^2 t + \sin^2 t = 1. \tag{3}$$

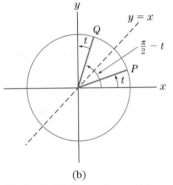

Figure 9 Diagrams for two identities.

The numbers t and $t \pm 2\pi$ determine the same point on the unit circle (because 2π represents a full revolution of the circle). But $t + 2\pi$ and $t - 2\pi$ correspond to the points $(\cos(t + 2\pi), \sin(t + 2\pi))$ and $(\cos(t - 2\pi), \sin(t - 2\pi))$, respectively. Hence,

$$\cos(t \pm 2\pi) = \cos t, \qquad \sin(t \pm 2\pi) = \sin t. \tag{4}$$

Fig. 9(a) illustrates another property of the sine and cosine:

$$\cos(-t) = \cos t, \qquad \sin(-t) = -\sin t. \tag{5}$$

Figure 9(b) shows that the points P and Q corresponding to t and to $\pi/2 - t$ are reflections of each other through the line $y = x$. Consequently, we obtain the coordinates of Q by interchanging the coordinates of P. This means that

$$\cos\left(\frac{\pi}{2} - t\right) = \sin t, \qquad \sin\left(\frac{\pi}{2} - t\right) = \cos t. \tag{6}$$

The equations in (3) to (6) are called *identities* because they hold for all values of t. Another identity that holds for all numbers s and t is

$$\sin(s + t) = \sin s \cos t + \cos s \sin t. \tag{7}$$

A proof of (7) may be found in any introductory book on trigonometry. There are a number of other identities concerning trigonometric functions, but we shall not need them here.

Graphs of sin *t* and cos *t*

Let us analyze what happens to $\sin t$ as t increases from 0 to π. When $t = 0$, the point $P = (\cos t, \sin t)$ is at $(1, 0)$, as in Fig. 10(a). As t increases, P moves counterclockwise around the unit circle. [See Fig. 10(b).] The y-coordinate of P—that is, $\sin t$—increases until $t = \pi/2$, where $P = (0, 1)$. [See Fig. 10(c).] As t increases from $\pi/2$ to π, the y-coordinate of P—that is, $\sin t$—decreases from 1 to 0. [See Figs. 10(d) and (e).]

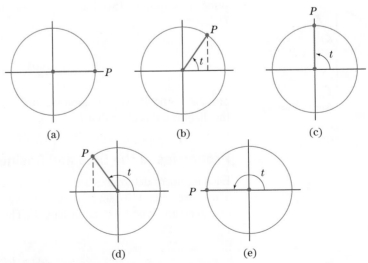

Figure 10 Movement along the unit circle.

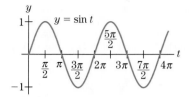

Figure 11

Part of the graph of $\sin t$ is sketched in Fig. 11. For t between 0 and π, notice that the values of $\sin t$ increase from 0 to 1 and then decrease back to 0, just as we predicted from Fig. 10. For t between π and 2π, the values of $\sin t$ are negative. The graph of $y = \sin t$ for t between 2π and 4π is exactly the same as the graph for t between 0 and 2π. This result follows from formula (4). We say that the sine function is *periodic with period 2π* because the graph repeats itself every 2π units. We can use this fact to make a quick sketch of part of the graph for negative values of t. (See Fig. 12.)

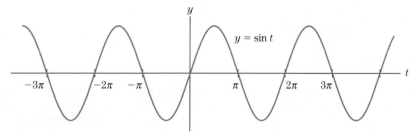

Figure 12 Graph of the sine function.

By analyzing what happens to the first coordinate of the point $(\cos t, \sin t)$ as t varies, we obtain the graph of $\cos t$. Note from Fig. 13 that the graph of the cosine function is also periodic with period 2π.

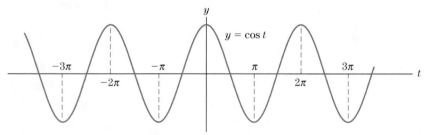

Figure 13 Graph of the cosine function.

NOTE The sine and cosine functions assign to each number t the values $\sin t$ and $\cos t$, respectively. There is nothing special, however, about the letter t. Although we chose to use the letters t, x, y, and r in the *definition* of the sine and cosine, other letters could have been used as well. Now that the sine and cosine of every number are defined, we are free to use *any* letter to represent the independent variable. «

INCORPORATING

TECHNOLOGY

Graphing Trigonometric Functions The ZTrig window setting is optimized for displaying graphs of trigonometric functions. This can be accessed via ZOOM 7, and it sets the window dimensions to $[-2\pi, 2\pi]$ by $[-4, 4]$ with an x-scale of $\pi/2$. Figure 14 is the graph of $y = \sin x$ in this setting.

Figure 14

Check Your Understanding 8.2

Solutions can be found following the section exercises.

1. Find cos t, where t is the radian measure of the angle shown in Fig. 15.

$(-3, -1)$

Figure 15

2. Assume that $\cos(1.17) = .390$. Use properties of the cosine and sine to determine $\sin(1.17)$, $\cos(1.17 + 4\pi)$, $\cos(-1.17)$, and $\sin(-1.17)$.

EXERCISES 8.2

In Exercises 1–12, give the values of sin t and cos t, where t is the radian measure of the angle shown.

1.

2

1

$\sqrt{3}$

2.
3

2

$\sqrt{5}$

3.

2

3

4.
4

1

5.

13

12

6.
5

3

7. $(-2, 1)$

t

8.
t

$(2, -3)$

9. $(-2, 2)$

t

10. $(.6, .8)$

t

11.

t

$(-.6, -.8)$

12.

t

$(.8, -.6)$

Exercises 13–20 refer to various right triangles whose sides and angles are labeled as in Fig. 16. Round off all lengths of sides to one decimal place.

c

b

t

a

Figure 16

13. Estimate t if $a = 12$, $b = 5$, and $c = 13$.

14. If $t = 1.1$ and $c = 10.0$, find b.

15. If $t = 1.1$ and $b = 3.2$, find c.

16. If $t = .4$ and $c = 5.0$, find a.

17. If $t = .4$ and $a = 10.0$, find c.

18. If $t = .9$ and $c = 20.0$, find a and b.

19. If $t = .5$ and $a = 2.4$, find b and c.

20. If $t = 1.1$ and $b = 3.5$, find a and c.

Find t such that $0 \le t \le \pi$ and t satisfies the stated condition.

21. $\cos t = \cos(-\pi/6)$ **22.** $\cos t = \cos(3\pi/2)$

23. $\cos t = \cos(5\pi/4)$ **24.** $\cos t = \cos(-4\pi/6)$

25. $\cos t = \cos(-5\pi/8)$ **26.** $\cos t = \cos(-3\pi/4)$

Find t such that $-\pi/2 \le t \le \pi/2$ and t satisfies the stated condition.

27. $\sin t = \sin(3\pi/4)$ **28.** $\sin t = \sin(7\pi/6)$

29. $\sin t = \sin(-4\pi/3)$ **30.** $\sin t = -\sin(3\pi/8)$

31. $\sin t = -\sin(\pi/6)$ **32.** $\sin t = -\sin(-\pi/3)$

33. $\sin t = \cos t$ **34.** $\sin t = -\cos t$

35. Refer to Fig. 10. Describe what happens to cos t as t increases from 0 to π.

36. Use the unit circle to describe what happens to sin t as t increases from π to 2π.

37. Determine the value of sin t when $t = 5\pi$, -2π, $17\pi/2$, $-13\pi/2$.

38. Determine the value of cos t when $t = 5\pi$, -2π, $17\pi/2$, $-13\pi/2$.

39. Assume that $\cos(.19) = .98$. Use properties of the cosine and sine to determine $\sin(.19)$, $\cos(.19 - 4\pi)$, $\cos(-.19)$, and $\sin(-.19)$.

40. Assume that sin(.42) = .41. Use properties of the cosine and sine to determine sin(−.42), sin(6π − .42), and cos(.42).

41. In any given locality, tap water temperature varies during the year. In Dallas, Texas, the tap water temperature (in degrees Fahrenheit) *t* days after the beginning of a year is given approximately by the formula

$$T = 59 + 14 \cos\left[\frac{2\pi}{365}(t - 208)\right], \qquad 0 \le t \le 365.$$

(*Source: Solar Energy.*)
(a) Graph the function in the window [0, 365] by [−10, 75].
(b) What is the temperature on February 14, that is, when *t* = 45 ?
(c) Use the fact that the value of the cosine function ranges from −1 to 1 to find the coldest and warmest tap water temperatures during the year.
(d) Use the TRACE feature or the MINIMUM command to estimate the day during which the tap water temperature is coldest. Find the exact day algebraically by using the fact that cos(−π) = −1.
(e) Use the TRACE feature or the MAXIMUM command to estimate the day during which the tap water temperature is warmest. Find the exact day algebraically by using the fact that cos(0) = 1.
(f) The average tap water temperature during the year is 59°. Find the two days during which the average

temperature is achieved. [*Note:* Answer this question both graphically and algebraically.]

42. In any given locality, the length of daylight varies during the year. In Des Moines, Iowa, the number of minutes of daylight in a day *t* days after the beginning of a year is given approximately by the formula

$$D = 720 + 200 \sin\left[\frac{2\pi}{365}(t - 79.5)\right], \qquad 0 \le t \le 365.$$

(*Source: School Science and Mathematics.*)
(a) Graph the function in the window [0, 365] by [−100, 940].
(b) How many minutes of daylight are there on February 14, that is, when *t* = 45?
(c) Use the fact that the value of the sine function ranges from −1 to 1 to find the shortest and longest amounts of daylight during the year.
(d) Use the TRACE feature or the MINIMUM command to estimate the day with the shortest amount of daylight. Find the exact day algebraically by using the fact that sin(3π/2) = −1.
(e) Use the TRACE feature or the MAXIMUM command to estimate the day with the longest amount of daylight. Find the exact day algebraically by using the fact that sin(π/2) = 1.
(f) Find the two days during which the amount of daylight equals the amount of darkness. (These days are called *equinoxes*.) [*Note:* Answer this question both graphically and algebraically.]

Solutions to Check Your Understanding 8.2

1. Here, $P = (x, y) = (-3, -1)$. The length of the line segment *OP* is

$$r = \sqrt{x^2 + y^2} = \sqrt{(-3)^2 + (-1)^2} = \sqrt{10}.$$

Then,

$$\cos t = \frac{x}{r} = \frac{-3}{\sqrt{10}} \approx -.94868.$$

2. Given cos(1.17) = .390, use the relation $\cos^2 t + \sin^2 t = 1$ with *t* = 1.17 to solve for sin(1.17):

$$\cos^2(1.17) + \sin^2(1.17) = 1$$
$$\sin^2(1.17) = 1 - \cos^2(1.17)$$
$$= 1 - (.390)^2 = .8479.$$

So,

$$\sin(1.17) = \sqrt{.8479} \approx .921.$$

Also, from properties (4) and (5),

$$\cos(1.17 + 4\pi) = \cos(1.17) = .390$$
$$\cos(-1.17) = \cos(1.17) = .390$$
$$\sin(-1.17) = -\sin(1.17) = -.921.$$

8.3 Differentiation and Integration of sin *t* and cos *t*

In this section, we study the two differentiation rules

$$\frac{d}{dt} \sin t = \cos t, \tag{1}$$

$$\frac{d}{dt} \cos t = -\sin t. \tag{2}$$

It is not difficult to see why these rules might be true. Formula (1) says that the slope of the curve $y = \sin t$ at a particular value of *t* is given by the corresponding value of cos *t*. To check it, we draw a careful graph of $y = \sin t$ and estimate the slope at various points. (See Fig. 1.) Let us plot the slope as a function of t. (See Fig. 2.) As can be seen, the "slope function" (the derivative) of $\sin t$ has a graph similar to the curve $y = \cos t$. Thus, formula (1) seems to be reasonable. A similar analysis of the graph of $y = \cos t$ would show why (2) might be true. Proofs of these differentiation rules are outlined in an appendix at the end of this section.

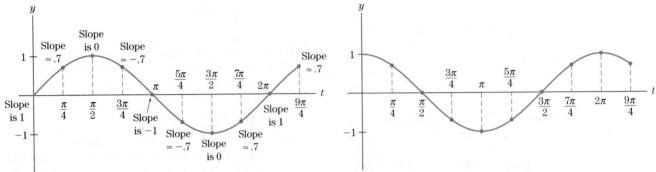

Figure 1 Slope estimates along the graph of $y = \sin t$.

Figure 2 Graph of the slope function for $y = \sin t$.

Combining (1), (2), and the chain rule, we obtain the following general rules:

$$\frac{d}{dt}(\sin g(t)) = [\cos g(t)]\, g'(t), \tag{3}$$

$$\frac{d}{dt}(\cos g(t)) = [-\sin g(t)]\, g'(t). \tag{4}$$

EXAMPLE 1 **Derivatives with the Sine Function** Differentiate
(a) $\sin 3t$ (b) $(t^2 + 3 \sin t)^5$.

SOLUTION (a) $\dfrac{d}{dt}(\sin 3t) = (\cos 3t)\dfrac{d}{dt}(3t) = (\cos 3t) \cdot 3 = 3 \cos 3t$ Use (3).

(b) $\dfrac{d}{dt}(t^2 + 3 \sin t)^5 = 5(t^2 + 3 \sin t)^4 \cdot \dfrac{d}{dt}(t^2 + 3 \sin t)$ Generalized power rule.

$$= 5(t^2 + 3 \sin t)^4 (2t + 3 \cos t)$$ Sum rule and (2).

≫ Now Try Exercise 1

EXAMPLE 2 **Derivatives with the Cosine Function** Differentiate
(a) $\cos(t^2 + 1)$ (b) $\cos^2 t$.

SOLUTION (a) $\dfrac{d}{dt}\cos(t^2 + 1) = -\sin(t^2 + 1)\dfrac{d}{dt}(t^2 + 1) = -\sin(t^2 + 1) \cdot (2t)$ Use (4).

$$= -2t \sin(t^2 + 1)$$

(b) Recall that the notation $\cos^2 t$ means $(\cos t)^2$.

$$\frac{d}{dt}\cos^2 t = \frac{d}{dt}(\cos t)^2$$

$$= 2(\cos t)\frac{d}{dt}\cos t = -2 \cos t \sin t$$ Generalized power rule and (4).

≫ Now Try Exercise 11

EXAMPLE 3   **Differentiating Sine and Cosine Functions** Differentiate
(a) $t^2 \cos 3t$ (b) $(\sin 2t)/t$.

SOLUTION   (a) From the product rule we have

$$\frac{d}{dt}(t^2 \cos 3t) = t^2 \frac{d}{dt}\cos 3t + (\cos 3t)\frac{d}{dt}t^2 \quad \text{Product rule.}$$
$$= t^2(-3 \sin 3t) + (\cos 3t)(2t) \quad \text{Use (4).}$$
$$= -3t^2 \sin 3t + 2t \cos 3t.$$

(b) From the quotient rule we have

$$\frac{d}{dt}\left(\frac{\sin 2t}{t}\right) = \frac{t\dfrac{d}{dt}\sin 2t - (\sin 2t)\cdot 1}{t^2} = \frac{2t \cos 2t - \sin 2t}{t^2}.$$

≫ Now Try Exercise 23

EXAMPLE 4   **Maximizing a Volume** A V-shaped trough is to be constructed with sides that are 200 centimeters long and 30 centimeters wide. (See Fig. 3.) Find the angle t between the sides that maximizes the capacity of the trough.

SOLUTION   The volume of the trough is its length times its cross-sectional area. Since the length is constant, it suffices to maximize the cross-sectional area. Let us rotate the diagram of a cross section so that one side is horizontal. (See Fig. 4.) Note that $h/30 = \sin t$, so $h = 30 \sin t$. Thus, the area A of the cross section is

$$A = \tfrac{1}{2} \cdot \text{base} \cdot \text{height}$$
$$= \tfrac{1}{2}(30)(h) = 15(30 \sin t) = 450 \sin t.$$

To find where A is a maximum, we set the derivative equal to zero and solve for t.

$$\frac{dA}{dt} = 0$$
$$450 \cos t = 0.$$

Physical considerations force us to consider only values of t between 0 and π. From the graph of $y = \cos t$, we see that $t = \pi/2$ is the only value of t between 0 and π that makes $\cos t = 0$. So, if we are to maximize the volume of the trough, the two sides should be perpendicular to one another.   **≪**

Figure 3

Figure 4

EXAMPLE 5   **Integrals of Sine Functions** Calculate the following indefinite integrals.

(a) $\displaystyle\int \sin t \, dt$ (b) $\displaystyle\int \sin 3t \, dt$

SOLUTION   (a) Since $\dfrac{d}{dt}(-\cos t) = \sin t$, we have

$$\int \sin t \, dt = -\cos t + C,$$

where C is an arbitrary constant.

(b) From part (a) we guess that an antiderivative of $\sin 3t$ should resemble the function $-\cos 3t$. However, if we differentiate, we find that

$$\frac{d}{dt}(-\cos 3t) = (\sin 3t)\cdot\frac{d}{dt}(3t) = 3 \sin 3t,$$

which is three times too much. So we multiply this last equation by $\tfrac{1}{3}$ to derive that

$$\frac{d}{dt}(-\tfrac{1}{3}\cos 3t) = \sin 3t,$$

so

$$\int \sin 3t \, dt = -\tfrac{1}{3}\cos 3t + C.$$ **≫ Now Try Exercise 35**

FOR REVIEW

The integral or antiderivative always has an arbitrary constant C. (Sec. 6.1, p. 295.)

Reversing the derivative formulas (1) and (2), we obtain the integral formulas

$$\int \cos t \, dt = \sin t + C \tag{5}$$

and

$$\int \sin t \, dt = -\cos t + C \tag{6}$$

More generally, if $a \neq 0$, we have

$$\int \cos at \, dt = \frac{1}{a} \sin at + C \tag{7}$$

and

$$\int \sin at \, dt = -\frac{1}{a} \cos at + C \tag{8}$$

To prove (7), we show that the derivative of the right side of (7) is equal to $\cos at$:

$$\frac{d}{dt} \left[\frac{1}{a} \sin at + C \right] = \frac{1}{a} \frac{d}{dt} [\sin at] + \frac{d}{dt} [C]$$

$$= \frac{1}{a} \cdot a \cos at + 0$$

$$= \cos at$$

The proof of (8) is done similarly.

EXAMPLE 6 **Area under a Sine Curve** Find the area under the curve $y = \sin 3t$ from $t = 0$ to $t = \pi/3$.

SOLUTION The area is shaded in Fig. 5.

Figure 5

$$[\text{shaded area}] = \int_0^{\pi/3} \sin 3t \, dt$$

$$= -\frac{1}{3} \cos 3t \Big|_0^{\pi/3} \qquad \text{By (8).}$$

$$= -\frac{1}{3} \cos \left(3 \cdot \frac{\pi}{3} \right) - \left(-\frac{1}{3} \cos 0 \right)$$

$$= -\frac{1}{3} \cos \pi + \frac{1}{3} \cos 0$$

$$= \frac{1}{3} + \frac{1}{3} = \frac{2}{3}. \qquad \cos \pi = -1, \cos 0 = 1.$$

≫ Now Try Exercise 45

As mentioned earlier, the trigonometric functions are required to model situations that are repetitive (or periodic). The next example illustrates such a situation.

EXAMPLE 7 **A Predators-and-Prey Model** In many mathematical models used to study the interaction between predators and prey, both the number of predators and the number of prey are described by periodic functions. Suppose that in one such model the number of predators (in a particular geographic region) at time t is given by the equation

$$N(t) = 5000 + 2000 \cos(2\pi t / 36),$$

where t is measured in months from June 1, 2010.

(a) At what rate was the number of predators changing on August 1, 2010?

(b) What is the average number of predators during the time interval from June 1, 2010, to June 1, 2022?

SOLUTION

(a) The date August 1, 2010, corresponds to $t = 2$. The rate of change of $N(t)$ is given by the derivative $N'(t)$:

$$N'(t) = \frac{d}{dt}\left[5000 + 2000\cos\left(\frac{2\pi t}{36}\right)\right]$$

$$= 2000\left[-\sin\left(\frac{2\pi t}{36}\right)\cdot\left(\frac{2\pi}{36}\right)\right] \qquad \text{Derivative of a constant is 0. Use (4).}$$

$$= -\frac{1000\pi}{9}\sin\left(\frac{2\pi t}{36}\right) \qquad \text{Simplify.}$$

$$N'(2) = -\frac{1000\pi}{9}\sin\left(\frac{\pi}{9}\right) \qquad \text{Evaluate for } t = 2.$$

$$\approx -119.$$

Thus, on August 1, 2010, the number of predators was decreasing at the rate of 119 per month.

(b) The time interval from June 1, 2010, to June 1, 2022, corresponds to $t = 0$ to $t = 144$. (There are 144 months in 12 years.) The average value of $N(t)$ over this interval is

$$\frac{1}{144 - 0}\int_0^{144} N(t)\,dt = \frac{1}{144}\int_0^{144}\left[5000 + 2000\cos\left(\frac{2\pi t}{36}\right)\right]dt$$

$$= \frac{1}{144}\left[5000t + \frac{2000}{2\pi/36}\sin\left(\frac{2\pi t}{36}\right)\right]\Bigg|_0^{144} \qquad \text{Use (7).}$$

$$= \frac{1}{144}\left[5000\cdot 144 + \frac{2000}{2\pi/36}\sin(8\pi)\right]$$

$$\quad - \frac{1}{144}\left[5000\cdot 0 + \frac{2000}{2\pi/36}\sin(0)\right]$$

$$= 5000.$$

The graph of $N(t)$ is sketched in Fig. 6. Note how $N(t)$ oscillates around 5000, the average value.

FOR REVIEW

The average of a function is its definite integral, divided by the length of the interval. (Sec. 6.5, p. 334.)

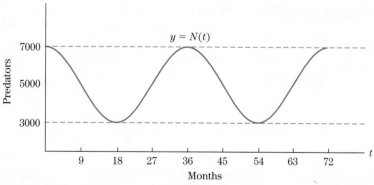

Figure 6 Periodic fluctuation of a predator population.   **» Now Try Exercise 51**

Informal Justification of the Differentiation Rules for sin t and cos t

First, let us examine the derivatives of $\cos t$ and $\sin t$ at $t = 0$. The function $\cos t$ has a maximum at $t = 0$; consequently, its derivative there must be zero. [See Fig. 7(a).] If we approximate the tangent line at $t = 0$ by a secant line, as in Fig. 7(b), the slope of the secant line must approach 0 as $h \to 0$. Since the slope of the secant line is $(\cos h - 1)/h$, we conclude that

$$\lim_{h \to 0} \frac{\cos h - 1}{h} = 0. \tag{9}$$

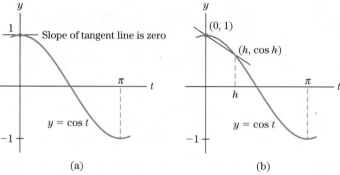

Figure 7

It appears from the graph of $y = \sin t$ that the tangent line at $t = 0$ has slope 1. [See Fig. 8(a).] If it does, the slope of the approximating secant line in Fig. 8(b) must approach 1, as h approaches 0. Since the slope of this line is $(\sin h)/h$, this implies that

$$\lim_{h \to 0} \frac{\sin h}{h} = 1. \tag{10}$$

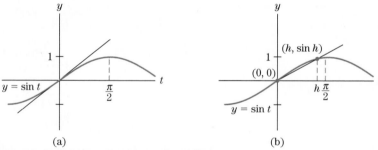

Figure 8 Slope of $\sin t$ at $t = 0$.

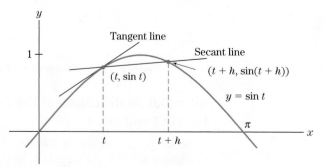

We can evaluate $\sin h/h$ for small values of h with a calculator. (See Fig. 9.) The numerical evidence does not prove (10), but should be sufficiently convincing for our purposes.

To obtain the differentiation formula for $\sin t$, we approximate the slope of a tangent line by the slope of a secant line. (See Fig. 10.) The slope of a secant line is

$$\frac{\sin(t + h) - \sin t}{h}.$$

Figure 9

Figure 10 Secant line approximation for $y = \sin t$.

From formula (7) of Section 8.2, we note that $\sin(t + h) = \sin t \cos h + \cos t \sin h$. Thus,

$$[\text{slope of secant line}] = \frac{(\sin t \cos h + \cos t \sin h) - \sin t}{h}$$

$$= \frac{\sin t (\cos h - 1) + \cos t \sin h}{h}$$

$$= (\sin t) \frac{\cos h - 1}{h} + (\cos t) \frac{\sin h}{h}.$$

From (9) and (10) it follows that

$$\frac{d}{dt} \sin t = \lim_{h \to 0} \left[(\sin t) \frac{\cos h - 1}{h} + (\cos t) \frac{\sin h}{h} \right]$$

$$= (\sin t) \lim_{h \to 0} \frac{\cos h - 1}{h} + (\cos t) \lim_{h \to 0} \frac{\sin h}{h}$$

$$= (\sin t) \cdot 0 + (\cos t) \cdot 1$$

$$= \cos t.$$

A similar argument may be given to verify the formula for the derivative of $\cos t$. Here is a shorter proof that uses the chain rule and the two identities

$$\cos t = \sin\left(\frac{\pi}{2} - t\right), \qquad \sin t = \cos\left(\frac{\pi}{2} - t\right).$$

[See formula (6) of Section 8.2.] We have

$$\frac{d}{dt} \cos t = \frac{d}{dt} \sin\left(\frac{\pi}{2} - t\right)$$

$$= \cos\left(\frac{\pi}{2} - t\right) \cdot \frac{d}{dt}\left(\frac{\pi}{2} - t\right)$$

$$= \cos\left(\frac{\pi}{2} - t\right) \cdot (-1)$$

$$= -\sin t.$$

Check Your Understanding 8.3

Solutions can be found following the section exercises.

1. Differentiate $y = 2 \sin[t^2 + (\pi/6)]$.

2. Differentiate $y = e^t \sin 2t$.

EXERCISES 8.3

Differentiate (with respect to t or x):

1. $y = \sin 4t$

2. $y = 2 \cos 2t$

3. $y = 4 \sin t$

4. $y = \cos(-4t)$

5. $y = 2 \cos 3t$

6. $y = -\dfrac{\sin 3t}{3}$

7. $y = t + \cos \pi t$

8. $y = t \cos t$

9. $y = \sin(\pi - t)$

10. $y = \dfrac{\cos(2x + 2)}{2}$

11. $y = \cos^3 t$

12. $y = \sin^3 t^2$

13. $y = \sin\sqrt{x - 1}$

14. $y = \cos(e^x)$

15. $y = \sqrt{\sin(x - 1)}$

16. $y = e^{\cos x}$

17. $y = (1 + \cos t)^8$

18. $y = \sqrt[3]{\sin \pi t}$

19. $y = \cos^2 x^3$

20. $y = \cos^2 x + \sin^2 x$

21. $y = e^x \sin x$

22. $y = (\cos x + \sin x)^2$

23. $y = \sin 2x \cos 3x$

24. $y = \dfrac{1 + x}{\cos x}$

25. $y = \dfrac{\sin t}{\cos t}$

26. $y = \cos(e^{2x+3})$

27. $y = \ln(\cos t)$

28. $y = \ln(\sin 2t)$

29. $y = \sin(\ln t)$

30. $y = (\cos t)\ln t$

31. Find the slope of the line tangent to the graph of $y = \cos 3x$ at $x = 13\pi/6$.

32. Find the slope of the line tangent to the graph of $y = \sin 2x$ at $x = 5\pi/4$.

33. Find the equation of the line tangent to the graph of $y = 3 \sin x + \cos 2x$ at $x = \pi/2$.

34. Find the equation of the line tangent to the graph of $y = 3 \sin 2x - \cos 2x$ at $x = 3\pi/4$.

Find the following indefinite integrals.

35. $\int \cos 2x \, dx$

36. $\int 3 \sin 3x \, dx$

37. $\int -\frac{1}{2} \cos \frac{x}{7} \, dx$

38. $\int 2 \sin \frac{x}{2} \, dx$

39. $\int (\cos x - \sin x) \, dx$

40. $\int \left(2 \sin 3x + \frac{\cos 2x}{2} \right) dx$

41. $\int (-\sin x + 3 \cos(-3x)) \, dx$

42. $\int \sin(-2x) \, dx$

43. $\int \sin(4x + 1) \, dx$

44. $\int \cos \frac{x-2}{2} \, dx$

45. Find the area under the curve $y = \cos t$ from $t = 0$ to $t = \frac{\pi}{2}$.

46. Find the area under the curve $y = \sin 2t$ from $t = 0$ to $t = \frac{\pi}{4}$.

47. A Model for Blood Pressure A person's blood pressure P at time t (in seconds) is given by $P = 100 + 20 \cos 6t$.

    (a) Find the maximum value of P (called the *systolic pressure*) and the minimum value of P (called the *diastolic pressure*) and give one or two values of t where these maximum and minimum values of P occur.

    (b) If time is measured in seconds, approximately how many heartbeats per minute are predicted by the equation for P?

48. Basal Metabolic Rate The *basal metabolism* (BM) of an organism over a certain time period may be described as the total amount of heat in kilocalories (kcal) that the organism produces during this period, assuming that the organism is at rest and not subject to stress. The *basal metabolic rate* (BMR) is the rate in kcal per hour at which the organism produces heat. The BMR of an animal such as a desert rat fluctuates in response to changes in temperature and other environmental factors. The BMR generally follows a *diurnal* cycle, rising at night during low temperatures and decreasing during the warmer daytime temperatures. Find the BM for 1 day if $\text{BMR}(t) = .4 + .2 \cos(\pi t/12)$ kcal per hour ($t = 0$ corresponds to 3 A.M.). (See Fig. 11.)

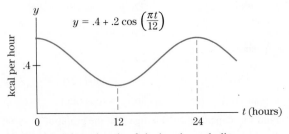

Figure 11 Diurnal cycle of the basal metabolic rate.

49. As h approaches 0, what value is approached by

$$\frac{\sin \left(\frac{\pi}{2} + h \right) - 1}{h} ?$$

$$\left[\textit{Hint: } \sin \frac{\pi}{2} = 1 \right]$$

50. As h approaches 0, what value is approached by

$$\frac{\cos(\pi + h) + 1}{h} ?$$

[*Hint:* $\cos \pi = -1$]

51. Average Temperature The average weekly temperature in Washington, D.C., t weeks after the beginning of the year is

$$f(t) = 54 + 23 \sin \left[\frac{2\pi}{52} (t - 12) \right].$$

The graph of this function is sketched in Fig. 12.

    (a) What is the average weekly temperature at week 18?

    (b) At week 20, how fast is the temperature changing?

    (c) When is the average weekly temperature 39 degrees?

    (d) When is the average weekly temperature falling at the rate of 1 degree per week?

    (e) When is the average weekly temperature greatest? Least?

    (f) When is the average weekly temperature increasing fastest? Decreasing fastest?

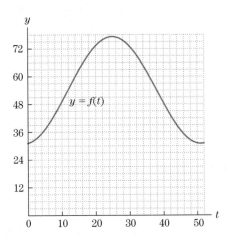

Figure 12

52. **Average Daylight Hours** The number of hours of daylight per day in Washington, D.C., t weeks after the beginning of the year is

$$f(t) = 12.18 + 2.725 \sin\left[\frac{2\pi}{52}(t - 12)\right].$$

The graph of this function is sketched in Fig. 13.

(a) How many hours of daylight are there after 42 weeks?
(b) After 32 weeks, how fast is the number of hours of daylight decreasing?
(c) When is there 14 hours of daylight per day?
(d) When is the number of hours of daylight increasing at the rate of 15 minutes per week?
(e) When are the days longest? Shortest?
(f) When is the number of hours of daylight increasing fastest? Decreasing fastest?

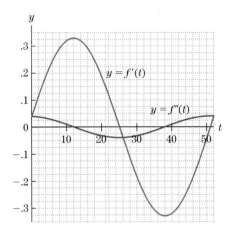

Figure 13

Solutions to Check Your Understanding 8.3

1. By the chain rule,

$$y' = 2\cos\left(t^2 + \frac{\pi}{6}\right) \cdot \frac{d}{dt}\left(t^2 + \frac{\pi}{6}\right)$$

$$= 2\cos\left(t^2 + \frac{\pi}{6}\right) \cdot 2t$$

$$= 4t\cos\left(t^2 + \frac{\pi}{6}\right).$$

2. By the product rule,

$$y' = e^t \frac{d}{dt}(\sin 2t) + (\sin 2t)\frac{d}{dt}e^t$$

$$= 2e^t \cos 2t + e^t \sin 2t.$$

8.4 The Tangent and Other Trigonometric Functions

Certain functions involving the sine and cosine functions occur so frequently in applications that they have been given special names. The **tangent** (tan), **cotangent** (cot), **secant** (sec), and **cosecant** (csc) are such functions and are defined as follows:

$$\tan t = \frac{\sin t}{\cos t}, \qquad \cot t = \frac{\cos t}{\sin t},$$

$$\sec t = \frac{1}{\cos t}, \qquad \csc t = \frac{1}{\sin t}.$$

They are defined only for t, such that the denominators in the preceding quotients are not zero. These four functions, together with the sine and cosine, are called the **trigonometric functions**. Our main interest in this section is in the tangent function. Some properties of the cotangent, secant, and cosecant are developed in the exercises.

Many identities involving the trigonometric functions can be deduced from the identities given in Section 8.2. We shall mention just one:

$$\tan^2 t + 1 = \sec^2 t. \tag{1}$$

[Here, $\tan^2 t$ means $(\tan t)^2$ and $\sec^2 t$ means $(\sec t)^2$.] This identity follows from the identity $\sin^2 t + \cos^2 t = 1$ when we divide everything by $\cos^2 t$.

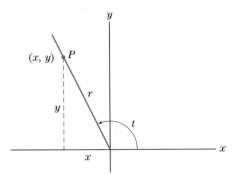

Figure 1

An important interpretation of the tangent function can be given in terms of the diagram used to define the sine and cosine. For a given t, let us construct an angle of t radians. (See Fig. 1.) Since $\sin t = y/r$ and $\cos t = x/r$, we have

$$\frac{\sin t}{\cos t} = \frac{y/r}{x/r} = \frac{y}{x},$$

where this formula holds provided that $x \neq 0$. Thus,

$$\tan t = \frac{y}{x}. \tag{2}$$

Three examples that illustrate this property of the tangent appear in Fig. 2.

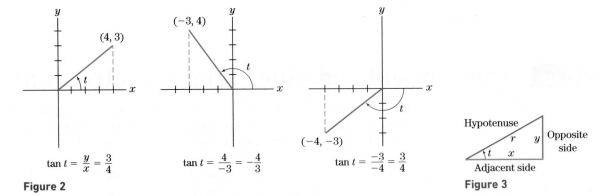

Figure 2

Figure 3

When $0 < t < \pi/2$, the value of $\tan t$ is a ratio of the lengths of the sides of a right triangle. In other words, if we are given a triangle as in Fig. 3, we would have

$$\tan t = \frac{\text{opposite}}{\text{adjacent}}. \tag{3}$$

EXAMPLE 1	**Determining the Height of a Building** The angle of elevation from an observer to the top of a building is 29°. (See Fig. 4.) If the observer is 100 meters from the base of the building, how high is the building?

SOLUTION   Let h denote the height of the building. Then, formula (3) implies that

$$\frac{h}{100} = \tan 29°$$

$$h = 100 \tan 29°.$$

We convert 29° into radians. We find that $29° = (\pi/180) \cdot 29$ radians $\approx .5$ radian, and $\tan .5 \approx .54630$. Hence,

$$h \approx 100(.54630) = 54.63 \text{ meters.}$$ **» Now Try Exercise 11**

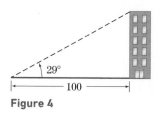

Figure 4

The Derivative of tan t

Since $\tan t$ is defined in terms of $\sin t$ and $\cos t$, we can compute the derivative of $\tan t$ from our rules of differentiation. That is, by applying the quotient rule for differentiation, we have

$$\frac{d}{dt}(\tan t) = \frac{d}{dt}\left(\frac{\sin t}{\cos t}\right) = \frac{(\cos t)(\cos t) - (\sin t)(-\sin t)}{(\cos t)^2}$$

$$= \frac{\cos^2 t + \sin^2 t}{\cos^2 t} = \frac{1}{\cos^2 t}.$$

Now

$$\frac{1}{\cos^2 t} = \frac{1}{(\cos t)^2} = \left(\frac{1}{\cos t}\right)^2 = (\sec t)^2 = \sec^2 t.$$

So the derivative of $\tan t$ can be expressed in two equivalent ways:

$$\frac{d}{dt}(\tan t) = \frac{1}{\cos^2 t} = \sec^2 t. \tag{4}$$

Combining (4) with the chain rule, we have

$$\frac{d}{dt}(\tan g(t)) = [\sec^2 g(t)]g'(t). \tag{5}$$

EXAMPLE 2	**Derivatives of Tangent Functions** Differentiate

(a) $\tan(t^3 + 1)$ (b) $\tan^3 t.$

SOLUTION   (a) From (5) we find that

$$\frac{d}{dt}[\tan(t^3 + 1)] = \sec^2(t^3 + 1) \cdot \frac{d}{dt}(t^3 + 1)$$
$$= 3t^2 \sec^2(t^3 + 1).$$

(b) We write $\tan^3 t$ as $(\tan t)^3$ and use the chain rule (in this case, the general power rule):

$$\frac{d}{dt}(\tan t)^3 = (3 \tan^2 t) \cdot \frac{d}{dt} \tan t = 3 \tan^2 t \sec^2 t.$$

» Now Try Exercise 17

The Graph of tan t

Recall that tan t is defined for all t except where cos $t = 0$. (We cannot have zero in the denominator of sin t/cos t.) The graph of tan t is sketched in Fig. 5. Note that tan t is periodic with period π.

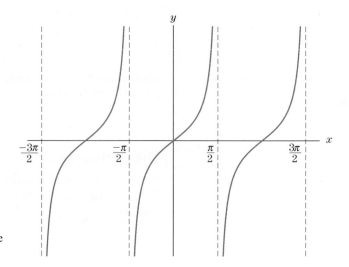

Figure 5 Graph of the tangent function.

Check Your Understanding 8.4

Solutions can be found following the section exercises.

1. Show that the slope of a straight line is equal to the tangent of the angle that the line makes with the x-axis.

2. Calculate $\int_0^{\pi/4} \sec^2 t \, dt$.

EXERCISES 8.4

1. If $0 < t < \pi/2$, use Fig. 3 to describe sec t as a ratio of the lengths of the sides of a right triangle.

2. Describe cot t for $0 < t < \pi/2$ as a ratio of the lengths of the sides of a right triangle.

In Exercises 3–10, give the values of tan t and sec t, where t is the radian measure of the angle shown.

3.

4.

5.
(−2, 1)

6.
(2, −3)

7.
(−2, 2)

8.
(.6, .8)

9.
(−.6, −.8)

10.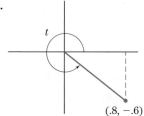
(.8, −.6)

11. Find the width of a river at points A and B if the angle BAC is 90°, the angle ACB is 40°, and the distance from A to C is 75 feet. See Fig. 6.

Figure 6

12. The angle of elevation from an observer to the top of a church is .3 radian, while the angle of elevation from the observer to the top of the church spire is .4 radian. If the observer is 70 meters from the church, how tall is the spire on top of the church?

Differentiate (with respect to t or x):

13. $f(t) = \sec t$

14. $f(t) = \csc t$

15. $f(t) = \cot t$

16. $f(t) = \cot 3t$

17. $f(t) = \tan 4t$

18. $f(t) = \tan \pi t$

19. $f(x) = 3 \tan(\pi - x)$

20. $f(x) = 5 \tan(2x + 1)$

21. $f(x) = 4 \tan(x^2 + x + 3)$

22. $f(x) = 3 \tan(1 - x^2)$

23. $y = \tan \sqrt{x}$

24. $y = 2 \tan \sqrt{x^2 - 4}$

25. $y = x \tan x$

26. $y = e^{3x} \tan 2x$

27. $y = \tan^2 x$

28. $y = \sqrt{\tan x}$

29. $y = (1 + \tan 2t)^3$

30. $y = \tan^4 3t$

31. $y = \ln(\tan t + \sec t)$

32. $y = \ln(\tan t)$

33. **(a)** Find the equation of the tangent line to the graph of $y = \tan x$ at the point $(\frac{\pi}{4}, 1)$.

    (b) Copy the portion of the graph of $y = \tan x$ for $-\frac{\pi}{2} < x < \frac{\pi}{2}$ from Fig. 5, then draw on this graph the tangent line that you found in part (a).

34. Repeat Exercise 33(a) and (b) using the point $(0, 0)$ on the graph of $y = \tan x$ instead of the point $(\frac{\pi}{4}, 1)$.

Evaluate the following integrals.

35. $\displaystyle\int \sec^2 3x \, dx$

36. $\displaystyle\int \sec^2(2x + 1) \, dx$

37. $\displaystyle\int_{-\pi/4}^{\pi/4} \sec^2 x \, dx$

38. $\displaystyle\int_{-\pi/8}^{\pi/8} \sec^2\left(x + \frac{\pi}{8}\right) dx$

39. $\displaystyle\int \frac{1}{\cos^2 x} \, dx$

40. $\displaystyle\int \frac{3}{\cos^2 2x} \, dx$

Solutions to Check Your Understanding 8.4

1. A line of positive slope m is shown in Fig. 7(a). Here, $\tan \theta = m/1 = m$. Suppose that $y = mx + b$ where the slope m is negative. The line $y = mx$ has the same slope and makes the same angle with the x-axis. [See Fig. 7(b).] We see that $\tan \theta = -m/-1 = m$.

2. $\displaystyle\int_0^{\pi/4} \sec^2 t \, dt = \tan t \Big|_0^{\pi/4} = \tan \frac{\pi}{4} - \tan 0 = 1 - 0 = 1.$

(a)

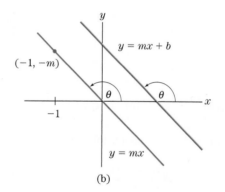

(b)

Figure 7

CHAPTER 8 Summary

KEY TERMS AND CONCEPTS	EXAMPLES

8.1 Radian Measure of Angles

To relate degrees and radians, we can use

$$d° = d \times \frac{\pi}{180} \text{ radians.}$$

In particular,

$$360° = 2\pi \text{ radians.}$$

An angle is in *standard position* on a coordinate system when the vertex of the angle is at $(0, 0)$ and one side, called the *initial side,* is along the positive x-axis. We measure such an angle from the initial side to the *terminal side,* where a *counterclockwise angle is positive* and a *clockwise angle is negative.*

1. Convert 35° to radian measure.

   Solution To convert degrees to radians, multiply the number by $\frac{\pi}{180}$. So,

   $$35° = 35 \times \frac{\pi}{180} \text{ radians} = \frac{7}{36}\pi \text{ radians} \approx .61 \text{ radians.}$$

2. Which of the following angles has the same terminal side as the angle with measure $\frac{\pi}{2}$?

   $$\frac{5\pi}{2}, \quad \frac{-7\pi}{2}$$

   Solution Observe that $\frac{5\pi}{2} = \frac{\pi}{2} + 2\pi$ and $\frac{-7\pi}{2} = \frac{\pi}{2} - 4\pi$. So, the angles $\frac{5\pi}{2}$ and $\frac{-7\pi}{2}$ differ from $\frac{\pi}{2}$ by an integer multiple of 2π, and they both have the same terminal side as $\frac{\pi}{2}$.

KEY TERMS AND CONCEPTS	EXAMPLES

8.2 The Sine and the Cosine

Let t represent the measure in radians of an angle as in the figure. We have

$$\sin t = \frac{y}{r} \quad \text{and} \quad \cos t = \frac{x}{r}.$$

The following are some of the important properties of the sine and cosine functions

$$\cos^2 t + \sin^2 t = 1$$
$$\cos(t \pm 2\pi) = \cos t, \quad \sin(t \pm 2\pi) = \sin t$$
$$\cos(-t) = \cos t, \quad \sin(-t) = -\sin t$$
$$\sin(s + t) = \sin s \cos t + \cos s \sin t$$

1. Let t be an angle between 0 and $\frac{\pi}{2}$ as in the figure. Determine $\sin t$ and $\cos t$.

Solution In the right triangle, the hypotenuse is equal to

$$r = \sqrt{1^2 + 3^2} = \sqrt{10}.$$

So,

$$\sin t = \frac{1}{\sqrt{10}} \quad \text{and} \quad \cos t = \frac{3}{\sqrt{10}}.$$

2. Given $\sin t = .3$, find $\cos t$
  (a) if t is an angle in the first quadrant.
  (b) if t is an angle in the second quadrant.

Solution (a) Using $\cos^2 t + \sin^2 t = 1$, we get

$$\cos^2 t + (.3)^2 = 1$$
$$\cos^2 t = 1 - .09 = .91$$
$$\cos t = \pm\sqrt{.91} \approx \pm.954.$$

If the angle is in the first quadrant, $\sin t$ and $\cos t$ are positive. In this case, $\cos t = \sqrt{.91} \approx .954$. (b) If the angle is in the second quadrant, $\sin t$ is positive while $\cos t$ is negative. In this case, $\cos t = -\sqrt{.91} \approx -.954$.

8.3 Differentiation and Integration of sin t and cos t

The two basic differentiation formulas are the following.

$$\frac{d}{dt}\sin t = \cos t \quad \frac{d}{dt}\cos t = -\sin t$$

Using the chain rule, we generalize these formulas as follows.

$$\frac{d}{dt}[\sin g(t)] = \cos g(t)\frac{d}{dt}[g(t)]$$

$$\frac{d}{dt}[\cos g(t)] = -\sin g(t)\frac{d}{dt}[g(t)]$$

Reversing the differentiation formulas, we obtain the following integration formulas.

$$\int \cos t \, dt = \sin t + C \quad \text{and} \quad \int \sin t \, dt = -\cos t + C$$

More generally, if $a \neq 0$,

$$\int \cos(at) \, dt = \frac{1}{a}\sin(at) + C$$

and

$$\int \sin(at) \, dt = -\frac{1}{a}\cos(at) + C.$$

1. Differentiate

  (a) $\sin(3t)$ **(b)** $\sin(3t^2)$ **(c)** $\sin^2(4t + 1)$.

  Solution

  (a) $\dfrac{d}{dt}[\sin(3t)] = \cos(3t)\dfrac{d}{dt}[3t] = 3\cos(3t)$

  (b) $\dfrac{d}{dt}[\sin(3t^2)] = \cos(3t^2)\dfrac{d}{dt}[3t^2] = 6t\cos(3t^2)$

  (c) $\dfrac{d}{dt}[\sin^2(4t + 1)] = 2\sin(4t + 1)\dfrac{d}{dt}[\sin(4t + 1)]$ Generalized power rule.

  $$= 2\sin(4t + 1)\cos(4t + 1)\frac{d}{dt}(4t + 1)$$

  $$= 8\sin(4t + 1)\cos(4t + 1)$$

2. Differentiate **(a)** $\cos t \sin t$ **(b)** $t^2\cos t^2$

  Solution

  (a) $\dfrac{d}{dt}[\cos t \sin t] = \cos t\dfrac{d}{dt}\sin t + \sin t\dfrac{d}{dt}\cos t$ Product rule.

  $$= \cos t \cos t + \sin t(-\sin t)$$

  $$= \cos^2 t - \sin^2 t$$

KEY TERMS AND CONCEPTS	EXAMPLES

EXAMPLES

(b) $\dfrac{d}{dt}[t^2\cos t^2] = t^2\dfrac{d}{dt}[\cos t^2] + \cos t^2\dfrac{d}{dt}(t^2)$ Product rule.

$\qquad\qquad\qquad = t^2(-\sin t^2)\dfrac{d}{dt}(t^2) + \cos t^2\,(2t)$

$\qquad\qquad\qquad = -t^2(2t)\sin t^2 + 2t\cos t^2$

$\qquad\qquad\qquad = 2t(-t^2\sin t^2 + \cos t^2)$

3. Evaluate the integrals

(a) $\displaystyle\int 3\cos 2t\,dt$ (b) $\displaystyle\int (\sin 4t - \cos t)\,dt.$

Solution

(a) $\displaystyle\int 3\cos 2t\,dt = 3\int\cos 2t\,dt = \dfrac{3}{2}\sin(2t) + C$

(b) $\displaystyle\int (\sin 4t - \cos t)\,dt = \int\sin 4t\,dt - \int\cos t\,dt$

$\qquad\qquad\qquad\qquad = -\dfrac{1}{4}\cos 4t - \sin t + C$

8.4 The Tangent and Other Trigonometric Functions

Other basic trigonometric functions involving the sine and the cosine are defined as follows.

$$\tan t = \dfrac{\sin t}{\cos t};\quad \cot t = \dfrac{\cos t}{\sin t}$$

$$\sec t = \dfrac{1}{\cos t};\quad \csc t = \dfrac{1}{\sin t}$$

We obtain the derivatives of these functions by using the derivative rules for the sine and the cosine and various differentiation formulas, such as the quotient rule. We have

$$\dfrac{d}{dt}\tan t = \sec^2 t;\quad \dfrac{d}{dt}\cot t = -\csc^2 t.$$

Reversing these formulas, we obtain

$$\int\sec^2 t\,dt = \tan t + C;\quad \int\csc^2 t\,dt = -\cot t + C.$$

More generally, if $a \neq 0$,

$$\int\sec^2 at\,dt = \dfrac{1}{a}\tan at + C$$

$$\int\csc^2 at\,dt = -\dfrac{1}{a}\cot at + C.$$

1. Differentiate

(a) $\tan(t^2 + 2t)$ (b) $t\tan^2 t.$

Solution

(a) $\dfrac{d}{dt}[\tan(t^2 + 2t)] = \sec^2(t^2 + 2t)\dfrac{d}{dt}[t^2 + 2t]$

$\qquad\qquad\qquad = (2t + 2)\sec^2(t^2 + 2t)$

(b) $\dfrac{d}{dt}[t\tan^2 t] = t\dfrac{d}{dt}[\tan^2 t] + \tan^2 t$

$\qquad\qquad\qquad = t(2\tan t)\dfrac{d}{dt}[\tan t] + \tan^2 t$

$\qquad\qquad\qquad = 2t\tan t\sec^2 t + \tan^2 t$

2. Evaluate the integrals

(a) $\displaystyle\int\sec^2 5t\,dt$ (b) $\displaystyle\int\dfrac{4}{\cos^2 3t}\,dt$

Solution

(a) $\displaystyle\int\sec^2 5t\,dt = \dfrac{1}{5}\tan 5t + C$

(b) $\displaystyle\int\dfrac{4}{\cos^2 3t}\,dt = \int 4\sec^2 3t\,dt = \dfrac{4}{3}\tan 3t + C$

CHAPTER 8   Fundamental Concept Check Exercises

1. Explain the radian measure of an angle.

2. Give the formula for converting degree measure to radian measure.

3. Give the triangle interpretation of $\sin t$, $\cos t$, and $\tan t$ for t between 0 and $\pi/2$.

4. Define $\sin t$, $\cos t$, and $\tan t$ for an angle of measure t for any t.

5. What does it mean when we say that the sine and cosine functions are periodic with period 2π?

6. Give verbal descriptions of the graphs of $\sin t$ and $\cos t$.

7. State as many identities involving the sine and cosine functions as you can recall.

8. Define $\cot t$, $\sec t$, and $\csc t$ for an angle of measure t.

9. State an identity involving $\tan t$ and $\sec t$.

10. What are the derivatives of $\sin g(t)$, $\cos g(t)$, and $\tan g(t)$?

CHAPTER 8 Review Exercises

Determine the radian measure of the angles shown in Exercises 1–3.

1.

2.

3.

Construct angles with the following radian measure.

4. $-\pi$

5. $\dfrac{5\pi}{4}$

6. $-\dfrac{9\pi}{2}$

In Exercises 7–10, the point with the given coordinates determines an angle of t radians, where $0 \le t \le 2\pi$. Find $\sin t$, $\cos t$, and $\tan t$.

7. $(3, 4)$

8. $(-.6, .8)$

9. $(-.6, -.8)$

10. $(3, -4)$

11. If $\sin t = \frac{1}{5}$, what are the possible values for $\cos t$?

12. If $\cos t = -\frac{2}{3}$, what are the possible values for $\sin t$?

13. Find the four values of t between -2π and 2π at which $\sin t = \cos t$.

14. Find the four values of t between -2π and 2π at which $\sin t = -\cos t$.

15. When $-\pi/2 < t < 0$, is $\tan t$ positive or negative?

16. When $\pi/2 < t < \pi$, is $\sin t$ positive or negative?

17. **Geometry of a Roof** A gabled roof is to be built on a house that is 30 feet wide so that the roof rises at a pitch of 23°. Determine the length of the rafters needed to support the roof.

18. **Determining the Height of a Tree** A tree casts a 60-foot shadow when the angle of elevation of the sun (measured from the horizontal) is 53°. How tall is the tree?

Differentiate (with respect to t or x):

19. $f(t) = 3 \sin t$

20. $f(t) = \sin 3t$

21. $f(t) = \sin \sqrt{t}$

22. $f(t) = \cos t^3$

23. $g(x) = x^3 \sin x$

24. $g(x) = \sin(-2x) \cos 5x$

25. $f(x) = \dfrac{\cos 2x}{\sin 3x}$

26. $f(x) = \dfrac{\cos x - 1}{x^3}$

27. $f(x) = \cos^3 4x$

28. $f(x) = \tan^3 2x$

29. $y = \tan(x^4 + x^2)$

30. $y = \tan e^{-2x}$

31. $y = \sin(\tan x)$

32. $y = \tan(\sin x)$

33. $y = \sin x \tan x$

34. $y = \ln x \cos x$

35. $y = \ln(\sin x)$

36. $y = \ln(\cos x)$

37. $y = e^{3x} \sin^4 x$

38. $y = \sin^4 e^{3x}$

39. $f(t) = \dfrac{\sin t}{\tan 3t}$

40. $f(t) = \dfrac{\tan 2t}{\cos t}$

41. $f(t) = e^{\tan t}$

42. $f(t) = e^t \tan t$

43. If $f(t) = \sin^2 t$, find $f''(t)$.

44. Show that $y = 3 \sin 2t + \cos 2t$ satisfies the differential equation $y'' = -4y$.

45. If $f(s, t) = \sin s \cos 2t$, find $\dfrac{\partial f}{\partial s}$ and $\dfrac{\partial f}{\partial t}$.

46. If $z = \sin wt$, find $\dfrac{\partial z}{\partial w}$ and $\dfrac{\partial z}{\partial t}$.

47. If $f(s, t) = t \sin st$, find $\dfrac{\partial f}{\partial s}$ and $\dfrac{\partial f}{\partial t}$.

48. The identity

$$\sin(s + t) = \sin s \cos t + \cos s \sin t$$

was given in Section 8.2. Compute the partial derivative of each side with respect to t, and obtain an identity involving $\cos(s + t)$.

49. Find the equation of the line tangent to the graph of $y = \tan t$ at $t = \pi/4$.

50. Sketch the graph of $f(t) = \sin t + \cos t$ for $-2\pi \le t \le 2\pi$, using the following steps:
 (a) Find all t between -2π and 2π such that $f'(t) = 0$. Plot the corresponding points on the graph of $y = f(t)$.
 (b) Check the concavity of $f(t)$ at the points in part (a). Make sketches of the graph near these points.
 (c) Determine any inflection points and plot them. Then, complete the sketch of the graph.

51. Sketch the graph of $y = t + \sin t$ for $0 \le t \le 2\pi$.

52. Find the area under the curve $y = 2 + \sin 3t$ from $t = 0$ to $t = \pi/2$.

53. Find the area of the region between the curve $y = \sin t$ and the t-axis from $t = 0$ to $t = 2\pi$.

54. Find the area of the region between the curve $y = \cos t$ and the t-axis from $t = 0$ to $t = 3\pi/2$.

55. Find the area of the region bounded by the curves $y = x$ and $y = \sin x$ from $x = 0$ to $x = \pi$.

Measuring Peak Flow A spirogram is a device that records on a graph the volume of air in a person's lungs as a function of time. If a person undergoes spontaneous hyperventilation, the spirogram trace will closely approximate a sine curve. A typical trace is given by

$$V(t) = 3 + .05 \sin\left(160\pi t - \frac{\pi}{2}\right),$$

where t is measured in minutes and $V(t)$ is the lung volume in liters. (See Fig. 1.) Exercises 56–58 refer to this function.

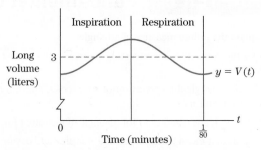

Figure 1 Spirogram trace.

56. (a) Compute $V(0)$, $V(\frac{1}{320})$, $V(\frac{1}{160})$, and $V(\frac{1}{80})$.

(b) What is the maximum lung volume?

57. (a) Find a formula for the rate of flow of air into the lungs at time t.

(b) Find the maximum rate of flow of air during inspiration (breathing in). This quantity is called the *peak inspiratory flow*.

(c) Inspiration occurs during the time from $t = 0$ to $t = 1/160$. Find the average rate of flow of air during inspiration. This quantity is called the *mean inspiratory flow*.

58. Minute Volume The **minute volume** is defined as the total amount of air inspired (breathed in) during 1 minute. According to a standard text on respiratory physiology, when a person undergoes spontaneous hyperventilation, the peak inspiratory flow equals π times the minute volume, and the mean inspiratory flow equals twice the minute volume. Verify these assertions using the data from Exercise 57. (*Source: Applied Respiratory Physiology.*)

Evaluate the following integrals.

59. $\displaystyle\int \sin(\pi - x)\,dx$

60. $\displaystyle\int (3\cos 3x - 2\sin 2x)\,dx$

61. $\displaystyle\int_{0}^{\pi/2} \cos 6x\,dx$

62. $\displaystyle\int \cos(6 - 2x)\,dx$

63. $\displaystyle\int_{0}^{\pi} (x - 2\cos(\pi - 2x))\,dx$

64. $\displaystyle\int_{-\pi}^{\pi} (\cos 3x + 2\sin 7x)\,dx$

65. $\displaystyle\int \sec^2 \frac{x}{2}\,dx$

66. $\displaystyle\int 2\sec^2 2x\,dx$

In Fig. 2:

67. Find the shaded area A_1.

68. Find the shaded area A_2.

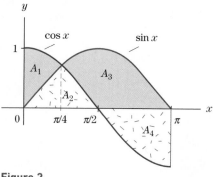

Figure 2

69. Find the shaded area A_3.

70. Find the shaded area A_4.

In Exercises 71–74, find the average of the function $f(t)$ over the given interval.

71. $f(t) = 1 + \sin 2t - \frac{1}{3}\cos 2t,\ 0 \le t \le 2\pi$

72. $f(t) = t - \cos 2t,\ 0 \le t \le \pi$

73. $f(t) = 1000 + 200\sin 2\left(t - \frac{\pi}{4}\right),\ 0 \le t \le \frac{3\pi}{4}$

74. $f(t) = \cos t + \sin t,\ -\pi \le t \le 0$

Evaluate the given integral. [*Hint:* Use identity (1), Section 8.4, to transform the integral before evaluating it.]

75. $\displaystyle\int \tan^2 x\,dx$

76. $\displaystyle\int \tan^2 3x\,dx$

77. $\displaystyle\int (1 + \tan^2 x)\,dx$

78. $\displaystyle\int (2 + \tan^2 x)\,dx$

79. $\displaystyle\int_{0}^{\pi/4} \tan^2 x\,dx$

80. $\displaystyle\int_{0}^{\pi/4} (2 + 2\tan^2 x)\,dx$

Techniques of Integration

In this chapter, we develop techniques for calculating integrals, both indefinite and definite. The need for these techniques has been justified in the preceding chapters. In addition to adding to our fund of applications, we will see even more clearly how the need to calculate integrals arises in physical problems.

Integration is the reverse process of differentiation. However, integration is much harder to carry out. If a function is an expression involving elementary functions (such as x^r, $\sin x$, e^x, . . .), so is its derivative. Moreover, we were able to develop methods of calculation that enable us to differentiate, with comparative ease, almost any function that we can write down. Although many integration problems have these characteristics, some do not. For some elementary functions (for example, e^{x^2}), an antiderivative cannot be expressed in terms of elementary functions. Even when an elementary antiderivative exists, the techniques for finding it are often complicated. For this reason, we must be prepared with a broad range of tools to cope with the problem of calculating integrals. Among the ideas to be discussed in this chapter are the following:

1. Techniques for evaluating indefinite integrals. We will concentrate on two methods: integration by substitution and integration by parts.

2. Evaluation of definite integrals.

3. Approximation of definite integrals. We will develop two new techniques for obtaining numerical approximations to $\int_a^b f(x)dx$. These techniques are especially useful in those cases in which we cannot find an antiderivative for $f(x)$.

Let us review the most elementary facts about integration. The indefinite integral

$$\int f(x)dx$$

is, by definition, a function whose derivative is $f(x)$. If $F(x)$ is one such function, then the most general function whose derivative is $f(x)$ is simply $F(x) + C$, where C is any constant. We write

$$\int f(x)dx = F(x) + C$$

to mean that all antiderivatives of $f(x)$ are precisely the functions $F(x) + C$, where C is any constant.

Each time we differentiate a function, we also derive an integration formula. For example, the fact that

$$\frac{d}{dx}(3x^2) = 6x$$

can be turned into the integration formula

$$\int 6x \, dx = 3x^2 + C.$$

Some of the formulas that follow immediately from differentiation formulas are reviewed in the following table.

Differentiation Formula	Corresponding Integration Formula
$\dfrac{d}{dx}(x^r) = rx^{r-1}$	$\displaystyle\int rx^{r-1} \, dx = x^r + C$ or
	$\displaystyle\int x^r \, dx = \dfrac{x^{r+1}}{r+1} + C, r \neq -1$
$\dfrac{d}{dx}(e^x) = e^x$	$\displaystyle\int e^x \, dx = e^x + C$
$\dfrac{d}{dx}(\ln|x|) = \dfrac{1}{x}$	$\displaystyle\int \frac{1}{x} dx = \ln|x| + C$
$\dfrac{d}{dx}(\sin x) = \cos x$	$\displaystyle\int \cos x \, dx = \sin x + C$
$\dfrac{d}{dx}(\cos x) = -\sin x$	$\displaystyle\int \sin x \, dx = -\cos x + C$
$\dfrac{d}{dx}(\tan x) = \sec^2 x$	$\displaystyle\int \sec^2 x \, dx = \tan x + C$

This table illustrates the need for techniques of integration. For although $\sin x$, $\cos x$, and $\sec^2 x$ occur as derivatives of simple trigonometric functions, the functions $\tan x$ and $\cot x$ are not on our list. In fact, if we experiment with various elementary combinations of the trigonometric functions, it is easy to convince ourselves that antiderivatives of $\tan x$ and $\cot x$ are not easy to compute. In this chapter, we develop techniques for calculating such antiderivatives (among others).

9.1 Integration by Substitution

Every differentiation formula can be turned into a corresponding integration formula. This point is true even for the chain rule. The resulting formula is called **integration by substitution** and is often used to transform a complicated integral into a simpler one.

Let $f(x)$ and $g(x)$ be two given functions, and let $F(x)$ be an antiderivative for $f(x)$. The chain rule asserts that

$$\frac{d}{dx}[F(g(x))] = F'(g(x))g'(x)$$

$$= f(g(x))g'(x) \quad \text{Since } F'(x) = f(x).$$

Turning this formula into an integration formula, we have

$$\int f(g(x))g'(x)dx = F(g(x)) + C, \tag{1}$$

where C is any constant.

FOR REVIEW

Do not leave out the arbitrary constant C in your answer. (See Sec. 6.1, p. 295.)

EXAMPLE 1 **Integration by Substitution** Determine

$$\int (x^2 + 1)^3 \cdot 2x \, dx.$$

FIRST SOLUTION If we set $f(x) = x^3$, $g(x) = x^2 + 1$, then $f(g(x)) = (x^2 + 1)^3$ and $g'(x) = 2x$. Therefore, we can apply formula (1). An antiderivative $F(x)$ of $f(x)$ is given by

$$F(x) = \frac{1}{4}x^4,$$

so that, by formula (1), we have

$$\int (x^2 + 1)^3 \cdot 2x \, dx = F(g(x)) + C = \frac{1}{4}(x^2 + 1)^4 + C.$$

» Now Try Exercise 1

   Formula (1) can be elevated from the status of a sometimes-useful formula to a technique of integration by the introduction of a simple mnemonic device. Suppose that we are faced with integrating a function of the form $f(g(x))g'(x)$. Of course, we know the answer from formula (1). However, let us proceed somewhat differently. Replace the expression $g(x)$ by a new variable u, and replace $g'(x)dx$ by du. Such a substitution has the advantage that it reduces the generally complex expression $f(g(x))$ to the simpler form $f(u)$. In terms of u, the integration problem may be written

$$\int f(g(x))g'(x)dx = \int f(u)du.$$

However, the integral on the right is easy to evaluate, since

$$\int f(u)du = F(u) + C.$$

Since $u = g(x)$, we obtain

$$\int f(g(x))g'(x)dx = F(u) + C = F(g(x)) + C,$$

which is the correct answer by (1). Remember, however, that replacing $g'(x)dx$ by du has status as a correct mathematical statement only, because doing so leads to the correct answers. We do not, in this book, seek to explain in any deeper way what this replacement means.

   Let us rework Example 1 using this method.

SECOND SOLUTION OF EXAMPLE 1   Set $u = x^2 + 1$. Then, $du = \dfrac{d}{dx}(x^2 + 1)dx = 2x\,dx$, and

$$\int (x^2 + 1)^3 \cdot 2x\,dx = \int u^3\,du$$

$$= \frac{1}{4}u^4 + C$$

$$= \frac{1}{4}(x^2 + 1)^4 + C. \quad u = x^2 + 1. \qquad \ll$$

EXAMPLE 2

Substitution with an Exponential Evaluate $\displaystyle\int 2xe^{x^2}\,dx$.

SOLUTION   Let $u = x^2$, so $du = \dfrac{d}{dx}(x^2)dx = 2x\,dx$. Therefore,

$$\int 2xe^{x^2}\,dx = \int e^{x^2} \cdot 2x\,dx$$

$$= \int e^u\,du$$

$$= e^u + C$$

$$= e^{x^2} + C. \qquad \text{➤➤ Now Try Exercise 5}$$

From Examples 1 and 2 we can deduce the following method for integration of functions of the form $f'(g(x))g'(x)$.

> **Integration by Substitution**
>
> 1. Define a new variable $u = g(x)$, where $g(x)$ is chosen in such a way that, when written in terms of u, the integrand is simpler than when written in terms of x.
> 2. Transform the integral with respect to x into an integral with respect to u by replacing $g(x)$ everywhere by u and $g'(x)dx$ by du.
> 3. Integrate the resulting function of u.
> 4. Rewrite the answer in terms of x by replacing u by $g(x)$.

Let us try a few more examples.

EXAMPLE 3

Substitution with a Radical Evaluate $\displaystyle\int 3x^2\sqrt{x^3 + 1}\,dx$.

SOLUTION   The first problem facing us is to find an appropriate substitution that will simplify the integral. An immediate possibility is offered by setting $u = x^3 + 1$. Then, $\sqrt{x^3 + 1}$ will become $\sqrt{u}$, a significant simplification. If $u = x^3 + 1$, then $du = \dfrac{d}{dx}(x^3 + 1)dx = 3x^2\,dx$, so

$$\int 3x^2\sqrt{x^3 + 1}\,dx = \int \sqrt{u}\,du$$

$$= \frac{2}{3}u^{3/2} + C$$

$$= \frac{2}{3}(x^3 + 1)^{3/2} + C. \qquad \text{➤➤ Now Try Exercise 7}$$

EXAMPLE 4

Substitution with a Logarithm Find $\int \dfrac{(\ln x)^2}{x}\,dx$.

SOLUTION Let $u = \ln x$. Then, $du = (1/x)dx$ and

$$\int \frac{(\ln x)^2}{x}\,dx = \int (\ln x)^2 \cdot \frac{1}{x}\,dx$$

$$= \int u^2\,du$$

$$= \frac{u^3}{3} + C$$

$$= \frac{(\ln x)^3}{3} + C. \quad \text{Since } u = \ln x.$$

》 Now Try Exercise 13

Knowing the correct substitution to make is a skill that develops through practice. Basically, we look for an occurrence of function composition, $f(g(x))$, where $f(x)$ is a function that we know how to integrate and where $g'(x)$ also appears in the integrand. Sometimes $g'(x)$ does not appear exactly but can be obtained by multiplication by a constant. Such a shortcoming is easily remedied, as is illustrated in Examples 5 and 6.

EXAMPLE 5

Substitution with an Exponential Find $\int x^2 e^{x^3}\,dx$.

SOLUTION Let $u = x^3$; then, $du = 3x^2\,dx$. The integrand involves $x^2\,dx$, not $3x^2\,dx$. To introduce the missing factor 3, we write

$$\int x^2 e^{x^3}\,dx = \int \frac{1}{3} \cdot 3x^2 e^{x^3}\,dx = \frac{1}{3}\int e^{x^3} 3x^2\,dx.$$

(Recall from Section 6.1 that constant multiples may be moved through the integral sign.) Substituting, we obtain

$$\int x^2 e^{x^3}\,dx = \frac{1}{3}\int e^u\,du = \frac{1}{3}e^u + C$$

$$= \frac{1}{3}e^{x^3} + C \quad u = x^3.$$

Another way to handle the missing factor 3 is to write

$$u = x^3, \quad du = 3x^2\,dx, \quad \text{and} \quad \frac{1}{3}\,du = x^2\,dx.$$

Then, substitution yields

$$\int x^2 e^{x^3}\,dx = \int e^{x^3} \cdot x^2\,dx = \int e^u \cdot \frac{1}{3}\,du = \frac{1}{3}\int e^u\,du$$

$$= \frac{1}{3}e^u + C = \frac{1}{3}e^{x^3} + C.$$

》 Now Try Exercise 11

EXAMPLE 6

Substitution with Fractional Powers Find $\int \dfrac{2 - x}{\sqrt{2x^2 - 8x + 1}}\,dx$.

SOLUTION Let $u = 2x^2 - 8x + 1$; then $du = (4x - 8)dx$. Observe that $4x - 8 = -4(2 - x)$. So, we multiply the integrand by -4 and compensate by placing a factor of $-\frac{1}{4}$ in front of the integral.

$$\int \frac{1}{\sqrt{2x^2 - 8x + 1}} \cdot (2 - x)dx = -\frac{1}{4}\int \frac{1}{\sqrt{2x^2 - 8x + 1}} \cdot (-4)(2 - x)dx$$

$$= -\frac{1}{4}\int \frac{1}{\sqrt{u}} du = -\frac{1}{4}\int u^{-1/2} du$$

$$= -\frac{1}{4} \cdot 2u^{1/2} + C = -\frac{1}{2}u^{1/2} + C$$

$$= -\frac{1}{2}(2x^2 - 8x + 1)^{1/2} + C.$$

» Now Try Exercise 17

EXAMPLE 7

Integral of a Rational Function Find $\int \frac{2x}{x^2 + 1} dx$.

SOLUTION

We note that the derivative of $x^2 + 1$ is $2x$. Thus, we make the substitution $u = x^2 + 1$, $du = 2x\, dx$ to derive

$$\int \frac{2x}{x^2 + 1} dx = \int \frac{1}{u} du = \ln|u| + C = \ln(x^2 + 1) + C.$$

» Now Try Exercise 21

FOR REVIEW

$$\int \frac{1}{u} du = \ln|u| + C$$

Do not leave out the absolute value. (See Sec. 6.1, p. 296.)

EXAMPLE 8

Substitution with Trigonometric Functions Evaluate $\int \tan x\, dx$.

SOLUTION

We have

$$\int \tan x\, dx = \int \frac{\sin x}{\cos x} dx. \quad \tan x = \frac{\sin x}{\cos x}$$

Let $u = \cos x$, so $du = -\sin x\, dx$. Then,

$$\int \frac{\sin x}{\cos x} dx = -\int \frac{-\sin x}{\cos x} dx$$

$$= -\int \frac{1}{u} du$$

$$= -\ln|u| + C$$

$$= -\ln|\cos x| + C.$$

Note that

$$-\ln|\cos x| = \ln\left|\frac{1}{\cos x}\right| = \ln|\sec x|.$$

So the preceding formula can be written

$$\int \tan x\, dx = \ln|\sec x| + C.$$

» Now Try Exercise 43

Check Your Understanding 9.1

Solutions can be found following the section exercises.

1. (*Review*) Differentiate the following functions:
   (a) $e^{(2x^3 + 3x)}$
   (b) $\ln x^5$
   (c) $\ln\sqrt{x}$
   (d) $\ln 5|x|$
   (e) $x \ln x$
   (f) $\ln(x^4 + x^2 + 1)$
   (g) $\sin x^3$
   (h) $\tan x$

2. Use the substitution $u = \frac{3}{x}$ to determine $\int \frac{e^{3/x}}{x^2} dx$.

EXERCISES 9.1

Determine the integrals in Exercises 1–36 by making appropriate substitutions.

1. $\displaystyle\int 2x(x^2 + 4)^5\, dx$

2. $\displaystyle\int 2(2x - 1)^7\, dx$

3. $\displaystyle\int \frac{2x + 1}{\sqrt{x^2 + x + 3}}\, dx$

4. $\displaystyle\int (x^2 + 2x + 3)^6(x + 1)\,dx$

5. $\displaystyle\int 3x^2 e^{(x^3 - 1)}\, dx$

6. $\displaystyle\int 2x e^{-x^2}\, dx$

7. $\displaystyle\int x\sqrt{4 - x^2}\, dx$

8. $\displaystyle\int \frac{(1 + \ln x)^3}{x}\, dx$

9. $\displaystyle\int \frac{1}{\sqrt{2x + 1}}\, dx$

10. $\displaystyle\int (x^3 - 6x)^7(x^2 - 2)\,dx$

11. $\displaystyle\int x e^{x^2}\, dx$

12. $\displaystyle\int \frac{e^{\sqrt{x}}}{\sqrt{x}}\, dx$

13. $\displaystyle\int \frac{\ln(2x)}{x}\, dx$

14. $\displaystyle\int \frac{\sqrt{\ln x}}{x}\, dx$

15. $\displaystyle\int \frac{x^4}{x^5 + 1}\, dx$

16. $\displaystyle\int \frac{x}{\sqrt{x^2 + 1}}\, dx$

17. $\displaystyle\int \frac{x - 3}{(1 - 6x + x^2)^2}\, dx$

18. $\displaystyle\int x^{-2}\left(\frac{1}{x} + 2\right)^5 dx$

19. $\displaystyle\int \frac{\ln\sqrt{x}}{x}\, dx$

20. $\displaystyle\int \frac{x^2}{3 - x^3}\, dx$

21. $\displaystyle\int \frac{x^2 - 2x}{x^3 - 3x^2 + 1}\, dx$

22. $\displaystyle\int \frac{\ln(3x)}{3x}\, dx$

23. $\displaystyle\int \frac{8x}{e^{x^2}}\, dx$

24. $\displaystyle\int \frac{3}{(2x + 1)^3}\, dx$

25. $\displaystyle\int \frac{1}{x \ln x^2}\, dx$

26. $\displaystyle\int \frac{2}{x(\ln x)^4}\, dx$

27. $\displaystyle\int (3 - x)(x^2 - 6x)^4\, dx$

28. $\displaystyle\int \frac{dx}{3 - 5x}$

29. $\displaystyle\int e^x(1 + e^x)^5\, dx$

30. $\displaystyle\int e^x\sqrt{1 + e^x}\, dx$

31. $\displaystyle\int \frac{e^x}{1 + 2e^x}\, dx$

32. $\displaystyle\int \frac{e^x + e^{-x}}{e^x - e^{-x}}\, dx$

33. $\displaystyle\int \frac{e^{-x}}{1 - e^{-x}}\, dx$

34. $\displaystyle\int \frac{(1 + e^{-x})^3}{e^x}\, dx$

[*Hint:* In Exercises 35 and 36, multiply the numerator and denominator by e^{-x}.]

35. $\displaystyle\int \frac{1}{1 + e^x}\, dx$

36. $\displaystyle\int \frac{e^{2x} - 1}{e^{2x} + 1}\, dx$

37. Figure 1 shows graphs of several functions $f(x)$ whose slope at each x is $x/\sqrt{x^2 + 9}$. Find the expression for the function $f(x)$ whose graph passes through $(4, 8)$.

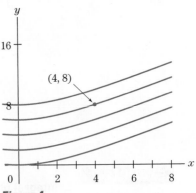

Figure 1

38. Figure 2 shows graphs of several functions $f(x)$ whose slope at each x is $(2\sqrt{x} + 1)/\sqrt{x}$. Find the expression for the function $f(x)$ whose graph passes through $(4, 15)$.

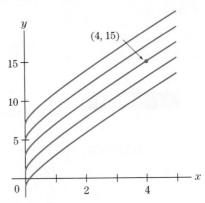

Figure 2

Determine the following integrals using the indicated substitution.

39. $\displaystyle\int (x + 5)^{-1/2} e^{\sqrt{x+5}}\, dx; \; u = \sqrt{x + 5}$

40. $\displaystyle\int \frac{x^4}{x^5 - 7} \ln(x^5 - 7)\,dx; \; u = \ln(x^5 - 7)$

41. $\displaystyle\int x\sec^2 x^2\, dx; \; u = x^2$

42. $\displaystyle\int (1 + \ln x)\sin(x \ln x)\,dx; \; u = x \ln x$

Determine the following integrals by making an appropriate substitution.

43. $\displaystyle\int \sin x \cos x\, dx$

44. $\displaystyle\int 2x \cos x^2\, dx$

45. $\displaystyle\int \frac{\cos\sqrt{x}}{\sqrt{x}}\, dx$

46. $\displaystyle\int \frac{\cos x}{(2 + \sin x)^3}\, dx$

47. $\displaystyle\int \cos^3 x \sin x\, dx$

48. $\displaystyle\int (\sin 2x)e^{\cos 2x}\, dx$

49. $\displaystyle\int \frac{\cos 3x}{\sqrt{2 - \sin 3x}}\, dx$

50. $\displaystyle\int \cot x\, dx$

51. $\displaystyle\int \frac{\sin x + \cos x}{\sin x - \cos x}\,dx$

52. $\displaystyle\int \tan x \sec^2 x\,dx$

53. Determine $\int 2x(x^2 + 5)dx$ by making a substitution. Then, determine the integral by multiplying out the integrand and antidifferentiating. Account for the difference in the two results.

Solutions to Check Your Understanding 9.1

1. (a) $\displaystyle\frac{d}{dx}e^{(2x^3+3x)} = e^{(2x^3+3x)} \cdot (6x^2 + 3)$ (chain rule)

(b) $\displaystyle\frac{d}{dx}\ln x^5 = \frac{d}{dx}5\ln x = 5 \cdot \frac{1}{x}$ (Logarithm Property LIV)

(c) $\displaystyle\frac{d}{dx}\ln\sqrt{x} = \frac{d}{dx}\frac{1}{2}\ln x = \frac{1}{2} \cdot \frac{1}{x} = \frac{1}{2x}$ (Logarithm Property LIV)

(d) $\displaystyle\frac{d}{dx}\ln 5|x| = \frac{d}{dx}[\ln 5 + \ln|x|] = 0 + \frac{1}{x} = \frac{1}{x}$
(Logarithm Property LI)

(e) $\displaystyle\frac{d}{dx}x\ln x = x \cdot \frac{1}{x} + (\ln x) \cdot 1 = 1 + \ln x$ (product rule)

(f) $\displaystyle\frac{d}{dx}\ln(x^4 + x^2 + 1) = \frac{4x^3 + 2x}{x^4 + x^2 + 1}$ (chain rule)

(g) $\displaystyle\frac{d}{dx}\sin x^3 = (\cos x^3) \cdot (3x^2)$ (chain rule)

(h) $\displaystyle\frac{d}{dx}\tan x = \sec^2 x$ (formula for the derivative of $\tan x$)

2. Let $u = 3/x$, $du = (-3/x^2)dx$. Then,

$$\int \frac{e^{3/x}}{x^2}\,dx = -\frac{1}{3}\int e^{3/x} \cdot \left(-\frac{3}{x^2}\right)dx$$

$$= -\frac{1}{3}\int e^u\,du = -\frac{1}{3}e^u + C = -\frac{1}{3}e^{3/x} + C.$$

9.2 Integration by Parts

In the preceding section, we developed the method of integration by substitution by turning the chain rule into an integration formula. Let us do the same for the product rule. Let $f(x)$ and $g(x)$ be any two functions, and let $G(x)$ be an antiderivative of $g(x)$. The product rule asserts that

$$\frac{d}{dx}[f(x)G(x)] = f(x)G'(x) + f'(x)G(x)$$

$$= f(x)g(x) + f'(x)G(x) \quad \text{[since } G'(x) = g(x)\text{]}.$$

Therefore,

$$f(x)G(x) = \int f(x)g(x)dx + \int f'(x)G(x)dx.$$

This last formula can be rewritten in the following more useful form.

> **Integration by Parts**
>
> $$\int f(x)g(x)dx = f(x)G(x) - \int f'(x)G(x)dx. \qquad (1)$$

Equation (1) is the principle of **integration by parts** and is one of the most important techniques of integration.

EXAMPLE 1 **Integration by Parts** Evaluate $\displaystyle\int xe^x\,dx$.

SOLUTION Set $f(x) = x$, $g(x) = e^x$. Then, $f'(x) = 1$, $G(x) = e^x$, and equation (1) yields

$$\int xe^x\,dx = xe^x - \int 1 \cdot e^x\,dx = xe^x - e^x + C. \quad \text{\textbf{»} Now Try Exercise 1}$$

The following principles underlie Example 1 and also illustrate general features of situations to which integration by parts may be applied:

1. The integrand is the product of two functions, $f(x) = x$ and $g(x) = e^x$.
2. It is easy to compute $f'(x)$ and $G(x)$. That is, we can differentiate $f(x)$ and integrate $g(x)$.
3. The integral $\int f'(x)G(x)dx$ can be calculated.

Let us consider another example to see how these three principles work.

EXAMPLE 2 **Integration by Parts** Evaluate $\int x(x + 5)^8 \, dx$.

SOLUTION Our calculations can be set up as follows:

$$f(x) = x, \qquad g(x) = (x + 5)^8,$$

$$f'(x) = 1, \qquad G(x) = \frac{1}{9}(x + 5)^9.$$

Then,

$$\int x(x + 5)^8 \, dx = x \cdot \frac{1}{9}(x + 5)^9 - \int 1 \cdot \frac{1}{9}(x + 5)^9 \, dx$$

$$= \frac{1}{9}x(x + 5)^9 - \frac{1}{9}\int (x + 5)^9 \, dx$$

$$= \frac{1}{9}x(x + 5)^9 - \frac{1}{9} \cdot \frac{1}{10}(x + 5)^{10} + C$$

$$= \frac{1}{9}x(x + 5)^9 - \frac{1}{90}(x + 5)^{10} + C.$$

≫ Now Try Exercise 3

We were led to try integration by parts because our integrand is the product of two functions. We choose $f(x) = x$ [and not $(x + 5)^8$] because $f'(x) = 1$, so the new integrand has only the factor $x + 5$, thereby simplifying the integral.

EXAMPLE 3 **Product of x and an Exponential Function** Evaluate $\int xe^{x/2} \, dx$.

SOLUTION Set

$$f(x) = x, \qquad g(x) = e^{x/2}$$
$$f'(x) = 1, \qquad G(x) = 2e^{x/2}.$$

To find $G(x)$, an antiderivative of $e^{x/2}$, we used the integral formula

$$\int e^{ax} \, dx = \frac{1}{a}e^{ax} + C \qquad (a \neq 0),$$

with $a = \frac{1}{2}$. Then,

$$\int xe^{x/2} \, dx = 2xe^{x/2} - \int 1 \cdot (2e^{x/2})dx$$

$$= 2xe^{x/2} - 2\int e^{x/2} \, dx$$

$$= 2xe^{x/2} - 4e^{x/2} + C,$$

where the last integral follows from the formula that we just recalled.

≫ Now Try Exercise 5

EXAMPLE 4

Product of x and a Trigonometric Function Evaluate $\displaystyle\int x \sin x \, dx$.

SOLUTION Let us set

$$f(x) = x, \qquad g(x) = \sin x,$$
$$f'(x) = 1, \qquad G(x) = -\cos x.$$

Then,

$$\int x \sin x \, dx = -x \cos x - \int 1 \cdot (-\cos x) dx$$

$$= -x \cos x + \int \cos x \, dx$$

$$= -x \cos x + \sin x + C. \qquad \text{» Now Try Exercise 17}$$

EXAMPLE 5

Product of a Power of x and a Logarithm Evaluate $\displaystyle\int x^2 \ln x \, dx$.

SOLUTION Set

$$f(x) = \ln x, \qquad g(x) = x^2$$
$$f'(x) = \frac{1}{x}, \qquad G(x) = \frac{x^3}{3}.$$

Then,

$$\int x^2 \ln x \, dx = \frac{x^3}{3}\ln x - \int \frac{1}{x} \cdot \frac{x^3}{3} \, dx$$

$$= \frac{x^3}{3}\ln x - \frac{1}{3}\int x^2 \, dx$$

$$= \frac{x^3}{3}\ln x - \frac{1}{9}x^3 + C. \qquad \text{» Now Try Exercise 19}$$

The next example shows how integration by parts can be used to compute a reasonably complicated integral.

EXAMPLE 6

Applying Integration by Parts Twice Find $\displaystyle\int x^2 \sin x \, dx$.

SOLUTION Let $f(x) = x^2$, $g(x) = \sin x$. Then, $f'(x) = 2x$ and $G(x) = -\cos x$. Applying our formula for integration by parts, we have

$$\int x^2 \sin x \, dx = -x^2 \cos x - \int 2x \cdot (-\cos x) \, dx$$

$$= -x^2 \cos x + 2\int x \cos x \, dx. \qquad (2)$$

The integral $\int x \cos x \, dx$ can itself be handled by integration by parts. Let $f(x) = x$, $g(x) = \cos x$. Then, $f'(x) = 1$ and $G(x) = \sin x$, so,

$$\int x \cos x \, dx = x \sin x - \int 1 \cdot \sin x \, dx$$

$$= x \sin x + \cos x + C. \qquad (3)$$

Combining (2) and (3), we see that

$$\int x^2 \sin x \, dx = -x^2 \cos x + 2(x \sin x + \cos x) + C$$

$$= -x^2 \cos x + 2x \sin x + 2 \cos x + C.$$

» *Now Try Exercise 23*

EXAMPLE 7   **A Logarithmic Function** Evaluate $\int \ln x \, dx$.

SOLUTION   Since $\ln x = 1 \cdot \ln x$, we may view $\ln x$ as a product $f(x)g(x)$, where $f(x) = \ln x$, $g(x) = 1$. Then,

$$f'(x) = \frac{1}{x}, \qquad G(x) = x.$$

Finally,

$$\int \ln x \, dx = x \ln x - \int \frac{1}{x} \cdot x \, dx$$

$$= x \ln x - \int 1 \, dx$$

$$= x \ln x - x + C. \qquad \text{» } \textit{Now Try Exercise 21}$$

Check Your Understanding 9.2

Solutions can be found following the section exercises.

Evaluate the following integrals.

1. $\int \dfrac{x}{e^{3x}} \, dx$

2. $\int \ln \sqrt{x} \, dx$

EXERCISES 9.2

Evaluate the following integrals:

1. $\int xe^{5x} \, dx$

2. $\int xe^{x/2} \, dx$

3. $\int x(x + 7)^4 \, dx$

4. $\int x(2x - 3)^2 \, dx$

5. $\int \dfrac{x}{e^x} \, dx$

6. $\int x^2 e^x \, dx$

7. $\int \dfrac{x}{\sqrt{x + 1}} \, dx$

8. $\int \dfrac{x}{\sqrt{3 + 2x}} \, dx$

9. $\int e^{2x}(1 - 3x) \, dx$

10. $\int (1 + x)^2 e^{2x} \, dx$

11. $\int \dfrac{6x}{e^{3x}} \, dx$

12. $\int \dfrac{x + 2}{e^{2x}} \, dx$

13. $\int x\sqrt{x + 1} \, dx$

14. $\int x\sqrt{2 - x} \, dx$

15. $\int \sqrt{x} \ln \sqrt{x} \, dx$

16. $\int x^5 \ln x \, dx$

17. $\int x \cos x \, dx$

18. $\int x \sin 8x \, dx$

19. $\int x \ln 5x \, dx$

20. $\int x^{-3} \ln x \, dx$

21. $\int \ln x^4 \, dx$

22. $\int \dfrac{\ln(\ln x)}{x} \, dx$

23. $\int x^2 e^{-x} \, dx$

24. $\int \ln \sqrt{x + 1} \, dx$

Evaluate the following integrals using techniques studied thus far.

25. $\int x(x + 5)^4 \, dx$

26. $\int 4x \cos(x^2 + 1) \, dx$

27. $\int x(x^2 + 5)^4 \, dx$

28. $\int 4x \cos(x + 1) \, dx$

29. $\int (3x + 1) e^{x/3} \, dx$

30. $\int \dfrac{(\ln x)^5}{x} \, dx$

31. $\int x \sec^2(x^2 + 1) \, dx$

32. $\int \dfrac{\ln x}{x^5} \, dx$

33. $\int (xe^{2x} + x^2) \, dx$

34. $\int (x^{3/2} + \ln 2x) \, dx$

35. $\int (xe^{x^2} - 2x) \, dx$

36. $\int (x^2 - x \sin 2x) \, dx$

37. Figure 1 shows graphs of several functions $f(x)$ whose slope at each x is $x/\sqrt{x+9}$. Find the expression for the function $f(x)$ whose graph passes through $(0, 2)$.

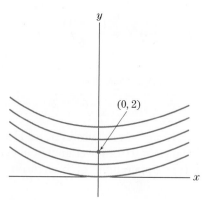

Figure 1

38. Figure 2 shows graphs of several functions $f(x)$ whose slope at each x is $\dfrac{x}{e^{x/3}}$. Find the expression for the function $f(x)$ whose graph passes through $(0, 6)$.

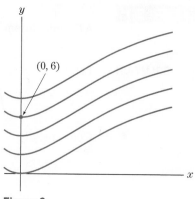

Figure 2

39. Evaluate

$$\int \frac{x\,e^x}{(x+1)^2}\,dx$$

using integration by parts. [*Hint:* $f(x) = xe^x$, $g(x) = \frac{1}{(x+1)^2}$]

40. Evaluate

$$\int x^7 e^{x^4}\,dx.$$

[*Hint:* First, make a substitution; then, use integration by parts.]

Solutions to Check Your Understanding 9.2

1. $\dfrac{x}{e^{3x}}$ is the same as xe^{-3x}, a product of two familiar functions. Set $f(x) = x$, $g(x) = e^{-3x}$. Then,

$$f'(x) = 1, \qquad G(x) = -\frac{1}{3}e^{-3x},$$

so,

$$\int \frac{x}{e^{3x}}\,dx = x \cdot \left(-\frac{1}{3}e^{-3x}\right) - \int 1 \cdot \left(-\frac{1}{3}e^{-3x}\right) dx$$

$$= -\frac{1}{3}xe^{-3x} + \frac{1}{3}\int e^{-3x}\,dx$$

$$= -\frac{1}{3}xe^{-3x} + \frac{1}{3}\left[-\frac{1}{3}e^{-3x}\right] + C$$

$$= -\frac{1}{3}xe^{-3x} - \frac{1}{9}e^{-3x} + C.$$

2. This problem is similar to Example 7, which asks for $\int \ln x\,dx$ and can be approached in the same way by letting $f(x) = \ln\sqrt{x}$ and $g(x) = 1$. Another approach is to use a property of logarithms to simplify the integrand:

$$\int \ln\sqrt{x}\,dx = \int \ln(x)^{1/2}\,dx$$

$$= \int \frac{1}{2}\ln x\,dx$$

$$= \frac{1}{2}\int \ln x\,dx.$$

Since we know $\int \ln x\,dx$ from Example 7,

$$\int \ln\sqrt{x}\,dx = \frac{1}{2}\int \ln x\,dx = \frac{1}{2}(x\ln x - x) + C.$$

9.3 Evaluation of Definite Integrals

Earlier, we discussed techniques for determining antiderivatives (indefinite integrals). One of the most important applications of such techniques concerns the computation of definite integrals. For if $F(x)$ is an antiderivative of $f(x)$, then

$$\int_a^b f(x)\,dx = F(b) - F(a).$$

Thus, the techniques of the previous sections can be used to evaluate definite integrals. Here, we will simplify the method of evaluating definite integrals in those cases where the antiderivative is found by integration by substitution or parts.

EXAMPLE 1

A Definite Integral with Substitution Evaluate $\int_0^1 2x(x^2 + 1)^5 \, dx$.

FIRST SOLUTION Let $u = x^2 + 1$, $du = 2x \, dx$. Then,

$$\int 2x(x^2 + 1)^5 \, dx = \int u^5 \, du = \frac{u^6}{6} + C = \frac{(x^2 + 1)^6}{6} + C.$$

Consequently,

$$\int_0^1 2x(x^2 + 1)^5 \, dx = \frac{(x^2 + 1)^6}{6} \bigg|_0^1 = \frac{2^6}{6} - \frac{1^6}{6} = \frac{21}{2}.$$

SECOND SOLUTION Again, we make the substitution $u = x^2 + 1$, $du = 2x \, dx$; however, we also apply the substitution to the limits of integration. When $x = 0$ (the lower limit of integration), we have $u = 0^2 + 1 = 1$; and when $x = 1$ (the upper limit of integration), we have $u = 1^2 + 1 = 2$. Therefore,

$$\int_0^1 2x(x^2 + 1)^5 \, dx = \int_1^2 u^5 \, du = \frac{u^6}{6} \bigg|_1^2 = \frac{2^6}{6} - \frac{1^6}{6} = \frac{21}{2}.$$

In utilizing the second method, notice that we did not need to reexpress the function $u^6/6$ in terms of x.

>> **Now Try Exercise 1**

The foregoing computation is an example of a general computational method, which can be expressed as follows:

Change of Limits Rule Suppose that the integral $\int f(g(x))g'(x)dx$ is subjected to the substitution $u = g(x)$ so that $\int f(g(x))g'(x)dx$ becomes $\int f(u)du$. Then,

$$\int_a^b f(g(x))g'(x) \, dx = \int_{g(a)}^{g(b)} f(u) \, du.$$

Justification of Change of Limits Rule If $F(x)$ is an antiderivative of $f(x)$, then

$$\frac{d}{dx}[F(g(x))] = F'(g(x))g'(x) = f(g(x))g'(x).$$

Therefore,

$$\int_a^b f(g(x))g'(x) \, dx = F(g(x)) \bigg|_a^b = F(g(b)) - F(ga)) = \int_{g(a)}^{g(b)} f(u) \, du. \qquad \ll$$

EXAMPLE 2

A Definite Integral with Substitution Evaluate $\int_3^5 x\sqrt{x^2 - 9} \, dx$.

SOLUTION Let $u = x^2 - 9$; then $du = 2x \, dx$. When $x = 3$, we have $u = 3^2 - 9 = 0$. When $x = 5$, we have $u = 5^2 - 9 = 16$. Thus,

$$\int_3^5 x\sqrt{x^2 - 9} \, dx = \frac{1}{2} \int_3^5 2x\sqrt{x^2 - 9} \, dx$$

$$= \frac{1}{2} \int_0^{16} \sqrt{u} \, du$$

$$= \frac{1}{2} \cdot \frac{2}{3} u^{3/2} \bigg|_0^{16}$$

$$= \frac{1}{3} \cdot [16^{3/2} - 0] = \frac{1}{3} \cdot 16^{3/2}$$

$$= \frac{1}{3} \cdot 64 = \frac{64}{3}.$$

» Now Try Exercise 3

EXAMPLE 3   **Area of an Ellipse** Determine the area of the ellipse $x^2/a^2 + y^2/b^2 = 1$. (See Fig. 1.)

SOLUTION   Owing to the symmetry of the ellipse, the area is equal to twice the area of the upper half of the ellipse. Solving for y,

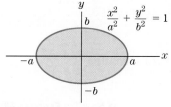

Figure 1

$$\frac{y^2}{b^2} = 1 - \frac{x^2}{a^2}$$

$$\frac{y}{b} = \pm \sqrt{1 - \left(\frac{x}{a}\right)^2}$$

$$y = \pm b \sqrt{1 - \left(\frac{x}{a}\right)^2}.$$

Since the area of the upper half-ellipse is the area under the curve

$$y = b\sqrt{1 - \left(\frac{x}{a}\right)^2},$$

the area of the ellipse is given by the integral

$$2 \int_{-a}^{a} b \sqrt{1 - \left(\frac{x}{a}\right)^2}\, dx.$$

Let $u = x/a$; then $du = 1/a\, dx$. When $x = -a$, we have $u = -a/a = -1$. When $x = a$, we have $u = a/a = 1$. So,

$$2 \int_{-a}^{a} b \sqrt{1 - \left(\frac{x}{a}\right)^2}\, dx = 2b \cdot a \int_{-a}^{a} \frac{1}{a} \sqrt{1 - \left(\frac{x}{a}\right)^2}\, dx$$

$$= 2ba \int_{-1}^{1} \sqrt{1 - u^2}\, du.$$

We cannot evaluate this integral using our existing techniques; we obtain its value immediately by recognizing that, since the area under the curve $y = \sqrt{1 - x^2}$ from $x = -1$ to $x = 1$ is the top half of a unit circle, and since we know that the area of a unit circle is π, the area of the ellipse is $2ba \cdot (\pi/2) = \pi ab$.   **«**

Integration by Parts in Definite Integrals

EXAMPLE 4

Definite Integral by Parts Evaluate $\displaystyle \int_{0}^{5} \frac{x}{\sqrt{x + 4}}\, dx$.

SOLUTION   Let $f(x) = x$, $g(x) = (x + 4)^{-1/2}$, $f'(x) = 1$, and $G(x) = 2(x + 4)^{1/2}$. Then,

$$\int_{0}^{5} \frac{x}{\sqrt{x + 4}}\, dx = 2x(x + 4)^{1/2}\Big|_{0}^{5} - \int_{0}^{5} 1 \cdot 2(x + 4)^{1/2}\, dx$$

$$= 2x(x + 4)^{1/2}\Big|_{0}^{5} - \frac{4}{3}(x + 4)^{3/2}\Big|_{0}^{5}$$

$$= [10(9)^{1/2} - 0] - \left[\frac{4}{3}(9)^{3/2} - \frac{4}{3}(4)^{3/2}\right]$$

$$= [30] - \left[36 - \frac{32}{3}\right] = 4\frac{2}{3}.$$

» Now Try Exercise 5

EXAMPLE 5

Definite Integral by Parts Evaluate $\displaystyle\int_0^{\pi/2} x \cos x \, dx$.

SOLUTION We use integration by parts to find an antiderivative of $x \cos x$. Let $f(x) = x$, $g(x) = \cos x$, $f'(x) = 1$, and $G(x) = \sin x$. Then,

$$\int x \cos x \, dx = x \sin x - \int 1 \cdot \sin x \, dx = x \sin x + \cos x + C.$$

Hence,

$$\int_0^{\pi/2} x \cos x \, dx = (x \sin x + \cos x)\Big|_0^{\pi/2}$$

$$= \left(\frac{\pi}{2} \sin \frac{\pi}{2} + \cos \frac{\pi}{2}\right) - (0 + \cos 0)$$

$$= \frac{\pi}{2} - 1.$$

≫ Now Try Exercise 19

Check Your Understanding 9.3

Solutions can be found following the section exercises.

Evaluate the following definite integrals:

1. $\displaystyle\int_0^1 (2x + 3)e^{x^2+3x+6} \, dx$

2. $\displaystyle\int_e^{e^{\pi/2}} \frac{\sin(\ln x)}{x} \, dx$

EXERCISES 9.3

Evaluate the following definite integrals.

1. $\displaystyle\int_{5/2}^3 2(2x - 5)^{14} \, dx$

2. $\displaystyle\int_2^6 \frac{1}{\sqrt{4x + 1}} \, dx$

3. $\displaystyle\int_0^2 4x(1 + x^2)^3 \, dx$

4. $\displaystyle\int_0^1 \frac{2x}{\sqrt{x^2 + 1}} \, dx$

5. $\displaystyle\int_0^3 \frac{x}{\sqrt{x + 1}} \, dx$

6. $\displaystyle\int_0^1 x(3 + x)^5 \, dx$

7. $\displaystyle\int_3^5 x\sqrt{x^2 - 9} \, dx$

8. $\displaystyle\int_0^1 \frac{1}{(1 + 2x)^4} \, dx$

9. $\displaystyle\int_{-1}^2 (x^2 - 1)(x^3 - 3x)^4 \, dx$

10. $\displaystyle\int_0^1 (2x - 1)(x^2 - x)^{10} \, dx$

11. $\displaystyle\int_0^1 \frac{x}{x^2 + 3} \, dx$

12. $\displaystyle\int_0^4 8x(x + 4)^{-3} \, dx$

13. $\displaystyle\int_1^3 x^2 e^{x^3} \, dx$

14. $\displaystyle\int_{-1}^1 2xe^x \, dx$

15. $\displaystyle\int_1^e \frac{\ln x}{x} \, dx$

16. $\displaystyle\int_1^e \ln x \, dx$

17. $\displaystyle\int_0^\pi e^{\sin x} \cos x \, dx$

18. $\displaystyle\int_0^{\pi/4} \tan x \, dx$

19. $\displaystyle\int_0^1 x \sin \pi x \, dx$

20. $\displaystyle\int_0^{\pi/2} \sin\left(2x - \frac{\pi}{2}\right) dx$

Use substitutions and the fact that a circle of radius r has area πr^2 to evaluate the following integrals.

21. $\displaystyle\int_{-\pi/2}^{\pi/2} \sqrt{1 - \sin^2 x} \cos x \, dx$

22. $\displaystyle\int_0^{\sqrt{2}} \sqrt{4 - x^4} \cdot 2x \, dx$

23. $\displaystyle\int_{-6}^0 \sqrt{-x^2 - 6x} \, dx$

[Complete the square: $-x^2 - 6x = 9 - (x + 3)^2$]

In Exercises 24 and 25, find the area of the shaded regions.

24.

$y = -x\sqrt{9 - x^2}$

25.

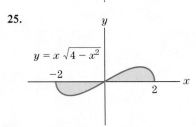

$y = x\sqrt{4 - x^2}$

Solutions to Check Your Understanding 9.3

1. Let $u = x^2 + 3x + 6$, so $du = (2x + 3)dx$. When $x = 0$, $u = 6$; when $x = 1$, $u = 10$. Thus,

$$\int_0^1 (2x + 3)e^{x^2 + 3x + 6} \, dx = \int_6^{10} e^u \, du = e^u r \Big|_6^{10}$$

$$= e^{10} - e^6.$$

2. Let $u = \ln x$, $du = (1/x)dx$. When $x = e$, $u = \ln e = 1$; when $x = e^{\pi/2}$, $u = \ln e^{\pi/2} = \pi/2$. Thus,

$$\int_e^{e^{\pi/2}} \frac{\sin(\ln x)}{x} \, dx = \int_1^{\pi/2} \sin u \, du = -\cos u \Big|_1^{\pi/2}$$

$$= -\cos \frac{\pi}{2} + \cos 1 \approx .54030.$$

9.4 Approximation of Definite Integrals

We cannot always evaluate the definite integrals that arise in practical problems by computing the net change in an antiderivative, as we did in the preceding section. Mathematicians have compiled extensive tables of antiderivatives. Moreover, many excellent software programs can be used to determine antiderivatives. However, the form of an antiderivative may be quite complex, and in some cases, there may actually be no way to express an antiderivative in terms of elementary functions. In this section, we discuss three methods for approximating the numerical value of the definite integral,

$$\int_a^b f(x)dx,$$

without computing an antiderivative.

Given a positive integer n, divide the interval $a \leq x \leq b$ into n equal subintervals, each of length $\Delta x = (b - a)/n$. Denote the endpoints of the subintervals by $a_0, a_1, \ldots, a_n$, and denote the midpoints of the subintervals by $x_1, x_2 \ldots, x_n$. (See Fig. 1.) Recall from Chapter 6 that the definite integral is the limit of Riemann sums. When the midpoints of the subintervals in Fig. 1 are used to construct a Riemann sum, the resulting approximation to $\int_a^b f(x)dx$ is called the **midpoint rule**.

Figure 1

Midpoint Rule

$$\int_a^b f(x)dx \approx f(x_1)\Delta x + f(x_2)\Delta x + \cdots + f(x_n)\Delta x$$

$$= [f(x_1) + f(x_2) + \cdots + f(x_n)]\Delta x. \tag{1}$$

EXAMPLE 1   **The Midpoint Rule** Use the midpoint rule with $n = 4$ to approximate

$$\int_0^2 \frac{1}{1 + e^x} \, dx.$$

SOLUTION   We have $\Delta x = (b - a)/n = (2 - 0)/4 = .5$. The endpoints of the four subintervals begin at $a = 0$ and are spaced .5 unit apart. The first midpoint is at $a + \Delta x/2 = .25$. The midpoints are also spaced .5 unit apart.

According to the midpoint rule, the integral is approximately equal to

$$\left[\frac{1}{1 + e^{0.25}} + \frac{1}{1 + e^{0.75}} + \frac{1}{1 + e^{1.25}} + \frac{1}{1 + e^{1.75}}\right](.5)$$

$$\approx .5646961 \quad \text{(to seven decimal places)}.$$

>> Now Try Exercise 7

A second method of approximation, the **trapezoidal rule**, uses the values of $f(x)$ at the endpoints of the subintervals of the interval $a \le x \le b$.

Trapezoidal Rule

$$\int_a^b f(x)dx \approx [f(a_0) + 2f(a_1) + \cdots + 2f(a_{n-1}) + f(a_n)]\frac{\Delta x}{2}. \quad (2)$$

We will discuss the origin of the trapezoidal rule and why we call it by that name later in this section.

EXAMPLE 2 **The Trapezoidal Rule** Use the trapezoidal rule with $n = 4$ to approximate

$$\int_0^2 \frac{1}{1 + e^x} dx.$$

SOLUTION As in Example 1, $\Delta x = .5$ and the endpoints of the subintervals are $a_0 = 0$, $a_1 = .5$, $a_2 = 1$, $a_3 = 1.5$, and $a_4 = 2$. The trapezoidal rule gives

$$\left[\frac{1}{1 + e^0} + 2 \cdot \frac{1}{1 + e^{0.5}} + 2 \cdot \frac{1}{1 + e^1} + 2 \cdot \frac{1}{1 + e^{1.5}} + \frac{1}{1 + e^2}\right] \cdot \frac{.5}{2}$$

$$\approx .5692545 \quad \text{(to seven decimal places)}.$$

>> Now Try Exercise 11

When the function $f(x)$ is given explicitly, either the midpoint rule or the trapezoidal rule may be used to approximate the definite integral. However, occasionally the values of $f(x)$ may be known only at the endpoints of the subintervals. This may happen, for instance, when the values of $f(x)$ are obtained from experimental data. In this case, the midpoint rule cannot be used.

EXAMPLE 3 **Measuring Cardiac Output** Five milligrams of dye is injected into a vein leading to the heart. The concentration of the dye in the aorta, an artery leading from the heart, is determined every 2 seconds for 22 seconds. (See Table 1.) Let $c(t)$ be the concentration in the aorta after t seconds. Use the trapezoidal rule to estimate $\int_0^{22} c(t)dt$.

Table 1 Concentration of Dye in the Aorta

Seconds after injection	0	2	4	6	8	10	12	14	16	18	20	22
Concentration (mg/liter)	0	0	.6	1.4	2.7	3.7	4.1	3.8	2.9	1.5	.9	.5

(*Source: Measuring Cardiac Output.*)

SOLUTION Let $n = 11$. Then, $a = 0$, $b = 22$, and $\Delta t = (22 - 0)/11 = 2$. The endpoints of the subintervals are $a_0 = 0$, $a_1 = 2$, $a_2 = 4, \ldots, a_{10} = 20$, and $a_{11} = 22$. By the trapezoidal rule,

$$\int_0^{22} c(t)\, dt \approx [c(0) + 2c(2) + 2c(4) + 2c(6) + \cdots + 2c(20) + c(22)]\left(\frac{2}{2}\right)$$

$$= [0 + 2(0) + 2(.6) + 2(1.4) + \cdots + 2(.9) + .5](1)$$

$$= 43.7 \text{ liters}.$$

Note that cardiac output is the rate (usually measured in liters per minute) at which the heart pumps blood, and it may be computed by the formula

$$R = \frac{60D}{\int_0^{22} c(t)\, dt},$$

where D is the quantity of dye injected. For the preceding data, $R = 60(5)/43.7 \approx 6.9$ liters per minute.   **» Now Try Exercise 27**

Let us return to the approximations to $\displaystyle\int_0^2 \frac{1}{1 + e^x}\, dx$ found in Examples 1 and 2. These numbers are shown in Fig. 2 along with the exact value of the definite integral to seven decimal places and the error of the two approximations. [The scale is greatly enlarged. The exact value of the integral is $\ln 2 - \ln(1 + e^{-2}) \approx .5662192$. To find it, see Exercise 35, Section 9.1.] It can be shown that, in general, the error from the midpoint rule is about one-half the error from the trapezoidal rule, and the estimates from these two rules are usually on opposite sides of the actual value of the definite integral. These observations suggest that we might improve our estimate of the value of a definite integral by using a "weighted average" of these two estimates. Let M and T denote the estimates from the midpoint and trapezoidal rules, respectively, and define

$$S = \frac{2}{3}M + \frac{1}{3}T = \frac{2M + T}{3}. \tag{3}$$

Error: .0015231   .0030353

.5646961   .5662192   .5692545
midpoint   actual   trapezoidal
rule   value   rule

Figure 2

The use of S as an estimate of the value of a definite integral is called **Simpson's rule**. If we use Simpson's rule to estimate the definite integral in Example 1, we find that

$$S = \frac{2(.5646961) + .5692545}{3} \approx .5662156.$$

The error here is only .0000036. The error from the trapezoidal rule, in this example, is over 800 times as large!

As the number n of subintervals increases, Simpson's rule becomes more accurate than both the midpoint rule and the trapezoidal rule. For a given definite integral, the error in the midpoint and trapezoidal rules is proportional to $1/n^2$, so doubling n will divide the error by 4. However, the error in Simpson's rule is proportional to $1/n^4$, so doubling n will divide the error by 16, and multiplying n by a factor of 10 will divide the error by 10,000.

It is possible to combine the formulas for the midpoint and trapezoidal rules into a single formula for Simpson's rule by using the fact that $S = (4M + 2T)/6$.

Simpson's Rule

$$\int_a^b f(x)\,dx \approx [f(a_0) + 4f(x_1) + 2f(a_1) + 4f(x_2) + 2f(a_2)$$

$$+ \cdots + 2f(a_{n-1}) + 4f(x_n) + f(a_n)]\frac{\Delta x}{3} \tag{4}$$

EXAMPLE 4

Distribution of IQs Psychologists use various standardized tests to measure intelligence. The method most commonly used to describe the results of such tests is an intelligence quotient (or IQ). An IQ is a positive number that, in theory, indicates how a person's mental age compares with the person's chronological age. The median IQ is arbitrarily set at 100, so half the population has an IQ less than 100 and half greater. IQs are distributed according to a bell-shaped curve called a **normal curve**, pictured in Fig. 3. The proportion of all people having IQs between A and B is given by the area under the curve from A to B, that is, by the integral

$$\frac{1}{16\sqrt{2\pi}} \int_{A}^{B} e^{-(1/2)[(x-100)/16]^2} \, dx.$$

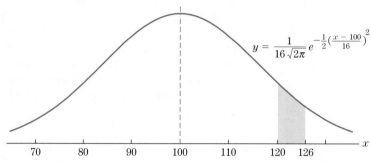

Figure 3 Proportion of IQs between 120 and 126.

Estimate the proportion of all people having IQs between 120 and 126.

SOLUTION

We have seen that this proportion is given by

$$\frac{1}{16\sqrt{2\pi}} \int_{120}^{126} f(x)dx, \quad \text{where } f(x) = e^{-(1/2)[(x-100)/16]^2}.$$

Let us approximate the definite integral by Simpson's rule with $n = 6$. Then, $\Delta x = (126 - 120)/6 = 1$. The endpoints of the subintervals are 120, 122, 124, and 126; the midpoints of these subintervals are 121, 123, and 125. Simpson's rule gives

$$[f(120) + 4f(121) + 2f(122) + 4f(123) + 2f(124) + 4f(125) + f(126)]\frac{\Delta x}{3}$$

$$\approx [.4578 + 1.6904 + .7771 + 1.4235 + .6493 + 1.1801 + .2671]\left(\frac{1}{3}\right)$$

$$\approx 2.1484.$$

Multiplying this estimate by $1/(16\sqrt{2\pi})$, the constant in front of the integral, we get .0536. Thus, approximately 5.36% of the population have IQs between 120 and 126.

>> Now Try Exercise 21

Geometric Interpretation of the Approximation Rules

Let $f(x)$ be a continuous nonnegative function on $a \leq x \leq b$. The approximation rules discussed previously may be interpreted as methods for estimating the area under the graph of $f(x)$. The midpoint rule arises from replacing this area by a collection of n rectangles, one lying over each subinterval, the first of height $f(x_1)$, the second of height $f(x_2)$, and so on. (See Fig. 4.)

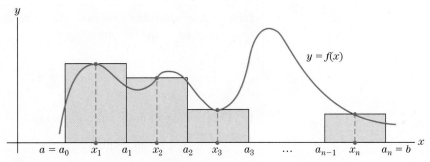

Figure 4 Approximation by rectangles.

If we approximate the area under the graph of $f(x)$ by trapezoids, as in Fig. 5, the total area of these trapezoids turns out to be the number given by the trapezoidal rule (hence the name of the rule).

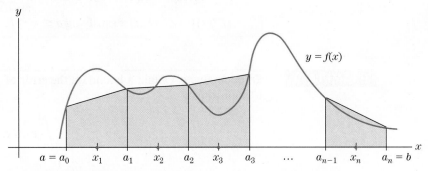

Figure 5 Approximation by trapezoids.

Simpson's rule corresponds to approximating the graph of $f(x)$ on each subinterval by a parabola instead of a straight line, as in the midpoint and trapezoidal rules. On each subinterval, the parabola is chosen so that it intersects the graph of $f(x)$ at the midpoint and both endpoints of the subinterval. (See Fig. 6.) It can be shown that the sum of the areas under these parabolas is the number given by Simpson's rule. We may obtain even more powerful approximation rules by approximating the graph of $f(x)$ on each subinterval by cubic curves, or graphs of higher-order polynomials.

Figure 6 Approximation by parabolas.

Error Analysis

A simple measure of the error of an approximation to a definite integral is the quantity

$$|\,[\text{approximate value}] - [\text{actual value}]\,|.$$

The following theorem gives an idea of how small this error must be for the various approximation rules. In a concrete example, the actual error of an approximation may be even substantially less than the "error bound" given in the theorem.

> **Error of Approximation Theorem** Let n be the number of subintervals used in an approximation of the definite integral
>
> $$\int_a^b f(x)\,dx.$$
>
> 1. The error for the midpoint rule is at most $\dfrac{A(b-a)^3}{24n^2}$, where A is a number such that $\left|f''(x)\right| \leq A$ for all x satisfying $a \leq x \leq b$.
>
> 2. The error for the trapezoidal rule is at most $\dfrac{A(b-a)^3}{12n^2}$, where A is a number such that $\left|f''(x)\right| \leq A$ for all x satisfying $a \leq x \leq b$.
>
> 3. The error for Simpson's rule is at most $\dfrac{A(b-a)^5}{2880n^4}$, where A is a number such that $\left|f''''(x)\right| \leq A$ for all x satisfying $a \leq x \leq b$.

EXAMPLE 5

Error Analysis Obtain a bound on the error of using the trapezoidal rule with $n = 20$ to approximate

$$\int_0^1 e^{x^2}\,dx.$$

SOLUTION

Here, $a = 0$, $b = 1$, and $f(x) = e^{x^2}$. Differentiating twice, we find that

$$f''(x) = (4x^2 + 2)e^{x^2}.$$

How large could $\left|f''(x)\right|$ be if x satisfies $0 \leq x \leq 1$? Since the function $(4x^2 + 2)e^{x^2}$ is clearly increasing on the interval from 0 to 1, its greatest value occurs at $x = 1$. (See Fig. 7.) Therefore, its greatest value is

$$(4 \cdot 1^2 + 2)e^{1^2} = 6e,$$

so we may take $A = 6e$ in the preceding theorem. The error of approximation using the trapezoidal rule is at most

$$\frac{6e(1 - 0)^3}{12(20)^2} = \frac{e}{800} \approx \frac{2.71828}{800} \approx .003398.$$

Figure 7

≫ Now Try Exercise 29

INCORPORATING
TECHNOLOGY

Approximating Integrals In the Incorporating Technology part of Section 6.3 we showed how to use the TI-83/84's **sum** and **seq** functions to evaluate Riemann sums. Here, we use that technique to demonstrate how to implement the approximations discussed in this section. In Fig. 8, we use the midpoint rule as in Example 1 and store the result in a variable M.

In Fig. 9 we implement the trapezoidal rule and store the result in a variable T. Here, we calculate the first term separately, use **sum(seq** to total the middle three terms, and then, we calculate the last term separately.

Figure 10 applies Simpson's rule to the same function and shows that it is very close to the highly accurate value given by **fnInt**.

We can also use our TI-83/84 to determine a suitable value for A in the error approximation theorem by finding the greatest value of either $f''(x)$ or $f''''(x)$. Graphing derivatives is explained in the Incorporating Technology section of Section 2.2. The method demonstrated there can be easily extended to graph the second and fourth derivatives of a function.

Figure 8 Calculation for Example 1.

Figure 9 Calculation for Example 2.

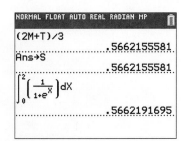

Figure 10 Simpson's rule.

Check Your Understanding 9.4

Solutions can be found following the section exercises.

Consider $\int_{1}^{3.4} (5x - 9)^2 \, dx$.

1. Divide the interval $1 \le x \le 3.4$ into three subintervals. List Δx and the endpoints and midpoints of the subintervals.

2. Approximate the integral by the midpoint rule with $n = 3$.

3. Approximate the integral by the trapezoidal rule with $n = 3$.

4. Approximate the integral by Simpson's rule with $n = 3$.

5. Find the exact value of the integral by integration.

EXERCISES 9.4

In Exercises 1 and 2, divide the given interval into n subintervals and list the value of Δx and the endpoints $a_0, a_1, \ldots, a_n$ of the subintervals.

1. $3 \le x \le 5; n = 5$
2. $-1 \le x \le 2; n = 5$

In Exercises 3 and 4, divide the interval into n subintervals and list the value of Δx and the midpoints $x_1, \ldots, x_n$ of the subintervals.

3. $-1 \le x \le 1; n = 4$
4. $0 \le x \le 3; n = 6$

5. Refer to the graph in Fig. 11. Draw the rectangles that approximate the area under the curve from 0 to 8 when using the midpoint rule with $n = 4$.

Figure 11

6. Refer to the graph in Fig. 11. Apply the trapezoidal rule with $n = 4$ to estimate the area under the curve.

Approximate the following integrals by the midpoint rule; then, find the exact value by integration. Express your answers to five decimal places.

7. $\int_{0}^{4} (x^2 + 5)dx; n = 2, 4$
8. $\int_{1}^{5} (x - 1)^2 \, dx; n = 2, 4$

9. $\int_{0}^{1} e^{-x} \, dx; n = 5$
10. $\int_{1}^{2} \frac{1}{x + 1} \, dx; n = 5$

Approximate the following integrals by the trapezoidal rule; then, find the exact value by integration. Express your answers to five decimal places.

11. $\int_{0}^{1} \left(x - \frac{1}{2}\right)^2 dx; n = 4$
12. $\int_{4}^{9} \frac{1}{x - 3} \, dx; n = 5$

13. $\int_{1}^{5} \frac{1}{x^2} \, dx; n = 3$
14. $\int_{-1}^{1} e^{2x} \, dx; n = 2, 4$

Approximate the following integrals by the midpoint rule, the trapezoidal rule, and Simpson's rule. Then, find the exact value by integration. Express your answers to five decimal places.

15. $\int_{1}^{4} (2x - 3)^3 \, dx; n = 3$
16. $\int_{10}^{20} \frac{\ln x}{x} \, dx; n = 5$

17. $\int_{0}^{2} 2xe^{x^2} \, dx; n = 4$
18. $\int_{0}^{3} x\sqrt{4 - x} \, dx; n = 5$

19. $\int_{2}^{5} xe^x \, dx; n = 5$
20. $\int_{1}^{5} (4x^3 - 3x^2)dx; n = 2$

The following integrals cannot be evaluated in terms of elementary antiderivatives. Find an approximate value by Simpson's rule. Express your answers to five decimal places.

21. $\int_{0}^{2} \sqrt{1 + x^3} \, dx; n = 4$
22. $\int_{0}^{1} \frac{1}{x^3 + 1} \, dx; n = 2$

23. $\int_{0}^{2} \sqrt{\sin x} \, dx; n = 5$
24. $\int_{-1}^{1} \sqrt{1 + x^4} \, dx; n = 4$

25. **Area** In a survey of a piece of oceanfront property, measurements of the distance to the water were made every 50 feet along a 200-foot side. (See Fig. 12.) Use the trapezoidal rule to estimate the area of the property.

Figure 12 Survey of an oceanfront property.

26. Area To determine the amount of water flowing down a certain 100-yard-wide river, engineers need to know the area of a vertical cross section of the river. Measurements of the depth of the river were made every 20 yards from one bank to the other. The readings in fathoms were 0, 1, 2, 3, 1, 0. (One fathom equals 2 yards.) Use the trapezoidal rule to estimate the area of the cross section.

27. Distance Traveled Upon takeoff, the velocity readings of a rocket noted every second for 10 seconds were 0, 30, 75, 115, 155, 200, 250, 300, 360, 420, and 490 feet per second. Use the trapezoidal rule to estimate the distance the rocket traveled during the first 10 seconds. [*Hint:* If $s(t)$ is the distance traveled by time t and $v(t)$ is the velocity at time t, then $s(10) = \int_0^{10} v(t)\,dt$.]

28. Distance Traveled In a drive along a country road, the speedometer readings are recorded each minute during a 5-minute interval.

Time (minutes)	0	1	2	3	4	5
Velocity (mph)	33	32	28	30	32	35

Use the trapezoidal rule to estimate the distance traveled during the 5 minutes. [*Hint:* If time is measured in minutes, velocity should be expressed in distance per minute. For example, 35 mph is $\frac{35}{60}$ miles per minute. Also, see the hint for Exercise 27.]

29. Consider $\int_0^2 f(x)\,dx$, where $f(x) = \frac{1}{12}x^4 + 3x^2$.
   (a) Make a rough sketch of the graph of $f''(x)$ for $0 \le x \le 2$.
   (b) Find a number A such that $|f''(x)| \le A$ for all x satisfying $0 \le x \le 2$.
   (c) Obtain a bound on the error of using the midpoint rule with $n = 10$ to approximate the definite integral.
   (d) The exact value of the definite integral (to four decimal places) is 8.5333, and the midpoint rule with $n = 10$ gives 8.5089. What is the error for the midpoint approximation? Does this error satisfy the bound obtained in part (c)?
   (e) Redo part (c) with the number of intervals doubled to $n = 20$. Is the bound on the error halved? Quartered?

30. Consider $\int_1^2 f(x)\,dx$, where $f(x) = 3 \ln x$.
   (a) Make a rough sketch of the graph of the fourth derivative of $f(x)$ for $1 \le x \le 2$.
   (b) Find a number A such that $|f''''(x)| \le A$ for all x satisfying $1 \le x \le 2$.
   (c) Obtain a bound on the error of using Simpson's rule with $n = 2$ to approximate the definite integral.
   (d) The exact value of the definite integral (to four decimal places) is 1.1589, and Simpson's rule with $n = 2$ gives

1.1588. What is the error for the approximation by Simpson's rule? Does this error satisfy the bound obtained in part (c)?
   (e) Redo part (c) with the number of intervals tripled to $n = 6$. Is the bound on the error divided by 3?

31. (a) Show that the area of the trapezoid in Fig. 13(a) is $\frac{1}{2}(h + k)l$. [*Hint:* Divide the trapezoid into a rectangle and a triangle.]
   (b) Show that the area of the first trapezoid on the left in Fig. 13(b) is $\frac{1}{2}[f(a_0) + f(a_1)]\Delta x$.
   (c) Derive the trapezoidal rule for the case $n = 4$.

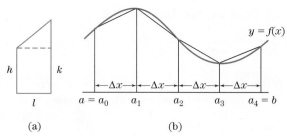

(a)   (b)

Figure 13 Derivation of the trapezoidal rule.

32. Approximate the value of $\int_a^b f(x)\,dx$, where $f(x) \ge 0$, by dividing the interval $a \le x \le b$ into four subintervals and constructing five rectangles. (See Fig. 14.) Note that the width of the three inside rectangles is Δx, while the width of the two outside rectangles is $\Delta x/2$. Compute the sum of the areas of these five rectangles and compare this sum with the trapezoidal rule for $n = 4$.

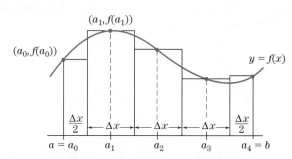

Figure 14 Another view of the trapezoidal rule.

33. (a) Suppose that the graph of $f(x)$ is above the x-axis and concave down on the interval $a_0 \le x \le a_1$. Let x_1 be the midpoint of this interval, let $\Delta x = a_1 - a_0$, and construct the line tangent to the graph of $f(x)$ at $(x_1, f(x_1))$, as in Fig. 15(a). Show that the area of the shaded trapezoid in Fig. 15(a) is the same as the area of the shaded rectangle in Fig. 15(c), that is, $f(x_1)\Delta x$. [*Hint:* Look at Fig. 15(b).] This shows that the area of the rectangle in Fig. 15(c) exceeds the area under the graph of $f(x)$ on the interval $a_0 \le x \le a_1$.
   (b) Suppose that the graph of $f(x)$ is above the x-axis and concave down for all x in the interval $a \le x \le b$. Explain why $T \le \int_a^b f(x)\,dx \le M$, where T and M are the approximations given by the trapezoidal and midpoint rules, respectively.

Figure 17

34. Riemann Sum Derivation of Formula for Cardiac Output; see Example 3. Subdivide the interval $0 \le t \le 22$ into n subintervals of length $\Delta t = 22/n$ seconds. Let t_i be a point in the ith subinterval.

(a) Show that $(R/60)\Delta t \approx$ [number of liters of blood flowing past the monitoring point during the ith time interval].

(b) Show that $c(t_i)(R/60)\Delta t \approx$ [quantity of dye flowing past the monitoring point during the ith time interval].

(c) Assume that basically all the dye will have flowed past the monitoring point during the 22 seconds. Explain why $D \approx (R/60)[c(t_1) + c(t_2) + \cdots + c(t_n)]\Delta t$, where the approximation improves as n gets large.

(d) Conclude that $D = \int_0^{22}(R/60)c(t)\,dt$, and solve for R.

TECHNOLOGY EXERCISES

35. In Fig. 16 a definite integral of the form $\int_a^b f(x)\,dx$ is approximated by the midpoint rule. Determine $f(x)$, a, b, and n.

NORMAL FLOAT AUTO REAL RADIAN CL

sum(seq(1/(25-X²),X,3.1,4.
9,.2))*.2
                    .4489645238
Ans→M
                    .4489645238

Figure 16

36. In Fig. 17 a definite integral of the form $\int_a^b f(x)\,dx$ is approximated by the trapezoidal rule. Determine $f(x)$, a, b, and n.

In Exercises 37–40, approximate the integrals by the midpoint rule, the trapezoidal rule, and Simpson's rule with $n = 10$. Then, find the exact value by integration and give the error for each approximation. Express your answers to the full accuracy given by the calculator or computer.

37. $\displaystyle\int_1^{11} \frac{1}{x}\,dx$

38. $\displaystyle\int_0^{\pi/2} \cos x\,dx$

39. $\displaystyle\int_0^{\pi/4} \sec^2 x\,dx$

40. $\displaystyle\int_0^1 2xe^{x^2}\,dx$

In Exercises 41 and 42, consider the definite integral $\displaystyle\int_0^1 \frac{4}{1 + x^2}\,dx$, which has the value π.

41. Suppose the midpoint rule with $n = 20$ is used to estimate π. Graph the second derivative of the function in the window $[0, 1]$ by $[-10, 10]$, and then use the graph to obtain a bound on the error of the estimate.

42. Suppose the trapezoidal rule with $n = 15$ is used to estimate π. Graph the second derivative of the function in the window $[0, 1]$ by $[-10, 10]$, and then use the graph to obtain a bound on the error of the estimate.

Solutions to Check Your Understanding 9.4

1. $\Delta x = (3.4 - 1)/3 = 2.4/3 = .8$. Each subinterval will have length .8. A good way to proceed is first to draw two hatchmarks that subdivide the interval into three equal subintervals. [See Fig. 18(a).] Then label the hatchmarks by successively adding .8 to the left endpoint. [See Fig. 18(b).] Obtain the first midpoint by adding one-half of .8 to the left endpoint. Then, add .8 to get the next midpoint, and so on. [See Fig. 18(c).]

2. The midpoint rule uses only the midpoints of the subintervals:

$$\int_1^{3.4} (5x - 9)^2\,dx \approx \{(5[1.4] - 9)^2 + (5[2.2] - 9)^2$$
$$+ (5[3] - 9)^2\}(.8)$$
$$= \{(-2)^2 + 2^2 + 6^2\}(.8)$$
$$= 35.2.$$

(a)   (b)   (c)

Figure 18

3. The trapezoidal rule uses only the endpoints of the subintervals:

$$\int_1^{3.4} (5x - 9)^2 \, dx \approx \{(5[1] - 9)^2 + 2(5[1.8] - 9)^2$$

$$+ 2(5[2.6] - 9)^2$$

$$+ (5[3.4] - 9)^2 \} \left(\frac{.8}{2}\right)$$

$$= \{(-4)^2 + 2(0)^2 + 2(4)^2 + 8^2 \}(.4)$$

$$= 44.8.$$

4. Using the formula $S = \dfrac{2M + T}{3}$, we obtain

$$\int_1^{3.4} (5x - 9)^2 \, dx \approx \frac{2(35.2) + 44.8}{3} = \frac{115.2}{3} = 38.4.$$

This approximation may also be obtained directly with formula (4), but the arithmetic requires about the same effort as calculating the midpoint and trapezoidal approximations separately and then combining them, as we do here.

5. $\displaystyle\int_1^{3.4} (5x - 9)^2 \, dx = \frac{1}{15} (5x - 9)^3 \Big|_1^{3.4}$

$$= \frac{1}{15} [8^3 - (-4)^3]$$

$$= 38.4.$$

(Notice that, here, Simpson's rule gives the exact answer. This is so since the function to be integrated is a quadratic polynomial. Actually, Simpson's rule gives the exact value of the definite integral of any polynomial of degree 3 or less. The reason can be discovered from the *error of approximation theorem*.)

9.5 Some Applications of the Integral

Recall that the integral

$$\int_a^b f(t) \, dt$$

can be approximated by a Riemann sum as follows: We divide the t-axis from a to b into n subintervals by adding intermediate points $t_0 = a, t_1, \ldots, t_{n-1}, t_n = b$.

We assume that the points are equally spaced so that each subinterval has length $\Delta t = (b - a)/n$. For large n, the integral is very closely approximated by the Riemann sum

$$\int_a^b f(t) \, dt \approx f(t_1)\Delta t + f(t_2)\Delta t + \cdots + f(t_n)\Delta t. \tag{1}$$

The approximation in (1) works both ways. If we encounter a Riemann sum like that in (1), we can approximate it by the corresponding integral. The approximation becomes better as the number of subintervals increases, that is, as n gets large. Thus, as n gets large, the sum approaches the value of the integral. This will be our approach in the examples that follow.

Our first two examples involve the concept of the present value of money. Suppose that we make an investment that promises to repay A dollars at time t (measuring the present as time 0). How much should we be willing to pay for such an investment? Clearly, we would not want to pay as much as A dollars. For if we had A dollars now, we could invest it at the current rate of interest and, at time t, we would get back our original A dollars plus the accrued interest. Instead, we should only be willing to pay an amount P that, if invested for t years, would yield A dollars. We call P the **present value of A dollars in t years**. We shall assume continuous compounding of interest. If the current (annual) rate of interest is r, then P dollars invested for t years will yield Pe^{rt} dollars (see Section 5.2). That is,

$$Pe^{rt} = A.$$

Thus, the formula for the present value of A dollars in t years at interest rate r is

$$P = Ae^{-rt}.$$

EXAMPLE 1   **Present Value of an Income Stream** Consider a small printing company that does most of its work on one printing press. The firm's profits are directly influenced by the amount of material that the press can produce (assuming that other factors, such as wages, are held constant). We may say that the press is producing a continuous stream of income for the company. Of course, the efficiency of the press may decline as it gets older. At time t, let $K(t)$ be the annual rate of income from the press. [This means that the press is producing $K(t) \cdot \frac{1}{365}$ dollars per day at time t.] Find a model for the present value of the income generated by the printing press over the next T years, assuming an interest rate r (with interest compounded continuously).

SOLUTION   We will divide the T-year period into n small subintervals of time, each of duration Δt years. (If each subinterval were 1 day, for example, Δt would equal $\frac{1}{365}$.)

We now consider the income produced by the printing press during a small time interval from t_{j-1} to t_j. Since Δt is small, the rate $K(t)$ of income production changes by only a negligible amount in that interval and can be considered approximately equal to $K(t_j)$. Since $K(t_j)$ gives an annual rate of income, the actual income produced during the period of Δt years is $K(t_j)\Delta t$. This income will be produced at approximately time t_j (t_j years from $t = 0$), so its present value is

$$[K(t_j)\Delta t]e^{-rt_j}.$$

The present value of the total income produced over the T-year period is the sum of the present values of the amounts produced during each time subinterval; that is,

$$K(t_1)e^{-rt_1}\Delta t + K(t_2)e^{-rt_2}\Delta t + \cdots + K(t_n)e^{-rt_n}\Delta t. \tag{2}$$

As the number of subintervals gets large, the length Δt of each subinterval becomes small, and the sum in (2) approaches the integral

$$\int_0^T K(t)e^{-rt}\, dt. \tag{3}$$

We call this quantity the **present value of the stream of income** produced by the printing press over the period from $t = 0$ to $t = T$ years. (The interest rate r used to compute present value is often called the company's *internal rate of return*.)

» Now Try Exercise 1

The concept of the present value of a continuous stream of income is an important tool in management decision processes involving the selection or replacement of equipment. It is also useful in the analysis of various investment opportunities. Even when $K(t)$ is a simple function, the evaluation of the integral in (3) usually requires special techniques, such as integration by parts, as we see in the next example.

EXAMPLE 2   **Present Value of an Income Stream** A company estimates that the rate of revenue produced by a machine at time t will be $5000 - 100t$ dollars per year. Find the present value of this continuous stream of income over the next 4 years at a 16% interest rate.

SOLUTION   We use formula (3) with $K(t) = 5000 - 100t$, $T = 4$, and $r = .16$. The present value of this income stream is

$$\int_0^4 (5000 - 100t)e^{-0.16t}\, dt.$$

Using integration by parts, with $f(t) = 5000 - 100t$ and $g(t) = e^{-0.16t}$, we find that the preceding integral equals

$$(5000 - 100t)\frac{1}{-.16}e^{-0.16t}\Big|_0^4 - \int_0^4 (-100)\frac{1}{-.16}e^{-0.16t}\, dt$$

$$\approx 16{,}090 - \frac{100}{.16} \cdot \frac{1}{-.16}e^{-0.16t}\Big|_0^4$$

$$\approx 16{,}090 - 1847 = \$14{,}243.$$ **≫ Now Try Exercise 5**

A slight modification of the discussion in Example 1 gives the following general result.

Present Value of a Continuous Stream of Income

$$[\text{present value}] = \int_{T_1}^{T_2} K(t)e^{-rt}\, dt,$$

where

1. $K(t)$ dollars per year is the annual rate of income at time t.

2. r is the annual interest rate of invested money.

3. T_1 to T_2 (years) is the time period of the income stream.

EXAMPLE 3   **A Demographic Model** It has been determined that in 1940 the population density t miles from the center of New York City was approximately $120e^{-0.2t}$ thousand people per square mile. Estimate the number of people who lived within 2 miles of the center of New York in 1940. (*Source: Journal of Royal Statistical Society* and *Journal of Urban Economics.*)

SOLUTION   Choose a fixed line emanating from the center of the city along which to measure distance. Subdivide this line from $t = 0$ to $t = 2$ into a large number of subintervals, each of length Δt. Each subinterval determines a ring. (See Fig. 1.)

Figure 1 A ring around the city center.

We will estimate the population of each ring and then add these populations together to obtain the total population. Suppose that j is an index ranging from 1 to n. If the outer circle of the jth ring is at a distance t_j from the center, the inner circle of this ring is at a distance $t_{j-1} = t_j - \Delta t$ from the center. The area of the jth ring is

$$\pi t_j^2 - \pi t_{j-1}^2 = \pi t_j^2 - \pi(t_j - \Delta t)^2$$
$$= \pi t_j^2 - \pi[t_j^2 - 2t_j\Delta t + (\Delta t)^2]$$
$$= 2\pi t_j\Delta t - \pi(\Delta t)^2.$$

Assume that Δt is very small. Then, $\pi(\Delta t)^2$ is much smaller than $2\pi t_j\Delta t$. Hence, the area of this ring is very close to $2\pi t_j\Delta t$.

Within the jth ring, the density of people is about $120e^{-0.2t_j}$ thousand people per square mile. So the number of people in this ring is approximately

$$[\text{population density}] \cdot [\text{area of ring}] \approx 120e^{-0.2t_j} \cdot 2\pi t_j\Delta t$$
$$= 240\pi t_j e^{-0.2t_j}\Delta t.$$

Adding up the populations of all the rings, we obtain a total of

$$240\pi t_1 e^{-0.2t_1}\Delta t + 240\pi t_2 e^{-0.2t_2}\Delta t + \cdots + 240\pi t_n e^{-0.2t_n}\Delta t,$$

which is a Riemann sum for the function $f(t) = 240t\pi e^{-0.2t}$ over the interval from $t = 0$ to $t = 2$. This approximation to the population improves as the number n increases. Thus, the number of people (in thousands) who lived within 2 miles of the center of the city was

$$\int_0^2 240\pi t e^{-0.2t}\,dt = 240\pi \int_0^2 t e^{-0.2t}\,dt.$$

The last integral can be computed with integration by parts, to obtain

$$240\pi \int_0^2 t e^{-0.2t}\,dt = 240\pi \frac{te^{-0.2t}}{-.2}\Big|_0^2 - 240\pi \int_0^2 \frac{e^{-0.2t}}{-.2}\,dt$$
$$= -2400\pi e^{-0.4} + 1200\pi\left(\frac{e^{-0.2t}}{-.2}\right)\Big|_0^2$$
$$= -2400\pi e^{-0.4} + (-6000\pi e^{-0.4} + 6000\pi)$$
$$\approx 1160.$$

Thus, in 1940, approximately 1,160,000 people lived within 2 miles of the center of the city.   **» Now Try Exercise 9**

An argument analogous to that in Example 3 leads to the following result.

$$\text{Total Population in a Ring around the City Center} = \int_a^b 2\pi t D(t)\,dt,$$

where
1. $D(t)$ is the density of population (in persons per square mile) at distance t miles from city center.
2. Ring includes all persons who live between a and b miles from the city center.

Check Your Understanding 9.5

Solutions can be found following the section exercises.

The integral formula

$$\int (at+b)e^{-rt}\,dt = e^{-rt}\left[\frac{-1}{r}(at+b) - \frac{a}{r^2}\right] + C \quad (r \neq 0)$$

is used in many applications in this section. Derive it using integration by parts.

EXERCISES 9.5

1. **Present Value** Find the present value of a continuous stream of income over 5 years when the rate of income is constant at $35,000 per year and the interest rate is 7%.

2. **Present Value** A continuous stream of income is being produced at the constant rate of $60,000 per year. Find the present value of the income generated during the time from $t = 2$ to $t = 6$ years, with a 6% interest rate.

3. **Present Value** Find the present value of a continuous stream of income over the time from $t = 1$ to $t = 5$ years when the interest rate is 10% and the income is produced at the rate of $12,000 per year.

4. **Present Value** Find the present value of a continuous stream of income over 4 years if the rate of income is $25e^{-0.02t}$ thousand dollars per year at time t and the interest rate is 8%.

5. **Present Value** Find the present value of a continuous stream of income over 3 years if the rate of income is $80e^{-0.08t}$ thousand dollars per year at time t and the interest rate is 11%.

6. **Present Value** A continuous stream of income is produced at the rate of $20e^{1-0.09t}$ thousand dollars per year at time t, and invested money earns 6% interest.
   (a) Write a definite integral that gives the present value of this stream of income over the time from $t = 2$ to $t = 5$ years.
   (b) Compute the present value described in part (a).

7. **A Company's Future Earnings** A growth company is one whose net earnings tend to increase each year. Suppose that the net earnings of a company at time t are being generated at the rate of $30 + 5t$ million dollars per year.
   (a) Write a definite integral that gives the present value of the company's earnings over the next 2 years using a 10% interest rate.
   (b) Compute the present value described in part (a).

8. **Present Value** Find the present value of a stream of earnings generated over the next 2 years at the rate of $50 + 7t$ thousand dollars per year at time t assuming a 10% interest rate.

9. **Demographic Model** In 2012, the population density of a city t miles from the city center was $120e^{-0.65t}$ thousand people per square mile.
   (a) Write a definite integral whose value equals the number of people (in thousands) who lived within 5 miles of the city center.
   (b) Calculate the definite integral in part (a).

10. **Demographic Model** Use the population density from Exercise 9 to calculate the number of people who lived between 3 and 5 miles from the city center.

11. **Population Model** The population density of Philadelphia in 1940 was given by the function $60e^{-0.4t}$. Calculate the number of people who lived within 5 miles of the city center. Sketch the graphs of the population densities for 1900 and 1940 (see Exercise 9) on a common graph. What trend do the graphs exhibit?

12. **Ecological Model** A volcano erupts and spreads lava in all directions. The density of the deposits at a distance t kilometers from the center is $D(t)$ thousand tons per square kilometer, where $D(t) = 11(t^2 + 10)^{-2}$. Find the tonnage of lava deposited between the distances of 1 and 10 kilometers from the center.

13. **Demographic Model** Suppose that the population density function for a city is $40e^{-0.5t}$ thousand people per square mile. Let $P(t)$ be the total population that lives within t miles of the city center, and let Δt be a small positive number.
   (a) Consider the ring about the city whose inner circle is at t miles and outer circle is at $t + \Delta t$ miles. The text shows that the area of this ring is approximately $2\pi t \Delta t$ square miles. Approximately how many people live within this ring? (Your answer will involve t and Δt.)
   (b) What does

$$\frac{P(t + \Delta t) - P(t)}{\Delta t}$$

   approach as Δt tends to zero?
   (c) What does the quantity $P(5 + \Delta t) - P(5)$ represent?
   (d) Use parts (a) and (c) to find a formula for

$$\frac{P(t + \Delta t) - P(t)}{\Delta t}$$

   and from that obtain an approximate formula for the derivative $P'(t)$. This formula gives the rate of change of total population with respect to the distance t from the city center.
   (e) Given two positive numbers a and b, find a formula involving a definite integral, for the number of people who live in the city between a miles and b miles of the city center. [*Hint:* Use part (d) and the fundamental theorem of calculus to compute $P(b) - P(a)$.]

Solutions to Check Your Understanding 9.5

Let $f(t) = at + b$, $g(t) = e^{-rt}$. Then, $f'(t) = a$ and $G(t) = -\frac{1}{r}e^{-rt}$. Using integration by parts, we have

$$\int (at + b)e^{-rt}\, dt = -\frac{1}{r}(at + b)e^{-rt} + \frac{a}{r}\int e^{-rt}\, dt$$

$$= -\frac{1}{r}(at + b)e^{-rt} + \frac{a}{r}\left(\frac{-1}{r}\right)e^{-rt} + C$$

$$= e^{-rt}\left[\frac{-1}{r}(at + b) - \frac{a}{r^2}\right] + C.$$

9.6 Improper Integrals

In applications of calculus, especially in statistics, it is often necessary to consider the area of a region that extends infinitely far to the right or left along the x-axis. We have drawn several such regions in Fig. 1. Note that the area under a curve that extends infinitely far along the x-axis is not necessarily infinite itself. The areas of such "infinite" regions may be computed with **improper integrals**.

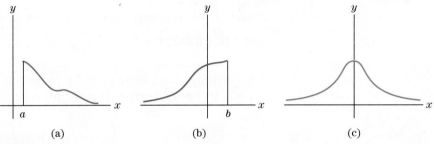

Figure 1

To motivate the idea of an improper integral, let us attempt to calculate the area under the curve $y = 3/x^2$ to the right of $x = 1$. (See Fig. 2.)

First, we compute the area under the graph of this function from $x = 1$ to $x = b$, where b is some number greater than 1. [See Fig. 3(a).] Then we examine how the area increases as we let b get larger. [See Figs. 3(b) and 3(c).] The area from 1 to b is given by

$$\int_1^b \frac{3}{x^2}\, dx = -\frac{3}{x}\Big|_1^b = \left(-\frac{3}{b}\right) - \left(-\frac{3}{1}\right) = 3 - \frac{3}{b}.$$

Figure 2

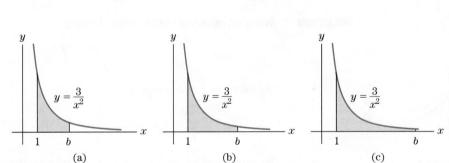

Figure 3

When b is large, $3/b$ is small, and the integral nearly equals 3. That is, the area under the curve from 1 to b nearly equals 3. (See Table 1.) In fact, the area gets arbitrarily close to 3 as b gets larger. Thus, it is reasonable to say that the region under the curve $y = 3/x^2$ for $x \geq 1$ has area 3.

Table 1 Value of an "Infinitely Long" Area as a Limit

b	Area $= \displaystyle\int_1^b \frac{3}{x^2}\, dx = 3 - \frac{3}{b}$
10	2.7000
100	2.9700
1,000	2.9970
10,000	2.9997

Recall from Chapter 1 that we write $b \to \infty$ as shorthand for "b gets arbitrarily large, without bound." Then, to express the fact that the value of

$$\int_1^b \frac{3}{x^2}\, dx$$

approaches 3 as $b \to \infty$, we write

$$\int_1^\infty \frac{3}{x^2}\, dx = \lim_{b \to \infty} \int_1^b \frac{3}{x^2}\, dx = 3.$$

We call $\int_1^\infty \frac{3}{x^2}\, dx$ an *improper* integral because the upper limit of the integral is ∞ (infinity), rather than a finite number.

> **DEFINITION** Let a be fixed and suppose that $f(x)$ is a nonnegative function for $x \geq a$. If $\lim_{b \to \infty} \int_a^b f(x)dx = L$, we define
>
> $$\int_a^\infty f(x)dx = \lim_{b \to \infty} \int_a^b f(x)dx = L.$$
>
> We say that the improper integral $\int_a^\infty f(x)dx$ is **convergent** and that the region under the curve $y = f(x)$ for $x \geq a$ has area L. (See Fig. 4.)

$y = f(x)$

Area $= \int_a^\infty f(x)\, dx$

Figure 4 Area defined by an improper integral.

It is possible to consider improper integrals in which $f(x)$ is both positive and negative. However, we shall consider only nonnegative functions, since this is the case occurring in most applications.

EXAMPLE 1   **A Convergent Improper Integral** Find the area under the curve $y = e^{-x}$ for $x \geq 0$. (See Fig. 5.)

SOLUTION   We must calculate the improper integral

$$\int_0^\infty e^{-x}\, dx.$$

We take $b > 0$ and compute

$$\int_0^b e^{-x}\, dx = -e^{-x}\Big|_0^b = (-e^{-b}) - (-e^0) = 1 - e^{-b} = 1 - \frac{1}{e^b}.$$

$y = e^{-x}$

Figure 5

We now consider the limit as $b \to \infty$ and note that $1/e^b$ approaches zero. Thus,

$$\int_0^\infty e^{-x}\, dx = \lim_{b \to \infty} \int_0^b e^{-x}\, dx = \lim_{b \to \infty}\left(1 - \frac{1}{e^b}\right) = 1.$$

Therefore, the region in Fig. 5 has area 1.   **» Now Try Exercise 15**

EXAMPLE 2

A Convergent Improper Integral Evaluate the improper integral $\int_7^\infty \frac{1}{(x-5)^2}\, dx$.

SOLUTION $\displaystyle \int_7^b \frac{1}{(x-5)^2}\, dx = -\frac{1}{x-5}\Big|_7^b = -\frac{1}{b-5} - \left(-\frac{1}{7-5}\right) = \frac{1}{2} - \frac{1}{b-5}$

As $b \to \infty$, the fraction $1/(b-5)$ approaches zero, so

$$\int_7^\infty \frac{1}{(x-5)^2}\, dx = \lim_{b \to \infty} \int_7^b \frac{1}{(x-5)^2}\, dx = \lim_{b \to \infty}\left(\frac{1}{2} - \frac{1}{b-5}\right) = \frac{1}{2}.$$

» Now Try Exercise 17

Not every improper integral is convergent. If the value $\int_a^b f(x)dx$ does not have a limit as $b \to \infty$, we cannot assign any numerical value to $\int_a^\infty f(x)dx$, and we say that the improper integral $\int_a^\infty f(x)dx$ is **divergent**.

EXAMPLE 3

A Divergent Improper Integral Show that $\int_1^\infty \frac{1}{\sqrt{x}}\,dx$ is divergent.

SOLUTION For $b > 1$, we have

$$\int_1^b \frac{1}{\sqrt{x}}\,dx = 2\sqrt{x}\,\Big|_1^b = 2\sqrt{b} - 2.$$

As $b \to \infty$, the quantity $2\sqrt{b} - 2$ increases without bound. That is, $2\sqrt{b} - 2$ can be made larger than any specific number. Therefore, $\int_1^b \frac{1}{\sqrt{x}}\,dx$ has no limit as $b \to \infty$, so $\int_1^\infty \frac{1}{\sqrt{x}}\,dx$ is divergent.

>> *Now Try Exercise 19*

In some cases, it is necessary to consider improper integrals of the form

$$\int_{-\infty}^b f(x)dx.$$

Let b be fixed and examine the value of $\int_a^b f(x)dx$ as $a \to -\infty$, that is, as a moves arbitrarily far to the left on the number line. If $\lim_{a \to -\infty} \int_a^b f(x)dx = L$, we say that the improper integral $\int_{-\infty}^b f(x)dx$ is *convergent*, and we write

$$\int_{-\infty}^b f(x)dx = L.$$

Otherwise, the improper integral is divergent. An integral of the form $\int_{-\infty}^b f(x)dx$ may be used to compute the area of a region such as that shown in Fig. 1(b).

EXAMPLE 4

A Convergent Improper Integral Determine if $\int_{-\infty}^0 e^{5x}\,dx$ is convergent. If convergent, find its value.

SOLUTION

$$\int_{-\infty}^0 e^{5x}\,dx = \lim_{a \to -\infty} \int_a^0 e^{5x}\,dx = \lim_{a \to -\infty} \frac{1}{5}e^{5x}\,\Big|_a^0 = \lim_{a \to -\infty} \left(\frac{1}{5} - \frac{1}{5}e^{5a} \right)$$

As $a \to -\infty$, e^{5a} approaches 0, so $\frac{1}{5} - \frac{1}{5}e^{5a}$ approaches $\frac{1}{5}$. Thus, the improper integral converges and has value $\frac{1}{5}$.

>> *Now Try Exercise 39*

Areas of regions that extend infinitely far to the left *and* right, such as the region in Fig. 1(c), are calculated using improper integrals of the form

$$\int_{-\infty}^\infty f(x)dx.$$

We define such an integral to have the value

$$\int_{-\infty}^0 f(x)dx + \int_0^\infty f(x)dx,$$

provided that both of the latter improper integrals are convergent.

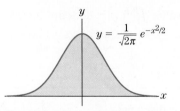

Figure 6 The standard normal curve.

An important area that arises in probability theory is the area under the so-called normal curve, whose equation is

$$y = \frac{1}{\sqrt{2\pi}}e^{-x^2/2}.$$

(See Fig. 6.) It is of fundamental importance for probability theory that this area be 1. In terms of an improper integral, this fact may be written as

$$\int_{-\infty}^{\infty} \frac{1}{\sqrt{2\pi}}e^{-x^2/2}\,dx = 1.$$

The proof of this result is beyond the scope of this book.

INCORPORATING TECHNOLOGY

Improper Integrals Although graphing calculators cannot tell you whether an improper integral converges, you can use the calculator to obtain a reliable indication of the behavior of the integral. Just look at values of $\int_a^b f(x)dx$ as b increases. Figures 7(a) and 7(b), which were created by setting $Y_1 = \mathbf{fnInt(e^(-X),X,0,X)}$, give convincing evidence that the value of the improper integral in Example 1 is 1. The final $\mathbf{X}$ in $\mathbf{fnInt(e^(-X),X,0,X)}$ is the upper limit of integration, that is, b.

(a)   (b)

Figure 7

Check Your Understanding 9.6

Solutions can be found following the section exercises.

1. Does $1 - 2(1 - 3b)^{-4}$ approach a limit as $b \to \infty$?

2. Evaluate $\int_1^{\infty} \frac{x^2}{x^3 + 8}\,dx$.

3. Evaluate $\int_{-\infty}^{-2} \frac{1}{x^4}\,dx$.

EXERCISES 9.6

In Exercises 1–12, determine if the given expression approaches a limit as $b \to \infty$, and find that number when it does.

1. $\dfrac{5}{b}$

2. b^2

3. $-3e^{2b}$

4. $\dfrac{1}{b} + \dfrac{1}{3}$

5. $\dfrac{1}{4} - \dfrac{1}{b^2}$

6. $\dfrac{1}{2}\sqrt{b}$

7. $2 - (b + 1)^{-1/2}$

8. $5 - (b - 1)^{-1}$

9. $5(b^2 + 3)^{-1}$

10. $4(1 - b^{-3/4})$

11. $e^{-b/2} + 5$

12. $2 - e^{-3b}$

13. Find the area under the graph of $y = 1/x^2$ for $x \geq 2$.

14. Find the area under the graph of $y = (x + 1)^{-2}$ for $x \geq 0$.

15. Find the area under the graph of $y = e^{-x/2}$ for $x \geq 0$.

16. Find the area under the graph of $y = 4e^{-4x}$ for $x \geq 0$.

17. Find the area under the graph of $y = (x + 1)^{-3/2}$ for $x \geq 3$.

18. Find the area under the graph of $y = (2x + 6)^{-4/3}$ for $x \geq 1$. (See Fig. 8.)

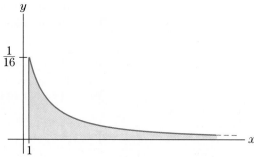

Figure 8

19. Show that the region under the graph of $y = (14x + 18)^{-4/5}$ for $x \geq 1$ cannot be assigned any finite number as its area. (See Fig. 9.)

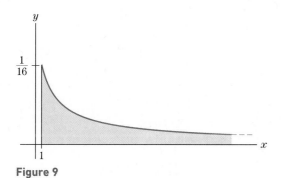

Figure 9

20. Show that the region under the graph of $y = (x - 1)^{-1/3}$ for $x \geq 2$ cannot be assigned any finite number as its area.

Evaluate the following improper integrals whenever they are convergent.

21. $\displaystyle\int_1^\infty \frac{1}{x^3}\, dx$

22. $\displaystyle\int_1^\infty \frac{2}{x^{3/2}}\, dx$

23. $\displaystyle\int_0^\infty \frac{1}{(2x + 3)^2}\, dx$

24. $\displaystyle\int_0^\infty e^{-3x}\, dx$

25. $\displaystyle\int_0^\infty e^{2x}\, dx$

26. $\displaystyle\int_0^\infty (x^2 + 1)\, dx$

27. $\displaystyle\int_2^\infty \frac{1}{(x - 1)^{5/2}}\, dx$

28. $\displaystyle\int_2^\infty e^{2-x}\, dx$

29. $\displaystyle\int_0^\infty .01 e^{-0.01x}\, dx$

30. $\displaystyle\int_0^\infty \frac{4}{(2x + 1)^3}\, dx$

31. $\displaystyle\int_0^\infty 6 e^{1-3x}\, dx$

32. $\displaystyle\int_1^\infty e^{-0.2x}\, dx$

33. $\displaystyle\int_3^\infty \frac{x^2}{\sqrt{x^3 - 1}}\, dx$

34. $\displaystyle\int_2^\infty \frac{1}{x \ln x}\, dx$

35. $\displaystyle\int_0^\infty x e^{-x^2}\, dx$

36. $\displaystyle\int_0^\infty \frac{x}{x^2 + 1}\, dx$

37. $\displaystyle\int_0^\infty 2x(x^2 + 1)^{-3/2}\, dx$

38. $\displaystyle\int_1^\infty (5x + 1)^{-4}\, dx$

39. $\displaystyle\int_{-\infty}^0 e^{4x}\, dx$

40. $\displaystyle\int_{-\infty}^0 \frac{8}{(x - 5)^2}\, dx$

41. $\displaystyle\int_{-\infty}^0 \frac{6}{(1 - 3x)^2}\, dx$

42. $\displaystyle\int_{-\infty}^0 \frac{1}{\sqrt{4 - x}}\, dx$

43. $\displaystyle\int_0^\infty \frac{e^{-x}}{(e^{-x} + 2)^2}\, dx$

44. $\displaystyle\int_{-\infty}^\infty \frac{e^{-x}}{(e^{-x} + 2)^2}\, dx$

45. If $k > 0$, show that $\displaystyle\int_0^\infty k e^{-kx}\, dx = 1$.

46. If $k > 0$, show that $\displaystyle\int_1^\infty \frac{k}{x^{k+1}}\, dx = 1$.

47. If $k > 0$, show that $\displaystyle\int_e^\infty \frac{k}{x(\ln x)^{k+1}}\, dx = 1$.

Capital Value of an Asset The *capital value* of an asset such as a machine is sometimes defined as the present value of all future net earnings. (See Section 9.5.) The actual lifetime of the asset may not be known, and since some assets may last indefinitely, the capital value of the asset may be written in the form

$$[\text{capital value}] = \int_0^\infty K(t) e^{-rt}\, dt,$$

where r is the annual rate of interest compounded continuously.

48. Find the capital value of an asset that generates income at the rate of $5000 per year, assuming an interest rate of 10%.

49. Construct a formula for the capital value of a rental property that will generate a fixed income at the rate of K dollars per year indefinitely, assuming an annual interest rate of r.

50. Suppose that a large farm with a known reservoir of gas beneath the ground sells the gas rights to a company for a guaranteed payment at the rate of $10{,}000 e^{0.04t}$ dollars per year. Find the present value of this perpetual stream of income, assuming an interest rate of 12%, compounded continuously.

Solutions to Check Your Understanding 9.6

1. The expression $1 - 2(1 - 3b)^{-4}$ may also be written in the form

$$1 - \frac{2}{(1 - 3b)^4}.$$

When b is large, $(1 - 3b)^4$ is very large, so $2/(1 - 3b)^4$ is very small. Thus, $1 - 2(1 - 3b)^{-4}$ approaches 1 as $b \to \infty$.

2. The first step is to find an antiderivative of $x^2/(x^3 + 8)$. Using the substitution $u = x^3 + 8$, $du = 3x^2\, dx$, we obtain

$$\int \frac{x^2}{x^3 + 8}\, dx = \frac{1}{3} \int \frac{1}{u}\, du$$

$$= \frac{1}{3} \ln |u| + C$$

$$= \frac{1}{3} \ln |x^3 + 8| + C.$$

Now,

$$\int_1^b \frac{x^2}{x^3 + 8} \, dx = \frac{1}{3} \ln |x^3 + 8| \Big|_1^b$$

$$= \frac{1}{3} \ln(b^3 + 8) - \frac{1}{3} \ln 9.$$

Finally, we examine what happens as $b \to \infty$. Certainly, $b^3 + 8$ gets arbitrarily large, so $\ln(b^3 + 8)$ must also get arbitrarily large. Hence,

$$\int_1^b \frac{x^2}{x^3 + 8} \, dx$$

has no limit as $b \to \infty$; so the improper integral

$$\int_1^\infty \frac{x^2}{x^3 + 8} \, dx$$

is divergent.

3. $\displaystyle\int_a^{-2} \frac{1}{x^4} \, dx = \int_a^{-2} x^{-4} \, dx$

$$= \frac{x^{-3}}{-3} \Big|_a^{-2} = \frac{1}{-3x^3} \Big|_a^{-2}$$

$$= \frac{1}{-3(-2)^3} - \left(\frac{1}{-3 \cdot a^3} \right)$$

$$= \frac{1}{24} + \frac{1}{3a^3}$$

$$\int_{-\infty}^{-2} \frac{1}{x^4} \, dx = \lim_{a \to -\infty} \int_a^{-2} \frac{1}{x^4} \, dx$$

$$= \lim_{a \to -\infty} \left(\frac{1}{24} + \frac{1}{3a^3} \right) = \frac{1}{24}.$$

CHAPTER 9 Summary

KEY TERMS AND CONCEPTS

9.1 Integration by Substitution

Every differentiation formula can be turned into a corresponding integration formula. If we apply this reasoning to the chain rule, we obtain the *integration by substitution formula*:

$$\int f(g(x))g'(x)dx = F(g(x)) + C,$$

where C is any constant and F is an antiderivative of f. This formula suggests a technique for evaluating integrals of products of functions by using a substitution $u = g(x)$ and $du = g'(x)dx$. This substitution will turn an integral of the form

$$\int f(g(x))g'(x)dx$$

into an integral that is easier to evaluate, since

$$\int f(u)du = F(u) + C.$$

The success of the method of integration by substitution depends on our choice of the substitution. Since there is no rule for selecting the "right" substitution, you can improve your skill in this area by considering as many examples as possible.

EXAMPLES

Evaluate the integrals by substitution:

(a) $\displaystyle\int 3x^2(x^3 + 7)dx$

(b) $\displaystyle\int \frac{x}{(x^2 + 1)^3} \, dx$

(c) $\displaystyle\int \frac{1}{x \ln x} \, dx$

(d) $\displaystyle\int e^x \sqrt{e^x + 1} \, dx$

Solution

(a) Let $u = x^3 + 7$, $du = 3x^2 \, dx$. Then,

$$\int 3x^2(x^3 + 7)dx = \int u \, du = \frac{1}{2}u^2 + C$$

$$= \frac{1}{2}(x^3 + 7)^2 + C.$$

(b) Let $u = x^2 + 1$, $du = 2x \, dx$. Since the integral does not contain the expression $2x \, dx$, we will use the equivalent substitution $x \, dx = \frac{1}{2}du$. Then,

$$\int \frac{x}{(x^2 + 1)^3} \, dx = \int \frac{\frac{1}{2}du}{u^3} = \frac{1}{2} \int u^{-3} \, du$$

$$= \frac{1}{2} \left(\frac{1}{-2} \right) u^{-2} + C$$

Since $\displaystyle\int u^r \, du = \frac{1}{r + 1} u^{r+1} + C \, (r \neq -1)$

$$= \frac{1}{4} \frac{1}{u^2} + C = -\frac{1}{4} \frac{1}{(x^2 + 1)^2} + C.$$

(c) Think of $\dfrac{1}{x \ln x}$ as a product of $\dfrac{1}{x}$ and $\dfrac{1}{\ln x}$. Let $u = \ln x$, $du = \dfrac{1}{x}dx$. Then,

$$\int \frac{1}{x \ln x} \, dx = \int \frac{1}{u} \, du = \ln |u| + C = \ln |\ln x| + C.$$

KEY TERMS AND CONCEPTS	EXAMPLES

(d) Let $u = e^x + 1$, $du = e^x\,dx$. Then,

$$\int e^x\sqrt{e^x + 1}\,dx = \int \sqrt{u}\,du = \int u^{1/2}\,du = \frac{2}{3}u^{3/2} + C$$

$$= \frac{2}{3}(e^x + 1)^{3/2} + C.$$

9.2 Integration by Parts

Integration by parts is an important integration technique based on reversing the product rule differentiation formula. Let $f(x)$ and $g(x)$ be any two functions and let $G(x)$ be an antiderivative of $g(x)$. The product rule asserts that

$$\frac{d}{dx}[f(x)G(x)] = f(x)G'(x) + f'(x)G(x)$$

$$= f(x)g(x) + f'(x)G(x)$$

Since $G'(x) = g(x)$.

Therefore,

$$f(x)G(x) = \int f(x)g(x)dx + \int f'(x)G(x)dx.$$

This last formula can be rewritten in the following more useful form:

$$\int f(x)g(x)dx = f(x)G(x) - \int f'(x)G(x)dx,$$

which is the basis for the **integration by parts technique**.

Evaluate the integrals by using integration by parts:

(a) $\int x(x + 7)^4\,dx$ **(b)** $\int (x + 1)e^{x/5}\,dx$ **(c)** $\int (x^3 + x)\ln x\,dx$

Solution

(a) Clearly the integral would be easier to evaluate if it were not for the presence of the factor x in the integrand. You can get rid of this factor by using integration by parts and setting $f(x) = x$. Then, $f'(x) = 1$. To complete the setup for integration by parts, we take $g(x) = (x + 7)^4$, and so, $G(x) = \frac{1}{5}(x + 7)^5$. Applying the integration by parts formula, we find

$$\int \overbrace{x \cdot (x + 7)^4}^{f(x)\cdot g(x)}\,dx = \overbrace{x \cdot \frac{1}{5}(x + 7)^5}^{f(x)\cdot G(x)} - \int \overbrace{\left(1 \cdot \frac{1}{5}(x + 7)^5\right)}^{f'(x)\cdot G(x)}\,dx$$

$$= \frac{x}{5}(x + 7)^5 - \int \frac{1}{5}(x + 7)^5\,dx$$

$$= \frac{x}{5}(x + 7)^5 - \frac{1}{5}\left(\frac{1}{6}\right)(x + 7)^6 + C$$

$$= \frac{x}{5}(x + 7)^5 - \frac{1}{30}(x + 7)^6 + C.$$

(b) This is a situation similar to that in (a), where we can simplify the integral by getting rid of the factor $(x + 1)$ in the integrand. Take $f(x) = x + 1$, $g(x) = e^{x/5}$. Then, $f'(x) = 1$ and $G(x) = 5e^{x/5}$. (In computing $G(x)$, we used the formula

$$\int e^{ax}\,dx = \frac{1}{a}e^{ax} + C, a \neq 0.$$

with $a = 1/5$.) So,

$$\int \overbrace{(x + 1) \cdot e^{x/5}}^{f(x)\cdot g(x)}\,dx = \overbrace{(x + 1) \cdot 5e^{x/5}}^{f(x)\cdot G(x)} - \int \overbrace{(1 \cdot 5e^{x/5})}^{f'(x)\cdot G(x)}\,dx$$

$$= 5(x + 1)e^{x/5} - 5\int e^{x/5}\,dx$$

$$= 5(x + 1)e^{x/5} - 25e^{x/5} + C$$

$$= (5x - 20)e^{x/5} + C.$$

(c) Take $f(x) = \ln x$, $g(x) = x^3 + x$. So, $f'(x) = \frac{1}{x}$ and $G(x) = \frac{1}{4}x^4 + \frac{1}{2}x^2$. Then,

$$\int (x^3 + x)\ln x\,dx = \overbrace{\ln x\left(\frac{1}{4}x^4 + \frac{1}{2}x^2\right)}^{f(x)\cdot G(x)} - \int \overbrace{\frac{1}{x} \cdot \left(\frac{1}{4}x^4 + \frac{1}{2}x^2\right)}^{f'(x)\cdot G(x)}\,dx$$

$$= \frac{x^2}{2}\left(\frac{1}{2}x^2 + 1\right)\ln x - \int \left(\frac{1}{4}x^3 + \frac{1}{2}x\right)\,dx$$

$$= \frac{x^2}{2}\left(\frac{1}{2}x^2 + 1\right)\ln x - \frac{1}{16}x^4 - \frac{1}{4}x^2 + C.$$

KEY TERMS AND CONCEPTS	EXAMPLES

9.3 Evaluation of Definite Integrals

When evaluating a definite integral that requires the application of a particular technique of integration, we carry out the computation of the indefinite integral first and then use the limits of the definite integral to complete the computation according to the formula

$$\int_a^b f(x)\,dx = F(b) - F(a).$$

If you use the method of substitution—say, $u = g(x)$—you can complete the evaluation of the definite integral in two equivalent ways: You can use the substitution with the limits and derive a definite integral with new limits on u, or you can write your indefinite integral with respect to x and use the limits on x.

1. Evaluate the integral

$$\int_0^1 2x(x^2 + 3)^4\,dx.$$

Solution: Method 1 We will use the substitution $u = x^2 + 3$, $du = 2x\,dx$ and use the substitution with the limits. If $x = 0$, then $u = 3$, and if $x = 1$, then $u = 4$. Making the substitution in the integral *and* the limits, we get

$$\int_0^1 2x(x^2 + 3)^4\,dx = \int_3^4 u^4\,du$$

$$= \frac{1}{5}u^5 \Big|_3^4 = \frac{1}{5}(4^5 - 3^5) = \frac{781}{5}.$$

Note that, in this method, we did not give the answer to the indefinite integral in terms of x: We kept it in terms of u and used the limits on u.

Solution: Method 2 Evaluate the integral using the same substitution and give your answer in terms of x. Then, evaluate the definite integral by using the limits on x. From the preceding computation, we found that

$$\int 2x(x^2 + 3)^4\,dx = \int u^4\,du = \frac{1}{5}u^5 + C = \frac{1}{5}(x^2 + 3)^5 + C.$$

So,

$$\int_0^1 2x(x^2 + 3)^4\,dx = \frac{1}{5}(x^2 + 3)^5 \Big|_0^1 = \frac{1}{5}(4^5 - 3^5) = \frac{781}{5},$$

which agrees with the previous answer.

2. Evaluate the integral

$$\int_{-1}^1 (2x + 3)e^x\,dx.$$

Solution Use integration by parts to evaluate the indefinite integral. Take $f(x) = 2x + 3$, $g(x) = e^x$. Then, $f'(x) = 2$, $G(x) = e^x$, and

$$\int (2x + 3)e^x\,dx = (2x + 3)e^x - \int 2e^x\,dx$$

$$= (2x + 3)e^x - 2e^x + C = (2x + 1)e^x + C.$$

We now use the limits on x to evaluate the definite integral:

$$\int_{-1}^1 (2x + 3)e^x\,dx = (2x + 1)e^x \Big|_{-1}^1$$

$$= (2 + 1)e^1 - (2(-1) + 1)e^{-1} = 3e + 1/e \approx 8.5.$$

9.4 Approximation of Definite Integrals

Midpoint Rule

$$\int_a^b f(x)\,dx \approx f(x_1)\Delta x + f(x_2)\Delta x + \cdots + f(x_n)\Delta x$$

$$= [f(x_1) + f(x_2) + \cdots + f(x_n)]\Delta x,$$

where $x_1, x_2, \ldots, x_n$ are the midpoints of the intervals. Another approximation that uses the endpoints of the subintervals is the trapezoidal rule.

Approximate the integral

$$\int_0^2 \frac{1}{x^2 + 1}\,dx$$

(a) By using the midpoint rule with $n = 4$.

(b) By using the trapezoidal rule with $n = 4$.

(c) By using Simpson's rule with $n = 4$.

KEY TERMS AND CONCEPTS	EXAMPLES

Trapezoidal Rule

$$\int_a^b f(x)\,dx$$

$$\approx [f(a_0) + 2f(a_1) + \cdots + 2f(a_{n-1}) + f(a_n)]\frac{\Delta x}{2}.$$

A third method, Simpson's rule, is based on taking a weighted average of the preceding two rules. Let M and T denote the estimates from the midpoint and trapezoidal rules, respectively.

Simpson's rule is based on the following approximation of the integral

$$S = \frac{2}{3}M + \frac{1}{3}T = \frac{2M + T}{3}.$$

You can express Simpson's rule in terms of the values of f at the endpoints and midpoints of the subintervals. See Section 9.4.

Solution

(a) We have $\Delta x = (b - a)/n = (2 - 0)/4 = .5$. The endpoints of the four subintervals begin at $a = 0$ and are spaced .5 unit apart. The first midpoint is at $a + \Delta x/2 = .25$. The midpoints are also spaced .5 unit apart. According to the midpoint rule, the integral is approximately equal to

$$\left[\frac{1}{1 + (.25)^2} + \frac{1}{1 + (.75)^2} + \frac{1}{1 + (1.25)^2} + \frac{1}{1 + (1.75)^2}\right](.5) \approx 1.10879.$$

(b) As in (a), $\Delta x = .5$ and the endpoints of the subintervals are $a_0 = 0$, $a_1 = .5$, $a_2 = 1$, $a_3 = 1.5$, and $a_4 = 2$. The trapezoidal rule gives

$$\left[\frac{1}{1 + 0^2} + 2\cdot\frac{1}{1 + (.5)^2} + 2\cdot\frac{1}{1 + 1^2} + 2\cdot\frac{1}{1 + (1.5)^2} + \frac{1}{1 + 2^2}\right]\cdot\frac{.5}{2}$$

$$\approx 1.10385.$$

(c) Simpson's rule yields the following approximation of the integral

$$\frac{(2)(1.10879) + 1.10385}{3} = 1.10714.$$

9.5 Some Applications of the Integral

Two interesting applications are presented based on modeling with sums that can be interpreted as Riemann sums. We then calculate the value of these sums using a definite integral and the fact that a definite integral approximates a Riemann sum. The first application is related to the present value of a continuous stream of income. The rule for computing such a value can be described as follows:

$$[\text{present value}] = \int_{T_1}^{T_2} K(t)e^{-rt}\,dt,$$

where $K(t)$ dollars per year is the annual rate of income at time t, r is the annual interest rate of invested money, T_1 to T_2 (years) is the time period of the income stream.

A second application is related to the calculation of the total population in a ring around the city center:

$$[\text{population}] = \int_a^b 2\pi t D(t)\,dt,$$

where $D(t)$ is the density of population (in persons per square mile) at distance t miles from the city center; the ring includes all persons who live between a and b miles from the city center.

A continuous stream of income is produced at the rate of $20e^{1-0.06t}$ thousand dollars per year at time t, and invested money earns 7% interest.

(a) Write a definite integral that gives the present value of this stream of income over the time from $t = 2$ to $t = 5$ years.

(b) Compute the present value described in part (a).

Solution

(a) Applying the formula, we find

$$[\text{present value}] = \int_2^5 20e^{1-0.06t}e^{-0.07t}\,dt = 20\int_2^5 e^{1-0.13t}\,dt.$$

(b) We compute the integral using the substitution $u = 1 - .13t$, $du = -.13\,dt$. When $t = 2$, $u = .74$, and when $t = 5$, $u = .35$, then,

$$20\int_2^5 e^{1-0.13t}\,dt = 20\int_{0.74}^{0.35} e^u\,\frac{du}{-.13}$$

$$= \frac{20}{-.13}e^u\Big|_{.74}^{.35} = \frac{20}{-.13}(e^{0.35} - e^{0.74}) \approx 104.134.$$

Thus, the present value of the stream of income over the given period is $104,134.

9.6 Improper Integrals

In this section, we consider improper integrals of the form

$$\int_a^\infty f(x)\,dx \quad \text{or} \quad \int_{-\infty}^b f(x)\,dx \quad \text{or} \quad \int_{-\infty}^\infty f(x)\,dx,$$

Evaluate the following improper integrals whenever they are convergent.

(a) $\displaystyle\int_1^\infty \frac{1}{x}\,dx$ (b) $\displaystyle\int_{-\infty}^0 e^{x+1}\,dx$ (c) $\displaystyle\int_0^\infty \frac{2x}{(x^2 + 1)^2}\,dx$

KEY TERMS AND CONCEPTS	EXAMPLES
where f is a continuous nonnegative function. An improper integral may *converge* or *diverge*. For example, the first improper integral is *convergent* if $$\lim_{b \to \infty} \int_a^b f(x)dx$$ has a limit L. In this case, we set $L = \int_a^\infty f(x)dx$. If the limit does not exist, we say that the integral is *divergent*. We similarly define the convergence and divergence of the second type of improper integrals. To work with the third integral, we write it as $$\int_{-\infty}^\infty f(x)dx = \int_{-\infty}^0 f(x)dx + \int_0^\infty f(x)dx$$ and consider it as the sum of two improper integrals.	**Solution** **(a)** We have $$\int_1^\infty \frac{1}{x}\,dx = \lim_{b \to \infty} \int_1^b \frac{1}{x}\,dx$$ $$= \lim_{b \to \infty} \ln x \Big|_1^b = \lim_{b \to \infty} (\ln b - \ln 1)$$ $$= \lim_{b \to \infty} \ln b$$ because $\ln 1 = 0$. As $b \to \infty$, $\ln b$ increases without bound. So the integral has no limit and is therefore divergent. **(b)** $$\int_{-\infty}^0 e^{x+1}\,dx = \lim_{a \to -\infty} \int_a^0 e^{x+1}\,dx$$ $$= \lim_{a \to -\infty} e^{x+1}\Big|_a^0$$ $$= \lim_{a \to -\infty} (e^1 - e^{a+1})$$ As $a \to -\infty$, $a+1$ tends to $-\infty$, and so e^{a+1} tends to 0. Thus, the integral is convergent and its value is $$\int_{-\infty}^0 e^{x+1}\,dx = \lim_{a \to -\infty} (e - e^{a+1}) = e.$$ **(c)** Use the substitution $u = x^2 + 1$, $du = 2x\,dx$. Then, when $x = 0$, $u = 1$, and when $x = b$, $u = b^2 + 1$. So, $$\int_0^\infty \frac{2x}{(x^2+1)^2}\,dx = \lim_{b \to \infty} \int_0^b \frac{2x}{(x^2+1)^2}\,dx$$ $$= \lim_{b \to \infty} \int_1^{b^2+1} \frac{du}{u^2}$$ $$= \lim_{b \to \infty} -\frac{1}{u}\Big|_1^{b^2+1}$$ $$= \lim_{b \to \infty} \left[-\frac{1}{b^2+1} - (-1) \right] = 1,$$ because $\lim_{b \to \infty} -\frac{1}{b^2+1} = 0$. Thus, the integral is convergent and its value is 1.

CHAPTER 9   Fundamental Concept Check Exercises

1. Describe integration by substitution in your own words.

2. Describe integration by parts in your own words.

3. Describe the change of limits rule for the integration by substitution of a definite integral.

4. State the formula for the integration by parts of a definite integral.

5. State the midpoint rule. (Include the meaning of all symbols used.)

6. State the trapezoidal rule. (Include the meaning of all symbols used.)

7. Explain the formula $S = \dfrac{2M + T}{3}$.

8. State the error of approximation theorem for each of the three approximation rules.

9. State the formula for each of the following quantities:
   (a) Present value of a continuous stream of income
   (b) Total population in a ring around the center of a city

10. How do you determine whether an improper integral is convergent?

CHAPTER 9 Review Exercises

Determine the following indefinite integrals:

1. $\displaystyle\int x \sin 3x^2 \, dx$

2. $\displaystyle\int \sqrt{2x + 1} \, dx$

3. $\displaystyle\int x(1 - 3x^2)^5 \, dx$

4. $\displaystyle\int \frac{(\ln x)^5}{x} \, dx$

5. $\displaystyle\int \frac{(\ln x)^2}{x} \, dx$

6. $\displaystyle\int \frac{1}{\sqrt{4x + 3}} \, dx$

7. $\displaystyle\int x\sqrt{4 - x^2} \, dx$

8. $\displaystyle\int x \sin 3x \, dx$

9. $\displaystyle\int x^2 e^{-x^3} \, dx$

10. $\displaystyle\int \frac{x \ln(x^2 + 1)}{x^2 + 1} \, dx$

11. $\displaystyle\int x^2 \cos 3x \, dx$

12. $\displaystyle\int \frac{\ln(\ln x)}{x \ln x} \, dx$

13. $\displaystyle\int \ln x^2 \, dx$

14. $\displaystyle\int x\sqrt{x + 1} \, dx$

15. $\displaystyle\int \frac{x}{\sqrt{3x - 1}} \, dx$

16. $\displaystyle\int x^2 \ln x^2 \, dx$

17. $\displaystyle\int \frac{x}{(1 - x)^5} \, dx$

18. $\displaystyle\int x(\ln x)^2 \, dx$

In Exercises 19–36, decide whether integration by parts or a substitution should be used to compute the indefinite integral. If substitution, indicate the substitution to be made. If by parts, indicate the functions $f(x)$ and $g(x)$ to be used in formula (1) of Section 9.2.

19. $\displaystyle\int xe^{2x} \, dx$

20. $\displaystyle\int (x - 3)e^{-x} \, dx$

21. $\displaystyle\int (x + 1)^{-1/2} e^{\sqrt{x+1}} \, dx$

22. $\displaystyle\int x^2 \sin(x^3 - 1) \, dx$

23. $\displaystyle\int \frac{x - 2x^3}{x^4 - x^2 + 4} \, dx$

24. $\displaystyle\int \ln\sqrt{5 - x} \, dx$

25. $\displaystyle\int e^{-x}(3x - 1)^2 \, dx$

26. $\displaystyle\int xe^{3 - x^2} \, dx$

27. $\displaystyle\int (500 - 4x)e^{-x/2} \, dx$

28. $\displaystyle\int x^{5/2} \ln x \, dx$

29. $\displaystyle\int \sqrt{x + 2} \ln(x + 2) \, dx$

30. $\displaystyle\int (x + 1)^2 e^{3x} \, dx$

31. $\displaystyle\int (x + 3)e^{x^2 + 6x} \, dx$

32. $\displaystyle\int \sin^2 x \cos x \, dx$

33. $\displaystyle\int x \cos(x^2 - 9) \, dx$

34. $\displaystyle\int (3 - x) \sin 3x \, dx$

35. $\displaystyle\int \frac{2 - x^2}{x^3 - 6x} \, dx$

36. $\displaystyle\int \frac{1}{x(\ln x)^{3/2}} \, dx$

Evaluate the following definite integrals:

37. $\displaystyle\int_0^1 \frac{2x}{(x^2 + 1)^3} \, dx$

38. $\displaystyle\int_0^{\pi/2} x \sin 8x \, dx$

39. $\displaystyle\int_0^2 xe^{-(1/2)x^2} \, dx$

40. $\displaystyle\int_{1/2}^1 \frac{\ln(2x + 3)}{2x + 3} \, dx$

41. $\displaystyle\int_1^2 xe^{-2x} \, dx$

42. $\displaystyle\int_1^2 x^{-3/2} \ln x \, dx$

Approximate the following definite integrals by the midpoint rule, the trapezoidal rule, and Simpson's rule.

43. $\displaystyle\int_1^9 \frac{1}{\sqrt{x}} \, dx; \, n = 4$

44. $\displaystyle\int_0^{10} e^{\sqrt{x}} \, dx; \, n = 5$

45. $\displaystyle\int_1^4 \frac{e^x}{x + 1} \, dx; \, n = 5$

46. $\displaystyle\int_{-1}^1 \frac{1}{1 + x^2} \, dx; \, n = 5$

Evaluate the following improper integrals whenever they are convergent.

47. $\displaystyle\int_0^\infty e^{6 - 3x} \, dx.$

48. $\displaystyle\int_1^\infty x^{-2/3} \, dx$

49. $\displaystyle\int_1^\infty \frac{x + 2}{x^2 + 4x - 2} \, dx$

50. $\displaystyle\int_0^\infty x^2 e^{-x^3} \, dx$

51. $\displaystyle\int_{-1}^\infty (x + 3)^{-5/4} \, dx$

52. $\displaystyle\int_{-\infty}^0 \frac{8}{(5 - 2x)^3} \, dx$

53. It can be shown that $\lim_{b \to \infty} be^{-b} = 0$. Use this fact to compute $\int_1^\infty xe^{-x} \, dx$.

54. Let k be a positive number. It can be shown that $\lim_{b \to \infty} be^{-kb} = 0$. Use this fact to compute $\int_0^\infty xe^{-kx} \, dx$.

55. Find the present value of a continuous stream of income over the next 4 years, where the rate of income is $50e^{-0.08t}$ thousand dollars per year at time t, and the interest rate is 12%.

56. **Property Tax** Suppose that t miles from the center of a certain city the property tax revenue is approximately $R(t)$ thousand dollars per square mile, where $R(t) = 50e^{-t/20}$. Use this model to predict the total property tax revenue that will be generated by property within 10 miles of the city center.

57. **Annual Rate of Maintenance** Suppose that a machine requires daily maintenance, and let $M(t)$ be the *annual* rate of maintenance expense at time t. Suppose that the interval $0 \leq t \leq 2$ is divided into n subintervals, with endpoints $t_0 = 0, t_1, \ldots, t_n = 2$.
  (a) Give a Riemann sum that approximates the total maintenance expense over the next 2 years. Then, write the integral that the Riemann sum approximates for large values of n.
  (b) Give a Riemann sum that approximates the present value of the total maintenance expense over the next 2 years using a 10% annual interest rate compounded continuously. Then, write the integral that the Riemann sum approximates for large values of n.

58. **Capitalized Cost** The *capitalized cost* of an asset is the total of the original cost and the present value of all future "renewals" or replacements. This concept is useful, for example, when you are selecting equipment that is manufactured by several different companies. Suppose that a corporation computes the present value of future expenditures using an annual interest rate r, with continuous compounding of interest. Assume that the original cost of an asset is $80,000 and the annual renewal expense will be $50,000, spread more-or-less evenly throughout each year, for a large but indefinite number of years. Find a formula involving an integral that gives the capitalized cost of the asset.

Differential Equations

A **differential equation** is an equation in which derivatives of an unknown function $y = f(t)$ occur. Examples of such equations are

$$y' = 6t + 3, \quad y' = 6y, \quad y'' = 3y' - x, \quad \text{and} \quad y' + 3y + t = 0.$$

As we shall see, many physical processes can be described by differential equations. In this chapter, we explore some topics in differential equations and use our resulting knowledge to study problems from many different fields.

10.1 Solutions of Differential Equations

A differential equation is an equation involving an unknown function y and one or more of the derivatives y', y'', y''', and so on. Suppose that y is a function of the variable t. A **solution** of a differential equation is any function $f(t)$ such that the differential equation becomes a true statement when y is replaced by $f(t)$, y' by $f'(t)$, y'' by $f''(t)$, and so forth.

EXAMPLE 1 **Verifying a Solution** Show that the function $f(t) = 5e^{-2t}$ is a solution of the differential equation

$$y' + 2y = 0. \tag{1}$$

SOLUTION The differential equation (1) says that $y' + 2y$ equals zero for all values of t. We must show that this result holds if y is replaced by $5e^{-2t}$ and y' is replaced by $(5e^{-2t})' = -10e^{-2t}$. Now,

$$\overbrace{(5e^{-2t})'}^{y'} + 2 \overbrace{(5e^{-2t})}^{y} = -10e^{-2t} + 10e^{-2t} = 0.$$

Therefore, $y = 5e^{-2t}$ is a solution of the differential equation (1). **▸▸ Now Try Exercise 1**

EXAMPLE 2

Verifying a Solution Show that the function $f(t) = \frac{1}{9}t + \sin 3t$ is a solution of the differential equation

$$y'' + 9y = t. \tag{2}$$

SOLUTION

If $f(t) = \frac{1}{9}t + \sin 3t$, then

$$f'(t) = \frac{1}{9} + 3\cos 3t,$$

$$f''(t) = -9\sin 3t.$$

Substituting $f(t)$ for y and $f''(t)$ for y'' in the left side of (2), we obtain

$$-9\sin 3t + 9\left(\tfrac{1}{9}t + \sin 3t\right) = -9\sin 3t + t + 9\sin 3t = t.$$

Therefore, $y'' + 9y = t$ if $y = \frac{1}{9}t + \sin 3t$, and hence, $y = \frac{1}{9}t + \sin 3t$ is a solution to $y'' + 9y = t$.

>> *Now Try Exercise 3*

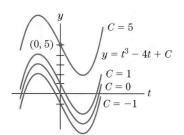

Figure 1 Typical solutions of $y' = 3t^2 - 4$.

The differential equation in Example 1 is said to be of **first order**, since it involves the first derivative of the unknown function y. The differential equation in Example 2 is of **second order**, since it involves the second derivative of y. In general, the **order** of a differential equation is the order of the highest derivative that appears in the equation.

The process of determining all the functions that are solutions of a differential equation is called *solving the differential equation*. The process of antidifferentiation amounts to solving a simple type of differential equation. For example, a solution of the differential equation

$$y' = 3t^2 - 4 \tag{3}$$

is a function y whose derivative is $3t^2 - 4$. Thus, solving (3) consists of finding all antiderivatives of $3t^2 - 4$. Clearly, y must be of the form $y = t^3 - 4t + C$, where C is a constant. The solutions of (3) corresponding to several values of C are sketched in Fig. 1. Each graph is called a **solution curve**.

We encountered differential equations such as

$$y' = 2y \tag{4}$$

in our discussion of exponential functions. Unlike (3), equation (4) does not give a specific formula for y' but instead describes a property of y': y' is proportional to y (with 2 as the constant of proportionality). At the moment, the only way we can "solve"(4) is simply to know in advance what the solutions are. Recall from Theorem 1, Section 5.1, that the solutions of (4) have the form $y = Ce^{2t}$ for any constant C. Some typical solutions of (4) are sketched in Fig. 2. In the next section we shall discuss a method for solving a class of differential equations that includes both (3) and (4) as special cases.

Figures 1 and 2 illustrate two important differences between differential equations and algebraic equations (such as $ax^2 + bx + c = 0$). First, a solution of a differential equation is a *function* rather than a number. Second, a differential equation usually has infinitely many solutions.

Sometimes, we want to find a particular solution that satisfies certain additional conditions called **initial conditions**. Initial conditions specify the values of a solution and a certain number of its derivatives at some specific value of t, often $t = 0$. If the solution to a differential equation is $y = f(t)$, we often write $y(0)$ for $f(0)$, $y'(0)$ for $f'(0)$, and so on. The problem of determining a solution of a differential equation that satisfies given initial conditions is called an **initial-value problem**.

FOR REVIEW

For additional examples of initial-value problems. See Section 5.1, pp. 256–257.

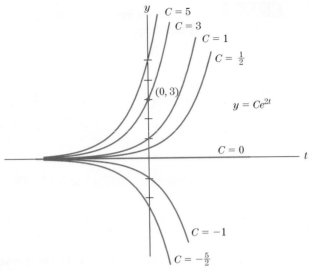

Figure 2 Typical solutions of $y' = 2y$.

EXAMPLE 3

Initial Value Problems
(a) Solve the initial-value problem $y' = 3t^2 - 4$, $y(0) = 5$.
(b) Solve the initial-value problem $y' = 2y$, $y(0) = 3$.

SOLUTION

(a) We have already noted that the general solution of $y' = 3t^2 - 4$ is $f(t) = t^3 - 4t + C$. We want the particular solution that satisfies $f(0) = 5$. Geometrically, we are looking for the curve in Fig. 1 that passes through the point $(0, 5)$. Using the general formula for $f(t)$, we have

$$5 = f(0) = (0)^3 - 4(0) + C = C$$
$$C = 5.$$

Thus, $f(t) = t^3 - 4t + 5$ is the desired solution.

(b) The general solution of $y' = 2y$ is $y = Ce^{2t}$. The condition $y(0) = 3$ means that y must be 3 when $t = 0$; that is, the point $(0, 3)$ must be on the graph of the solution to $y' = 2y$. (See Fig. 2.) We have

$$3 = y(0) = Ce^{2(0)} = C \cdot 1 = C$$
$$C = 3.$$

Thus, $y = 3e^{2t}$ is the desired solution. **≫ Now Try Exercise 5**

A constant function that satisfies a differential equation is called a **constant solution** of the differential equation. Constant solutions occur in many of the applied problems considered later in the chapter.

EXAMPLE 4

Constant Solutions Find a constant solution of $y' = 3y - 12$.

SOLUTION

Let $f(t) = c$ for all t. Then, $f'(t)$ is zero for all t. If $f(t)$ satisfies the differential equation

$$f'(t) = 3 \cdot f(t) - 12,$$

then

$$0 = 3 \cdot c - 12,$$

and so $c = 4$. This is the only possible value for a constant solution. Substitution shows that the function $f(t) = 4$ is indeed a solution of the differential equation.

≫ Now Try Exercise 9

Modeling with Differential Equations

Equations describing conditions in a physical process are often referred to as **mathematical models**, and discovering these equations is called **modeling**. In our next example, we show how we can use a differential equation to model a physical process. The example should be studied carefully, for it contains the key to understanding many similar problems that will appear in exercises and in later sections.

EXAMPLE 5

Newton's Law of Cooling Suppose that a red-hot steel rod is plunged into a bath of cool water. Let $f(t)$ be the temperature of the rod at time t, and suppose that the water is maintained at a constant temperature of 10°C. According to Newton's law of cooling, the rate of change of $f(t)$ is proportional to the difference between the two temperatures 10° and $f(t)$. Find a differential equation that describes this physical law.

SOLUTION

FOR REVIEW

A good example of modeling is presented in Sec. 5.2, where a differential equation describes continuous compounding (p. 265).

The two key ideas are *rate of change* and *proportional*. The rate of change of $f(t)$ is the derivative $f'(t)$. Since this is proportional to the difference $10 - f(t)$, there exists a constant k such that

$$f'(t) = k[10 - f(t)]. \tag{5}$$

The term *proportional* does not tell us whether k is positive or negative (or zero). We must decide this, if possible, from the context of the problem. In the present situation, the steel rod is hotter than the water, so $10 - f(t)$ is negative. Also, $f(t)$ will decrease as time passes, so $f'(t)$ should be negative. Thus, to make $f'(t)$ negative in equation (5), k must be a positive number. From (5), we see that $y = f(t)$ satisfies a differential equation of the form

$$y' = k(10 - y), \quad k \text{ a positive constant.} \quad \text{\textbf{\textgg} Now Try Exercise 13}$$

EXAMPLE 6

Newton's Law of Cooling Suppose that the constant of proportionality in Example 5 is $k = .2$ and time is measured in seconds. How fast is the temperature of the steel rod changing when the temperature is 110°C?

SOLUTION

The relationship between the temperature and the rate of change of the temperature is given by the differential equation $y' = .2(10 - y)$, where $y = f(t)$ is the temperature after t seconds. When $y = 110$, the rate of change is

$$y' = .2(10 - 110) = .2(-100) = -20.$$

The temperature is decreasing at the rate of 20 degrees per second.   **\textless\textless**

The differential equation in Examples 5 and 6 is a special case of the differential equation

$$y' = k(M - y).$$

Differential equations of this type describe not only cooling, but also many important applications in economics, medicine, population dynamics, and engineering. Some of these applications are discussed in the exercises and in Sections 10.4 and 10.6.

Geometric Meaning of a Differential Equation: Slope Fields

As observed earlier, a differential equation such as

$$y' = t - y \tag{6}$$

FOR REVIEW

The derivative is a slope formula. See Sec. 1.3.

does not give a specific formula for y' in terms of the variable t; instead, it describes a property of y'. The key to understanding this property is to remember the geometric interpretation of the derivative as a slope formula. So, if $y = f(t)$ is a solution of (6) and (t, y) is a point on the graph of this solution, equation (6) is telling us that the slope of

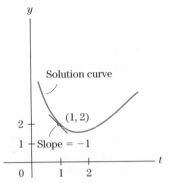

Figure 3 Solution curve of $y' = t - y$ and a tangent line through the point $(1, 2)$.

the graph at the point (t, y) [that is, $y'(t)$] is just $t - y$. For example, if a solution curve goes through the point $(1, 2)$, without knowing the formula for this solution we can say that the slope of the solution through the point $(1, 2)$ is

$$y'(1) = 1 - 2 = -1.$$

This information gives us the direction of the solution curve through the point $(1, 2)$. Figure 3 shows a portion of the solution curve through the point $(1, 2)$, along with the tangent line at that point. Note how the slope of the tangent lines matches with the value $y'(1) = -1$.

Imagine now repeating the previous construction of the tangent line at many points in the ty-plane, not just the point $(1, 2)$. This laborious process is usually done with the help of a computer or a graphing calculator that can generate a collection of small line segments, called a **slope field** or **direction field** of the differential equation. [See Fig. 4(a).] Since each line segment in the slope field is tangent to a solution curve, we can visualize a solution curve by following the flow of the slope field.

(a) Slope field of $y' = t - y$.

Start at the point $(0, 1)$.

(b) Hand-drawn approximation of the solution curve through the point $(0, 1)$.

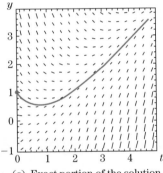

(c) Exact portion of the solution curve through $(0, 1)$.

Figure 4

EXAMPLE 7   **Slope Field** Use the slope field of the differential equation $y' = t - y$, shown in Fig. 4(a), to draw by hand a portion of the solution curve that passes through the point $(0, 1)$.

SOLUTION   We are given one point on the solution curve and the slope field. To draw the desired curve, we start at the point $(0, 1)$ and trace a curve in the ty-plane tangent to the line segments in the slope field. The result is the curve shown in Fig. 4(b). Figure 4(c) shows a portion of the exact solution curve for the purpose of comparison. As you can see, the hand-drawn curve in Fig. 4(b) is a pretty good approximation of the exact solution curve.

>> Now Try Exercise 27

In general, we can construct a slope field for any first-order differential equation of the form

$$y' = g(t, y), \tag{7}$$

where $g(t, y)$ is a function of t and y. The idea is that at any given point (a, b) of a solution curve the slope $y'(a)$ is given by $g(a, b)$.

As our next example shows, slope fields are useful in deriving qualitative properties of solutions.

EXAMPLE 8   **Using a Slope Field to Analyze a Solution** Let $f(t)$ denote the number of people who have contracted a certain strain of flu after t days. The function $f(t)$ satisfies the initial-value problem $y' = .0002y(5000 - y)$, $y(0) = 1000$. A slope field of the differential equation is shown in Fig. 5(a). Based on the slope field, do you think that the number of infected people will ever exceed 5000?

SOLUTION The initial condition $f(0) = 1000$ tells us that the point $(0, 1000)$ is on the graph of the solution curve $y = f(t)$. Starting at this point in the ty-plane and tracing a curve tangent to the line segments in the slope field, we obtain an approximation to a portion of the solution curve through the point $(0, 1000)$. [See Fig. 5(b).] According to this curve, we can conclude that the solution curve will come very close to, but will not exceed, the value 5000. Hence, the number of infected people will not exceed 5000.

>> Now Try Exercise 29

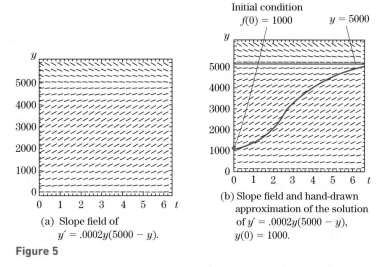

(a) Slope field of
$y' = .0002y(5000 - y)$.

(b) Slope field and hand-drawn approximation of the solution of $y' = .0002y(5000 - y)$, $y(0) = 1000$.

Figure 5

The differential equation in Example 8 is a special case of the logistic differential equation, which arises frequently in the study of population growth in a restricted environment. This important equation will be studied in greater detail in later sections.

Throughout this section we have assumed that our solution functions y are functions of the variable t, which in applications usually stands for time. In most of this chapter, we will continue to use the variable t. However, if we use another variable occasionally, such as x, we will make the variable explicit by writing $\frac{dy}{dx}$ instead of y'.

INCORPORATING
TECHNOLOGY

Figure 6

Slope Fields Drawing slope fields with your TI-83/84 Plus calculator requires a program that is freely available from the Texas Instruments Web site education.ti.com. Point your web browser to education.ti.com and search for SLPFLD.8xp. Download this file to your computer, connect your calculator to your computer, and use the TI-Connect© program to transfer the file to the calculator.

To demonstrate the usage of the SLPFLD program, we consider the differential equation $y' = t - y$. The SLPFLD program requires that the independent variable be X (rather than t), so we begin by pressing ⟨Y=⟩ and setting $Y_1 = X - Y$. Now, return to the home screen and press ⟨PRGM⟩. From the **EXEC** menu, scroll down to **SLPFLD** and press ⟨ENTER⟩. You will then see **prgmSLPFLD** on the home screen. Press ⟨ENTER⟩. Figure 6 shows the result.

Check Your Understanding 10.1

Solutions can be found following the section exercises.

1. Show that any function of the form $y = Ae^{t^3/3}$, where A is a constant, is a solution of the differential equation $y' - t^2y = 0$.

2. If the function $f(t)$ is a solution of the initial-value problem $y' = (t + 2)y$, $y(0) = 3$, find $f(0)$ and $f'(0)$.

3. Let $f(t)$ be the size of a population after t days. Suppose that $y = f(t)$ satisfies $y' = .06y$, $y(0) = 1000$. Describe this initial-value problem in words. (Include the phrase "is proportional to.") How fast is the population growing when it contains 3000 members?

EXERCISES 10.1

1. Show that the function $f(t) = \frac{3}{2}e^{t^2} - \frac{1}{2}$ is a solution of the differential equation $y' - 2ty = t$.

2. Show that the function $f(t) = t^2 - \frac{1}{2}$ is a solution of the differential equation $(y')^2 - 4y = 2$.

3. Show that the function $f(t) = 5e^{2t}$ satisfies $y'' - 3y' + 2y = 0$, $y(0) = 5$, $y'(0) = 10$.

4. Show that the function $f(t) = (e^{-t} + 1)^{-1}$ satisfies $y' + y^2 = y$, $y(0) = \frac{1}{2}$.

In Exercises 5 and 6, state the order of the differential equation and verify that the given function is a solution.

5. $(1 - t^2)y'' - 2ty' + 2y = 0$, $y(t) = t$

6. $(1 - t^2)y'' - 2ty' + 6y = 0$, $y(t) = \frac{1}{2}(3t^2 - 1)$

7. Is the constant function $f(t) = 3$ a solution of the differential equation $y' = 6 - 2y$?

8. Is the constant function $f(t) = -4$ a solution of the differential equation $y' = t^2(y + 4)$?

9. Find a constant solution of $y' = t^2y - 5t^2$.

10. Find two constant solutions of $y' = 4y(y - 7)$.

11. If the function $f(t)$ is a solution of the initial-value problem $y' = 2y - 3$, $y(0) = 4$, find $f(0)$ and $f'(0)$.

12. If the function $f(t)$ is a solution of the initial-value problem $y' = e^t + y$, $y(0) = 0$, find $f(0)$ and $f'(0)$.

13. **Velocity of a Skydiver** Let $y = v(t)$ be the downward speed (in feet per second) of a skydiver after t seconds of free fall. This function satisfies the differential equation $y' = .2(160 - y)$, $y(0) = 0$. What is the skydiver's acceleration when her downward speed is 60 feet per second? [*Note:* Acceleration is the derivative of speed.]

14. **Population of Fish in a Lake** A lake is stocked with 100 fish. Let $f(t)$ be the number of fish after t months, and suppose that $y = f(t)$ satisfies the differential equation $y' = .0004y(1000 - y)$. Figure 7 shows the graph of the solution to this differential equation. The graph is asymptotic to the line $y = 1000$, the maximum number of fish that the lake can support. How fast is the fish population growing when it reaches one-half of its maximum population?

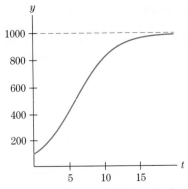

Figure 7 Growth of a fish population.

15. **Savings Account** Let $f(t)$ be the balance in a savings account at the end of t years, and suppose that $y = f(t)$ satisfies the differential equation $y' = .05y - 10,000$.
    (a) If after 1 year the balance is \$150,000, is it increasing or decreasing at that time? At what rate is it increasing or decreasing at that time?
    (b) Write the differential equation in the form $y' = k(y - M)$.
    (c) Describe this differential equation in words.

16. **Savings Account** Let $f(t)$ be the balance in a savings account at the end of t years. Suppose that $y = f(t)$ satisfies the differential equation $y' = .04y + 2000$.
    (a) If after 1 year the balance is \$10,000, is it increasing or decreasing at that time? At what rate is it increasing or decreasing at that time?
    (b) Write the differential equation in the form $y' = k(y + M)$.
    (c) Describe this differential equation in words.

17. **Spread of News** A certain piece of news is being broadcast to a potential audience of 200,000 people. Let $f(t)$ be the number of people who have heard the news after t hours. Suppose that $y = f(t)$ satisfies

$$y' = .07(200,000 - y), \qquad y(0) = 10.$$

Describe this initial-value problem in words.

18. **Paramecium Growth** Let $f(t)$ be the size of a paramecium population after t days. Suppose that $y = f(t)$ satisfies the differential equation

$$y' = .003y(500 - y), \qquad y(0) = 20.$$

Describe this initial-value problem in words.

19. **Rate of Net Investment** Let $f(t)$ denote the amount of capital invested by a certain business firm at time t. The rate of change of invested capital, $f'(t)$, is sometimes called the *rate of net investment*. The management of the firm decides that the optimum level of investment should be C dollars and that, at any time, the rate of net investment should be proportional to the difference between C and the total capital invested. Construct a differential equation that describes this situation.

20. **Newton's Law of Cooling** A cool object is placed in a room that is maintained at a constant temperature of 20°C. The rate at which the temperature of the object rises is proportional to the difference between the room temperature and the temperature of the object. Let $y = f(t)$ be the temperature of the object at time t; give a differential equation that describes the rate of change of $f(t)$.

21. **Carbon Dioxide Diffusion in Lungs during Breath Holding** When the breath is held, carbon dioxide (CO_2) diffuses from the blood into the lungs at a steadily decreasing rate. Let P_0 and P_b denote the pressure of CO_2 in the lungs, respectively, in the blood at the moment when the breath is held. Suppose that P_b is constant during breath holding, and let $P(t)$ denote the pressure of CO_2 in the lungs at time $t > 0$. Experiments

show that the rate of change of $P(t)$ is proportional to the difference between the two pressures $P(t)$ and P_b. Find an initial-value problem that describes the diffusion of CO_2 in the lungs during breath holding.

22. **Slope Field** The slope field in Fig. 4(a) suggests that the solution curve of the differential equation $y' = t - y$ through the point $(0, -1)$ is a straight line.
    (a) Assuming that this is true, find the equation of the line. [*Hint:* Use the differential equation to get the slope of the line through the point $(0, -1)$.]
    (b) Verify that the function that you found in part (a) is a solution by plugging its formula into the differential equation.

23. Verify that the function $f(t) = 2e^{-t} + t - 1$ is a solution of the initial-value problem $y' = t - y$, $y(0) = 1$. [This is the function shown in Fig. 4(c). In Section 10.3, you will learn how to derive this solution.]

24. On the slope field in Fig. 5(a) or a copy of it, draw the solution of the initial-value problem $y' = .0002y(5000 - y)$, $y(0) = 500$.

25. **Flu Epidemic** The health officials that studied the flu epidemic in Example 8 made an error in counting the initial number of infected people. They are now claiming that $f(t)$ (the number of infected people after t days) is a solution of the initial-value problem $y' = .0002y(5000 - y)$, $y(0) = 1500$. Under this new assumption, can $f(t)$ exceed 5000? [*Hint:* Since the differential equation is the same as the one in Example 8, you can use the slope field in Fig. 5(a) to answer the question.]

26. On the slope field in Fig. 4(a) or a copy of it, draw an approximation of a portion of the solution curve of the differential equation $y' = t - y$ that goes through the point $(0, 2)$. In your opinion, based on the slope field, can this solution pass through the point $(.5, 2.2)$?

27. Figure 8 shows a slope field of the differential equation $y' = 2y(1 - y)$. With the help of this figure, determine the constant solutions, if any, of the differential equation. Verify your answer by substituting back into the equation.

28. Figure 8 shows a portion of the solution curve of the differential equation $y' = 2y(1 - y)$ through the point $(0, 2)$. On Fig. 8

Figure 8 Slope field of $y' = 2y(1 - y)$.

or a copy of it, draw an approximation of the solution curve of the differential equation $y' = 2y(1 - y)$ through the point $(0, 3)$. Use the slope field to guide your graph.

29. If $y_0 > 1$, is the solution $y = f(t)$ of the initial-value problem $y' = 2y(1 - y)$, $y(0) = y_0$, decreasing for all $t > 0$? Answer this question based on the slope field shown in Fig. 8.

30. Answer the question in Exercise 29 by using the differential equation to determine the sign of $f'(t)$.

TECHNOLOGY EXERCISES

31. Consider the differential equation $y' = .2(10 - y)$ from Example 6. If the initial temperature of the steel rod is $510°$, the function $f(t) = 10 + 500e^{-0.2t}$ is the solution of the differential equation.
    (a) Graph the function in the window $[0, 30]$ by $[-75, 550]$.
    (b) In the home screen, compute $.2(10 - f(5))$ and compare this value with $f'(5)$.

32. The function $f(t) = \dfrac{5000}{1 + 49e^{-t}}$ is the solution of the differential equation $y' = .0002y(5000 - y)$ from Example 8.
    (a) Graph the function in the window $[0, 10]$ by $[-750, 5750]$.
    (b) In the home screen, compute $.0002 f(3)(5000 - f(3))$, and compare this value with $f'(3)$.

Solutions to Check Your Understanding 10.1

1. If $y = Ae^{t^3/3}$, then

$$\overbrace{(Ae^{t^3/3})'}^{y'} - t^2 \overbrace{(Ae^{t^3/3})}^{y} = At^2 e^{t^3/3} - t^2 Ae^{t^3/3} = 0.$$

Therefore, $y' - t^2 y = 0$ if $y = Ae^{t^3/3}$.

2. The initial condition $y(0) = 3$ says that $f(0) = 3$. Since $f(t)$ is a solution to $y' = (t + 2)y$,

$$f'(t) = (t + 2)f(t),$$

and hence,

$$f'(0) = (0 + 2)f(0) = 2 \cdot 3 = 6.$$

3. Initially, the population has 1000 members. At any time, the rate of growth of the population is proportional to the size of the population at that time, and the constant of proportionality is .06. When $y = 3000$,

$$y' = .06y = .06 \cdot 3000 = 180.$$

Therefore, the population is growing at the rate of 180 members per day.

10.2   Separation of Variables

Here, we describe a technique for solving an important class of differential equations, those of the form

$$y' = p(t)q(y),$$

where $p(t)$ is a function of t only and $q(y)$ is a function of y only. Two equations of this type are

$$y' = \frac{3t^2}{y^2} \qquad \left[p(t) = 3t^2, q(y) = \frac{1}{y^2} \right], \tag{1}$$

$$y' = e^{-y}(2t + 1) \qquad [p(t) = 2t + 1, q(y) = e^{-y}]. \tag{2}$$

The main feature of such equations is that we may *separate the variables*; that is, we may rewrite the equations so that y occurs only on one side of the equation and t on the other. For example, if we multiply both sides of equation (1) by y^2, the equation becomes

$$y^2 y' = 3t^2;$$

if we multiply both sides of equation (2) by e^y, the equation becomes

$$e^y y' = 2t + 1.$$

It should be pointed out that the differential equation

$$y' = 3t^2 - 4$$

is of the preceding type. Here, $p(t) = 3t^2 - 4$ and $q(y) = 1$. The variables are already separated, however. Similarly, the differential equation

$$y' = 5y$$

is of the preceding type, with $p(t) = 5$, $q(y) = y$. We can separate the variables by writing the equation as

$$\frac{1}{y} y' = 5.$$

In the next example, we present a procedure for solving differential equations in which the variables are separated.

EXAMPLE 1   **Separation of Variables**   Find all solutions of the differential equation $y^2 y' = 3t^2$.

SOLUTION   (a)   Write y' as $\dfrac{dy}{dt}$: $\qquad y^2 \dfrac{dy}{dt} = 3t^2.$

(b)   Integrate both sides with respect to t:

$$\int y^2 \frac{dy}{dt} \, dt = \int 3t^2 \, dt.$$

(c)   Rewrite the left side, "canceling the dt":

$$\int y^2 \, dy = \int 3t^2 \, dt.$$

(See the following discussion for an explanation of exactly what this means.)

(d)   Calculate the antiderivatives:

$$\frac{1}{3} y^3 + C_1 = t^3 + C_2.$$

(e) Solve for y in terms of t:

$$y^3 = 3(t^3 + C_2 - C_1)$$
$$y = \sqrt[3]{3t^3 + C}, \qquad C \text{ a constant.}$$

We can check that this method works by showing that $y = \sqrt[3]{3t^3 + C}$ is a solution to $y^2 y' = 3t^2$. Since $y = (3t^3 + C)^{1/3}$, we have

$$y' = \frac{1}{3}(3t^3 + C)^{-2/3} \cdot 3 \cdot 3t^2 = 3t^2(3t^3 + C)^{-2/3}$$
$$y^2 y' = [(3t^3 + C)^{1/3}]^2 \cdot 3t^2 \, (3t^3 + C)^{-2/3}$$
$$= 3t^2.$$

Figure 1 Solution curves in Example 1.

Figure 1 shows solution curves for various values of C. Note the linear solution that corresponds to $C = 0$.

Discussion of Step (c) Suppose that $y = f(t)$ is a solution of the differential equation $y^2 y' = 3t^2$. Then,

$$[f(t)]^2 f'(t) = 3t^2.$$

Integrating, we have

$$\int [f(t)]^2 f'(t) \, dt = \int 3t^2 \, dt.$$

Make the substitution $y = f(t)$, $dy = f'(t) \, dt$ in the left side to get

$$\int y^2 \, dy = \int 3t^2 \, dt.$$

This is just the result of step (c). The process of "canceling the dt" and integrating with respect to y is just equivalent to making the substitution $y = f(t)$, $dy = f'(t) \, dt$.

» Now Try Exercise 1

The technique used in Example 1 can be used for any differential equation with separated variables. Suppose that we are given such an equation:

$$h(y)y' = p(t),$$

where $h(y)$ is a function of y only and $p(t)$ is a function of t only. Our method of solution can be summarized as follows:

(a) Write y' as $\dfrac{dy}{dt}$:

$$h(y)\frac{dy}{dt} = p(t).$$

(b) Integrate both sides with respect to t:

$$\int h(y)\frac{dy}{dt} \, dt = \int p(t) \, dt.$$

(c) Rewrite the left side by "canceling the dt":

$$\int h(y) \, dy = \int p(t) \, dt.$$

(d) Calculate the antiderivatives $H(y)$ for $h(y)$ and $P(t)$ for $p(t)$:

$$H(y) = P(t) + C.$$

(e) Solve for y in terms of t:

$$y = \ldots.$$

NOTE In step (d), there is no need to write two constants of integration (as we did in Example 1), since they will be combined into one in step (e). «

EXAMPLE 2   **Separation of Variables** Solve $e^y y' = 2t + 1$, $y(0) = 1$.

SOLUTION   (a) $e^y \dfrac{dy}{dt} = 2t + 1$

Figure 2 Solution curves and the solution of the initial value problem in Example 2.

(b) $\displaystyle\int e^y \dfrac{dy}{dt}\,dt = \int (2t+1)\,dt$

(c) $\displaystyle\int e^y \,dy = \int (2t+1)\,dt$

(d) $e^y = t^2 + t + C$

(e) $y = \ln(t^2 + t + C)$ Take logarithms of both sides of the equation in step (d).

If $y = \ln(t^2 + t + C)$ is to satisfy the initial condition $y(0) = 1$, then

$$1 = y(0) = \ln(0^2 + 0 + C) = \ln C,$$

so that $C = e$ and $y = \ln(t^2 + t + e)$. Figure 2 shows solutions of the differential equation for various values of C. The curve that goes through the point $(0, 1)$ is the solution of the initial-value problem.   **» Now Try Exercise 5**

EXAMPLE 3   **Separation of Variables** Solve $y' = t^3 y^2 + y^2$.

SOLUTION   As the equation is given, the right side is not in the form $p(t)q(y)$. However, we may rewrite the equation in the form $y' = (t^3 + 1)y^2$. Now, we may separate the variables, dividing both sides by y^2, to get

$$\frac{1}{y^2} y' = t^3 + 1. \tag{3}$$

Then, we apply our method of solution:

(a) $\dfrac{1}{y^2}\dfrac{dy}{dt} = t^3 + 1$

(b) $\displaystyle\int \dfrac{1}{y^2}\dfrac{dy}{dt}\,dt = \int (t^3 + 1)\,dt$

(c) $\displaystyle\int \dfrac{1}{y^2}\,dy = \int (t^3 + 1)\,dt$

(d) $-\dfrac{1}{y} = \dfrac{1}{4}t^4 + t + C$, C a constant

(e) $y = -\dfrac{1}{\frac{1}{4}t^4 + t + C}.$

Our method yields all the solutions of equation (3). However, we have ignored an important point. We wish to solve $y' = y^2(t^3 + 1)$ and not equation (3). Do the two equations have precisely the same solutions? We obtained (3) from the given equation by dividing by y^2. This is a permissible operation, provided that y is not equal to zero for all t. (Of course, if y is zero for some t, the resulting differential equation is understood to hold only for some limited range of t.) Thus, in dividing by y^2, we must assume that y is not the zero function. However, note that $y = 0$ is a solution of the original equation because

$$0 = (0)' = t^3 \cdot 0^2 + 0^2.$$

So, when we divided by y^2, we "lost" the solution $y = 0$. Finally, we see that the solutions of the differential equation $y' = t^3 y^2 + y^2$ are

$$y = -\frac{1}{\frac{1}{4}t^4 + t + C}, \quad C \text{ a constant,} \quad \text{and} \quad y = 0.$$

Figure 3 shows two solutions corresponding to different values of the constant C, and the solution $y = 0$. It is interesting to note how a great diversity of solutions arises from a relatively simple differential equation.   **» Now Try Exercise 11**

CAUTION ▶ If the equation in Example 3 had been

$$y' = t^3 y^2 + 1,$$

we would not have been able to use the method of separation of variables, because the expression $t^3 y^2 + 1$ cannot be written in the form $p(t)q(y)$. **«**

EXAMPLE 4   **Initial Value Problem** Solve the initial-value problem $3y' + y^4 \cos t = 0$, $y\left(\frac{\pi}{2}\right) = \frac{1}{2}$.

SOLUTION   Write the equation in the form

$$3y' = -y^4 \cos t. \tag{4}$$

Clearly, the constant function $y = 0$ is a solution of the differential equation because it makes both sides of (4) zero for all t. Now, supposing that $y \neq 0$, we may divide by y^4 and get

$$\frac{3}{y^4} y' = -\cos t$$

$$\int \frac{3}{y^4} \frac{dy}{dt} \, dt = -\int \cos t \, dt$$

$$-y^{-3} = -\sin t + C, \quad C \text{ a constant}$$

$$y^{-3} = \sin t + C.$$

Figure 3 Solution curves in Example 3.

In the last step, there is no need to change the sign of C, since it stands for an arbitrary constant. Solving for y in terms of t, we find

$$y^3 = \frac{1}{\sin t + C}$$

$$y = \frac{1}{\sqrt[3]{\sin t + C}}.$$

If y is to satisfy the initial condition $y\left(\frac{\pi}{2}\right) = \frac{1}{2}$, we must have

$$\frac{1}{2} = y\left(\frac{\pi}{2}\right) = \frac{1}{\sqrt[3]{\sin\left(\frac{\pi}{2}\right) + C}} = \frac{1}{\sqrt[3]{1 + C}} \qquad \text{Recall that } \sin\left(\frac{\pi}{2}\right) = 1.$$

$$2 = \sqrt[3]{1 + C} \qquad \text{Take inverses.}$$

$$2^3 = 1 + C$$

$$C = 7.$$

Hence, the desired solution is

$$y = \frac{1}{\sqrt[3]{\sin t + 7}}.$$

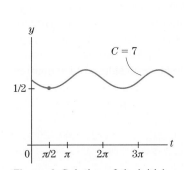

Figure 4 Solution of the initial-value problem in Example 4.

The graph of y in Fig. 4 goes through the point $\left(\frac{\pi}{2}, \frac{1}{2}\right)$.   **» Now Try Exercise 21**

EXAMPLE 5   **Initial Value Problem**   Solve $y' = te^t/y$, $y(0) = -5$.

SOLUTION   Separating the variables, we have

$$yy' = te^t$$

$$\int y \frac{dy}{dt}\, dt = \int te^t\, dt$$

$$\int y\, dy = \int te^t\, dt.$$

The integral $\int te^t\, dt$ may be found by integration by parts:

$$\int te^t\, dt = te^t - \int 1 \cdot e^t\, dt = te^t - e^t + C.$$

Therefore,

$$\tfrac{1}{2}y^2 = te^t - e^t + C$$

$$y^2 = 2te^t - 2e^t + C_1$$

$$y = \pm\sqrt{2te^t - 2e^t + C_1}.$$

Note that the $\pm$ appears because there are two square roots of $2te^t - 2e^t + C_1$ that differ from one another by a minus sign. Thus, the solutions are of two sorts:

$$y = +\sqrt{2te^t - 2e^t + C_1}$$

$$y = -\sqrt{2te^t - 2e^t + C_1}.$$

Two of these solutions are shown in Fig. 5 with different signs in front of the square root. We must choose C_1 so that $y(0) = -5$. Since the values of y for the first solution are always positive, the given initial condition must correspond to the second solution, and we must have

$$-5 = y(0) = -\sqrt{2\cdot 0 \cdot e^0 - 2e^0 + C_1} = -\sqrt{-2 + C_1}$$

$$-2 + C_1 = 25$$

$$C_1 = 27.$$

Hence, the desired solution is

$$y = -\sqrt{2te^t - 2e^t + 27}.$$

Its graph goes through the point $(0, -5)$ in Fig. 5.   **» Now Try Exercise 19**

$y = \sqrt{2te^t - 2e^t + 11}$

$y = -\sqrt{2te^t - 2e^t + 27}$

Figure 5 Solution curves in Example 5.

When working the exercises at the end of the section, it is a good practice first to find the constant solution(s), if any. A constant function $y = c$ is a solution of $y' = p(t)q(y)$ if, and only if, $q(c) = 0$. [For $y = c$ implies that $y' = (c)' = 0$, and $p(t)q(y)$ will be zero for all t if, and only if, $q(y) = 0$—that is, $q(c) = 0$.] After listing the constant solutions, one may assume that $q(y) \neq 0$ and go on to divide both sides of the equation $y' = p(t)q(y)$ by $q(y)$ to separate the variables.

Check Your Understanding 10.2

Solutions can be found following the section exercises.

1. Solve the initial-value problem $y' = 5y$, $y(0) = 2$, by separation of variables.

2. Solve $y' = \sqrt{ty}$, $y(1) = 4$.

EXERCISES 10.2

Solve the following differential equations:

1. $\dfrac{dy}{dt} = \dfrac{5-t}{y^2}$

2. $\dfrac{dy}{dt} = te^{2y}$

3. $\dfrac{dy}{dt} = \dfrac{e^t}{e^y}$

4. $\dfrac{dy}{dt} = -\dfrac{1}{t^2 y^2}$

5. $\dfrac{dy}{dt} = t^{1/2} y^2$

6. $\dfrac{dy}{dt} = \dfrac{t^2 y^2}{t^3 + 8}$

7. $y' = \left(\dfrac{t}{y}\right)^2 e^{t^3}$

8. $y' = e^{4y} t^3 - e^{4y}$

9. $y' = \sqrt{\dfrac{y}{t}}$

10. $y' = \left(\dfrac{e^t}{y}\right)^2$

11. $y' = 3t^2 y^2$

12. $(1 + t^2) y' = ty^2$

13. $y' e^y = te^{t^2}$

14. $y' = \dfrac{1}{ty + y}$

15. $y' = \dfrac{\ln t}{ty}$

16. $y^2 y' = \tan t$

17. $y' = (y - 3)^2 \ln t$

18. $yy' = t \sin(t^2 + 1)$

Solve the following differential equations with the given initial conditions.

19. $y' = 2te^{-2y} - e^{-2y}$, $y(0) = 3$

20. $y' = y^2 - e^{3t} y^2$, $y(0) = 1$

21. $y^2 y' = t \cos t$, $y(0) = 2$

22. $y' = t^2 e^{-3y}$, $y(0) = 2$

23. $3y^2 y' = -\sin t$, $y(\frac{\pi}{2}) = 1$

24. $y' = -y^2 \sin t$, $y(\frac{\pi}{2}) = 1$

25. $\dfrac{dy}{dt} = \dfrac{t+1}{ty}$, $t > 0$, $y(1) = -3$

26. $\dfrac{dy}{dt} = \left(\dfrac{1+t}{1+y}\right)^2$, $y(0) = 2$

27. $y' = 5ty - 2t$, $y(0) = 1$

28. $y' = \dfrac{t^2}{y}$, $y(0) = -5$

29. $\dfrac{dy}{dx} = \dfrac{\ln x}{\sqrt{xy}}$, $y(1) = 4$

30. $\dfrac{dN}{dt} = 2tN^2$, $N(0) = 5$

31. Relationship between Price and Sales A model that describes the relationship between the price and the weekly sales of a product might have a form such as

$$\frac{dy}{dp} = -\frac{1}{2}\left(\frac{y}{p+3}\right),$$

where y is the volume of sales and p is the price per unit. That is, at any time, the rate of decrease of sales with respect to price is directly proportional to the sales level

and inversely proportional to the sales price plus a constant. Solve this differential equation. (Figure 6 shows several typical solutions.)

Figure 6 Demand curves.

32. Model in Psychology One problem in psychology is to determine the relation between some physical stimulus and the corresponding sensation or reaction produced in a subject. Suppose that, measured in appropriate units, the strength of a stimulus is s and the intensity of the corresponding sensation is some function of s, say, $f(s)$. Some experimental data suggest that the rate of change of intensity of the sensation with respect to the stimulus is directly proportional to the intensity of the sensation and inversely proportional to the strength of the stimulus; that is, $f(s)$ satisfies the differential equation

$$\frac{dy}{ds} = k\frac{y}{s}$$

for some positive constant k. Solve this differential equation. (Figure 7 shows several solutions corresponding to $k = .4$.)

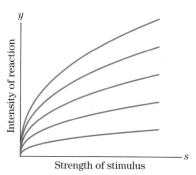

Figure 7 Reaction to stimuli.

33. Probability of Accidents Let t represent the total number of hours that a truck driver spends during a year driving on a certain highway connecting two cities, and let $p(t)$ represent the probability that the driver will have at least one accident during these t hours. Then, $0 \le p(t) \le 1$, and $1 - p(t)$ represents the probability of not having an accident. Under ordinary conditions, the rate of increase in the probability of an accident (as a function of t) is proportional to the probability of not having an accident. Construct and solve a differential equation for this situation.

34. **Amount of Information Learned** In certain learning situations a maximum amount, M, of information can be learned, and at any time, the rate of learning is proportional to the amount yet to be learned. Let $y = f(t)$ be the amount of information learned up to time t. Construct and solve a differential equation that is satisfied by $f(t)$.

35. **Volume of a Mothball** Mothballs tend to evaporate at a rate proportional to their surface area. If V is the volume of a mothball, then its surface area is roughly a constant times $V^{2/3}$. So the mothball's volume decreases at a rate proportional to $V^{2/3}$. Suppose that initially a mothball has a volume of 27 cubic centimeters and 4 weeks later has a volume of 15.625 cubic centimeters. Construct and solve a differential equation satisfied by the volume at time t. Then, determine if and when the mothball will vanish ($V = 0$).

36. **Construction Cost Index** Some homeowner's insurance policies include automatic inflation coverage based on the U.S. Commerce Department's construction cost index (CCI). Each year, the property insurance coverage is increased by an amount based on the change in the CCI. Let $f(t)$ be the CCI at time t years since January 1, 1990, and let $f(0) = 100$. Suppose that the construction cost index is rising at a rate proportional to the CCI and the index was 115 on January 1, 1992. Construct and solve a differential equation satisfied by $f(t)$. Then, determine when the CCI will reach 200.

37. **Gompertz Equation** The Gompertz growth equation is

$$\frac{dy}{dt} = -ay \ln \frac{y}{b},$$

where a and b are positive constants. This equation is used in biology to describe the growth of certain populations. Find the general form of solutions to this differential equation. (Figure 8 shows several solutions corresponding to $a = .04$ and $b = 90$.)

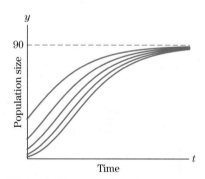

Figure 8 Gompertz growth curves.

38. **Rate of Decomposition** When a certain liquid substance A is heated in a flask, it decomposes into a substance B at such a rate (measured in units of A per hour) that at any time t is proportional to the square of the amount of substance A present. Let $y = f(t)$ be the amount of substance A present at time t. Construct and solve a differential equation that is satisfied by $f(t)$.

39. **Population of Fish** Let $f(t)$ denote the number (in thousands) of fish in a lake after t years, and suppose that $f(t)$ satisfies the differential equation

$$y' = 0.1y(5 - y).$$

The slope field for this equation is shown in Fig. 9.
(a) With the help of the slope field, discuss what happens to an initial population of 6000 fish. Does it increase or decrease?
(b) How about an initial population of 1000 fish? Does it increase or decrease?
(c) On the slope field in Fig. 9 or a copy of it, draw the solution of the initial-value problem

$$y' = 0.1y(5 - y), \quad y(0) = 1.$$

What does this solution represent?

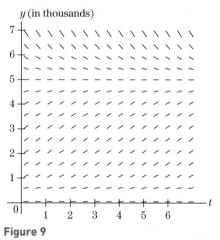

Figure 9

40. Refer to the differential equation in Exercise 39.
(a) Obviously, if you start with zero fish, $f(t) = 0$ for all t. Confirm this on the slope field. Are there any other constant solutions?
(b) Describe the population of fish if the initial population is greater than 5000; less than 5000. On the slope field in Fig. 9 or a copy of it, draw solution curves to illustrate your description.

Solutions to Check Your Understanding 10.2

1. The constant function $y = 0$ is a solution of $y' = 5y$. If $y \neq 0$, we may divide by y and obtain

$$\frac{1}{y} y' = 5$$

$$\int \frac{1}{y} \frac{dy}{dt} \, dt = \int 5 \, dt$$

$$\int \frac{1}{y} \, dy = \int 5 \, dt$$

$$\ln |y| = 5t + C$$

$$|y| = e^{5t + C} = e^C \cdot e^{5t}$$

$$y = \pm e^C \cdot e^{5t}.$$

These two types of solutions and the constant solution may all be written in the form

$$y = Ae^{5t},$$

where A is an arbitrary constant (positive, negative, or zero). The initial condition $y(0) = 2$ implies that

$$2 = y(0) = Ae^{5(0)} = A.$$

Hence, the solution of the initial-value problem is $y = 2e^{5t}$.

2. We rewrite $y' = \sqrt{ty}$ as $y' = \sqrt{t} \cdot \sqrt{y}$. The constant function $y = 0$ is one solution. To find the others, we suppose that $y \ne 0$ and divide by $\sqrt{y}$ to obtain

$$\frac{1}{\sqrt{y}} y' = \sqrt{t}$$

$$\int y^{-1/2} \frac{dy}{dt} dt = \int t^{1/2} dt$$

$$\int y^{-1/2} dy = \int t^{1/2} dt$$

$$2y^{1/2} = \frac{2}{3} t^{3/2} + C$$

$$y^{1/2} = \frac{1}{3} t^{3/2} + C_1 \tag{5}$$

$$y = \left(\frac{1}{3} t^{3/2} + C_1 \right)^2. \tag{6}$$

We must choose C_1 so that $y(1) = 4$. The quickest method is to use (5) instead of (6). We have $y = 4$ when $t = 1$, so

$$4^{1/2} = \frac{1}{3}(1)^{3/2} + C_1$$

$$2 = \frac{1}{3} + C_1$$

$$C_1 = \frac{5}{3}.$$

Hence, the desired solution is

$$y = \left(\frac{1}{3} t^{3/2} + \frac{5}{3} \right)^2.$$

10.3 First-Order Linear Differential Equations

In this section, we study first-order differential equations of the form

$$y' + a(t)y = b(t), \tag{1}$$

where $a(t)$ and $b(t)$ are continuous functions on a given interval. Equation (1) is called a **first-order linear differential equation in standard form**.

The following are examples of first-order linear differential equations

$$y' - 2ty = 0 \qquad [a(t) = -2t, b(t) = 0]$$
$$y' + y = 2 \qquad [a(t) = 1, b(t) = 2]$$
$$ty' = ty + t^2 + 1 \qquad \left[y' - y = \frac{t^2 + 1}{t}, a(t) = -1, b(t) = \frac{t^2 + 1}{t} \right]$$
$$e^t y' + e^t y = 5 \qquad [y' + y = 5e^{-t}, a(t) = 1, b(t) = 5e^{-t}]$$

In the last two examples, we first put the equation in standard form $y' + a(t)y = b(t)$ before determining the functions $a(t)$ and $b(t)$.

Given equation (1), form the function $e^{A(t)}$, called an **integrating factor**, where $A(t) = \int a(t)\, dt$. Observe that, as a consequence of the chain rule and the product rule,

$$\frac{d}{dt}\left[e^{A(t)} \right] = e^{A(t)} \frac{d}{dt} A(t) = e^{A(t)} a(t) = a(t)e^{A(t)},$$

and

$$\frac{d}{dt}\left[e^{A(t)} y \right] = e^{A(t)} y' + a(t)e^{A(t)} y$$

$$\overbrace{\phantom{= e^{A(t)} [y' + a(t)y].}}^{\text{left side of (1)}}$$
$$= e^{A(t)} \overbrace{[y' + a(t)y]}. \tag{2}$$

Returning to equation (1), multiply both sides by $e^{A(t)}$, and then simplify the resulting equation with the help of (2) as follows:

$$e^{A(t)}[y' + a(t)y] = e^{A(t)}b(t)$$

$$\frac{d}{dt}\left[e^{A(t)}y\right] = e^{A(t)}b(t) \quad \text{[by (2)]}. \tag{3}$$

Equation (3) is equivalent to equation (1). Integrate both sides to get rid of the derivative on the left side of (3) and get

$$e^{A(t)}y = \int e^{A(t)}b(t)\, dt + C.$$

Solve for y by multiplying both sides by $e^{-A(t)}$:

$$y = e^{-A(t)}\left[\int e^{A(t)}b(t)\, dt + C\right], \quad C \text{ a constant.} \tag{4}$$

This formula gives all the solutions of equation (1). It is called the **general solution** of (1). As our next examples illustrate, to solve a first-order linear differential equation, you can appeal directly to (4), or you can use an integrating factor and repeat the steps leading to (4).

EXAMPLE 1   **A First-Order Differential Equation** Solve $y' = 3 - 2y$.

SOLUTION   **Step 1** Put the equation in standard form: $y' + 2y = 3$.

Step 2 Find an integrating factor $e^{A(t)}$. We have $a(t) = 2$, so

$$A(t) = \int a(t)\, dt = \int 2\, dt = 2t.$$

Note how we picked one antiderivative of $a(t)$ by setting the constant of integration equal to zero. Hence, the integrating factor is $e^{A(t)} = e^{2t}$.

Step 3 Multiply both sides of the differential equation by the integrating factor e^{2t}:

$$\overbrace{e^{2t}y' + 2e^{2t}y}^{\frac{d}{dt}[e^{2t}y]} = 3e^{2t}.$$

Recognizing the terms on the left side as a derivative of the product $e^{2t}y$, we get

$$\frac{d}{dt}\left[e^{2t}y\right] = 3e^{2t}.$$

Step 4 Integrating both sides and solving for y, we obtain

$$e^{2t}y = \int 3e^{2t}\, dt = \frac{3}{2}e^{2t} + C$$

$$y = e^{-2t}\left[\frac{3}{2}e^{2t} + C\right] = \frac{3}{2} + Ce^{-2t}.$$

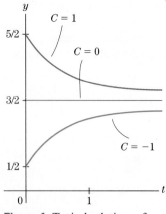

Figure 1 Typical solutions of the differential equation in Example 1: $y' + 2y = 3$.

Figure 1 shows solution curves for various values of C. Note the constant solution, $y = \frac{3}{2}$, corresponding to $C = 0$.   **» Now Try Exercise 7**

From the preceding example, we can state a step-by-step process for solving a first-order linear differential equation.

> **Solving a First-Order Linear Differential Equation**
>
> **Step 1**   Put the equation in the standard form $y' + a(t)y = b(t)$.
>
> **Step 2**   Compute an antiderivative of $a(t)$, $A(t) = \int a(t)\, dt$. [When evaluating $\int a(t)\, dt$, it is customary to choose 0 for the constant of integration.] Form the integrating factor $e^{A(t)}$.
>
> **Step 3**   Multiply the differential equation by the integrating factor $e^{A(t)}$. This transforms the terms on the left side of the equation into the derivative of a product, $\frac{d}{dt}\left[e^{A(t)}y\right]$, as in equation (3).
>
> **Step 4**   Integrate to get rid of the derivative, and then solve for y.

EXAMPLE 2   **A First-Order Differential Equation**   Solve $\dfrac{1}{3t^2}y' + y = 4$, $\quad t > 0$.

SOLUTION   **Step 1**   Multiply through by $3t^2$ and get

$$y' + 3t^2 y = 12t^2.$$

Hence, $a(t) = 3t^2$.

Step 2   An antiderivative of $a(t)$ is

$$A(t) = \int a(t)\, dt = \int 3t^2\, dt = t^3.$$

So the integrating factor is $e^{A(t)} = e^{t^3}$.

Step 3   Multiply both sides of the differential equation by e^{t^3} and get

$$e^{t^3}y' + 3t^2 e^{t^3} y = 12t^2 e^{t^3}$$

$$\frac{d}{dt}\left[e^{t^3}y\right] = 12t^2 e^{t^3}.$$

Step 4   Integrating both sides, we obtain

$$e^{t^3}y = \int 12t^2 e^{t^3}\, dt = 4e^{t^3} + C$$

$$y = e^{-t^3}\left[4e^{t^3} + C\right] = 4 + Ce^{-t^3}.$$

In evaluating the integral $\int 12t^2 e^{t^3}\, dt$, we used integration by substitution (Section 9.1): Let $u = t^3$, $du = 3t^2\, dt$. Then,

$$\int 12t^2\, e^{t^3}\, dt = 4\int \overbrace{e^{t^3}}^{e^u}\, \overbrace{3t^2\, dt}^{du} = 4\int e^u\, du = 4e^u + C = 4e^{t^3} + C.$$

Figure 2 shows solution curves for various values of C. Note the constant solution, $y = 4$, which corresponds to $C = 0$.   **» Now Try Exercise 13**

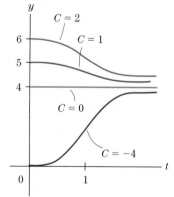

Figure 2 Typical solutions of the differential equation in Example 2.

EXAMPLE 3   **Initial-Value Problem**   Solve the initial-value problem $t^2 y' + ty = 2$, $y(1) = 1$, $t > 0$.

SOLUTION   **Step 1**   Divide by t^2 to put the equation in standard form:

$$y' + \frac{1}{t}y = \frac{2}{t^2}.$$

Step 2   An antiderivative of $a(t) = \frac{1}{t}$ is

$$A(t) = \int \frac{1}{t}\, dt = \ln t.$$

Hence, the integrating factor is $e^{A(t)} = e^{\ln t} = t$.

Step 3   Multiply both sides of the differential equation by t and get

$$ty' + y = \frac{2}{t}$$

$$\frac{d}{dt}[ty] = \frac{2}{t}.$$

Step 4   Integrating both sides and solving for y, we obtain

$$ty = \int \frac{2}{t}\, dt = 2 \ln t + C$$

$$y = \frac{2 \ln t + C}{t}.$$

To satisfy the initial condition, we must have

$$1 = y(1) = \frac{2 \ln(1) + C}{1} = C \quad [\ln(1) = 0].$$

Hence, the solution of the initial-value problem is

$$y = \frac{2 \ln t + 1}{t}, \quad t > 0.$$

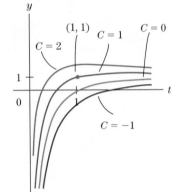

Figure 3 Solution of the initial-value problem $t^2 y' + ty = 2$, $y(1) = 1, t > 0$.

Among all the solution curves that are shown in Fig. 3, the solution of the initial-value problem is the curve that goes through the point $(1, 1)$.   **>> Now Try Exercise 21**

In the following section, we present several interesting applications of first-order linear differential equations.

Check Your Understanding 10.3

Solutions can be found following the section exercises.

1. Using an integrating factor, solve $y' + y = 1 + e^{-t}$.

2. Find an integrating factor for the differential equation $y' = -\dfrac{y}{1 + t} + 1, t \geq 0$.

EXERCISES 10.3

In Exercises 1–6, find an integrating factor for each equation. Take $t > 0$.

1. $y' - 2y = t$

2. $y' + ty = 6t$

3. $t^3 y' + y = 0$

4. $y' + \sqrt{t}\, y = 2(t + 1)$

5. $y' - \dfrac{y}{10 + t} = 2$

6. $y' = t^2(y + 1)$

In Exercises 7–20, solve the given equation using an integrating factor. Take $t > 0$.

7. $y' + y = 1$

8. $y' + 2ty = 0$

9. $y' - 2ty = -4t$

10. $y' = 2(20 - y)$

11. $y' = .5(35 - y)$

12. $y' + y = e^{-t} + 1$

13. $y' + \dfrac{y}{10 + t} = 0$

14. $y' - 2y = e^{2t}$

15. $(1 + t)y' + y = -1$

16. $y' = e^{-t}(y + 1)$

17. $6y' + ty = t$

18. $e^t y' + y = 1$

19. $y' + y = 2 - e^t$

20. $\dfrac{1}{\sqrt{t + 1}} y' + y = 1$

In Exercises 21–28, solve the initial-value problem.

21. $y' + 2y = 1, y(0) = 1$

22. $ty' + y = \ln t, y(e) = 0, t > 0$

23. $y' + \dfrac{y}{1 + t} = 20, y(0) = 10, t \geq 0$

24. $y' = 2(10 - y), y(0) = 1$

25. $y' + y = e^{2t}, y(0) = -1$

26. $ty' - y = -1, y(1) = 1, t > 0$

27. $y' + 2y \cos(2t) = 2 \cos(2t), y(\frac{\pi}{2}) = 0$

28. $ty' + y = \sin t, y(\frac{\pi}{2}) = 0, t > 0$

TECHNOLOGY EXERCISES

29. Consider the initial-value problem

$$y' = -\frac{y}{1 + t} + 10, \quad y(0) = 50.$$

(a) Is the solution increasing or decreasing when $t = 0$? [*Hint:* Compute $y'(0)$.]

(b) Find the solution and plot it for $0 \leq t \leq 4$.

Solutions to Check Your Understanding 10.3

1. We follow the step-by-step method outlined in this section. The equation $y' + y = 1 + e^{-t}$ is already in standard form. We have $a(t) = 1$, $A(t) = \int 1\, dt = t$, and so the integrating factor is $e^{A(t)} = e^t$. Multiplying the equation by the integrating factor, we transform the terms on its left side into the derivative of a product, as follows:

$e^t(y' + y) = e^t(1 + e^{-t})$ [Multiply the equation by the integrating factor e^t.]

$e^t y' + e^t y = e^t + 1$ [The left side is the derivative of the product $e^t y$.]

$\dfrac{d}{dt}[e^t y] = e^t + 1$

Integrating both sides, we get rid of the derivative and obtain

$$e^t y = \int (e^t + 1)\, dt = e^t + t + C.$$

To solve for y, we multiply both sides by e^{-t} and get

$$y = e^{-t}(e^t + t + C) = 1 + te^{-t} + Ce^{-t}.$$

2. In standard form, the equation becomes

$$y' + \frac{y}{1 + t} = 1.$$

We have $a(t) = \dfrac{1}{1 + t}$. An antiderivative of $a(t)$ is

$$A(t) = \int \frac{1}{1 + t}\, dt = \ln|1 + t|.$$

But since $t \geq 0$, it follows that $1 + t \geq 0$; hence, $\ln|1 + t| = \ln(1 + t)$. So the integrating factor is

$$e^{A(t)} = e^{\ln(1+t)} = 1 + t.$$

10.4 Applications of First-Order Linear Differential Equations

This section discusses several interesting applications that lead to first-order linear differential equations. Modeling or setting up the mathematical equations is a crucial part of the solution. You can develop your modeling skills by studying carefully the details leading to the mathematical equations in each of the following examples. Look for the key expression "rate of change" and translate it as a derivative. Then, continue your description of the rate of change to obtain a differential equation.

Our first examples are related to the topics of compound interest from Section 5.2 and future value of an income stream from Section 6.5.

EXAMPLE 1   **A Retirement Account** You invest in a retirement account that pays 6% interest per year. You make an initial deposit of $1000 and plan on making future deposits at the rate of $2400 per year. Assume that the deposits are made continuously and that interest is compounded continuously. Let $P(t)$ denote the amount of money in the account t years after the initial deposit.
(a) Set up an initial-value problem that is satisfied by $P(t)$.
(b) Find $P(t)$.

SOLUTION   (a) If no deposits or withdrawals are made, only interest is being added to the account at a rate proportional to the amount in the account with constant of proportionality $k = .06$ or 6%. Since in this case the growth of $P(t)$ comes from interest only, it follows that $P(t)$ satisfies the equation

$$\underbrace{y'}_{\begin{bmatrix}\text{rate of} \\ \text{change of } y\end{bmatrix}} = \underbrace{.06}_{\begin{bmatrix}\text{constant of} \\ \text{proportionality}\end{bmatrix}} \times \underbrace{y}_{\begin{bmatrix}\text{amount in} \\ \text{the account}\end{bmatrix}}.$$

Taking into consideration the deposits, which are adding money to the account at the rate of $2400 per year, we see that there are two influences on the way the amount of money in the account changes: the rate at which interest is added and the rate at which deposits are made. The rate of change of $P(t)$ is the *net effect* of these two influences. That is, $P(t)$ now satisfies the first-order linear differential equation

$$\underbrace{y'}_{\begin{bmatrix}\text{rate of} \\ \text{change of } y\end{bmatrix}} = \underbrace{.06y}_{\begin{bmatrix}\text{rate at which} \\ \text{interest is added}\end{bmatrix}} + \underbrace{2400}_{\begin{bmatrix}\text{rate at which} \\ \text{money is deposited}\end{bmatrix}}.$$

Since the initial deposit in the account was $1000, it follows that $P(t)$ satisfies the initial condition $y(0) = 1000$. Putting the equation in standard form [see Section 10.3, equation (1)], we conclude that $P(t)$ satisfies the initial-value problem

$$y' - .06y = 2400, \qquad y(0) = 1000.$$

(b) To solve the equation, use the step-by-step method from the previous section. We have $a(t) = -.06$. So $A(t) = -\int .06\, dt = -.06t$; hence, the integrating factor is $e^{A(t)} = e^{-0.06t}$. Multiplying both sides of the equation by the integrating factor, we obtain

$$e^{-0.06t}[y' - .06y] = 2400e^{-0.06t}$$

$$e^{-0.06t}y' - .06e^{-0.06t}y = 2400e^{-0.06t} \quad \text{[The left side is a derivative of a product.]}$$

$$\frac{d}{dt}[e^{-0.06t}y] = 2400e^{-0.06t}.$$

Integrating both sides and solving for y, we get

$$e^{-0.06t}y = \int 2400e^{-0.06t}\, dt$$

$$e^{-0.06t}y = -40{,}000e^{-0.06t} + C \qquad \int 2400e^{-0.06t}\, dt = -\frac{2400}{.06}e^{-0.06t} + C$$

$$y = e^{0.06t}\left[-40{,}000e^{-0.06t} + C.\right] \quad \text{Solve for } y.$$

$$y = -40{,}000 + Ce^{0.06t}. \quad \text{Simplify.}$$

Thus, the solution is $P(t) = -40{,}000 + Ce^{0.06t}$. To satisfy the initial condition, we must have

$$1000 = P(0) = -40{,}000 + Ce^{0.06 \cdot (0)} = -40{,}000 + C$$

$$C = 41{,}000$$

Hence, the amount of money in the account at time t is given by

$$P(t) = -40{,}000 + 41{,}000e^{0.06t}.$$

The graph of $P(t)$ is shown in Fig. 1. ≫ **Now Try Exercise 1**

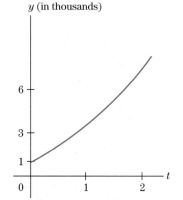

Figure 1 The account $P(t)$ in Example 1.

Let us quickly examine the modeling ideas in Example 1. The quantity that we were looking for was the amount of money in the account at time t. Its rate of change was affected by two influences: the interest rate and the rate at which money was deposited. To describe the rate of change of the money in the account, we added both rates of change and obtained the differential equation $y' = .06y + 2400$.

We can model many interesting situations by refining these ideas. First, identify the quantity of concern (amount of money in the account). Then, identify the various influences that affect the rate of change of this quantity (interest rate, rate of deposits). Finally, derive a differential equation by expressing the rate of change of the quantity of interest in terms of the rates of change of the various influences. In Example 1, we added two rates of change to describe y'. As our next example illustrates, to model problems involving amortization of consumer loans and mortgages, you may have to subtract two rates of change.

EXAMPLE 2   **Paying Off a Car Loan** You took a loan of $25,000 to pay for a new car. The interest rate on the loan is 5%. You arranged through your online banking to make daily payments totaling $4800 per year. This allows you to assume that your payments are flowing continuously into your account. Let $P(t)$ denote the amount that you owe on the loan at time t (in years). Assume that interest is compounded continuously. Set up an initial-value problem that is satisfied by $P(t)$.

SOLUTION   As in Example 1, there are two influences on the way the amount of money owed changes: the rate at which interest is added to the amount owed and the rate at which payments are subtracted from the amount owed. We know that interest is being added at a rate proportional to the amount owed, with constant of proportionality $k = .05$. The effect of the payments is to subtract from the amount owed at the rate of $4800 per year. Since the rate of change of $P(t)$ is the *net effect* of these two influences, we see that $P(t)$ satisfies the first-order linear differential equation

$$y' \quad = \quad .05y \quad - \quad 4800$$
$$\begin{bmatrix} \text{rate of} \\ \text{change of } y \end{bmatrix} = \begin{bmatrix} \text{rate at which} \\ \text{interest is added} \end{bmatrix} - \begin{bmatrix} \text{rate of} \\ \text{payments} \end{bmatrix}.$$

Recalling the initial condition and rewriting the equation in standard form, we find that $P(t)$ satisfies the initial-value problem

$$y' - .05y = -4800, \quad y(0) = 25{,}000. \qquad \text{\textbf{\textgg Now Try Exercise 3}}$$

Using an integrating factor and proceeding in a manner very similar to that in Example 1, we find the solution of the initial-value problem in Example 2 to be

$$P(t) = 96{,}000 - 71{,}000e^{0.05t}.$$

The graph of $P(t)$ is shown in Fig. 2. From the graph, we see that $P(t) = 0$ when $t \approx 6$ years. This is how long it will take to pay off the entire loan.

In the previous examples, the annual rates of deposits or loan payments were constant. If you expect your annual income to increase (which is a reasonable expectation), you may want to increase the rate of your savings deposits or you may consider paying off your loan at a faster rate. The modeling involved in these situations is similar to that in the previous examples but may lead to more complicated differential equations. See the exercises for several illustrations.

In Chapter 5, we learned that simple population models are based on the assumption that the growth rate of a population is proportional to the size of the population at time t. The constant of proportionality is called the **growth constant** and is specific to the population. In reality, the rate of growth of a population may be affected by several other factors. In the next example, we consider the effect of emigration on the size of a population.

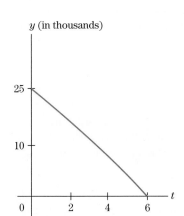

y (in thousands)

Figure 2 $P(t)$ is the amount owed at time t. The loan is paid off in approximately 6 years.

EXAMPLE 3   **A Population Model with Emigration** In 2005, people in a country suffering from economic problems started to emigrate to other countries. Let $P(t)$ denote the population of the country in millions t years after 2005. Sociologists studying this population determined that for the next 30 years the number of people emigrating would gradually increase as the news of better prospects outside the country spread. Suppose that the rate of emigration is given by $.004e^{0.04t} + .04$ millions per year t years after 2005. Suppose further that the growth constant of the population is $\frac{3}{125}$. Find a differential equation satisfied by $P(t)$.

SOLUTION   In our model, for the next 30 years (from 2005), the rate of growth of the population is affected by two influences: the rate at which the population is growing and the rate at which the population is emigrating. The rate of change of the population is the net effect of these two influences. Thus, $P(t)$ satisfies the differential equation

$$y' \quad = \quad \tfrac{3}{125}y \quad - \quad (.004e^{0.04t} + .04)$$
$$\begin{bmatrix} \text{rate of} \\ \text{change of } y \end{bmatrix} = \begin{bmatrix} \text{rate at which} \\ \text{population is growing} \end{bmatrix} - \begin{bmatrix} \text{rate at which} \\ \text{population is emigrating} \end{bmatrix}.$$

Putting the equation in standard form, we obtain

$$y' - \frac{3}{125}y = -.004e^{0.04t} - .04.$$

$\text{\textbf{\textgg Now Try Exercise 9}}$

Figure 3 Population with emigration.

To obtain a formula for the population in Example 3, you need to know the initial population. For example, if in 2005 the size of the population was 2 million, then $P(0) = 2$. Solving the differential equation, subject to this initial condition, we obtain

$$P(t) = \frac{7}{12}e^{\frac{3}{125}t} - \frac{1}{4}e^{\frac{t}{25}} + \frac{5}{3}.$$

(The details are straightforward and are left to Exercise 16.) Figure 3 shows the size of the population since 2005. Starting in 2005, the population continued to grow at a decreasing rate. According to our model, and assuming that the economical conditions remain the same, the population will peak at about 2.055 million between the years 2025 and 2030, and then, it will start to decrease.

Our final topic deals with Newton's law of cooling from Example 5, Section 10.1. The modeling ideas involved are useful in many interesting applications, such as determining the time of death of a person and the study of the concentration of waste products in the body and their clearance by artificial kidneys (dialysis).

EXAMPLE 4   **Newton's Law of Cooling** A hungry college student, in a rush to eat, turns the oven on and puts a frozen pizza in it without preheating the oven. Let $f(t)$ denote the temperature of the pizza and $T(t)$ the oven's temperature t minutes after the oven was turned on. According to Newton's law of cooling, the rate of change of $f(t)$ is proportional to the difference between the oven's temperature and the temperature of the pizza. Find a differential equation that is satisfied by $f(t)$.

SOLUTION   We reason as we did in Example 5, Section 10.1. The rate of change of the temperature of the pizza is the derivative of $f(t)$. This derivative is proportional to the difference $T(t) - f(t)$. Thus, there exists a constant k such that

$$f'(t) = k[T(t) - f(t)].$$

Is k positive or negative? While the pizza is heating, its temperature is rising. So $f'(t)$ is positive. Also, the temperature of the oven is always higher than the temperature of the pizza. So $T(t) - f(t)$ is positive. Thus, to make $f'(t)$ positive, k must be a positive number. Consequently, the differential equation satisfied by $f(t)$ is

$$y' = k[T(t) - y],$$

where k is a positive constant.   **≪**

Example 5 treats an interesting case of the model in Example 4. In the solution, we will appeal to the integral

$$\int (at + b)e^{ct}\, dt = \frac{1}{c^2}e^{ct}(act + bc - a) + C \quad (c \neq 0). \tag{1}$$

You do not need to memorize this formula, but you should be able to evaluate integrals of this form by using integration by parts, as we now illustrate. Set $f(t) = at + b$ and $g(t) = e^{ct}$. Then, $f'(t) = a$ and $G(t) = \frac{1}{c}e^{ct}$. Integrating by parts, we get

$$\int (at + b)e^{ct}\, dt = \frac{1}{c}(at + b)e^{ct} - \int \frac{a}{c}e^{ct}\, dt$$

$$= \frac{1}{c}(at + b)e^{ct} - \frac{a}{c^2}e^{ct} + C$$

$$= \frac{1}{c^2}e^{ct}(c(at + b) - a) + C,$$

and equation (1) follows upon simplifying.

EXAMPLE 5   **Newton's Law of Cooling** Suppose that the oven's temperature in Example 4 is given by $T(t) = 70 + 50t$ for $0 \leq t \leq 8$. [So when the student turned on the oven ($t = 0$), the oven's temperature was 70°, and then it started to rise at the rate of 50° per minute for the next 8 minutes. Presumably, the student set the oven's temperature at 470°.] Suppose further that the constant of proportionality is $k = .1$ and that the initial temperature of the frozen pizza was 27°. Determine the temperature of the pizza during the first 8 minutes of heating. What is the temperature of the pizza after 8 minutes of heating?

SOLUTION   Substituting $T(t) = 70 + 50t$ and $k = .1$ into the differential equation in Example 4, we see that $f(t)$ satisfies

$$y' = .1[70 + 50t - y].$$

Putting the equation in standard form and recalling the initial condition $f(0) = 27$, we obtain the initial-value problem

$$y' + .1y = 5t + 7, \quad y(0) = 27.$$

Multiplying both sides of the equation by the integrating factor $e^{0.1t}$ and combining terms, we obtain

$$\frac{d}{dt}[e^{0.1t}y] = (5t + 7)e^{0.1t}.$$

Integrating both sides, we get

$$e^{0.1t}y = \int (5t + 7)e^{0.1t}\, dt$$

$$= 100e^{0.1t}(.5t + .7 - 5) + C,$$

where we have used equation (1) with $a = 5$, $b = 7$, and $c = .1$. Multiplying both sides by $e^{-0.1t}$, we find

$$y = 100(.5t - 4.3) + Ce^{-0.1t} = 50t - 430 + Ce^{-0.1t}.$$

To satisfy the initial condition, we must have

$$f(0) = 27 = -430 + C.$$

Hence, $C = 457$, and the temperature of the pizza at time t is

$$f(t) = 50t - 430 + 457e^{-0.1t}.$$

After 8 minutes of heating, the temperature of the pizza is

$$f(8) = 50(8) - 430 + 457e^{-0.1(8)} = -30 + 457e^{-0.8} \approx 175°. \quad \text{>> \emph{Now Try Exercise 13}}$$

The modeling techniques of this section have many interesting applications in finance, biology, medicine, and sociology. Some of these applications are presented in the exercises.

Check Your Understanding 10.4

Solutions can be found following the section exercises.

A savings account earns 4% interest per year, compounded continuously, and continuous withdrawals are made from the account at the rate of $1200 per year. Set up a differential equation that is satisfied by the amount $f(t)$ of money in the account at time t.

EXERCISES 10.4

1. **A Retirement Account** Refer to Example 1.
   (a) How fast was the amount in the account growing when it reached $30,000?
   (b) How much money was in the account when it was growing at twice the rate of your annual contribution?
   (c) How long do you have to wait for the money in the account to reach $40,000?

2. Refer to Example 2. Answer questions (a) and (b) in Exercise 1 if the interest rate is 7%. How long will it take to pay off the $25,000 loan in this case?

3. **A Retirement Account** A person planning for her retirement arranges to make continuous deposits into a savings account at the rate of $3600 per year. The savings account earns 5% interest compounded continuously.
   (a) Set up a differential equation that is satisfied by $f(t)$, the amount of money in the account at time t.
   (b) Solve the differential equation in part (a), assuming that $f(0) = 0$, and determine how much money will be in the account at the end of 25 years.

4. **A Savings Account** A person deposits $10,000 in a bank account and decides to make additional deposits at the rate of A dollars per year. The bank compounds interest continuously at the annual rate of 6%, and the deposits are made continuously into the account.
   (a) Set up a differential equation that is satisfied by the amount $f(t)$ in the account at time t.
   (b) Determine $f(t)$ (as a function of A).
   (c) Determine A if the initial deposit is to double in 5 years.

5. **Comparing Growth of Saving Accounts** Twenty years ahead of her retirement, Kelly opened a savings account that earns 5% interest rate compounded continuously, and she contributed to this account at the annual rate of $1200 per year for 20 years. Ten years ahead of his retirement, John opened a similar savings account that earns 5% interest rate compounded continuously and decided to double the annual rate of contribution to $2400 per year for 10 years. Who has more money in his or her savings account at retirement? (Assume that the contributions are made continuously into the accounts.)

6. Answer the question in Exercise 5 if John contributed to his savings account at the annual rate of $3000 per year for 10 years.

7. A person took out a loan of $100,000 from a bank that charges 7.5% interest compounded continuously. What should be the annual rate of payments if the loan is to be paid in full in exactly 10 years? (Assume that the payments are made continuously throughout the year.)

8. **Car Prices in 2012** The National Automobile Dealers Association reported that the average retail selling price of a new vehicle was $30,303 in 2012. A person purchased a new car at the average price and financed the entire amount. Suppose that the person can only afford to pay $500 per month. Assume that the payments are made at a continuous annual rate and that interest is compounded continuously at the rate of 3.5%. (*Source: The National Automobile Dealers Association, www.nada.com.*)
   (a) Set up a differential equation that is satisfied by the amount $f(t)$ of money owed on the car loan at time t.
   (b) How long will it take to pay off the car loan?

9. **New Home Prices in 2012** The Federal Housing Finance Board reported that the national average price of a new one-family house in 2012 was $278,900. At the same time, the average interest rate on a conventional 30-year fixed-rate mortgage was 3.1%. A person purchased a home at the average price, paid a down payment equal to 10% of the purchase price, and financed the remaining balance with a 30-year fixed-rate mortgage. Assume that the person makes payments continuously at a constant annual rate A and that interest is compounded continuously at the rate of 3.1%. (*Source: The Federal Housing Finance Board, www.fhfb.gov.*)
   (a) Set up a differential equation that is satisfied by the amount $f(t)$ of money owed on the mortgage at time t.
   (b) Determine A, the rate of annual payments that is required to pay off the loan in 30 years. What will the monthly payments be?
   (c) Determine the total interest paid during the 30-year term mortgage.

10. Answer parts (a), (b), and (c) of Exercise 9 if the person takes a 15-year fixed-rate mortgage with a 6% interest rate and intends to pay off the entire loan in 15 years.

11. **Elasticity of Demand** Let $q = f(p)$ be the demand function for a certain commodity, where q is the demand quantity and p the price of 1 unit. In Section 5.3, we defined the elasticity of demand as
$$E(p) = \frac{-pf'(p)}{f(p)}.$$
   (a) Find a differential equation satisfied by the demand function if the elasticity of demand is a linear function of price given by $E(p) = p + 1$.
   (b) Find the demand function in part (a), given $f(1) = 100$.

12. Find the demand function if the elasticity of demand is a linear function of price given by $E(p) = ap + b$, where a and b are constants.

13. **Temperature of a Steel Rod** When a red-hot steel rod is plunged in a bath of water that is kept at a constant temperature 10°C, the temperature of the rod at time t, $f(t)$, satisfies the differential equation
$$y' = k[10 - y],$$
where $k > 0$ is a constant of proportionality. Determine $f(t)$ if the initial temperature of the rod is $f(0) = 350$°C and $k = .1$.

14. **Temperature of a Steel Rod** Rework Exercise 13 for a metal with a constant of proportionality $k = .2$. Which rod cools faster, the rod with a constant of proportionality $k = .1$ or the rod with a constant of proportionality $k = .2$? What can you say about the effect of varying the constant of proportionality in a cooling problem?

15. **Determining the Time of Death** A body was found in a room when the room's temperature was 70°F. Let $f(t)$ denote the temperature of the body t hours from the time of death. According to Newton's law of cooling, f satisfies a differential equation of the form
$$y' = k(T - y).$$
   (a) Find T.
   (b) After several measurements of the body's temperature, it was determined that when the temperature of the body was 80 degrees, it was decreasing at the rate of 5 degrees per hour. Find k.
   (c) Suppose that, at the time of death, the body's temperature was about normal, say, 98°F. Determine $f(t)$.
   (d) When the body was discovered, its temperature was 85°F. Determine how long ago the person died.

16. Derive the formula for the population in Example 3, if the population in 1995 was 2 million. (The formula is given following the solution of Example 3.)

17. **A Bacteria Culture** In an experiment, a certain type of bacteria was being added to a culture at the rate of $e^{0.03t} + 2$ thousand bacteria per hour. Suppose that the bacteria grow at a rate proportional to the size of the culture at time t, with constant of proportionality $k = .45$. Let $P(t)$ denote the number of bacteria in the culture at time t. Find a differential equation satisfied by $P(t)$.

18. Find a formula for $P(t)$ in Exercise 17 if, initially, 10,000 bacteria were present in the culture.

19. **Dialysis and Creatinine Clearance** According to the National Kidney Foundation, in 1997 more than 260,000 Americans

suffered from chronic kidney failure and needed an artificial kidney (dialysis) to stay alive. (*Source: The National Kidney Foundation, www.kidney.org.*) When the kidneys fail, toxic waste products such as creatinine and urea build up in the blood. One way to remove these wastes is to use a process known as peritoneal dialysis, in which the patient's peritoneum, or lining of the abdomen, is used as a filter. When the abdominal cavity is filled with a certain dialysate solution, the waste products in the blood filter through the peritoneum into the solution. After a waiting period of several hours, the dialysate solution is drained out of the body along with the waste products.

In one dialysis session, the abdomen of a patient with an elevated concentration of creatinine in the blood equal to 110 grams per liter was filled with two liters of a dialysate (containing no creatinine). Let $f(t)$ denote the concentration of creatinine in the dialysate at time t. The rate of change of $f(t)$ is proportional to the difference between 110 (the maximum concentration that can be attained in the dialysate) and $f(t)$. Thus, $f(t)$ satisfies the differential equation

$$y' = k(110 - y).$$

(a) Suppose that, at the end of a 4-hour dialysis session, the concentration in the dialysate was 75 grams per liter and it was rising at the rate of 10 grams per liter per hour. Find k.

(b) What is the rate of change of the concentration at the beginning of the dialysis session? By comparing with the rate at the end of the session, can you give a (simplistic) justification for draining and replacing the dialysate with a fresh solution after 4 hours of dialysis? [*Hint:* You do not need to solve the differential equation.]

20. **Radioactive Decay** Radium 226 is a radioactive substance with a decay constant .00043. Suppose that radium 226 is being continuously added to an initially empty container at a constant rate of 3 milligrams per year. Let $P(t)$ denote the number of grams of radium 226 remaining in the container after t years.

(a) Find an initial-value problem satisfied by $P(t)$.

(b) Solve the initial-value problem for $P(t)$.

(c) What is the limit of the amount of radium 226 in the container as t tends to infinity?

In Exercises 21–25, solving the differential equations that arise from modeling may require using integration by parts. [See formula (1).]

21. **Growth of a Savings Account** A person deposits an inheritance of $100,000 in a savings account that earns 4% interest compounded continuously. This person intends to make withdrawals that will increase gradually in size with time. Suppose that the annual rate of withdrawals is $2000 + 500t$ dollars per year, t years from the time the account was opened.

(a) Assume that the withdrawals are made at a continuous rate. Set up a differential equation that is satisfied by the amount $f(t)$ in the account at time t.

(b) Determine $f(t)$.

(c) With the help of your calculator, plot $f(t)$ and approximate the time it will take before the account is depleted.

22. **A Savings Account** You make an initial deposit of $500 in a savings account and plan on making future deposits at a gradually increasing annual rate given by $90t + 810$ dollars

per year, t years after the initial deposit. Assume that the deposits are made continuously and that interest is compounded continuously at the rate of 6%. Let $P(t)$ denote the amount of money in the account.

(a) Set up an initial-value problem that is satisfied by $P(t)$.

(b) Find $P(t)$.

23. **A Savings Account with Withdrawals** After depositing an initial amount of $10,000 in a savings account that earns 4% interest compounded continuously, a person continued to make deposits for a certain period of time and then started to make withdrawals from the account. The annual rate of deposits was given by $3000 - 500t$ dollars per year, t years from the time the account was opened. (Here, negative rates of deposits correspond to withdrawals.)

(a) How many years did the person contribute to the account before starting to withdraw money from it?

(b) Let $P(t)$ denote the amount of money in the account, t years after the initial deposit. Find an initial-value problem satisfied by $P(t)$. (Assume that the deposits and withdrawals were made continuously.)

24. Figure 4 contains the solution of the initial-value problem in Exercise 23.

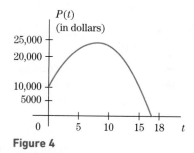

Figure 4

(a) With the help of the graph, approximate how long it will take before the account is depleted.

(b) Solve the initial-value problem to determine $P(t)$.

(c) Use the formula for $P(t)$ to verify your answer in part (a) with the help of a calculator.

25. **Morphine Infusion** Morphine is a drug that is widely used for pain management. However, morphine can cause fatal respiratory arrest. Since pain perception and drug tolerance vary with patients, morphine is gradually administered with small increments until pain is controlled or side effects begin to appear.

In one intravenous infusion, morphine was injected continuously at an increasing rate of t milligrams per hour. Suppose that the body removes the drug at a rate proportional to the amount of the drug present in the body, with constant of proportionality $k = .35$. Let $f(t)$ denote the amount of morphine in the body, t hours from the beginning of the infusion.

(a) Find a differential equation satisfied by $f(t)$.

(b) Assuming that the infusion lasted 8 hours, determine the amount of morphine in the body during the infusion if at the beginning of the infusion the body was free of morphine.

TECHNOLOGY EXERCISES

26. **Therapeutic Level of a Drug** A certain drug is administered intravenously to a patient at the continuous rate of r milligrams per hour. The patient's body removes the drug from the

bloodstream at a rate proportional to the amount of the drug in the blood, with constant of proportionality $k = .5$.

(a) Write a differential equation that is satisfied by the amount $f(t)$ of the drug in the blood at time t (in hours).

(b) Find $f(t)$ assuming that $f(0) = 0$. (Give your answer in terms of r.)

(c) In a therapeutic 2-hour infusion, the amount of drug in the body should reach 1 milligram within 1 hour of

administration and stay above this level for another hour. However, to avoid toxicity, the amount of drug in the body should not exceed 2 milligrams at any time. Plot the graph of $f(t)$ on the interval $1 \le t \le 2$, as r varies between 1 and 2 by increments of $.1$. That is, plot $f(t)$ for $r = 1, 1.1, 1.2, 1.3, \ldots, 2$. By looking at the graphs, pick the values of r that yield a therapeutic and nontoxic 2-hour infusion.

Solution to Check Your Understanding 10.5

We reason as in Example 1. There are two influences on the way the savings account changes: the rate at which interest is added to the account and the rate at which money is withdrawn from the account. We know that interest is being added at a rate proportional to the amount in the account and that withdrawals are made at a rate of $1200 per year. Since the rate of change of $f(t)$ is the net effect of these two influences, we see that $f(t)$ satisfies the first-order differential equation

$$
\begin{array}{ccccc}
y' & = & .04\,y & - & 1200 \\
\begin{bmatrix} \text{rate of} \\ \text{change of } y \end{bmatrix} & = & \begin{bmatrix} \text{rate at which} \\ \text{interest is added} \end{bmatrix} & - & \begin{bmatrix} \text{rate at which} \\ \text{money is withdrawn} \end{bmatrix}.
\end{array}
$$

The standard form of this first-order equation is $y' - .04y = -1200$.

10.5 Graphing Solutions of Differential Equations

In this section, we present a technique for sketching solutions to differential equations of the form $y' = g(y)$ *without having to solve the differential equation.* This technique is based on the geometric interpretation of a differential equation that we introduced in Section 10.1 and used to construct slope fields. The technique is valuable for three reasons. First, for many differential equations explicit solutions cannot be written down. Second, even when an explicit solution is available, we still face the problem of determining its behavior. For example, does the solution increase or decrease? If it increases, does it approach an asymptote or does it grow arbitrarily large? Third, and probably most significant, in many applications the explicit formula for a solution is unnecessary; only a general knowledge of the behavior of the solution is needed. That is, a qualitative understanding of the solution is sufficient.

The theory introduced in this section is part of what is called the *qualitative theory of differential equations.* We limit our attention to differential equations of the form $y' = g(y)$. Such differential equations are called **autonomous**. The term *autonomous* here means "independent of time" and refers to the fact that the right side of $y' = g(y)$ depends only on y and not on t. All applications studied in the next section involve autonomous differential equations.

Throughout this section, we consider the values of each solution $y = f(t)$ only for $t \ge 0$. To introduce the qualitative theory, let us examine the graphs of the various typical solutions of the differential equation $y' = \frac{1}{2}(1 - y)(4 - y)$. The solution curves in Fig. 1 illustrate the following properties.

Property I Corresponding to each zero of $g(y)$, there is a constant solution of the differential equation. Specifically, if $g(c) = 0$, the constant function $y = c$ is a solution. (The constant solutions in Fig. 1 are $y = 1$ and $y = 4$.)

Property II The constant solutions divide the ty-plane into horizontal strips. Each nonconstant solution lies completely in one strip.

Property III Each nonconstant solution is either strictly increasing or decreasing.

Property IV Each nonconstant solution either is asymptotic to a constant solution or else increases or decreases without bound.

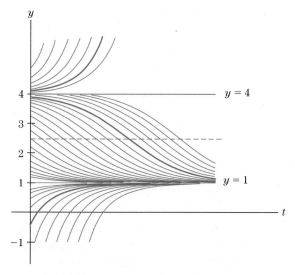

Figure 1 Solutions of $y' = \frac{1}{2}(1 - y)(4 - y)$.

It can be shown that Properties I through IV are valid for the solutions of any autonomous differential equation $y' = g(y)$, provided that $g(y)$ is a "sufficiently well-behaved" function. We shall assume these properties in this chapter.

Using Properties I through IV, we can sketch the general shape of any solution curve by looking at the graph of the function $g(y)$ and the behavior of that graph near $y(0)$. The procedure for doing this is illustrated in the following example.

EXAMPLE 1   **An Autonomous Differential Equation** Sketch the solution to $y' = e^{-y} - 1$ that satisfies $y(0) = -2$.

SOLUTION   Here $g(y) = e^{-y} - 1$. On a yz-coordinate system we draw the graph of the function $z = g(y) = e^{-y} - 1$. [See Fig. 2(a).] The function $g(y) = e^{-y} - 1$ has a zero when $y = 0$. Therefore, the differential equation $y' = e^{-y} - 1$ has the constant solution $y = 0$. We indicate this constant solution on a ty-coordinate system in Fig. 2(b). To begin the sketch of the solution satisfying $y(0) = -2$, we locate this initial value of y on the (horizontal) y-axis in Fig. 2(a) and on the (vertical) y-axis in Fig. 2(b).

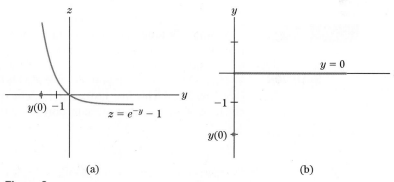

Figure 2

To determine whether the solution increases or decreases when it leaves the initial point $y(0)$ on the ty-graph, we look at the yz-graph and note that $z = g(y)$ is positive at $y = -2$. [See Fig. 3(a).] Consequently, since $y' = g(y)$, the derivative of the solution is positive, which implies that the solution is increasing. We indicate this by an arrow at the initial point in Fig. 3(b). Moreover, the solution y will increase asymptotically to the constant solution $y = 0$, by Properties III and IV of autonomous differential equations.

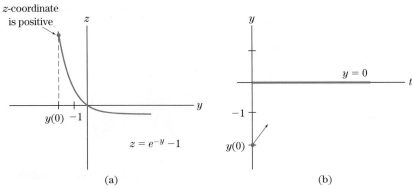

Figure 3

Next, we place an arrow in Fig. 4(a) to remind us that y will move from $y = -2$ toward $y = 0$. As y moves to the right toward $y = 0$ in Fig. 4(a), the z-coordinate of points on the graph of $g(y)$ becomes less positive; that is, $g(y)$ becomes less positive. Consequently, since $y' = g(y)$, the slope of the solution curve becomes less positive. Thus, the solution curve is concave down. [See Fig. 4(b).]   **» Now Try Exercise 7**

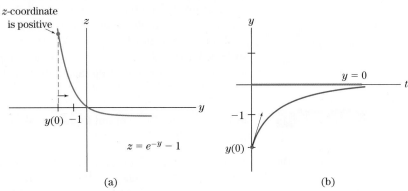

Figure 4

An important point to remember when sketching solutions is that z-coordinates on the yz-graph are values of $g(y)$, and since $y' = g(y)$, a z-coordinate gives the *slope* of the solution curve at the corresponding point on the ty-graph.

EXAMPLE 2   **Graphing a Solution** Sketch the graphs of the solutions to $y' = y + 2$ satisfying
          (a) $y(0) = 1$ **(b)** $y(0) = -3$.

SOLUTION   Here, $g(y) = y + 2$. The graph of $z = g(y)$ is a straight line of slope 1 and z-intercept 2. [See Fig. 5(a).] This line crosses the y-axis only where $y = -2$. Thus, the differential equation $y' = y + 2$ has one constant solution, $y = -2$. [See Fig. 5(b).]

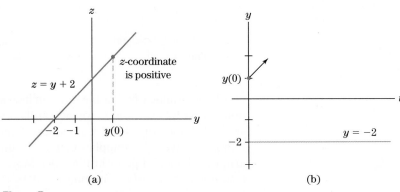

Figure 5

(a) We locate the initial value $y(0) = 1$ on the y-axes of both graphs in Fig. 5. The corresponding z-coordinate on the yz-graph is positive; therefore, the solution on the ty-graph has positive slope and is increasing as it leaves the initial point. We indicate this by an arrow in Fig. 5(b). Now, Property IV of autonomous differential equations implies that y will increase without bound from its initial value. As we let y increase from 1 in Fig. 6(a), we see that the z-coordinates [the values of $g(y)$] increase. Consequently, y' is increasing, so the graph of the solution must be concave up. We have sketched the solution in Fig. 6(b).

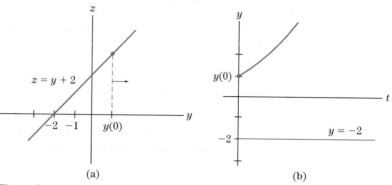

Figure 6

(b) Next, we graph the solution for which $y(0) = -3$. From the graph of $z = y + 2$, we see that z is negative when $y = -3$. This implies that the solution is decreasing as it leaves the initial point. (See Fig. 7.) It follows that the values of y will continue to decrease without bound and become more and more negative. This means that, on the yz-graph, y must move to the *left*. [See Fig. 8(a).] We now examine what happens to $g(y)$ as y moves to the left. (This is the opposite of the ordinary way to read a graph.) The z-coordinate becomes more negative; hence, the slopes on the solution curve will become more negative. Thus, the solution curve must be concave down, as in Fig. 8(b).

≫ Now Try Exercise 7

Figure 7

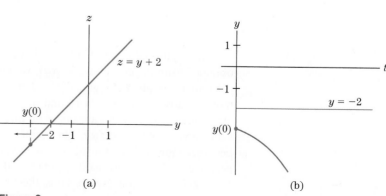

Figure 8

From the preceding examples, we can state a few rules for sketching a solution to $y' = g(y)$ with $y(0)$ given:

1. Sketch the graph of $z = g(y)$ on a yz-coordinate system. Find and label the zeros of $g(y)$.

2. For each zero c of $g(y)$, draw the constant solution $y = c$ on the ty-coordinate system.

3. Plot $y(0)$ on the y-axes of the two coordinate systems.

4. Determine whether the value of $g(y)$ is positive or negative when $y = y(0)$. This tells us whether the solution is increasing or decreasing. On the ty-graph, indicate the direction of the solution through $y(0)$.

5. On the yz-graph, indicate in which direction y should move. (*Note:* If y is moving *down* on the ty-graph, y moves to the *left* on the yz-graph.) As y moves in the proper direction on the yz-graph, determine whether $g(y)$ becomes more positive, less positive, more negative, or less negative. This tells us the concavity of the solution.

6. Beginning at $y(0)$ on the ty-graph, sketch the solution, being guided by the principle that the solution will grow (positively or negatively) without bound unless it encounters a constant solution. In this case, it will approach the constant solution asymptotically.

EXAMPLE 3   **Constant Solutions and Graphing** Sketch the solutions to $y' = y^2 - 4y$ satisfying $y(0) = 4.5$ and $y(0) = 3$.

SOLUTION   Refer to Fig. 9. Since $g(y) = y^2 - 4y = y(y - 4)$, the zeros of $g(y)$ are 0 and 4; hence, the constant solutions are $y = 0$ and $y = 4$. The solution satisfying $y(0) = 4.5$ is increasing, because the z-coordinate is positive when $y = 4.5$ on the yz-graph. This solution continues to increase without bound. The solution satisfying $y(0) = 3$ is decreasing because the z-coordinate is negative when $y = 3$ on the yz-graph. This solution will decrease and approach asymptotically the constant solution $y = 0$.

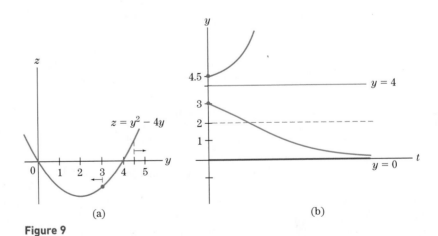

Figure 9

An additional piece of information about the solution satisfying $y(0) = 3$ may be obtained from the graph of $z = g(y)$. We know that y decreases from 3 and approaches 0. From the graph of $z = g(y)$ in Fig. 9, it appears that, at first, the z-coordinates become more negative until y reaches 2 and then become less negative as y moves on toward 0. Since these z-coordinates are slopes on the solution curve, we conclude that, as the solution moves downward from its initial point on the ty-coordinate system, its slope becomes more negative until the y-coordinate is 2, and then, the slope becomes less negative as the y-coordinate approaches 0. Hence, the solution is concave down until $y = 2$ and then is concave up. Thus, there is an inflection point at $y = 2$, where the concavity changes.   **>> Now Try Exercise 9**

We saw in Example 3 that the inflection point at $y = 2$ was produced by the fact that $g(y)$ had a minimum at $y = 2$. A generalization of the argument in Example 3 shows that inflection points of solution curves occur at each value of y where $g(y)$ has a nonzero relative maximum or minimum point. Thus, we may formulate an additional rule for sketching a solution of $y' = g(y)$.

7. On the ty-coordinate system, draw dashed horizontal lines at all values of y at which $g(y)$ has a *nonzero* relative maximum or minimum point. A solution curve will have an inflection point whenever it crosses such a dashed line.

It is useful to note that when $g(y)$ is a quadratic function, as in Example 3, its maximum or minimum point occurs at a value of y halfway between the zeros of $g(y)$. This is because the graph of a quadratic function is a parabola, which is symmetric about a vertical line through its vertex.

EXAMPLE 4   **Concavity of a Solution Curve**   Sketch a solution to $y' = e^{-y}$ with $y(0) > 0$.

SOLUTION   Refer to Fig. 10. Since $g(y) = e^{-y}$ is always positive, there are no constant solutions to the differential equation, and every solution will increase without bound. When drawing solutions that asymptotically approach a horizontal line, we have no choice as to whether to draw it concave up or concave down. This decision will be obvious from its increasing or decreasing nature and from knowledge of inflection points. However, for solutions that grow without bound, we must look at $g(y)$ to determine concavity. In this example, as t increases, the values of y increase. As y increases, $g(y)$ becomes less positive. Since $g(y) = y'$, we deduce that the slope of the solution curve becomes less positive; therefore, the solution curve is concave down.   **» Now Try Exercise 33**

   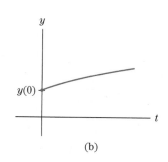

(a)   (b)

Figure 10

Check Your Understanding 10.5

Solutions can be found following the section exercises.

Consider the differential equation $y' = g(y)$, where $g(y)$ is the function whose graph is drawn in Fig. 11.

1. How many constant solutions are there to the differential equation $y' = g(y)$?

2. For what initial values $y(0)$ will the corresponding solution of the differential equation be an increasing function?

3. If the initial value $y(0)$ is near 4, will the corresponding solution be asymptotic to the constant solution $y = 4$?

4. For what initial values $y(0)$ will the corresponding solution of the differential equation have an inflection point?

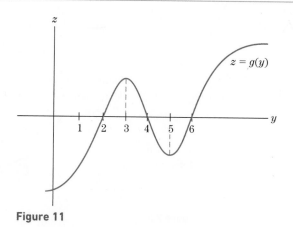

Figure 11

EXERCISES 10.5

Exercises 1–6 review concepts that are important in this section. In each exercise, sketch the graph of a function with the stated properties.

1. Domain: $0 \le t \le 3$; $(0, 1)$ is on the graph; the slope is always positive, and the slope becomes less positive (as t increases).

2. Domain: $0 \le t \le 4$; $(0, 2)$ is on the graph; the slope is always positive, and the slope becomes more positive (as t increases).

3. Domain: $0 \le t \le 5$; $(0, 3)$ is on the graph; the slope is always negative, and the slope becomes less negative.

4. Domain: $0 \le t \le 6$; $(0, 4)$ is on the graph; the slope is always negative, and the slope becomes more negative.

5. Domain: $0 \le t \le 7$; $(0, 2)$ is on the graph; the slope is always positive, the slope becomes more positive as t increases from 0 to 3, and the slope becomes less positive as t increases from 3 to 7.

6. Domain: $0 \le t \le 8$; $(0, 6)$ is on the graph; the slope is always negative, the slope becomes more negative as t increases from 0 to 3, and the slope becomes less negative as t increases from 3 to 8.

One or more initial conditions are given for each differential equation in the following exercises. Use the qualitative theory of autonomous differential equations to sketch the graphs of the corresponding solutions. Include a yz-graph if one is not already provided. Always indicate the constant solutions on the ty-graph whether they are mentioned or not.

7. $y' = 3 - \frac{1}{2}y$, $y(0) = 4$, $y(0) = 8$. [The graph of $z = g(y)$ is drawn in Fig. 12.]

8. $y' = \frac{2}{3}y - 3$, $y(0) = 3$, $y(0) = 6$. (See Fig. 13.)

9. $y' = y^2 - 5$, $y(0) = -4$, $y(0) = 2$, $y(0) = 3$. (See Fig. 14.)

10. $y' = 6 - y^2$, $y(0) = -3$, $y(0) = 3$. (See Fig. 15.)

11. $y' = -\frac{1}{3}(y + 2)(y - 4)$, $y(0) = -3$, $y(0) = -1$, $y(0) = 6$. (See Fig. 16.)

12. $y' = y^2 - 6y + 5$ or $y' = (y - 1)(y - 5)$, $y(0) = -2$, $y(0) = 2$, $y(0) = 4$, $y(0) = 6$. (See Fig. 17.)

13. $y' = y^3 - 9y$ or $y' = y(y^2 - 9)$, $y(0) = -4$, $y(0) = -1$, $y(0) = 2$, $y(0) = 4$. (See Fig. 18.)

14. $y' = 9y - y^3$, $y(0) = -4$, $y(0) = -1$, $y(0) = 2$, $y(0) = 4$. (See Fig. 19.)

15. Use the graph in Fig. 20 to sketch the solutions to the Gompertz growth equation
$$\frac{dy}{dt} = -\frac{1}{10}y \ln \frac{y}{100}$$
satisfying $y(0) = 10$ and $y(0) = 150$.

16. The graph of $z = -\frac{1}{2}y \ln(y/30)$ has the same general shape as the graph in Fig. 20 with relative maximum point at $y \approx 11.0364$ and y-intercept at $y = 30$; $y(0) = 1$, $y(0) = 20$, and $y(0) = 40$. Sketch the solutions to the Gompertz growth equation
$$\frac{dy}{dt} = -\frac{1}{2}y \ln \frac{y}{30}.$$

17. $y' = g(y)$, $y(0) = -.5$, $y(0) = .5$, where $g(y)$ is the function whose graph is given in Fig. 21.

Figure 12

Figure 13

Figure 14

Figure 15

Figure 16

Figure 17

Figure 18

Figure 19

Figure 20

Figure 21

18. $y' = g(y)$, $y(0) = 0$, $y(0) = 4$, where the graph of $g(y)$ is given in Fig. 22.

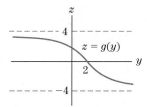

Figure 22

19. $y' = g(y)$, $y(0) = 0$, $y(0) = 1.2$, $y(0) = 5$, $y(0) = 7$, where the graph of $g(y)$ is given in Fig. 23.

Figure 23

20. $y' = g(y)$, $y(0) = 1$, $y(0) = 3$, $y(0) = 11$, where the graph of $g(y)$ is given in Fig. 24.

Figure 24

21. $y' = \frac{3}{4}y - 3$, $y(0) = 2$, $y(0) = 4$, $y(0) = 6$

22. $y' = -\frac{1}{2}y$, $y(0) = -2$, $y(0) = 0$, $y(0) = 2$

23. $y' = 5y - y^2$, $y(0) = 1$, $y(0) = 7$

24. $y' = -y^2 + 10y - 21$, $y(0) = 1$, $y(0) = 4$

25. $y' = y^2 - 3y - 4$, $y(0) = 0$, $y(0) = 3$

26. $y' = \frac{1}{2}y^2 - 3y$, $y(0) = 3$, $y(0) = 6$, $y(0) = 9$

27. $y' = y^2 + 2$, $y(0) = -1$, $y(0) = 1$

28. $y' = y - \frac{1}{4}y^2$, $y(0) = -1$, $y(0) = 1$

29. $y' = \sin y$, $y(0) = -\pi/6$, $y(0) = \pi/6$, $y(0) = 7\pi/4$

30. $y' = 1 + \sin y$, $y(0) = 0$, $y(0) = \pi$

31. $y' = 1/y$, $y(0) = -1$, $y(0) = 1$

32. $y' = y^3$, $y(0) = -1$, $y(0) = 1$

33. $y' = ky^2$, where k is a negative constant, $y(0) = -2$, $y(0) = 2$

34. $y' = ky(M - y)$, where $k > 0$, $M > 10$, and $y(0) = 1$

35. $y' = ky - A$, where k and A are positive constants. Sketch solutions where $0 < y(0) < A/k$ and $y(0) > A/k$.

36. $y' = k(y - A)$, where $k < 0$ and $A > 0$. Sketch solutions where $y(0) < A$ and $y(0) > A$.

37. Growth of a Plant Suppose that, once a sunflower plant has started growing, the rate of growth at any time is proportional to the product of its height and the difference between its height at maturity and its current height. Give a differential equation that is satisfied by $f(t)$, the height at time t, and sketch the solution.

38. Terminal Velocity A parachutist has a terminal velocity of -176 feet per second. That is, no matter how long a person falls, his or her speed will not exceed 176 feet per second, but it will get arbitrarily close to that value. The velocity in feet per second, $v(t)$, after t seconds satisfies the differential equation $v'(t) = 32 - k \cdot v(t)$. What is the value of k?

TECHNOLOGY EXERCISES

39. Draw the graph of $g(x) = (x - 2)^2 (x - 6)^2$, and use the graph to sketch the solutions of the differential equation $y' = (y - 2)^2 (y - 6)^2$ with initial conditions $y(0) = 1$, $y(0) = 3$, $y(0) = 5$, and $y(0) = 7$ on a ty-coordinate system.

40. Draw the graph of $g(x) = e^x - 100x^2 - 1$, and use the graph to sketch the solution of the differential equation $y' = e^y - 100y^2 - 1$ with initial condition $y(0) = 4$ on a ty-coordinate system.

Solutions to Check Your Understanding 10.5

1. Three. The function $g(y)$ has zeros when y is 2, 4, and 6. Therefore, $y' = g(y)$ has the constant functions $y = 2$, $y = 4$, and $y = 6$ as solutions.

2. For $2 < y(0) < 4$ and $y(0) > 6$. Since nonconstant solutions are either strictly increasing or strictly decreasing, a solution is an increasing function, provided that it is increasing at time $t = 0$. This is the case when the first derivative is positive at $t = 0$. When $t = 0$, $y' = g(y(0))$. Therefore, the solution corresponding to $y(0)$ is increasing whenever $g(y(0))$ is positive.

3. Yes. If $y(0)$ is slightly to the right of 4, then $g(y(0))$ is negative, so the corresponding solution will be a decreasing function with values moving to the left closer and closer to 4. If $y(0)$ is slightly to the left of 4, then $g(y(0))$ is positive, so the corresponding solution will be an increasing function with

values moving to the right closer and closer to 4. (The constant solution $y = 4$ is referred to as a *stable constant solution*. The solution with initial value 4 stays at 4, and solutions with initial values near 4 move toward 4. The constant solution $y = 2$ is *unstable*. Solutions with initial values near 2 move away from 2.)

4. For $2 < y(0) < 3$ and $5 < y(0) < 6$. Inflection points of solutions correspond to relative maximum and relative minimum points of the function $g(y)$. If $2 < y(0) < 3$, the corresponding solution will be an increasing function. The values of y will move to the right (toward 4) and therefore will cross 3, a place at which $g(y)$ has a relative maximum point. Similarly, if $5 < y(0) < 6$, the corresponding solution will be decreasing. The values of y on the yz-graph will move to the left and cross 5.

10.6   Applications of Differential Equations

In this section, we study real-life situations that may be modeled by an autonomous differential equation $y' = g(y)$. Here, y will represent some quantity that is changing with time, and the equation $y' = g(y)$ will be obtained from a description of the rate of change of y.

We have already encountered many situations where the rate of change of y is *proportional* to some quantity. For example,

1. $y' = ky$: "The rate of change of y is proportional to y'" (exponential growth or decay).
2. $y' = k(M - y)$: "The rate of change of y is proportional to the difference between M and y'" (Newton's law of cooling, for example).

Both situations involve *linear* first-order differential equations. The following example gives rise to an equation that is not linear. It concerns the rate at which a technological innovation may spread through an industry, a subject of interest to both sociologists and economists.

EXAMPLE 1   **Setting up a Differential Equation**   The by-product coke oven was first introduced into the iron and steel industry in 1894. It took about 30 years before all the major steel producers had adopted this innovation. Let $f(t)$ be the percentage of the producers that had installed the new coke ovens by time t. Then, a reasonable model for the way $f(t)$ increased is given by the assumption that the rate of change of $f(t)$ at time t was proportional to the product of $f(t)$ and the percentage of firms that had not yet installed the new coke ovens at time t. Write a differential equation that is satisfied by $f(t)$. (*Source: Econometrica.*)

SOLUTION   Since $f(t)$ is the *percentage* of firms that have the new coke oven, $100 - f(t)$ is the percentage of firms that still have not installed any new coke ovens. We are told that the rate of change of $f(t)$ is proportional to the product of $f(t)$ and $100 - f(t)$. Hence, there is a constant of proportionality k such that

$$f'(t) = kf(t)[100 - f(t)].$$

Replacing $f(t)$ by y and $f'(t)$ by y', we obtain the desired differential equation,

$$y' = ky(100 - y).$$

Note that both y and $100 - y$ are nonnegative quantities. Clearly, y' must be positive, because $y = f(t)$ is an increasing function. Hence, the constant k must be positive.

» Now Try Exercise 13

The differential equation obtained in Example 1 is a special case of the **logistic differential equation**,

$$y' = ky(a - y), \tag{1}$$

where k and a are positive constants. This equation is used as a simple mathematical model of a wide variety of physical phenomena. In Section 5.4, we described applications of the logistic equation to restricted population growth and to the spread of an epidemic. Let us use the qualitative theory of differential equations to gain more insight into this important equation.

The first step in sketching solutions of (1) is to draw the yz-graph. Rewriting the equation $z = ky(a - y)$ in the form

$$z = -ky^2 + kay,$$

we see that the equation is quadratic in y, and hence, its graph will be a parabola. The parabola is concave down because the coefficient of y^2 is negative (since k is a positive

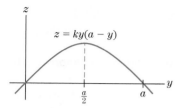

Figure 1 yz-graph for a logistic differential equation.

Figure 2

(a)

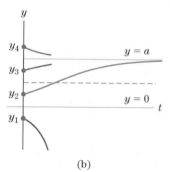

(b)

Figure 3

constant). The zeros of the quadratic expression $ky(a - y)$ occur where $y = 0$ and $y = a$. Since a represents some positive constant, we select an arbitrary point on the positive y-axis and label it "a." With this information we can sketch a representative graph. (See Fig. 1.) Note that the vertex of the parabola occurs at $y = a/2$, halfway between the y-intercepts. (You should review how we obtained this graph, given only that k and a are positive constants. Similar situations will arise in the exercises.)

We begin the ty-graph showing the constant solutions and placing a dashed line at $y = a/2$, where certain solution curves will have an inflection point. (See Fig. 2.) On either side of the constant solutions we choose initial values for y—say, y_1, y_2, y_3, y_4. Then, we use the yz-graph to sketch the corresponding solution curves. (See Fig. 3.)

The solution in Fig. 3(b) beginning at y_2 has the general shape usually referred to as a *logistic curve*. This is the type of solution that would model the situation described in Example 1. The solution in Fig. 3(b) beginning at y_1 usually has no physical significance. The other solutions shown in Fig. 3(b) can occur in practice, particularly in the study of population growth.

In ecology, the growth of a population is often described by a logistic equation written in the form

$$\frac{dN}{dt} = rN\frac{K - N}{K} \tag{2}$$

or, equivalently,

$$\frac{dN}{dt} = \frac{r}{K}N(K - N),$$

where N is used instead of y to denote the size of the population at time t. Typical solutions of this equation are sketched in Fig. 4. The constant K is called the *carrying capacity* of the environment. When the initial population is close to zero, the population curve has the typical S-shaped appearance, and N approaches the carrying capacity asymptotically. When the initial population is greater than K, the population decreases in size, again asymptotically approaching the carrying capacity.

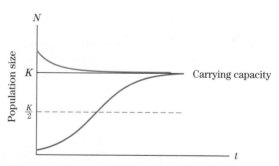

Figure 4 A logistic model for population change.

The quantity $(K - N)/K$ in equation (2) is a fraction between 0 and 1. It reflects the limiting effect of the environment on the population and is close to 1 when N is close to 0. If this fraction were replaced by the constant 1, then (2) would become

$$\frac{dN}{dt} = rN.$$

This is the equation for ordinary exponential growth, where r is the growth rate. For this reason, the parameter r in (2) is called the **intrinsic rate of growth** of the population. It expresses how the population would grow if the environment were to permit unrestricted exponential growth.

We now consider a concrete situation that gives rise to a logistic equation.

EXAMPLE 2   **Fish Population** A pond on a fish farm has a carrying capacity of 1000 fish. The pond was originally stocked with 100 fish. Let $N(t)$ denote the number of fish in the pond after t months.

(a) Set up a logistic differential equation satisfied by $N(t)$, and plot an approximate graph of the fish population.

(b) Find the size of the population of fish with the highest rate of growth. Find this rate, given that the intrinsic rate of growth is .3.

SOLUTION   (a) We are told that the equation is a logistic equation with carrying capacity $K = 1000$. Hence, from (2), the equation is

$$\frac{dN}{dt} = rN\frac{1000 - N}{1000} = \frac{r}{1000}N(1000 - N).$$

The fish population at time t is given by the solution of this differential equation with the initial condition $N(0) = 100$. Even though we do not have a numerical value for the intrinsic rate r, we can still estimate the shape of the solution by using qualitative techniques. First, sketch the constant solutions, $N = 0$ and $N = 1000$; then, place a dashed line at $N = 500$, where certain solutions will have inflection points. The solution starting at $N = 100$ is a typical logistic curve. It is increasing, with a horizontal asymptote $N = 1000$ and inflection point at $N = 500$, where the graph changes concavity. A solution curve satisfying these properties is shown in Fig. 5(b).

(a)   (b)

Figure 5

(b) Since the question concerns the rate of growth, we should look at the equation itself for answers. The equation tells us that the rate of growth is given by the quadratic function

$$\frac{dN}{dt} = \frac{r}{1000}N(1000 - N),$$

whose graph is an inverted parabola with intercepts at $N = 0$ and $N = 1000$. [See Fig. 5(a).] Since the parabola is concave down, it has a maximum at $N = 500$, halfway between 0 and 1000. Thus, the size of the population with the highest rate of growth is 500. To find the numerical value of the fastest growth rate, given that $r = .3$, we substitute $r = .3$ and $N = 500$ into the equation and get

$$\left.\frac{dN}{dt}\right|_{N=500} = \frac{.3}{1000}(500)(1000 - 500) = 75 \text{ fish per month}.$$

This is the maximum rate of growth of the fish population. It is attained when 500 fish are in the pond. Note that 500 is not the maximum size of the population. In fact, we know that the fish population will approach 1000 asymptotically. [See Fig. 5(b).]   **»» Now Try Exercise 17**

We now turn to applications that involve a different sort of autonomous differential equation. The main idea is illustrated in the following familiar example of the savings account that we discussed in Section 10.4.

EXAMPLE 3	**Savings Account with Withdrawals** A savings account earns 6% interest per year, compounded continuously, and continuous withdrawals are made from the account at the rate of $900 per year. Set up a differential equation that is satisfied by the amount $f(t)$ of money in the account at time t. Sketch typical solutions of the differential equation.

SOLUTION   At first, let us ignore the withdrawals from the account. In Section 5.2, we discussed continuous compounding of interest and showed that, if no deposits or withdrawals are made, $f(t)$ satisfies the equation

$$y' = .06y.$$

That is, the savings account grows at a rate proportional to the size of the account. Since this growth comes from the interest, we conclude that *interest is being added to the account at a rate proportional to the amount in the account.*

Now, suppose that continuous withdrawals are made from this same account at the rate of $900 per year. Then, there are two influences on the way the amount of money in the account changes: the rate at which interest is added and the rate at which money is withdrawn. The rate of change of $f(t)$ is the *net effect* of these two influences. That is, $f(t)$ now satisfies the equation

$$y' = .06y - 900$$

$$\begin{bmatrix} \text{rate of} \\ \text{change of } y \end{bmatrix} = \begin{bmatrix} \text{rate at which} \\ \text{interest is added} \end{bmatrix} - \begin{bmatrix} \text{rate at which} \\ \text{money is withdrawn} \end{bmatrix}.$$

The qualitative sketches for this differential equation are given in Fig. 6. We find the constant solution by solving $.06y - 900 = 0$, which gives $y = 900/.06 = 15,000$. If the initial amount $y(0)$ in the account is \$15,000, the balance in the account will always be \$15,000. If the initial amount is greater than \$15,000, the savings account will accumulate money without bound. If the initial amount is less than \$15,000, the account balance will decrease. Presumably, the bank will stop withdrawals when the account balance reaches zero.   **» Now Try Exercise 19**

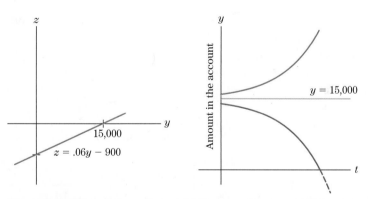

Figure 6 A differential equation model for a savings account: $y' = .06y - 900$.

We may think of the savings account in Example 3 as a compartment or container into which money (interest) is being steadily added and also from which money is being steadily withdrawn. (See Fig. 7.)

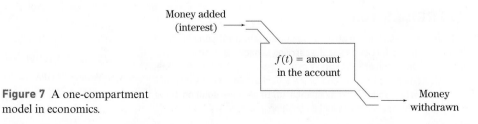

Figure 7 A one-compartment model in economics.

A similar situation arises frequently in physiology in what are called "one-compartment problems." Typical examples of a compartment are a person's lungs, the digestive system, and the cardiovascular system. A common problem is to study the rate at which the amount of some substance in the compartment is changing when two or more processes act on the substance in the compartment. In many important cases, each of these processes changes the substance, either at a constant rate or at a rate proportional to the amount in the compartment. (*Source: Mathematical Techniques for Physiology and Medicine.*)

An earlier example of such a one-compartment problem, discussed in Section 5.4, concerned the continuous infusion of glucose into a patient's bloodstream. A similar situation is discussed in the next example.

EXAMPLE 4   **A One-Compartment Mixing Process** Consider a flask that contains 3 liters of salt water. Suppose that water containing 25 grams per liter of salt is pumped into the flask at the rate of 2 liters per hour, and the mixture, being steadily stirred, is pumped out of the flask at the same rate. Find a differential equation satisfied by the amount of salt $f(t)$ in the flask at time t.

SOLUTION   Let $f(t)$ be the amount of salt measured in grams. Since the volume of the mixture in the flask is being held constant at 3 liters, the concentration of salt in the flask at time t is

$$[\text{concentration}] = \frac{[\text{amount of salt}]}{[\text{volume of mixture}]} = \frac{f(t)\text{ grams}}{3\text{ liters}} = \frac{1}{3}f(t)\frac{\text{grams}}{\text{liter}}.$$

Next, we compute the rates at which salt enters and leaves the flask at time t:

$$[\text{rate of salt entering}] = [\text{entering concentration}] \times [\text{flow rate}]$$

$$= \left[25\frac{\text{grams}}{\text{liter}}\right] \times \left[2\frac{\text{liters}}{\text{hour}}\right]$$

$$= 50\frac{\text{grams}}{\text{hour}}.$$

$$[\text{rate of salt leaving}] = [\text{concentration}] \times [\text{flow rate}]$$

$$= \left[\frac{1}{3}f(t)\frac{\text{grams}}{\text{liter}}\right] \times \left[2\frac{\text{liters}}{\text{hour}}\right]$$

$$= \frac{2}{3}f(t)\frac{\text{grams}}{\text{hour}}.$$

The *net* rate of change of salt (in grams per hour) at time t is $f'(t) = 50 - \frac{2}{3}f(t)$. Hence, the desired differential equation is

$$y' = 50 - \tfrac{2}{3}y.$$

» Now Try Exercise 21

Check Your Understanding 10.6

Solutions can be found following the section exercises.

1. Refer to Example 4, involving the flow of saltwater through a flask. Will $f(t)$ be an increasing or a decreasing function?

2. **Rate of Litter Accumulation** In a certain tropical forest, litter (mainly dead vegetation such as leaves and vines) forms on the ground at the rate of 10 grams per square centimeter per year. At the same time, however, the litter is decomposing at the rate of 80% per year. Let $f(t)$ be the amount of litter (in grams per square centimeter) present at time t. Find a differential equation satisfied by $f(t)$.

EXERCISES 10.6

In Exercises 1–4, you are given a logistic equation with one or more initial conditions. (a) Determine the carrying capacity and intrinsic rate. (b) Sketch the graph of $\frac{dN}{dt}$ versus N in an Nz-plane. (c) In the tN-plane, plot the constant solutions and place a dashed line where the concavity of certain solutions may change. (d) Sketch the solution curve corresponding to each given initial condition.

1. $dN/dt = N(1 - N)$, $N(0) = .75$

2. $dN/dt = .3N(100 - N)$, $N(0) = 25$

3. $dN/dt = -.01N^2 + N$, $N(0) = 5$

4. $dN/dt = -N^2 + N$, $N(0) = .5$

5. Answer part (a) in Example 2 if the pond was originally stocked with 600 fish and all other data are unchanged, how does the graph of the fish population in this case differ from the one in Example 2?

6. Answer parts (a) and (b) in Example 2 if the pond has a carrying capacity of 2000 fish and all other data are unchanged.

7. Social Diffusion For information being spread by mass media, rather than through individual contact, the rate of spread of the information at any time is proportional to the percentage of the population not having the information at that time. Give the differential equation that is satisfied by $y = f(t)$, the percentage of the population having the information at time t. Assume that $f(0) = 1$. Sketch the solution.

8. Gravity At one point in his study of a falling body starting from rest, Galileo conjectured that its velocity at any time is proportional to the distance it has dropped. Using this hypothesis, set up the differential equation whose solution is $y = f(t)$, the distance fallen by time t. By making use of the initial value, show why Galileo's original conjecture is invalid.

9. Autocatalytic Reaction In an autocatalytic reaction, one substance is converted into a second substance in such a way that the second substance catalyzes its own formation. This is the process by which trypsinogen is converted into the enzyme trypsin. The reaction starts only in the presence of some trypsin, and each molecule of trypsinogen yields 1 molecule of trypsin. The rate of formation of trypsin is proportional to the product of the amounts of the two substances present. Set up the differential equation that is satisfied by $y = f(t)$, the amount (number of molecules) of trypsin present at time t. Sketch the solution. For what value of y is the reaction proceeding the fastest? [*Note:* Letting M be the total amount of the two substances, the amount of trypsinogen present at time t is $M - f(t)$.]

10. Drying A porous material dries outdoors at a rate that is proportional to the moisture content. Set up the differential equation whose solution is $y = f(t)$, the amount of water at time t in a towel on a clothesline. Sketch the solution.

11. Movement of Solutes through a Cell Membrane Let c be the concentration of a solute outside a cell that we assume to be constant throughout the process, that is, unaffected by the small influx of the solute across the membrane due to a difference in concentration. The rate of change of the concentration of the solute inside the cell at any time t is proportional to the difference between the outside concentration and the inside concentration. Set up the differential equation whose solution is $y = f(t)$, the concentration of the solute inside the cell at time t. Sketch a solution.

12. Bacteria Growth An experimenter reports that a certain strain of bacteria grows at a rate proportional to the square of the size of the population. Set up a differential equation that describes the growth of the population. Sketch a solution.

13. Chemical Reaction Suppose that substance A is converted into substance B at a rate that, at any time t, is proportional to the square of the amount of A. This situation occurs, for instance, when it is necessary for two molecules of A to collide to create one molecule of B. Set up the differential equation that is satisfied by $y = f(t)$, the amount of substance A at time t. Sketch a solution.

14. War Fever L. F. Richardson proposed the following model to describe the spread of war fever. If $y = f(t)$ is the percentage of the population advocating war at time t, the rate of change of $f(t)$ at any time is proportional to the product of the percentage of the population advocating war and the percentage not advocating war. Set up a differential equation that is satisfied by $y = f(t)$, and sketch a solution. (*Source: Psychometrica.*)

15. Capital Investment Model In economic theory, the following model is used to describe a possible capital investment policy. Let $f(t)$ represent the total invested capital of a company at time t. Additional capital is invested whenever $f(t)$ is below a certain equilibrium value E, and capital is withdrawn whenever $f(t)$ exceeds E. The rate of investment is proportional to the difference between $f(t)$ and E. Construct a differential equation whose solution is $f(t)$, and sketch two or three typical solution curves.

16. Evans Price Adjustment Model Consider a certain commodity that is produced by many companies and purchased by many other firms. Over a relatively short period, there tends to be an equilibrium price p_0 per unit of the commodity that balances the supply and the demand. Suppose that, for some reason, the price is different from the equilibrium price. The Evans price adjustment model says that the rate of change of price with respect to time is proportional to the difference between the actual market price p and the equilibrium price. Write a differential equation that expresses this relation. Sketch two or more solution curves.

17. Fish Population with Harvesting The fish population in a pond with carrying capacity 1000 is modeled by the logistic equation

$$\frac{dN}{dt} = \frac{.4}{1000}N(1000 - N).$$

Here, $N(t)$ denotes the number of fish at time t in years. When the number of fish reached 275, the owner of the pond decided to remove 75 fish per year.

(a) Modify the differential equation to model the population of fish from the time it reached 275.

(b) Plot several solution curves of the new equation, including the solution curve with $N(0) = 275$.

(c) Is the practice of catching 75 fish per year sustainable, or will it deplete the fish population in the pond? Will the size of the fish population ever come close to the carrying capacity of the pond?

18. Continuous Annuity A *continuous annuity* is a steady stream of money that is paid to some person. Such an annuity may be established, for example, by making an initial deposit in a savings account and then making steady withdrawals to pay the continuous annuity. Suppose that an initial deposit of $5400 is made into a savings account that earns $5\frac{1}{2}\%$ interest compounded continuously, and immediately continuous withdrawals are begun at the rate of $300 per year. Set up the differential equation that is satisfied by the amount $f(t)$ of money in the account at time t. Sketch the solution.

19. Savings Account with Deposits A company wishes to set aside funds for future expansion and so arranges to make continuous *deposits* into a savings account at the rate of $10,000 per year. The savings account earns 5% interest compounded continuously.

(a) Set up the differential equation that is satisfied by the amount $f(t)$ of money in the account at time t.

(b) Solve the differential equation in part (a), assuming that $f(0) = 0$, and determine how much money will be in the account at the end of 5 years.

20. Savings Account A company arranges to make continuous deposits into a savings account at the rate of P dollars per year. The savings account earns 5% interest compounded continuously. Find the approximate value of P that will make the savings account balance amount to $50,000 in 4 years.

21. Amount of CO_2 in a Room The air in a crowded room full of people contains .25% carbon dioxide (CO_2). An air conditioner is turned on that blows fresh air into the room at the rate of 500 cubic feet per minute. The fresh air mixes with the stale air, and the mixture leaves the room at the rate of 500 cubic feet per minute. The fresh air contains .01% CO_2, and the room has a volume of 2500 cubic feet.

(a) Find a differential equation satisfied by the amount $f(t)$ of CO_2 in the room at time t.

(b) The model developed in part (a) ignores the CO_2 produced by the respiration of the people in the room. Suppose that the people generate .08 cubic foot of CO_2 per minute. Modify the differential equation in part (a) to take into account this additional source of CO_2.

22. Elimination of a Drug from the Bloodstream A certain drug is administered intravenously to a patient at the continuous rate of 5 milligrams per hour. The patient's body removes the drug from the bloodstream at a rate proportional to the amount of the drug in the blood. Write a differential equation that is satisfied by the amount $f(t)$ of the drug in the blood at time t. Sketch a typical solution.

23. Elimination of a Drug A single dose of iodine is injected intravenously into a patient. The iodine mixes thoroughly in the blood before any is lost as a result of metabolic processes (ignore the time required for this mixing process). Iodine will leave the blood and enter the thyroid gland at a rate proportional to the amount of iodine in the blood. Also, iodine will leave the blood and pass into the urine at a (different) rate proportional to the amount of iodine in the blood. Suppose that the iodine enters the thyroid at the rate of 4% per hour, and the iodine enters the urine at the rate of 10% per hour. Let $f(t)$ denote the amount of iodine in the blood at time t. Write a differential equation satisfied by $f(t)$.

24. Litter in a Forest Show that the mathematical model in Check Your Understanding 2 predicts that the amount of litter in the forest will eventually stabilize. What is the "equilibrium level" of litter in that problem? [*Note:* Today, most forests are close to their equilibrium levels. This was not so during the Carboniferous period when the great coal deposits were formed.]

25. Population Model In the study of the effect of natural selection on a population, we encounter the differential equation

$$\frac{dq}{dt} = -.0001q^2(1 - q),$$

where q is the frequency of a gene a and the selection pressure is against the recessive genotype aa. Sketch a solution of this equation when $q(0)$ is close to but slightly less than 1.

Solutions to Check Your Understanding 10.6

1. The nature of the function $f(t)$ depends on the initial amount of saltwater in the flask. Fig. 8 contains solutions for three different initial amounts, $y(0)$. If the initial amount is less than 75 grams, the amount of salt in the flask will increase asymptotically to 75. If the initial concentration is greater than 75 grams, the amount of salt in the flask will decrease asymptotically to 75. Of course, if the initial concentration is exactly 75 grams, the amount of salt in the flask will remain constant.

2. This problem resembles a one-compartment problem, where the forest floor is the compartment. We have

$$\begin{bmatrix} \text{rate of change} \\ \text{of litter} \end{bmatrix} = \begin{bmatrix} \text{rate of litter} \\ \text{formation} \end{bmatrix} - \begin{bmatrix} \text{rate of litter} \\ \text{decomposition} \end{bmatrix}.$$

If $f(t)$ is the amount of litter (in grams per square centimeter) at time t, the 80% decomposition rate means that at time t the litter is decaying at the rate of $.80f(t)$ grams per square centimeter per year. Thus, the net rate of change of litter is $f'(t) = 10 - .80f(t)$. The desired differential equation is $y' = 10 - .80y$.

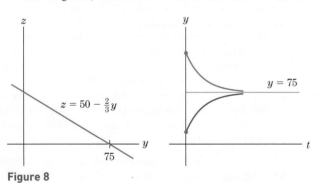

Figure 8

10.7 Numerical Solution of Differential Equations

Many differential equations that arise in real-life applications cannot be solved by *any* known method. However, approximate solutions may be obtained by several different numerical techniques. In this section we describe what is known as **Euler's method** for approximating solutions to initial-value problems of the form

$$y' = g(t, y), \qquad y(a) = y_0 \tag{1}$$

for values of t in some interval $a \le t \le b$ Here, $g(t, y)$ is some reasonably well-behaved function of two variables. Equations of the form $y' = p(t)q(y)$ (studied in Section 10.2) and linear equations of the form $y' = -a(t)y + b(t)$ (studied in Sections 10.3 and 10.4) are special cases of equation (1).

In the following discussion we assume that $f(t)$ is a solution of (1) for $a \le t \le b$. The basic idea on which Euler's method rests is the following: *If the graph of $y = f(t)$ passes through some given point (t, y), the slope of the graph (that is, the value of y') at that point is just $g(t, y)$, because $y' = g(t, y)$.* This is the same idea that we used in Section 10.1 in our discussion of slope fields. Euler's method uses this observation to approximate the graph of $f(t)$ by a polygonal path. (See Fig. 1.)

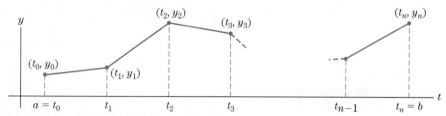

Figure 1 A polygonal path.

The t-axis from a to b is subdivided by the equally spaced points $t_0, t_1, \ldots, t_n$. Each of the n subinterval has length $h = (b - a)/n$. The initial condition $y(a) = y_0$ in (1) implies that the graph of the solution $f(t)$ passes through the point (t_0, y_0). As noted previously, the slope of this graph at (t_0, y_0) must be $g(t_0, y_0)$. Thus, on the first subinterval, Euler's method approximates the graph of $f(t)$ by the straight line

$$y = y_0 + g(t_0, y_0) \cdot (t - t_0),$$

which passes through (t_0, y_0) and has slope $g(t_0, y_0)$. (See Fig. 2.) When $t = t_1$, the y-coordinate on this line is

$$y_1 = y_0 + g(t_0, y_0) \cdot (t_1 - t_0) = y_0 + g(t_0, y_0) \cdot h.$$

Since the graph of $f(t)$ is close to the point (t_1, y_1) on the line, the slope of the graph of $f(t)$ when $t = t_1$ will be close to $g(t_1, y_1)$. So we draw the straight line

$$y = y_1 + g(t_1, y_1) \cdot (t - t_1) \tag{2}$$

through (t_1, y_1) with slope $g(t_1, y_1)$, and we use this line to approximate $f(t)$ on the second subinterval. From (2) we determine an estimate y_2 for the value of $f(t)$ at $t = t_2$:

$$y_2 = y_1 + g(t_1, y_1) \cdot h.$$

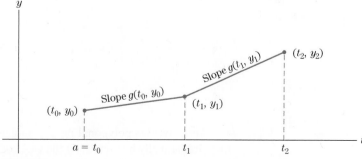

Figure 2

The slope of the graph of $f(t)$ at t_2 is now estimated by $g(t_2, y_2)$, and so on. Let us summarize this procedure:

Euler's Method The endpoints $(t_0, y_0), \ldots, (t_n, y_n)$ of the line segments approximating the solution of equation (1) on the interval $a \leq t \leq b$ are given by the following formulas, where $h = (b - a)/n$:

$$t_0 = a \text{ (given)}, \qquad y_0 \text{ (given)}$$
$$t_1 = t_0 + h, \qquad y_1 = y_0 + g(t_0, y_0) \cdot h,$$
$$t_2 = t_1 + h, \qquad y_2 = y_1 + g(t_1, y_1) \cdot h,$$
$$\vdots \qquad\qquad \vdots$$
$$t_n = t_{n-1} + h. \qquad y_n = y_{n-1} + g(t_{n-1}, y_{n-1}) \cdot h.$$

EXAMPLE 1

Euler's Method Use Euler's method with $n = 4$ to approximate the solution $f(t)$ to $y' = 2t - 3y$, $y(0) = 4$, for t in the interval $0 \leq t \leq 2$. In particular, estimate $f(2)$.

SOLUTION

Here $g(t, y) = 2t - 3y$, $a = 0$, $b = 2$, $y_0 = 4$, and $h = (2 - 0)/4 = \frac{1}{2}$. Starting with $(t_0, y_0) = (0, 4)$, we find that $g(0, 4) = -12$. Thus,

$$t_1 = \frac{1}{2}, \qquad y_1 = 4 + (-12) \cdot \frac{1}{2} = -2.$$

Next, $g\left(\frac{1}{2}, -2\right) = 7$, so

$$t_2 = 1, \qquad y_2 = -2 + 7 \cdot \frac{1}{2} = \frac{3}{2}.$$

Next, $g\left(1, \frac{3}{2}\right) = -\frac{5}{2}$, so

$$t_3 = \frac{3}{2}, \qquad y_3 = \frac{3}{2} + \left(-\frac{5}{2}\right) \cdot \frac{1}{2} = \frac{1}{4}.$$

Finally, $g\left(\frac{3}{2}, \frac{1}{4}\right) = \frac{9}{4}$, so

$$t_4 = 2, \qquad y_4 = \frac{1}{4} + \frac{9}{4} \cdot \frac{1}{2} = \frac{11}{8}.$$

Thus, the approximation to the solution $f(t)$ is given by the polygonal path shown in Fig. 3. The last point $\left(2, \frac{11}{8}\right)$ is close to the graph of $f(t)$ at $t = 2$, so $f(2) \approx \frac{11}{8}$.

» Now Try Exercise 5

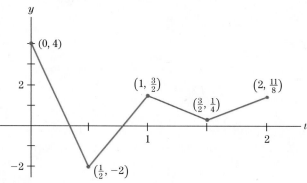

Figure 3

Actually, this polygonal path is somewhat misleading. We can improve the accuracy dramatically by increasing the value of n. Figure 4 shows the Euler approximations for $n = 8$ and $n = 20$. The graph of the exact solution is shown for comparison.

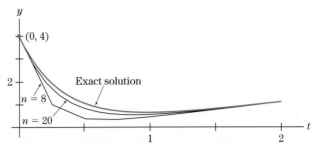

Figure 4 Approximating an exact solution by polygonal paths.

For many purposes, we can obtain satisfactory graphs by running Euler's method on a computer with large values of n. There is a limit to the accuracy obtainable, however, because each computer calculation involves a slight round-off error. When n is very large, the cumulative round-off error can become significant.

INCORPORATING

TECHNOLOGY

Euler's Method Here, we demonstrate how to implement Euler's method on a TI-83/84 to approximate the solution to the differential equation from Example 1. In particular, we will approximate the solution to $y' = 2t - 3y$ with $y(0) = 4$ for t in the interval $0 \le t \le 2$. As indicated in Example 1, the accuracy of the approximation obtained from Euler's method can be improved with large values of n, and for this, a computer or calculator is necessary. In this example, we will implement Euler's method with $n = 100$.

The method we are demonstrating requires that the calculator be in sequence mode. To invoke the sequence mode, press MODE, move the cursor down to the fourth line, move the cursor right to **Seq**, and press ENTER. Now, press Y= to obtain the sequence editor.

With the calculator in sequence mode, the values for $t_0, t_1, t_2, \ldots$ are stored as the sequence values $u(0), u(1), u(2), \ldots$, and the values for $y_0, y_1, y_2, \ldots$ are stored as the sequence values $v(0), v(1), v(2), \ldots$.

We begin counting our sequences with $n = 0$, so cursor up and set nMin $= 0$.

Recall that, in Euler's method, we obtain each successive value of t by adding the step size h to the previous value of t. To implement this, we set $u(n) = u(n - 1) + 0.02$. [In Seq mode, pressing 2nd [u] (the second function of the 7 key) generates u, and pressing X,T,Θ,n generates n. In this example, $h = (2 - 0)/100 = 0.02$.]

In our example, we have $t_0 = 0$, so we set $u(\text{nMin}) = 0$.

In Euler's method, the formula for computing successive values of the dependent variable is $y_n = y_{n-1} + g(t_{n-1}, y_{n-1})h$, and in our example, $g(t, y) = 2t - 3y$. To implement this, we set $v(n) = v(n - 1) + (2u(n - 1) - 3v(n - 1))0.02$. [In Seq mode, pressing 2nd [v] (the second function of the 8 key) generates v.]

In our example, we have $y_0 = 4$, so we set $v(\text{nMin}) = 4$.

Figure 5

We are now nearly ready to graph the approximation, but first, we must set up the calculator to display the graph properly. To begin, we set up the calculator to graph u on the horizontal axis and v on the vertical axis. Press 2nd [FORMAT], move on the first row to **uv**, and press ENTER.

Now press WINDOW and set nMin $= 0$, nMax $= 100$. Our variable t is in [0, 2], so we set Xmin $= 0$ and Xmax $= 2$. Finally, set Ymin $= 0$ and Ymax $= 4$, and leave Xscl, Yscl, PlotStart, and PlotStep set at their default values of 1.

Now, to display a graph of the solution, press GRAPH. (See Fig. 5.)

To display a table of the points on the solution given by Euler's method, first press 2nd [TBLSET], and set TblStart $= 0$, ΔTbl $= 1$, and the other items to Auto. Then, press 2nd [TABLE]. The successive values of t and y are contained in the $u(n)$ and $v(n)$ columns, respectively. (See Fig. 6.)

Note: After using Euler's method, reset the calculator to function mode by pressing MODE, moving to **Func** in the fourth line, and pressing ENTER.

NORMAL FLOAT AUTO REAL RADIAN MP		
PRESS + FOR ΔTbl		
n	$u(n)$	$v(n)$
0	0	4
1	.02	3.76
2	.04	3.5352
3	.06	3.3247
4	.08	3.1276
5	.1	2.9432
6	.12	2.7706
7	.14	2.6091
8	.16	2.4582
9	.18	2.3171
10	.2	2.1853
$n=0$		

Figure 6

Check Your Understanding 10.7

Solutions can be found following the section exercises.

Let $f(t)$ be the solution of $y' = \sqrt{ty}$, $y(1) = 4$.

1. Use Euler's method with $n = 2$ on the interval $1 \le t \le 2$ to estimate $f(2)$.

2. Draw the polygonal path corresponding to the application of Euler's method in Problem 1.

EXERCISES 10.7

1. Suppose that $f(t)$ is a solution of the differential equation $y' = ty - 5$ and the graph of $f(t)$ passes through the point $(2, 4)$. What is the slope of the graph at this point?

2. Suppose that $f(t)$ is a solution of $y' = t^2 - y^2$ and the graph of $f(t)$ passes through the point $(2, 3)$. Find the slope of the graph when $t = 2$.

3. Suppose that $f(t)$ satisfies the initial-value problem $y' = y^2 + ty - 7$, $y(0) = 3$. Is $f(t)$ increasing or decreasing at $t = 0$?

4. Suppose that $f(t)$ satisfies the initial-value problem $y' = y^2 + ty - 7$, $y(0) = 2$. Is the graph of $f(t)$ increasing or decreasing at $t = 0$?

5. Use Euler's method with $n = 2$ on the interval $0 \le t \le 1$ to approximate the solution $f(t)$ to $y' = t^2 y$, $y(0) = -2$. In particular, estimate $f(1)$.

6. Use Euler's method with $n = 2$ on the interval $2 \le t \le 3$ to approximate the solution $f(t)$ to $y' = t - 2y$, $y(2) = 3$. Estimate $f(3)$.

7. Use Euler's method with $n = 4$ to approximate the solution $f(t)$ to $y' = 2t - y + 1$, $y(0) = 5$ for $0 \le t \le 2$. Estimate $f(2)$.

8. Let $f(t)$ be the solution of $y' = y(2t - 1)$, $y(0) = 8$. Use Euler's method with $n = 4$ to estimate $f(1)$.

9. Let $f(t)$ be the solution of $y' = -(t + 1)y^2$, $y(0) = 1$. Use Euler's method with $n = 5$ to estimate $f(1)$. Then, solve the differential equation, find an explicit formula for $f(t)$, and compute $f(1)$. How accurate is the estimated value of $f(1)$?

10. Let $f(t)$ be the solution of $y' = 10 - y$, $y(0) = 1$. Use Euler's method with $n = 5$ to estimate $f(1)$. Then, solve the differential equation and find the exact value of $f(1)$.

11. Suppose that the Consumer Products Safety Commission issues new regulations that affect the toy-manufacturing industry. Every toy manufacturer will have to make certain changes in its manufacturing process. Let $f(t)$ be the fraction of manufacturers that have complied with the regulations within t months. Note that $0 \le f(t) \le 1$. Suppose that the rate at which new companies comply with the regulations is proportional to the fraction of companies who have not yet complied, with constant of proportionality $k = .1$.
   (a) Construct a differential equation satisfied by $f(t)$.
   (b) Use Euler's method with $n = 3$ to estimate the fraction of companies that comply with the regulations within the first 3 months.
   (c) Solve the differential equation in part (a) and compute $f(3)$.
   (d) Compare the answers in parts (b) and (c) and approximate the error in using Euler's method.

12. **Rate of Evaporation** The Los Angeles Zoo plans to transport a California sea lion to the San Diego Zoo. The animal will be wrapped in a wet blanket during the trip. At any time t, the blanket will lose water (due to evaporation) at a rate proportional to the amount $f(t)$ of water in the blanket, with constant of proportionality $k = -.3$. Initially, the blanket will contain 2 gallons of seawater.
   (a) Set up the differential equation satisfied by $f(t)$.
   (b) Use Euler's method with $n = 2$ to estimate the amount of moisture in the blanket after 1 hour.
   (c) Solve the differential equation in part (a) and compute $f(1)$.
   (d) Compare the answers in parts (b) and (c) and approximate the error in using Euler's method.

TECHNOLOGY EXERCISES

13. The differential equation $y' = .5(1 - y)(4 - y)$ has five types of solutions labeled A–E. For each of the following initial values, graph the solution of the differential equation and identify the type of solution. Use a small value of h, let t range from 0 to 4, and let y range from -1 to 5. Use the technique of Section 10.6 to verify your answer.
   (a) $y(0) = -1$ (b) $y(0) = 1$ (c) $y(0) = 2$
   (d) $y(0) = 3.9$ (e) $y(0) = 4.1$

   A. Constant solution.
   B. Decreasing, has an inflection point, and asymptotic to the line $y = 1$.
   C. Increasing, concave down, and asymptotic to the line $y = 1$.
   D. Concave up and increasing indefinitely.
   E. Decreasing, concave up, and asymptotic to the line $y = 1$.

14. The differential equation $y' = .5(y - 1)(4 - y)$ has five types of solutions labeled A–E. For each of the following initial values, graph the solution of the differential equation and identify the type of solution. Use a small value of h, let t range from 0 to 4, and let y range from -1 to 5. Use the technique of Section 10.6 to verify your answer.
   (a) $y(0) = .9$ (b) $y(0) = 1.1$ (c) $y(0) = 3$
   (d) $y(0) = 4$ (e) $y(0) = 5$

   A. Constant solution.
   B. Decreasing, concave up, and asymptotic to the line $y = 4$.
   C. Increasing, has an inflection point, and asymptotic to the line $y = 4$.
   D. Increasing, concave down, and asymptotic to the line $y = 4$.
   E. Concave down and decreasing indefinitely.

15. The differential equation $y' = e^t - 2y$, $y(0) = 1$, has solution $y = \frac{1}{3}(2e^{-2t} + e^t)$. In the following table, fill in the second row with the values obtained from the use of a numerical method and the third row with the actual values calculated from the solution. What is the greatest difference between corresponding values in the second and third rows?

t_i	0	.25	.50	.75	1	1.25	1.5	1.75	2
y_i	1								
y	1	.8324							2.4752

16. The differential equation $y' = 2ty + e^{t^2}$, $y(0) = 5$, has solution $y = (t + 5)e^{t^2}$. In the following table, fill in the second row with the values obtained from the use of a numerical method and the third row with the actual values calculated from the solution. What is the greatest difference between corresponding values in the second and third rows?

t_i	0	.2	.4	.6	.8	1	1.2	1.4	1.6	1.8	2
y_i	5										
y	5	5.412									382.2

Solutions to Check Your Understanding 10.7

1. Here $g(t, y) = \sqrt{ty}$, $a = 1$, $b = 2$, $y_0 = 4$, and $h = (2 - 1)/2 = \frac{1}{2}$. We have

$$t_0 = 1, \quad y_0 = 4, \qquad g(1, 4) = \sqrt{1 \cdot 4}$$
$$= 2,$$

$$t_1 = \tfrac{3}{2}, \quad y_1 = 4 + 2\left(\tfrac{1}{2}\right) = 5, \quad g(\tfrac{3}{2}, 5) = \sqrt{\tfrac{3}{2} \cdot 5}$$
$$\approx 2.7386,$$

$$t_2 = 2, \quad y_2 = 5 + (2.7386)\left(\tfrac{1}{2}\right) = 6.3693.$$

Hence, $f(2) \approx y_2 = 6.3693$. [In Practice Problems 10.2, we found the solution of $y' = \sqrt{ty}$, $y(1) = 4$, to be $f(t) = \left(\tfrac{1}{3}t^{3/2} + \tfrac{5}{3}\right)^2$. We find that $f(2) = 6.8094$ (to four decimal places). The error $6.8094 - 6.3693 = .4401$ in the preceding approximation is about 6.5%.]

2. To find the polygonal path, plot the points (t_0, y_0), (t_1, y_1), and (t_2, y_2) and join them by line segments. See Fig. 7.

Figure 7

CHAPTER 10 Summary

KEY TERMS AND CONCEPTS	EXAMPLES

10.1 Solutions of Differential Equations

A differential equation is an equation involving an unknown function y and one or more of the derivatives y', y'', y''', and so on. A **solution** of a differential equation is any function $f(t)$ such that the differential equation becomes a true statement when y is replaced by $f(t)$, y' by $f'(t)$, y'' by $f''(t)$, and so forth. The problem of determining a solution of a differential equation that satisfies given initial conditions is called an **initial-value problem**.

The *order* of a differential equation is the order of the highest derivative that appears in the equation.

Suppose that (t_1, y_1) is a point in the ty-plane on the graph of a solution curve of the differential equation $y' = g(t, y)$. The equation tells us that, at this point, the derivative y' is equal to $y' = g(t_1, y_1)$. Using this information, we can draw a small line segment through the point (t_1, y_1), with slope $y' = g(t_1, y_1)$, that represents a small portion of the tangent line to the solution curve through the point (t_1, y_1). If we repeat this process at many points in the ty-plane by drawing small line segments at each point (t, y) with slope $y' = g(t, y)$, we obtain a *slope field* of the differential equation $y' = g(t, y)$. Since each line segment represents a small portion of a tangent line to a solution curve, we can use the slope field to visualize solution curves by following the flow of the tangent line segments.

1. Show that the function $y = t^2 + 2t$ is a solution of the initial value problem $ty' - 2y = -2t$, $y(0) = 0$.

Solution If $y = t^2 + 2t$, then $y' = 2t + 2$ and so

$$ty' - 2y = t(2t + 2) - 2(t^2 + 2t) = -2t,$$

which shows that $y = t^2 + 2t$ does satisfy the differential equation. Also, $y(0) = 0^2 + 2(0) = 0$, and so the initial condition is satisfied. Hence, $y = t^2 + 2t$ is a solution of the initial value problem.

2. The slope field of the equation $y' = 2y(1 - y)$ is shown in Figure 1. Draw a solution curve through the point $(0, \frac{1}{2})$.

Solution Start at the point $(0, \frac{1}{2})$ and trace a curve that is tangent to the line segments in the slope field. The resulting curve is shown in Fig. 1. It has the line $y = 1$ as an asymptote.

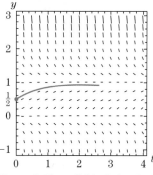

Figure 1 Slope field of $y' = 2y(1 - y)$.

KEY TERMS AND CONCEPTS	EXAMPLES

10.2 Separation of Variables

The separation-of-variables method is used to solve an important class of differential equations of the form $\frac{dy}{dt} = p(t)q(y)$, where $p(t)$ is a function of t only and $q(y)$ is a function of y only. To solve the equation, we first separate the variables and multiply by dt both sides of the equation to get

$$\frac{1}{q(y)}\frac{dy}{dt}\,dt = p(t)\,dt.$$

We may solve this equation by integrating both sides of the equal sign, as illustrated by the example.

Solve the differential equation $y' = (t + 1)e^y$ using the method of separation of variables.

Solution

$$\frac{dy}{dt} = (t + 1)e^y \qquad\qquad \left[\text{Write } y' \text{ as } \frac{dy}{dt}.\right]$$

$$e^{-y}\frac{dy}{dt} = (t + 1) \qquad [\text{Multiply both sides by } e^{-y}.]$$

$$\int e^{-y}\frac{dy}{dt}\,dt = \int (t + 1)\,dt$$

$$[\text{Integrate both sides with respect to } t.]$$

$$\int e^{-y}\,dy = \frac{1}{2}t^2 + t + C$$

$$[\text{Rewrite the left-hand side, "canceling the } dt."]$$

$$-e^{-y} = \frac{1}{2}t^2 + t + C \qquad [\text{Integrate the right side.}]$$

$$e^{-y} = -\left(\frac{1}{2}t^2 + t + C\right)$$

To solve for y in terms of t, we apply the logarithm to both sides of the equation and get

$$\ln(e^{-y}) = \ln\left(-\left(\frac{1}{2}t^2 + t + C\right)\right)$$

$$-y = \ln\left(-\left(\frac{1}{2}t^2 + t + C\right)\right)$$

$$y = -\ln\left(-\left(\frac{1}{2}t^2 + t + C\right)\right).$$

10.3 First-Order Linear Differential Equations

The equation $y' + a(t)y = b(t)$ is called a *first-order linear differential equation in standard form*. Its general solution is given by the formula

$$y = e^{-A(t)}\left[\int e^{A(t)}b(t)\,dt + C\right], \quad C \text{ a constant.}$$

where $A(t)$ is an antiderivative of $a(t)$; that is, $A(t) = \int a(t)\,dt$, where it is customary to choose the constant of integration equal to 0. The expression $e^{A(t)}$ is called an *integrating factor*.

Solve the differential equation $y' + 2ty = 4t$.

Solution The equation is a first-order linear differential equation in standard form, with $a(t) = 2t$ and $b(t) = 4t$. Applying the outlined method for solving such equations, we have $A(t) = \int a(t)\,dt = \int 2t\,dt = t^2$, the integrating factor is $e^{A(t)} = e^{t^2}$, and the solution is

$$y = e^{-A(t)}\left[\int e^{A(t)}b(t)\,dt + C\right]$$

$$= e^{-t^2}\left[\int e^{t^2}4t\,dt + C\right]$$

$$= e^{-t^2}\left[2\int e^u\,du + C\right]$$

$$[\text{In the integral, use the substitution } u = t^2, du = 2t\,dt.]$$

$$= e^{-t^2}[2e^u + C] = e^{-t^2}[2e^{t^2} + C]$$

$$= 2 + Ce^{-t^2}.$$

It is straightforward to verify that $y = 2 + Ce^{-t^2}$ is indeed a solution of $y' + 2ty = 4t$. Moreover, every solution of the equation is of this form, and so, $y = 2 + Ce^{-t^2}$ is the *general* solution of the equation.

KEY TERMS AND CONCEPTS	EXAMPLES

10.4 Applications of First-Order Linear Differential Equations

Many interesting applications in economics, business and finance, population dynamics, and physics are modeled by first-order linear differential equations. In this section, we considered a selection of such applications that illustrates the important ideas behind modeling with differential equations and solving applied problems.

You opened a retirement account, with an initial deposit of $1000, that pays 6% interest per year, and planned on making future deposits at the rate of $100 per month. Assume that the deposits are made continuously and that interest is compounded continuously. Let $y = P(t)$ denote the amount of money in the account t years after the initial deposit.

(a) Set up an initial value problem that is satisfied by $P(t)$.

(b) Find $P(t)$.

Solution

(a) The amount of money in the account changes under two influences: Interest is being added to the account at a rate proportional to y, with constant of proportionality $k = .06$ or 6%; and deposits are made at the rate of $100 per month or $1200 per year. So, the interest is being added at a rate equal to $.06y$. The rate of change of y is the sum of the rates of change due to interest and deposits. Thus, $y' = .06y + 1200$ or, in standard form, $y' - .06y = 1200$. Since the initial deposit in the account was $1000, we have $y(0) = 1000$. Thus, $P(t)$ satisfies the initial value problem

$$y' - .06y = 1200, \qquad y(0) = 1000.$$

(b) To find y, we appeal to the general solution of the first-order linear differential equation with $a(t) = -.06$ and $b(t) = 1200$. Then, $A(t) = \int a(t)dt = -.06t$,

$$y = e^{-A(t)}\left[\int e^{A(t)}b(t)\,dt + C\right]$$

$$= e^{0.06t}\left[\int e^{-0.06t}1200\,dt + C\right]$$

$$= e^{0.06t}\left[\frac{1200}{-.06}e^{-0.06t} + C\right]$$

$$= -20{,}000 + Ce^{0.06t}.$$

Setting $t = 0$ and using the initial condition $y(0) = 1000$, we find $1000 = -20{,}000 + C$; so, $C = 21{,}000$, and $y = -20{,}000 + 21{,}000e^{0.06t}$.

10.5 Graphing Solutions of Differential Equations

This section and the next deal with an area known as the **qualitative theory of differential equations**. Its purpose is to study the solution curves of a differential equation without having to solve the equation explicitly. The theory applies to a class of differential equations of the form $y' = g(y)$, known as *autonomous differential equations*. The study is based on the following properties of autonomous equations.

Property I Corresponding to each zero of $g(y)$, there is a constant solution of the differential equation. Specifically, if $g(c) = 0$, the constant function $y = c$ is a solution. (The constant solutions in Fig. 2 are $y = 1$ and $y = 3$.)

Property II The constant solutions divide the ty-plane into horizontal strips. Each nonconstant solution lies completely in one strip.

Property III Each nonconstant solution is either strictly increasing or decreasing.

Sketch the solutions to $y' = (y - 1)(y - 3)$ satisfying $y(0) = 4$ and $y(0) = 2$.

Solution Let $z = g(y) = (y - 1)(y - 3)$ (Fig. 2). The zeros of $g(y)$ are 1 and 3; hence, the constant solutions are $y = 1$ and $y = 3$. The solution satisfying $y(0) = 4$ is increasing, because the z-coordinate is positive when $y = 4$ on the yz-graph. This solution continues to increase without bound. The solution satisfying $y(0) = 2$ is decreasing because the z-coordinate is negative when $y = 2$ on the yz-graph. This solution will decrease and approach asymptotically the constant solution $y = 1$. (See Fig. 3.)

KEY TERMS AND CONCEPTS	EXAMPLES

Property IV Each nonconstant solution either is asymptotic to a constant solution or else increases or decreases without bound.

Figure 2

Figure 3

10.6 Applications of Differential Equations

The qualitative theory of the previous section was used in this section to study applications involving an important class of (nonlinear) differential equations, known as the *logistic differential equation*, $y' = ky(a - y)$. More specifically, the logistic equation

$$\frac{dN}{dt} = rN\left(\frac{K - N}{K}\right) \quad \text{or, equivalently,}$$

$$\frac{dN}{dt} = \frac{r}{K}N(K - N),$$

where N is used instead of y to denote the size of the population at time t, K is the *carrying capacity* of the environment, and r denotes the *intrinsic rate of growth*.

Interesting applications from business and finance, population dynamics, ecology, and population genetics were presented.

A new branch of BK Bank opened with 1500 customers. The branch is equipped to handle a maximum number of 20,000 customers. Suppose that the number of customers grows according to a logistic equation with intrinsic rate of growth $r = .2$.

(a) Set up the differential equation that is satisfied by $N(t)$, the number of bank customers, t years from the time the branch opened.
(b) Plot an approximate graph of $N(t)$.
(c) Find the number of customers with the largest rate of growth.

Solution
(a) The equation is a logistic equation with carrying capacity $K = 20{,}000$ and intrinsic rate of growth $r = .2$. Hence,

$$\frac{dN}{dt} = (.2)N\left(\frac{20{,}000 - N}{20{,}000}\right) = \frac{.1}{10{,}000}N(20{,}000 - N).$$

(See Fig. 4.) The number of customers at time t is given by the solution of this differential equation with the initial condition $N(0) = 1500$.

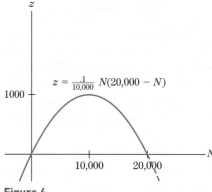

Figure 4

| KEY TERMS AND CONCEPTS | EXAMPLES |

Figure 5

(b) Using the qualitative techniques from the previous section, we first sketch the constant solutions, $N = 0$ and $N = 20{,}000$. The solution starting at $N = 1500$ is a typical logistic curve. It is increasing, with a horizontal asymptote $N = 20{,}000$ and inflection point at $N = 10{,}000$, where the graph changes concavity (see Fig. 5).

(c) Recall that the rate of growth of $N(t)$ is dN/dt and is given by the differential equation

$$\frac{dN}{dt} = \frac{.1}{10{,}000} N(20{,}000 - N).$$

The graph of dN/dt is an inverted parabola with intercepts at $N = 0$ and $N = 20{,}000$, as shown in Fig. 4. The parabola has a maximum at $N = 10{,}000$. The maximum value of dN/dt when $N = 10{,}000$ is

$$\frac{dN}{dt} = \frac{.1}{10{,}000}(10{,}000)(20{,}000 - 10{,}000) = 1000.$$

Thus, the maximum rate of growth is 1000 customers per year and is attained when the branch reaches 10,000 customers. (See Fig. 4)

10.7 Numerical Solution of Differential Equations

Euler's method is used to approximate the solution of the initial-value problem $y' = g(t, y)$, $y(a) = y_0$, where $a \le t \le b$. Let $y = f(t)$ be the solution. Then, Euler's method uses the tangent line approximation of a function to generate a polygonal line that approximates the solution curve, $y = f(t)$.

The endpoints of the line segments approximating the solution, $(t_0, y_0), \ldots, (t_n, y_n)$, are given by the following formulas, where $h = (b - a)/n$:

$$
\begin{aligned}
t_0 &= a \text{ (given)}, & y_0 &\text{ (given)} \\
t_1 &= t_0 + h, & y_1 &= y_0 + g(t_0, y_0) \cdot h, \\
t_2 &= t_1 + h, & y_2 &= y_1 + g(t_1, y_1) \cdot h, \\
&\ \ \vdots & &\ \ \vdots \\
t_n &= t_{n-1} + h. & y_n &= y_{n-1} + g(t_{n-1}, y_{n-1}) \cdot h.
\end{aligned}
$$

Use Euler's method with $n = 4$ to approximate the solution $f(t)$ to $y' = ty$, $y(0) = 1$, for t in the interval $0 \le t \le 2$. In particular, estimate $f(2)$.

Solution Here, $g(t, y) = ty$, $a = 0$, $b = 2$, $y_0 = 1$ and $h = (2 - 0)/4 = \frac{1}{2}$. Starting with $(t_0, y_0) = (0, 1)$, we find that $g(0, 1) = 0$. Thus,

$$t_1 = \frac{1}{2}, \qquad y_1 = 1 + 0 \cdot \frac{1}{2} = 1.$$

Next, $g(\frac{1}{2}, 1) = \frac{1}{2}$, so

$$t_2 = 1, \qquad y_2 = 1 + \frac{1}{2} \cdot \frac{1}{2} = \frac{5}{4} = 1.25.$$

Next, $g(1, \frac{5}{4}) = \frac{5}{4}$, so

$$t_3 = \frac{3}{2}, \qquad y_3 = \frac{5}{4} + \left(\frac{5}{4}\right) \cdot \frac{1}{2} = \frac{5}{4} + \frac{5}{8} = \frac{15}{8} = 1.875.$$

Finally, $g(\frac{3}{2}, \frac{15}{8}) = \frac{45}{16}$, so

$$t_4 = 2, \qquad y_4 = \frac{15}{8} + \frac{45}{16} \cdot \frac{1}{2} = \frac{15}{8} + \frac{45}{32} = \frac{105}{32} \approx 3.281.$$

Thus, the approximation to the solution $f(t)$ is given by the polygonal path shown in the figure. The last point $\left(2, \frac{105}{32}\right)$ is close to the graph of $f(t)$ at $t = 2$, so $f(2) \approx \frac{105}{32}$.

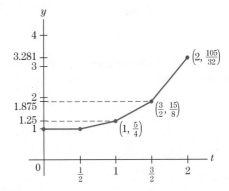

CHAPTER 10 Fundamental Concept Check Exercises

1. What is a differential equation?

2. What does it mean for a function to be a solution to a differential equation?

3. What is a solution curve?

4. What is a constant solution to a differential equation?

5. What is the slope field?

6. Describe the separation-of-variables technique for obtaining the solution to a differential equation.

7. What is a first-order linear differential equation?

8. What is an integrating factor and how does it help you solve a first-order linear differential equation?

9. What is an autonomous differential equation?

10. How do you recognize an autonomous differential equation from its slope field?

11. Outline the procedure for sketching a solution of an autonomous differential equation.

12. What is the logistic differential equation?

13. Describe Euler's method for approximating the solution of a differential equation.

CHAPTER 10 Review Exercises

Solve the differential equations in Exercises 1–10.

1. $y^2 y' = 4t^3 - 3t^2 + 2$

2. $\dfrac{y'}{t+1} = y + 1$

3. $y' = \dfrac{y}{t} - 3y, \; t > 0$

4. $(y')^2 = t$

5. $y = 7y' + ty', \; y(0) = 3$

6. $y' = te^{t+y}, \; y(0) = 0$

7. $yy' + t = 6t^2, \; y(0) = 7$

8. $y' = 5 - 8y, \; y(0) = 1$

9. $y' - \dfrac{2}{1-t}y = (1-t)^4$

10. $y' - \dfrac{1}{2(1+t)}y = 1 + t, \; t \geq 0$

11. Find a curve in the xy-plane passing through the origin and whose slope at the point (x, y) is $x + y$.

12. Let $P(t)$ denote the price in dollars of a certain commodity at time t in days. Suppose that the rate of change of P is proportional to the difference $D - S$ of the demand D and supply S at any time t. Suppose further that the demand and supply are related to the price by $D = 10 - .3P$ and $S = -2 + 3P$.
    (a) Find a differential equation that is satisfied by P, given that the price was falling at the rate of one dollar a day when $D = 10$ and $S = 20$.
    (b) Find P, given that $P(0) = 1$.

13. If $f(t)$ is a solution of $y' = (2 - y)e^{-y}$, is $f(t)$ increasing or decreasing at some value of t where $f(t) = 3$?

14. Solve the initial-value problem
$$y' = e^{y^2}(\cos y)(1 - e^{y-1}), \qquad y(0) = 1.$$

Sketch the solutions of the differential equations in Exercises 15–24. In each case, also indicate the constant solutions.

15. $y' = 2\cos y, \; y(0) = 0$

16. $y' = 5 + 4y - y^2, \; y(0) = 1$

17. $y' = y^2 + y, \; y(0) = -\frac{1}{3}$

18. $y' = y^2 - 2y + 1, \; y(0) = -1$

19. $y' = \ln y, \; y(0) = 2$

20. $y' = 1 + \cos y, \; y(0) = -\frac{3}{4}$

21. $y' = \dfrac{1}{y^2 + 1}, \; y(0) = -1$

22. $y' = \dfrac{3}{y + 3}, \; y(0) = 2$

23. $y' = .4y^2(1 - y), \; y(0) = -1, \; y(0) = .1, \; y(0) = 2$

24. $y' = y^3 - 6y^2 + 9y, \; y(0) = -\frac{1}{4}, \; y(0) = \frac{1}{4}, \; y(0) = 4$

25. The birth rate in a certain city is 3.5% per year, and the death rate is 2% per year. Also, there is a net movement of population out of the city at a steady rate of 3000 people per year. Let $N = f(t)$ be the city's population at time t.
    (a) Write a differential equation satisfied by N.
    (b) Use a qualitative analysis of the equation to determine if there is a size at which the population would remain constant. Is it likely that a city would have such a constant population?

26. Suppose that in a chemical reaction, each gram of substance A combines with 3 grams of substance B to form 4 grams of substance C. The reaction begins with 10 grams of A, 15 grams of B, and 0 grams of C present. Let $y = f(t)$ be the amount of C present at time t. The rate at which substance C is formed is proportional to the product of the unreacted amounts of A and B present. That is, $f(t)$ satisfies the differential equation.
$$y' = k\left(10 - \tfrac{1}{4}y\right)\left(15 - \tfrac{3}{4}y\right), \qquad y(0) = 0,$$
where k is a constant.
    (a) What do the quantities $10 - \frac{1}{4}f(t)$ and $15 - \frac{3}{4}f(t)$ represent?
    (b) Should the constant k be positive or negative?
    (c) Make a qualitative sketch of the solution of the preceding differential equation.

27. A bank account has $20,000 earning 5% interest compounded continuously. A pensioner uses the account to pay himself an annuity, drawing continuously at a $2000 annual rate. How long will it take for the balance in the account to drop to zero?

28. A continuous annuity of $12,000 per year is to be funded by steady withdrawals from a savings account that earns 6% interest compounded continuously.
    (a) What is the smallest initial amount in the account that will fund such an annuity forever?
    (b) What initial amount will fund such an annuity for exactly 20 years (at which time the savings account balance will be zero)?

29. Let $f(t)$ be the solution to $y' = 2e^{2t-y}$, $y(0) = 0$. Use Euler's method with $n = 4$ on $0 \le t \le 2$ to estimate $f(2)$. Then show that Euler's method gives the exact value of $f(2)$ by solving the differential equation.

30. Let $f(t)$ be the solution to $y' = (t + 1)/y$, $y(0) = 1$. Use Euler's method with $n = 3$ on $0 \le t \le 1$ to estimate $f(1)$. Then, show that Euler's method gives the exact value of $f(1)$ by solving the differential equation.

31. Use Euler's method with $n = 6$ on the interval $0 \le t \le 3$ to approximate the solution $f(t)$ to

$$y' = .1y(20 - y), \qquad y(0) = 2.$$

32. Use Euler's method with $n = 5$ on the interval $0 \le t \le 1$ to approximate the solution $f(t)$ to

$$y' = \frac{1}{2}y(y - 10), \qquad y(0) = 9.$$

Taylor Polynomials and Infinite Series

I n earlier chapters, we introduced the functions e^x, $\ln x$, $\sin x$, $\cos x$, and $\tan x$. Whenever we needed the value of one of these functions for a particular value of x, such as $e^{0.023}$, $\ln 5.8$, or $\sin .25$, we had to use a calculator. Now, we shall take up the problem of numerically computing the values of such functions for particular choices of the variable x. The computational methods developed have many applications, for example, to differential equations and probability theory.

11.1 Taylor Polynomials

A polynomial of degree n is a function of the form

$$p(x) = a_0 + a_1 x + \cdots + a_n x^n,$$

where $a_0, a_1, \ldots, a_n$ are given numbers and $a_n \neq 0$. In many instances in mathematics and its applications, calculations are much simpler for polynomials than for other functions. In this section, we show how to approximate a given function, $f(x)$, by a polynomial, $p(x)$, for all values of x near some specified number, say, $x = a$. To simplify matters, we begin by considering values of x near $x = 0$.

Figure 1 shows the graph of the function $f(x) = e^x$ together with the tangent line through $(0, f(0)) = (0, 1)$. Since $f'(x) = e^x$, the slope of the tangent line is $f'(0) = e^0 = 1$.

So, the equation of the tangent line is

$$y - f(0) = f'(0)(x - 0)$$
$$y = f(0) + f'(0)x$$
$$y = 1 + x.$$

FOR REVIEW

The exponential function $f(x) = e^x$ has base $e \approx 2.71828$, p. 226.

Figure 1 A linear approximation of e^x at $x = 0$.

From our discussion of the derivative, we know that the tangent line at $x = 0$ closely approximates the graph of $y = e^x$ for values of x near 0. Thus, if we let $p_1(x) = 1 + x$, the values of $p_1(x)$ are close to the corresponding values of $f(x) = e^x$ for x near 0.

In general, a given function, $f(x)$, may be approximated for values of x near 0 by the polynomial

$$p_1(x) = f(0) + f'(0)x,$$

which is called the **first Taylor polynomial of** $f(x)$ **at** $x = 0$. The graph of $p_1(x)$ is just the tangent line to $y = f(x)$ at $x = 0$.

The first Taylor polynomial "resembles" $f(x)$ near $x = 0$ in the sense that

$p_1(0) = f(0)$ Both graphs go through the same point at $x = 0$.

$p_1'(0) = f'(0)$ Both graphs have the same slope at $x = 0$.

That is, $p_1(x)$ coincides with $f(x)$ in both its value at $x = 0$ and the value of its first derivative at $x = 0$. This suggests that to approximate $f(x)$ even more closely at $x = 0$, we look for a polynomial that coincides with $f(x)$ in its value at $x = 0$ and in the values of its first *and second* derivatives at $x = 0$. We can obtain a further approximation by going out to the third derivative, and so on.

EXAMPLE 1

A Polynomial Approximation Given a function $f(x)$, suppose that $f(0) = 1$, $f'(0) = -2$, $f''(0) = 7$, and $f'''(0) = -5$. Find a polynomial of degree 3,

$$p(x) = a_0 + a_1 x + a_2 x^2 + a_3 x^3,$$

such that $p(x)$ coincides with $f(x)$ up to the third derivative at $x = 0$; that is,

$p(0) = f(0) = 1$ Same value at $x = 0$.

$p'(0) = f'(0) = -2$ Same first derivative at $x = 0$.

$p''(0) = f''(0) = 7$ Same second derivative at $x = 0$.

$p'''(0) = f'''(0) = -5$ Same third derivative at $x = 0$.

SOLUTION   To find the coefficients $a_0, \ldots, a_3$ of $p(x)$, we first compute the values of $p(x)$ and its derivatives at $x = 0$:

$$p(x) = a_0 + a_1 x + a_2 x^2 + a_3 x^3, \qquad p(0) = a_0,$$
$$p'(x) = 0 + a_1 + 2a_2 x + 3a_3 x^2, \qquad p'(0) = a_1,$$
$$p''(x) = 0 + 0 + 2a_2 + 2 \cdot 3a_3 x, \qquad p''(0) = 2a_2,$$
$$p'''(x) = 0 + 0 + 0 + 2 \cdot 3a_3 \qquad p'''(0) = 2 \cdot 3a_3.$$

Since we want $p(x)$ and its derivatives to coincide with the given values of $f(x)$ and its derivatives, we must have

$$a_0 = 1, \quad a_1 = -2, \quad 2a_2 = 7, \quad \text{and} \quad 2 \cdot 3a_3 = -5.$$

So,

$$a_0 = 1, \quad a_1 = -2, \quad a_2 = \frac{7}{2}, \quad \text{and} \quad a_3 = \frac{-5}{2 \cdot 3}.$$

Rewriting the coefficients slightly, we have

$$p(x) = 1 + \frac{(-2)}{1}x + \frac{7}{1 \cdot 2}x^2 + \frac{-5}{1 \cdot 2 \cdot 3}x^3.$$

The form in which we have written $p(x)$ clearly exhibits the values $1, -2, 7, -5$ of $f(x)$ and its derivatives at $x = 0$. In fact, we could also write this formula for $p(x)$ in the form

$$p(x) = f(0) + \frac{f'(0)}{1}x + \frac{f''(0)}{1 \cdot 2}x^2 + \frac{f'''(0)}{1 \cdot 2 \cdot 3}x^3. \qquad \ll$$

Given a function $f(x)$, we may use the formula in Example 1 to find a polynomial that coincides with $f(x)$ up to the third derivative at $x = 0$. To describe the general formula for higher-order polynomials, we let $f^{(n)}(x)$ denote the nth derivative of $f(x)$, and we let $n!$ (read "n factorial") denote the product of all the integers from 1 to n, so that $n! = 1 \cdot 2 \cdot \cdots \cdot (n-1) \cdot n$. (Thus, $1! = 1, 2! = 1 \cdot 2 = 2, 3! = 1 \cdot 2 \cdot 3 = 6$, and so forth.)

Given a function $f(x)$, the **nth Taylor polynomial of $f(x)$ at $x = 0$** is the polynomial $p_n(x)$ defined by

$$p_n(x) = f(0) + \frac{f'(0)}{1!}x + \frac{f''(0)}{2!}x^2 + \cdots + \frac{f^{(n)}(0)}{n!}x^n.$$

This polynomial coincides with $f(x)$ up to the nth derivative at $x = 0$ in the sense that

$$p_n(0) = f(0), \quad p_n'(0) = f'(0), \ldots, p_n^{(n)}(0) = f^{(n)}(0).$$

The next example shows how Taylor polynomials are used to approximate values of e^x for x near 0. The choice of which polynomial to use depends on how accurate we want the values of e^x to be.

EXAMPLE 2   **Taylor Polynomial Approximation of e^x** Determine the first three Taylor polynomials of $f(x) = e^x$ at $x = 0$ and sketch their graphs.

SOLUTION   Since all derivatives of e^x are e^x, we see that

$$f(0) = f'(0) = f''(0) = f'''(0) = e^0 = 1.$$

Thus, the desired Taylor polynomials are

$$p_1(x) = 1 + \frac{1}{1!} \cdot x = 1 + x,$$

$$p_2(x) = 1 + \frac{1}{1!}x + \frac{1}{2!}x^2 = 1 + x + \frac{1}{2}x^2,$$

$$p_3(x) = 1 + \frac{1}{1!}x + \frac{1}{2!}x^2 + \frac{1}{3!}x^3 = 1 + x + \frac{1}{2}x^2 + \frac{1}{6}x^3.$$

The relative accuracy of these approximations to e^x may be seen from the graphs in Fig. 2.

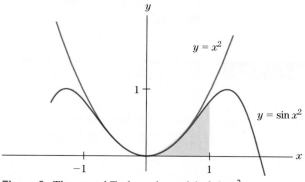

Figure 2 Taylor polynomials of e^x at $x = 0$.

» **Now Try Exercise 3**

EXAMPLE 3   **A Taylor Polynomial Approximation** Determine the nth Taylor polynomial of

$$f(x) = \frac{1}{1 - x}$$

at $x = 0$.

SOLUTION

$$
\begin{aligned}
f(x) &= (1 - x)^{-1}, & f(0) &= 1, \\
f'(x) &= 1(1 - x)^{-2}, & f'(0) &= 1, \\
f''(x) &= 1 \cdot 2(1 - x)^{-3} = 2!(1 - x)^{-3}, & f''(0) &= 2!, \\
f'''(x) &= 1 \cdot 2 \cdot 3(1 - x)^{-4} = 3!(1 - x)^{-4}, & f'''(0) &= 3!, \\
f^{(4)}(x) &= 1 \cdot 2 \cdot 3 \cdot 4(1 - x)^{-5} = 4!(1 - x)^{-5}, & f^{(4)}(0) &= 4!.
\end{aligned}
$$

From the pattern of calculations, it is clear that $f^{(k)}(0) = k!$ for each k. Therefore,

$$
\begin{aligned}
p_n(x) &= 1 + \frac{1}{1!}x + \frac{2!}{2!}x^2 + \frac{3!}{3!}x^3 + \cdots + \frac{n!}{n!}x^n \\
&= 1 + x + x^2 + x^3 + \cdots + x^n.
\end{aligned}
$$

» **Now Try Exercise 13**

We have already mentioned the possibility of using a polynomial to approximate the values of a function near $x = 0$. Here is another way to use a polynomial approximation.

EXAMPLE 4   **Using Taylor Polynomials to Approximate an Integral** It can be shown that the second Taylor polynomial of $\sin x^2$ at $x = 0$ is $p_2(x) = x^2$. Use this polynomial to approximate the area under the graph of $y = \sin x^2$ from $x = 0$ to $x = 1$. (See Fig. 3.)

Figure 3 The second Taylor polynomial of $\sin x^2$ at $x = 0$.

SOLUTION Since the graph of $p_2(x)$ is very close to the graph of $\sin x^2$ for x near 0, the areas under the two graphs should be close. The area under the graph of $p_2(x)$ is

$$\int_0^1 p_2(x)\,dx = \int_0^1 x^2\,dx = \tfrac{1}{3}x^3 \Big|_0^1 = \tfrac{1}{3}.$$

>> *Now Try Exercise 15*

In Example 4, the exact area under the graph of $\sin x^2$ is given by

$$\int_0^1 \sin x^2\,dx.$$

However, this integral cannot be computed by the usual method because there is no way to construct an antiderivative of $\sin x^2$ consisting of elementary functions. Using an approximation technique from Chapter 9, we can find that the value of the integral is .3103 to four decimal places. Thus, the error in using $p_2(x)$ as an approximation for $\sin x^2$ is about .023. (The error can be reduced further if we use a Taylor polynomial of higher degree. In this particular example, the effort involved is far less than that required by the approximation methods of Chapter 9.)

Taylor Polynomials at $x = a$

Suppose now that we wish to approximate a given function $f(x)$ by a polynomial for values of x near some number a. Since the behavior of $f(x)$ near $x = a$ is determined by the values of $f(x)$ and its derivatives at $x = a$, we should try to approximate $f(x)$ by a polynomial $p(x)$ for which the values of $p(x)$ and its derivatives at $x = a$ are the same as those of $f(x)$. This is easily done if we use a polynomial that has the form

$$p(x) = a_0 + a_1(x - a) + a_2(x - a)^2 + \cdots + a_n(x - a)^n.$$

We call this a **polynomial in** $x - a$. In this form, it is easy to compute $p(a), p'(a)$, and so on, because setting $x = a$ in $p(x)$ or one of its derivatives makes most of the terms equal to zero. The following result is easily verified.

Given a function $f(x)$, the **nth Taylor polynomial of $f(x)$ at $x = a$** is the polynomial $p_n(x)$ defined by

$$p_n(x) = f(a) + \frac{f'(a)}{1!}(x - a) + \frac{f''(a)}{2!}(x - a)^2 + \cdots + \frac{f^{(n)}(a)}{n!}(x - a)^n.$$

This polynomial coincides with $f(x)$ up to the nth derivative at $x = a$ in the sense that

$$p_n(a) = f(a), \quad p_n'(a) = f'(a), \ldots, p_n^{(n)}(a) = f^{(n)}(a).$$

When $a = 0$, of course, these Taylor polynomials are just the same as those introduced earlier.

EXAMPLE 5 **Centering a Taylor Polynomial** Calculate the second Taylor polynomial of $f(x) = \sqrt{x}$ at $x = 1$, and use this polynomial to estimate $\sqrt{1.02}$.

SOLUTION Here, $a = 1$. Since we want the second Taylor polynomial, we must calculate the values of $f(x)$ and of its first two derivatives at $x = 1$:

$$f(x) = x^{1/2}, \qquad f'(x) = \tfrac{1}{2}x^{-1/2}, \qquad f''(x) = -\tfrac{1}{4}x^{-3/2},$$
$$f(1) = 1, \qquad f'(1) = \tfrac{1}{2}, \qquad f''(1) = -\tfrac{1}{4}.$$

Therefore, the desired Taylor polynomial is

$$p_2(x) = 1 + \frac{1/2}{1!}(x-1) + \frac{-1/4}{2!}(x-1)^2$$

$$= 1 + \tfrac{1}{2}(x-1) - \tfrac{1}{8}(x-1)^2.$$

Since 1.02 is close to 1, $p_2(1.02)$ gives a good approximation to $f(1.02)$, that is, to $\sqrt{1.02}$.

$$p_2(1.02) = 1 + \tfrac{1}{2}(1.02 - 1) - \tfrac{1}{8}(1.02 - 1)^2$$

$$= 1 + \tfrac{1}{2}(.02) - \tfrac{1}{8}(.02)^2$$

$$= 1 + .01 - .00005$$

$$= 1.00995. \qquad \text{➤➤ \textit{Now Try Exercise 21}}$$

Accuracy of the Approximation

The solution to Example 5 is incomplete in a practical sense, for it offers no information about how close 1.00995 is to the true value of $\sqrt{1.02}$. In general, when we obtain an approximation of some quantity, we also want an indication of the quality of the approximation.

To measure the accuracy of an approximation to a function $f(x)$ by its Taylor polynomial at $x = a$, we define

$$R_n(x) = f(x) - p_n(x).$$

This difference between $f(x)$ and $p_n(x)$ is called the **nth remainder of $f(x)$ at $x = a$**. The following formula is derived in advanced texts:

The Remainder Formula   Suppose that the function $f(x)$ can be differentiated $n + 1$ times on an interval containing the number a. Then, for each x in this interval, there exists a number c between a and x such that

$$R_n(x) = \frac{f^{(n+1)}(c)}{(n+1)!}(x-a)^{n+1}. \qquad (1)$$

Usually the precise value of c is unknown. However, if we can find a number M such that $|f^{(n+1)}(c)| \le M$ for all c between a and x, we do not need to know which c appears in (1), because we have

$$|f(x) - p_n(x)| = |R_n(x)| \le \frac{M}{(n+1)!}|x-a|^{n+1}.$$

EXAMPLE 6   **Using the Remainder Formula** Determine the accuracy of the estimate in Example 5.

SOLUTION   The second remainder for a function $f(x)$ at $x = 1$ is

$$R_2(x) = \frac{f^{(3)}(c)}{3!}(x-1)^3,$$

where c is between 1 and x (and where c depends on x). Here, $f(x) = \sqrt{x}$, and therefore, $f^{(3)}(c) = \tfrac{3}{8}c^{-5/2}$. We are interested in $x = 1.02$, and so, $1 \le c \le 1.02$. We observe that since $c^{5/2} \ge 1^{5/2} = 1$, we have $c^{-5/2} \le 1$. Thus,

$$|f^{(3)}(c)| \le \tfrac{3}{8} \cdot 1 = \tfrac{3}{8},$$

and

$$|R_2(1.02)| \le \frac{3/8}{3!}(1.02 - 1)^3$$
$$= \tfrac{3}{8} \cdot \tfrac{1}{6}(.02)^3$$
$$= .0000005.$$

Thus, the error in using $p_2(1.02)$ as an approximation of $f(1.02)$ is at most .0000005.

≫ Now Try Exercise 29

INCORPORATING TECHNOLOGY

Taylor Polynomials Graphing calculators can be used to determine how well a function is approximated by a Taylor polynomial. Figure 4 shows the (heavily shaded) graph of $\mathbf{Y_1} = \sin(x^2)$ and the graph of its sixth Taylor polynomial $\mathbf{Y_2} = x^2 - \dfrac{x^6}{6}$. The two graphs appear identical on the screen for x between -1.1 and 1.1. Figure 5 shows that the distance between the two functions at $x = 1.1$ is about .02. As x increases beyond 1.1, the goodness of fit deteriorates. For instance, the two functions are about 5.9 units apart when $x = 2$.

Figure 4 $\sin(x^2)$ and its sixth Taylor polynomial.

Figure 5 Two differences between the function and its Taylor polynomial.

Check Your Understanding 11.1

Solutions can be found following the section exercises.

1. **(a)** Determine the third Taylor polynomial of $f(x) = \cos x$ at $x = 0$.
   (b) Use the result of part (a) to estimate $\cos .12$.

2. Determine all Taylor polynomials of $f(x) = 3x^2 - 17$ at $x = 3$.

EXERCISES 11.1

In Exercises 1–8, determine the third Taylor polynomial of the given function at $x = 0$.

1. $f(x) = \sin x$

2. $f(x) = e^{-x/2}$

3. $f(x) = 5e^{2x}$

4. $f(x) = \cos(\pi - 5x)$

5. $f(x) = \sqrt{4x + 1}$

6. $f(x) = \dfrac{1}{x + 2}$

7. $f(x) = xe^{3x}$

8. $f(x) = \sqrt{1 - x}$

9. Determine the fourth Taylor polynomial of $f(x) = e^x$ at $x = 0$, and use it to estimate $e^{0.01}$.

10. Determine the fourth Taylor polynomial of $f(x) = \ln(1 - x)$ at $x = 0$, and use it to estimate $\ln(.9)$.

11. Sketch the graphs of $f(x) = \dfrac{1}{1 - x}$ and its first three Taylor polynomials at $x = 0$.

12. Sketch the graphs of $f(x) = \sin x$ and its first three Taylor polynomials at $x = 0$.

13. Determine the nth Taylor polynomial for $f(x) = e^x$ at $x = 0$.

14. Determine all Taylor polynomials for $f(x) = x^2 + 2x + 1$ at $x = 0$.

15. Use a second Taylor polynomial at $x = 0$ to estimate the area under the curve $y = \ln(1 + x^2)$ from $x = 0$ to $x = \frac{1}{2}$.

16. Use a second Taylor polynomial at $x = 0$ to estimate the area under the curve $y = \sqrt{\cos x}$ from $x = -1$ to $x = 1$. (The exact answer to three decimal places is 1.828.)

17. Determine the third Taylor polynomial of $\dfrac{1}{5 - x}$ at $x = 4$.

18. Determine the fourth Taylor polynomial of $\ln x$ at $x = 1$.

19. Determine the third and fourth Taylor polynomials of $\cos x$ at $x = \pi$.

20. Determine the third and fourth Taylor polynomials of $x^3 + 3x - 1$ at $x = -1$.

21. Use the second Taylor polynomial of $f(x) = \sqrt{x}$ at $x = 9$ to estimate $\sqrt{9.3}$.

22. Use the second Taylor polynomial of $f(x) = \ln x$ at $x = 1$ to estimate $\ln .8$.

23. Determine all Taylor polynomials of $f(x) = x^4 + x + 1$ at $x = 2$.

24. Determine the nth Taylor polynomial of $f(x) = 1/x$ at $x = 1$.

25. If $f(x) = 3 + 4x - \dfrac{5}{2!}x^2 + \dfrac{7}{3!}x^3$, what are $f''(0)$ and $f'''(0)$?

26. If $f(x) = 2 - 6(x - 1) + \dfrac{3}{2!}(x - 1)^2 - \dfrac{5}{3!}(x - 1)^3 +$
$\dfrac{1}{4!}(x - 1)^4$, what are $f''(1)$ and $f'''(1)$?

27. The third remainder for $f(x)$ at $x = 0$ is
$$R_3(x) = \frac{f^{(4)}(c)}{4!}x^4,$$
where c is a number between 0 and x. Let $f(x) = \cos x$, as in Check Your Understanding Problem 11.1.
   (a) Find a number M such that $|f^{(4)}(c)| \leq M$ for all values of c.
   (b) In Check Your Understanding Problem 11.1, the error in using $p_3(.12)$ as an approximation to $f(.12) = \cos .12$ is given by $R_3(.12)$. Show that this error does not exceed 8.64×10^{-6}.

28. Let $p_4(x)$ be the fourth Taylor polynomial of $f(x) = e^x$ at $x = 0$. Show that the error in using $p_4(.1)$ as an approximation for $e^{0.1}$ is at most 2.5×10^{-7}. [*Hint:* Observe that if $x = .1$ and if c is a number between 0 and $.1$, then $|f^{(5)}(c)| \leq f^{(5)}(.1) = e^{0.1} \leq e^1 \leq 3$.]

29. Let $p_2(x)$ be the second Taylor polynomial of $f(x) = \sqrt{x}$ at $x = 9$, as in Exercise 21.
   (a) Give the second remainder for $f(x)$ at $x = 9$.
   (b) Show that $|f^{(3)}(c)| \leq \dfrac{1}{648}$ if $c \geq 9$.
   (c) Show that the error in using $p_2(9.3)$ as an approximation for $\sqrt{9.3}$ is at most $\dfrac{1}{144} \times 10^{-3} < 7 \times 10^{-6}$.

30. Let $p_2(x)$ be the second Taylor polynomial of $f(x) = \ln x$ at $x = 1$, as in Exercise 22.
   (a) Show that $|f^{(3)}(c)| < 4$ if $c \geq .8$.
   (b) Show that the error in using $p_2(.8)$ as an approximation for $\ln .8$ is at most $\dfrac{16}{3} \times 10^{-3} < .0054$.

TECHNOLOGY EXERCISES

31. Graph the function $\mathbf{Y_1} = \dfrac{1}{1 - x}$ and its fourth Taylor polynomial in the window $[-1, 1]$ by $[-1, 5]$. Find a number b such that graphs of the two functions appear identical on the screen for x between 0 and b. Calculate the difference between the function and its Taylor polynomial at $x = b$.

32. Repeat Exercise 31 for the function $\mathbf{Y_1} = \dfrac{1}{1 - x}$ and its seventh Taylor polynomial.

33. Graph the function $\mathbf{Y_1} = e^x$ and its fourth Taylor polynomial in the window $[0, 3]$ by $[-2, 20]$. Find a number b such that graphs of the two functions appear identical on the screen for x between 0 and b. Calculate the difference between the function and its Taylor polynomial at $x = b$ and at $x = 3$.

34. Graph the function $\mathbf{Y_1} = \cos x$ and its second Taylor polynomial in the window ZDECIMAL. Find an interval of the form $[-b, b]$ over which the Taylor polynomial is a good fit to the function. What is the greatest difference between the two functions on this interval?

Solution to Check Your Understanding 11.1

1. (a) $f(x) = \cos x, \qquad f(0) = 1$
   $f'(x) = -\sin x, \quad f'(0) = 0$
   $f''(x) = -\cos x, \quad f''(0) = -1$
   $f'''(x) = \sin x, \qquad f'''(0) = 0$

   Therefore,
   $$p_3(x) = 1 + \frac{0}{1!}x + \frac{-1}{2!}x^2 + \frac{0}{3!}x^3 = 1 - \frac{1}{2}x^2.$$

   [Notice that, here the third Taylor polynomial is actually a polynomial of degree 2. The important thing about $p_3(x)$ is not its degree but the fact that it agrees with $f(x)$ at $x = 0$ up to its third derivative.]

   (b) By part (a), $\cos x \approx 1 - \frac{1}{2}x^2$ when x is near 0. Therefore,
   $$\cos .12 \approx 1 - \tfrac{1}{2}(.12)^2 = .9928.$$

   (*Note:* To five decimal places, $\cos .12 = .99281$.)

2. $f(x) = 3x^2 - 17, \qquad f(3) = 10$
   $f'(x) = 6x, \qquad\qquad f'(3) = 18$
   $f''(x) = 6, \qquad\qquad f''(3) = 6$
   $f^{(3)}(x) = 0, \qquad\qquad f^{(3)}(3) = 0$

The derivatives $f^{(n)}(x)$ for $n \geq 3$ are all the zero constant function. In particular, $f^{(n)}(3) = 0$ for $n \geq 3$. Therefore,

$$p_1(x) = 10 + 18(x - 3),$$
$$p_2(x) = 10 + 18(x - 3) + \frac{6}{2!}(x - 3)^2,$$
$$p_3(x) = 10 + 18(x - 3) + \frac{6}{2!}(x - 3)^2 + \frac{0}{3!}(x - 3)^3.$$

For $n \geq 3$, we have

$$p_n(x) = p_2(x) = 10 + 18(x - 3) + 3(x - 3)^2.$$

[This is the appropriate form of the Taylor polynomial at $x = 3$. However, it is instructive to multiply out the terms in $p_2(x)$ and collect the like powers of x:

$$p_2(x) = 10 + 18x - 18 \cdot 3 + 3(x^2 - 6x + 9)$$
$$= 10 + 18x - 54 + 3x^2 - 18x + 27 = 3x^2 - 17.$$

That is, $p_2(x)$ is $f(x)$, but written in a different form. This is not too surprising, since $f(x)$ itself is a polynomial that agrees with $f(x)$ and all its derivatives at $x = 3$.]

11.2 The Newton–Raphson Algorithm

Figure 1

Many applications of mathematics involve the solution of equations. Often, we have a function $f(x)$ and must find a value of x—say, $x = r$—such that $f(r) = 0$. Such a value of x is called a **zero** of the function or, equivalently, a **root** of the equation $f(x) = 0$. Graphically, a zero of $f(x)$ is a value of x where the graph of $y = f(x)$ crosses the x-axis. (See Fig. 1.) When $f(x)$ is a polynomial, it is sometimes possible to factor $f(x)$ and quickly discover the zeros of $f(x)$. Unfortunately, in most realistic applications there is no simple way to locate zeros. However, there are several methods for finding an approximate value of a zero to any desired degree of accuracy. We shall describe one such method, the Newton–Raphson algorithm.

Suppose that we know that a zero of $f(x)$ is approximately x_0. The idea of the Newton–Raphson algorithm is to obtain an even better approximation of the zero by replacing $f(x)$ by its first Taylor polynomial at x_0, that is, by

$$p(x) = f(x_0) + \frac{f'(x_0)}{1}(x - x_0).$$

Since $p(x)$ closely resembles $f(x)$ near $x = x_0$, the zero of $f(x)$ should be close to the zero of $p(x)$. But solving the equation $p(x) = 0$ for x gives

$$f(x_0) + f'(x_0)(x - x_0) = 0$$
$$xf'(x_0) = f'(x_0)x_0 - f(x_0)$$
$$x = x_0 - \frac{f(x_0)}{f'(x_0)}.$$

That is, if x_0 is an approximation to the zero r, the number

$$x_1 = x_0 - \frac{f(x_0)}{f'(x_0)} \tag{1}$$

generally provides an improved approximation.

We may visualize the situation geometrically as in Fig. 2. The first Taylor polynomial $p(x)$ at x_0 has as its graph the tangent line to $y = f(x)$ at the point $(x_0, f(x_0))$. The value of x for which $p(x) = 0$—that is, $x = x_1$—corresponds to the point where the tangent line crosses the x-axis.

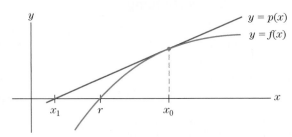

Figure 2 Obtaining x_1 from x_0.

Now, let us use x_1 in place of x_0 as an approximation to the zero r. We obtain a new approximation x_2 from x_1 in the same way that we obtained x_1 from x_0:

$$x_2 = x_1 - \frac{f(x_1)}{f'(x_1)}.$$

We may repeat this process over and over. At each stage, a new approximation x_{new} is obtained from the old approximation x_{old} by the following formula:

Newton–Raphson Method

$$x_{\text{new}} = x_{\text{old}} - \frac{f(x_{\text{old}})}{f'(x_{\text{old}})}.$$

In this way, we obtain a sequence of approximations $x_0, x_1, x_2, \ldots$, which usually approach as close to r as desired. (See Fig. 3.)

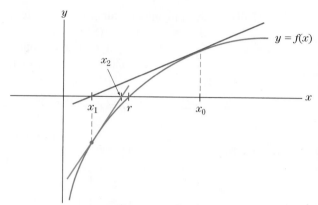

Figure 3 A sequence of approximations to r.

EXAMPLE 1

Approximation Using Newton–Raphson The polynomial $f(x) = x^3 - x - 2$ has a zero between 1 and 2. Let $x_0 = 1$, and find the next three approximations of the zero of $f(x)$ using the Newton–Raphson algorithm.

SOLUTION

Since $f'(x) = 3x^2 - 1$, formula (1) becomes

$$x_1 = x_0 - \frac{x_0^3 - x_0 - 2}{3x_0^2 - 1}.$$

With $x_0 = 1$, we have

$$x_1 = 1 - \frac{1^3 - 1 - 2}{3(1)^2 - 1} = 1 - \frac{-2}{2} = 2,$$

$$x_2 = 2 - \frac{2^3 - 2 - 2}{3(2)^2 - 1} = 2 - \frac{4}{11} = \frac{18}{11},$$

$$x_3 = \frac{18}{11} - \frac{\left(\frac{18}{11}\right)^3 - \frac{18}{11} - 2}{3\left(\frac{18}{11}\right)^2 - 1} \approx 1.530.$$

The actual value of r to three decimal places is 1.521.   **» Now Try Exercise 1**

EXAMPLE 2

Iterating the Newton–Raphson Algorithm Use four repetitions of the Newton–Raphson algorithm to approximate $\sqrt{2}$.

SOLUTION

$\sqrt{2}$ is a zero of the function $f(x) = x^2 - 2$. Since $\sqrt{2}$ clearly lies between 1 and 2, let us take our initial approximation as $x_0 = 1$. ($x_0 = 2$ would do just as well.) Since $f'(x) = 2x$, we have

$$x_1 = x_0 - \frac{x_0^2 - 2}{2x_0} = 1 - \frac{1^2 - 2}{2(1)} = 1 - \left(-\frac{1}{2}\right) = 1.5,$$

$$x_2 = 1.5 - \frac{(1.5)^2 - 2}{2(1.5)} \approx 1.4167,$$

$$x_3 = 1.4167 - \frac{(1.4167)^2 - 2}{2(1.4167)} \approx 1.41422,$$

$$x_4 = 1.41422 - \frac{(1.41422)^2 - 2}{2(1.41422)} \approx 1.41421.$$

This approximation to $\sqrt{2}$ is correct to five decimal places.   **» Now Try Exercise 3**

EXAMPLE 3 **Approximating Zeros of a Polynomial** Approximate the zeros of the polynomial $x^3 + x + 3$.

SOLUTION By applying our curve-sketching techniques, we can make a rough sketch of the graph of $y = x^3 + x + 3$. (See Fig. 4.) The graph crosses the x-axis between $x = -2$ and $x = -1$. So, the polynomial has one zero lying between -2 and -1. Let us therefore set $x_0 = -1$. Since $f'(x) = 3x^2 + 1$, we have

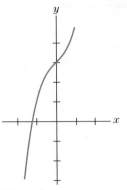

$$x_1 = x_0 - \frac{x_0^3 + x_0 + 3}{3x_0^2 + 1} = -1 - \frac{(-1)^3 + (-1) + 3}{3(-1)^2 + 1} = -1.25,$$

$$x_2 = -1.25 - \frac{(-1.25)^3 + (-1.25) + 3}{3(-1.25)^2 + 1} \approx -1.21429,$$

$$x_3 \approx -1.21341,$$

$$x_4 \approx -1.21341.$$

Figure 4

Therefore, the zero of the given polynomial is approximately -1.21341.

» Now Try Exercise 9

EXAMPLE 4 **Approximating Solutions** Approximate the positive solution of $e^x - 4 = x$.

SOLUTION The rough sketches of the two graphs in Fig. 5 indicate that the solution lies near 2. Let $f(x) = e^x - 4 - x$. Then, the solution of the original equation will be a zero of $f(x)$. We apply the Newton–Raphson algorithm to $f(x)$ with $x_0 = 2$. Since $f'(x) = e^x - 1$,

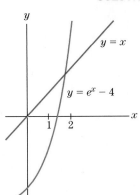

$$x_1 = x_0 - \frac{e^{x_0} - 4 - x_0}{e^{x_0} - 1} = 2 - \frac{e^2 - 4 - 2}{e^2 - 1} \approx 2 - \frac{1.38906}{6.38906} \approx 1.78,$$

$$x_2 = 1.78 - \frac{e^{1.78} - 4 - (1.78)}{e^{1.78} - 1} \approx 1.78 - \frac{.14986}{4.92986} \approx 1.75,$$

$$x_3 = 1.75 - \frac{e^{1.75} - 4 - (1.75)}{e^{1.75} - 1} \approx 1.75 - \frac{.0046}{4.7546} \approx 1.749.$$

Figure 5

Therefore, an approximate solution is $x = 1.749$.

» Now Try Exercise 11

EXAMPLE 5 **Internal Rate of Return** Suppose that an investment of $100 yields the following returns:

$2 at the end of the first month
$15 at the end of the second month
$45 at the end of the third month
$50 at the end of the fourth (and last) month

The total of these returns is $112. This represents the initial investment of $100, plus earnings of $12 during the 4 months. The **internal rate of return** on this investment is the interest rate (per month) for which the sum of the present values of the returns equals the initial investment, $100. Determine the internal rate of return.

SOLUTION Let i be the *monthly* rate of interest. The present value of an amount A to be received in k months is $A(1 + i)^{-k}$. Therefore, we must solve

$$\begin{bmatrix} \text{amount of initial} \\ \text{investment} \end{bmatrix} = \begin{bmatrix} \text{sum of present} \\ \text{values of returns} \end{bmatrix}$$

$$100 = 2(1 + i)^{-1} + 15(1 + i)^{-2} + 45(1 + i)^{-3} + 50(1 + i)^{-4}.$$

Multiplying both sides of the equation by $(1 + i)^4$ and taking all terms to the left, we obtain

$$100(1 + i)^4 - 2(1 + i)^3 - 15(1 + i)^2 - 45(1 + i) - 50 = 0.$$

Let $x = 1 + i$, and solve the resulting equation by the Newton–Raphson algorithm with $x_0 = 1.1$.

$$100x^4 - 2x^3 - 15x^2 - 45x - 50 = 0$$

$$f(x) = 100x^4 - 2x^3 - 15x^2 - 45x - 50$$

$$f'(x) = 400x^3 - 6x^2 - 30x - 45$$

$$x_1 = x_0 - \frac{100x_0^4 - 2x_0^3 - 15x_0^2 - 45x_0 - 50}{400x_0^3 - 6x_0^2 - 30x_0 - 45}$$

$$= 1.1 - \frac{100(1.1)^4 - 2(1.1)^3 - 15(1.1)^2 - 45(1.1) - 50}{400(1.1)^3 - 6(1.1)^2 - 30(1.1) - 45}$$

$$= 1.1 - \frac{26.098}{447.14} \approx 1.042$$

$$x_2 \approx 1.035$$

$$x_3 \approx 1.035$$

Therefore, an approximate solution is $x = 1.035$. Hence, $i = .035$ and the investment had an internal rate of return of 3.5% per month.   **» Now Try Exercise 13**

In general, if an investment of P dollars produces the returns

R_1 at the end of the first period,
R_2 at the end of the second period,
$\vdots$
R_N at the end of the Nth (and last) period,

then we obtain the internal rate of return, i, by solving the equation

$$P(1 + i)^N - R_1(1 + i)^{N-1} - R_2(1 + i)^{N-2} - \cdots - R_N = 0$$

for its positive root. We are assuming that all the returns are nonnegative and add up to at least P.

When a loan of P dollars is paid back with N equal periodic payments of R dollars at interest rate i per period, the equation to be solved for i becomes

$$P(1 + i)^N - R(1 + i)^{N-1} - R(1 + i)^{N-2} - \cdots - R = 0.$$

This equation can be simplified to

$$Pi + R[(1 + i)^{-N} - 1] = 0.$$

(See Exercise 41 in Section 11.3.)

EXAMPLE 6   **Monthly Interest Rate of a Loan** A mortgage of $104,880 is repaid in 360 monthly payments of $755. Use two iterations of the Newton–Raphson algorithm to estimate the monthly rate of interest.

SOLUTION   Here, $P = 104,880$, $R = 755$, and $N = 360$. Therefore, we must solve the equation

$$104,880i + 755[(1 + i)^{-360} - 1] = 0.$$

Let $f(i) = 104,880i + 755[(1 + i)^{-360} - 1]$. Then,

$$f'(i) = 104,880 - 271,800(1 + i)^{-361}.$$

Apply the Newton–Raphson algorithm to $f(i)$ with $i_0 = .01$:

$$i_1 = i_0 - \frac{104,880i_0 + 755[(1 + i_0)^{-360} - 1]}{104,880 - 271,800(1 + i_0)^{-361}} \approx .00677$$

$$i_2 \approx .00650.$$

Therefore, the monthly interest rate is approximately .65%.   **» Now Try Exercise 15**

NOTE 1. The values of successive approximations in the Newton–Raphson algorithm depend on the extent of round-off used during the calculation.

2. The Newton–Raphson algorithm is an excellent computational tool. However, in some cases, it will not work. For instance, if $f'(x_n) = 0$ for some approximation x_n, there is no way to compute the next approximation. Other instances in which the algorithm fails are presented in Exercises 25 and 26.

3. It can be shown that, if $f(x)$, $f'(x)$, and $f''(x)$ are continuous near r [a zero of $f(x)$] and $f'(r) \neq 0$, the Newton–Raphson algorithm will definitely work provided that the initial approximation x_0 is not too far away. ≪

INCORPORATING

TECHNOLOGY

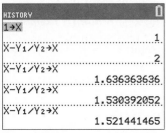

Figure 6 Successive approximations for Example 1.

▦ **The Newton–Raphson Algorithm**   The TI-83/84 can generate a new approximation for the Newton–Raphson algorithm with each press of the ENTER key. To illustrate, let us use the polynomial $f(x) = x^3 - x - 2$ from Example 1. Begin by pressing Y=, and assign $Y_1 = X^3 - X - 2$. Assign to Y_2 the derivative of Y_1 so that $Y_2 = 3X^2 - 1$. Return to the home screen. Our initial approximation in Example 1 was $x_0 = 1$, so we begin the Newton–Raphson algorithm by assigning the value 1 to the variable X. We accomplish this by pressing 1 STO▶ X,T,Θ,n and then ENTER.

Now that we have initialized the algorithm, we compute the next approximation by calculating the value of $X - Y_1/Y_2$. As indicated in Fig. 6, after typing in $X - Y_1/Y_2$, we immediately press STO▶ X,T,Θ,n to assign the value of that calculation to X. Then, each time we press the ENTER key, another approximation is displayed. Note that, when assigning to Y_2 the derivative of Y_1, we could just as well have set in classic mode $Y_2 = \text{nDeriv}(Y_1, X, X)$. In this case, the successive approximations will differ slightly from those obtained with Y_2 equal to the exact derivative.

Check Your Understanding 11.2

Solutions can be found following the section exercises.

1. Use three repetitions of the Newton–Raphson algorithm to estimate $\sqrt[3]{7}$.

2. Use three repetitions of the Newton–Raphson algorithm to estimate the zeros of $f(x) = 2x^3 + 3x^2 + 6x - 3$.

EXERCISES 11.2

In Exercises 1–8, use three repetitions of the Newton–Raphson algorithm to approximate the following:

1. $\sqrt{5}$ 2. $\sqrt{7}$ 3. $\sqrt[3]{6}$ 4. $\sqrt[3]{11}$

5. The zero of $x^2 - x - 5$ between 2 and 3

6. The zero of $x^2 + 3x - 11$ between -5 and -6

7. The zero of $\sin x + x^2 - 1$ near $x_0 = 0$

8. The zero of $e^x + 10x - 3$ near $x_0 = 0$

9. Sketch the graph of $y = x^3 + 2x + 2$, and use the Newton–Raphson algorithm (three repetitions) to approximate all x-intercepts.

10. Sketch the graph of $y = x^3 + x - 1$, and use the Newton–Raphson algorithm (three repetitions) to approximate all x-intercepts.

11. Use the Newton–Raphson algorithm to find an approximate solution to $e^{-x} = x^2$.

12. Use the Newton–Raphson algorithm to find an approximate solution to $e^{5-x} = 10 - x$.

13. **Internal Rate of Return** Suppose that an investment of \$500 yields returns of \$100, \$200, and \$300 at the end of the first, second, and third months, respectively. Determine the internal rate of return on this investment.

14. **Internal Rate of Return** An investor buys a bond for \$1000. She receives \$10 at the end of each month for 2 months and then sells the bond at the end of the second month for \$1040. Determine the internal rate of return on this investment.

15. **Monthly Interest Rate** A \$663 flat-screen TV is purchased with a down payment of \$100 and a loan of \$563 to be repaid in five monthly installments of \$116. Determine the monthly rate of interest on the loan.

16. **Monthly Interest Rate** A mortgage of \$100,050 is repaid in 240 monthly payments of \$900. Determine the monthly rate of interest.

17. **Estimating Roots of a Function** A function $f(x)$ has the graph given in Fig. 7. Let x_1 and x_2 be the estimates of a root of $f(x)$ obtained by applying the Newton–Raphson algorithm using an initial approximation of $x_0 = 5$. Draw the appropriate tangent lines and estimate the numerical values of x_1 and x_2.

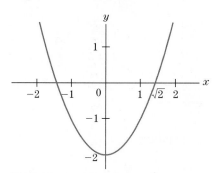

Figure 7

18. Redo Exercise 17 with $x_0 = 1$.

19. Estimating the Root of a Function Suppose that the line $y = 4x + 5$ is tangent to the graph of the function $f(x)$ at $x = 3$. If the Newton–Raphson algorithm is used to find a root of $f(x) = 0$ with the initial guess $x_0 = 3$, what is x_1?

20. Estimating the Root of a Function Suppose that the graph of the function $f(x)$ has slope -2 at the point $(1, 2)$. If the Newton–Raphson algorithm is used to find a root of $f(x) = 0$ with the initial guess $x_0 = 1$, what is x_1?

21. Figure 8 contains the graph of the function $f(x) = x^2 - 2$. The function has zeros at $x = \sqrt{2}$ and $x = -\sqrt{2}$. When the Newton–Raphson algorithm is applied to find a zero, what values of x_0 lead to the zero $\sqrt{2}$?

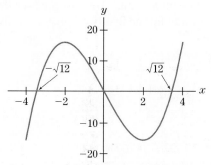

Figure 8 Graph of $f(x) = x^2 - 2$.

22. Figure 9 contains the graph of the function $f(x) = x^3 - 12x$. The function has zeros at $x = -\sqrt{12}$, 0, and $\sqrt{12}$. Which zero of $f(x)$ will be approximated by the Newton–Raphson method starting with $x_0 = 4$? Starting with $x_0 = 1$? Starting with $x_0 = -1.8$?

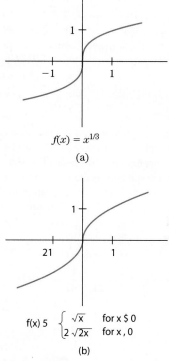

Figure 9 Graph of $f(x) = x^3 - 12x$.

23. What special occurrence takes place when the Newton–Raphson algorithm is applied to the linear function $f(x) = mx + b$ with $m \neq 0$?

24. What happens when the first approximation, x_0, is actually a zero of $f(x)$?

Exercises 25 and 26 present two examples in which successive repetitions of the Newton–Raphson algorithm do not approach a root.

25. Apply the Newton–Raphson algorithm to the function $f(x) = x^{1/3}$ whose graph is drawn in Fig. 10(a). Use $x_0 = 1$.

$f(x) = x^{1/3}$

(a)

$f(x) = \begin{cases} \sqrt{x} & \text{for } x \leq 0 \\ 2\sqrt{2x} & \text{for } x > 0 \end{cases}$

(b)

Figure 10

26. Apply the Newton–Raphson algorithm to the function whose graph is drawn in Fig. 10(b). Use $x_0 = 1$.

TECHNOLOGY EXERCISES

27. The functions $f(x) = x^2 - 4$ and $g(x) = (x - 2)^2$ both have a zero at $x = 2$. Apply the Newton–Raphson algorithm to each function with $x_0 = 3$, and determine the value of n for which x_n appears on the screen as exactly 2. Graph the two functions and explain why the sequence for $f(x)$ converges so quickly to 2, whereas the sequence for $g(x)$ converges so slowly.

28. Apply the Newton–Raphson algorithm to the function $f(x) = x^3 - 5x$ with $x_0 = 1$. After observing the behavior, graph the function along with the tangent lines at $x = 1$ and $x = -1$, and explain geometrically what is happening.

29. Draw the graph of $f(x) = x^4 - 2x^2$, $[-2, 2]$ by $[-2, 2]$. The function has zeros at $x = -\sqrt{2}$, $x = 0$, and $x = \sqrt{2}$. By looking at the graph, guess which zero will be approached when you apply the Newton–Raphson algorithm to each of the following initial approximations:

(a) $x_0 = 1.1$ **(b)** $x_0 = .95$ **(c)** $x_0 = .9$

Then, test your guesses by actually carrying out the computations.

30. Graph the function $f(x) = \dfrac{x^2}{1 + x^2}$, $[-2, 2]$ by $[-.5, 1]$. The function has 0 as a zero. By looking at the graph, guess at a value of x_0 for which x_1 will be exactly 0 when the Newton–Raphson algorithm is invoked. Then, test your guess by carrying out the computation.

Solutions to Check Your Understanding 11.2

1. We wish to approximate a zero of $f(x) = x^3 - 7$. Since $f(1) = -6 < 0$ and $f(2) = 1 > 0$, the graph of $f(x)$ crosses the x-axis somewhere between $x = 1$ and $x = 2$. Take $x_0 = 2$ as the initial approximation to the zero. Since $f'(x) = 3x^2$, we have

$$x_1 = x_0 - \frac{x_0^3 - 7}{3x_0^2} = 2 - \frac{2^3 - 7}{3(2)^2} = \frac{23}{12} \approx 1.9167,$$

$$x_2 = 1.9167 - \frac{(1.9167)^3 - 7}{3(1.9167)^2} \approx 1.91294,$$

$$x_3 = 1.91294 - \frac{(1.91294)^3 - 7}{3(1.91294)^2} \approx 1.91293.$$

2. As a preliminary step, we use the methods of Chapter 2 to sketch the graph of $f(x)$. (See Fig. 11.) We see that $f(x)$ has a zero that occurs for a positive value of x. Since $f(0) = -3$ and $f(1) = 8$, the graph crosses the x-axis between 0 and 1. Let us choose $x_0 = 0$ as our initial approximation to the zero of $f(x)$. Since $f'(x) = 6x^2 + 6x + 6$, we then have

$$x_1 = x_0 - \frac{2x_0^3 + 3x_0^2 + 6x_0 - 3}{6x_0^2 + 6x_0 + 6} = 0 - \frac{-3}{6} = \frac{1}{2},$$

$$x_2 = \frac{1}{2} - \frac{1}{\frac{21}{2}} = \frac{1}{2} - \frac{2}{21} = \frac{17}{42} \approx .40476.$$

Continuing, we find that $x_3 \approx .39916$.

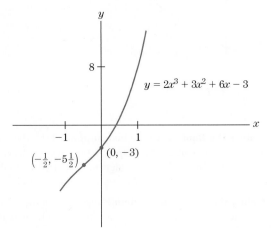

Figure 11

11.3 Infinite Series

An **infinite series** is an infinite addition of numbers

$$a_1 + a_2 + a_3 + a_4 + \cdots.$$

Here are some examples:

$$1 + \tfrac{1}{2} + \tfrac{1}{4} + \tfrac{1}{8} + \tfrac{1}{16} + \cdots, \tag{1}$$

$$1 + 1 + 1 + 1 + \cdots, \tag{2}$$

$$1 - 1 + 1 - 1 + \cdots. \tag{3}$$

Some infinite series can be associated with a "sum." To illustrate how this is done, let us consider the infinite series (1). If we add up the first two, three, four, five, and six terms of the infinite series (1), we obtain

$$1 + \tfrac{1}{2} = 1\tfrac{1}{2},$$

$$1 + \tfrac{1}{2} + \tfrac{1}{4} = 1\tfrac{3}{4},$$

$$1 + \tfrac{1}{2} + \tfrac{1}{4} + \tfrac{1}{8} = 1\tfrac{7}{8},$$

$$1 + \tfrac{1}{2} + \tfrac{1}{4} + \tfrac{1}{8} + \tfrac{1}{16} = 1\tfrac{15}{16},$$

$$1 + \tfrac{1}{2} + \tfrac{1}{4} + \tfrac{1}{8} + \tfrac{1}{16} + \tfrac{1}{32} = 1\tfrac{31}{32}.$$

Each total lies halfway between the preceding total and the number 2. It appears from these calculations that, by increasing the number of terms, we bring the total arbitrarily close to 2. Indeed, this is supported by further calculation. For example,

$$\underbrace{1 + \frac{1}{2} + \frac{1}{4} + \cdots + \frac{1}{2^9}}_{10 \text{ terms}} = 2 - \frac{1}{2^9} \approx 1.998047,$$

$$\underbrace{1 + \frac{1}{2} + \frac{1}{4} + \cdots + \frac{1}{2^{19}}}_{20 \text{ terms}} = 2 - \frac{1}{2^{19}} \approx 1.999998,$$

$$\underbrace{1 + \frac{1}{2} + \frac{1}{4} + \cdots + \frac{1}{2^{n-1}}}_{n \text{ terms}} = 2 - \frac{1}{2^{n-1}}.$$

Therefore, it seems reasonable to assign the infinite series (1) the "sum" 2:

$$1 + \tfrac{1}{2} + \tfrac{1}{4} + \tfrac{1}{8} + \tfrac{1}{16} + \cdots = 2. \tag{4}$$

The sum of the first n terms of an infinite series is called its **nth partial sum** and is denoted S_n. In series (1), we were very fortunate that the partial sums approached a limiting value, 2. This is not always the case. For example, consider the infinite series (2). If we form the first few partial sums, we get

$$\begin{aligned} S_2 &= 1 + 1 &&= 2, \\ S_3 &= 1 + 1 + 1 &&= 3, \\ S_4 &= 1 + 1 + 1 + 1 &&= 4. \end{aligned}$$

We see that these sums do not approach any limit. Rather, they become larger and larger, eventually exceeding any specified number.

The partial sums need not grow without bound for an infinite series not to have a sum. For example, consider the infinite series (3). Here, the sums of initial terms are

$$\begin{aligned} S_2 &= 1 - 1 &&= 0, \\ S_3 &= 1 - 1 + 1 &&= 1, \\ S_4 &= 1 - 1 + 1 - 1 &&= 0, \\ S_5 &= 1 - 1 + 1 - 1 + 1 &&= 1, \end{aligned}$$

and so forth. The partial sums alternate between 0 and 1 and do not approach a limit. So the infinite series (3) has no sum.

An infinite series whose partial sums approach a limit is called **convergent**. The limit is then called the **sum** of the infinite series. An infinite series whose partial sums do not approach a limit is called **divergent**. From our preceding discussion, we know that the infinite series (1) is convergent, whereas (2) and (3) are divergent.

It is often extremely difficult to determine whether a given infinite series is convergent, and intuition is not always an accurate guide. For example, we might at first suspect that the infinite series

$$1 + \tfrac{1}{2} + \tfrac{1}{3} + \tfrac{1}{4} + \tfrac{1}{5} + \cdots$$

(the **harmonic series**) is convergent. However, it is not. The sums of its initial terms increase without bound, although they do so very slowly. For example, it takes about 12,000 terms before the sum exceeds 10 and about 2.7×10^{43} terms before the sum exceeds 100. Nevertheless, the sum eventually exceeds any prescribed number. (See Exercise 42 and Section 11.4.)

There is an important type of infinite series whose convergence or divergence is easily determined. Let a and r be given nonzero numbers. A series of the form

$$a + ar + ar^2 + ar^3 + ar^4 + \cdots$$

is called a **geometric series with ratio** r. (The "ratio" of consecutive terms is r.)

The Geometric Series The infinite series

$$a + ar + ar^2 + ar^3 + ar^4 + \cdots$$

converges if and only if $|r| < 1$. When $|r| < 1$, the sum of the series is

$$\frac{a}{1 - r}. \tag{5}$$

For example, if $a = 1$ and $r = \frac{1}{2}$, we obtain the infinite series (1). In this case,

$$\frac{a}{1 - r} = \frac{1}{1 - \frac{1}{2}} = \frac{1}{\frac{1}{2}} = 2,$$

in agreement with our previous observation. Also, series (2) and (3) are divergent geometric series, with $r = 1$ and $r = -1$, respectively. A proof of the foregoing result is outlined in Exercise 41.

EXAMPLE 1   **Geometric Series** Calculate the sums of the following geometric series:

(a) $1 + \dfrac{1}{5} + \dfrac{1}{5^2} + \dfrac{1}{5^3} + \dfrac{1}{5^4} + \cdots$

(b) $\dfrac{2}{3^2} + \dfrac{2}{3^4} + \dfrac{2}{3^6} + \dfrac{2}{3^8} + \dfrac{2}{3^{10}} + \cdots$

(c) $\dfrac{5}{2^2} - \dfrac{5^2}{2^5} + \dfrac{5^3}{2^8} - \dfrac{5^4}{2^{11}} + \dfrac{5^5}{2^{14}} - \cdots$

SOLUTION   (a) Here, $a = 1$ and $r = \frac{1}{5}$. The sum of the series is

$$\frac{a}{1 - r} = \frac{1}{1 - \frac{1}{5}} = \frac{1}{\frac{4}{5}} = \frac{5}{4}.$$

(b) We find r by dividing any term by the preceding term. So,

$$r = \frac{\dfrac{2}{3^4}}{\dfrac{2}{3^2}} = \frac{2}{3^4} \cdot \frac{3^2}{2} = \frac{1}{3^2} = \frac{1}{9}.$$

Since the series is a geometric series, we obtain the same result using any other pair of successive terms. For instance,

$$\frac{\dfrac{2}{3^8}}{\dfrac{2}{3^6}} = \frac{2}{3^8} \cdot \frac{3^6}{2} = \frac{1}{3^2} = \frac{1}{9}.$$

The first term of the series is $a = \dfrac{2}{3^2} = \dfrac{2}{9}$, so the sum of the series is

$$\frac{a}{1 - r} = a \cdot \frac{1}{1 - r} = \frac{2}{9} \cdot \frac{1}{1 - \frac{1}{9}} = \frac{2}{9} \cdot \frac{9}{8} = \frac{1}{4}.$$

(c) We may find r as in part (b), or we may observe that the numerator of each fraction in the series (c) is increasing by a factor of 5, while the denominator is increasing by a factor of $2^3 = 8$. So the ratio of successive fractions is $\frac{5}{8}$. However, the terms in the series are alternately positive and negative, so the ratio of successive terms must be negative. Hence, $r = -\frac{5}{8}$. Next, $a = \frac{5}{2^2} = \frac{5}{4}$, so the sum of series (c) is

$$a \cdot \frac{1}{1-r} = \frac{5}{4} \cdot \frac{1}{1 - \left(-\frac{5}{8}\right)} = \frac{5}{4} \cdot \frac{1}{\frac{13}{8}} = \frac{5}{4} \cdot \frac{8}{13} = \frac{10}{13}.$$

>> Now Try Exercise 1

Sometimes, a rational number is expressed as an infinite repeating decimal, such as $.1212\overline{12}$. The value of such a "decimal expansion" is, in fact, the sum of an infinite series.

EXAMPLE 2   **A Rational Number as a Geometric Series** What rational number has the decimal expansion $.1212\overline{12}$?

SOLUTION   This number denotes the infinite series

$$.12 + .0012 + .000012 + \cdots = \frac{12}{100} + \frac{12}{100^2} + \frac{12}{100^3} + \cdots,$$

a geometric series with $a = \frac{12}{100}$ and $r = \frac{1}{100}$. The sum of the geometric series is

$$\frac{a}{1-r} = a \cdot \frac{1}{1-r} = \frac{12}{100} \cdot \frac{1}{1 - \frac{1}{100}} = \frac{12}{100} \cdot \frac{100}{99} = \frac{12}{99} = \frac{4}{33}.$$

Hence, $.1212\overline{12} = \frac{4}{33}$.

>> Now Try Exercise 15

EXAMPLE 3   **The Multiplier Effect in Economics** Suppose that the federal government enacts an income tax cut of $10 billion. Assume that each person will spend 93% of all resulting extra income and save the rest. Estimate the total effect of the tax cut on economic activity.

SOLUTION   Express all amounts of money in billions of dollars. Of the increase in income created by the tax cut, $(.93)(10)$ billion dollars will be spent. These dollars become extra income to someone and, hence, 93% will be spent again and 7% saved, so additional spending of $(.93)(.93)(10)$ billion dollars is created. The recipients of those dollars will spend 93% of them, creating again additional spending of

$$(.93)(.93)(.93)(10) = 10(.93)^3$$

billion dollars, and so on. The total amount of new spending created by the tax cut is thus given by the infinite series

$$10(.93) + 10(.93)^2 + 10(.93)^3 + \cdots.$$

This is a geometric series with initial term $10(.93)$ and ratio $.93$. Its sum is

$$\frac{a}{1-r} = \frac{10(.93)}{1 - .93} = \frac{9.3}{.07} \approx 132.86.$$

Thus, a $10-billion tax cut creates new spending of about $132.86 billion.

>> Now Try Exercise 23

Example 3 illustrates the **multiplier effect**. The proportion of each extra dollar that a person will spend is called the **marginal propensity to consume**, denoted MPC. In Example 3, MPC $= .93$. As we observed, the total new spending generated by the tax cut is

$$[\text{total new spending}] = 10 \cdot \frac{.93}{1 - .93} = [\text{tax cut}] \cdot \frac{\text{MPC}}{1 - \text{MPC}}.$$

The tax cut is multiplied by the "multiplier" $\dfrac{\text{MPC}}{1 - \text{MPC}}$ to obtain its true effect.

EXAMPLE 4   **Elimination of a Drug**   Patients with certain heart problems are often treated with digitoxin, a derivative of the digitalis plant. The rate at which a person's body eliminates digitoxin is proportional to the amount of digitoxin present. In 1 day (24 hours) about 10% of any given amount of the drug will be eliminated. Suppose that a maintenance dose of .05 milligram (mg) is given daily to a patient. Estimate the total amount of digitoxin that should be present in the patient after several months of treatment.

SOLUTION   For a moment, let us consider what happens to the initial dose of .05 mg and disregard the subsequent doses. After 1 day, 10% of the .05 mg will have been eliminated and $(.90)(.05)$ mg will remain. By the end of the second day, this smaller amount will be reduced 10% to $(.90)(.90)(.05)$ mg, and so on, until after n days only $(.90)^n(.05)$ mg of the original dose will remain. (See Fig. 1.) To determine the cumulative effect of all the doses of digitoxin, we observe that at the time of the second dose (1 day after the first dose), the patient's body will contain the second dose of .05 mg plus $(.90)(.05)$ mg of the first dose. A day later there will be the third dose of .05 mg, plus $(.90)(.05)$ mg of the second dose, plus $(.90)^2(.05)$ of the first dose. At the time of any new dose, the patient's body will contain that dose plus the amounts that remain of earlier doses. Let us tabulate this.

	Total Amount (mg) of Digitoxin
0	.05
1	$.05 + (.90)(.05)$
2	$.05 + (.90)(.05) + (.90)^2(.05)$
$\vdots$	
n	$.05 + (.90)(.05) + (.90)^2(.05) + \cdots + (.90)^n(.05)$

Days after Initial Dose

Figure 1 Exponential decrease of the initial dose.

We can see that the amounts present at the time of each new dose correspond to the partial sums of the geometric series

$$.05 + (.90)(.05) + (.90)^2(.05) + (.90)^3(.05) + \cdots,$$

where $a = .05$ and $r = .90$. The sum of this series is

$$\frac{.05}{1 - .90} = \frac{.05}{.10} = .5.$$

Since the partial sums of the series approach the sum of .5, we may conclude that a daily maintenance dose of .05 mg will eventually raise the level of digitoxin in the patient to a plateau of .5 mg. Between doses the level will drop 10% down to $(.90)(.5) = .45$ mg. The use of a regular maintenance dose to sustain a certain level of a drug in a patient is an important technique in drug therapy.   **»» Now Try Exercise 29**

Sigma Notation   When studying series, it is often convenient to use the Greek capital letter sigma to indicate summation. For example, the sum

$$a_2 + a_3 + \cdots + a_{10}$$

is denoted by

$$\sum_{k=2}^{10} a_k$$

(read "the sum of a sub k from k equals 2 to 10"). The nth partial sum of a series, $a_1 + a_2 + \cdots + a_n$, is written as $\sum_{k=1}^{n} a_k$. In these examples the letter k is called the **index of summation**. Sometimes, we require the index of summation to begin with 0 and sometimes with 1, but any integer value may be used for k. Any letter not already in use may be used as the index of summation. For instance, both

$$\sum_{i=0}^{4} a_i \quad \text{and} \quad \sum_{j=0}^{4} a_j$$

indicate the sum $a_0 + a_1 + a_2 + a_3 + a_4$.

Finally, a formal infinite series

$$a_1 + a_2 + a_3 + \cdots$$

is written as

$$\sum_{k=1}^{\infty} a_k \quad \text{or} \quad \sum_{1}^{\infty} a_k.$$

We will also write $\sum_{k=1}^{\infty} a_k$ as the symbol for the numerical value of the series when it is convergent. With this notation (and writing ar^0 in place of a), the main result about the geometric series may be written as follows:

$$\sum_{k=0}^{\infty} ar^k = \frac{a}{1-r} \qquad \text{if } |r| < 1,$$

$$\sum_{k=0}^{\infty} ar^k \text{ is divergent} \qquad \text{if } |r| \geq 1.$$

EXAMPLE 5   **Geometric Series and Sigma Notation**   Determine the sums of the following infinite series.

(a) $\displaystyle\sum_{k=0}^{\infty} \left(\frac{2}{3}\right)^k$ **(b)** $\displaystyle\sum_{j=0}^{\infty} 4^{-j}$ **(c)** $\displaystyle\sum_{i=3}^{\infty} \frac{2}{7^i}$

SOLUTION   In each case, the first step is to write out the first few terms of the series.

(a) $\displaystyle\sum_{k=0}^{\infty} \left(\frac{2}{3}\right)^k = \underset{[k=0]}{1} + \underset{[k=1]}{\frac{2}{3}} + \underset{[k=2]}{\left(\frac{2}{3}\right)^2} + \underset{[k=3]}{\left(\frac{2}{3}\right)^3} + \cdots$

This is a geometric series with initial term $a = 1$ and ratio $r = \frac{2}{3}$; its sum is

$$\frac{1}{1 - \frac{2}{3}} = \frac{1}{\frac{1}{3}} = 3.$$

(b) $\displaystyle\sum_{j=0}^{\infty} 4^{-j} = 4^0 + 4^{-1} + 4^{-2} + 4^{-3} + \cdots = 1 + \frac{1}{4} + \left(\frac{1}{4}\right)^2 + \left(\frac{1}{4}\right)^3 + \cdots = \frac{1}{1 - \frac{1}{4}} = \frac{4}{3}.$

(c) $\displaystyle\sum_{i=3}^{\infty} \frac{2}{7^i} = \frac{2}{7^3} + \frac{2}{7^4} + \frac{2}{7^5} + \frac{2}{7^6} + \cdots$

This is a geometric series with $a = \dfrac{2}{7^3}$ and $r = \dfrac{1}{7}$; its sum is

$$a \cdot \frac{1}{1-r} = \frac{2}{7^3} \cdot \frac{1}{1 - \frac{1}{7}} = \frac{2}{7^3} \cdot \frac{7}{6} = \frac{1}{147}.$$

≫ Now Try Exercise 35

Figure 2

Figure 3

INCORPORATING

TECHNOLOGY

Finite Sums Graphing calculators can compute finite sums. The variable **x** can be used as the index of summation. If $f(x)$ is an expression involving x, the sum

$$\sum_{x=m}^{n} f(x)$$

can be computed on TI calculators as

$$\textbf{sum(seq(f(X), X, m, n, 1)).}$$

For instance,

$$\sum_{x=1}^{99} \frac{1}{x} = 1 + \frac{1}{2} + \frac{1}{3} + \cdots + \frac{1}{99}$$

is calculated in classic mode as **sum(seq(1/X, X, 1, 99, 1))**, and

$$\sum_{x=1}^{10} \frac{2}{3^{2x}} = \frac{2}{3^2} + \frac{2}{3^4} + \cdots + \frac{2}{3^{20}}$$

from Example 1(b) is calculated as **sum(seq(2/3^(2X), X, 1, 10, 1))**. (See Fig. 2.) You access both the expressions **sum(** and **seq(** by pressing [2nd] [LIST]. Then, **sum(** is the fifth option under the **MATH** menu, and **seq(** is the fifth option under the **OPS** menu. Figure 2 shows the result of this calculation and also the sum from Example 1(b).

Also, you can generate successive partial sums by repeatedly pressing the ENTER key. In Fig. 3 the variable **s** holds the current partial sum for the series in Example 5(a).

Check Your Understanding 11.3

Solutions can be found following the section exercises.

1. Determine the sum of the geometric series

$$8 - \frac{8}{3} + \frac{8}{9} - \frac{8}{27} + \frac{8}{81} - \cdots.$$

2. Find the value of $\displaystyle\sum_{k=0}^{\infty} (.7)^{-k+1}$.

EXERCISES 11.3

Determine the sums of the following geometric series when they are convergent.

1. $1 + \dfrac{1}{6} + \dfrac{1}{6^2} + \dfrac{1}{6^3} + \dfrac{1}{6^4} \cdots$

2. $1 + \dfrac{3}{4} + \left(\dfrac{3}{4}\right)^2 + \left(\dfrac{3}{4}\right)^3 + \left(\dfrac{3}{4}\right)^4 + \cdots$

3. $1 - \dfrac{1}{3^2} + \dfrac{1}{3^4} - \dfrac{1}{3^6} + \dfrac{1}{3^8} - \cdots$

4. $1 + \dfrac{1}{2^3} + \dfrac{1}{2^6} + \dfrac{1}{2^9} + \dfrac{1}{2^{12}} + \cdots$

5. $2 + \dfrac{2}{3} + \dfrac{2}{9} + \dfrac{2}{27} + \dfrac{2}{81} + \cdots$

6. $3 + \dfrac{6}{5} + \dfrac{12}{25} + \dfrac{24}{125} + \dfrac{48}{625} + \cdots$

7. $\dfrac{1}{5} + \dfrac{1}{5^4} + \dfrac{1}{5^7} + \dfrac{1}{5^{10}} + \dfrac{1}{5^{13}} + \cdots$

8. $\dfrac{1}{3^2} - \dfrac{1}{3^3} + \dfrac{1}{3^4} - \dfrac{1}{3^5} + \dfrac{1}{3^6} - \cdots$

9. $3 - \dfrac{3^2}{7} + \dfrac{3^3}{7^2} - \dfrac{3^4}{7^3} + \dfrac{3^5}{7^4} - \cdots$

10. $6 - 1.2 + .24 - .048 + .0096 - \cdots$

11. $\dfrac{2}{5^4} - \dfrac{2^4}{5^5} + \dfrac{2^7}{5^6} - \dfrac{2^{10}}{5^7} + \dfrac{2^{13}}{5^8} - \cdots$

12. $\dfrac{3^2}{2^5} + \dfrac{3^4}{2^8} + \dfrac{3^6}{2^{11}} + \dfrac{3^8}{2^{14}} + \dfrac{3^{10}}{2^{17}} + \cdots$

13. $5 + 4 + 3.2 + 2.56 + 2.048 + \cdots$

14. $\dfrac{5^3}{3} - \dfrac{5^5}{3^4} + \dfrac{5^7}{3^7} - \dfrac{5^9}{3^{10}} + \dfrac{5^{11}}{3^{13}} - \cdots$

Sum an appropriate infinite series to find the rational number whose decimal expansion is given.

15. $.27272\overline{27}$

16. $.173\overline{173}$

17. $.22\overline{2}$

18. $.1515\overline{15}$

19. $4.011\overline{011}$ $(= 4 + .011\overline{011})$

20. $5.44\overline{4}$

21. Show that $.99\overline{9} = 1$.

22. **A Rational Number** Compute the value of $.1212\overline{1212}$ as a geometric series with $a = .1212$ and $r = .0001$. Compare your answer with the result of Example 2.

23. **Marginal Propensity to Consume and the Multiplier Effect** Compute the total new spending created by a $10-billion federal income tax cut when the population's marginal

propensity to consume is 95%. Compare your result with that of Example 3, and note how a small change in the MPC makes a dramatic change in the total spending generated by the tax cut.

24. **The Multiplier Effect** Compute the effect of a $20-billion federal income tax cut when the population's marginal propensity to consume is 98%. What is the "multiplier" in this case?

A *perpetuity* is a periodic sequence of payments that continues forever. The *capital value* of the perpetuity is the sum of the present values of all future payments.

25. **Perpetuity** Consider a perpetuity that promises to pay $100 at the beginning of each month. If the interest rate is 12% compounded monthly, the present value of $100 in k months is $100(1.01)^{-k}$.
    (a) Express the capital value of the perpetuity as an infinite series.
    (b) Find the sum of the infinite series.

26. **Perpetuity** Consider a perpetuity that promises to pay P dollars at the end of each month. (The first payment will be received in 1 month.) If the interest rate per month is r, the present value of P dollars in k months is $P(1 + r)^{-k}$. Find a simple formula for the capital value of the perpetuity.

27. **Bonus plus Taxes on Taxes** A generous corporation not only gives its CEO a $1,000,000 bonus, but gives her enough money to cover the taxes on the bonus, the taxes on the additional taxes, the taxes on the taxes on the additional taxes, and so on. If she is in the 39.6% tax bracket, how large is her bonus?

28. **Total Distance Traveled by a Bouncing Ball** The *coefficient of restitution* of a ball, a number between 0 and 1, specifies how much energy is conserved when the ball hits a rigid surface. A coefficient of .9, for instance, means a bouncing ball will rise to 90% of its previous height after each bounce. The coefficients of restitution for a tennis ball, basketball, super ball, and softball are .7, .75, .9, and .3, respectively. Find the total distance traveled by a tennis ball dropped from a height of 6 feet.

29. **Elimination of a Drug** A patient receives 6 mg of a certain drug daily. Each day the body eliminates 30% of the amount of the drug present in the system. After extended treatment, estimate the total amount of the drug that should be present immediately after a dose is given.

30. **Elimination of a Drug** A patient receives 2 mg of a certain drug each day. Each day the body eliminates 20% of the amount of drug present in the system. After extended treatment, estimate the total amount of the drug present immediately *before* a dose is given.

31. **Drug Dosage** A patient receives M mg of a certain drug each day. Each day the body eliminates 25% of the amount of drug present in the system. Determine the value of the maintenance dose M such that after many days approximately 20 mg of the drug is present immediately after a dose is given.

32. **Drug Dosage** A patient receives M mg of a certain drug daily. Each day, the body eliminates a fraction q of the amount of the drug present in the system. After extended treatment, estimate the total amount of the drug that should be present immediately after a dose is given.

33. The infinite series $a_1 + a_2 + a_3 + \cdots$ has partial sums given by $S_n = 3 - \dfrac{5}{n}$.

(a) Find $\sum_{k=1}^{10} a_k$.
(b) Does the infinite series converge? If so, to what value does it converge?

34. The infinite series $a_1 + a_2 + a_3 + \cdots$ has partial sums given by $S_n = n - \dfrac{1}{n}$.

(a) Find $\sum_{k=1}^{10} a_k$.
(b) Does the infinite series converge? If so, to what value does it converge?

Determine the sums of the following infinite series:

35. $\displaystyle\sum_{k=0}^{\infty} \left(\frac{5}{6}\right)^k$ 36. $\displaystyle\sum_{k=0}^{\infty} \frac{7}{10^k}$

37. $\displaystyle\sum_{j=1}^{\infty} 5^{-2j}$ 38. $\displaystyle\sum_{j=0}^{\infty} \frac{(-1)^j}{3^j}$

39. $\displaystyle\sum_{k=0}^{\infty} (-1)^k \frac{3^{k+1}}{5^k}$ 40. $\displaystyle\sum_{k=1}^{\infty} \left(\frac{1}{3}\right)^{2k}$

41. Let a and r be given nonzero numbers.
    (a) Show that
    $$(1 - r)(a + ar + ar^2 + \cdots + ar^n) = a - ar^{n+1},$$
    and from this conclude that, for $r \neq 1$,
    $$a + ar + ar^2 + \cdots + ar^n = \frac{a}{1 - r} - \frac{ar^{n+1}}{1 - r}.$$
    (b) Use the result of part (a) to explain why the geometric series $\sum_0^{\infty} ar^k$ converges to $\dfrac{a}{1 - r}$ when $|r| < 1$.
    (c) Use the result of part (a) to explain why the geometric series diverges for $|r| > 1$.
    (d) Explain why the geometric series diverges for $r = 1$ and $r = -1$.

42. Show that the infinite series
    $$1 + \frac{1}{2} + \frac{1}{3} + \frac{1}{4} + \frac{1}{5} + \cdots$$
    diverges. [*Hint*: $\frac{1}{3} + \frac{1}{4} > \frac{1}{2}$; $\frac{1}{5} + \frac{1}{6} + \frac{1}{7} + \frac{1}{8} > \frac{1}{2}$; $\frac{1}{9} + \cdots + \frac{1}{16} > \frac{1}{2}$; etc.]

TECHNOLOGY EXERCISES

43. What is the exact value of the infinite geometric series whose partial sum appears at the first entry in Fig. 3?

44. What is the exact value of the infinite geometric series whose partial sum appears at the second entry in Fig. 2?

In Exercises 45 and 46, the calculator screen computes a partial sum of an infinite series. Write out the first five terms of the series and determine the exact value of the infinite series.

45.
```
NORMAL FLOAT AUTO REAL RADIAN CL
sum(seq((-1)^(2X)/2^(X+1),
X,1,20,1))
                     .4999995232
```

46.

47. Verify the formula

$$\sum_{x=1}^{n} x = \frac{n(n+1)}{2}$$

for $n = 10$, 50, and 100.

48. The sum of the first n odd numbers is n^2; that is,

$$\sum_{x=1}^{n} (2x - 1) = n^2.$$

Verify this formula for $n = 5$, 10, and 25.

In Exercises 49 and 50, convince yourself that the equation is correct by summing up the first 999 terms of the infinite series and comparing the sum with the value on the right.

49. $\displaystyle\sum_{x=1}^{\infty} \frac{1}{x^2} = \frac{\pi^2}{6}$

50. $\displaystyle\sum_{x=1}^{\infty} \frac{(-1)^{x+1}}{x} = \ln 2$

Solutions to Check Your Understanding 11.3

1. Answer: 6. To obtain the sum of a geometric series, identify a and r, and (provided that $|r| < 1$) substitute these values into the formula $\dfrac{a}{1 - r}$. The initial term a is just the first term of the series: $a = 8$. The ratio r is often obvious by inspection. However, if in doubt, divide any term by the *preceding* term. Here, the second term divided by the first term is $\frac{-8}{3}/8 = -\frac{1}{3}$, so $r = -\frac{1}{3}$. Since $|r| = \frac{1}{3}$, the series is convergent and the sum is

$$\frac{a}{1 - r} = \frac{8}{1 - \left(-\frac{1}{3}\right)} = \frac{8}{\frac{4}{3}} = 8 \cdot \frac{3}{4} = 6.$$

2. Write out the first few terms of the series and then proceed as in Problem 1.

$$\sum_{k=0}^{\infty} (.7)^{-k+1} = \underset{[k=0]}{(.7)^1} + \underset{[k=1]}{(.7)^0} + \underset{[k=2]}{(.7)^{-1}}$$
$$+ \underset{[k=3]}{(.7)^{-2}} + \underset{[k=4]}{(.7)^{-3}}$$
$$= .7 + 1 + \frac{1}{.7} + \frac{1}{(.7)^2} + \frac{1}{(.7)^3} + \cdots.$$

Here, $a = .7$ and $r = 1/.7 = \frac{10}{7}$. Since $|r| = \frac{10}{7} > 1$, the series is divergent and has no sum. (The formula $\dfrac{a}{1 - r}$ yields $-\frac{49}{30}$; however, this value is meaningless. The formula applies only to the case in which the series is convergent.)

11.4 Series with Positive Terms

Very often, it is difficult to determine the sum of an infinite series. As a fallback to determining the sum, we can at least check that the series converges and so has a sum (even if we cannot determine its exact value). Calculus includes many tests for telling if an infinite series converges. This section presents two convergence tests for infinite series consisting of positive terms. The tests are derived from geometric models of the series.

Throughout this section, we consider only series for which each term a_k is positive (or zero). Suppose that $\sum_{k=1}^{\infty} ar^k$ is such a series. Consider the corresponding collection of rectangles in Fig. 1. Each rectangle is 1 unit wide, and the height of the kth rectangle is a_k. Hence, the area of the kth rectangle is a_k, and the area of the region consisting of the first n rectangles is the nth partial sum $S_n = a_1 + a_2 + \cdots + a_n$. As n increases, the partial sums increase and approach the area of the region consisting of all the rectangles. If this area is finite, the infinite series converges to this area. If this area is infinite, the series is divergent.

Figure 1 Representation of an infinite series by rectangles.

FOR REVIEW

Improper integrals are discussed in Section 9.6. See the introduction in See. 9.6.

This geometric "picture" of an infinite series provides a convergence test that relates the convergence of the series to the convergence of an improper integral. For example, consider

$$\sum_{k=1}^{\infty} \frac{1}{k^2} \quad \text{and} \quad \int_1^{\infty} \frac{1}{x^2}\, dx.$$

Notice that the series and the integral have a similar form:

$$\sum_{k=1}^{\infty} f(k) \quad \text{and} \quad \int_1^{\infty} f(x)\, dx, \tag{1}$$

where $f(x) = 1/x^2$.

The techniques of Section 9.6 show that the integral is convergent. Hence, the area under the graph of $y = 1/x^2$ for $x \geq 1$ is finite. Figure 2 shows the graph of $y = 1/x^2$ ($x \geq 1$) superimposed on a geometric model of the series

$$\sum_{k=1}^{\infty} \frac{1}{k^2} = 1 + \frac{1}{4} + \frac{1}{9} + \frac{1}{16} + \cdots.$$

The area of the first rectangle is 1. The region consisting of all the remaining rectangles has finite area since it is contained in the region under the graph of $y = 1/x^2$ ($x \geq 1$), which has finite area. Hence, the total area of all the rectangles is finite and the series $\sum_{k=1}^{\infty} \frac{1}{k^2}$ is convergent.

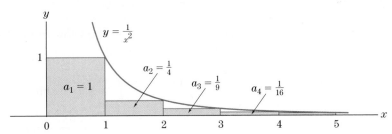

Figure 2 A convergent series.

As a second example, consider

$$\sum_{k=1}^{\infty} \frac{1}{k} \quad \text{and} \quad \int_1^{\infty} \frac{1}{x}\, dx.$$

Once again, the series and the integral have the forms in (1), where $f(x) = 1/x$. It is easy to check that the integral is divergent, and hence, the area under the graph of $y = 1/x$ for $x \geq 1$ is infinite. Figure 3 shows the graph of $y = 1/x$ superimposed on a geometric model of the series

$$\sum_{k=1}^{\infty} \frac{1}{k} = 1 + \frac{1}{2} + \frac{1}{3} + \frac{1}{4} + \cdots.$$

Figure 3 A divergent series.

Since the area of the region formed by the rectangles clearly exceeds the infinite area of the region under the graph of $y = 1/x$ $(x \geq 1)$, the series $\sum_{k=1}^{\infty} \dfrac{1}{k}$ is divergent.

The reasoning for these two examples can be used to derive the following important test:

The Integral Test   Let $f(x)$ be continuous, decreasing, and positive for $x \geq 1$. Then, the infinite series

$$\sum_{k=1}^{\infty} f(k)$$

is convergent if the improper integral

$$\int_{1}^{\infty} f(x)\, dx$$

is convergent, and the infinite series is divergent if the improper integral is divergent.

EXAMPLE 1   **Applying the Integral Test** Use the integral test to determine whether the infinite series

$$\sum_{k=1}^{\infty} \frac{1}{e^k}$$

is convergent or divergent.

SOLUTION   Here, $f(x) = 1/e^x = e^{-x}$. We know from Chapter 4 that $f(x)$ is a positive, decreasing, continuous function. Also,

$$\int_{1}^{\infty} e^{-x}\, dx = \lim_{b \to \infty} \int_{1}^{b} e^{-x}\, dx = \lim_{b \to \infty} -e^{-x} \Big|_{1}^{b}$$

$$= \lim_{b \to \infty} (-e^{-b} + e^{-1}) = e^{-1} = \frac{1}{e}.$$

Since the improper integral is convergent, so is the infinite series.   **▸▸ Now Try Exercise 1**

NOTE▸ The integral test does not provide the exact value of the sum of a convergent infinite series. It only verifies convergence. Other techniques, sometimes quite advanced, must be used to find the sum. (The sum of the series in Example 1 is readily found since the series happens to be a geometric series with ratio $1/e$.) ◂

Instead of starting our infinite series with $k = 1$, we may start with $k = N$, where N is any positive integer. To test the convergence of such a series, we determine the convergence (or divergence) of the improper integral

$$\int_{N}^{\infty} f(x)\, dx.$$

EXAMPLE 2   **Applying the Integral Test** Determine whether the series

$$\sum_{k=3}^{\infty} \frac{\ln k}{k}$$

is convergent.

SOLUTION   We take $f(x) = (\ln x)/x$. Note that $f(x)$ is continuous and positive for $x \geq 3$. Moreover,

$$f'(x) = \frac{x \cdot \dfrac{1}{x} - \ln x \cdot 1}{x^2} = \frac{1 - \ln x}{x^2}.$$

Since $\ln x > 1$ for $x \geq 3$, we conclude that $f'(x)$ is negative and hence that $f(x)$ is a decreasing function.

To antidifferentiate $(\ln x)/x$, make the substitution $u = \ln x$, $du = (1/x)\,dx$:

$$\int \frac{\ln x}{x}\,dx = \int u\,du = \frac{u^2}{2} + C = \frac{(\ln x)^2}{2} + C.$$

Hence,

$$\int_3^\infty \frac{\ln x}{x}\,dx = \lim_{b \to \infty} \int_3^b \frac{\ln x}{x}\,dx = \lim_{b \to \infty} \left.\frac{(\ln x)^2}{2}\right|_3^b$$

$$= \lim_{b \to \infty} \frac{(\ln b)^2}{2} - \frac{(\ln 3)^2}{2} = \infty.$$

Thus, the series $\sum_{k=3}^\infty \dfrac{\ln k}{k}$ is divergent.   **» Now Try Exercise 3**

When using the integral test, we tested the convergence of an infinite series by relating it to an improper integral. In many situations, it is possible to accomplish the same result by comparing the series with another infinite series whose convergence or divergence is already known.

Suppose that $\sum_{k=1}^\infty a_k$ and $\sum_{k=1}^\infty b_k$ have the property that $0 \leq a_k \leq b_k$ for all k. In Fig. 4, we have superimposed the geometric models for the two series. Each rectangle for the series $\sum_{k=1}^\infty a_k$ lies inside (or coincides with) the corresponding rectangle for the series $\sum_{k=1}^\infty b_k$. Clearly, if the region formed by all the rectangles for $\sum_{k=1}^\infty b_k$ has finite

Figure 4 Term-by-term comparison of two series.

area, so does the region for $\sum_{k=1}^\infty a_k$. On the other hand, if the region for $\sum_{k=1}^\infty a_k$ has infinite area, so does the region for $\sum_{k=1}^\infty b_k$. These geometric conclusions may be stated in terms of the infinite series as follows:

Comparison Test   Suppose that $0 \leq a_k \leq b_k$ for $k = 1, 2, \ldots$.

$$\text{If } \sum_{k=1}^\infty b_k \text{ converges, so does } \sum_{k=1}^\infty a_k.$$

$$\text{If } \sum_{k=1}^\infty a_k \text{ diverges, so does } \sum_{k=1}^\infty b_k.$$

The comparison test also applies to two series whose terms eventually satisfy $0 \leq a_k \leq b_k$, say, for $k \geq N$, where N is some positive integer. This is so because the convergence or divergence of $\sum_{k=1}^\infty a_k$ and $\sum_{k=1}^\infty b_k$ is not affected by the removal of some terms from the beginning of the series.

EXAMPLE 3   **A Comparison Test** Determine the convergence or divergence of the series

$$\sum_{k=1}^{\infty} \frac{3}{1+5^k} = \frac{3}{6} + \frac{3}{26} + \frac{3}{126} + \frac{3}{626} + \cdots.$$

SOLUTION   Compare the series with the convergent geometric series

$$\sum_{k=1}^{\infty} \frac{3}{5^k} = \frac{3}{5} + \frac{3}{25} + \frac{3}{125} + \frac{3}{625} + \cdots.$$

This series converges since the ratio of successive terms is $r = \frac{1}{5}$. The kth terms of these two series satisfy

$$\frac{3}{1+5^k} < \frac{3}{5^k}$$

because the denominator of the left fraction is greater than the denominator of the right fraction. Since the series $\sum_{k=1}^{\infty} \frac{3}{5^k}$ converges, the comparison test implies that $\sum_{k=1}^{\infty} \frac{3}{1+5^k}$ also converges.

➤ *Now Try Exercise 21*

This section has considered convergence tests only for series of positive terms. Here is a version of the comparison test that works even if one of the two series has some negative terms:

> Suppose that $\sum_{k=1}^{\infty} b_k$ is a convergent series of positive terms and that $|a_k| \leq b_k$ for $k = 1, 2, 3, \ldots$. Then, $\sum_{k=1}^{\infty} a_k$ is convergent.

Check Your Understanding 11.4

Solutions can be found following the section exercises.

1. (a) What is the improper integral associated with the infinite series $\sum_{k=1}^{\infty} \frac{k^2}{(k^3+6)^2}$?

   (b) Is the improper integral found in part (a) convergent or divergent?

   (c) Is the infinite series $\sum_{k=1}^{\infty} \frac{k^2}{(k^3+6)^2}$ convergent or divergent?

2. The two series

$$\sum_{k=1}^{\infty} \frac{1}{4k} \quad \text{and} \quad \sum_{k=1}^{\infty} \frac{1}{4k+3}$$

   are both divergent. (This is easily established by the integral test.) Which of these series can be used in the comparison test to show that the series $\sum_{k=1}^{\infty} \frac{1}{4k+1}$ is divergent?

EXERCISES 11.4

In Exercises 1–16, use the integral test to determine whether the infinite series is convergent or divergent. (You may assume that the hypotheses of the integral test are satisfied.)

1. $\sum_{k=1}^{\infty} \frac{3}{\sqrt{k}}$

2. $\sum_{k=1}^{\infty} \frac{5}{k^{3/2}}$

3. $\sum_{k=2}^{\infty} \frac{1}{(k-1)^3}$

4. $\sum_{k=0}^{\infty} \frac{7}{k+100}$

5. $\sum_{k=1}^{\infty} \frac{2}{5k-1}$

6. $\sum_{k=2}^{\infty} \frac{1}{k\sqrt{\ln k}}$

7. $\sum_{k=2}^{\infty} \frac{k}{(k^2+1)^{3/2}}$

8. $\sum_{k=1}^{\infty} \frac{1}{(2k+1)^3}$

9. $\sum_{k=2}^{\infty} \frac{1}{k(\ln k)^2}$

10. $\sum_{k=1}^{\infty} \frac{1}{(3k)^2}$

11. $\sum_{k=1}^{\infty} e^{3-k}$

12. $\sum_{k=1}^{\infty} \frac{1}{e^{2k+1}}$

13. $\sum_{k=1}^{\infty} ke^{-k^2}$

14. $\sum_{k=1}^{\infty} k^{-3/4}$

15. $\sum_{k=1}^{\infty} \frac{2k+1}{k^2+k+2}$

16. $\sum_{k=2}^{\infty} \frac{k+1}{(k^2+2k+1)^2}$

17. It can be shown that

$$\int_0^{\infty} \frac{3}{9+x^2}\,dx$$

is convergent. Use this fact to show that an appropriate infinite series converges. Give the series, and show that the hypotheses of the integral test are satisfied.

18. Use the integral test to determine if $\sum_{k=1}^{\infty} \dfrac{e^{1/k}}{k^2}$ is convergent. Show that the hypotheses of the integral test are satisfied.

19. It can be shown that $\lim_{b \to \infty} b e^{-b} = 0$. Use this fact and the integral test to show that $\sum_{k=1}^{\infty} \dfrac{k}{e^k}$ is convergent.

20. Is the series $\sum_{k=1}^{\infty} \dfrac{3^k}{4^k}$ convergent? What is the easiest way to answer this question? Can you tell if

$$\int_1^{\infty} \frac{3^x}{4^x} \, dx$$

is convergent?

In Exercises 21–26, use the comparison test to determine whether the infinite series is convergent or divergent.

21. $\displaystyle\sum_{k=2}^{\infty} \frac{1}{k^2 + 5}$ $\left[\text{Compare with } \displaystyle\sum_{k=2}^{\infty} \frac{1}{k^2}.\right]$

22. $\displaystyle\sum_{k=2}^{\infty} \frac{1}{\sqrt{k^2 - 1}}$ $\left[\text{Compare with } \displaystyle\sum_{k=2}^{\infty} \frac{1}{k}.\right]$

23. $\displaystyle\sum_{k=1}^{\infty} \frac{1}{2^k + k}$ $\left[\text{Compare with } \displaystyle\sum_{k=1}^{\infty} \frac{1}{2^k}.\right]$

24. $\displaystyle\sum_{k=1}^{\infty} \frac{1}{k3^k}$ $\left[\text{Compare with } \displaystyle\sum_{k=1}^{\infty} \frac{1}{3^k}.\right]$

25. $\displaystyle\sum_{k=1}^{\infty} \frac{1}{5^k} \cos^2\left(\frac{k\pi}{4}\right)$ $\left[\text{Compare with } \displaystyle\sum_{k=1}^{\infty} \frac{1}{5^k}.\right]$

26. $\displaystyle\sum_{k=0}^{\infty} \frac{1}{\left(\frac{3}{4}\right)^k + \left(\frac{5}{4}\right)^k}$ $\left[\text{Compare with } \displaystyle\sum_{k=0}^{\infty} \left(\tfrac{3}{4}\right)^{-k} \text{ or } \displaystyle\sum_{k=0}^{\infty} \left(\tfrac{5}{4}\right)^{-k}.\right]$

27. Can the comparison test be used with $\sum_{k=2}^{\infty} \dfrac{1}{k \ln k}$ and $\sum_{k=2}^{\infty} \dfrac{1}{k}$ to deduce anything about the first series?

28. Can the comparison test be used with $\sum_{k=1}^{\infty} \dfrac{1}{k^2 \ln k}$ and $\sum_{k=2}^{\infty} \dfrac{1}{k^2}$ to deduce anything about the first series?

29. The following property is true for any two series (with possibly some negative terms): Let $\sum_{k=1}^{\infty} a_k$ and $\sum_{k=1}^{\infty} b_k$ be convergent series whose sums are S and T, respectively. Then, $\sum_{k=1}^{\infty} (a_k + b_k)$ is a convergent series whose sum is $S + T$. Make a geometric picture to illustrate why this property is true when the terms a_k and b_k are all positive.

30. Let $\sum_{k=1}^{\infty} a_k$ be a convergent series with sum S, and let c be a constant. Then, $\sum_{k=1}^{\infty} c a_k$ is a convergent series whose sum is $c \cdot S$. Make a geometric picture to illustrate why this is true when $c = 2$ and the terms a_k are all positive.

31. Use Exercise 29 to show that the series $\sum_{k=0}^{\infty} \dfrac{8^k + 9^k}{10^k}$ is convergent, and determine its sum.

32. Use Exercise 30 to show that the series $\sum_{k=1}^{\infty} \dfrac{3}{k^2}$ is convergent. Then, use the comparison test to show that the series $\sum_{k=1}^{\infty} \dfrac{e^{1/k}}{k^2}$ is convergent.

Solutions to Check Your Understanding 11.4

1. (a) $\displaystyle\int_1^{\infty} \frac{x^2}{(x^3 + 6)^2} \, dx$. In general, to find the function $f(x)$, replace each occurrence of k by x; then, replace the summation sign by an integral sign and adjoin dx.

(b) A substitution must first be made to antidifferentiate the function. Let $u = x^3 + 6$. Then, $du = 3x^2 \, dx$ and

$$\int \frac{x^2}{(x^3 + 6)^2} \, dx = \frac{1}{3} \int \frac{3x^2}{(x^3 + 6)^2} \, dx$$

$$= \frac{1}{3} \int \frac{1}{u^2} \, du$$

$$= -\frac{1}{3} u^{-1} + C$$

$$= -\frac{1}{3} \cdot \frac{1}{x^3 + 6} + C.$$

Hence,

$$\int_1^{\infty} \frac{x^2}{(x^3 + 6)^2} \, dx = \lim_{b \to \infty} \left[-\frac{1}{3} \cdot \frac{1}{x^3 + 6} \right]\Big|_1^b$$

$$= \lim_{b \to \infty} \left[-\frac{1}{3} \cdot \frac{1}{b^3 + 6} + \frac{1}{3} \cdot \frac{1}{7} \right]$$

$$= \frac{1}{21}.$$

Therefore, the improper integral is convergent.

(c) Convergent, since the infinite series is convergent if, and only if, the associated improper integral is convergent.

2. To show that a series is divergent, we need to show that its terms are *greater* than the corresponding terms of some divergent series. Since

$$\frac{1}{4k + 3} < \frac{1}{4k + 1},$$

the comparison should be made with

$$\sum_{k=1}^{\infty} \frac{1}{4k+3}.$$

(*Note:* If instead we had been trying to establish the *convergence* of an infinite series, we would try to show that its terms are *less* than the corresponding terms of a convergent series.)

11.5　Taylor Series

Consider the infinite series $1 + x + x^2 + x^3 + x^4 + \cdots$. This series is of a different type from those discussed in the preceding two sections. Its terms are not numbers but are powers of x. However, for some specific values of x, the series is convergent. In fact, for any value of x between -1 and 1, the series is a convergent geometric series with ratio x and sum $\dfrac{1}{1-x}$. We write

$$\frac{1}{1-x} = 1 + x + x^2 + x^3 + x^4 + \cdots, \qquad |x| < 1. \tag{1}$$

Looking at (1) from a different point of view, we see that the function $f(x) = \dfrac{1}{1-x}$ is represented as a series involving the powers of x. This representation is not valid throughout the entire domain of the function $\dfrac{1}{1-x}$, but just for values of x with $-1 < x < 1$.

In many important cases, a function $f(x)$ may be represented by a series of the form

$$f(x) = a_0 + a_1 x + a_2 x^2 + a_3 x^3 + \cdots, \tag{2}$$

where $a_0, a_1, a_2, \ldots$ are suitable constants and where x ranges over values that make the series converge to $f(x)$. The series is called a **power series** (because it involves powers of x). It may be shown that when a function $f(x)$ has a representation by a power series as in (2), the coefficients $a_0, a_1, a_2, \ldots$ are uniquely determined by $f(x)$ and its derivatives at $x = 0$. In fact, $a_0 = f(0)$, and $a_k = f^{(k)}(0)/k!$ for $k = 1, 2, \ldots,$ so

$$f(x) = f(0) + \frac{f'(0)}{1!}x + \frac{f''(0)}{2!}x^2 + \cdots + \frac{f^{(k)}(0)}{k!}x^k + \cdots. \tag{3}$$

The series in (3) is often called the **Taylor series** of $f(x)$ at $x = 0$ because the partial sums of the series are the Taylor polynomials of $f(x)$ at $x = 0$. The entire equation (3) is called the **Taylor series expansion** of $f(x)$ at $x = 0$.

EXAMPLE 1　**A Taylor Series Centered at 0**　Find the Taylor series expansion of

$$\frac{1}{1-x} \quad \text{at } x = 0.$$

SOLUTION　We already know how to represent $\dfrac{1}{1-x}$ as a power series for $|x| < 1$. However, let us use the formula for the Taylor series to see if we get the same result.

$$f(x) = \frac{1}{1-x} = (1-x)^{-1}, \qquad f(0) = 1,$$

$$f'(x) = (1-x)^{-2}, \qquad\qquad f'(0) = 1,$$

$$f''(x) = 2(1-x)^{-3}, \qquad\qquad f''(0) = 2,$$

$$f'''(0) = 3 \cdot 2(1-x)^{-4}, \qquad f'''(0) = 3 \cdot 2,$$

$$f^{(4)}(x) = 4 \cdot 3 \cdot 2(1-x)^{-5}, \qquad f^{(4)}(0) = 4 \cdot 3 \cdot 2,$$

$$\vdots \qquad\qquad\qquad \vdots$$

Therefore,

$$\frac{1}{1-x} = 1 + \frac{1}{1!}x + \frac{2}{2!}x^2 + \frac{3 \cdot 2}{3!}x^3 + \frac{4 \cdot 3 \cdot 2}{4!}x^4 + \cdots$$

$$= 1 + x + x^2 + x^3 + x^4 + \cdots.$$

We have verified that the Taylor series for $\dfrac{1}{1-x}$ is the familiar geometric power series. The Taylor series expansion is valid for $|x| < 1$. **» Now Try Exercise 1**

EXAMPLE 2

A Taylor Series Centered at 0 Find the Taylor series at $x = 0$ for $f(x) = e^x$.

SOLUTION

$$f(x) = e^x, \quad f'(x) = e^x, \quad f''(x) = e^x, \quad f'''(x) = e^x, \ldots$$
$$f(0) = 1, \quad f'(0) = 1, \quad f''(0) = 1, \quad f'''(0) = 1, \ldots.$$

Therefore,

$$e^x = 1 + x + \frac{1}{2!}x^2 + \frac{1}{3!}x^3 + \frac{1}{4!}x^4 + \cdots.$$

It can be shown that this Taylor series expansion of e^x is valid for all x. (*Note:* A Taylor polynomial of e^x gives only an approximation of e^x, but the infinite Taylor series actually *equals* e^x for all x, in the sense that for any given x the sum of the series is the same as the value of e^x.) **» Now Try Exercise 3**

Operations on Taylor Series

It is often helpful to think of a Taylor series as a polynomial of infinite degree. Many operations on polynomials are also legitimate for Taylor series, provided that we restrict attention to values of x within an appropriate interval. For example, if we have a Taylor series expansion of $f(x)$, we may differentiate the series term by term to obtain the Taylor series expansion of $f'(x)$. An analogous result holds for antiderivatives. Other permissible operations that produce Taylor series include multiplying a Taylor series expansion by a constant or a power of x, replacing x by a power of x or by a constant times a power of x, and adding or subtracting two Taylor series expansions. The use of such operations often makes it possible to find the Taylor series of a function without directly using the formal definition of a Taylor series. (The process of computing higher-order derivatives can become quite laborious when the product or quotient rule is involved.) Once a power series expansion of a function $f(x)$ is found, that series *must* be the Taylor series of the function, since the coefficients of the series are uniquely determined by $f(x)$ and its derivatives at $x = 0$.

EXAMPLE 3

Differentiation and Integration Term by Term Use the Taylor series at $x = 0$ for

$$\frac{1}{1-x}$$

to find the Taylor series at $x = 0$ for the following functions

(a) $\dfrac{1}{(1-x)^2}$ **(b)** $\dfrac{1}{(1-x)^3}$ **(c)** $\ln(1-x)$

SOLUTION We begin with the series expansion

$$\frac{1}{1-x} = 1 + x + x^2 + x^3 + x^4 + x^5 + \cdots, \quad |x| < 1.$$

(a) When we differentiate both sides of this equation, we obtain

$$\frac{1}{(1-x)^2} = 1 + 2x + 3x^2 + 4x^3 + 5x^4 + \cdots, \quad |x| < 1.$$

(b) Differentiating the series in part (a), we find that

$$\frac{2}{(1-x)^3} = 2 + 3\cdot 2x + 4\cdot 3x^2 + 5\cdot 4x^3 + \cdots, \quad |x| < 1.$$

We may multiply a convergent series by a constant. Multiplying by $\frac{1}{2}$, we have

$$\frac{1}{(1-x)^3} = 1 + 3x + 6x^2 + 10x^3 + \cdots + \frac{(n+2)(n+1)}{2}x^n + \cdots$$

for $|x| < 1$.

(c) For $|x| < 1$, we have

$$\int \frac{1}{1-x}\,dx = \int (1 + x + x^2 + x^3 + \cdots)\,dx,$$

$$-\ln(1-x) + C = x + \tfrac{1}{2}x^2 + \tfrac{1}{3}x^3 + \tfrac{1}{4}x^4 + \cdots,$$

where C is the constant of integration. If we set $x = 0$ in both sides, we obtain

$$0 + C = 0,$$

so $C = 0$. Thus,

$$\ln(1-x) = -x - \tfrac{1}{2}x^2 - \tfrac{1}{3}x^3 - \tfrac{1}{4}x^4 - \cdots, \quad |x| < 1.$$

>> *Now Try Exercise 9*

EXAMPLE 4	**Evaluating a Taylor Series** Use the results of Example 3(c) to compute ln 1.1.

SOLUTION   Take $x = -.1$ in the Taylor series expansion of $\ln(1-x)$. Then,

$$\ln(1-(-.1)) = -(-.1) - \tfrac{1}{2}(-.1)^2 - \tfrac{1}{3}(-.1)^3 - \tfrac{1}{4}(-.1)^4 - \cdots$$

$$\ln 1.1 = .1 - \frac{.01}{2} + \frac{.001}{3} - \frac{.0001}{4} + \frac{.00001}{5} - \cdots.$$

This infinite series may be used to compute ln 1.1 to any degree of accuracy required. For instance, the fifth partial sum gives $\ln 1.1 \approx .09531$, which is correct to five decimal places.   ◀◀

EXAMPLE 5	**New Taylor Series from Known Ones** Use the Taylor series at $x = 0$ for e^x to find the Taylor series at $x = 0$ for

(a) $x(e^x - 1)$ **(b)** e^{x^2}

SOLUTION   **(a)** If we subtract 1 from the Taylor series for e^x, we obtain a series that converges to $e^x - 1$:

$$e^x - 1 = \left(1 + x + \frac{1}{2!}x^2 + \frac{1}{3!}x^3 + \frac{1}{4!}x^4 + \cdots\right) - 1$$

$$= x + \frac{1}{2!}x^2 + \frac{1}{3!}x^3 + \frac{1}{4!}x^4 + \cdots.$$

Now we multiply this series by x, term by term:

$$x(e^x - 1) = x^2 + \frac{1}{2!}x^3 + \frac{1}{3!}x^4 + \frac{1}{4!}x^5 + \cdots.$$

(b) To obtain the Taylor series for e^{x^2}, we replace every occurrence of x with x^2 in the Taylor series for e^x,

$$e^{x^2} = 1 + (x^2) + \frac{1}{2!}(x^2)^2 + \frac{1}{3!}(x^2)^3 + \frac{1}{4!}(x^2)^4 + \cdots$$

$$= 1 + x^2 + \frac{1}{2!}x^4 + \frac{1}{3!}x^6 + \frac{1}{4!}x^8 + \cdots.$$

>> *Now Try Exercise 21*

EXAMPLE 6   **New Taylor Series from Known Ones** Find the Taylor series at $x = 0$ for

(a) $\dfrac{1}{1 + x^3}$ (b) $\dfrac{x^2}{1 + x^3}$

SOLUTION   (a) In the Taylor series at $x = 0$ for $\dfrac{1}{1 - x}$, we replace x by $-x^3$, to obtain

$$\frac{1}{1 - (-x^3)} = 1 + (-x^3) + (-x^3)^2 + (-x^3)^3 + (-x^3)^4 + \cdots$$

$$\frac{1}{1 + x^3} = 1 - x^3 + x^6 - x^9 + x^{12} - \cdots.$$

(b) If we multiply the series in part (a) by x^2, we obtain

$$\frac{x^2}{1 + x^3} = x^2 - x^5 + x^8 - x^{11} + x^{14} - \cdots.$$

» Now Try Exercise 25

Definite Integrals   The **standard normal curve** of statistics has the equation

$$y = \frac{1}{\sqrt{2\pi}} e^{-x^2/2}.$$

Areas under the curve cannot be found by direct integration since there is no simple formula for an antiderivative of $e^{-x^2/2}$. However, Taylor series can be used to calculate these areas with a high degree of accuracy.

EXAMPLE 7   **Integration Term by Term** Find the area under the standard normal curve from $x = 0$ to $x = .8$; that is, calculate

$$\frac{1}{\sqrt{2\pi}} \int_0^{0.8} e^{-x^2/2}\, dx.$$

SOLUTION   A Taylor expansion for e^x was obtained in Example 2:

$$e^x = 1 + x + \frac{1}{2!}x^2 + \frac{1}{3!}x^3 + \frac{1}{4!}x^4 + \cdots.$$

Replace each occurrence of x by $-x^2/2$. Then,

$$e^{-x^2/2} = 1 + \left(-\frac{x^2}{2}\right) + \frac{1}{2!}\left(-\frac{x^2}{2}\right)^2 + \frac{1}{3!}\left(-\frac{x^2}{2}\right)^3 + \frac{1}{4!}\left(-\frac{x^2}{2}\right)^4 + \cdots,$$

$$e^{-x^2/2} = 1 - \frac{1}{2 \cdot 1!}x^2 + \frac{1}{2^2 \cdot 2!}x^4 - \frac{1}{2^3 \cdot 3!}x^6 + \frac{1}{2^4 \cdot 4!}x^8 - \cdots.$$

Integrating, we obtain

$$\frac{1}{\sqrt{2\pi}} \int_0^{0.8} e^{-x^2/2}\, dx$$

$$= \frac{1}{\sqrt{2\pi}}\left(x - \frac{1}{3 \cdot 2 \cdot 1!}x^3 + \frac{1}{5 \cdot 2^2 \cdot 2!}x^5 - \frac{1}{7 \cdot 2^3 \cdot 3!}x^7 + \frac{1}{9 \cdot 2^4 \cdot 4!}x^9 - \cdots\right)\Bigg|_0^{0.8}$$

$$= \frac{1}{\sqrt{2\pi}}\left[.8 - \frac{1}{6}(.8)^3 + \frac{1}{40}(.8)^5 - \frac{1}{336}(.8)^7 + \frac{1}{3456}(.8)^9 - \cdots\right].$$

The infinite series on the right converges to the value of the definite integral. Summing up the five terms displayed gives the approximation .28815, which is accurate to four places. We can make this approximation arbitrarily accurate by summing additional terms.

» Now Try Exercise 39

Convergence of Power Series When we differentiate, integrate, or perform algebraic operations on Taylor series, we are using the fact that Taylor series are *functions*. In fact, any power series in x is a function of x, whether or not its coefficients are obtained from the derivatives of some function. The **domain of a power series function** is the set of all x for which the series converges. The function value at a specific x in its domain is the number to which the series converges.

For instance, the geometric series $\sum_{k=0}^{\infty} x^k$ defines a function whose domain is the set of all x for which $|x| < 1$. The familiar Taylor series expansion

$$\frac{1}{1-x} = \sum_{k=0}^{\infty} x^k$$

simply states that the functions $\frac{1}{1-x}$ and $\sum_{k=0}^{\infty} x^k$ have the same value for each x such that $|x| < 1$.

Given any power series $\sum_{k=0}^{\infty} a_k x^k$, one of three possibilities must occur:

(i) There is a positive constant R such that the series converges for $|x| < R$ and diverges for $|x| > R$.

(ii) The series converges for all x.

(iii) The series converges only for $x = 0$.

In case (i), we call R the **radius of convergence** of the series. The series converges for all x in the interval $-R < x < R$ and may or may not converge at one or both of the endpoints of this interval. In case (ii), we say that the radius of convergence is ∞, and in case (iii) we say that the radius of convergence is 0.

When a power series with a positive radius of convergence is differentiated term by term, the new series will have the same radius of convergence. An analogous result holds for antiderivatives. Other operations, such as replacing x by a constant times a power of x, may affect the radius of convergence.

Suppose that we begin with a function that has derivatives of all orders, and we write down its formal Taylor series at $x = 0$. Can we conclude that the Taylor series and the function have the same values for every x within the radius of convergence of the series? For all the functions that we consider, the answer is yes. However, it is possible for the two values to differ. In this case, we say that the function does not admit a power series expansion. To show that a function admits a power series expansion, it is necessary to show that the partial sums of the Taylor series converge to the function. The nth partial sum of the series is the nth Taylor polynomial p_n. Recall from Section 11.1 that we considered the nth remainder of $f(x)$,

$$R_n(x) = f(x) - p_n(x).$$

For a fixed x, the Taylor series converges to $f(x)$ if, and only if, $R_n(x) \to 0$ as $n \to \infty$. Exercises 45 and 46 illustrate how convergence can be verified by use of the remainder formula from Section 11.1.

Taylor Series at $x = a$ To simplify the discussion in this section, we have restricted our attention to series that involve powers of x, rather than powers of $x - a$. However, Taylor series, just like Taylor polynomials, can be formed as sums of powers of $x - a$. The Taylor expansion of $f(x)$ at $x = a$ is

$$f(x) = f(a) + \frac{f'(a)}{1!}(x - a) + \frac{f''(a)}{2!}(x - a)^2 + \frac{f'''(a)}{3!}(x - a)^3 + \cdots$$
$$+ \frac{f^{(n)}(a)}{n!}(x - a)^n + \cdots.$$

Check Your Understanding 11.5

Solutions can be found following the section exercises.

1. Find the Taylor series expansion of $\sin x$ at $x = 0$.
2. Find the Taylor series expansion of $\cos x$ at $x = 0$.
3. Find the Taylor series expansion of $x^3 \cos 7x$ at $x = 0$.

4. If $f(x) = x^3 \cos 7x$, find $f^{(5)}(0)$. [*Hint:* How are the coefficients in the Taylor series of $f(x)$ related to $f(x)$ and its derivatives at $x = 0$?]

EXERCISES 11.5

In Exercises 1–4, find the Taylor series at $x = 0$ of the given function by computing three or four derivatives and using the definition of the Taylor series.

1. $\dfrac{1}{2x + 3}$

2. $\ln(1 - 3x)$

3. $\sqrt{1 + x}$

4. $(1 + x)^3$

In Exercises 5–20, find the Taylor series at $x = 0$ of the given function. Use suitable operations (differentiation, substitution, etc.) on the Taylor series at $x = 0$ of $\dfrac{1}{1 - x}$, e^x, or $\cos x$. These series are derived in Examples 1 and 2 and Check Your Understanding Problem 2.

5. $\dfrac{1}{1 - 3x}$

6. $\dfrac{1}{1 + x}$

7. $\dfrac{1}{1 + x^2}$

8. $\dfrac{x}{1 + x^2}$

9. $\dfrac{1}{(1 + x)^2}$

10. $\dfrac{x}{(1 - x)^3}$

11. $5e^{x/3}$

12. $x^3 e^{x^2}$

13. $1 - e^{-x}$

14. $3(e^{-2x} - 2)$

15. $\ln(1 + x)$

16. $\ln(1 + x^2)$

17. $\cos 3x$

18. $\cos x^2$

19. $\sin 3x$

20. $x \sin x^2$

21. Find the Taylor series of xe^{x^2} at $x = 0$.

22. Show that $\ln\left(\dfrac{1 + x}{1 - x}\right) = 2x + \tfrac{2}{3}x^3 + \tfrac{2}{5}x^5 + \tfrac{2}{7}x^7 + \cdots$, $|x| < 1$.

[*Hint:* Use Exercise 15 and Example 3.] This series converges much more quickly than the series for $\ln(1 - x)$ in Example 3, particularly for x close to zero. The series gives a formula for $\ln y$, where y is any number and $x = \dfrac{y - 1}{y + 1}$.

23. The **hyperbolic cosine** of x, denoted by $\cosh x$, is defined by
$$\cosh x = \tfrac{1}{2}(e^x + e^{-x}).$$
This function occurs often in physics and probability theory. The graph of $y = \cosh x$ is called a *catenary*.
   (a) Use differentiation and the definition of a Taylor series to compute the first four nonzero terms in the Taylor series of $\cosh x$ at $x = 0$.
   (b) Use the known Taylor series for e^x to obtain the Taylor series for $\cosh x$ at $x = 0$.

24. The **hyperbolic sine** of x is defined by
$$\sinh x = \tfrac{1}{2}(e^x - e^{-x}).$$
Repeat parts (a) and (b) of Exercise 23 for $\sinh x$.

25. Given the Taylor series expansion
$$\frac{1}{\sqrt{1 + x}} = 1 - \frac{1}{2}x + \frac{1 \cdot 3}{2 \cdot 4}x^2 - \frac{1 \cdot 3 \cdot 5}{2 \cdot 4 \cdot 6}x^3 + \frac{1 \cdot 3 \cdot 5 \cdot 7}{2 \cdot 4 \cdot 6 \cdot 8}x^4 - \cdots,$$
find the first four terms in the Taylor series of $\dfrac{1}{\sqrt{1 - x}}$ at $x = 0$.

26. Find the first four terms in the Taylor series of $\dfrac{1}{\sqrt{1 - x^2}}$ at $x = 0$. (See Exercise 25.)

27. Use Exercise 25 and the fact that
$$\int \frac{1}{\sqrt{1 + x^2}} \, dx = \ln(x + \sqrt{1 + x^2}) + C$$
to find the Taylor series of $\ln(x + \sqrt{1 + x^2})$ at $x = 0$.

28. Use the Taylor series expansion for $\dfrac{x}{(1 - x)^2}$ to find the function whose Taylor series is
$$1 + 4x + 9x^2 + 16x^3 + 25x^4 + \cdots.$$

29. Use the Taylor series for e^x to show that $\dfrac{d}{dx} e^x = e^x$.

30. Use the Taylor series for $\cos x$ (see Problem 2 in Check Your Understanding) to show that $\cos(-x) = \cos x$.

31. The Taylor series at $x = 0$ for
$$f(x) = \ln\left(\frac{1 + x}{1 - x}\right)$$
is given in Exercise 22. Find $f^{(5)}(0)$.

32. The Taylor series at $x = 0$ for $f(x) = \sec x$ is
$1 + \tfrac{1}{2}x^2 + \tfrac{5}{24}x^4 + \tfrac{61}{720}x^6 + \cdots$. Find $f^{(4)}(0)$.

33. The Taylor series at $x = 0$ for $f(x) = \tan x$ is
$x + \tfrac{1}{3}x^3 + \tfrac{2}{15}x^5 + \tfrac{17}{315}x^7 + \cdots$. Find $f^{(4)}(0)$.

34. The Taylor series at $x = 0$ for $\dfrac{1 + x^2}{1 - x}$ is
$1 + x + 2x^2 + 2x^3 + 2x^4 + \cdots$. Find $f^{(4)}(0)$, where
$f(x) = \dfrac{1 + x^4}{1 - x^2}$.

In Exercises 35–37, find the Taylor series expansion at $x = 0$ of the given antiderivative.

35. $\displaystyle\int e^{-x^2}\,dx$ **36.** $\displaystyle\int xe^{x^3}\,dx$ **37.** $\displaystyle\int \frac{1}{1+x^3}\,dx$

In Exercises 38–40, find an infinite series that converges to the value of the given definite integral.

38. $\displaystyle\int_0^1 \sin x^2\,dx$ **39.** $\displaystyle\int_0^1 e^{-x^2}\,dx$ **40.** $\displaystyle\int_0^1 xe^{x^3}\,dx$

41. (a) Use the Taylor series for e^x at $x = 0$ to show that $e^x > x^2/2$ for $x > 0$.
   (b) Deduce that $e^{-x} < 2/x^2$ for $x > 0$.
   (c) Show that xe^{-x} approaches 0 as $x \to \infty$.

42. Let k be a positive constant.

   (a) Show that $e^{kx} > \dfrac{k^2x^2}{2}$, for $x > 0$.

   (b) Deduce that $e^{-kx} < \dfrac{2}{k^2x^2}$, for $x > 0$.

   (c) Show that xe^{-kx} approaches 0 as $x \to \infty$.

43. Show that $e^x > x^3/6$ for $x > 0$, and from this, deduce that x^2e^{-x} approaches 0 as $x \to \infty$.

44. If k is a positive constant, show that x^2e^{-kx} approaches 0 as $x \to \infty$.

Exercises 45 and 46 rely on the fact that

$$\lim_{n \to \infty} \frac{|x|^{n+1}}{(n+1)!} = 0.$$

The proof of this fact is omitted.

45. Let $R_n(x)$ be the nth remainder of $f(x) = \cos x$ at $x = 0$. (See Section 11.1.) Show that, for any fixed value of x, $|R_n(x)| \le |x|^{n+1}/(n+1)!$, and hence, conclude that $|R_n(x)| \to 0$ as $n \to \infty$. This shows that the Taylor series for $\cos x$ converges to $\cos x$ for every value of x.

46. Let $R_n(x)$ be the nth remainder of $f(x) = e^x$ at $x = 0$. (See Section 11.1.) Show that, for any fixed value of x, $|R_n(x)| \le e^{|x|}\cdot|x|^{n+1}/(n+1)!$, and hence, conclude that $|R_n(x)| \to 0$ as $n \to \infty$. This shows that the Taylor series for e^x converges to e^x for every value of x.

Solutions to Check Your Understanding 11.5

1. Use the definition of the Taylor series as an extended Taylor polynomial:

$$f(x) = \sin x, \qquad f'(x) = \cos x,$$
$$f(0) = 0, \qquad f'(0) = 1,$$
$$f''(x) = -\sin x, \qquad f'''(x) = -\cos x,$$
$$f''(0) = 0, \qquad f'''(0) = -1,$$
$$f^{(4)}(x) = \sin x, \qquad f^{(5)}(x) = \cos x,$$
$$f^{(4)}(0) = 0, \qquad f^{(5)}(0) = 1.$$

Therefore,

$$\sin x = 0 + 1\cdot x + \frac{0}{2!}x^2 + \frac{-1}{3!}x^3 + \frac{0}{4!}x^4 + \frac{1}{5!}x^5 + \cdots$$
$$= x - \frac{1}{3!}x^3 + \frac{1}{5!}x^5 - \cdots.$$

2. Differentiate the Taylor series in the previous problem:

$$\frac{d}{dx}\sin x = \frac{d}{dx}\left(x - \frac{1}{3!}x^3 + \frac{1}{5!}x^5 - \cdots\right),$$
$$\cos x = 1 - \frac{1}{2!}x^2 + \frac{1}{4!}x^4 - \cdots.$$

$\left[\, \textit{Note:} \text{ We use the fact that } \dfrac{3}{3!} = \dfrac{3}{1\cdot 2\cdot 3} = \dfrac{1}{1\cdot 2} = \dfrac{1}{2!} \text{ and}\right.$

$\left.\dfrac{5}{5!} = \dfrac{1}{4!}.\right]$

3. Replace x by $7x$ in the Taylor series for $\cos x$ and then multiply by x^3:

$$\cos x = 1 - \frac{1}{2!}x^2 + \frac{1}{4!}x^4 - \cdots$$

$$\cos 7x = 1 - \frac{1}{2!}(7x)^2 + \frac{1}{4!}(7x)^4 - \cdots$$

$$= 1 - \frac{7^2}{2!}x^2 + \frac{7^4}{4!}x^4 - \cdots$$

$$x^3\cos 7x = x^3\left(1 - \frac{7^2}{2!}x^2 + \frac{7^4}{4!}x^4 - \cdots\right)$$

$$= x^3 - \frac{7^2}{2!}x^5 + \frac{7^4}{4!}x^7 - \cdots.$$

4. The coefficient of x^5 in the Taylor series of $f(x)$ is $\dfrac{f^{(5)}(0)}{5!}$. By Problem 3, this coefficient is $-\dfrac{7^2}{2!}$. Therefore,

$$\frac{f^{(5)}(0)}{5!} = -\frac{7^2}{2!}$$

$$f^{(5)}(0) = -\frac{7^2}{2!}\cdot 5! = -\frac{49}{2}\cdot 120$$
$$= -(49)(60) = -2940.$$

CHAPTER 11 Summary

KEY TERMS AND CONCEPTS	EXAMPLES

11.1 Taylor Polynomials

To approximate the values of a function $f(x)$ near a point $x = a$, we can use a *Taylor polynomial*, $p_n(x)$, of degree n. The polynomial is defined in terms of the values of f and its derivatives at $x = a$, by the following formula:

$$p_n(x) = f(a) + \frac{f'(a)}{1!}(x - a) + \frac{f''(a)}{2!}(x - a)^2 + \cdots$$
$$+ \frac{f^{(n)}(a)}{n!}(x - a)^n.$$

This polynomial coincides with $f(x)$ up to the nth derivative at $x = a$ in the sense that

$$p_n(a) = f(a), \quad p_n'(a) = f'(a), \ldots, p_n^{(n)}(a) = f^{(n)}(a).$$

For all values of x near a, $p_n(x)$ approximates $f(x)$.

(a) Find the second Taylor polynomial of $f(x) = e^{x^2}$ at $x = 0$.

(b) Use the second Taylor polynomial to approximate $e^{0.25}$.

Solution

(a) The second Taylor polynomial has degree 2 and is given by

$$p_2(x) = f(0) + f'(0)x + \frac{f''(0)}{2}x^2$$

We have

$$f(x) = ex^2, \quad f(0) = 1$$
$$f'(x) = 2xe^{x^2}, \quad f'(0) = 0$$
$$f''(x) = 2x(2x)e^{x^2} + 2e^{x^2}, \quad f''(0) = 2.$$

Thus, $p_2(x) = 1 + \frac{2}{2}x^2 = 1 + x^2$.

(b) Note that $.25 = (.5)^2$. Using the approximation $e^{x^2} \approx p_2(x)$, we find

$$e^{0.25} = e^{(0.5)^2} \approx p_2(.5) = 1 + (.5)^2 = 1.25.$$

(With the help of a calculator, we found $e^{0.25} \approx 1.28$, which is close to our Taylor polynomial approximation.)

11.2 The Newton–Raphson Algorithm

Suppose that r is a zero of the function $f(x)$; that is, $f(r) = 0$. In many situations, it is difficult to find the exact value of r. The *Newton–Raphson algorithm* is one of several methods that can be used to find an approximate value of r to any desired degree of accuracy. The algorithm is defined by the following formula

$$x_{\text{new}} = x_{\text{old}} - \frac{f(x_{\text{old}})}{f'(x_{\text{old}})}$$

and works as follows.

Step 1 Based on the graph of f or by estimating, guess a value x_0 that is close to r. This is your first approximation of r.

Step 2 Use x_0 in the Newton–Raphson algorithm to generate the value x_1. So,

$$x_1 = x_0 - \frac{f(x_0)}{f'(x_0)}.$$

Step 3 Using x_1 in the formula, compute x_2. Continue in this manner to obtain a sequence of approximations $x_0, x_1, x_2, \ldots$, which usually approach as close to r as desired.

Approximate the first positive solution of $e^{0.2x} - 3x = 0$.

Solution The graph of $f(x) = e^{0.2x} - 3x$ indicates that there are two positive solutions: a first one that lies between 0 and 2; and a second one that lies between 20 and 25 (see Fig. 1). To approximate the first one, let us apply the Newton–Raphson algorithm to $f(x)$ with $x_0 = 1$. Since $f'(x) = .2e^{0.2x} - 3$,

$$x_1 = x_0 - \frac{e^{0.2x_0} - 3x_0}{0.2e^{0.2x_0} - 3} = 1 - \frac{e^{0.2} - 3}{.2e^{0.2} - 3} \approx 1 - \frac{-1.7786}{-2.75572} \approx .35458,$$

$$x_2 = .35458 - \frac{e^{(0.2)(0.35458)} - 3(.35458)}{.2e^{(0.2)(0.35458)} - 3} \approx .358081$$

$$x_3 = .358081 - \frac{e^{(0.2)(0.358081)} - 3(.358081)}{.2e^{(0.2)(0.358081)} - 3} \approx .358081.$$

Therefore, an approximate solution is $x = .358081$.

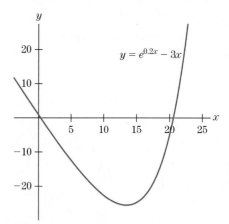

Figure 1

KEY TERMS AND CONCEPTS	EXAMPLES

11.3 Infinite Series

An *infinite series* is an infinite addition of numbers

$$a_1 + a_2 + a_3 + a_4 + \cdots.$$

The sum of the first n terms of an infinite series is called its *nth partial sum* and is denoted by S_n. If S_n approaches a limit as n increases, then the series is called *convergent*. The limit of S_n is then called the *sum* of the infinite series. An infinite series whose partial sums do not approach a limit is called *divergent*.

A very important series is the *geometric series with ratio r*:

$$a + ar + ar^2 + ar^3 + ar^4 + \cdots.$$

The geometric series converges if, and only if, $|r| < 1$. When $|r| < 1$, the sum of the series is

$$\frac{a}{1-r}.$$

A convenient notation for series and sums is the sigma notation. In this notation, the geometric series may be written as

$$\sum_{j=0}^{\infty} ar^j.$$

Determine the sums of the following geometric series, when they are convergent.

(a) $7 + \dfrac{7}{6} + \dfrac{7}{6^2} + \dfrac{7}{6^3} + \dfrac{7}{6^4} + \cdots$ **(b)** $\left(\dfrac{3}{4}\right)^2 + \left(\dfrac{3}{4}\right)^3 + \left(\dfrac{3}{4}\right)^4 + \cdots$

Solution

(a) This is a geometric series with $a = 7$ and $r = 1/6$. Since $|r| < 1$, the series is convergent and its sum is

$$\frac{a}{1-r} = \frac{7}{1-\frac{1}{6}} = \frac{42}{5}.$$

(b) We have $a = \left(\frac{3}{4}\right)^2$ and, to find r, we compute the ratio of two consecutive terms in this geometric series. Let us consider the first two terms

$$r = \frac{\left(\frac{3}{4}\right)^3}{\left(\frac{3}{4}\right)^2} = \frac{3}{4}.$$

Since $|r| < 1$, the series is convergent and its sum is

$$\frac{a}{1-r} = \frac{\left(\frac{3}{4}\right)^2}{1-\frac{3}{4}} = \left(\frac{3}{4}\right)^2 (4) = \frac{9}{4}.$$

11.4 Series with Positive Terms

The Integral Test Let $f(x)$ be continuous, decreasing, and positive for $x \geq 1$. Then, the infinite series

$$\sum_{k=1}^{\infty} f(k)$$

is convergent if the improper integral

$$\int_1^x f(x)\, dx$$

is convergent, and the infinite series is divergent if the improper integral is divergent.

The Comparison Test Suppose that $0 \leq a_k \leq b_k$ for $k = 1, 2, \ldots$. If $\sum_{k=1}^{\infty} b_k$ converges, so does $\sum_{k=1}^{\infty} a_k$. If $\sum_{k=1}^{\infty} a_k$ diverges, so does $\sum_{k=1}^{\infty} b_k$.

We also have the following version of the comparison test for series with positive and negative terms.

The Comparison Test Suppose that $0 \leq |a_k| \leq b_k$ for $k = 1, 2, \ldots$, and $\sum_{k=1}^{\infty} b_k$ converges, then $\sum_{k=1}^{\infty} a_k$ converges.

Use the integral test to determine whether the infinite series $\sum_{k=1}^{\infty} \frac{1}{k\sqrt{k}}$ is convergent or divergent (You may assume that the hypotheses of the integral test are satisfied.)

Solution The series converges if the improper integral

$$\int_1^{\infty} \frac{1}{x\sqrt{x}}\, dx$$

converges. We have

$$\int_1^{\infty} \frac{1}{x\sqrt{x}}\, dx = \lim_{b \to \infty} \int_1^b \frac{1}{x\sqrt{x}}\, dx = \lim_{b \to \infty} \int_1^b \frac{1}{x^{3/2}}\, dx$$

$$= \lim_{b \to \infty} \int_1^b x^{-3/2}\, dx = \lim_{b \to \infty} \frac{1}{-\frac{3}{2}+1} x^{-\frac{3}{2}+1} \Big|_1^b$$

$$= \lim_{b \to \infty} -2x^{-\frac{1}{2}} \Big|_1^b = \lim_{b \to \infty} \left[-2b^{-\frac{1}{2}} - \left(-2(1)^{-\frac{1}{2}} \right) \right]$$

$$= \lim_{b \to \infty} \left[-\frac{2}{\sqrt{b}} + 2 \right] = 2.$$

Since the integral converges, the series converges.

KEY TERMS AND CONCEPTS	EXAMPLES

11.5 Taylor Series

A *power series* is a series that is written in terms of powers of x and is of the form

$$a_0 + a_1 x + \cdots + a_n x^n + \cdots .$$

Roughly speaking, a power series is a polynomial of infinite degree in x. A *Taylor series expansion* of $f(x)$ at $x = 0$ is the series

$$f(x) = f(0) + \frac{f'(0)}{1!} x + \frac{f''(0)}{2!} x^2 + \cdots + \frac{f^{(k)}(0)}{k!} x^k + \cdots .$$

A Taylor series is a particular power series used to represent $f(x)$ near $x = 0$. The series representation is either valid for all x; or is valid for $x = 0$ only; or is valid in an interval $(-R, R)$, centered at $x = 0$, with radius $R > 0$, called the *radius of convergence* of the Taylor series. In the latter case, the series diverges (and thus is not valid) outside the interval $[-R, R]$.

A Taylor series representation may be centered at a point $x = a$, where a may be different from 0. In this case, we have the Taylor series representation

$$f(x) = f(a) + \frac{f'(a)}{1!} (x - a) + \frac{f''(a)}{2!} (x - a)^2 + \cdots$$
$$+ \frac{f^{(k)}(a)}{k!} (x - a)^k + \cdots .$$

1. Find the Taylor series of $f(x) = e^{2x}$ at $x = 0$ by computing three derivatives and using the definition of the Taylor series.

   Solution

   $$\begin{aligned} f(x) &= e^{2x}, & f(0) &= 1, \\ f'(x) &= 2e^{2x}, & f'(0) &= 2, \\ f''(x) &= 4e^{2x}, & f''(0) &= 4, \\ f'''(x) &= 8e^{2x}, & f'''(0) &= 8 \end{aligned}$$

   Therefore,

   $$\begin{aligned} e^{2x} &= f(0) + \frac{f'(0)}{1!} x + \frac{f''(0)}{2!} x^2 + \frac{f^{(3)}(0)}{3!} x^3 + \cdots \\ &= 1 + 2x + \frac{4}{2!} x^2 + \frac{8}{3!} x^3 + \cdots \\ &= 1 + 2x + 2x^2 + \frac{4}{3} x^3 + \cdots . \end{aligned}$$

2. Find the Taylor series of $f(x) = e^{2x}$ at $x = 0$ by using the Taylor series expansion of e^x.

   Solution In the Taylor series expansion of e^u at $u = 0$,

   $$e^u = 1 + u + \frac{u^2}{2!} + \frac{u^3}{3!} + \cdots ,$$

   we substitute $u = 2x$ and get

   $$e^{2x} = 1 + 2x + \frac{(2x)^2}{2!} + \frac{(2x)^3}{3!} + \cdots = 1 + 2x + 2x^2 + \frac{4}{3} x^3 + \cdots .$$

   The series is precisely the one that we found in Example 1. Since the Taylor series of e^u is valid for all u, we conclude that the series expansion of e^{2x} is valid for all x.

CHAPTER 11 Fundamental Concept Check Exercises

1. Define the nth Taylor polynomial of $f(x)$ at $x = a$.

2. In what way is the nth Taylor polynomial of $f(x)$ at $x = a$ like $f(x)$ at $x = a$?

3. State the remainder formula for the nth Taylor polynomial of $f(x)$ at $x = a$.

4. Explain how the Newton–Raphson algorithm is used to approximate a zero of a function.

5. What is the nth partial sum of an infinite series?

6. What is a convergent infinite series? Divergent?

7. What is meant by the sum of a convergent infinite series?

8. What is a geometric series and when does it converge?

9. What is the sum of a convergent geometric series?

10. Define the Taylor series of $f(x)$ at $x = 0$.

11. Discuss the three possibilities for the radius of convergence of a Taylor series.

CHAPTER 11 Review Exercises

1. Find the second Taylor polynomial of $x(x + 1)^{3/2}$ at $x = 0$.

2. Find the fourth Taylor polynomial of $(2x + 1)^{3/2}$ at $x = 0$.

3. Find the fifth Taylor polynomial of $x^3 - 7x^2 + 8$ at $x = 0$.

4. Find the nth Taylor polynomial of $\dfrac{2}{2 - x}$ at $x = 0$.

5. Find the third Taylor polynomial of x^2 at $x = 3$.

6. Find the third Taylor polynomial of e^x at $x = 2$.

7. Use a second Taylor polynomial at $t = 0$ to estimate the area under the graph of $y = -\ln(\cos 2t)$ between $t = 0$ and $t = \frac{1}{2}$.

8. Use a second Taylor polynomial at $x = 0$ to estimate the value of $\tan(.1)$.

9. (a) Find the second Taylor polynomial of $\sqrt{x}$ at $x = 9$.
   (b) Use part (a) to estimate $\sqrt{8.7}$ to six decimal places.
   (c) Use the Newton–Raphson algorithm with $n = 2$ and $x_0 = 3$ to approximate the solution of the equation $x^2 - 8.7 = 0$. Express your answer to six decimal places.

10. (a) Use the third Taylor polynomial of $\ln(1 - x)$ at $x = 0$ to approximate $\ln 1.3$ to four decimal places.
    (b) Find an approximate solution of the equation $e^x = 1.3$ using the Newton–Raphson algorithm with $n = 2$ and $x_0 = 0$. Express your answer to four decimal places.

11. Use the Newton–Raphson algorithm with $n = 2$ to approximate the zero of $x^2 - 3x - 2$ near $x_0 = 4$.

12. Use the Newton–Raphson algorithm with $n = 3$ to approximate the solution of the equation $e^{2x} = 1 + e^{-x}$.

In Exercises 13–20, find the sum of the given infinite series if it is convergent.

13. $1 - \dfrac{3}{4} + \dfrac{9}{16} - \dfrac{27}{64} + \dfrac{81}{256} - \cdots$

14. $\dfrac{5^2}{6} + \dfrac{5^3}{6^2} + \dfrac{5^4}{6^3} + \dfrac{5^5}{6^4} + \dfrac{5^6}{6^5} + \cdots$

15. $\dfrac{1}{8} + \dfrac{1}{8^2} + \dfrac{1}{8^3} + \dfrac{1}{8^4} + \dfrac{1}{8^5} + \cdots$

16. $\dfrac{2^2}{7} - \dfrac{2^5}{7^2} + \dfrac{2^8}{7^3} - \dfrac{2^{11}}{7^4} + \dfrac{2^{14}}{7^5} - \cdots$

17. $\dfrac{1}{m + 1} + \dfrac{m}{(m + 1)^2} + \dfrac{m^2}{(m + 1)^3} + \dfrac{m^3}{(m + 1)^4} + \cdots$, where m is a positive number

18. $\dfrac{1}{m} - \dfrac{1}{m^2} + \dfrac{1}{m^3} - \dfrac{1}{m^4} + \dfrac{1}{m^5} - \cdots$, where m is a positive number

19. $1 + 2 + \dfrac{2^2}{2!} + \dfrac{2^3}{3!} + \dfrac{2^4}{4!} + \cdots$

20. $1 + \dfrac{1}{3} + \dfrac{1}{2!}\left(\dfrac{1}{3}\right)^2 + \dfrac{1}{3!}\left(\dfrac{1}{3}\right)^3 + \dfrac{1}{4!}\left(\dfrac{1}{3}\right)^4 + \cdots$

21. Use properties of convergent series to find $\displaystyle\sum_{k=0}^{\infty} \dfrac{1 + 2^k}{3^k}$.

22. Find $\displaystyle\sum_{k=0}^{\infty} \dfrac{3^k + 5^k}{7^k}$.

In Exercises 23–26, determine if the given series is convergent.

23. $\displaystyle\sum_{k=1}^{\infty} \dfrac{1}{k^3}$

24. $\displaystyle\sum_{k=1}^{\infty} \dfrac{1}{3^k}$

25. $\displaystyle\sum_{k=1}^{\infty} \dfrac{\ln k}{k}$

26. $\displaystyle\sum_{k=0}^{\infty} \dfrac{k^3}{(k^4 + 1)^2}$

27. For what values of p is $\displaystyle\sum_{k=1}^{\infty} \dfrac{1}{k^p}$ convergent?

28. For what values of p is $\displaystyle\sum_{k=1}^{\infty} \dfrac{1}{p^k}$ convergent?

In Exercises 29–32, find the Taylor series at $x = 0$ of the given function. Use suitable operations on the Taylor series at $x = 0$ of $\dfrac{1}{1 - x}$ and e^x.

29. $\dfrac{1}{1 + x^3}$

30. $\ln(1 + x^3)$

31. $\dfrac{1}{(1 - 3x)^2}$

32. $\dfrac{e^x - 1}{x}$

33. (a) Find the Taylor series of $\cos 2x$ at $x = 0$, either by direct calculation or by using the known series for $\cos x$.
    (b) Use the trigonometric identity
    $$\sin^2 x = \tfrac{1}{2}(1 - \cos 2x)$$
    to find the Taylor series of $\sin^2 x$ at $x = 0$.

34. (a) Find the Taylor series of $\cos 3x$ at $x = 0$.
    (b) Use the trigonometric identity
    $$\cos^3 x = \tfrac{1}{4}(\cos 3x + 3 \cos x)$$
    to find the fourth Taylor polynomial of $\cos^3 x$ at $x = 0$.

35. Use the decomposition
    $$\frac{1 + x}{1 - x} = \frac{1}{1 - x} + \frac{x}{1 - x}$$
    to find the Taylor series of $\dfrac{1 + x}{1 - x}$ at $x = 0$.

36. Find an infinite series that converges to
    $$\int_0^{1/2} \frac{e^x - 1}{x}\, dx.$$

    [*Hint:* Use Exercise 32.]

37. It can be shown that the sixth Taylor polynomial of $f(x) = \sin x^2$ at $x = 0$ is $x^2 - \tfrac{1}{6}x^6$. Use this fact in parts (a), (b), and (c).
    (a) What is the fifth Taylor polynomial of $f(x)$ at $x = 0$?
    (b) What is $f'''(0)$?
    (c) Estimate the area under the graph of $y = \sin x^2$ between $x = 0$ and $x = 1$. Use four decimal places, and compare your answer with the values given in Example 4 of Section 11.1.

38. Let $f(x) = \ln|\sec x + \tan x|$. It can be shown that $f'(0) = 1$, $f''(0) = 0$, $f'''(0) = 1$, and $f^{(4)}(0) = 0$. What is the fourth Taylor polynomial of $f(x)$ at $x = 0$?

39. Let $f(x) = 1 + x^2 + x^4 + x^6 + \cdots$.
    (a) Find the Taylor series expansion of $f'(x)$ at $x = 0$.
    (b) Find the simple formula for $f'(x)$ not involving a series. [*Hint:* First find a simple formula for $f(x)$.]

40. Let $f(x) = x - 2x^3 + 4x^5 - 8x^7 + 16x^9 - \cdots$.
    (a) Find the Taylor series expansion of $\int f(x)\, dx$ at $x = 0$.
    (b) Find a simple formula for $\int f(x)\, dx$ not involving a series. [*Hint:* First find a simple formula for $f(x)$.]

41. **Fractional Reserve Banking** Suppose that the Federal Reserve (the Fed) buys $100 million of government debt obligations from private owners. This creates $100 million of new money and sets off a chain reaction because of the "fractional reserve" banking system. When the $100 million is deposited into private bank accounts, the banks keep only 15% in reserve and may loan out the remaining 85%, creating more

new money: $(.85)(100)$ million dollars. The companies that borrow this money turn around and spend it, and the recipients deposit the money in their bank accounts. Assuming that all the $(.85)(100)$ million is redeposited, the banks may again loan out 85% of this amount, creating $(.85)^2(100)$ million additional dollars. This process may be repeated indefinitely. Compute the total amount of new money that can be created theoretically by this process, beyond the original $100 million. (In practice, only about an additional $300 million is created, usually within a few weeks of the action of the Fed.)

42. **Federal Reserve Banking** Suppose that the Federal Reserve creates $100 million of new money, as in Exercise 41, and the banks lend 85% of all new money they receive. However, suppose that out of each loan, only 80% is redeposited into the banking system. Thus, whereas the first set of loans totals $(.85)(100)$ million dollars, the second set is only 85% of $(.80)(.85)(100)$, or $(.80)(.85)^2(100)$ million, and the next set is 85% of $(.80)^2(.85)^2(100)$, or $(.80)^2(.85)^3(100)$ million dollars, and so on. Compute the total theoretical amount that may be loaned in this situation.

Perpetuity Suppose that, when you die, the proceeds of a life insurance policy will be deposited into a trust fund that will earn 8% interest, compounded continuously. According to the terms of your will, the trust fund must pay to your descendants and their heirs c_1 dollars (total) at the end of the first year, c_2 dollars at the end of the second year, c_3 dollars at the end of the third year, and so on, forever. The amount that must be in the trust fund initially to make the kth payment is $c_k e^{-0.08k}$, the present value of the amount to be paid in k years. So the life insurance policy should pay a total of $\sum_{k=1}^{\infty} c_k e^{-0.08k}$ dollars into the trust fund to provide for all the payments.

43. How large must the insurance policy be if $c_k = 10{,}000$ for all k? (Find the sum of the series.)

44. How large must the insurance policy be if $c_k = 10{,}000(.9)^k$ for all k?

45. Suppose that $c_k = 10{,}000(1.08)^k$ for all k. Find the sum of the series above if the series converges.

Probability and Calculus

In this chapter, we shall survey a few applications of calculus to the theory of probability. Since we do not intend this chapter to be a self-contained course in probability, we shall select only a few salient ideas to present a taste of probability theory and provide a starting point for further study.

12.1 Discrete Random Variables

We will motivate the concepts of mean, variance, standard deviation, and random variable by analyzing examination grades.

Suppose that the grades on an exam taken by 10 people are 50, 60, 60, 70, 70, 90, 100, 100, 100, 100. This information is displayed in a frequency table in Fig. 1.

Grade	50	60	70	90	100
Frequency	1	2	2	1	4

Figure 1

One of the first things we do when looking over the results of an exam is to compute the **mean** or **average** of the grades. We do this by totaling the grades and dividing by the number of people. This is the same as multiplying each distinct grade by the frequency with which it occurs, adding up those products, and dividing by the sum of the frequencies:

$$[\text{mean}] = \frac{50 \cdot 1 + 60 \cdot 2 + 70 \cdot 2 + 90 \cdot 1 + 100 \cdot 4}{10} = \frac{800}{10} = 80.$$

To get an idea of how spread out the grades are, we can compute the difference between each grade and the average grade. We have tabulated these differences in Fig. 2. For example, if a person received a 50, then [grade] − [mean] is $50 - 80 = -30$. As a measure of the spread of the grades, statisticians compute the average of the squares of these differences and call it the **variance** of the grade distribution. We have

[Grade] − [Mean]	−30	−20	−10	10	20
Frequency	1	2	2	1	4

Figure 2

$$[\text{variance}] = \frac{(-30)^2 \cdot 1 + (-20)^2 \cdot 2 + (-10)^2 \cdot 2 + (10)^2 \cdot 1 + (20)^2 \cdot 4}{10}$$

$$= \frac{900 + 800 + 200 + 100 + 1600}{10} = \frac{3600}{10} = 360.$$

The square root of the variance is called the **standard deviation** of the grade distribution. In this case, we have

$$[\text{standard deviation}] = \sqrt{360} \approx 18.97.$$

There is another way of looking at the grade distribution and its mean and variance. This new point of view is useful because it can be generalized to other situations. We begin by converting the frequency table to a relative frequency table. (See Fig. 3.) Below each grade we list the fraction of the class receiving that grade. The grade of 50 occurred $\frac{1}{10}$ of the time, the grade of 60 occurred $\frac{2}{10}$ of the time, and so on. Note that the relative frequencies add up to 1, because they represent the various fractions of the class grouped by test scores.

Grade	50	60	70	90	100
Relative frequency	$\frac{1}{10}$	$\frac{2}{10}$	$\frac{2}{10}$	$\frac{1}{10}$	$\frac{4}{10}$

Figure 3

It is sometimes helpful to display the data in the relative frequency table by constructing a **relative frequency histogram**. (See Fig. 4.) Over each grade we place a rectangle whose height equals the relative frequency of that grade.

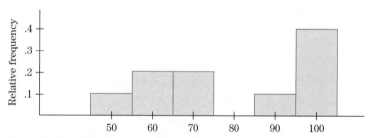

Figure 4 A relative frequency histogram.

An alternative way to compute the mean grade is

$$[\text{mean}] = \frac{50 \cdot 1 + 60 \cdot 2 + 70 \cdot 2 + 90 \cdot 1 + 100 \cdot 4}{10}$$

$$= 50 \cdot \frac{1}{10} + 60 \cdot \frac{2}{10} + 70 \cdot \frac{2}{10} + 90 \cdot \frac{1}{10} + 100 \cdot \frac{4}{10}$$

$$= 5 + 12 + 14 + 9 + 40 = 80.$$

Looking at the second line of this computation, we see that the mean is a sum of the various grades times their relative frequencies. We say that the mean is the **weighted sum** of the grades. (Grades are weighted by their relative frequencies.)

In a similar manner we see that the variance is also a weighted sum.

$$[\text{variance}] = [(50 - 80)^2 \cdot 1 + (60 - 80)^2 \cdot 2 + (70 - 80)^2 \cdot 2$$

$$+ (90 - 80)^2 \cdot 1 + (100 - 80)^2 \cdot 4] \frac{1}{10}$$

$$= (50 - 80)^2 \cdot \frac{1}{10} + (60 - 80)^2 \cdot \frac{2}{10} + (70 - 80)^2 \cdot \frac{2}{10}$$

$$+ (90 - 80)^2 \cdot \frac{1}{10} + (100 - 80)^2 \cdot \frac{4}{10}$$

$$= 90 + 80 + 20 + 10 + 160 = 360.$$

The relative frequency table shown in Fig. 3 is also called a **probability table**. The reason for this terminology is as follows. Suppose that we perform an **experiment** that consists of picking an exam paper at random from among the 10 papers. If the experiment is repeated many times, we expect the grade of 50 to occur about $\frac{1}{10}$ of the time, the grade of 60 about $\frac{2}{10}$ of the time, and so on. We say that the *probability* of the grade of 50 being chosen is $\frac{1}{10}$, the probability of the grade of 60 being chosen is $\frac{2}{10}$, and so on. In other words, the probability associated with a given grade measures the likelihood that an exam having that grade is chosen.

In this section, we consider various experiments described by probability tables similar to the one in Fig. 3. The results of these experiments will be numbers (such as the preceding exam scores) called the **outcomes** of the experiment. We will also be given the probability of each outcome, indicating the relative frequency with which the given outcome is expected to occur if the experiment is repeated very often. If the outcomes of an experiment are $a_1, a_2, \ldots, a_n$, with respective probabilities $p_1, p_2, \ldots, p_n$, we describe the experiment by a probability table. (See Fig. 5.) Since the probabilities indicate relative frequencies, we see that

$$0 \leq p_i \leq 1$$

and

$$p_1 + p_2 + \cdots + p_n = 1.$$

Outcome	a_1	a_2	a_3	$\cdots$	a_n
Probability	p_1	p_2	p_3	$\cdots$	p_n

Figure 5

The last equation indicates that the outcomes $a_1, \ldots, a_n$ comprise all possible results of the experiment. We will usually list the outcomes of our experiments in ascending order, so $a_1 < a_2 < \cdots < a_n$.

We may display the data of a probability table in a histogram that has a rectangle of height p_i over the outcome a_i. (See Fig. 6.)

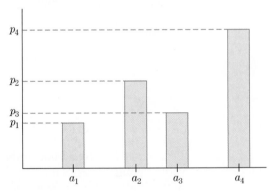

Figure 6

Let us define the **expected value** (or **mean**) of the probability table of Fig. 5 to be the weighted sum of the outcomes $a_1, \ldots, a_n$, each outcome weighted by the probability of its occurrence. That is,

$$[\text{expected value}] = a_1 p_1 + a_2 p_2 + \cdots + a_n p_n.$$

Similarly, let us define the **variance** of the probability table to be the weighted sum of the squares of the differences between each outcome and the expected value. That is, if m denotes the expected value, then

$$[\text{variance}] = (a_1 - m)^2 p_1 + (a_2 - m)^2 p_2 + \cdots + (a_n - m)^2 p_n.$$

To keep from writing the "outcome" so many times, we shall abbreviate by X the outcome of our experiment. That is, X is a variable that takes on the values $a_1, a_2, \ldots, a_n$ with respective probabilities $p_1, p_2, \ldots, p_n$. We will assume that our experiment is performed many times, being repeated in an unbiased (or random) way. Then, X is a variable whose value depends on chance, and for this reason, we say that X is a **random variable**. Instead of speaking of the expected value (mean) and the variance of a probability table, let us speak of the **expected value** and the **variance of the random variable X** that is associated with the probability table. We shall denote the expected value of X by $\mathrm{E}(X)$ and the variance of X by $\mathrm{Var}(X)$. The **standard deviation** of X is defined to be $\sqrt{\mathrm{Var}(X)}$.

EXAMPLE 1   **Expected Value and Variance** One possible bet in roulette is to wager $1 on red. The two possible outcomes are lose $1 and win $1. These outcomes and their probabilities are given in Fig. 7. (*Note:* A roulette wheel in Las Vegas has 18 red numbers, 18 black numbers, and two green numbers.) Compute the expected value and the variance of the amount won.

SOLUTION   Let X be the random variable "amount won." Then,

Amount won	-1	1
Probability	$\frac{20}{38}$	$\frac{18}{38}$

Figure 7 Las Vegas roulette.

$$\mathrm{E}(X) = -1 \cdot \frac{20}{38} + 1 \cdot \frac{18}{38} = -\frac{2}{38} \approx -.0526,$$

$$\mathrm{Var}(X) = \left[-1 - \left(-\frac{2}{38}\right)\right]^2 \cdot \frac{20}{38} + \left[1 - \left(-\frac{2}{38}\right)\right]^2 \cdot \frac{18}{38}$$

$$= \left(-\frac{36}{38}\right)^2 \cdot \frac{20}{38} + \left(\frac{40}{38}\right)^2 \cdot \frac{18}{38} \approx .997.$$

The expected value of the amount won is approximately $-5\frac{1}{4}$ cents. In other words, sometimes we will win $1 and sometimes we will lose $1, but in the long run we can expect to lose an average of about $5\frac{1}{4}$ cents for each time we bet.   **» Now Try Exercise 1**

EXAMPLE 2   **Expected Value and Variance** An experiment consists of selecting a number at random from the set of integers $\{1, 2, 3\}$. The probabilities are given by the table in Fig. 8. Let X designate the outcome. Find the expected value and the variance of X.

SOLUTION

Number	1	2	3
Probability	$\frac{1}{3}$	$\frac{1}{3}$	$\frac{1}{3}$

Figure 8

$$\mathrm{E}(X) = 1 \cdot \frac{1}{3} + 2 \cdot \frac{1}{3} + 3 \cdot \frac{1}{3} = 2,$$

$$\mathrm{Var}(X) = (1-2)^2 \cdot \frac{1}{3} + (2-2)^2 \cdot \frac{1}{3} + (3-2)^2 \cdot \frac{1}{3}$$

$$= (-1)^2 \cdot \frac{1}{3} + 0 + (1)^2 \cdot \frac{1}{3} = \frac{2}{3}.$$

» Now Try Exercise 3

EXAMPLE 3   **Decision Making Based on Expected Value** A cement company plans to bid on a contract for constructing the foundations of new homes in a housing development. The company is considering two bids: a high bid that will produce $75,000 profit (if the bid is accepted) and a low bid that will produce $40,000 profit. From past experience, the company estimates that the high bid has a 30% chance of acceptance and the low bid a 50% chance. Which bid should the company make?

SOLUTION   The standard method of decision making is to choose the bid that has the higher expected value. Let X be the amount the company makes if it submits the high bid, and let Y be the amount it makes if it submits the low bid. Then, the company must analyze the situation using the probability table shown in Table 1. The expected values are

$$\mathrm{E}(X) = (75{,}000)(.30) + 0(.70) = 22{,}500,$$
$$\mathrm{E}(Y) = (40{,}000)(.50) + 0(.50) = 20{,}000.$$

If the cement company has many opportunities to bid on similar contracts, a high bid each time will be accepted sufficiently often to produce an average profit of $22,500 per bid.

A consistently low bid will produce an average profit of $20,000 per bid. Thus, the company should submit the high bid.

Table 1 Bids on a Cement Contract

	High Bid			Low Bid	
	Accepted	Rejected		Accepted	Rejected
Value of X	75,000	0	Value of Y	40,000	0
Probability	.30	.70	Probability	.50	.50

>> *Now Try Exercise 9*

When a probability table contains a large number of possible outcomes of an experiment, the associated histogram for the random variable X becomes a valuable aid for visualizing the data in the table. Look at Fig. 9, for example. Since the rectangles that make up the histogram all have the same width, their areas are in the same ratios as their heights. By an appropriate change of scale on the y-axis, we may assume that the *area* (instead of the height) of each rectangle gives the associated probability of X. Such a histogram is sometimes referred to as a **probability density histogram**.

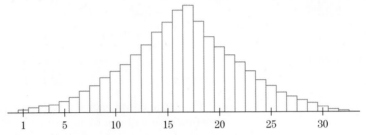

Figure 9 Probabilities displayed as areas.

A histogram that displays probabilities as areas is useful when we wish to visualize the probability that X has a value between two specified numbers. For example, in Fig. 9, suppose that the probabilities associated with $X = 5, X = 6, \ldots, X = 10$ are $p_5, p_6, \ldots, p_{10}$, respectively. Then, the probability that X lies between 5 and 10 inclusive is $p_5 + p_6 + \cdots + p_{10}$. In terms of areas, this probability is just the total area of those rectangles over the values $5, 6, \ldots, 10$. (See Fig. 10.) We will consider analogous situations in the next section.

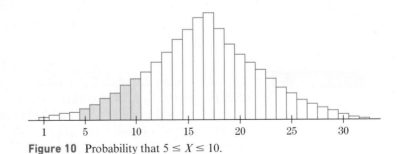

Figure 10 Probability that $5 \leq X \leq 10$.

Check Your Understanding 12.1

Solutions can be found following the section exercises.

1. Compute the expected value and the variance of the random variable X with Table 2 as its probability table.

Table 2

Value of X	−1	0	1	2
Probability	$\frac{1}{8}$	$\frac{1}{8}$	$\frac{3}{8}$	$\frac{3}{8}$

2. The production department at a radio factory sends CB radios to the inspection department in lots of 100. There an inspector examines three radios at random from each lot. If at least one of the three radios is defective and needs adjustment, the entire lot is sent back to the production department. Records of the inspection department show that the number X of defective radios in a sample of three radios has Table 3 as its probability table.

Table 3 Quality Control Data

Defectives	0	1	2	3
Probability	.7265	.2477	.0251	.0007

(a) What percentage of the lots does the inspection department reject?

(b) Find the mean number of defective radios in the samples of three radios.

(c) Based on the evidence in part (b), estimate the average number of defective radios in each lot of 100 radios.

EXERCISES 12.1

1. Table 4 is the probability table for a random variable X. Find $E(X)$, $Var(X)$, and the standard deviation of X.

Table 4

Outcome	0	1
Probability	$\frac{1}{5}$	$\frac{4}{5}$

2. Find $E(X)$, $Var(X)$, and the standard deviation of X, where X is the random variable whose probability table is given in Table 5.

Table 5

Outcome	1	2	3
Probability	$\frac{4}{9}$	$\frac{4}{9}$	$\frac{1}{9}$

3. Compute the variances of the three random variables whose probability tables are given in Table 6. Relate the sizes of the variances to the spread of the values of the random variable.

Table 6

	Outcome	Probability
(a)	4	.5
	6	.5
(b)	3	.5
	7	.5
(c)	1	.5
	9	.5

4. Compute the variances of the two random variables whose probability tables are given in Table 7. Relate the sizes of the variances to the spread of the values of the random variables.

Table 7

	Outcome	Probability
(a)	2	.1
	4	.4
	6	.4
	8	.1
(b)	2	.3
	4	.2
	6	.2
	8	.3

5. **Expected Value** The number of accidents per week at a busy intersection was recorded for a year. There were 11 weeks with no accidents, 26 weeks with one accident, 13 weeks with two accidents, and 2 weeks with three accidents. A week is to be selected at random and the number of accidents noted. Let X be the outcome. Then, X is a random variable taking on the values 0, 1, 2, and 3.
   (a) Write out a probability table for X.
   (b) Compute $E(X)$.
   (c) Interpret $E(X)$.

6. **Probability Table, Expected Value** The number of phone calls coming into a telephone switchboard during each minute was recorded during an entire hour. During 30 of the 1-minute intervals there were no calls, during 20 intervals there was one call, and during 10 intervals there were two calls. A 1-minute interval is to be selected at random and the number of calls noted. Let X be the outcome. Then, X is a random variable taking on the values 0, 1, and 2.
   (a) Write out a probability table for X.
   (b) Compute $E(X)$.
   (c) Interpret $E(X)$.

7. Consider a circle with radius 1.
   (a) What percentage of the points lies within $\frac{1}{2}$ unit of the center?
   (b) Let c be a constant with $0 < c < 1$. What percentage of the points lies within c unit of the center?

8. Consider a circle with circumference 1. An arrow (or spinner) is attached at the center so that, when flicked, it spins freely. Upon stopping, it points to a particular point on the circumference of the circle. Determine the likelihood that the point is
   (a) On the top half of the circumference.
   (b) On the top quarter of the circumference.
   (c) On the top one-hundredth of the circumference.
   (d) Exactly at the top of the circumference.

9. **Decision Making Based on Expected Value** A citrus grower anticipates a profit of $100,000 this year if the nightly temperatures remain mild. Unfortunately, the weather forecast indicates a 25% chance that the temperatures will drop below freezing during the next week. Such freezing weather will destroy 40% of the crop and reduce the profit to $60,000. However, the grower can protect the citrus fruit against the possible freezing (using smudge pots, electric fans, and so on) at a cost of $5000. Should the grower spend the $5000 and thereby reduce the profit to $95,000? [*Hint:* Compute $E(X)$, where X is the profit the grower will get if he does nothing to protect the fruit.]

10. Suppose that the weather forecast in Exercise 9 indicates a 10% chance that cold weather will reduce the citrus grower's profit from $100,000 to $85,000 and a 10% chance that cold weather will reduce the profit to $75,000. Should the grower spend $5000 to protect the citrus fruit against the possible bad weather?

Solutions to Check Your Understanding 12.1

1. $E(X) = (-1) \cdot \frac{1}{8} + 0 \cdot \frac{1}{8} + 1 \cdot \frac{3}{8} + 2 \cdot \frac{3}{8} = 1,$

$\text{Var}(X) = (-1-1)^2 \cdot \frac{1}{8} + (0-1)^2 \cdot \frac{1}{8}$

$\qquad + (1-1)^2 \cdot \frac{3}{8} + (2-1)^2 \cdot \frac{3}{8}$

$\qquad = 4 \cdot \frac{1}{8} + 1 \cdot \frac{1}{8} + 0 + 1 \cdot \frac{3}{8} = 1.$

2. (a) In three cases a lot will be rejected: $X = 1, 2,$ or 3. Adding the corresponding probabilities, we find that the probability of rejecting a lot is $.2477 + .0251 + .0007 = .2735,$ or 27.35%. (An alternative method of solution uses the fact that the sum of the probabilities for *all* possible cases must be 1. From the table, we see that the probability of

accepting a lot is .7265, so the probability of rejecting a lot is $1 - .7265 = .2735.$)

(b) $E(X) = 0(.7265) + 1(.2477)$

$\qquad + 2(.0251) + 3(.0007)$

$\qquad = .3000.$

(c) In part (b) we found that an average of .3 radio in every sample of three radios is defective. Thus, about 10% of the radios in the sample are defective. Since the samples are chosen at random, we may assume that about 10% of *all* the radios are defective. Thus, we estimate that, on the average, 10 out of each lot of 100 radios will be defective.

12.2 Continuous Random Variables

Consider a cell population that is growing vigorously. When a cell is T days old, it divides and forms two new daughter cells. If the population is sufficiently large, it will contain cells of many different ages between 0 and T. It turns out that the proportion of cells of various ages remains constant. That is, if a and b are any two numbers between 0 and T, with $a < b$, the proportion of cells whose ages lie between a and b is essentially constant from one moment to the next, even though individual cells are aging and new cells are being formed all the time. In fact, biologists have found that under the ideal circumstances described, the proportion of cells whose ages are between a and b is given by the area under the graph of the function $f(x) = 2ke^{-kx}$ from $x = a$ to $x = b$, where $k = (\ln 2)/T$. (See Fig. 1.)

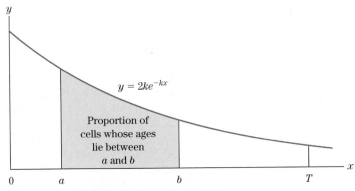

Figure 1 Age distribution in a cell population.

Now consider an experiment in which we select a cell at random from the population and observe its age, X. Then, the probability that X lies between a and b is given by the area under the graph of $f(x) = 2ke^{-kx}$ from a to b. (See Fig. 1.) Let us denote this probability by $\Pr(a \le X \le b)$. Using the fact that the area under the graph of $f(x)$ is given by a definite integral, we have

$$\Pr(a \le X \le b) = \int_a^b f(x)\,dx = \int_a^b 2ke^{-kx}\,dx. \tag{1}$$

Since X can assume any one of the (infinitely many) numbers in the continuous interval from 0 to T, we say that X is a **continuous random variable**. The function $f(x)$ that determines the probability in (1) for each a and b is called the **(probability) density function** of X (or of the experiment whose outcome is X).

More generally, consider an experiment whose outcome may be any value between A and B. The outcome of the experiment, denoted X, is called a **continuous random variable**.

For the cell population described previously, $A = 0$ and $B = T$. Another typical experiment might consist of choosing a number X at random between $A = 5$ and $B = 6$. Or we could observe the duration X of a random telephone call passing through a given telephone switchboard. If we have no way of knowing how long a call might last, X might be any nonnegative number. In this case it is convenient to say that X lies between 0 and ∞ and to take $A = 0$ and $B = \infty$. On the other hand, if the possible values of X for some experiment include rather large negative numbers, we sometimes takes $A = -\infty$.

Given an experiment whose outcome is a continuous random variable X, the probability $\Pr(a \leq X \leq b)$ is a measure of the likelihood that an outcome of the experiment will lie between a and b. If the experiment is repeated many times, the proportion of times that X has a value between a and b should be close to $\Pr(a \leq X \leq b)$. In experiments of practical interest involving a continuous random variable X, it is usually possible to find a function $f(x)$ such that

$$\Pr(a \leq X \leq b) = \int_a^b f(x)\,dx \tag{2}$$

for all a and b in the range of possible values of X. Such a function $f(x)$ is called a **probability density function** and satisfies the following properties:

(I) $f(x) \geq 0$ for $A \leq x \leq B$.

(II) $\displaystyle\int_A^B f(x)\,dx = 1.$

Indeed, Property I means that, for x between A and B, the graph of $f(x)$ must lie on or above the x-axis. Property II simply says that there is probability 1 that X has a value between A and B. (Of course, if $B = \infty$ and/or $A = -\infty$, the integral in Property II is an improper integral.) Properties I and II characterize probability density functions, in the sense that any function $f(x)$ satisfying I and II is the probability density function for some continuous random variable X. Moreover, $\Pr(a \leq X \leq b)$ can then be calculated with equation (2).

Unlike a probability table for a discrete random variable, a density function $f(x)$ does *not* give the probability that X has a certain value. Instead, $f(x)$ can be used to find the probability that X is *near* a specific value in the following sense. If x_0 is a number between A and B and if Δx is the width of a small interval centered at x_0, the probability that X is between $x_0 - \frac{1}{2}\Delta x$ and $x_0 + \frac{1}{2}\Delta x$ is approximately $f(x_0)\Delta x$, that is, the area of the rectangle shown in Fig. 2.

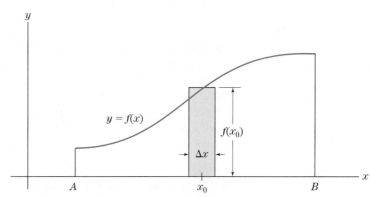

Figure 2 Area of rectangle gives approximate probability that X is near x_0.

EXAMPLE 1

Probability Density Consider the cell population described earlier. Let $f(x) = 2ke^{-kx}$, where $k = (\ln 2)/T$. Show that $f(x)$ is indeed a probability density function on $0 \leq x \leq T$.

SOLUTION   Clearly, $f(x) \geq 0$, since ln 2 is positive and the exponential function is never negative. Thus, Property I is satisfied. For Property II, we check that

$$\int_0^T f(x)\, dx = \int_0^T 2ke^{-kx}\, dx = -2e^{-kx}\Big|_0^T = -2e^{-kT} + 2e^0$$

$$= -2e^{-[(\ln 2)/T]T} + 2 = -2e^{-\ln 2} + 2$$

$$= -2e^{\ln(1/2)} + 2 = -2\left(\frac{1}{2}\right) + 2 = 1.$$ **≫ Now Try Exercise 1**

EXAMPLE 2   **Determining a Probability Density** Let $f(x) = kx^2$.
(a) Find the value of k that makes $f(x)$ a probability density function on $0 \leq x \leq 4$.
(b) Let X be a continuous random variable whose density function is $f(x)$. Compute $\Pr(1 \leq X \leq 2)$.

SOLUTION   (a) We must have $k \geq 0$ so that Property I is satisfied. For Property II, we calculate

$$\int_0^4 f(x)\, dx = \int_0^4 kx^2\, dx = \frac{1}{3}kx^3\Big|_0^4 = \frac{1}{3}k(4)^3 - 0 = \frac{64}{3}k.$$

Figure 3

To satisfy Property II, we must have $\frac{64}{3}k = 1$, or $k = \frac{3}{64}$. Thus, $f(x) = \frac{3}{64}x^2$.

(b) $\Pr(1 \leq X \leq 2) = \int_1^2 f(x)\, dx = \int_1^2 \frac{3}{64}x^2\, dx = \frac{1}{64}x^3\Big|_1^2 = \frac{8}{64} - \frac{1}{64} = \frac{7}{64}.$

The area corresponding to this probability is shown in Fig. 3.

≫ Now Try Exercise 7

The density function in the next example is a special case of what statisticians sometimes call a *beta* **probability density**.

EXAMPLE 3   **A Beta Probability Density** The parent corporation for a franchised chain of fast-food restaurants claims that the proportion of their new restaurants that makes a profit during their first year of operation has the probability density function

$$f(x) = 12x(1-x)^2, \qquad 0 \leq x \leq 1.$$

(a) What is the probability that less than 40% of the restaurants opened this year will make a profit during their first year of operation?
(b) What is the probability that more than 50% of the restaurants will make a profit during their first year of operation?

SOLUTION   Let X be the proportion of new restaurants opened this year that make a profit during their first year of operation. Then, the possible values of X range between 0 and 1.

(a) The probability that X is less than .4 equals the probability that X is between 0 and .4. We note that $f(x) = 12x(1 - 2x + x^2) = 12x - 24x^2 + 12x^3$, and therefore,

$$\Pr(0 \leq X \leq .4) = \int_0^{.4} f(x)\, dx = \int_0^{.4}(12x - 24x^2 + 12x^3)\, dx$$

$$= (6x^2 - 8x^3 + 3x^4)\Big|_0^{.4} = .5248.$$

(b) The probability that X is greater than .5 equals the probability that X is between .5 and 1. Thus,

$$\Pr(.5 \leq X \leq 1) = \int_{.5}^1 (12x - 24x^2 + 12x^3)\, dx$$

$$= (6x^2 - 8x^3 + 3x^4)\Big|_{.5}^1 = .3125.$$

≫ Now Try Exercise 15

Each probability density function is closely related to another important function called a *cumulative distribution function*. To describe this relationship, let us consider an experiment whose outcome is a continuous random variable X, with values between A and B, and let $f(x)$ be the associated density function. For each number x between A and B, let $F(x)$ be the probability that X is less than or equal to the number x. Sometimes, we write $F(x) = \Pr(X \leq x)$; however, since X is never less than A, we may also write

$$F(x) = \Pr(A \leq X \leq x). \tag{3}$$

Graphically, $F(x)$ is the area under the graph of the probability density function $f(x)$ from A to x. (See Fig. 4.) The function $F(x)$ is called the **cumulative distribution function** of the random variable X (or of the experiment whose outcome is X). Note that $F(x)$ also has the properties

$$F(A) = \Pr(A \leq X \leq A) = 0, \tag{4}$$

$$F(B) = \Pr(A \leq X \leq B) = 1. \tag{5}$$

Since $F(x)$ is an area function that gives the area under the graph of $f(x)$ from A to x, we know from Section 6.4 that $F(x)$ is an antiderivative of $f(x)$. That is,

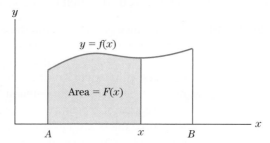

Figure 4 The cumulative distribution function $F(x)$.

$$F'(x) = f(x), \qquad A \leq x \leq B. \tag{6}$$

It follows that we may use $F(x)$ to compute probabilities, since

FOR REVIEW

Compare (7) with
Theorem I, Sec. 6.3, p. 310.

$$\Pr(a \leq X \leq b) = \int_a^b f(x)\,dx = F(b) - F(a), \tag{7}$$

for any a and b between A and B.

The relation (6) between $F(x)$ and $f(x)$ makes it possible to find one of these functions when the other is known, as we see in the following two examples.

EXAMPLE 4   **Age of a Cell** Let X be the age of a cell selected at random from the cell population described earlier. The probability density function for X is $f(x) = 2ke^{-kx}$, where $k \doteq (\ln 2)/T$. (See Fig. 5.) Find the cumulative distribution function $F(x)$ for X.

SOLUTION   Since $F(x)$ is an antiderivative of $f(x) = 2ke^{-kx}$, we have $F(x) = -2e^{-kx} + C$ for some constant C. Now, $F(x)$ is defined for $0 \leq x \leq T$. Thus, (4) implies that $F(0) = 0$. Setting $F(0) = -2e^0 + C = 0$, we find that $C = 2$, so

$$F(x) = -2e^{-kx} + 2.$$

(See Fig. 6.)

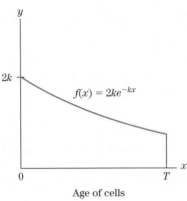

Figure 5   Probability density function.

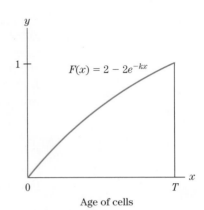

Figure 6   Cumulative distribution function.

>> *Now Try Exercise 25*

EXAMPLE 5   **Random Selection**   Let X be the random variable associated with the experiment that consists of selecting a point at random from the region inside a circle of radius 1 and observing its distance from the center. Find the probability density function $f(x)$ and cumulative distribution function $F(x)$ of X.

SOLUTION   The distance of a point from the center of the unit circle is a number between 0 and 1. Suppose that $0 \leq x \leq 1$. Let us first compute the cumulative distribution function $F(x) = \Pr(0 \leq X \leq x)$. That is, let us find the probability that a point selected at random lies within x units of the center of the circle, in other words, lies inside the circle of radius x. See the shaded region in Fig. 7(b). Since the area of this shaded region is πx^2 and the area of the entire unit circle is $\pi \cdot 1^2 = \pi$, the proportion of points inside the shaded region is $\pi x^2 / \pi = x^2$. Thus, the probability is x^2 that a point selected at random will be in this shaded region. Hence,

$$F(x) = x^2.$$

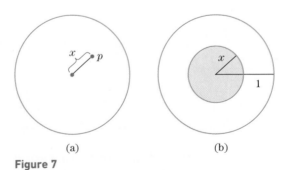

(a)   (b)

Figure 7

Differentiating, we find that the probability density function for X is

$$f(x) = F'(x) = 2x. \qquad \text{>> } \textit{Now Try Exercise 27}$$

Our final example involves a continuous random variable X whose possible values lie between $A = 1$ and $B = \infty$; that is, X is any number greater than or equal to 1.

EXAMPLE 6   **Working with a Probability Density**   Let $f(x) = 3x^{-4}$, $x \geq 1$.
   (a) Show that $f(x)$ is the probability density function of some random variable X.
   (b) Find the cumulative distribution function $F(x)$ of X.
   (c) Compute $\Pr(X \leq 4)$, $\Pr(4 \leq X \leq 5)$, and $\Pr(4 \leq X)$.

SOLUTION (a) It is clear that $f(x) \geq 0$ for $x \geq 1$. Thus, Property I holds. To check for Property II, we must compute

$$\int_1^\infty 3x^{-4}\, dx.$$

But

$$\int_1^b 3x^{-4}\, dx = -x^{-3}\Big|_1^b = -b^{-3} + 1,$$

which has limit 1 as $b \to \infty$ because b^{-3} tends to 0 as $b \to \infty$. Thus,

$$\int_1^\infty 3x^{-4}\, dx = 1,$$

and Property II holds.

(b) Since $F(x)$ is an antiderivative of $f(x) = 3x^{-4}$, we have

$$F(x) = \int 3x^{-4}\, dx = -x^{-3} + C.$$

Since X has values greater than or equal to 1, we must have $F(1) = 0$. Setting $F(1) = -1 + C = 0$, we find that $C = 1$, so

$$F(x) = 1 - x^{-3}.$$

(c) $\Pr(X \leq 4) = F(4) = 1 - 4^{-3} = 1 - \frac{1}{64} = \frac{63}{64}$

Since we know $F(x)$, we may use it to compute $\Pr(4 \leq X \leq 5)$, as follows:

$$\Pr(4 \leq X \leq 5) = F(5) - F(4) = (1 - 5^{-3}) - (1 - 4^{-3})$$

$$= \frac{1}{4^3} - \frac{1}{5^3} \approx .0076.$$

We may compute $\Pr(4 \leq X)$ directly by evaluating the improper integral

$$\int_4^\infty 3x^{-4}\, dx.$$

However, there is a simpler method. We know that

$$\int_1^4 3x^{-4}\, dx + \int_4^\infty 3x^{-4}\, dx = \int_1^\infty 3x^{-4}\, dx = 1. \tag{8}$$

In terms of probabilities, (8) may be written as

$$\Pr(X \leq 4) + \Pr(4 \leq X) = 1.$$

Hence,

$$\Pr(4 \leq X) = 1 - \Pr(X \leq 4) = 1 - \frac{63}{64} = \frac{1}{64}.$$

» Now Try Exercise 37

Check Your Understanding 12.2

Solutions can be found following the section exercises.

1. In a certain farming region and in a certain year, the number of bushels of wheat produced per acre is a random variable X with a density function

$$f(x) = \frac{x - 30}{50}, \qquad 30 \leq x \leq 40.$$

(a) What is the probability that an acre selected at random produced less than 35 bushels of wheat?

(b) If the farming region had 20,000 acres of wheat, how many acres produced less than 35 bushels of wheat per acre?

2. The density function for a continuous random variable X on the interval $1 \leq x \leq 2$ is $f(x) = 8/(3x^3)$. Find the corresponding cumulative distribution function for X.

EXERCISES 12.2

Verify that each of the following functions is a probability density function.

1. $f(x) = \frac{1}{18}x,\ 0 \le x \le 6$

2. $f(x) = 2(x - 1),\ 1 \le x \le 2$

3. $f(x) = \frac{1}{4},\ 1 \le x \le 5$

4. $f(x) = \frac{8}{9}x,\ 0 \le x \le \frac{3}{2}$

5. $f(x) = 5x^4,\ 0 \le x \le 1$

6. $f(x) = \frac{3}{2}x - \frac{3}{4}x^2,\ 0 \le x \le 2$

In Exercises 7–12, find the value of k that makes the given function a probability density function on the specified interval.

7. $f(x) = kx,\ 1 \le x \le 3$

8. $f(x) = kx^2,\ 0 \le x \le 2$

9. $f(x) = k,\ 5 \le x \le 20$

10. $f(x) = k/\sqrt{x},\ 1 \le x \le 4$

11. $f(x) = kx^2(1 - x),\ 0 \le x \le 1$

12. $f(x) = k(3x - x^2),\ 0 \le x \le 3$

13. The density function of a continuous random variable X is $f(x) = \frac{1}{8}x,\ 0 \le x \le 4$. Sketch the graph of $f(x)$ and shade in the areas corresponding to (a) $\Pr(X \le 1)$; (b) $\Pr(2 \le X \le 2.5)$; (c) $\Pr(3.5 \le X)$.

14. The density function of a continuous random variable X is $f(x) = 3x^2,\ 0 \le x \le 1$. Sketch the graph of $f(x)$ and shade in the areas corresponding to (a) $\Pr(X \le .3)$; (b) $\Pr(.5 \le X \le .7)$; (c) $\Pr(.8 \le X)$.

15. Find $\Pr(1 \le X \le 2)$ when X is a random variable whose density function is given in Exercise 1.

16. Find $\Pr(1.5 \le X \le 1.7)$ when X is a random variable whose density function is given in Exercise 2.

17. Find $\Pr(X \le 3)$ when X is a random variable whose density function is given in Exercise 3.

18. Find $\Pr(1 \le X)$ when X is a random variable whose density function is given in Exercise 4.

19. Lifetime of a Battery Suppose that the lifetime X (in hours) of a certain type of flashlight battery is a random variable on the interval $30 \le x \le 50$ with density function $f(x) = \frac{1}{20}$, $30 \le x \le 50$. Find the probability that a battery selected at random will last at least 35 hours.

20. Waiting Time At a certain supermarket, the amount of wait time at the express lane is a random variable with density function $f(x) = 11/[10(x + 1)^2]$, $0 \le x \le 10$. (See Fig. 8.) Find the probability of having to wait less than 4 minutes at the express lane.

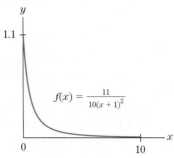

$$f(x) = \frac{11}{10(x + 1)^2}$$

Figure 8 A density function.

21. The cumulative distribution function for a random variable X on the interval $1 \le x \le 5$ is $F(x) = \frac{1}{2}\sqrt{x - 1}$. (See Fig. 9.) Find the corresponding density function.

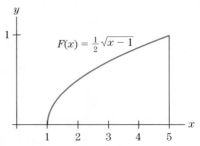

$$F(x) = \frac{1}{2}\sqrt{x - 1}$$

Figure 9 A cumulative distribution function.

22. The cumulative distribution function for a random variable X on the interval $1 \le x \le 2$ is $F(x) = \frac{4}{3} - 4/(3x^2)$. Find the corresponding density function.

23. Compute the cumulative distribution function corresponding to the density function $f(x) = \frac{1}{5},\ 2 \le x \le 7$.

24. Compute the cumulative distribution function corresponding to the density function $f(x) = \frac{1}{2}(3 - x),\ 1 \le x \le 3$.

25. The time (in minutes) required to complete a certain subassembly is a random variable X with the density function $f(x) = \frac{1}{21}x^2,\ 1 \le x \le 4$.
   (a) Use $f(x)$ to compute $\Pr(2 \le X \le 3)$.
   (b) Find the corresponding cumulative distribution function $F(x)$.
   (c) Use $F(x)$ to compute $\Pr(2 \le X \le 3)$.

26. The density function for a continuous random variable X on the interval $1 \le x \le 4$ is $f(x) = \frac{4}{9}x - \frac{1}{9}x^2$.
   (a) Use $f(x)$ to compute $\Pr(3 \le X \le 4)$.
   (b) Find the corresponding cumulative distribution function $F(x)$.
   (c) Use $F(x)$ to compute $\Pr(3 \le X \le 4)$.

An experiment consists of selecting a point at random from the region inside the square in Fig. 10(a). Let X be the maximum of the coordinates of the point.

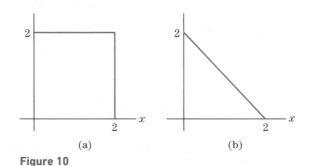

(a)   **(b)**

Figure 10

27. Show that the cumulative distribution function of X is $F(x) = x^2/4,\ 0 \le x \le 2$.

28. Find the corresponding density function of X.

An experiment consists of selecting a point at random from the region inside the triangle in Fig. 10(b). Let X be the sum of the coordinates of the point.

29. Show that the cumulative distribution function of X is $F(x) = x^2/4$, $0 \le x \le 2$.

30. Find the corresponding density function of X.

In a certain cell population, cells divide every 10 days, and the age of a cell selected at random is a random variable X with the density function $f(x) = 2ke^{-kx}$, $0 \le x \le 10$, $k = (\ln 2)/10$.

31. Find the probability that a cell is at most 5 days old.

32. Upon examination of a slide, 10% of the cells are found to be undergoing mitosis (a change in the cell leading to division). Compute the length of time required for mitosis; that is, find the number M such that

$$\int_{10-M}^{10} 2ke^{-kx}\, dx = .10.$$

33. A random variable X has a density function $f(x) = \frac{1}{3}$, $0 \le x \le 3$. Find b such that $\Pr(0 \le X \le b) = .6$.

34. A random variable X has a density function $f(x) = \frac{2}{3}x$ on $1 \le x \le 2$. Find a such that $\Pr(a \le X) = \frac{1}{3}$.

35. A random variable X has a cumulative distribution function $F(x) = \frac{1}{4}x^2$ on $0 \le x \le 2$. Find b such that $\Pr(X \le b) = .09$.

36. A random variable X has a cumulative distribution function $F(x) = (x-1)^2$ on $1 \le x \le 2$. Find b such that $\Pr(X \le b) = \frac{1}{4}$.

37. Let X be a continuous random variable with values between $A = 1$ and $B = \infty$, and with the density function $f(x) = 4x^{-5}$.
   (a) Verify that $f(x)$ is a probability density function for $x \ge 1$.
   (b) Find the corresponding cumulative distribution function $F(x)$.
   (c) Use $F(x)$ to compute $\Pr(1 \le X \le 2)$ and $\Pr(2 \le X)$.

38. Let X be a continuous random variable with the density function $f(x) = 2(x+1)^{-3}$, $x \ge 0$.
   (a) Verify that $f(x)$ is a probability density function for $x \ge 0$.
   (b) Find the cumulative distribution function for X.
   (c) Compute $\Pr(1 \le X \le 2)$ and $\Pr(3 \le X)$.

Solutions to Check Your Understanding 12.2

1. (a) $\Pr(X \le 35) = \displaystyle\int_{30}^{35} \frac{x-30}{50}\, dx = \left. \frac{(x-30)^2}{100} \right|_{30}^{35}$

$= \dfrac{5^2}{100} - 0 = .25$

   (b) Using part (a), we see that 25% of the 20,000 acres, or 5000 acres, produced less than 35 bushels of wheat per acre.

2. The cumulative distribution function $F(x)$ is an antiderivative of $f(x) = 8/(3x^3) = \frac{8}{3}x^{-3}$. Thus, $F(x) = -\frac{4}{3}x^{-2} + C$ for some constant C. Since X varies over the interval $1 \le x \le 2$, we must have $F(1) = 0$; that is, $-\frac{4}{3}(1)^{-2} + C = 0$. Thus, $C = \frac{4}{3}$, and

$$F(x) = \frac{4}{3} - \frac{4}{3}x^{-2}.$$

12.3 Expected Value and Variance

When studying the cell population described in Section 12.2, we might reasonably ask for the average age of the cells. In general, given an experiment described by a random variable X and a probability density function $f(x)$, it is often important to know the average outcome of the experiment and the degree to which the experimental outcomes are spread out around the average. To provide this information in Section 12.1, we introduced the concepts of expected value and variance of a discrete random variable. Let us now examine the analogous definition for a continuous random variable.

> **DEFINITION** Let X be a continuous random variable whose possible values lie between A and B, and let $f(x)$ be the probability density function for X. Then, the **expected value** (or **mean**) of X is the number $E(X)$ defined by
>
> $$E(X) = \int_A^B xf(x)\, dx. \tag{1}$$
>
> The **variance** of X is the number $\mathrm{Var}(X)$ defined by
>
> $$\mathrm{Var}(X) = \int_A^B [x - E(X)]^2 f(x)\, dx. \tag{2}$$

The expected value of X has the same interpretation as in the discrete case: If the experiment whose outcome is X is performed many times, the average of all the outcomes will approximately equal $E(X)$. As in the case of a discrete random variable, the variance of X is a quantitative measure of the likely spread of the values of X about the mean $E(X)$ when the experiment is performed many times.

To explain why definition (1) of $E(X)$ is analogous to the definition in Section 12.1, let us approximate the integral in (1) by a Riemann sum of the form

$$x_1 \, f(x_1)\Delta x + x_2 \, f(x_2)\Delta x + \cdots + x_n \, f(x_n)\Delta x. \tag{3}$$

Here, $x_1, \ldots, x_n$ are the midpoints of subintervals of the interval from A to B, each subinterval of width $\Delta x = (B - A)/n$. (See Fig. 1.) Now recall from Section 12.2 that, for $i = 1, \ldots, n$, the quantity $f(x_i)\Delta x$ is approximately the probability that X is close to x_i, that is, the probability that X lies in the subinterval centered at x_i. If we write $\Pr(X \approx x_i)$ for this probability, then (3) is nearly the same as

$$x_1 \cdot \Pr(X \approx x_1) + x_2 \cdot \Pr(X \approx x_2) + \cdots + x_n \cdot \Pr(X \approx x_n). \tag{4}$$

As the number of subintervals increases, the sum becomes closer and closer to the integral in equation (1) defining $E(X)$. Furthermore, each approximating sum in (4) resembles the sum in the definition of the expected value of a discrete random variable, where we compute the weighted sum over all possible outcomes, with each outcome weighted by the probability of its occurrence.

FOR REVIEW

The definite integral is a limit of Riemann sums. See Sec. 6.3, Theorem II, p. 315.

Figure 1

A similar analysis will show that definition (2) of variance is analogous to the definition for the discrete case.

EXAMPLE 1   **Cumulative Distribution, Expected Value, and Variance**   Let us consider the experiment of selecting a number at random from among the numbers between 0 and B. Let X denote the associated random variable. Determine the cumulative distribution function of X, the density function of X, and the mean and variance of X.

SOLUTION

$$F(x) = \frac{[\text{length of the interval from 0 to } x]}{[\text{length of the interval from 0 to } B]} = \frac{x}{B}.$$

Since $f(x) = F'(x)$, we see that $f(x) = 1/B$. Thus, we have

$$E(X) = \int_0^B x \cdot \frac{1}{B} \, dx = \frac{1}{B} \int_0^B x \, dx = \frac{1}{B} \cdot \frac{B^2}{2} = \frac{B}{2},$$

$$Var(X) = \int_0^B \left(x - \frac{B}{2}\right)^2 \cdot \frac{1}{B} \, dx = \frac{1}{B} \int_0^B \left(x - \frac{B}{2}\right)^2 dx$$

$$= \frac{1}{B} \cdot \frac{1}{3}\left(x - \frac{B}{2}\right)^3 \Big|_0^B = \frac{1}{3B}\left[\left(\frac{B}{2}\right)^3 - \left(-\frac{B}{2}\right)^3\right] = \frac{B^2}{12}.$$

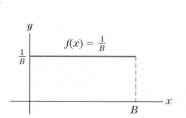

Figure 2   A uniform probability density function.

The graph of the density function $f(x)$ is shown in Fig. 2. Since the density function has a flat graph, the random variable X is called the **uniform random variable** on the interval from 0 to B.   **▶▶ Now Try Exercise 1**

EXAMPLE 2   **Average Age of a Cell Population**   Let X be the age of a cell chosen at random from the population described in Section 12.2, where the density function for X was given as

$$f(x) = 2ke^{-kx}, \quad 0 \le x \le T,$$

and $k = (\ln 2)/T$. Find the average age, $E(X)$, of the cell population.

SOLUTION   By definition,

$$E(X) = \int_0^T x \cdot 2ke^{-kx}\, dx.$$

To calculate this integral, we need integration by parts, with $f(x) = 2x$, $g(x) = ke^{-kx}$, $f'(x) = 2$, and $G(x) = -e^{-kx}$. We have

$$\int_0^T 2xke^{-kx}\, dx = -2xe^{-kx}\Big|_0^T - \int_0^T -2e^{-kx}\, dx$$

$$= -2Te^{-kT} - \left(\frac{2}{k}e^{-kx}\right)\Big|_0^T$$

$$= -2Te^{-kT} - \frac{2}{k}e^{-kT} + \frac{2}{k}.$$

We may simplify this formula for $E(X)$ by noting that $e^{-kT} = e^{-\ln 2} = \frac{1}{2}$. Thus,

$$E(X) = -2T\left(\frac{1}{2}\right) - \frac{2}{k}\left(\frac{1}{2}\right) + \frac{2}{k} = \frac{1}{k} - T$$

$$= \frac{T}{\ln 2} - T = \left(\frac{1}{\ln 2} - 1\right)T$$

$$\approx .4427T. \qquad\qquad \text{**»** Now Try Exercise 9}$$

EXAMPLE 3   **Random Selection in a Circle** Consider the experiment of selecting a point at random in a circle of radius 1, and let X be the distance from this point to the center. Compute the expected value and variance of the random variable X.

SOLUTION   We showed in Example 5 of Section 12.2 that the density function for X is given by $f(x) = 2x$, $0 \le x \le 1$. Therefore, we see that

$$E(X) = \int_0^1 x \cdot 2x\, dx = \int_0^1 2x^2\, dx = \frac{2x^3}{3}\Big|_0^1 = \frac{2}{3}$$

and

$$\text{Var}(X) = \int_0^1 \left(x - \frac{2}{3}\right)^2 \cdot 2x\, dx \quad \text{Since } E(X) = \frac{2}{3}$$

$$= \int_0^1 \left(x^2 - \frac{4}{3}x + \frac{4}{9}\right) \cdot 2x\, dx$$

$$= \int_0^1 \left(2x^3 - \frac{8}{3}x^2 + \frac{8}{9}x\right) dx$$

$$= \left(\frac{1}{2}x^4 - \frac{8}{9}x^3 + \frac{4}{9}x^2\right)\Big|_0^1$$

$$= \frac{1}{2} - \frac{8}{9} + \frac{4}{9} = \frac{1}{18}.$$

From our first calculation, we see that if a large number of points is chosen randomly from a circle of radius 1, their average distance to the center should be about $\frac{2}{3}$.

» *Now Try Exercise 11*

The following alternative formula for the variance of a random variable is usually easier to use than the actual definition of $\text{Var}(X)$.

Let X be a continuous random variable whose values lie between A and B, and let $f(x)$ be the density function for X. Then,

$$\text{Var}(X) = \int_A^B x^2 f(x)\, dx - \text{E}(X)^2. \qquad (5)$$

To prove (5), we let $m = \text{E}(X) = \int_A^B x f(x)\, dx$. Then,

$$
\begin{aligned}
\text{Var}(X) &= \int_A^B (x - m)^2 f(x)\, dx = \int_A^B (x^2 - 2xm + m^2) f(x)\, dx \\
&= \int_A^B x^2 f(x)\, dx - 2m \int_A^B x f(x)\, dx + m^2 \int_A^B f(x)\, dx \\
&= \int_A^B x^2 f(x)\, dx - 2m \cdot m + m^2 \cdot 1 \qquad \text{By Property II.} \\
&= \int_A^B x^2 f(x)\, dx - m^2.
\end{aligned}
$$

EXAMPLE 4 **Proportion of Students Using a Library** A college library has found that, in any given month during a school year, the proportion of students who make some use of the library is a random variable X with the cumulative distribution function

$$F(x) = 4x^3 - 3x^4, \qquad 0 \le x \le 1.$$

(a) Compute $\text{E}(X)$ and give an interpretation of this quantity.
(b) Compute $\text{Var}(X)$.

SOLUTION (a) To compute $\text{E}(X)$, we first find the probability density function $f(x)$. From Section 12.2, we know that

$$f(x) = F'(x) = 12x^2 - 12x^3.$$

Hence,

$$
\begin{aligned}
\text{E}(X) &= \int_0^1 x f(x)\, dx = \int_0^1 (12x^3 - 12x^4)\, dx \\
&= \left(3x^4 - \frac{12}{5}x^5 \right)\Big|_0^1 = 3 - \frac{12}{5} = \frac{3}{5}.
\end{aligned}
$$

The meaning of $\text{E}(X)$ in this example is that, over a period of many months (during school years), the average proportion of students each month who make some use of the library should be close to $\frac{3}{5}$.

(b) We first compute

$$
\begin{aligned}
\int_0^1 x^2 f(x)\, dx &= \int_0^1 (12x^4 - 12x^5)\, dx = \left(\frac{12}{5}x^5 - 2x^6 \right)\Big|_0^1 \\
&= \frac{12}{5} - 2 = \frac{2}{5}.
\end{aligned}
$$

Then, from the alternative formula (5) for the variance, we find that

$$\text{Var}(X) = \frac{2}{5} - \text{E}(X)^2 = \frac{2}{5} - \left(\frac{3}{5} \right)^2 = \frac{1}{25}.$$

>> *Now Try Exercise 15*

Check Your Understanding 12.3

Solutions can be found following the section exercises.

1. Find the expected value and variance of the random variable X whose density function is $f(x) = 1/(2\sqrt{x})$, $1 \le x \le 4$.

2. An insurance company finds that the proportion X of its salespeople who sell more than \$25,000 worth of insurance in a given week is a random variable with the beta probability

density function
$$f(x) = 60x^3(1-x)^2, \qquad 0 \le x \le 1.$$

(a) Compute $E(X)$ and give an interpretation of this quantity.

(b) Compute $\text{Var}(X)$.

EXERCISES 12.3

Find the expected value and variance for each random variable whose probability density function is given. When computing the variance, use formula (5).

1. $f(x) = \frac{1}{18}x$, $0 \le x \le 6$

2. $f(x) = 2(x-1)$, $1 \le x \le 2$

3. $f(x) = \frac{1}{4}$, $1 \le x \le 5$

4. $f(x) = \frac{8}{9}x$, $0 \le x \le \frac{3}{2}$

5. $f(x) = 5x^4$, $0 \le x \le 1$

6. $f(x) = \frac{3}{2}x - \frac{3}{4}x^2$, $0 \le x \le 2$

7. $f(x) = 12x(1-x)^2$, $0 \le x \le 1$

8. $f(x) = \dfrac{3\sqrt{x}}{16}$, $0 \le x \le 4$

9. **A Beta Probability Density** A newspaper publisher estimates that the proportion X of space devoted to advertising on a given day is a random variable with the beta probability density $f(x) = 30x^2(1-x)^2$, $0 \le x \le 1$.
   (a) Find the cumulative distribution function for X.
   (b) Find the probability that less than 25% of the newspaper's space on a given day contains advertising.
   (c) Find $E(X)$ and give an interpretation of this quantity.
   (d) Compute $\text{Var}(X)$.

10. **Proportion of Successful Restaurants** Let X be the proportion of new restaurants in a given year that make a profit during their first year of operation, and suppose that the density function for X is $f(x) = 20x^3(1-x)$, $0 \le x \le 1$.
    (a) Find $E(X)$ and give an interpretation of this quantity.
    (b) Compute $\text{Var}(X)$.

11. **Expected Life of a Machine** The useful life (in hundreds of hours) of a certain machine component is a random variable X with the cumulative distribution function $F(x) = \frac{1}{9}x^2$, $0 \le x \le 3$.
    (a) Find $E(X)$, and give an interpretation of this quantity.
    (b) Compute $\text{Var}(X)$.

12. **Expected Assembly Time** The time (in minutes) required to complete an assembly on a production line is a random variable X with the cumulative distribution function $F(x) = \frac{1}{125}x^3$, $0 \le x \le 5$.
    (a) Find $E(X)$ and give an interpretation of this quantity.
    (b) Compute $\text{Var}(X)$.

13. **Expected Reading Time** The amount of time (in minutes) that a person spends reading the editorial page of the newspaper is a random variable with the density function $f(x) = \frac{1}{72}x$, $0 \le x \le 12$. Find the average time spent reading the editorial page.

14. **Time between Buses** At a certain bus stop the time between buses is a random variable X with the density function $f(x) = 6x(10 - x)/1000$, $0 \le x \le 10$. Find the average time between buses.

15. **Time to Complete a Job** When preparing a bid on a large construction project, a contractor analyzes how long each phase of the construction will take. Suppose that the contractor estimates that the time required for the electrical work will be X hundred worker-hours, where X is a random variable with density function $f(x) = x(6 - x)/18$, $3 \le x \le 6$. (See Fig. 3.)
    (a) Find the cumulative distribution function $F(x)$.
    (b) What is the likelihood that the electrical work will take less than 500 worker-hours?
    (c) Find the mean time to complete the electrical work.
    (d) Find $\text{Var}(X)$.

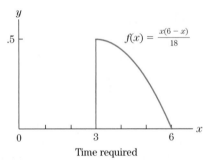

Figure 3 Density function for construction bid.

16. **Amount of Milk Sold** The amount of milk (in thousands of gallons) that a dairy sells each week is a random variable X with the density function $f(x) = 4(x - 1)^3$, $1 \le x \le 2$. (See Fig. 4.)
    (a) What is the likelihood that the dairy will sell more than 1500 gallons?
    (b) What is the average amount of milk that the dairy sells each week?

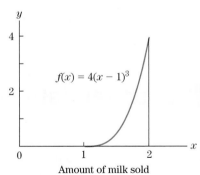

Figure 4 Density function for sales of milk.

17. Let X be a continuous random variable with values between $A = 1$ and $B = \infty$, and with the density function $f(x) = 4x^{-5}$. Compute $E(X)$ and $\text{Var}(X)$.

18. Let X be a continuous random variable with density function $f(x) = 3x^{-4}$, $x \geq 1$. Compute $E(X)$ and $Var(X)$.

If X is a random variable with density function $f(x)$ on $A \leq x \leq B$, the **median** of X is that number M such that

$$\int_A^M f(x)\, dx = \frac{1}{2}.$$

In other words, $Pr(X \leq M) = \frac{1}{2}$.

19. Find the median of the random variable whose density function is $f(x) = \frac{1}{18}x$, $0 \leq x \leq 6$.

20. Find the median of the random variable whose density function is $f(x) = 2(x - 1)$, $1 \leq x \leq 2$.

21. The machine component described in Exercise 11 has a 50% chance of lasting at least how long?

22. In Exercise 12, find the length of time T such that half of the assemblies are completed in T minutes or less.

23. In Exercise 20 of Section 12.2, find the length of time T such that about half of the time you wait only T minutes or less in the express lane at the supermarket.

24. Find the number M such that, half of the time, the dairy in Exercise 16 sells M thousand gallons of milk or less.

25. Show that $E(X) = B - \int_A^B F(x)\, dx$, where $F(x)$ is the cumulative distribution function for X on $A \leq x \leq B$.

26. Use the formula in Exercise 25 to compute $E(X)$ for the random variable X in Exercise 12.

Solutions to Check Your Understanding 12.3

1. $E(X) = \int_1^4 x \cdot \frac{1}{2\sqrt{x}}\, dx = \int_1^4 \frac{1}{2} x^{1/2}\, dx = \frac{1}{3} x^{3/2} \Big|_1^4$

$= \frac{1}{3}(4)^{3/2} - \frac{1}{3} = \frac{8}{3} - \frac{1}{3} = \frac{7}{3}$

To find $Var(X)$, we first compute

$\int_1^4 x^2 \cdot \frac{1}{2\sqrt{x}}\, dx = \int_1^4 \frac{1}{2} x^{3/2}\, dx = \frac{1}{5} x^{5/2} \Big|_1^4$

$= \frac{1}{5}(4)^{5/2} - \frac{1}{5} = \frac{32}{5} - \frac{1}{5} = \frac{31}{5}.$

Then, from formula (5),

$$Var(X) = \frac{31}{5} - \left(\frac{7}{3}\right)^2 = \frac{34}{45}.$$

2. (a) First note that $f(x) = 60x^3(1 - x)^2 = 60x^3(1 - 2x + x^2) = 60x^3 - 120x^4 + 60x^5$. Then,

$E(X) = \int_0^1 xf(x)\, dx$

$= \int_0^1 (60x^4 - 120x^5 + 60x^6)\, dx$

$= \left(12x^5 - 20x^6 + \frac{60}{7} x^7\right) \Big|_0^1$

$= 12 - 20 + \frac{60}{7} = \frac{4}{7}.$

Thus, in an average week, about four-sevenths of the salespeople sell more than \$25,000 worth of insurance. More precisely, over a period of many weeks, we expect an average of four-sevenths of the salespeople each week to sell more than \$25,000 worth of insurance.

(b) $\int_0^1 x^2 f(x)\, dx = \int_0^1 (60x^5 - 120x^6 + 60x^7)\, dx$

$= \left(10x^6 - \frac{120}{7} x^7 + \frac{60}{8} x^8\right) \Big|_0^1$

$= 10 - \frac{120}{7} + \frac{60}{8} = \frac{5}{14}.$

Hence,

$$Var(X) = \frac{5}{14} - \left(\frac{4}{7}\right)^2 = \frac{3}{98}.$$

12.4 Exponential and Normal Random Variables

This section is devoted to the two most important types of probability density functions, the exponential and normal density functions. These functions are associated with random variables that arise in a wide variety of applications. We will describe some typical examples.

Exponential Density Functions

Let k be a positive constant. Then, the function

$$f(x) = ke^{-kx}, \qquad x \geq 0,$$

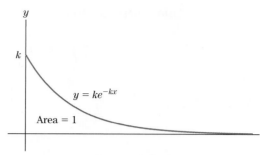

Figure 1 Exponential density function.

is called an **exponential density function**. (See Fig. 1.) This function is indeed a probability density function. First, $f(x)$ is clearly greater than or equal to 0. Second,

$$\int_0^b ke^{-kx}\,dx = -e^{-kx}\Big|_0^b = 1 - e^{-kb} \to 1 \quad \text{as} \quad b \to \infty,$$

so,

$$\int_0^\infty ke^{-kx}\,dx = 1.$$

A random variable X with an exponential density function is called an **exponential random variable**, and the values of X are said to be **exponentially distributed**. Exponential random variables are used in reliability calculations to represent the lifetime (or time to failure) of electronic components such as computer chips. They are used to describe the length of time between two successive random events, such as the interarrival times between successive telephone calls at a switchboard. Also, exponential random variables can arise in the study of service times, such as the length of time a person spends in a doctor's office or at a gas station.

Let us compute the expected value of an exponential random variable X:

$$E(X) = \int_0^\infty xf(x)\,dx = \int_0^\infty xke^{-kx}\,dx.$$

We may approximate this improper integral by a definite integral and use integration by parts to find that

$$\int_0^b xke^{-kx}\,dx = -xe^{-kx}\Big|_0^b - \int_0^b -e^{-kx}\,dx$$

$$= (-be^{-kb} + 0) - \frac{1}{k}e^{-kx}\Big|_0^b$$

$$= -be^{-kb} - \frac{1}{k}e^{-kb} + \frac{1}{k}. \tag{1}$$

As $b \to \infty$, this quantity approaches $1/k$, because the numbers $-be^{-kb}$ and $-(1/k)e^{-kb}$ both approach 0. (See Section 11.5, Exercise 42.) Thus,

$$E(X) = \int_0^\infty xke^{-kx}\,dx = \frac{1}{k}.$$

Let us now compute the variance of X. From the alternative formula for $\text{Var}(X)$ given in Section 12.3, we have

$$\text{Var}(X) = \int_0^\infty x^2 f(x)\,dx - E(X)^2$$

$$= \int_0^\infty x^2 ke^{-kx}\,dx - \frac{1}{k^2}. \tag{2}$$

Using integration by parts, we obtain

$$\int_0^b x^2 k e^{-kx}\,dx = x^2(-e^{-kx})\Big|_0^b - \int_0^b 2x(-e^{-kx})\,dx$$

$$= (-b^2 e^{-kb} + 0) + 2\int_0^b x e^{-kx}\,dx$$

$$= -b^2 e^{-kb} + \frac{2}{k}\int_0^b x k e^{-kx}\,dx. \tag{3}$$

Now let $b \to \infty$. We know from our calculation (1) of $E(X)$ that the integral in the second term of (3) approaches $1/k$; also, it can be shown that $-b^2 e^{-kb}$ approaches 0 (see Section 11.5, Exercise 44). Therefore,

$$\int_0^\infty x^2 k e^{-kx}\,dx = \frac{2}{k}\cdot\frac{1}{k} = \frac{2}{k^2}.$$

And by equation (2), we have

$$\text{Var}(X) = \frac{2}{k^2} - \frac{1}{k^2} = \frac{1}{k^2}.$$

Let us summarize our results:

> Let X be a random variable with an exponential density function $f(x) = k e^{-kx}$ ($x \geq 0$). Then,
>
> $$E(X) = \frac{1}{k} \quad \text{and} \quad \text{Var}(X) = \frac{1}{k^2}.$$

EXAMPLE 1 **Exponential Density** Suppose that the number of days of continuous use provided by a certain brand of lightbulb is an exponential random variable X with expected value 100 days.
(a) Find the density function of X.
(b) Find the probability that a randomly chosen bulb will last between 80 and 90 days.
(c) Find the probability that a randomly chosen bulb will last for more than 40 days.

SOLUTION (a) Since X is an exponential random variable, its density function must be of the form $f(x) = k e^{-kx}$ for some $k > 0$. Since the expected value of such a density function is $1/k$ and is equal to 100 in this case, we see that

$$\frac{1}{k} = 100,$$

$$k = \frac{1}{100} = .01.$$

Thus, $f(x) = .01 e^{-0.01x}$.

(b) $\Pr(80 \leq X \leq 90) = \int_{80}^{90} .01 e^{-0.01x}\,dx = -e^{-0.01x}\Big|_{80}^{90} = -e^{-0.9} + e^{-0.8} \approx .04276$

(c) $\Pr(X > 40) = \int_{40}^\infty .01 e^{-0.01x}\,dx = 1 - \int_0^{40} .01 e^{-0.01x}\,dx$

[since $\int_0^\infty f(x)\,dx = 1$], so

$$\Pr(X > 40) = 1 + (e^{-0.01x})\Big|_0^{40} = 1 + (e^{-0.4} - 1) = e^{-0.4} \approx .67032.$$

>> *Now Try Exercise 3*

| EXAMPLE 2 | **Time between Phone Calls** During a certain part of the day, the interarrival time between successive phone calls at a central telephone exchange is an exponential random variable X with expected value $\frac{1}{3}$ second. |

(a) Find the density function of X.

(b) Find the probability that between $\frac{1}{3}$ and $\frac{2}{3}$ of a second elapses between consecutive phone calls.

(c) Find the probability that the time between successive phone calls is more than 2 seconds.

SOLUTION **(a)** Since X is an exponential random variable, its density function is $f(x) = ke^{-kx}$ for some $k > 0$. Since the expected value of X is $1/k = \frac{1}{3}$, we have $k = 3$ and $f(x) = 3e^{-3x}$.

(b) $\Pr\left(\frac{1}{3} \le X \le \frac{2}{3}\right) = \int_{1/3}^{2/3} 3e^{-3x}\, dx = -e^{-3x}\Big|_{1/3}^{2/3} = -e^{-2} + e^{-1} \approx .23254$

(c) $\Pr(X > 2) = \int_{2}^{\infty} 3e^{-3x}\, dx = 1 - \int_{0}^{2} 3e^{-3x}\, dx = 1 + (e^{-3x})\Big|_{0}^{2} = e^{-6} \approx .00248.$

In other words, about .25% of the time, the waiting time between consecutive calls is at least 2 seconds. **» Now Try Exercise 5**

Normal Density Functions

Let μ and σ be given numbers, with $\sigma > 0$. Then, the function

$$f(x) = \frac{1}{\sigma\sqrt{2\pi}}\, e^{-(1/2)[(x-\mu)/\sigma]^2} \tag{4}$$

is called a **normal density function**. A random variable X whose density function has this form is called a **normal random variable**, and the values of X are said to be **normally distributed**. Many random variables in applications are approximately normal. For example, errors that occur in physical measurements and various manufacturing processes, as well as many human physical and mental characteristics, are all conveniently modeled by normal random variables.

The graph of the density function in definition (4) is called a **normal curve** (Fig. 2). A normal curve is symmetric about the line $x = \mu$ and has inflection points at $\mu - \sigma$ and $\mu + \sigma$. Figure 3 shows three normal curves corresponding to different values of σ. The parameters μ and σ determine the shape of the curve. The value of μ determines the point where the curve reaches its maximum height, and the value of σ determines how sharp a peak the curve has.

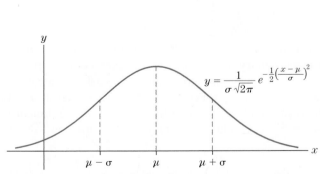

Figure 2 A normal density function.

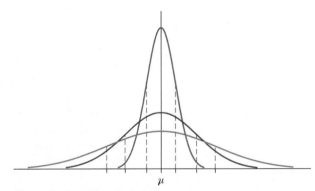

Figure 3 Several normal curves.

It can be shown that the constant $1/(\sigma\sqrt{2\pi})$ in definition (4) of a normal density function $f(x)$ is needed to make the area under the normal curve equal to 1, that is, to make $f(x)$ a probability density function. The theoretical values of a normal random variable X include all positive and negative numbers, but the normal curve approaches the horizontal axis so rapidly beyond the inflection points that the probabilities associated with intervals on the x-axis far to the left or right of $x = \mu$ are negligible.

Using techniques outside the scope of this book, we can verify the following basic facts about a normal random variable.

Let X be a random variable with the normal density function

$$f(x) = \frac{1}{\sigma\sqrt{2\pi}}\, e^{-(1/2)[(x-\mu)/\sigma]^2}.$$

Then, the expected value (mean), variance, and standard deviation of X are given by

$$\mathrm{E}(X) = \mu, \quad \mathrm{Var}(X) = \sigma^2, \quad \text{and} \quad \sqrt{\mathrm{Var}(X)} = \sigma.$$

A normal random variable with expected value $\mu = 0$ and standard deviation $\sigma = 1$ is called a **standard normal random variable** and is often denoted by the letter Z. Using these values for μ and σ in (4) and writing z in place of the variable x, we see that the density function for Z is

$$f(z) = \frac{1}{\sqrt{2\pi}}\, e^{-(1/2)z^2}.$$

The graph of this function is called the **standard normal curve**. (See Fig. 4.)

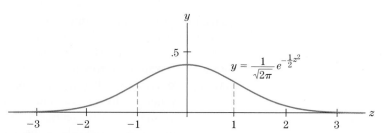

Figure 4 The standard normal curve.

Probabilities involving a standard normal random variable Z may be written in the form

$$\Pr(a \le Z \le b) = \int_a^b \frac{1}{\sqrt{2\pi}}\, e^{-(1/2)z^2}\, dz.$$

Such an integral cannot be evaluated directly, because the density function for Z cannot be antidifferentiated in terms of elementary functions. However, tables of such probabilities have been compiled using numerical approximations to the definite integrals. For $z \ge 0$, let $A(z) = \Pr(0 \le Z \le z)$ and $A(-z) = \Pr(-z \le Z \le 0)$. That is, let $A(z)$ and $A(-z)$ be the areas of the regions shown in Fig. 5. From the symmetry of the standard normal curve, it is clear that $A(-z) = A(z)$. Values of $A(z)$ for $z \ge 0$ are listed in Table 1 of the Appendix.

Figure 5 Areas under the standard normal curve.

EXAMPLE 3 **A Standard Normal Random Variable** Let Z be a standard normal random variable. Use Table 1 of the Appendix to compute the following probabilities:
(a) $\Pr(0 \le Z \le 1.84)$ (b) $\Pr(-1.65 \le Z \le 0)$
(c) $\Pr(.7 \le Z)$ (d) $\Pr(.5 \le Z \le 2)$ (e) $\Pr(-.75 \le Z \le 1.46)$

SOLUTION

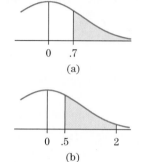

0   .7
(a)

0   .5   2
(b)

−.75 0   1.46
(c)

Figure 6

(a) $\Pr(0 \le Z \le 1.84) = A(1.84)$. In Table 1 we move down the column under z until we reach 1.8; then, we move right in the same row to the column with the heading .04. There, we find that $A(1.84) = .4671$.
(b) $\Pr(-1.65 \le Z \le 0) = A(-1.65) = A(1.65) = .4505$ (from Table 1)
(c) Since the area under the normal curve is 1, the symmetry of the curve implies that the area to the right of the y-axis is .5. Now $\Pr(.7 \le Z)$ is the area under the curve to the right of .7, and so, we can find this area by subtracting from .5 the area between 0 and .7. [See Fig. 6(a).] Thus,

$$\Pr(.7 \le Z) = .5 - \Pr(0 \le Z \le .7)$$
$$= .5 - A(.7) = .5 - .2580 \quad \text{From Table 1}$$
$$= .2420.$$

(d) The area under the standard normal curve from .5 to 2 equals the area from 0 to 2 minus the area from 0 to .5. [See Fig. 6(b).] Thus, we have

$$\Pr(.5 \le Z \le 2) = A(2) - A(.5)$$
$$= .4772 - .1915 = .2857.$$

(e) The area under the standard normal curve from −.75 to 1.46 equals the area from −.75 to 0 plus the area from 0 to 1.46. [See Fig. 6(c).] Thus,

$$\Pr(-.75 \le Z \le 1.46) = A(-.75) + A(1.46)$$
$$= A(.75) + A(1.46)$$
$$= .2734 + .4279 = .7013.$$

» Now Try Exercise 23

When X is an *arbitrary* normal random variable, with mean μ and standard deviation σ, we may compute a probability such as $\Pr(a \le X \le b)$ by making the change of variable $z = (x - \mu)/\sigma$. This converts the integral for $\Pr(a \le X \le b)$ into an integral involving the standard normal density function. The following example illustrates this procedure.

EXAMPLE 4 **Length of a Metal Flange** A metal flange on a truck must be between 92.1 and 94 millimeters long to fit properly. The lengths of the flanges supplied to the truck manufacturer are normally distributed with mean $\mu = 93$ millimeters and standard deviation $\sigma = .4$ millimeter.
(a) What percentage of the flanges have an acceptable length?
(b) What percentage of the flanges are too long?

SOLUTION Let X be the length of a metal flange selected at random from the supply of flanges.
(a) We have

$$\Pr(92.1 \le X \le 94) = \int_{92.1}^{94} \frac{1}{(.4)\sqrt{2\pi}} e^{-(1/2)[(x-93)/0.4]^2} \, dx.$$

Using the substitution $z = (x - 93)/.4$, $dz = (1/.4) \, dx$, we note that, if $x = 92.1$, then $z = (92.1 - 93)/.4 = -.9/.4 = -2.25$, and if $x = 94$, then $z = (94 - 93)/.4 = 1/.4 = 2.5$. Hence,

$$\Pr(92.1 \le X \le 94) = \int_{-2.25}^{2.5} \frac{1}{\sqrt{2\pi}} e^{-(1/2)z^2} \, dz.$$

The value of this integral is the area under the standard normal curve from -2.25 to 2.5, which equals the area from -2.25 to 0 plus the area from 0 to 2.5. Thus,

$$\Pr(92.1 \le X \le 94) = A(-2.25) + A(2.5)$$
$$= A(2.25) + A(2.5)$$
$$= .4878 + .4938 = .9816.$$

From this probability, we conclude that about 98% of the flanges will have an acceptable length.

(b) $$\Pr(94 \le X) = \int_{94}^{\infty} \frac{1}{(4)\sqrt{2\pi}} e^{-(1/2)[(x-93)/0.4]^2} \, dx$$

This integral is approximated by an integral from $x = 94$ to $x = b$, where b is large. If we substitute $z = (x - 93)/.4$, we find that

$$\int_{94}^{b} \frac{1}{(.4)\sqrt{2\pi}} e^{-(1/2)[(x-93)/0.4]^2} \, dx = \int_{2.5}^{(b-93)/0.4} \frac{1}{\sqrt{2\pi}} e^{-(1/2)z^2} \, dz. \tag{5}$$

Now, when $b \to \infty$, the quantity $(b - 93)/.4$ also becomes arbitrarily large. Since the left integral in (5) approaches $\Pr(94 \le X)$, we conclude that

$$\Pr(94 \le X) = \int_{2.5}^{\infty} \frac{1}{\sqrt{2\pi}} e^{-(1/2)z^2} \, dz.$$

To calculate this integral, we use the method of Example 3(c). The area under the standard normal curve to the right of 2.5 equals the area to the right of 0 minus the area from 0 to 2.5. That is,

$$\Pr(94 \le X) = .5 - A(2.5)$$
$$= .5 - .4938 = .0062.$$

Approximately .6% of the flanges exceed the maximum acceptable length.

》 Now Try Exercise 27

INCORPORATING

TECHNOLOGY

The TI-83 graphing calculator can easily calculate normal probabilities with the **normalcdf** function from the DISTR menu. The area under the normal curve with mean μ and standard deviation σ from $x = a$ to $x = b$ is given by **normalcdf (a,b,μ, σ)**. The normal probabilities from Example 4 are calculated in Fig. 7. (*Note:* E99, which equals 10^{99}, is generated when you press **1** [2nd] [EE] **99** and is used in place of infinity. Minus infinity is represented by $-$E99.)

Figure 7 Probabilities from Example 4.

Check Your Understanding 12.4

Solutions can be found following the section exercises.

1. The emergency flasher on an automobile is warranted for the first 12,000 miles that the car is driven. During that period a defective flasher will be replaced free. If the time before failure of the emergency flasher (measured in thousands of miles) is an exponential random variable X with mean 50 (thousand miles), what percentage of the flashers will have to be replaced during the warranty period?

2. The lead time between ordering furniture from a certain company and receiving delivery is a normal random variable with $\mu = 18$ weeks and $\sigma = 5$ weeks. Find the likelihood that a customer will have to wait more than 16 weeks.

EXERCISES 12.4

Find (by inspection) the expected values and variances of the exponential random variables with the density functions given in Exercises 1–4.

1. $3e^{-3x}$

2. $\frac{1}{4}e^{-x/4}$

3. $.2e^{-0.2x}$

4. $1.5e^{-1.5x}$

In a large factory there is an average of two accidents per day, and the time between accidents has an exponential density function with an expected value of $\frac{1}{2}$ day.

5. Find the probability that the time between two accidents will be more than $\frac{1}{2}$ day and less than 1 day.

6. Find the probability that the time between accidents will be less than 8 hours ($\frac{1}{3}$ day).

The amount of time required to serve a customer at a bank has an exponential density function with mean 3 minutes.

7. Find the probability that a customer is served in less than 2 minutes.

8. Find the probability that serving a customer will require more than 5 minutes.

During a certain part of the day, the time between arrivals of automobiles at the tollgate on a turnpike is an exponential random variable with expected value 20 seconds.

9. Find the probability that the time between successive arrivals is more than 60 seconds.

10. Find the probability that the time between successive arrivals is greater than 10 seconds and less than 30 seconds.

In a study of the vacancies occurring in the U.S. Supreme Court, it has been determined that the time elapsed between successive resignations is an exponential random variable with expected value 2 years.

11. Exponential Distribution and the Supreme Court A new president takes office at the same time that a justice retires. Find the probability that the next vacancy on the court will take place during the president's 4-year term.

12. Find the probability that the composition of the U.S. Supreme Court will remain unchanged for a period of 5 years or more.

13. Reliability of Electronic Components Suppose that the average life span of an electronic component is 72 months and that the life spans are exponentially distributed.
 (a) Find the probability that a component lasts for more than 24 months.
 (b) The *reliability function* $r(t)$ gives the probability that a component will last for more than t months. Compute $r(t)$ in this case.

14. Survival Function Consider a group of patients who have been treated for an acute disease such as cancer, and let X be the number of years a person lives after receiving the treatment (the survival time). Under suitable conditions, the density function for X will be $f(x) = ke^{-kx}$ for some constant k.
 (a) The *survival function* $S(x)$ is the probability that a person chosen at random from the group of patients survives until at least time x. Explain why $S(x) = 1 - F(x)$, where $F(x)$ is the cumulative distribution function for X, and compute $S(x)$.

(b) Suppose that the probability is .90 that a patient will survive at least 5 years [$S(5) = .90$]. Find the constant k in the exponential density function $f(x)$.

Find the expected values and the standard deviations (by inspection) of the normal random variables with the density functions given in Exercises 15–18.

15. $\frac{1}{\sqrt{2\pi}} e^{-(1/2)(x-4)^2}$

16. $\frac{1}{\sqrt{2\pi}} e^{-(1/2)(x+5)^2}$

17. $\frac{1}{3\sqrt{2\pi}} e^{-(1/18)x^2}$

18. $\frac{1}{5\sqrt{2\pi}} e^{-(1/2)[(x-3)/5]^2}$

19. Show that the function $f(x) = e^{-x^2/2}$ has a relative maximum at $x = 0$.

20. Show that the function $f(x) = e^{-(1/2)[(x-\mu)/\sigma]^2}$ has a relative maximum at $x = \mu$.

21. Show that the function $f(x) = e^{-x^2/2}$ has inflection points at $x = \pm 1$.

22. Show that the function $f(x) = e^{-(1/2)[(x-\mu)/\sigma]^2}$ has inflection points at $x = \mu \pm \sigma$.

23. Let Z be a standard normal random variable. Calculate
 (a) $\Pr(-1.3 \leq Z \leq 0)$ **(b)** $\Pr(.25 \leq Z)$
 (c) $\Pr(-1 \leq Z \leq 2.5)$ **(d)** $\Pr(Z \leq 2)$

24. Calculate the area under the standard normal curve for values of z
 (a) between .5 and 1.5,
 (b) between $-.75$ and .75,
 (c) to the left of $-.3$,
 (d) to the right of -1.

25. Time of Birth The gestation period (length of pregnancy) of a certain species is approximately normally distributed with a mean of 6 months and standard deviation of $\frac{1}{2}$ month.
 (a) Find the percentage of births that occur after a gestation period of between 6 and 7 months.
 (b) Find the percentage of births that occur after a gestation period of between 5 and 6 months.

26. Normal Distribution and Life of a Tire Suppose that the life span of a certain automobile tire is normally distributed, with $\mu = 25{,}000$ miles and $\sigma = 2000$ miles.
 (a) Find the probability that a tire will last between 28,000 and 30,000 miles.
 (b) Find the probability that a tire will last more than 29,000 miles.

27. Amount of Milk in a Container If the amount of milk in a gallon container is a normal random variable, with $\mu = 128.2$ ounces and $\sigma = .2$ ounce, find the probability that a random container of milk contains less than 128 ounces.

28. Breaking Weight The amount of weight required to break a certain brand of twine has a normal density function, with $\mu = 43$ kilograms and $\sigma = 1.5$ kilograms. Find the probability that the breaking weight of a piece of the twine is less than 40 kilograms.

29. Time of a Commute A student with an eight o'clock class at the University of Maryland commutes to school by car. She has discovered that along each of two possible routes her traveling time to school (including the time to get to class) is approximately a normal random variable. If she uses the Capital Beltway for most of her trip, $\mu = 25$ minutes and $\sigma = 5$ minutes. If she drives a longer route over local city streets, $\mu = 28$ minutes and $\sigma = 3$ minutes. Which route should the student take if she leaves home at 7:30 A.M.? (Assume that the best route is one that minimizes the probability of being late to class.)

30. Which route should the student in Exercise 29 take if she leaves home at 7:26 A.M.?

31. **Diameter of a Bolt** A certain type of bolt must fit through a 20-millimeter test hole or else it is discarded. If the diameters of the bolts are normally distributed, with $\mu = 18.2$ millimeters and $\sigma = .8$ millimeters, what percentage of the bolts will be discarded?

32. **SAT Scores Distribution** The Math SAT scores of a recent freshman class at a university were normally distributed, with $\mu = 535$ and $\sigma = 100$.
    (a) What percentage of the scores were between 500 and 600?
    (b) Find the minimum score needed to be in the top 10% of the class.

33. **Life of a Computer Chip** Let X be the time to failure (in years) of a computer chip, and suppose that the chip has been operating properly for a years. Then, it can be shown that the probability that the chip will fail within the next b years is

$$\frac{\Pr(a \le X \le a + b)}{\Pr(a \le X)}. \qquad (6)$$

Compute this probability for the case when X is an exponential random variable with density function $f(x) = ke^{-kx}$, and show that this probability equals $\Pr(0 \le X \le b)$. This means that the probability given by (6) does not depend on how long the chip has already been operating. Exponential random variables are therefore said to be *memoryless*.

34. Recall that the *median* of an exponential density function is that number M such that $\Pr(X \le M) = \frac{1}{2}$. Show that $M = (\ln 2)/k$. (We see that the median is less than the mean.)

35. **Life of a Lightbulb** If the lifetime (in weeks) of a certain brand of lightbulb has an exponential density function and 80% of all lightbulbs fail within the first 100 weeks, find the average lifetime of a lightbulb.

TECHNOLOGY EXERCISES

36. The computations of the expected value and variance of an exponential random variable relied on the fact that, for any positive number k, be^{-kb} and $b^2 e^{-kb}$ approach 0 as b gets large. That is,

$$\lim_{x \to \infty} \frac{x}{e^{kx}} = 0 \quad \text{and} \quad \lim_{x \to \infty} \frac{x^2}{e^{kx}} = 0.$$

The validity of these limits for the case $k = 1$ is shown in Figs. 8 and 9. Convince yourself that these limits hold for all positive values of k by generating graphs for $k = .1$, $k = .5$, and $k = 2$.

Figure 8 Graph of $\mathbf{Y_1} = \dfrac{x}{e^x}$.

Figure 9 Graph of $\mathbf{Y_1} = \dfrac{x^2}{e^x}$.

37. Use the integral routine to convince yourself that $\int_{-\infty}^{\infty} x^2 f(x)\, dx = 1$, where $f(x)$ is the standard normal density function. [*Note:* Since $f(x)$ approaches zero so rapidly as x gets large in magnitude, the value of the improper integral is nearly the same as the definite integral of $x^2 f(x)$ from $x = -8$ to $x = 8$.] Conclude that the standard deviation of the standard normal random variable is 1.

Solutions to Check Your Understanding 12.4

1. The density function for X is $f(x) = ke^{-kx}$, where $1/k = 50$ (thousand miles) and $k = 1/50 = .02$. Then,

$$\Pr(X \le 12) = \int_0^{12} .02 e^{-0.02x}\, dx = -e^{-0.02x} \Big|_0^{12}$$
$$= 1 - e^{-0.24} \approx .21337.$$

About 21% of the flashers will have to be replaced during the warranty period.

2. Let X be the time between ordering and receiving the furniture. Since $\mu = 18$ and $\sigma = 5$, we have

$$\Pr(16 < X) = \int_{16}^{\infty} \frac{1}{5\sqrt{2\pi}} e^{-(1/2)[(x-18)/5]^2}\, dx.$$

If we substitute $z = (x - 18)/5$, then $dz = \frac{1}{5} dx$, and $z = -.4$ when $x = 16$:

$$\Pr(16 < X) = \int_{-.4}^{\infty} \frac{1}{\sqrt{2\pi}} e^{-(1/2)z^2}\, dz.$$

[A similar substitution was made in Example 4(b).] The preceding integral gives the area under the standard normal curve to the right of $-.4$. Since the area between $-.4$ and 0 is $A(-.4) = A(.4)$, and the area to the right of 0 is .5, we have

$$\Pr(16 < X) = A(.4) + .5 = .1554 + .5 = .6554.$$

12.5 Poisson and Geometric Random Variables

Probability theory is widely applied in business, biology and the social sciences to situations that involve counting. The probability models in this section involve a random variable X whose values are the discrete numbers $0, 1, 2, \ldots$. Usually, there is no specific upper limit to the value of X, even though extremely large values of X are highly unlikely. Here are typical examples of such experiments. In each case, X represents the outcome of the experiment.

1. At an insurance company, count the number of fire insurance claims submitted in any one month (selected at random).
2. In a microbiological study of a pond, count the number of protozoa in a drop-size random sample of water.
3. In a factory, count the number of times per month that a certain type of machine breaks down.

Suppose X denotes a random variable for an experiment whose outcome is one of the values $0, 1, 2, \ldots$, and for each possible value n, let p_n be the associated probability of occurrence. That is,

$$p_0 = \Pr(X = 0),$$
$$p_1 = \Pr(X = 1),$$
$$\vdots$$
$$p_n = \Pr(X = n)$$
$$\vdots$$

Notice that, since $p_0, p_1, p_2, \ldots$ are probabilities, each lies between 0 and 1. Also, the sum of these probabilities must be 1. (One of the outcomes $0, 1, 2, \ldots$ always occurs.) That is,

$$p_0 + p_1 + p_2 + \cdots + p_n + \cdots = 1.$$

Unlike the situation in Section 12.1, this sum is an infinite series such as those studied in Sections 11.3 and 11.5.

In analogy with the case of experiments having a finite number of possible outcomes, we may define the **expected value** (or average value) of the random variable X (or of the experiment whose outcome is X) to be the number $E(X)$ given by the following formula:

$$E(X) = 0 \cdot p_0 + 1 \cdot p_1 + 2 \cdot p_2 + 3 \cdot p_3 + \cdots$$

(provided that the infinite series converges). That is, the expected value $E(X)$ is formed by addition of the products of the possible outcomes by their respective probabilities of occurrence.

In a similar fashion, letting m denote $E(X)$, we define the *variance* of X by

$$\mathrm{Var}(x) = (0 - m)^2 \cdot p_0 + (1 - m)^2 \cdot p_1 + (2 - m)^2 \cdot p_2 + (3 - m)^2 \cdot p_3 + \cdots.$$

Poisson Random Variables

In many experiments, the probabilities p_n involve a parameter λ (depending on the particular experiment), and they have the following special form:

$$p_0 = e^{-\lambda},$$

$$p_1 = \frac{\lambda}{1} e^{-\lambda},$$

$$p_2 = \frac{\lambda^2}{2 \cdot 1} e^{-\lambda},$$

$$p_3 = \frac{\lambda^3}{3 \cdot 2 \cdot 1} e^{-\lambda},$$

$$\vdots$$

$$p_n = \frac{\lambda^n}{n!} e^{-\lambda}. \tag{1}$$

FOR REVIEW

The factorial
$n! = n(n-1)\cdots 2.1$
(See p. 514.)

The constant $e^{-\lambda}$ in each probability is necessary to make the sum of all the probabilities equal to 1. A random variable X whose probabilities are given by formula (1) is called a **Poisson random variable**, and the probabilities for X are said to form a **Poisson distribution** with parameter λ. The histograms in Fig. 1 show Poisson distributions for $\lambda = 1.5$, 3, and 5.

(a) $\lambda = 1.5$

(b) $\lambda = 3$

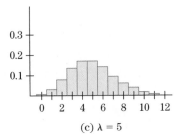
(c) $\lambda = 5$

Figure 1

EXAMPLE 1

A Poisson Distribution The annual number of deaths in a certain town due to a particular disease has a Poisson distribution with parameter $\lambda = 3$. Verify that the associated probabilities sum to 1.

SOLUTION

The first few probabilities are (to four decimal places)

$$p_0 = e^{-3} \approx .0498,$$

$$p_1 = \frac{3}{1} e^{-3} = 3 \cdot p_0 \approx .1494,$$

$$p_2 = \frac{3^2}{2 \cdot 1} e^{-3} = \frac{3 \cdot 3}{2 \cdot 1} e^{-3} = \frac{3}{2} p_1 \approx .2240,$$

$$p_3 = \frac{3^3}{3 \cdot 2 \cdot 1} e^{-3} = \frac{3 \cdot 3 \cdot 3}{3 \cdot 2 \cdot 1} e^{-3} = \frac{3}{3} p_2 \approx .2240,$$

$$p_4 = \frac{3^4}{4 \cdot 3 \cdot 2 \cdot 1} e^{-3} = \frac{3 \cdot 3 \cdot 3 \cdot 3}{4 \cdot 3 \cdot 2 \cdot 1} e^{-3} = \frac{3}{4} p_3 \approx .1680,$$

$$p_5 = \frac{3^5}{5 \cdot 4 \cdot 3 \cdot 2 \cdot 1} e^{-3} = \frac{3 \cdot 3 \cdot 3 \cdot 3 \cdot 3}{5 \cdot 4 \cdot 3 \cdot 2 \cdot 1} e^{-3} = \frac{3}{5} p_4 \approx .1008.$$

Notice how each probability p_n for $n \geq 1$ is computed from the preceding probability p_{n-1}. In general, $p_n = (\lambda/n)p_{n-1}$.

To sum the probabilities for all n, use the exact values, not the decimal approximations:

$$e^{-3} + \frac{3}{1}e^{-3} + \frac{3^2}{2 \cdot 1}e^{-3} + \frac{3^3}{3 \cdot 2 \cdot 1}e^{-3} + \frac{3^4}{4 \cdot 3 \cdot 2 \cdot 1}e^{-3} + \cdots$$

$$= e^{-3}\left(1 + 3 + \frac{1}{2!}3^2 + \frac{1}{3!}3^3 + \frac{1}{4!}3^4 + \cdots\right).$$

From Section 11.5, you should recognize that the series inside the parentheses is the power series for e^x evaluated at $x = 3$. The preceding sum is $e^{-3} \cdot e^3$, which equals 1.

» Now Try Exercise 1

The following facts about Poisson random variables provide an interpretation for the parameter λ.

Let X be a random variable whose probabilities are Poisson distributed with parameter λ; that is,

$$p_0 = e^{-\lambda},$$

$$p_n = \frac{\lambda^n}{n!}e^{-\lambda} \quad (n = 1, 2, \ldots).$$

Then, the expected value and variance of X are given by

$$E(X) = \lambda, \qquad \text{Var}(X) = \lambda.$$

We shall verify only the statement about $E(X)$. The argument uses the Taylor series for e^λ. We have

$$E(X) = 0 \cdot p_0 + 1 \cdot p_1 + 2 \cdot p_2 + 3 \cdot p_3 + 4 \cdot p_4 + \cdots$$

$$= 0 \cdot e^{-\lambda} + \cancel{1} \cdot \frac{\lambda}{\cancel{1}}e^{-\lambda} + \cancel{2} \cdot \frac{\lambda^2}{1 \cdot \cancel{2}}e^{-\lambda}$$

$$+ \cancel{3} \cdot \frac{\lambda^3}{1 \cdot 2 \cdot \cancel{3}}e^{-\lambda} + \cancel{4} \cdot \frac{\lambda^4}{1 \cdot 2 \cdot 3 \cdot \cancel{4}}e^{-\lambda} + \cdots$$

$$= \lambda e^{-\lambda} + \frac{\lambda^2}{1}e^{-\lambda} + \frac{\lambda^3}{1 \cdot 2}e^{-\lambda} + \frac{\lambda^4}{1 \cdot 2 \cdot 3}e^{-\lambda} + \cdots$$

$$= \lambda e^{-\lambda}\left(1 + \frac{\lambda}{1} + \frac{\lambda^2}{1 \cdot 2} + \frac{\lambda^3}{1 \cdot 2 \cdot 3} + \cdots\right)$$

$$= \lambda e^{-\lambda} \cdot e^\lambda$$

$$= \lambda.$$

The next two examples illustrate some applications of Poisson random variables.

EXAMPLE 2　**Computing Probabilities with a Poisson Distribution** Suppose that we observe the number X of calls received by a telephone switchboard during a 1-minute interval. Experience suggests that X is Poisson distributed with $\lambda = 5$.

(a) Determine the probability that zero, one, or two calls arrive during a particular minute.

(b) Determine the probability that three or more calls arrive during a particular minute.

(c) Determine the average number of calls received per minute.

SOLUTION **(a)** The probability that zero, one, or two calls arrive during a given minute is $p_0 + p_1 + p_2$. Moreover,

$$p_0 = e^{-\lambda} = e^{-5} \approx .00674,$$

$$p_1 = \frac{\lambda}{1} e^{-\lambda} = 5e^{-5} \approx .03369,$$

$$p_2 = \frac{\lambda^2}{1 \cdot 2} e^{-\lambda} = \frac{5}{2} p_1 \approx .0842.$$

Thus, $p_0 + p_1 + p_2 \approx .12465$. That is, during approximately 12% of the minutes, either zero, one, or two calls are received.

(b) The probability of receiving three or more calls is the same as the probability of *not* receiving zero, one, or two calls and so is equal to

$$1 - (p_0 + p_1 + p_2) = 1 - .12465 = .87535.$$

(c) The average number of calls received per minute is equal to λ. That is, on average the switchboard receives five calls per minute. **» Now Try Exercise 7**

EXAMPLE 3 **Modeling with a Poisson Distribution** Drop-size water samples are drawn from a New England pond. The numbers of protozoa in many different samples are counted, and the average number is found to be about 8.3. What is the probability that a sample chosen at random contains at most four protozoa?

SOLUTION Under the assumption that the protozoa are thoroughly spread throughout the pond, without clumping, the number of protozoa per drop is a Poisson random variable, call it X. From the experimental data, we assume that $E(X) = 8.3$. Since $\lambda = E(X)$, the probabilities for X are given by

$$p_n = \frac{8.3^n}{n!} e^{-8.3}.$$

The probability of "at most four" is $\Pr(X \le 4)$. Using a calculator to generate the probabilities, we find that

$$\Pr(X \le 4) = p_0 + p_1 + p_2 + p_3 + p_4$$
$$\approx .00025 + .00206 + .00856 + .02368 + .04914$$
$$= .08369.$$

The probability of at most four protozoa is about 8.4%. **» Now Try Exercise 9**

Geometric Random Variables

The following two experiments give rise to discrete random variables whose values are $0, 1, \ldots,$ but whose distributions are not Poisson.

- Toss a coin until a head appears and count the number of tails that precede it.
- As part of a quality-control procedure, test the items coming off an assembly line. Count the number of acceptable items before the first defective item is found.

Each of these experiments involves a trial having two results (tail, head) or (acceptable, defective). In general, the two results are named **success** and **failure**, and the trial is repeated until a failure occurs. The outcome of the experiment is the number X of successes $(0, 1, 2, \ldots)$ that precede the first failure. If for some number p between 0 and 1 the probabilities for X have the form

$$p_0 = 1 - p,$$
$$p_1 = p(1 - p),$$
$$p_2 = p^2(1 - p),$$
$$\vdots$$
$$p_n = p^n(1 - p), \qquad \qquad (2)$$
$$\vdots$$

then X is called a **geometric random variable**, and the probabilities for X are said to form a **geometric distribution** with parameter p. In this case, each trial of the experiment has the same probability p of success. (The probability of failure is $1 - p$.) Also, the outcome of each trial is independent of the other trials. The histograms in Fig. 2 display geometric distributions for $p = .6$ and $.8$.

The term *geometric* is associated with formula (2) because the probabilities form a geometric series, with initial term $a = 1 - p$ and ratio $r = p$. The sum of the series is

$$p_0 + p_1 + p_2 + \cdots = \frac{a}{1 - r} = \frac{1 - p}{1 - p} = 1.$$

This is just what you would expect: The probabilities add up to 1.

EXAMPLE 4	**A Geometric Distribution in Quality Control** An assembly line produces a small mechanical toy, and about 2% of the toys are defective. A quality-control inspector selects a toy at random, inspects it, and repeats this process until a defective toy is found. The probability of success (passing the test) is .98 for each inspection (trial).

(a) Find the probability that exactly three toys pass the test before a defective toy is found.

(b) Find the probability that at most three toys pass the test before a defective toy is found.

(c) What is the probability that at least four toys will pass the test before the first defective toy is found?

SOLUTION   Let X be the number of acceptable toys found before the first defective toy. A reasonable assumption is that the random variable X has a geometric distribution with parameter $p = .98$.

(a) The probability that $X = 3$ is $p_3 = (.98)^3(1 - .98) \approx .0188$.

(b) $\Pr(X \le 3) = p_0 + p_1 + p_2 + p_3$

$$= .02 + (.98)(.02) + (.98)^2(.02) + (.98)^3(.02)$$

$$\approx .02 + .0196 + .0192 + .0188 = .0776.$$

(c) $\Pr(X \ge 4) = 1 - \Pr(X \le 3) \approx 1 - .0776 = .9224.$ **» Now Try Exercise 17**

The following properties of a geometric distribution can be established with facts about power series.

(a) $p = .6$

(b) $p = .8$

Figure 2

> Let X be a geometric random variable with parameter p, that is,
>
> $$p_n = p^n(1 - p) \qquad (n = 0, 1, \ldots).$$
>
> Then, the expected value and variance of X are given by
>
> $$\mathrm{E}(X) = \frac{p}{1 - p}, \qquad \mathrm{Var}(X) = \frac{p}{(1 - p)^2}.$$

EXAMPLE 5	**Expected Number of Defective Toys** For the situation described in Example 4, what is the average number of toys that pass inspection before the first defective toy is found?

SOLUTION   Since $p = .98$, $\mathrm{E}(X) = .98/(1 - .98) = 49$. If many inspections are made, then we can expect on average that 49 toys will pass inspection before a defective toy is found.

» Now Try Exercise 11

INCORPORATING
TECHNOLOGY

Poisson Probabilities The TI-83 graphing calculator has two functions that calculate Poisson probabilities. The value of **poissonpdf(λ,n)** is $p_n = \frac{\lambda^n}{n!}e^{-\lambda}$, and the value of **poissoncdf(λ,n)** is $p_0 + p_1 + p_2 + \cdots + p_n$. (These two functions are invoked from the DISTR menu.) In Fig. 3, the functions are used to calculate two probabilities from Example 2(a). With other graphing calculators, you can find the sum of successive Poisson probabilities with **sum(seq**, as shown in the last computation of Fig. 3. In Fig. 4 **sum(seq** is used to compute the geometric probabilities from parts (b) and (c) of Example 4.

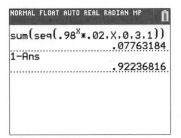

Figure 3 Two probabilities from Example 2.

Figure 4 Two probabilities from Example 4.

Check Your Understanding 12.5

Solutions can be found following the section exercises.

A public health officer is tracking down the source of a bacterial infection in a certain city. She analyzes the reported incidence of the infection in each city block and finds an average of three cases per block. A certain block is found to have seven cases. What is the probability that a randomly chosen block has at least seven cases, assuming that the number of cases per block is Poisson distributed?

EXERCISES 12.5

1. Suppose that a random variable X has a Poisson distribution with $\lambda = 3$, as in Example 1. Compute the probabilities p_6, p_7, p_8.

2. Let X be a Poisson random variable with parameter $\lambda = 5$. Compute the probabilities $p_0, \ldots, p_6$ to four decimal places.

3. Repeat Exercise 2 with $\lambda = .75$ and make a histogram.

4. Repeat Exercise 2 with $\lambda = 2.5$ and make a histogram.

5. **Number of Insurance Claims** The monthly number of fire insurance claims filed with the Firebug Insurance Company is Poisson distributed with $\lambda = 10$.
   (a) What is the probability that in a given month no claims are filed?
   (b) What is the probability that in a given month no more than two claims are filed? (The number of claims is either zero, one, or two.)
   (c) What is the probability that in a given month at least three claims are filed?

6. **Waiting Time in an Emergency Room** On a typical weekend evening at a local hospital, the number of persons waiting for treatment in the emergency room is Poisson distributed with $\lambda = 6.5$.
   (a) What is the likelihood that either no one or only one person is waiting for treatment?
   (b) What is the likelihood that no more than four persons are waiting?
   (c) What is the likelihood that at least five persons are waiting?

7. **Distribution of Typos** The number of typographical errors per page of a certain newspaper has a Poisson distribution, and there is an average of 1.5 errors per page.
   (a) What is the probability that a randomly selected page is error-free?
   (b) What is the probability that a page has either two or three errors?
   (c) What is the probability that a page has at least four errors?

8. **Number of Cars at a Tollgate** During a certain part of the day, an average of five automobiles arrives every minute at the tollgate on a turnpike. Let X be the number of automobiles that arrive in any 1-minute interval selected at random. Let Y be the interarrival time between any two successive arrivals. (The average interarrival time is $\frac{1}{5}$ minute.) Assume that X is a Poisson random variable and that Y is an exponential random variable.
   (a) Find the probability that at least five cars arrive during a given 1-minute interval.
   (b) Find the probability that the time between any two successive cars is less than $\frac{1}{5}$ minute.

9. **Poisson Distribution in a Mixing Problem** A bakery makes gourmet cookies. For a batch of 4800 oatmeal and raisin cookies, how many raisins should be used so that the probability of a cookie having no raisins is .01? [*Note:* A reasonable assumption is that the number of raisins in a random cookie has a Poisson distribution.]

10. If X is a geometric random variable with parameter $p = .9$, compute the probabilities $p_0, \ldots, p_5$ and make a histogram.

11. Repeat Exercise 10 with $p = .6$.

12. **Quality Control** The quality-control department at a sewing machine factory has determined that 1 out of 40 machines does not pass inspection. Let X be the number of machines on an assembly line that pass inspection before a machine is found that fails inspection.
   (a) Write the formula for $\Pr(X = n)$.
   (b) What is the probability that, of five machines coming off the assembly line, the first four pass inspection and the fifth does not?

13. **Two Competing Companies** In a certain town, there are two competing taxicab companies, Red Cab and Blue Cab. The taxis mix with downtown traffic in a random manner. There are three times as many Red taxis as there are Blue taxis. Suppose that you stand on a downtown street corner and count the number X of Red taxis that appear before the first Blue taxi appears.

(a) Determine the formula for $\Pr(X = n)$.

(b) What is the likelihood of observing at least three Red taxis before the first Blue taxi?

(c) What is the average number of consecutive Red taxis prior to the appearance of a Blue taxi?

14. At a certain junior high school, two-thirds of the students have at least one tooth cavity. A dental survey is made of the students. What is the probability that the first student to have a cavity is the third student examined?

15. Let X be a geometric random variable with parameter $p < 1$. Find a formula for $\Pr(X < n)$, for $n > 0$. [*Note:* The partial sum of a geometric series with ratio r is given by

$$1 + r + \cdots + r^{n-1} = \frac{1 - r^n}{1 - r}.]$$

16. Paper Jam in Copy Machines Whenever a document is fed into a high-speed copy machine, there is a .5% chance that a paper jam will stop the machine.

(a) What is the expected number of documents that can be copied before a paper jam occurs?

(b) Determine the probability that at least 100 documents can be copied before a paper jam occurs. [*Hint:* See Exercise 15.]

17. Probability of an Infection Suppose that a large number of persons become infected by a particular strain of staphylococcus that is present in food served by a fast-food restaurant and that the germ usually produces a certain symptom in 5% of the persons infected. What is the probability that, when customers are examined, the first person to have the symptom is the fifth customer examined?

18. Coin Toss Suppose that you toss a fair coin until a head appears and count the number X of consecutive tails that precede it.

(a) Determine the probability that exactly n consecutive tails occur.

(b) Determine the average number of consecutive tails that occur.

(c) Write down the infinite series that gives the variance for the number of consecutive tails. Use Exercise 28 in Section 11.5 to show that the variance equals 2.

Maximum Likelihood Exercises 19 and 20 illustrate a technique from statistics (called the *method of maximum likelihood*) that estimates a parameter for a probability distribution.

19. In a production process, a box of fuses is examined and found to contain two defective fuses. Suppose that the probability

of having two defective fuses in a box selected at random is $(\lambda^2/2)e^{-\lambda}$ for some λ. Take first and second derivatives to determine the value of λ for which the probability has its maximum value.

20. A person shooting at a target has five successive hits and then a miss. If x is the probability of success on each shot, the probability of having five successive hits followed by a miss is $x^5(1 - x)$. Take first and second derivatives to determine the value of x for which the probability has its maximum value.

21. Let X be a geometric random variable with parameter p. Derive the formula for $E(X)$ by using the power series formula (see Example 3 in Section 11.5):

$$1 + 2x + 3x^2 + \cdots = \frac{1}{(1 - x)^2} \quad \text{for } |x| < 1.$$

22. Let X be a Poisson random variable with parameter λ. Use Exercise 23 in Section 11.5 to show that the probability that X is an even integer (including 0) is $e^{-\lambda} \cosh \lambda$.

TECHNOLOGY EXERCISES

23. The number of times a printing press breaks down each month is Poisson distributed with $\lambda = 4$. What is the probability that the printing press breaks down between 2 and 8 times during a particular month?

24. The number of people arriving during a 5-minute interval at a supermarket checkout counter is Poisson distributed with $\lambda = 8$.

(a) What is the probability that exactly eight people arrive during a particular 5-minute period?

(b) What is the probability that at most eight people arrive during a particular 5-minute period?

25. The number of babies born each day in a certain hospital is Poisson distributed with $\lambda = 6.9$.

(a) During a particular day, are 7 babies more likely to be born than 6 babies?

(b) What is the probability that at most 15 babies will be born during a particular day?

26. The number of accidents occurring each month at a certain intersection is Poisson distributed with $\lambda = 4.8$.

(a) During a particular month, are five accidents more likely to occur than four accidents?

(b) What is the probability that more than eight accidents will occur during a particular month?

Solutions to Check Your Understanding 12.5

The number of cases per block is Poisson distributed with $\lambda = 3$. So the probability of having at least seven cases in a given block is

$$p_7 + p_8 + p_9 + \cdots$$
$$= 1 - (p_0 + p_1 + p_2 + p_3 + p_4 + p_5 + p_6).$$

However,

$$p_n = \frac{3^n}{1 \cdot 2 \cdots n} e^{-3}$$

so

$$p_0 = .04979, \quad p_1 = .14936, \quad p_2 = .22404$$
$$p_3 = .22404, \quad p_4 = .16803, \quad p_5 = .10082,$$
$$p_6 = .05041.$$

Therefore, the probability of at least seven cases in a given block is $1 - (.04979 + .14936 + .22404 + .22404 + .16803 + .10082 + .05041) = .03351$. (See Fig. 5.)

Figure 5

CHAPTER 12 Summary

KEY TERMS AND CONCEPTS	EXAMPLES

12.1 Discrete Random Variables

Suppose that X is a discrete random variable. Let $a_1, a_2, \ldots, a_n$ denote the outcomes of X with respective probabilities $p_1, p_2, \ldots, p_n$. The *mean value* or *expected value* of X is

$$E(X) = a_1 p_1 + a_2 p_2 + \cdots + a_n p_n.$$

To define the *variance* of X, $\mathrm{Var}(X)$, let $m = E(X)$; then,

$$\mathrm{Var}(X) = (a_1 - m)^2 p_1 + (a_2 - m)^2 p_2 + \cdots + (a_n - m)^2 p_n.$$

The *standard deviation* of X is $\sqrt{\mathrm{Var}(X)}$.

EXAMPLES

The outcomes from throwing a skewed die are 1, 2, 3, ..., 6, with respective probabilities .3, .2, .2, .1, .1, .1. Let X denote the random variable representing the outcomes of the die. Compute the mean and variance of X.

Solution

$$E(X) = (.3)(1) + (.2)(2) + (.2)(3) + (.1)(4) + (.1)(5) + (.1)(6)$$
$$= 2.8$$

$$\mathrm{Var}(X) = (.3)(1 - 2.8)^2 + (.2)(2 - 2.8)^2 + (.2)(3 - 2.8)^2$$
$$+ (.1)(4 - 2.8)^2 + (.1)(5 - 2.8)^2 + (.1)(6 - 2.8)^2$$
$$= 2.76.$$

[standard deviation] $= \sqrt{2.76} \approx 1.66.$

12.2 Continuous Random Variables

Let X denote a continuous random variable whose possible values lie in the interval $[A, B]$. The *probability density function* of X is a function $f(x)$, defined on $[A, B]$, with the following properties:

(I) $f(x) \geq 0$ for $A \leq x \leq B$. (II) $\int_A^B f(x)\,dx = 1$.

The probability that the outcomes of X lie in an interval $[a, b]$ is given by

$$\mathrm{Pr}(a \leq X \leq b) = \int_a^b f(x)\,dx.$$

The *cumulative distribution* of X is the antiderivative of $f(x)$ given by

$$F(x) = \int_A^x f(t)\,dt.$$

Note that

$$F(x) = \mathrm{Pr}(A \leq X \leq x) = \int_A^x f(t)\,dt$$

and

$$\mathrm{Pr}(a \leq X \leq b) = F(b) - F(a).$$

(If $B = \infty$ and/or $A = -\infty$, the integrals are improper integrals.)

EXAMPLES

1. Let $f(x) = \frac{2}{x^2}$ if $1 \leq x \leq 2$.
   (a) Show that $f(x)$ is a probability density function on $1 \leq x \leq 2$.
   (b) Let X be a random variable with probability density function $f(x)$. Compute $\mathrm{Pr}(1 \leq X \leq 1.5)$.

   Solution

   (a) Since $f(x) \geq 0$ for all $1 \leq x \leq 2$, it suffices to show that $\int_1^2 f(x)\,dx = 1$. We have

   $$\int_1^2 \frac{2}{x^2}\,dx = -\frac{2}{x}\Big|_1^2$$
   $$= -1 - (-2) = 1.$$

   (b) $$\mathrm{Pr}(1 \leq X \leq 1.5) = \int_1^{1.5} \frac{2}{x^2}\,dx$$
   $$= -\frac{2}{x}\Big|_1^{1.5} = -\frac{2}{1.5} - (-2)$$
   $$= -\frac{4}{3} + 2 = \frac{2}{3}$$

2. Find $F(x)$, the cumulative distribution of the random variable in the previous example, and then calculate $\mathrm{Pr}(1 \leq X \leq 1.5)$ using $F(x)$.

   Solution For $1 \leq x \leq 2$, we have

   $$F(x) = \int_1^x f(t)\,dt = \int_1^x \frac{2}{t^2}\,dt$$
   $$= -\frac{2}{t}\Big|_1^x = -\frac{2}{x} - (-2) = 2 - \frac{2}{x}.$$

   So,

   $$\mathrm{Pr}(1 \leq X \leq 1.5) = F(1.5) - F(1) = 2 - \frac{2}{1.5} - \left(2 - \frac{2}{1}\right) = \frac{2}{3},$$

   which agrees with the answer of the previous example.

KEY TERMS AND CONCEPTS	EXAMPLES

12.3 Expected Value and Variance

Suppose that X is a continuous random variable whose possible values lie in the interval $[A, B]$, and let $f(x)$ denote the probability density function of X. The *expected value* or *mean* of X is given by

$$E(X) = \int_A^B x f(x)\, dx.$$

The *variance* of X is

$$\text{Var}(X) = \int_A^B [x - E(X)]^2 f(x)\, dx.$$

We also have

$$\text{Var}(X) = \int_A^B x^2 f(x)\, dx - E(X)^2.$$

Let X be a random variable with probability density function $f(x) = \frac{2}{x^2}$ for $1 \le x \le 2$. Compute $E(X)$ and $\text{Var}(X)$.

Solution We have

$$E(X) = \int_1^2 x f(x)\, dx = \int_1^2 x\left(\frac{2}{x^2}\right) dx$$
$$= \int_1^2 \frac{2}{x}\, dx = 2\ln x \Big|_1^2$$
$$= 2(\ln 2 - \ln 1) = 2\ln 2,$$

because $\ln 1 = 0$, and

$$\text{Var}(X) = \int_1^2 x^2 f(x)\, dx - E(X)^2$$
$$= \int_1^2 2\, dx - (2\ln 2)^2$$
$$= 2x \Big|_1^2 - (2\ln 2)^2 = 2 - 4(\ln 2)^2.$$

12.4 Exponential and Normal Random Variables

A random variable with an exponential density function of the form

$$f(x) = k e^{-kx}, \qquad x \ge 0,$$

where $k > 0$, is called an *exponential random variable*. For an exponential random variable, we have

$$E(X) = \frac{1}{k} \quad \text{and} \quad \text{Var}(X) = \frac{1}{k^2}.$$

A *normal random variable* is a random variable with a normal density function of the form

$$f(x) = \frac{1}{\sigma\sqrt{2\pi}} e^{-(1/2)[(x-\mu)/\sigma]^2},$$

where $\mu > 0$ and $\sigma > 0$. For a normal random variable, we have

$$E(X) = \mu \quad \text{and} \quad \text{Var}(X) = \sigma^2.$$

Suppose that X is normally distributed with mean $\mu = 5$ and standard deviation $\sigma = 2$. Compute $\Pr(6 \le X \le 8)$.

Solution Using $z = \dfrac{x - \mu}{\sigma} = \dfrac{x - 5}{2}$,

when $x = 6$, $z = \frac{1}{2}$, and when $x = 8$, $z = \frac{3}{2}$. Thus,

$$\Pr(6 \le X \le 8) = \Pr\left(\frac{1}{2} \le Z \le \frac{3}{2}\right) = A\left(\frac{3}{2}\right) - A\left(\frac{1}{2}\right).$$

With the help of Table 1 in the Appendix, we find $A(.5) = .192$ and $A(1.5) = .433$. Thus, the desired probability is $.433 - .192 = .241$.

12.5 Poisson and Geometric Random Variables

A *Poisson random variable* with parameter $\lambda > 0$ has outcomes $0, 1, 2, 3, \dots$ and respective probabilities $e^{-\lambda}, \frac{\lambda}{1} e^{-\lambda}, \frac{\lambda^2}{2 \cdot 1} e^{-\lambda}, \frac{\lambda^3}{3 \cdot 2 \cdot 1} e^{-\lambda}, \dots$. For a Poisson random variable, we have

$$E(X) = \lambda \quad \text{and} \quad \text{Var}(X) = \lambda.$$

A *geometric random variable* with parameter $0 < p < 1$ has outcomes $0, 1, 2, 3, \dots$ and respective probabilities $1 - p$, $p(1-p)$, $p^2(1-p)$, $p^3(1-p), \dots$. For a geometric random variable, we have

$$E(X) = \frac{p}{1-p} \quad \text{and} \quad \text{Var}(X) = \frac{p}{(1-p)^2}.$$

Suppose that X has a Poisson distribution with mean $\mu = 3$. Find the probability that $X = 3$.

Solution Since the mean is 3, it follows that $\lambda = 3$, and so the probability that $X = 3$ is

$$p_3 = \frac{\lambda^3}{3 \cdot 2 \cdot 1} e^{-\lambda} = \frac{3^2}{2} e^{-3} = .224.$$

CHAPTER 12 Fundamental Concept Check Exercises

1. What is a probability table?

2. What is a discrete random variable?

3. Make a small probability table for a discrete random variable X and use it to define $E(X)$, $Var(X)$, and the standard deviation of X.

4. Explain how to create a probability density histogram.

5. What is the difference between a discrete random variable and a continuous random variable?

6. What are the two properties of a probability density function?

7. How is a probability density function used to calculate probabilities?

8. What is a cumulative distribution function, and how is it related to the corresponding probability density function?

9. How is the expected value of a continuous random variable computed?

10. Give two ways to compute the variance of a continuous random variable.

11. What is an exponential density function? Give an example.

12. What is the expected value of an exponential random variable?

13. What is the density function for a normal random variable with mean μ and standard deviation σ?

14. What is a standard normal random variable? Write the density function.

15. How is an integral involving a normal density function converted to an integral involving a standard normal density function?

16. What is $\Pr(X = n)$ for a Poisson random variable with parameter λ? What is $E(X)$ in this case?

17. What is $\Pr(X = n)$ for a geometric random variable with parameter p (the probability of success)? What is $E(X)$ in this case?

CHAPTER 12 Review Exercises

1. Let X be a continuous random variable on $0 \le x \le 2$, with the density function $f(x) = \frac{3}{8} x^2$.
   (a) Calculate $\Pr(X \le 1)$ and $\Pr(1 \le X \le 1.5)$.
   (b) Find $E(X)$ and $Var(X)$.

2. Let X be a continuous random variable on $3 \le x \le 4$, with the density function $f(x) = 2(x - 3)$.
   (a) Calculate $\Pr(3.2 \le X)$ and $\Pr(3 \le X)$.
   (b) Find $E(X)$ and $Var(X)$.

3. For any number A, verify that $f(x) = e^{A-x}$, $x \ge A$, is a density function. Compute the associated cumulative distribution for X.

4. **Pareto Distribution** For any positive constants k and A, verify that the function $f(x) = kA^k/x^{k+1}$, $x \ge A$, is a density function. The associated cumulative distribution function $F(x)$ is called a *Pareto distribution*. Compute $F(x)$.

5. **Chi-Square Density** For any positive integer n, the function $f_n(x) = c_n x^{(n-2)/2} e^{-x/2}$, $x \ge 0$, where c_n is an appropriate constant, is called the *chi-square density function* with n degrees of freedom. Find c_2 and c_4 such that $f_2(x)$ and $f_4(x)$ are probability density functions.

6. For any positive number k, verify that $f(x) = 1/(2k^3) x^2 e^{-x/k}$, $x \ge 0$, is a density function.

7. **Blood Test Probabilities** A medical laboratory tests many blood samples for a certain disease that occurs in about 5% of the samples. The lab collects samples from 10 persons and mixes together some blood from each sample. If a test on the mixture is positive, an additional 10 tests must be run, one on each individual sample. But if the test on the mixture is negative, no other tests are needed. It can be shown that the test of the mixture will be negative with probability $(.95)^{10} = .599$, because each of the 10 samples has a 95% chance of being free of the disease. If X is the total number of tests required, X has the probability table shown in Table 1.

Table 1 Blood Test Probabilities for Batches of 10 Samples

	Test of Mixture	
	Negative	Positive
Total tests	1	11
Probability	.599	.401

(a) Find $E(X)$.
(b) If the laboratory uses the procedure described on 200 blood samples (that is, 20 batches of 10 samples), about how many tests can it expect to run?

8. If the laboratory in Exercise 7 uses batches of 5 instead of 10 samples, the probability of a negative test on the mixture of 5 samples is $(.95)^5 = .774$. Thus, Table 2 gives the probabilities for the number X of tests required.
(a) Find $E(X)$.
(b) If the laboratory uses this procedure on 200 blood samples (that is, 40 batches of 5 samples), about how many tests can it expect to run?

Table 2 Blood Test Probabilities for Batches of 5 Samples

	Test of Mixture	
	Negative	Positive
Total tests	1	6
Probability	.774	.226

9. **Probability of Gasoline Sales** A certain gas station sells X thousand gallons of gas each week. Suppose that the cumulative distribution function for X is $F(x) = 1 - \frac{1}{4}(2 - x)^2$, $0 \leq x \leq 2$.
   (a) If the tank contains 1.6 thousand gallons at the beginning of the week, find the probability that the gas station will have enough gas for its customers throughout the week.
   (b) How much gas must be in the tank at the beginning of the week to have a probability of .99 that there will be enough gasoline for the week?
   (c) Compute the density function for X.

10. **Expected Value of a Service Contract** A service contract on a computer costs $100 per year. The contract covers all necessary maintenance and repairs on the computer. Suppose that the actual cost to the manufacturer for providing this service is a random variable X (measured in hundreds of dollars) whose probability density function is $f(x) = (x - 5)^4/625$, $0 \leq x \leq 5$. Compute E(X) and determine how much money the manufacturer expects to make on each service contract on average.

11. A random variable X has a uniform density function $f(x) = \frac{1}{5}$ on $20 \leq x \leq 25$.
   (a) Find E(X) and Var(X).
   (b) Find b such that Pr$(X \leq b) = .3$.

12. A random variable X has a cumulative distribution function $F(x) = (x^2 - 9)/16$ on $3 \leq x \leq 5$.
   (a) Find the density function for X.
   (b) Find a such that Pr$(a \leq X) = \frac{1}{4}$.

13. **Income Distribution** The annual incomes of the households in a certain community range between 5 and 25 thousand dollars. Let X represent the annual income (in thousands of dollars) of a household chosen at random in this community, and suppose that the probability density function for X is $f(x) = kx$, $5 \leq x \leq 25$.
   (a) Find the value of k that makes $f(x)$ a density function.
   (b) Find the fraction of households whose income exceeds $20,000.
   (c) Find the mean annual income of the households in the community.

14. **Lifetime of a Battery** The density function $f(x)$ for the lifetime of a certain battery is shown in Fig. 1. Each battery lasts between 3 and 10 hours.
   (a) Sketch the graph of the corresponding cumulative distribution function $F(x)$.
   (b) What is the meaning of the number $F(7) - F(5)$?
   (c) Formulate the number in part (b) in terms of $f(x)$.

Figure 1

15. **Random Selection of Points** A point is selected at random from the rectangle of Fig. 2; call its coordinates (θ, y). Find the probability that $y \leq \sin \theta$.

Figure 2

16. **Buffon Needle Problem** A needle of length 1 unit is dropped on a floor that is ruled with parallel lines, 1 unit apart. [See Fig. 3.] Let P be the lowest point of the needle, y the distance of P from the ruled line above it, and θ the angle the needle makes with a line parallel to the ruled lines. Show that the needle touches a ruled line if, and only if, $y \leq \sin \theta$. Conclude that the probability of the needle touching a ruled line is the probability found in Exercise 15.

Figure 3

17. **Expected Revenue from Sales of Monitors** The lifetime of a certain computer monitor is an exponential random variable with an expected value of 5 years. The manufacturer sells the monitor for $100, but will give a complete refund if the monitor burns out within 3 years. Then, the revenue that the manufacturer receives on each monitor is a discrete random variable Y, with values 100 and 0. Determine the expected revenue per monitor.

18. **Deciding on a Service Contract** The condenser motor in an air conditioner costs $300 to replace, but a home air-conditioning service will guarantee to replace it free when it burns out if you will pay an annual insurance premium of $25. The life span of the motor is an exponential random variable with an expected life of 10 years. Should you take out the insurance for the first year? [*Hint:* Consider the random variable Y such that $Y = 300$ if the motor burns out during the year and $Y = 0$ otherwise. Compare E(Y) with the cost of 1 year's insurance.]

19. **Relief Time for Arthritic Patients** An exponential random variable X has been used to model the relief times (in minutes) of arthritic patients who have taken an analgesic for their pain. Suppose that the density function for X is $f(x) = ke^{-kx}$ and that a certain analgesic provides relief within 4 minutes for 75% of a large group of patients. Then, we may estimate that Pr$(X \leq 4) = .75$. Use this estimate to

find an approximate value for k. [*Hint:* First show that $\Pr(X \le 4) = 1 - e^{-4k}$.]

20. **Earnings from a Machine** A piece of new equipment has a useful life of X thousand hours, where X is a random variable with the density function $f(x) = .01xe^{-x/10}$, $x \ge 0$. A manufacturer expects the machine to generate $\$5000$ of additional income for every thousand hours of use, but the machine costs $\$60,000$. Should the manufacturer purchase the new equipment? [*Hint:* Compute the expected value of the additional earnings generated by the machine.]

21. **Life Span of a Product** Extensive records are kept of the life spans (in months) of a certain product, and a relative frequency histogram is constructed from the data, using areas to represent relative frequencies (as in Fig. 4 in Section 12.1). It turns out that the upper boundary of the relative frequency histogram is approximated closely by the graph of the function

$$f(x) = \frac{1}{8\sqrt{2\pi}} e^{-(1/2)[(x-50)/8]^2}.$$

Determine the probability that the life span of such a product is between 30 and 50 months.

22. **Tolerance Limits** A certain machine part has a nominal length of 80 millimeters, with a tolerance of $\pm .05$ millimeter. Suppose that the actual length of the parts supplied is a normal random variable with mean 79.99 millimeters and standard deviation .02 millimeter. How many parts in a lot of 1000 should you expect to lie outside the tolerance limits?

23. **Height Requirements by a Police Department** The men hired by a certain city police department must be at least 69 inches tall. If the heights of adult men in the city are normally distributed, with $\mu = 70$ inches and $\sigma = 2$ inches, what percentage of the men are tall enough to be eligible for recruitment by the police department?

24. Suppose that the police force in Exercise 23 maintains the same height requirements for women as men and that the heights of women in the city are normally distributed, with $\mu = 65$ inches and $\sigma = 1.6$ inches. What percentage of the women are eligible for recruitment?

25. Let Z be a standard normal random variable. Find the number a such that $\Pr(a \le Z) = .40$.

26. **Scores on an Entrance Exam** Scores on a school's entrance exam are normally distributed, with $\mu = 500$ and $\sigma = 100$. If the school wishes to admit only the students in the top 40%, what should be the cutoff grade?

27. **Area under the Normal Curve** It is useful in some applications to know that about 68% of the area under the standard normal curve lies between -1 and 1.
   (a) Verify this statement.
   (b) Let X be a normal random variable with expected value μ and variance σ^2. Compute
   $$\Pr(\mu - \sigma \le X \le \mu + \sigma).$$

28. (a) Show that about 95% of the area under the standard normal curve lies between -2 and 2.
   (b) Let X be a normal random variable with expected value μ and variance σ^2. Compute
   $$\Pr(\mu - 2\sigma \le X \le \mu + 2\sigma).$$

29. **Chebyshev's Inequality** The Chebyshev inequality says that for any random variable X with expected value μ and standard deviation σ,
   $$\Pr(\mu - n\sigma \le X \le \mu + n\sigma) \ge 1 - \frac{1}{n^2}.$$
   (a) Take $n = 2$. Apply the Chebyshev inequality to an exponential random variable.
   (b) By integrating, find the exact value of the probability in part (a).

30. Do the same as in Exercise 29 with a normal random variable.

White Blood Cell Count A small volume of blood is selected and examined under a microscope, and the number of white blood cells is counted. Suppose that for healthy people the number of white blood cells in such a specimen is Poisson distributed with $\lambda = 4$.

31. What is the probability that a specimen from a healthy person has exactly four white blood cells?

32. What is the probability that a specimen from a healthy person has eight or more white blood cells?

33. What is the average number of white blood cells per specimen from a healthy person?

Rolling Dice A pair of dice is rolled until a 7 or an 11 appears, and the number of rolls preceding the final roll is observed. The probability of rolling 7 or 11 is $\frac{2}{9}$.

34. Determine the formula for p_n, the probability of exactly n consecutive rolls preceding the final roll.

35. Determine the average number of consecutive rolls preceding the final roll.

36. What is the probability that at least three consecutive rolls precede the final roll?

Appendix

Areas under the Standard Normal Curve

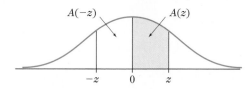

TABLE 1 Areas under the Standard Normal Curve

z	.00	.01	.02	.03	.04	.05	.06	.07	.08	.09
0.0	.0000	.0040	.0080	.0120	.0160	.0199	.0239	.0279	.0319	.0359
0.1	.0398	.0438	.0478	.0517	.0557	.0596	.0636	.0675	.0714	.0754
0.2	.0793	.0832	.0871	.0910	.0948	.0987	.1026	.1064	.1103	.1141
0.3	.1179	.1217	.1255	.1293	.1331	.1368	.1406	.1443	.1480	.1517
0.4	.1554	.1591	.1628	.1664	.1700	.1736	.1772	.1808	.1844	.1879
0.5	.1915	.1950	.1985	.2019	.2054	.2088	.2123	.2157	.2190	.2224
0.6	.2258	.2291	.2324	.2357	.2389	.2422	.2454	.2486	.2518	.2549
0.7	.2580	.2612	.2642	.2673	.2704	.2734	.2764	.2794	.2823	.2852
0.8	.2881	.2910	.2939	.2967	.2996	.3023	.3051	.3078	.3106	.3133
0.9	.3159	.3186	.3212	.3238	.3264	.3289	.3315	.3340	.3365	.3389
1.0	.3413	.3438	.3461	.3485	.3508	.3531	.3554	.3577	.3599	.3621
1.1	.3643	.3665	.3686	.3708	.3729	.3749	.3770	.3790	.3810	.3820
1.2	.3849	.3869	.3888	.3907	.3925	.3944	.3962	.3980	.3997	.4015
1.3	.4032	.4049	.4066	.4082	.4099	.4115	.4131	.4147	.4162	.4177
1.4	.4192	.4207	.4222	.4236	.4251	.4265	.4279	.4292	.4306	.4319
1.5	.4332	.4345	.4357	.4370	.4382	.4394	.4406	.4418	.4429	.4441
1.6	.4452	.4463	.4474	.4484	.4495	.4505	.4515	.4525	.4535	.4545
1.7	.4554	.4564	.4573	.4582	.4591	.4599	.4608	.4616	.4625	.4633
1.8	.4641	.4649	.4656	.4664	.4671	.4678	.4686	.4693	.4699	.4706
1.9	.4713	.4719	.4726	.4732	.4738	.4744	.4750	.4756	.4761	.4767
2.0	.4772	.4778	.4783	.4788	.4793	.4798	.4803	.4808	.4812	.4817
2.1	.4821	.4826	.4830	.4834	.4838	.4842	.4846	.4850	.4854	.4857
2.2	.4861	.4864	.4868	.4871	.4875	.4878	.4881	.4884	.4887	.4890
2.3	.4893	.4896	.4898	.4901	.4904	.4906	.4909	.4911	.4913	.4916
2.4	.4918	.4920	.4922	.4925	.4927	.4929	.4931	.4932	.4934	.4936
2.5	.4938	.4940	.4941	.4943	.4945	.4946	.4948	.4949	.4951	.4952
2.6	.4953	.4955	.4956	.4957	.4959	.4960	.4961	.4962	.4963	.4964
2.7	.4965	.4966	.4967	.4968	.4969	.4970	.4971	.4972	.4973	.4974
2.8	.4974	.4975	.4976	.4977	.4977	.4978	.4979	.4979	.4980	.4981
2.9	.4981	.4982	.4982	.4983	.4984	.4984	.4985	.4985	.4986	.4986
3.0	.4987	.4987	.4987	.4988	.4988	.4989	.4989	.4989	.4990	.4990
3.1	.4990	.4991	.4991	.4991	.4992	.4992	.4992	.4992	.4993	.4993
3.2	.4993	.4993	.4994	.4994	.4994	.4994	.4994	.4995	.4995	.4995
3.3	.4995	.4995	.4995	.4996	.4996	.4996	.4996	.4996	.4996	.4997
3.4	.4997	.4997	.4997	.4997	.4997	.4997	.4997	.4997	.4997	.4998
3.5	.4998	.4998	.4998	.4998	.4998	.4998	.4998	.4998	.4998	.4998

Sources

Chapter 0

Section 0.2

1. Exercise 20, from Adair, R. K. (1990). *The physics of baseball*. New York, NY: Harper & Row.

Section 0.4

1. Exercise 46, from Burghes, D., Huntley, I., & McDonald, J. (1982). *Applying mathematics: A course in mathematical modelling*. New York, NY: Halstead Press, pp. 57–60.

Chapter 1

Section 1.8

1. Technology Exercise 33, from Anderson, J. R. (1992, Summer). Automaticity and the ACT theory. *American Journal of Psychology, 105*(2), 165–180.

Chapter 2

Section 2.3

1. Exercise 45, from Fidelity slashes index-fund fees. *The Boston Globe*, September 1, 2004.

Section 2.4

1. Technology Exercise 37, from Johnson, J. D., Wogenrich, W. J., Hsi, K. C., Skipper, B. J., & Greenberg, R. E. (1991). Growth retardation during the suckling period in expanded litters of rats; observations of growth patterns and protein turnover. *Growth, Development, and Aging, 55*, 263–273.

2. Technology Exercise 38, from Woodward & Prine. (1993). Crop quality and utilization. *Crop Science, 33*, 818–824.

Section 2.6

1. Example 2, from Van Horne, J. C. (1983). *Financial management and policy* (6th ed., pp. 416–420). Englewood Cliffs, NJ: Prentice-Hall.

Chapter 3

Section 3.1

1. Technology Exercise 69, from Crawford, B. H. (1937). The dependence of pupil size on the external light stimulus under static and variable conditions. *Proceedings of the Royal Society, Series B, 121*, 376–395.

Section 3.2

1. Exercise 50, from Platt, D. R. (1969). Natural history of the hognose snakes *Heterodon platyrhinos* and *Heterodon nasicus*. University of Kansas Publications, *Museum of Natural History, 18*(4), 253–420.

Chapter 4

Section 4.3

1. Technology Exercise 43, from Baker, G. M., Goddard, H. L., Clarke, M. B., & Whimster, W. F. (1990). Proportion of necrosis in transplanted murine adenocarcinomas and its relationship to tumor growth. *Growth, Development, and Aging, 54*, 85–93.

Section 4.4

1. Exercise 50, from Cox, G., Collier, B., Johnson, A., & Miller, P. (1973). *Dynamic ecology*. Englewood Cliffs, NJ: Prentice-Hall, pp. 113–115.

Section 4.5

1. Technology Exercise 38, from Rao, K. M., Prakash, S., Kumar, S., Suryanarayana, M. V. S., Bhagwat, M. M., Gharia, M. M., & Bhavsar, R. B. (1991, January). N-diethylphenylacetamide in treated fabrics, as a repellent against *Aedes aegypti* and *Culex quinquefasclatus* (Deptera: Culiciae). *Journal of Medical Entomology, 28*(1).

Section 4.6

1. Exercise 51, from Batschelet, E. (1971). *Introduction to mathematics for life scientists*. New York, NY: Springer-Verlag, pp. 305–307.

Chapter 5

Section 5.4

1. Example 1, from Coleman, J. (1964). *Introduction to mathematical sociology*. New York, NY: Free Press, p. 43.

2. Check Your Understanding 1 from Coleman, J. S., Katz, E., & Menzel, H. (1957). The diffusion of an innovation among physicians. *Sociometry, 20*, 253–270.

Chapter 7

Section 7.2

1. Exercise 32, from Stone, R. (1945). The analysis of market demand. *Journal of the Royal Statistical Society, 108*, 286–391.

2. Exercise 38, from Routh, J. (1971). *Mathematical preparation for laboratory technicians*. Philadelphia, PA: W. B. Saunders, p. 92.

Section 7.3

1. Example 3, from March, L. (1972). Elementary models of built forms. In L. Martin & L. March (Eds.), *Urban space and structures*. New York, NY: Cambridge University Press.

Section 7.5

1. Exercise 11, from U.S. Health Care Financing Administration, *Health Care Financing Review*, Spring 2012.

Chapter 8

Section 8.2

1. Technology Exercise 41, from Rapp, D. (1981). *Solar energy*. Upper Saddle River, NJ: Prentice-Hall, p. 171.

2. Technology Exercise 42, from Duncan, D. R., et al. (1976, January). Climate curves. *School Science and Mathematics, 76*, 41–49.

Review Exercise

3. Exercise 58, from Nunn, J. F. (1977). *Applied respiratory physiology* (2nd ed.). London, England: Butterworths, p. 122.

Chapter 9

Section 9.4

1. Example 3, from Horelick, B., & Koont, S. (1978). Project UMAP, *Measuring cardiac output.* Newton, MA: Educational Development Center.

Section 9.5

1. Example 3, from Clark, C. (1951). Urban population densities. *Journal of the Royal Statistical Society, Series A, 114,* 490–496.

2. Example 3, from White, M. J. (1977). On cumulative urban growth and urban density functions. *Journal of Urban Economics, 4,* 104–112.

Chapter 10

Section 10.4

1. Exercise 8, from the National Automobile Dealers Association, www.nada.com

2. Exercise 9, from the Federal Housing Finance Board, www.fhfb.gov

3. Exercise 19, from the National Kidney Foundation, www.kidney.org

Section 10.6

1. Example 1, from Mansfield, E. (1961). Technical change and the rate of imitation. *Econometrica, 29,* 741–766.

2. Example 4, from Simon, W. (1972). *Mathematical techniques for physiology and medicine.* New York, NY: Academic Press, Chap. 5.

3. Example 4, from Li, C. C. (1955). *Population genetics.* Chicago, IL: University of Chicago Press, pp. 240–263, 283–286.

4. Exercise 14, from Richardson, L. F. (1948). War moods I. *Psychometrica,* p. 13.

Chapter 11

Section 11.2

1. Example 5, from Paley, H., Colwell, P., & Cannaday, R. (1984). *Internal rates of return,* UMAP Module 640. Lexington, MA: COMAP.

Section 11.3

1. Example 4, from Oates, J. A., & Wilkinson, G. R. (1977). Principles of drug therapy (pp. 334–346). In T. R. Harrison (Ed.), *Principles of internal medicine* (8th ed.). New York, NY: McGraw-Hill.

Chapter 12

Section 12.2

2. From Cook, J. R., & James, T. W. (1964). Age distribution of cells in logarithmically growing cell populations (pp. 485–495). In E. Zeuthen (Ed.), *Synchrony in cell division and growth.* New York, NY: Wiley.

Section 12.4

3. From Wallis, W. A. (1936). The poisson distribution and the Supreme Court. *Journal of the American Statistical Association, 31,* 376–380.

Selected Answers

PREREQUISITE SKILLS DIAGNOSTIC TEST

1. 1 **2.** 9 **3.** $\dfrac{1}{3}$ **4.** 2 **5.** x^6 **6.** $1 + x^4$ **7.** $\dfrac{y^2}{x^3}$ **8.** x^4 **9.** $\dfrac{x^2 + 3x + 1}{x + 1}$ **10.** x **11.** $\dfrac{x^2 + 2x + 1}{x}$ **12.** $-\dfrac{1}{x + 1}$ **13.** $\dfrac{t^2}{(t + 1)^2}$

14. $\dfrac{t^2}{t^2 + 1}$ **15.** $\dfrac{t^4}{(t + 1)^4}$ **16.** $\dfrac{(t + 1)^2}{(t + 2)^2}$ **17.** $2x + 2 + h$ **18.** $-\dfrac{1}{x(x + h)}$ **19.** $f(x) = \dfrac{1}{\sqrt{x + h} + \sqrt{x}}$ **20.** $3x^2 + 3xh + h^2$

21. **22.** **23.** **24.** **25.** $\left(-\dfrac{3}{4}, -\dfrac{5}{4}\right)$

26. $(6, 3)$

27. $\left(\dfrac{7}{4}, 0\right)$

28. None **29.** $(x + 7)(x - 2)$ **30.** $(x + 4)(x + 1)$ **31.** $x(x + 2)(x - 1)$ **32.** $x(x + 1)(x - 3)$ **33.** ± 12 **34.** -2 **35.** $0, -3, -5$

36. $0, -\dfrac{3}{2}, -\dfrac{1}{3}$ **37.** $\dfrac{-3 \pm \sqrt{17}}{4}$ **38.** $\dfrac{-1 \pm \sqrt{5}}{2}$ **39.** No real solutions **40.** $-2 \pm 2\sqrt{2}$ **41.** $2, 5$ **42.** $-\dfrac{3}{2}, \dfrac{1}{2}$

43. $0, -2$ **44.** $\dfrac{3 \pm \sqrt{5}}{2}$

CHAPTER 0

Exercises 0.1, page 9

1. **3.** **5.** **7.** $[2, 3)$ **9.** $[-1, 0)$ **11.** $(-\infty, 3)$

13. $f(0) = 0, f(5) = 10, f(3) = 0, f(-7) = 70$ **15.** $a^2 - 1, a^2 + 2a$ **17.** 3 **19. (a)** 5660 **(b)** 10220 **21.** $x \neq 1, 2$ **23.** $x < 3$

25. **27.**

29. Function **31.** Not a function
33. Not a function **35.** $f(0) = 1, f(7) = -1$
37. Positive **39.** $[-1, 3]$ **41.** $(-\infty, -1], [5, 9]$
43. $f(1) \approx .03, f(5) \approx .037$ **45.** $[0, .05]$
47. No **49.** Yes **51.** $(a + 1)^3$ **53.** $1, 3, 4$
55. $\pi, 3, 12$

57. $f(x) = \begin{cases} .06x & \text{for } 50 \le x \le 3000 \\ .02x + 15 & \text{for } 3000 < x \end{cases}$

$f(3000) = .06(3000) = 180, f(4500) = .02(4500) + 15 = 105$. So $180 and $105

59. **61.** You need parentheses $Y_1 = 1/(X + 1)$. **63.**

Exercises 0.2, page 18

1.   **3.**   **5.**   **7.**   **9.**

11. $\left(-\frac{1}{3}, 0\right), (0, 3)$ **13.** y-intercept $(0, 5)$ **15.** $(0, 0)$ **17. (a)** \$114   **(b)** $f(x) = .45x + 24$ **19.** $700x + 1900, x =$ number of days

21. The cost for another 5% is \$25 million. The cost for the final 5% is 21 times as much.   **23. (a)** $K = \frac{1}{250}, V = \frac{1}{50}$ **(b)** $\left(-\frac{1}{K}, 0\right), \left(0, \frac{1}{V}\right)$

25. $a = 3, b = -4, c = 0$ **27.** $a = -2, b = 3, c = 1$ **29.** $a = -1, b = 0, c = 1$

31.   **33.**   **35.**

37. 1   **39.** 10^{-2} **41.** 2.5   **43.** $-3985, 3008$ **45.** $-4.60569, 231.499$

Exercises 0.3, page 23

1. $x^2 + 9x + 1$ **3.** $9x^3 + 9x$ **5.** $\dfrac{t^2 + 1}{9t}$ **7.** $\dfrac{3x + 1}{x^2 - x - 6}$ **9.** $\dfrac{4x}{x^2 - 12x + 32}$ **11.** $\dfrac{2x^2 + 5x + 50}{x^2 - 100}$ **13.** $\dfrac{2x^2 - 2x + 10}{x^2 + 3x - 10}$

15. $\dfrac{-x^2 + 5x}{x^2 + 3x - 10}$ **17.** $\dfrac{x^2 + 5x}{-x^2 + 7x - 10}$ **19.** $\dfrac{-x^2 + 3x + 4}{x^2 + 5x - 6}$ **21.** $\dfrac{-x^2 - 3x}{x^2 + 15x + 50}$ **23.** $\dfrac{5u - 1}{5u + 1}, u \neq 0$ **25.** $\left(\dfrac{x}{1 - x}\right)^6$

27. $\left(\dfrac{x}{1 - x}\right)^3 - 5\left(\dfrac{x}{x - 1}\right)^2 + 1$ **29.** $\dfrac{t^3 - 5t^2 + 1}{-t^3 + 5t^2}$ **31.** $2xh + h^2$ **33.** $4 - 2t - h$ **35. (a)** $C(A(t)) = 3000 + 1600t - 40t^2$

(b) \$6040   **37.** $h(x) = x + \dfrac{1}{8}$; $h(x)$ converts from British sizes to U.S. sizes.

39. The graph of $f(x) + c$ is the graph of $f(x)$ shifted up (if $c > 0$) or down (if $c < 0$) by $|c|$ units.

41. 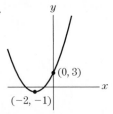   **43.** $f(f(x)) = x, x \neq 1$

Exercises 0.4, page 31
1. $2, \frac{3}{2}$ **3.** $\frac{3}{2}$ **5.** No zeros   **7.** $1, -\frac{1}{5}$ **9.** $5, 4$ **11.** $2 + \frac{\sqrt{6}}{3}, 2 - \frac{\sqrt{6}}{3}$ **13.** $(x + 5)(x + 3)$ **15.** $(x - 4)(x + 4)$ **17.** $3(x + 2)^2$
19. $-2(x - 3)(x + 5)$ **21.** $x(3 - x)$ **23.** $-2x(x - \sqrt{3})(x + \sqrt{3})$ **25.** $(x - 1)(x^2 + x + 1)$ **27.** $(2x + 3)(4x^2 - 6x + 9)$
29. $(x - 7)^2$ **31.** $(-1, 1), (5, 19)$ **33.** $(-1, 9), (4, 4)$ **35.** $(0, 0), (2, -2)$ **37.** $(0, 5), (2 - \sqrt{3}, 25 - 23\sqrt{3}/2), (2 + \sqrt{3}, 25 + 23\sqrt{3}/2)$
39. $-7, 3$ **41.** $-2, 3$ **43.** -7 **45.** 16,667 and 78,571 subscribers   **47.** $-1, 2$ **49.** ≈ 4.56 **51.** $\approx (-.41, -1.83), (2.41, 3.83)$
53. $\approx (2.14, -25.73), (4.10, -21.80)$ **55.** $[-5, 22]$ by $[-1400, 100]$ **57.** $[-20, 4]$ by $[-500, 2500]$

Exercises 0.5, page 39
1. 27   **3.** 1   **5.** .0001   **7.** -16 **9.** 4   **11.** .01   **13.** $\frac{1}{6}$ **15.** 100   **17.** 16   **19.** 125   **21.** 1   **23.** 4   **25.** $\frac{1}{2}$ **27.** 1000   **29.** 10

31. 6   **33.** 16   **35.** 18   **37.** $\frac{4}{9}$ **39.** 7   **41.** $x^6 y^6$ **43.** $x^3 y^3$ **45.** $\dfrac{1}{x^{1/2}}$ **47.** $\dfrac{x^{12}}{y^6}$ **49.** $x^{12} y^{20}$ **51.** $x^2 y^6$ **53.** $16x^4$ **55.** x^2

57. $\dfrac{1}{x^7}$ **59.** x **61.** $\dfrac{27x^6}{8y^3}$ **63.** $2\sqrt{x}$ **65.** $\dfrac{1}{8x^6}$ **67.** $\dfrac{1}{32x^2}$ **69.** $9x^3$ **71.** $\dfrac{1}{x^{5/3}}$ **73.** $\dfrac{1}{x^{7/3}}$ **75.** $\dfrac{1}{x^5}$ **77.** $\dfrac{1}{x^{5/6}}$ **79.** $\dfrac{1}{x^{2/3}}$ **81.** $x^{1/9}$

83. $x - 1$ **85.** $1 + 6x^{1/2}$ **87.** $a^{1/2} \cdot b^{1/2} = (ab)^{1/2}$ (Law 5) **89.** 16 **91.** $\frac{1}{4}$ **93.** 8 **95.** $\frac{1}{32}$ **97.** \$709.26 **99.** \$127,857.61

101. \$164.70 **103.** \$1592.75 **105.** \$3268.00 **107.** $\frac{500}{256}(256 + 256r + 96r^2 + 16r^3 + r^4) = A$ **109.** $\frac{1}{20}(2x)^2 = \frac{1}{20}(4x^2) = 4(\frac{1}{20}x^2)$

111. .0008103 **113.** .00000823

Exercises 0.6, page 47

1.   **3.**   **5.**   **7.** $P = 8x; 3x^2 = 25$ **9.** $A = \pi r^2; 2\pi r = 15$

11. $V = x^2 h; x^2 + 4xh = 65$ **13.** $\pi r^2 h = 100; C = 11\pi r^2 + 14\pi rh$ **15.** $2x + 3h = 5000; A = xh$
17. $C = 36l + 20h$ **19.** $75\,\text{cm}^2$ **21.** (a) 38 (b) \$40 **23.** (a) 200 (b) 275 (c) 25
25. (a) $P(x) = 12x - 800$ (b) \$640 (c) \$3150 **27.** 270 cents
29. A 100-in.^{3} cylinder of radius 3 in. costs \$1.62 to construct. **31.** \$1.08

33. $R(30) = 1800; C(30) = 1200$ **35.** 40 **37.** $C(1000) = 4000$ **39.** Find the y-coordinate of the point on the graph whose x-coordinate is 400. **41.** The greatest profit, \$52,500, occurs when 2500 units of goods are produced. **43.** Find the x-coordinate of the point on the graph whose y-coordinate is 30,000. **45.** Find $h(3)$. Find the y-coordinate of the point on the graph whose t-coordinate is 3. **47.** Find the maximum value of $h(t)$. Find the y-coordinate of the highest point of the graph. **49.** Solve $h(t) = 100$. Find the t-coordinates of the points whose y-coordinate is 100.

51. (a)   (b) 96 feet
(c) 1 and 4 seconds
(d) 5 seconds
(e) 2.5 seconds; 100 feet

53. (a)   (b) 350 bicycles per year
(c) \$68,000
(d) \$5000
(e) No, revenue would only increase by \$4000.

Chapter 0: Chapter Review Exercises, page 54

1. $2, 27\frac{1}{3}, -2, -2\frac{1}{8}, \frac{5\sqrt{2}}{2}$ **2.** $f(0) = 0, f(-\frac{1}{4}) = -\frac{5}{16}, f(\frac{1}{\sqrt{2}}) = \frac{2\sqrt{2} + 3}{2}$ **3.** $a^2 - 4a + 2$ **4.** $f(a + 1) = \frac{1}{a + 2} - (a + 1)^2$
5. $x \ne 0, -3$ **6.** $x \ge 1$ **7.** All x **8.** $x > 0$ **9.** yes **10.** No **11.** $5x(x - 1)(x + 4)$ **12.** $3(x - 5)(x + 4)$ **13.** $(-1)(x - 6)(x + 3)$
14. $x^3(x - 2)(x + 1)$ **15.** $-\frac{2}{5}, 1$ **16.** $x = \dfrac{-1 \pm \sqrt{17}}{4}$ **17.** $\left(\dfrac{5 + 3\sqrt{5}}{10}, \dfrac{3\sqrt{5}}{5}\right), \left(\dfrac{5 - 3\sqrt{5}}{10}, \dfrac{-3\sqrt{5}}{5}\right)$
18. $(-\sqrt{6}, -\sqrt{6} - 5), (\sqrt{6}, \sqrt{6} - 5)$ **19.** $x^2 + x - 1$ **20.** $x^2 - 5x + 1$ **21.** $x^{5/2} - 2x^{3/2}$ **22.** $3x^3 - 7x^2 + 2x$ **23.** $x^{3/2} - 2x^{1/2}$
24. $3x^{3/2} - x^{1/2}$ **25.** $\dfrac{x^2 - x + 1}{x^2 - 1}$ **26.** $\dfrac{x^3 + x^2 + x}{(x + 2)(x^2 - 1)}$ **27.** $-\dfrac{3x^2 + 1}{3x^2 + 4x + 1}$ **28.** $\dfrac{5x^2 + x - 2}{(x^2 - 1)(3x + 1)}$ **29.** $\dfrac{-3x^2 + 9x - 10}{3x^2 - 5x - 8}$
30. $\dfrac{-x^2 + 3x - 1}{x^2 - 1}$ **31.** $\dfrac{1}{x^4} - \dfrac{2}{x^2} + 4$ **32.** $\dfrac{1}{(x^2 - 2x + 4)^2}$ **33.** $(\sqrt{x} - 1)^2$ **34.** $\dfrac{|x|}{1 - |x|}$ **35.** $\dfrac{1}{(\sqrt{x} - 1)^2} - \dfrac{2}{\sqrt{x} - 1} + 4$
36. $(\sqrt{x^2 - 2x + 4} - 1)^{-1}$ **37.** 27, 32, 4 **38.** 1000, 0.1 **39.** $301 + 10t + .04t^2$ **40.** $R(d) = 30\left(1 - \dfrac{200}{d + 200}\right) - 36\left(1 - \dfrac{200}{d + 200}\right)^2$
41. $x^2 + 2x + 1$ **42.** x^6/y^3 **43.** x **44.** $8x$ **45.** (a) $A(t) = 15,000\left(1 + \frac{.04}{12}\right)^{12t} = 15,000(1.0333)^{12t}$
(b) $A(2) = \$16,247.1, A(5) = \$18,314.9$ **46.** (a) $A(t) = 7000\left(1 + \frac{.09}{2}\right)^{2t} = 7000(1.045)^{2t}$ (b) $A(10) = \$16,882, A(20) = \$40,714.6$
47. (a) $A(r) = 15,000(1 + r)^{10}$ (b) $A(.04) \approx \$22,203.7, A(.06) \approx \$26,862.7$ **48.** (a) $A(r) = 7000(1 + r)^{20}$
(b) $A(.07) \approx \$27,087.8, A(.12) \approx \$67,524.1$

CHAPTER 1

Exercises 1.1, page 63
1. $m = -7, b = 3$ **3.** $m = \frac{1}{2}, b = \frac{3}{2}$ **5.** $m = \frac{1}{7}, b = -5$ **7.** $y - 1 = -(x - 7)$ **9.** $y - 1 = \frac{1}{2}(x - 2)$ **11.** $y - 5 = \frac{63}{10}(x - \frac{5}{7})$
13. $y = 0$ **15.** $y = 9$ **17.** $-\frac{x}{\pi} + y = 1$ or $y = \frac{x}{\pi} + 1$ **19.** $y = -2x - 4$ **21.** $y = x - 2$ **23.** $y = 3x - 6$ **25.** $y = x - 2$

27. Start at $(1, 0)$. To get back to the line, move one unit to the right and then one unit up.

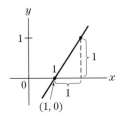

29. Start at $(1, -1)$. To get back to the line, move 3 units to the right and then 1 unit down.

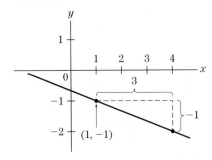

31. (a) C **(b)** B **(c)** D **(d)** A   **33.** 1   **35.** $-\frac{3}{4}$
37. $(2, 5)$; $(3, 7)$; $(0, 1)$ **39.** $f(3) = 2$ **41.** l_1

43.

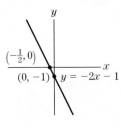

45. $a = 2$, $f(2) = 2$ **47. (a)** $C(10) = \$1220$.   **(b)** Marginal cost is $12 per item.   **(c)** $C(11) - C(10) = $ marginal cost $ = \$12$.
49. $P(x) = 2.19 - .04x$ dollars, where $P(x)$ is the price of 1 gallon x months since January 1. Price of 1 gallon on April 1 is $P(3) = \$2.07$. Price of 15 gallons is $31.05. Price of 1 gallon on September 1 is $P(8) = \$1.87$. Price of 15 gallons is $28.05.
51. $C(x) = .03x + 5$.   **53.** $G(x) = -\frac{5000}{3}x + 5000$, $G(2.34) = 1100$ gallons.   **55. (a)** $C(x) = 7x + 1500$.   **(b)** Marginal cost is $7 per rod.
(c) $7   **57.** If the manufacturer wants to sell one more unit of goods, then the price per unit must be lowered by 2 cents. No one is willing to pay $7 or more for a unit of goods.   **59.** Let $A(x)$ be the amount in ml of the drug in the blood after x minutes; then, $A(x) = 6x + 1.5$.   **61.** $y(t) = 2t - 212$ **63. (a)** $C(x) = 7x + 230$ dollars   **(b)** $R(x) = 12x$ dollars   **67. (a)** $y = \frac{1}{3}x + 54$.
(b) Every year since 2014, .33% more of the world becomes urban.   **(c)** 56%.   **(d)** 2068.

Exercises 1.2, page 69

1. $-\frac{4}{3}$ **3.** 1   **5.** 1   **7.** -2 **9.** Small positive, large positive   **11.** Zero slope, small negative   **13.** $-.8$, $y - .16 = -.8(x + .4)$
15. $\frac{2}{3}$; $y - \frac{1}{9} = \frac{2}{3}(x - \frac{1}{3})$ **17.** $m = -\frac{1}{2}$ **19.** $y - 6.25 = 5(x - 2.5)$ **21.** $(\frac{7}{4}, \frac{49}{16})$ **23.** $(-\frac{1}{3}, \frac{1}{9})$ **25.** Approximately $17.5. The price was rising on both days.   **27.** $27.25; rising at the rate of $0.05 per day.   **29.** 12   **31.** $\frac{3}{4}$ **33.** $a = 1$, $f(1) = 1$, slope is 2.   **35.** $(\frac{1}{\sqrt{2}}, \frac{1}{2\sqrt{2}})$, $(-\frac{1}{\sqrt{2}}, -\frac{1}{2\sqrt{2}})$.   **37. (a)** $3, 9$ **(b)** increase.   **39.** -3 **41.** .25

Exercises 1.3, page 79

1. 3   **3.** $\frac{3}{4}$ **5.** $7x^6$ **7.** $\frac{2}{3}x^{-1/3} = \frac{2}{3\sqrt[3]{x}}$ **9.** $-\frac{5}{2}x^{-7/2}$ **11.** $\frac{1}{3}x^{-2/3}$ **13.** $f(x) = x^2$; $f'(x) = 2x$ **15.** 0   **17.** $\frac{3}{4}$ **19.** $-\frac{9}{4}$ **21.** 1
23. 2   **25.** 32   **27.** $f(-5) = -125$, $f'(-5) = 75$ **29.** $f(8) = 2$, $f'(8) = \frac{1}{12}$ **31.** $f(-2) = -\frac{1}{32}$, $f'(-2) = -\frac{5}{64}$ **33.** $y + 8 = 12(x + 2)$
35. $y = 3x + 1$ **37.** $y - \frac{1}{3} = \frac{3}{2}(x - \frac{1}{9})$ **39.** $y - 1 = -\frac{1}{2}(x - 1)$ **41.** $f(x) = x^4$, $f'(x) = 4x^3$, $f(1) = 1$, $f'(1) = 4$; then, (6) (with $a = 1$) becomes $y - 1 = 4(x - 1)$, which is the given equation of the tangent line.   **43.** $P = (\frac{1}{16}, \frac{1}{4})$, $b = \frac{1}{8}$

45. (a) $(16, 4)$ **(b)**

47. No, such a point would have to have an x-coordinate satisfying $3x^2 = -1$, which has no real solutions.   **49.** $8x^7$ **51.** $\frac{3}{4}x^{-1/4}$ **53.** 0   **55.** $\frac{1}{5}x^{-4/5}$ **57.** $4, \frac{1}{3}$ **59.** $a = 4$, $b = 1$ **61.** $1, 1.5; 2$ **63.** $y - 5 = \frac{1}{2}(x - 4)$ **65.** $4x + 2h$ **67.** $-2x + 2 - h$
69. $3x^2 + 3xh + h^2$ **71.** $f'(x) = -2x$ **73.** $f'(x) = 14x + 1$ **75.** $f'(x) = 3x^2$

77. (a) and **(b)**

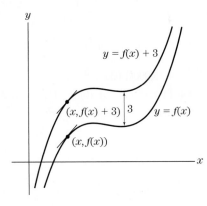

(c) Parallel lines have equal slopes: Slope of the graph of $y = f(x)$ at the point $(x, f(x))$ is equal to the slope of the graph of $y = f(x) + 3$ at the point $(x, f(x) + 3)$, which implies the given equation.

79. .69315   **81.** .70711   **83.** .11111

Exercises 1.4, page 89

1. No limit **3.** 1 **5.** No limit **7.** -5 **9.** 5 **11.** $\frac{13}{5}$ **13.** 288 **15.** $\frac{\sqrt{14}}{23}$ **17.** 3 **19.** -4 **21.** -8 **23.** $\frac{6}{7}$ **25.** No limit

27. (a) 0 **(b)** $-\frac{3}{2}$ **(c)** $-\frac{1}{4}$ **(d)** -1 **29.** 6 **31.** 3 **33.** Step 1. $\frac{f(x+h)-f(x)}{h} = \frac{(x+h)^2+1-(x^2+1)}{h}$ Step 2. $\frac{f(x+h)-f(x)}{h} = 2x + h$

Step 3. $f'(x) = 2x$ **35.** Step 1. $\frac{f(x+h)-f(x)}{h} = \frac{(x+h)^3-1-(x^3-1)}{h}$ Step 2. $\frac{f(x+h)-f(x)}{h} = 3x^2 + 3xh + h^2$ Step 3. $f'(x) = 3x^2$

37. Steps 1, 2: $\frac{f(3+h)-f(3)}{h} = 3$. Step 3: $f'(x) = \lim\limits_{h\to 0} 3 = 3$ **39.** Steps 1, 2: $\frac{f(x+h)-f(x)}{h} = 1 + \frac{-1}{x(x+h)}$.

Step 3: $f'(x) = \lim\limits_{h\to 0} 1 + \frac{-1}{x(x+h)} = 1 - \frac{1}{x^2}$ **41.** Steps 1, 2: $\frac{f(x+h)-f(x)}{h} = \frac{1}{(x+1)(x+h+1)}$. Step 3: $f'(x) = \lim\limits_{h\to 0} \frac{1}{(x+1)(x+h+1)} = \frac{1}{(x+1)^2}$

43. Steps 1, 2: $\frac{f(x+h)-f(x)}{h} = \frac{-2x-h}{((x+h)^2+1)(x^2+1)}$. Step 3: $f'(x) = \lim\limits_{h\to 0} \frac{-2x-h}{((x+h)^2+1)(x^2+1)} = \frac{-2x}{(x^2+1)^2}$ **45.** Steps 1, 2: $\frac{f(x+h)-f(x)}{h} = \frac{1}{\sqrt{x+h+2}+\sqrt{x+2}}$.

Step 3: $f'(x) = \lim\limits_{h\to 0} \frac{1}{\sqrt{x+h+2}+\sqrt{x+2}} = \frac{1}{2\sqrt{x+2}}$ **47.** Steps 1, 2: $\frac{f(x+h)-f(x)}{h} = \frac{-1}{\sqrt{x}\sqrt{x+h}(\sqrt{x}+\sqrt{x+h})}$.

Step 3: $f'(x) = \lim\limits_{h\to 0} \frac{-1}{\sqrt{x}\sqrt{x+h}(\sqrt{x}+\sqrt{x+h})} = \frac{-1}{2x\sqrt{x}} = \frac{-1}{2x^{3/2}}$ **49.** $f(x) = x^2; a = 1$ **51.** $f(x) = \frac{1}{x}; a = 10$ **53.** $f(x) = \sqrt{x}; a = 9$

55. 0 **57.** $\frac{5}{3}$ **59.** 0 **61.** $\frac{3}{4}$ **63.** 0 **65.** 0 **67.** 0 **69.** .5

Exercises 1.5, page 96

1. No **3.** Yes **5.** No **7.** No **9.** Yes **11.** No **13.** Continuous, differentiable **15.** Continuous, not differentiable
17. Continuous, not differentiable **19.** Not continuous, not differentiable **21.** $f(5) = 3$ **23.** Not possible **25.** $f(0) = 12$

27. (a) $T(x) = \begin{cases} .15x & \text{for } 0 < x \le 27{,}050 \\ .275x - 3381.25 & \text{for } 27{,}050 < x \le 65{,}550 \\ .305x - 5347.75 & \text{for } 65{,}550 < x \le 136{,}750 \end{cases}$

(b)

$T(x)$

36,361

14,645

4057.5

0   27,050 65,550 100,000 136,750 x
Taxable income in dollars

29. (a) $R(x) = \begin{cases} .07x + 2.5 & \text{for } 0 \le x \le 100 \\ .04x + 5.5 & \text{for } 100 < x \end{cases}$

(b) $P(x) = \begin{cases} .04x + 2.5 & \text{for } 0 \le x \le 100 \\ .01x + 5.5 & \text{for } 100 < x \end{cases}$

31. (a) $3000 per hour **(b)** $3000 per hour, between 8 A.M. and 10 A.M. **33.** $a = 1$

(c) $T(65{,}550) - T(27{,}050) = 10{,}587.5$ dollars

Exercises 1.6, page 102

1. $18x^2$ **3.** $3\left(\frac{1}{3}\right)x^{-2/3}$ or $\frac{1}{x^{2/3}}$ **5.** $\frac{1}{2} - 2(-1)x^{-2}$ or $\frac{1}{2} + \frac{2}{x^2}$ **7.** $4x^3 + 3x^2 + 1$ **9.** $3(2x+4)^2(2)$ or $6(2x+4)^2$

11. $7(x^3 + x^2 + 1)^6(3x^2 + 2x)$ **13.** $-\frac{8}{x^3}$ **15.** $3\left(\frac{1}{3}\right)(2x^2 + 1)^{-\frac{2}{3}}(4x)$ or $4x(2x^2 + 1)^{-\frac{2}{3}}$ **17.** $2 + 3(x+2)^2$ **19.** $\frac{1}{5}(-5)x^{-6}$ or $-\frac{1}{x^6}$

21. $(-1)(x^3 + 1)^{-2}(3x^2)$ or $-\frac{3x^2}{(x^3+1)^2}$ **23.** $1 - (x+1)^{-2}$ **25.** $\frac{45x^2+5}{2\sqrt{3x^3+x}}$ **27.** 3 **29.** $\frac{1}{2}(1 + x + x^2)^{-\frac{1}{2}}(1 + 2x)$ **31.** $10(1-5x)^{-2}$

33. $-45(1 + x + \sqrt{x})^{-2}(1 + \frac{1}{2}x^{-\frac{1}{2}})$ **35.** $1 + \frac{1}{2}(x+1)^{-\frac{1}{2}}$ **37.** $\frac{3}{2}\left(\frac{\sqrt{x}}{2} + 1\right)^{1/2}\left(\frac{1}{4}x^{-1/2}\right)$ or $\frac{3}{8\sqrt{x}}\left(\frac{\sqrt{x}}{2} + 1\right)^{1/2}$ **39.** 4 **41.** 15

43. $f'(4) = 48, y = 48x - 191$ **45. (a)** $y' = 2(3x^2 + x - 2) \cdot (6x + 1) = 36x^3 + 18x^2 - 22x - 4$

(b) $y = 9x^4 + 6x^3 - 11x^2 - 4x + 4, y' = 36x^3 + 18x^2 - 22x - 4$ **47.** 4.8; 1.8 **49.** 14; 11 **51.** 10; $\frac{15}{4}$ **53.** $\left(5, \frac{161}{3}\right)$; (3, 49)

55. $f(4) = 5, f'(4) = \frac{1}{2}$

Exercises 1.7, page 110

1. $10t(t^2 + 1)^4$ **3.** $8t + \frac{11}{2}t^{-\frac{1}{2}}$ **5.** $5T^4 - 16T^3 + 6T - 1$ **7.** $6P - \frac{1}{2}$ **9.** $2a^2t + b^2$ **11.** $y' = 1, y'' = 0$ **13.** $y' = \frac{1}{2}x^{-1/2}$,

$y'' = -\frac{1}{4}x^{-3/2}$ **15.** $y' = \frac{1}{2}(x+1)^{-\frac{1}{2}}, y'' = -\frac{1}{4}(x+1)^{-\frac{3}{2}}$ **17.** $f'(r) = 2\pi r, f''(r) = 2\pi$ **19.** $f'(P) = 15(3P+1)^4, f''(P) = 180(3P+1)^3$

21. 36 **23.** 0 **25.** 34 **27.** $f'(1) = -\frac{1}{9}, f''(1) = \frac{2}{27}$ **29.** $\frac{d}{dt}\left(\frac{dx}{dt}\right) = 18t + \frac{8}{t^3}$; 37. **31.** 20 **33. (a)** $\frac{ds}{dP} = T$ **(b)** $\frac{ds}{dT} = P$

35. (a) $2Tx + 3P$ **(b)** $3x$ **(c)** $x^2 + 2T$ **37.** When 50 bicycles are manufactured, the cost is $5000. For every additional bicycle manufactured, there is an additional cost of $45. **39. (a)** $2.60 per unit **(b)** 100 or 200 more units **41. (a)** $R(12) = 22$, $R'(12) = .075$ **(b)** $P(x) = R(x) - C(x)$ so $P'(x) = R'(x) - C'(x)$. When 1200 chips are produced, the marginal profit is $.75 - 1.5 = -.75$ dollars per chip. **43. (a)** $S(1) = 120.560, S'(1) = 1.5$ **(b)** $S(3) = 80, S'(3) = -6$ **45. (a)** $S(10) = \frac{372}{121} \approx 3.074$ thousand dollars; $S'(10) = -\frac{18}{11^3} \approx -.0135$ thousand dollars per day. **(b)** $S(11) \approx S(10) + S'(10) \approx 3.061$ thousand dollars.

$S(11) = \frac{49}{16} = 3.0625$ thousand dollars. **47. (a)** $A(8) = 12$, $A'(8) = .5$ **(b)** $A(9) \approx A(8) + A'(8) = 12.5$. If the company spends $9000, it should expect to sell 1250 computers. **49. (a)** $f'''(x) = 60x^2 - 24x$ **(b)** $f'''(x) = \frac{15}{2\sqrt{x}}$ **51.**

Exercises 1.8, page 119

1. (a) $1\frac{f(2)-f(1)}{2-1} = 6$; **(b)** $\frac{f(1.5)-f(1)}{1.5-1} = 5.5$; **(c)** $\frac{f(1.1)-f(1)}{1.1-1} = 5.1$ **3. (a)** 12; 10; 8.4 **(b)** 8 **5. (a)** 14 **(b)** 13
7. (a) 28 km/h **(b)** 96 km **(c)** $\frac{1}{2}$ hr **9.** 63 units/h **11. (a)** $v(t) = -12t + 72$ feet per sec. **(b)** $a(t) = -12$ feet per sec per sec.
(c) $t = 6$ sec. **(d)** $s(6) = 216$ feet **13. (a)** 160 ft/sec **(b)** 96 ft/sec **(c)** -32 ft/sec^2 **(d)** 10 sec **(e)** -160 ft/sec
15. A–b; B–d; C–f; D–e; E–a; F–c; G–g **17. (a)** 15 ft/sec. **(b)** No; positive velocity indicates the object is moving
away from the reference point. **(c)** 5 ft/sec. **19. (a)** 5010 **(b)** 5005 **(c)** 4990 **(d)** 4980 **(e)** 4997.5
21. Four minutes after it has been poured, the coffee is 120°. At that time, its temperature is decreasing by 5°/min; 119.5°.
23. When the price of a car is $10,000, 200,000 cars are sold. At that price, the number of cars sold decreases by 3 for each dollar
increase in price. **25.** When the price is set at $1200, 60,000 computers are sold. For every $100 increase in price, the sales decrease by
2000 computers. 59,000 computers. **27.** The profit from manufacturing and selling 100 cars is $90,000. Each additional car made and
sold creates an additional profit of $1200; $88,800. **29. (a)** $C'(5) = \$74,000$ per unit. **(b)** $C(5.25) \approx C(5) + C'(5)(.25) = 256.5$
thousand dollars. **(c)** $x = 4$ **(d)** $C'(4) = 62$, $R'(4) = 29$. If the production is increased by one unit, cost will rise by $62,000, and
revenue will increase by $29,000. The company should not increase production beyond the breakeven point. **31. (a)** $500 billion
(b) $50 billion/yr **(c)** 1994 **(d)** 1994
33. (a) **(b)** .85 sec
(c) 5 days
(d) $-.05$ sec/day
(e) 3 days

Chapter 1: Review Exercises, page 129

1. $y = -2x + 3$ **2.**   **3.** $y = 5x - 10$ **4.** $y - 4 = -\frac{1}{3}(x - 1)$

   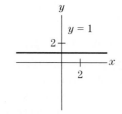

5. $y = -2x + 11$ **6.** $y - 1 = \frac{2}{3}x$ **7.** $y = \frac{3}{4}x + \frac{19}{4}$ **8.** $y = 1$

   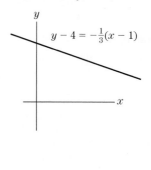

9. $y = -\frac{1}{3}x + \frac{7}{3}$

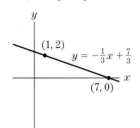

10. $y - 7 = \frac{4}{3}(x - 6)$

11. $y = 3$

12. $x = 4$

13. $x = 0$

14. $y = 0$

15. $7x^6 + 3x^2$
16. $40x^7$
17. $\frac{3}{\sqrt{x}}$
18. $7x^6 + 15x^4$
19. $-\frac{3}{x^2}$
20. $4x^3 + \frac{4}{x^2}$
21. $48x(3x^2 - 1)^7$

22. $x^{1/3} + x^{-1/4}$
23. $-\frac{5}{(5x - 1)^2}$
24. $5(x^3 + x^2 + 1)^4(3x^2 + 2x)$
25. $\frac{x}{\sqrt{x^2+1}}$
26. $-\frac{70x}{(7x^2+1)^2}$
27. $-\frac{1}{4x^{5/4}}$
28. $6(2x + 1)^2$

29. 0 **30.** $\frac{5}{2} + \frac{2}{5x^2}$ **31.** $10[x^5 - (x - 1)^5]^9[5x^4 - 5(x - 1)^4]$ **32.** $10t^9 - 90t^8$ **33.** $\frac{3}{2}t^{-1/2} + \frac{3}{2}t^{-3/2}$ **34.** 0
35. $2(9t^2 - 1)/(t - 3t^3)^2$ **36.** $2.8^{-0.3}$ **37.** $\frac{9}{4}x^{1/2} - 4x^{-1/3}$ **38.** $\frac{1}{2\sqrt{x+\sqrt{x}}} \cdot (1 + \frac{1}{2\sqrt{x}})$ **39.** 28 **40.** 10π **41.** 14; 3
42. $-1/2; 0$ **43.** $15/2$ **44.** 12 **45.** 33 **46.** -5 **47.** $4x^3 - 4x$ **48.** $\frac{5}{2}t^{3/2} + 3t^{1/2} - \frac{1}{2}t^{-1/2}$ **49.** $\frac{-3}{2\sqrt{1-3P}}$ **50.** $-5n^{-6}$
51. 29 **52.** 320 **53.** $300(5x + 1)^2$ **54.** $-\frac{1}{2}t^{-3/2}$ **55.** -2 **56.** 0 **57.** $3x^{-1/2}$ **58.** $\frac{2}{3}t^{-3}$ **59.** Slope -4; tangent $y = -4x + 6$
60. $y + 1/2 = -\frac{3}{4}(x - 1)$ **61.** $y = 3x - \frac{9}{4}$

62. $y - 4 = -4(x + 2)$ **63.** $y = 2$ **64.** $y - 8 = 60(x - 2)$
65. $f(2) = 3, f'(2) = -1$ **66.** $a = -1$ **67.** 96 ft/sec.
68. 45 tons/hr **69.** 11 feet **70.** $v_{\text{ave}} = \frac{6-1}{4-1} = \frac{5}{3}$ feet/second
71. $\frac{5}{3}$ feet/second **72.** faster at $t = 6$ **73.** (a) \$16.10 (b) \$16
74. (a) 4400 (b) 4700 (c) 4100 (d) 4900
75. $\frac{3}{4}$ inch **76.** \$12.53 **77.** 4 **78.** Does not exist
79. Does not exist **80.** 0 **81.** $-\frac{1}{50}$ **82.** 4
83. The slope of a secant line at $(3, 9)$ **84.** $-\frac{1}{4}$

CHAPTER 2

Exercises 2.1, page 138

1. (a), (e), (f) **3.** (b), (c), (d) **5.** Increasing for $x < \frac{1}{2}$, relative maximum point at $x = \frac{1}{2}$, maximum value $= 1$, decreasing for $x > \frac{1}{2}$, concave down, y-intercept $(0, 0)$, x-intercepts $(0, 0)$ and $(1, 0)$. **7.** Decreasing for $x < 0$, relative minimum point at $x = 0$, relative minimum value $= 2$, increasing for $0 < x < 2$, relative maximum point at $x = 2$, relative maximum value $= 4$, decreasing for $x > 2$, concave up for $x < 1$, concave down for $x > 1$, inflection point at $(1, 3)$, y-intercept $(0, 2)$, x-intercept $(3.6, 0)$. **9.** Decreasing for $x < 2$, relative minimum at $x = 2$, minimum value $= 3$, increasing for $x > 2$, concave up for all x, no inflection point, defined for $x > 0$, the line $y = x$ is an asymptote, the y-axis is an asymptote. **11.** Decreasing for $1 \le x < 3$, relative minimum point at $x = 3$, increasing for $x > 3$, maximum value $= 6$ (at $x = 1$), minimum value $= .9$ (at $x = 3$), inflection point at $x = 4$, concave up for $1 \le x < 4$, concave down for $x > 4$, the line $y = 4$ is an asymptote. **13.** Slope decreases for all x. **15.** Slope decreases for $x < 1$, increases for $x > 1$. Minimum slope occurs at $x = 1$. **17.** (a) C, F (b) A, B, F (c) C

19.

21.

23.

25.

27.

29. 1960 **31.** The parachutist's velocity levels off to 15 ft/sec.

33.

35.

$y = \frac{3}{4}x + 5$

37. (a) Yes **(b)** Yes

39. $x = 2$ **41.** $y = x$

Exercises 2.2, page 145

1. (e) **3.** (a), (b), (d), (e) **5.** (d)

7.

(2, 1)

9.

(3, 5)

11.

(0, 6)

(2, 3)

(4, 0)

13.

(3, 4)

15.

(3, 1)

17.

(3, −2)

19.

	f	f'	f''
A	POS	POS	NEG
B	0	NEG	0
C	NEG	0	POS

21. $t = 1$ **23. (a)** Decreasing **(b)** The function $f(x)$ is increasing for $1 \le x < 2$ because the values of $f'(x)$ are positive. The function $f(x)$ is decreasing for $2 < x \le 3$ because the values of $f'(x)$ are negative. Therefore, $f(x)$ has a relative maximum at $x = 2$. Coordinates: (2, 9) **(c)** The function $f(x)$ is decreasing for $9 \le x < 10$ because the values of $f'(x)$ are negative. The function $f(x)$ is increasing for $10 < x \le 11$ because the values of $f'(x)$ are positive. Therefore, $f(x)$ has a relative minimum at $x = 10$. **(d)** Concave down **(e)** At $x = 6$; coordinates: (6, 5) **(f)** $x = 15$ **25.** The slope is positive because $f'(6) = 2$, a positive number. **27.** The slope is 0 because $f'(3) = 0$. Also, $f'(x)$ is positive for x slightly less than 3, and $f'(x)$ is negative for x slightly greater than 3. Hence, $f(x)$ changes from increasing to decreasing at $x = 3$. **29.** $f'(x)$ is increasing at $x = 0$, so the graph of $f(x)$ is concave up. **31.** At $x = 1$, $f'(x)$ changes from increasing to decreasing, so the concavity of the graph of $f(x)$ changes from concave up to concave down. **33.** $y - 3 = 2(x - 6)$ **35.** 3.25 **37. (a)** $\frac{1}{6}$ in. **(b)** (ii), Because the water level is falling. **39.** II. The derivative is positive for $x > 0$, so the function should be increasing. **41.** I **43. (a)** 2 million **(b)** 30,000 farms per year **(c)** 1940 **(d)** 1945 and 1978 **(e)** 1960 **45.** Rel. max: $x \approx -2.34$; rel. min: $x \approx 2.34$; inflection point: $x = 0$, $x \approx \pm 1.41$

Exercises 2.3, page 156

1. $f'(x) = 3(x + 3)(x - 3)$; relative maximum point (−3, 54); relative minimum point (3, −54)

Critical Values		−3		3	
	$x < -3$		$-3 < x < 3$		$3 < x$
$f'(x)$	+	0	−	0	+
$f(x)$	Increasing on $(-\infty, -3)$	54	Decreasing on $(-3, 3)$	−54	Increasing on $(3, \infty)$
		Local maximum (−3, 54)		Local minimum (3, −54)	

3. $f'(x) = -3(x - 1)(x - 3)$; relative maximum point (3, 1); relative minimum point (1, −3)

Critical Values		1		3	
	$x < 1$		$1 < x < 3$		$3 < x$
$f'(x)$	−	0	+	0	−
$f(x)$	Decreasing on $(-\infty, 1)$	−3	Increasing on $(1, 3)$	1	Decreasing on $(3, \infty)$
		Local minimum (1, −3)		Local maximum (3, 1)	

5. $f'(x) = x(x - 2)$; relative maximum point $(0, 1)$; relative minimum point $(2, -1/3)$

Critical Values		0		2	
	$x < 0$		$0 < x < 2$		$2 < x$
$f'(x)$	+	0	−	0	+
$f(x)$	Increasing on $(-\infty, 0)$		Decreasing on $(0, 2)$		Increasing on $(2, \infty)$

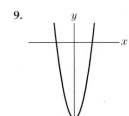

Local maximum $(0, 1)$ Local minimum $(2, -1/3)$

7. $f'(x) = -3x(x + 8)$; relative maximum point $(0, -2)$; relative minimum point $(-8, -258)$

Critical Values		−8		0	
	$x < -8$		$-8 < x < 0$		$0 < x$
$f'(x)$	−	0	+	0	−
$f(x)$	Decreasing on $(-\infty, -8)$		Increasing on $(-8, 0)$		Decreasing on $(0, \infty)$

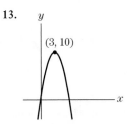

Local minimum $(-8, -258)$ Local maximum $(0, -2)$

9.

11.

13.

15.

17.

19.

21.

23.

25.

27.

29.

31.

33. No, $f''(x) = 2a \neq 0$.   **35.** $(4, 3)$ min.   **37.** $(1, 5)$ max.   **39.** $(-.1, -3.05)$ min.   **41.** $f'(x) = g(x)$ **43.** (a) f has a relative min.
(b) f has an inflection point.   **45.** (a) $A(x) = -893.103x + 460.759$ (billion dollars).   (b) Revenue is $x\%$ of assets or
$R(x) = \frac{xA(x)}{100} = \frac{x}{100}(-893.103x + 460.759)$. $R(.3) \approx .578484$ billion dollars or \$578.484 million $R(.1) \approx .371449$ billion dollars or
\$371.449 million.   (c) Maximum revenue when $R'(x) = 0$ or $x \approx .258$. Maximum revenue $R(.258) \approx .594273$ or \$594.273 million.
47. $f'(x)$ is always nonnegative.   **49.** They both have a minimum point. The parabola does not have a vertical asymptote.

Exercises 2.4, page 162

1. $\left(\frac{3 \pm \sqrt{5}}{2}, 0\right)$ **3.** $(-2, 0), \left(-\frac{1}{2}, 0\right)$ **5.** $\left(\frac{1}{2}, 0\right)$ **7.** The derivative $x^2 - 4x + 5$ has no zeros; no relative extreme points.

9.

11.

13.

15.

17.   **19.**   **21.**   **23.**

25.   **27.**   **29.**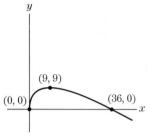

31. $g(x) = f'(x)$ **33.** $f(2) = 0$ implies $4a + 2b + c = 0$. Local maximum at $(0, 1)$ implies $f'(0) = 0$ and $f(0) = 1$. $a = -1/4, b = 0, c = 1, f(x) = -1/4x^2 + 1$.   **35.** If $f'(a) = 0$ and f' is increasing at $x = a$, then $f'(x) < 0$ for $x < a$ and $f'(x) > 0$ for $x > a$. By the first-derivative test (case (b)), f has a local minimum at $x = a$.

37. (a)   **(b)** 15.0 g   **(c)** after 12.0 days   **(d)** 1.6 g/day   **(e)** after 6.0 days and after 17.6 days   **(f)** after 11.8 days

Exercises 2.5, page 168

1. 20   **3.** $t = 4$, $f(4) = 8$ **5.** $x = 1, y = 1$, maximum $= 1$ **7.** $x = 3, y = 3$, minimum $= 18$ **9.** $x = 6, y = 6$, minimum $= 12$ **11. (a)** Objective: $A = xy$; constraint: $8x + 4y = 320$ **(b)** $A = -2x^2 + 80x$ **(c)** $x = 20$ ft, $y = 40$ ft
13. (a) 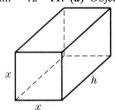   **(b)** $h + 4x$ **(c)** Objective: $V = x^2h$; constraint: $h + 4x = 84$ **(d)** $V = -4x^3 + 84x^2$
(e) $x = 14$ in., $h = 28$ in.   **15.** Let x be the length of the fence and y the other dimension. Objective: $C = 15x + 20y$; constraint: $xy = 75$; $x = 10$ ft, $y = 7.5$ ft.   **17.** Let x be the length of each edge of the base and h the height. Objective: $A = 2x^2 + 4xh$; constraint: $x^2h = 8000$; 20 cm by 20 cm by 20 cm
19. Let x be the length of the fence parallel to the river and y the length of each section perpendicular to the river. Objective: $A = xy$; constraint: $6x + 15y = 1500$; $x = 125$ ft, $y = 50$ ft
21. Objective: $P = xy$; constraint: $x + y = 100$; $x = 50, y = 50$ **23.** Objective: $A = \frac{\pi x^2}{2} + 2xh$; constraint: $(2 + \pi)x + 2h = 14$; $x = \frac{14}{4 + \pi}$ ft

25. $w = 20$ ft, $x = 10$ ft   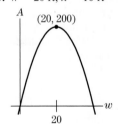   **27.** $C(x) = 6x + 10\sqrt{(20 - x)^2 + 24^2}$; $C'(x) = 6 - \frac{10(20 - x)}{\sqrt{(20 - x)^2 + 24^2}}$; $C'(x) = 0$ ($0 \le x \le 20$) implies $x = 2$. Use the first-derivative test to conclude that the minimum cost is $C(2) = \$312$.   **29.** $\left(\frac{3}{2}, \sqrt{\frac{3}{2}}\right)$
31. $x = 2, y = 1$

Exercises 2.6, page 175

1. (a) 90   **(b)** 180   **(c)** 6   **(d)** 1080 pounds   **3. (a)** $C = 16r + 2x$ **(b)** Constraint $rx = 800$ **(c)** $x = 80, r = 10$, minimum inventory cost $= \$320$ **5.** Let x be the number of cases per order and r the number of orders per year. Objective: $C = 80r + 5x$; constraint: $rx = 10{,}000$ (a) $\$4100$ (b) 400 cases   **7.** Let r be the number of production runs and x the number of microscopes manufactured per run. Objective: $C = 2500r + 25x$; constraint: $rx = 1600$; 4 runs   **11.** Objective: $A = (100 + x)w$; constraint: $2x + 2w = 300$; $x = 25$ ft, $w = 125$ ft   **13.** Objective: $F = 2x + 3w$; constraint: $xw = 54$; $x = 9$ m, $w = 6$ m   **15. (a)** $A(x) = 100x + 1000$
(b) $R(x) = A(x) \cdot (\text{Price}) = (100x + 1000)(18 - x)$ ($0 \le x \le 18$). The graph of $R(x)$ is a parabola looking downward, with a maximum at $x = 4$.   **(c)** $A(x)$ does not change, $R(x) = (100x + 1000)(9 - x)$ ($0 \le x \le 9$). Maximum value when $x = 0$.
17. Let x be the length of each edge of the base and h the height. Objective: $C = 6x^2 + 10xh$; constraint: $x^2h = 150$; 5 ft by 5 ft by 6 ft

19. Let x be the length of each edge of the end and h the length. Objective: $V = x^2h$; constraint: $2x + h = 120$; 40 cm by 40 cm by 40 cm
21. objective: $V = w^2x$; constraint: $2x + w = 16$; $\frac{8}{3}$ in.   **23.** After 20 days   **25.** $2\sqrt{3}$ by 6   **27.** 10 in. by 10 in. by 4 in.   **29.** ≈ 3.77 cm

Exercises 2.7, page 183

1. $1   **3.** 32   **5.** 5   **7.** $x = 20$ units, $p = \$133.33$.   **9.** 2 million tons, $156 per ton   **11. (a)** $3.00   **(b)** $3.30   **13.** Let x be the number of prints and p the price per print. Demand equation: $p = 650 - 5x$; revenue: $R(x) = (650 - 5x)x$; 65 prints   **15.** Let x be the number of tables and p the profit per table. $p = 16 - .5x$; profit from the café: $R = (16 - .5x)x$; 16 tables.   **17. (a)** $x = 15 \cdot 10^5$, $p = \$45$.
(b) No. Profit is maximized when price is increased to $50.   **19.** 5%   **21. (a)** $75,000   **(b)** $3200 per unit   **(c)** 15 units
(d) 32.5 units   **(e)** 35 units

Chapter 2: Chapter Review Exercises, page 190

1. (a) increasing: $-3 < x < 1, x > 5$; decreasing: $x < -3, 1 < x < 5$ **(b)** concave up: $x < -1, x > 3$; concave down: $-1 < x < 3$
2. (a) $f(3) = 2$ **(b)** $f'(3) = \frac{1}{2}$ **(c)** $f''(3) = 0$

3.   **4.**   **5.**   **6.**

7. d, e   **8.** b   **9.** c, d   **10.** a   **11.** e   **12.** b   **13.** Graph goes through (1, 2), increasing at $x = 1$.   **14.** Graph goes through (1, 5), decreasing at $x = 1$.   **15.** Increasing and concave up at $x = 3$.   **16.** Decreasing and concave down at $x = 2$.   **17.** (10, 2) is a relative minimum point.   **18.** Graph goes through (4, −2), increasing and concave down at $x = 4$.   **19.** Graph goes through (5, −1), decreasing at $x = 5$.   **20.** (0, 0) is a relative minimum.   **21. (a)** after 2 hours   **(b)** .8   **(c)** after 3 hours   **(d)** −.02 unit per hour
22. (a) 400 trillion kilowatt-hours   **(b)** 35 trillion kilowatt-hours per year   **(c)** 1995   **(d)** 10 trillion kilowatt-hours per year in 1935
(e) 1600 trillion kilowatt-hours in 1970

23.   **24.**   **25.**   **26.**

27.   **28.**   **29.**   **30.**

31.   **32.**   **33.**   **34.**

35.   **36.**   **37.**   **38.**

39.

$(-2, \frac{14}{3})$

40.

$(-1, 58)$ $y = x^3 - 6x^2 - 15x + 50$ $(2, 4)$ $(5, -50)$

41.

$(-1, -1)$ $(1, -1)$

42.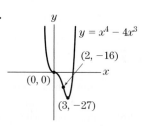

$y = x^4 - 4x^3$ $(2, -16)$ $(0, 0)$ $(3, -27)$

43.

$(10, 7)$ $y = \frac{x}{5} + 3$

44.

$y = \frac{1}{2x} + 2x + 1$ $(\frac{1}{2}, 3)$ $y = 2x + 1$

45. $f'(x) = 3x(x^2 + 2)^{1/2}$, $f'(0) = 0$
46. $f'(x) = 6x(2x^2 + 3)^{1/2}$, $f'(x) > 0$ if $x > 0$ and $f'(x) < 0$ if $x < 0$.
47. $f''(x) = -2x(1 + x^2)^{-2}$, $f''(x)$ is positive for $x < 0$ and negative for $x > 0$.   **48.** $f''(x) = \frac{1}{2}(5x^2 + 1)^{-1/2}(10x)$, so $f''(0) = 0$. Since $f'(x) > 0$ for all x, $f''(x)$ is positive for $x > 0$ and negative for $x < 0$, and it follows that 0 must be an inflection point.   **49.** A–c, B–e, C–f, D–b, E–a, F–d   **50.** A–c, B–e, C–f, D–b, E–a, F–d   **51. (a)** the number of people living between $10 + h$ and 10 mi from the center of the city

(b) If so, $f(x)$ would be decreasing at $x = 10$.   **52.** $x = 8$ **53.** The endpoint maximum value of 2 occurs at $x = 0$.
54. $g(3) = 0$ **55.** Let x be the width and h the height. Objective: $A = 4x + 2xh + 8h$; constraint: $4xh = 200$; 4 ft by 10 ft by 5 ft
56. 2 ft × 2 ft × 4 ft   **57.** $\frac{15}{2}$ in.   **58.** 45 trees   **59.** Let r be the number of production runs and x the number of books manufactured per run. Objective: $C = 1000r + (.25)x$; constraint: $rx = 400,000$; $x = 40,000$ **60.** $x = 3500$ **61.** A to P, $\frac{8}{3}$ miles from C **62.** Let x be the number of people and c the cost. Objective: $R = xc$; constraint: $c = 1040 - 20x$; 25 people.

CHAPTER 3

Exercises 3.1, page 199

1. $(x + 1) \cdot (3x^2 + 5) + (x^3 + 5x + 2) \cdot 1$, or $4x^3 + 3x^2 + 10x + 7$ **3.** $(2x^4 - x + 1) \cdot (-5x^4) + (-x^5 + 1) \cdot (8x^3 - 1)$, or $-18x^8 + 6x^5 - 5x^4 + 8x^3 - 1$ **5.** $x(4)(x^2 + 1)^3(2x) + (x^2 + 1)^4(1)$ or $(x^2 + 1)^3(9x^2 + 1)$ **7.** $(x^2 + 3) \cdot 10(x^2 - 3)^9(2x) + (x^2 - 3)^{10} \cdot 2x$, or $2x(x^2 - 3)^9(11x^2 + 27)$ **9.** $(5x + 1) \cdot 2x + (x^2 - 1) \cdot 5 + \frac{2}{3}$, or $15x^2 + 2x - \frac{13}{3}$ **11.** $\dfrac{(x + 1) \cdot 1 - (x - 1) \cdot 1}{(x + 1)^2}$, or $\dfrac{2}{(x + 1)^2}$

13. $\dfrac{(x^2 + 1) \cdot 2x - (x^2 - 1) \cdot 2x}{(x^2 + 1)^2}$, or $\dfrac{4x}{(x^2 + 1)^2}$ **15.** $\dfrac{(2x + 1)^2 - 2(2x + 1)(2)(x + 3)}{(2x + 1)^4}$, or $\dfrac{-2x - 11}{(2x + 1)^3}$ **17.** $\dfrac{-4x}{(x^2 + 1)^2}$

19. $\dfrac{2x(x^2 + 1)^2 - x^2 2(x^2 + 1)(2x)}{(x^2 + 1)^4}$, or $\dfrac{-2x^3 + 2x}{(x^2 + 1)^3}$ **21.** $2[(3x^2 + 2x + 2)(x - 2)][(3x^2 + 2x + 2) \cdot 1 + (x - 2) \cdot (6x + 2)]$,

or $2(3x^2 + 2x + 2)(x - 2)(9x^2 - 8x - 2)$ **23.** $\dfrac{-1}{2\sqrt{x}(\sqrt{x} + 1)^2}$ **25.** $-42\dfrac{(x + 11)^2}{(x - 3)^4}$ **27.** $\sqrt{x + 2}\, 2(2x + 1)(2) + (2x + 1)^2 \dfrac{1}{2\sqrt{x + 2}}$,

or $\dfrac{(2x + 1)(10x + 17)}{2\sqrt{x + 2}}$ **29.** $y - 16 = 88(x - 3)$ **31.** 2, 7   **33.** 0, ± 2, $\pm\dfrac{5}{4}$ **35.** $\left(\dfrac{1}{2}, \dfrac{3}{2}\right)$, $\left(-\dfrac{1}{2}, \dfrac{9}{2}\right)$ **37.** $\dfrac{dy}{dx} = 8x(x^2 + 1)^3$,

$\dfrac{d^2y}{dx^2} = 8(x^2 + 1)^2(7x^2 + 1)$ **39.** $\dfrac{dy}{dx} = \dfrac{x}{2\sqrt{x + 1}} + \sqrt{x + 1}$ or $\dfrac{dy}{dx} = \dfrac{3x + 2}{2\sqrt{x + 1}}$, $\dfrac{d^2y}{dx^2} = \dfrac{3x + 4}{4(x + 1)^{3/2}}$ **41.** $x \cdot f'(x) + f(x)$

43. $\dfrac{(x^2 + 1) \cdot f'(x) - 2x \cdot f(x)}{(x^2 + 1)^2}$ **45.** $2 \times 3 \times 1$ **47.** 150; $AC(150) = 35 = C'(150)$ **49.** AR is maximized where

$0 = \dfrac{d}{dx}(AR) = \dfrac{x \cdot R'(x) - R(x) \cdot 1}{x^2}$. This happens when the production level, x, satisfies $xR'(x) - R(x) = 0$, and hence,

$R'(x) = R(x)/x = AR$.   **51.** 38 in.²/s   **53.** 150, 853, 600 gal/yr   **55.** (2, 10)   **57.** 10   **59.** 18   **61.** 9

63. (b) $f'(x) \cdot g'(x) = -\dfrac{1}{x^2}(3x^2) = -3$; $[f(x)g(x)]' = 2x$ **67.** $\dfrac{h(t)w'(t) - 2w(t)h'(t)}{[h(t)]^3}$

69. (a)

(b) 10.8 mm²   **(c)** 2.61 units of light   **(d)** $-.55$ mm²/unit of light

Exercises 3.2, page 206

1. $\dfrac{x^3}{x^3+1}$ **3.** $\sqrt{x}(x+1)$ **5.** $f(x)=x^5,\, g(x)=x^3+8x-2$ **7.** $f(x)=\sqrt{x},\, g(x)=4-x^2$ **9.** $f(x)=\dfrac{1}{x},\, g(x)=x^3-5x^2+1$

11. $30x(x^2+5)^{14}$ **13.** $6x^2\cdot 3(x-1)^2(1)+(x-1)^3\cdot 12x,$ or $6x(x-1)^2(5x-2)$ **15.** $2(x^3-1)\cdot 4(3x^2+1)^3(6x)+(3x^2+1)^4\cdot 2(3x^2),$

or $6x(3x^2+1)^3(11x^3+x-8)$ **17.** 8   **19.** 25   **21.** $2xf'(x^2)$ **23.** $f'(-x)$ **25.** $\dfrac{2x^2f'(x^2)-f(x^2)}{x^2}$

27.   **29.** $30(6x-1)^4$ **31.** $-(1-x^2)^{-2}\cdot(-2x),$ or $2x(1-x^2)^{-2}$ **33.** $[4(x^2-4)^3-2(x^2-4)]\cdot(2x),$ or

$8x(x^2-4)^3-4x(x^2-4)$

35. $2[(x^2+5)^3+1]3(x^2+5)^2\cdot(2x),$ or $12x[(x^2+5)^3+1](x^2+5)^2$ **37.** $6(4x+1)^{1/2}$

39. $\dfrac{dy}{dx}=\left(\dfrac{1}{2}-\dfrac{2}{u^2}\right)(1-2x)$ or $\dfrac{dy}{dx}=\left(\dfrac{1}{2}-\dfrac{2}{(x-x^2)^2}\right)(1-2x)$ **41.** $\dfrac{dy}{dt}=(2x-3)(2t)$

when $t=0,\, x=3,\, \dfrac{dy}{dt}=0$ **43.** $\dfrac{dy}{dt}=\dfrac{-2}{(x-1)^2}\cdot\dfrac{t}{2}$ when $t=3,\, x=\dfrac{9}{4},\, \dfrac{dy}{dt}=-\dfrac{48}{25}$ **45.** $y=62x-300$

47. 1; 2; 3   **49.** (a) $\dfrac{dV}{dt}=\dfrac{dV}{dx}\cdot\dfrac{dx}{dt}$ (b) 2   **51.** (a) (i) $\dfrac{dy}{dt},$ (ii) $\dfrac{dP}{dy},$ (iii) $\dfrac{dP}{dt}$ (b) $\dfrac{dP}{dt}=\dfrac{dP}{dy}\cdot\dfrac{dy}{dt}$ **53.** (a) $\dfrac{200(100-x^2)}{(100+x^2)^2}$

(b) $\dfrac{400[100-(4+2t)^2]}{[100+(4+2t)^2]^2}$ (c) falling at the rate of \$480 per week   **55.** (a) $.4+.0002x$

(b) increasing at the rate of 25,000 persons per year   (c) rising at the rate of 14 ppm per year

57. x^3+1 **59.** 24   **61.** (a) $t=1.5,\, x=40,\, W\approx 77.209$ million dollars; $t=3.5,\, x=30,\, W\approx \$76.364$ million   (b) $\dfrac{dx}{dt}\Big|_{t=1.5}=20;$

the price of one share is \$40 and is rising at the rate of \$20 per month. $\dfrac{dx}{dt}\Big|_{t=3.5}=0;$ the price of one share is steady at \$30 a share.

63. (a) $t=2.5,\, x=40,\, \dfrac{dx}{dt}\Big|_{t=2.5}=-20;$ the price of one share is \$40 and is falling at the rate of \$20 per month. $t=4,\, x=30,\, \dfrac{dx}{dt}\Big|_{t=4}=0;$

the price of one share is steady at \$30 a share.   (b) $\dfrac{dW}{dt}\Big|_{t=2.5}=\dfrac{dW}{dx}\Big|_{x=40}\dfrac{dx}{dt}\Big|_{t=2.5}=\dfrac{120}{1849}\cdot(-20)\approx -1.3;$ the total value of the

company is falling at the rate of \$1.3 million per month. $\dfrac{dW}{dt}=\dfrac{dW}{dx}\dfrac{dx}{dt}=0;$ when $x=30,$ the total value of the company is steady at

$W\approx \$76.364$ million.   **65.** The derivative of the composite function $f(g(x))$ is the derivative of the outer function evaluated at the inner function and then multiplied by the derivative of the inner function.

Exercises 3.3, page 215

1. $\dfrac{x}{y}$ **3.** $\dfrac{1+6x}{5y^4}$ **5.** $\dfrac{2x^3-x}{2y^3-y}$ **7.** $\dfrac{1-6x^2}{1-6y^2}$ **9.** $-\dfrac{y}{x}$ **11.** $-\dfrac{y+2}{5x}$ **13.** $\dfrac{8-3xy^2}{2x^2y}$ **15.** $\dfrac{x^2(y^3-1)}{y^2(1-x^3)}$ **17.** $-\dfrac{y^2+2xy}{x^2+2xy}$ **19.** $\dfrac{1}{2}$

21. $-\dfrac{8}{3}$ **23.** $-\dfrac{2}{15}$ **25.** $y-\dfrac{1}{2}=-\dfrac{1}{16}(x-4),\, y+\dfrac{1}{2}=\dfrac{1}{16}(x-4)$ **27.** (a) $\dfrac{2x-x^3-xy^2}{2y+y^3+x^2y}$ (b) 0   **29.** (a) $\dfrac{-y}{2x}$ (b) $\left|\dfrac{-27}{16}\right|=\dfrac{27}{16}$

31. $-\dfrac{x^3}{y^3}\dfrac{dx}{dt}$ **33.** $\dfrac{2x-y}{x}\dfrac{dx}{dt}$ **35.** $\dfrac{2x+2y}{3y^2-2x}\dfrac{dx}{dt}$ **37.** $-\dfrac{15}{8}$ units per second   **39.** Rising at 3,000 units per week

41. Increasing at \$20,000 per month   **43.** (a) $x^2+y^2=100$ (b) $\dfrac{dy}{dt}=-4,$ so the top of the ladder is falling at the rate of 4 feet

per second.   **45.** Decreasing at $\dfrac{22}{\sqrt{5}}$ feet per second (or 9.84 feet per second)

Chapter 3: Review Exercises, page 219

1. $(4x-1)\cdot 4(3x+1)^3(3)+(3x+1)^4\cdot 4,$ or $4(3x+1)^3(15x-2)$ **2.** $-6(5-x)^2(6x-1)+12(5-x)^3$ or $6(5-x)^2(11-8x)$

3. $x\cdot 3(x^5-1)^2\cdot 5x^4+(x^5-1)^3\cdot 1,$ or $(x^5-1)^2(16x^5-1)$ **4.** $5(2x+1)^{3/2}(4x-1)^{3/2}+6(2x+1)^{5/2}(4x-1)^{1/2},$

or $(2x+1)^{3/2}(4x-1)^{1/2}(32x+1)$ **5.** $5(x^{1/2}-1)^4\cdot 2(x^{1/2}-2)\left(\dfrac{1}{2}x^{-1/2}\right)+(x^{1/2}-2)^2\cdot 20(x^{1/2}-1)^3\left(\dfrac{1}{2}x^{-1/2}\right),$

or $5x^{-1/2}(x^{1/2} - 1)^3(x^{1/2} - 2)(3x^{1/2} - 5)$ **6.** $\dfrac{2}{\sqrt{x}(\sqrt{x} + 4)^2}$ **7.** $3(x^2 - 1)^3 \cdot 5(x^2 + 1)^4(2x) + (x^2 + 1)^5 \cdot 9(x^2 - 1)^2(2x)$,

or $12x(x^2 - 1)^2(x^2 + 1)^4(4x^2 - 1)$ **8.** $\dfrac{-12x - 30}{(x^2 + 5x + 1)^7}$ **9.** $\dfrac{(x - 2) \cdot (2x - 6) - (x^2 - 6x) \cdot 1}{(x - 2)^2}$, or $\dfrac{x^2 - 4x + 12}{(x - 2)^2}$ **10.** $\dfrac{4}{(2 - 3x)^2}$

11. $2\left(\dfrac{3 - x^2}{x^3}\right) \cdot \dfrac{x^3 \cdot (-2x) - (3 - x^2) \cdot 3x^2}{x^6}$, or $\dfrac{2(3 - x^2)(x^2 - 9)}{x^7}$ **12.** $\dfrac{x(x - 1)(3x^2 + 1) - x(x^2 + 1)(2x - 1)}{(x^2 - x)^2}$ **13.** $-\dfrac{1}{3}, 3, \dfrac{31}{27}$

14. $x = 0$ **15.** $y + 32 = 176(x + 1)$ **16.** $y = \dfrac{1}{2}x - \dfrac{3}{2}$ **17.** $x = 44$ m, $y = 22$ m **18.** $x = (4 + 10\sqrt{8})$ meters, $y = (4 + 10\sqrt{8})$ meters

19. $\dfrac{dC}{dt} = \dfrac{dC}{dx} \cdot \dfrac{dx}{dt} = 40 \cdot 3 = 120$. Costs are rising \$120 per day. **20.** $\dfrac{dy}{dt} = \dfrac{dy}{dP} \cdot \dfrac{dP}{dt}$ **21.** $\dfrac{3x^2}{x^6 + 1}$ **22.** $\dfrac{d}{dx}f(g(x)) = -\dfrac{1}{1 + x^2}, x \neq 0$

23. $\dfrac{2x}{(x^2 + 1)^2 + 1}$ **24.** $\dfrac{d}{dx}f(g(x)) = 2x^3\sqrt{1 - x^4}$ **25.** $\dfrac{1}{2}\sqrt{1 - x}$ **26.** $\dfrac{d}{dx}f(g(x)) = \dfrac{3}{2}x^2\sqrt{1 - x^3}$ **27.** $\dfrac{3x^2}{2(x^3 + 1)}$

28. $\dfrac{dy}{dx} = \dfrac{2x(x^2 + 1)}{x^4 + 2x^2 + 2}$ **29.** $-\dfrac{25}{x(25 + x^2)}$ **30.** $\dfrac{dy}{dx} = \dfrac{2x^3}{\sqrt{1 + x^8}}$ **31.** $\dfrac{x^{1/2}}{(1 + x^2)^{1/2}} \cdot \dfrac{1}{2}(x^{-1/2})$, or $\dfrac{1}{2\sqrt{1 + x^2}}$ **32.** $\dfrac{dy}{dx} = \dfrac{-4}{x^3\sqrt{1 + \dfrac{16}{x^4}}}$

33. $0; -\dfrac{7}{2}$ **34.** $h(1) = 6; h'(1) = 2f'(1) + 3g'(1) = 2\left(\dfrac{1}{2}\right) + 3\left(\dfrac{3}{2}\right) = \dfrac{11}{2}$ or $5\dfrac{1}{2}$ **35.** $\dfrac{3}{2}; -\dfrac{7}{8}$ **36.** $h(1) = 9$ $h'(1) = 3$ **37.** $1; -\dfrac{3}{2}$

38. $h(1) = g(3) = 1; h'(1) = g'(f(1))f'(1) = -\dfrac{1}{4}$ **39. (a) (i)** $\dfrac{dR}{dA}$, **(ii)** $\dfrac{dA}{dt}$, **(iii)** $\dfrac{dR}{dx}$, **(iv)** $\dfrac{dx}{dA}$ **(b)** $\dfrac{dR}{dt} = \dfrac{dR}{dx}\dfrac{dx}{dA}\dfrac{dA}{dt}$

40. (a) (i) $\dfrac{dP}{dt}$; **(ii)** $\dfrac{dA}{dP}$; **(iii)** $\dfrac{dS}{dP}$; **(iv)** $\dfrac{dA}{dS}$ **(b)** $\dfrac{dA}{dt} = \dfrac{dA}{dS} \cdot \dfrac{dS}{dP} \cdot \dfrac{dP}{dt}$ **41. (a)** $-y^{1/3}/x^{1/3}$ **(b)** 1 **42. (a)** $\dfrac{x^2 - 3y}{3x - y^2}$ **(b)** $\dfrac{4}{5}$

43. -3 **44.** $-\dfrac{1}{6}$ **45.** $\dfrac{3}{5}$ **46.** $\dfrac{1}{4}$ **47. (a)** $\dfrac{dy}{dx} = \dfrac{15x^2}{2y}$ **(b)** $\dfrac{20}{3}$ thousand dollars per thousand-unit increase in production

(c) $\dfrac{dy}{dt} = \dfrac{15x^2}{2y}\dfrac{dx}{dt}$ **(d)** \$2,000 per week **48. (a)** $\dfrac{dy}{dx} = \dfrac{16,000x}{3y^2}$ **(b)** $4.\overline{4}$ **(c)** $\dfrac{dy}{dt} = \dfrac{16,000x}{3y^2} \cdot \dfrac{dx}{dt}$ **(d)** 8; 8000 books/year

49. Increasing at the rate of 2.5 units per unit time **50.** $\dfrac{40}{\pi}$ meters per hour, or approximately 12.73 m per hour.

51. 1.89 m^2/year **52.** 200 dishwashers per month

CHAPTER 4

Exercises 4.1, page 225

1. $2^{2x}, 3^{(1/2)x}, 3^{-2x}$ **3.** $2^{2x}, 3^{3x}, 2^{-3x}$ **5.** $2^{-4x}, 2^{9x}, 3^{-2x}$ **7.** $2^x, 3^x, 3^x$ **9.** $3^{2x}, 2^{6x}, 3^{-x}$ **11.** $2^{(1/2)x}, 3^{(4/3)x}$ **13.** $2^{-2x}, 3^x$ **15.** $\dfrac{1}{9}$

17. 1 **19.** 2 **21.** -1 **23.** $\dfrac{1}{5}$ **25.** $\dfrac{5}{2}$ **27.** -1 **29.** 4 **31.** 1 **33.** 1 or 2 **35.** 1 or 2 **37.** 2^h **39.** $2^h - 1$ **41.** $3^x + 1$

43.

$\approx .6931$

45. 2.7

Exercises 4.2, page 231

1. 1.16, 1.10, 1.10 **3. (a)** $2m$ **(b)** $m/4$ where $m \approx .693$ **5. (a)** e **(b)** $1/e$ **7.** e^{2x}, e^{-x} **9.** e^{-6x}, e^{2x} **11.** e^{6x}, e^{2x}

13. $x = 4$ **15.** $x = 4, -2$ **17.** 1 or -1 **19.** $y - \dfrac{1}{e} = \dfrac{1}{e}(x + 1)$ or $y = .37x + .74$ **21.** $f'(x) = e^x > 0$, $f(x)$ is always increasing, no

maximum or minimum point, $f''(x) = e^x > 0$, $f(x)$ concave up. **23.** b **25.** $3e^x - 7$ **27.** $xe^x + e^x = (x + 1)e^x$

29. $8e^x(2)(1 + 2e^x)(2e^x) + 8e^x(1 + 2e^x)^2$ or $8e^x(6e^x + 1)(1 + 2e^x)$ **31.** $\dfrac{e^x(x + 1) - e^x}{(x + 1)^2}$ or $\dfrac{xe^x}{(x + 1)^2}$ **33.** $\dfrac{e^x(e^x + 1) - e^x(e^x - 1)}{(e^x + 1)^2}$ or

$\dfrac{2e^x}{(e^x + 1)^2}$ **35.** Maximum at $(0, -1)$ **37.** $y' = e^x(1 + x)^2$, $y' = 0$ when $x = -1$. This point is $(-1, 2/e)$. **39.** 1 **41.** $y - \dfrac{1}{3} = \dfrac{1}{9}x$

43. $f'(x) = e^x(x^2 + 4x + 3)$, $f''(x) = e^x(x^2 + 6x + 7)$ **45.** $5e^x$ **(b)** $10e^{10x}$ **(c)** e^{2+x}

47.   **49.**   **51.**   **53.** $y = x + 1$
55.

Exercises 4.3, page 235

1. $2e^{2x+3}$ **3.** $(8x - 1)e^{4x^2 - x}$ **5.** $e^{e^x}e^x$ **7.** $\dfrac{1}{2\sqrt{x}}e^{\sqrt{x}}$ **9.** $-e^{\frac{x}{7}}$ **11.** $.2e^{0.05\,t} - .23e^{0.01t}$

13. $(t^2 + 2e^t)e^{t-1} + e^{t-1}(2t + 2e^t) = e^{t-1}(t^2 + 2t + 4e^t)$ **15.** $2\left(x + \dfrac{1}{x}\right)e^{2x} + e^{2x}\left(1 - \dfrac{1}{x^2}\right) = e^{2x}\left(1 + 2x + \dfrac{2}{x} - \dfrac{1}{x^2}\right)$

17. $\dfrac{(e^x - e^{-x})(e^x - e^{-x}) - (e^x + e^{-x})(e^x + e^{-x})}{(e^x - e^{-x})^2} = \dfrac{-4}{(e^x - e^{-x})^2}$ **19.** $\dfrac{e^x}{2\sqrt{e^x + 1}}$ **21.** $f(x) = e^{15x}, f'(x) = 15e^{15x}$

23. $f(x) = e^{-x/2}, f'(x) = -\dfrac{1}{2}e^{-x/2} = -\dfrac{1}{2\sqrt{e^x}}$ **25.** $f(x) = 1 + 5e^x, f'(x) = 5e^x$ **27.** Max at $x = -2/3$ **29.** Min at $x = 5/4$

31. Max at $x = 9/10$ **33.** (a) \$5000 (b) \$6,259.92 (c) $\approx$ \$287.18 per year **35.** (a) $f(18) \approx 179.408$ million dollars
(b) $f'(18) \approx 17.223$ million dollars per year (c) $f(23) \approx 289.937$ million dollars, $f'(23) \approx 27.834$ million dollars per year
37. (a) 45 m/sec (b) 10 m/sec^2 (c) 4 sec (d) 4 sec **39.** 2 in./ week **41.** $.02e^{-2e^{-0.01x}}e^{-0.01x}$
43. (a) Volume seems to stabilize near 6 ml (b) 3.2 ml (c) 7.7 wk (d) .97 ml/wk (e) 3.7 wk (f) 1.13 ml/wk

Exercises 4.4, page 240

1. $\frac{1}{2}$ **3.** $\ln 5$ **5.** $\dfrac{1}{e}$ **7.** -3 **9.** e **11.** 0 **13.** x^2 **15.** $\frac{1}{49}$ **17.** $2x$ **19.** $\frac{1}{2}\ln 5$ **21.** $4 - e^{1/2}$ **23.** $\pm e^{9/2}$ **25.** $\dfrac{-\ln .5}{.00012}$ **27.** $\frac{5}{3}$

29. $\dfrac{e}{3}$ **31.** $3 \ln \frac{9}{2}$ **33.** $\frac{1}{2}e^{8/5}$ **35.** $\frac{1}{2}\ln 4$ **37.** $-\ln \frac{3}{2}$ **39.** $x = \ln 5, y = 5(1 - \ln 5) \approx -3.047$ **41.** (a) $-5 + e^x = 3, x = \ln 8 \approx 2.08$
(b) No, because $-5 + e^x = -7$ or $e^x = -2$ has no solution **43.** (a) $y' = -e^{-x}, y' = -2$ when $x = -\ln 2 = \ln\left(\frac{1}{2}\right)$.
The point is $(\ln(\frac{1}{2}), 2)$. (b)

45. $(-\ln 3, 3 - 3 \ln 3)$, minimum **47.** $t = \ln 4/.11 \approx 12.6$
49. Max at $t = 2 \ln 51$ **51.** $\ln 2$
53. The graph of $y = e^{\ln x}$ is the same as the graph of $y = x$ for $x > 0$. **55.**

$\frac{1}{5}e^2 \approx 1.4778$

Exercises 4.5, page 244

1. $\dfrac{3}{x}$ **3.** $x \ln x + \dfrac{x}{2} = x(\ln x + \frac{1}{2})$ **5.** $\dfrac{e^x}{x} + e^x \ln x$ **7.** $\dfrac{1}{x\sqrt{x}} - \dfrac{1}{2x\sqrt{x}}\ln x = \dfrac{1}{x\sqrt{x}}\left(1 - \dfrac{\ln x}{2}\right)$ **9.** $\dfrac{2x}{x^2} = \dfrac{2}{x}$ **11.** $x \cdot \dfrac{-1}{x^2} = \dfrac{-1}{x}$

13. $\dfrac{12x^3 - 2x}{3x^4 - x^2} = \dfrac{12x^2 - 2}{3x^3 - x}$ **15.** $y = (\ln x)^{-1}, y' = \dfrac{-1}{(\ln x)^2} \cdot \dfrac{1}{x}$ **17.** $\dfrac{\ln 2x - \ln x}{x(\ln 2x)^2}$ **19.** $3x^2 \ln(x^3 + 1) + 3x^2 = 3x^2(1 + \ln(x^3 + 1))$

21. $3 + 2 \ln t$ **23.** $\left(e^2, \dfrac{2}{e}\right)$ **25.** $y = 1$ **27.** (a) $t > 1$. (b) $t > e$. **29.** $y' = x(1 + 2 \ln x), y'' = 3 + 2 \ln x, y' = 0$ when
$x = e^{-1/2}, y''(e^{-1/2}) > 0$, relative minimum at $x = e^{-1/2}, y = -e^{-1}/2$. **31.** (a) The derivatives $1 + 1/x$ and $1/x$ are always positive for

$x > 0$. (b) $\ln 2$ **33.** $\left(\dfrac{1}{\sqrt{2}}, \dfrac{1}{2} + \dfrac{1}{2}\ln 2\right)$ **34.** $\frac{1}{2}$ **35.** $R'(x) = \dfrac{45(\ln x - 1)}{(\ln x)^2}$; $R'(20) \approx 10$. **37.** Maximum area when $x = e^{-1} \approx .3679$.

Exercises 4.6, page 248

1. $\ln 5x$ **3.** $\ln 3$ **5.** $\ln 2$ **7.** x^2 **9.** $\ln \dfrac{x^5 z^3}{y^{1/2}}$ **11.** $3 \ln x$ **13.** $3 \ln 3$ **15.** (a) $2 \ln 2 = 2(.69) = 1.38$ (b) $\ln 2 + \ln 3 = 1.79$

(c) $3 \ln 3 + \ln 2 = 3.99$ **17.** (a) $-\ln 2 - \ln 3 = -1.79$ (b) $\ln 2 - 2 \ln 3 = -1.51$ (c) $-\frac{1}{2} \ln 2 = -.345$ **19.** (d)   **21.** (d)

23. 3   **25.** $\sqrt{e}$ **27.** e or e^{-1} **29.** 1 or e^4 **31.** $\dfrac{1 + 2e}{e - 1}$ **33.** $\dfrac{1}{x+5} + \dfrac{2}{2x-1} - \dfrac{1}{4-x}$ **35.** $\dfrac{2}{1+x} + \dfrac{3}{2+x} + \dfrac{4}{3+x}$ **37.** $\dfrac{1}{2x} + x$

39. $\dfrac{4}{x+1} - 1$ **41.** $\ln(3x+1)\dfrac{5}{5x+1} + \ln(5x+1)\dfrac{3}{3x+1}$ **43.** $(x+1)^4(4x-1)^2\left(\dfrac{4}{x+1} + \dfrac{8}{4x-1}\right)$

45. $\dfrac{(x+1)(2x+1)(3x+1)}{\sqrt{4x+1}}\left(\dfrac{1}{x+1} + \dfrac{2}{2x+1} + \dfrac{3}{3x+1} - \dfrac{2}{4x+1}\right)$ **47.** $2^x \ln 2$ **49.** $x^x[1 + \ln x]$ **51.** $y = cx^k$ **53.** $h = 3, k = \ln 2$

Chapter 4: Review Exercises, page 252

1. 81   **2.** 8   **3.** $\frac{1}{25}$ **4.** $\frac{1}{2}$ **5.** 4   **6.** 4   **7.** 9   **8.** 2   **9.** e^{3x^2} **10.** e^{7x} **11.** e^{2x} **12.** 6^x **13.** $e^{11x} + 7e^x$ **14.** $e^{2x} - e^{2.5x}$ **15.** $x = 4$

16. $-1, 2$ **17.** $x = -5$ **18.** $\frac{3}{5}$ **19.** $70e^{7x}$ **20.** $\dfrac{e^{\sqrt{x}}}{2\sqrt{x}}$ **21.** $e^{x^2} + 2x^2 e^{x^2}$ **22.** $\dfrac{xe^x - 2e^x - 1}{(x-1)^2}$ **23.** $e^x \cdot e^{e^x} = e^{x + e^x}$

24. $e^{-2x}\left(\dfrac{1}{2\sqrt{x}} - 2\sqrt{x} - 2\right)$ **25.** $\dfrac{(e^{3x} + 3)(2x - 1) - (x^2 - x + 5)3e^{3x}}{(e^{3x} + 3)^2}$ **26.** exe^{-1} **27.** $x = 0, \pm \sqrt{\ln 4}$

28. $f''(x) = e^{x^2}(4x^2 + 2) - 8, f''(0) = -6 < 0$, relative maximum   **29.** $t = \dfrac{\ln 2}{0.03} \approx 23.1$ **30.** $t = \dfrac{\ln 8}{.98} \approx 2.12$

31. $x = \dfrac{\ln 4}{1 - \ln 2} \approx 4.52$ **32.** $x = \dfrac{\ln 2}{\ln 3 - 1} \approx 7.03$ **33.** $\ln 4$ **34.** $\dfrac{\ln 2}{3}$ **35.** Increasing when $x > 0$, decreasing when $x < 0$

36. Increasing when $x > 1/e$, decreasing when $0 < x < 1/e$ **37.** $y = \frac{1}{4}x + \frac{1}{2}$ **38.** $y'(1) = \dfrac{4}{(e + e^{-1})^2}, y'(-1) = \dfrac{4}{(e + e^{-1})^2}$ same slope

39. $\sqrt{5}$ **40.** x^2 **41.** $\frac{2}{3}$ **42.** 4   **43.** 1   **44.** x^2 **45.** $e, \dfrac{1}{e}$ **46.** $\dfrac{e}{3}$ **47.** $\frac{1}{2} \ln 5$ **48.** $2 \ln 4$ **49.** $e^{5/2}$ **50.** $\dfrac{\ln 2}{.3} = \dfrac{10 \ln 2}{3}$

51. $\dfrac{6x^5 + 12x^3}{x^6 + 3x^4 + 1}$ **52.** $\dfrac{\ln x - 1}{(\ln x)^2}$ **53.** $\dfrac{5}{5x - 7}$ **54.** $\dfrac{1}{x}$ **55.** $\dfrac{2 \ln x}{x}$ **56.** $3(\ln x + 1)(x \ln x)^2$ **57.** $\dfrac{1}{x} + 1 - \dfrac{1}{2(1+x)}$

58. $6 + \dfrac{10x}{x^2 + 3} - \dfrac{12x^2}{x^3 + 1}$ **59.** $\ln x$ **60.** $2(x+1)$ **61.** $\dfrac{1}{x \ln x}$ **62.** $\dfrac{-1}{x(\ln x)^2}$ **63.** $\dfrac{e^x}{x} + e^x \ln x$ **64.** $\dfrac{2x + e^x}{x^2 + e^x}$ **65.** $\dfrac{x}{x^2 + 1} - \dfrac{1}{2x + 3}$

66. $\dfrac{2}{2x - 1}$ **67.** $2x - \dfrac{1}{x}$ **68.** $\dfrac{x^2 + 1}{x^3 + 3x - 2}$ **69.** $\ln 2$ **70.** $\ln 3$ **71.** $\dfrac{1}{x - 1}$ **72.** $4(2x + 1)$ **73.** $-\frac{1}{2}x^{-\frac{1}{2}}$ or $\dfrac{-1}{2\sqrt{x}}$ **74.** $\dfrac{e^x - 3e^{-x}}{e^x + 3e^{-x}}$

75. $\sqrt[5]{\dfrac{x^5 + 1}{x^5 + 5x + 1}}\left[\dfrac{x^4}{x^5 + 1} - \dfrac{x^4 + 1}{x^5 + 5x + 1}\right]$ **76.** $2^x \ln 2$ **77.** $x^{\sqrt{x} - \frac{1}{2}}\left[1 + \frac{1}{2} \ln x\right]$ **78.** $b^x \ln b$

79. $(x^2 + 5)^6(x^3 + 7)^8(x^4 + 9)^{10}\left[\dfrac{12x}{x^2 + 5} + \dfrac{24x^2}{x^3 + 7} + \dfrac{40x^3}{x^4 + 9}\right]$ **80.** $x^{1+x}\left(\ln x + \dfrac{1}{x} + 1\right)$ **81.** $10^x \ln 10$ **82.** $\sqrt{x^2 + 5}\, e^{x^2}\left(\dfrac{x}{x^2 + 5} + 2x\right)$

83. $\frac{1}{2}\sqrt{\dfrac{xe^x}{x^3 + 3}}\left[\dfrac{1}{x} + 1 - \dfrac{3x^2}{x^3 + 3}\right]$ **84.** $\dfrac{e^x \sqrt{x + 1}(x^2 + 2x + 3)^2}{4x^2}\left[1 + \dfrac{1}{2(x+1)} + \dfrac{2(2x + 2)}{x^2 + 2x + 3} - \dfrac{2}{x}\right]$ **85.** $e^{x+1}(x^2 + 1)x\left[1 + \dfrac{2x}{x^2 + 1} + \dfrac{1}{x}\right]$

or $e^{x+1}(x^3 + 3x^2 + x + 1)$ **86.** $e^x x^2 2^x\left(1 + \dfrac{2}{x} + \ln 2\right)$ **87.** (a) 800 g/cm^2 (b) 14 km   (c) -50 g/cm^2 per km   (d) 2 km

88. (a) $\approx \$181.969$ billion   (b) $\approx \$10.212$ billion/year   (c) ≈ 2004 (d) ≈ 2008

CHAPTER 5

Exercises 5.1, page 262

1. $k = 1, y = Ce^t$ **3.** $k = 1.7, y = Ce^{1.7t}$ **5.** $k = \frac{1}{2}, y = Ce^{0.5t}$ **7.** $k = \frac{1}{4}, y = Ce^{0.25t}$ **9.** $k = \frac{1}{12}, y = Ce^{\frac{1}{12}t}$ **11.** $y = e^{3t}$

13. $y = 2e^{2t}$ **15.** $y = 5e^{0.6t}$ **17.** $y = 12e^{\frac{1}{6}t}$ **19.** (a) $P(t) = 2e^{0.01t}$,   (b) 2 million,   (c) $P(4) = 2e^{0.01(4)} \approx 2.082$ million.

21. (a) $P(t) = P_0 e^{kt}$, $P(2) = 4P_0, k = \ln 2$,   (b) $P(.5) = 20000e^{\ln 2(.5)} \approx 28,284$ bacteria.   **23.** (a) $P(t) = 5$,

$P'(t) = .03P(t) = .03(5) = .15$ million per year   (b) $P'(t) = 400,000 = .4$ million per year. $P(t) = P'(t)/(.03) = .4/.03 \approx 13.33$ million.

(c) $P(t) = 4e^{0.03t}$ **25.** (a) 5000   (b) $P'(t) = .2P(t)$ (c) 3.5 h   (d) 6.9 h   **27.** .017   **29.** 22 yr   **31.** 27 million cells

33. 34.0 million   **35.** (a) $P(t) = 8e^{-0.021t}$ (b) 8 g   (c) .021   (d) 6.5 g   (e) .021 g/yr   (f) 5 g   (g) 4 g; 2 g; 1 g

37. (a) $f'(t) = -.6f(t)$ (b) 14.9 mg   (c) 1.2 h   **39.** 30.1 yr   **41.** 176 days   **43.** $f(t) = 8e^{-0.014t}$

45. (a) 8 g   (b) 3.5 h   (c) .6 g/h   (d) 8 h   **47.** 13,412 years   **49.** 58.3%   **51.** 10,900 years ago   **53.** a–D, b–G, c–E, d–B, e–H,

f–F, g–A, h–C   **55.** (a) $y - 10 = -5t$ or $y = -5t + 10$ (b) $P(t) = 10e^{-0.5t}$ (c) $T = 2$.

Exercises 5.2, page 269

1. (a) $5000 **(b)** 4% **(c)** $7459.12 **(d)** $A'(t) = .04A(t)$ **(e)** $298.36 per year **(f)** $7000 **3. (a)** $A(t) = 4000e^{0.035t}$
(b) $A'(t) = .035A(t)$ **(c)** $4290.03 **(d)** 6.4 yr **(e)** $175 per year **5.** $378 per year **7.** 15.3 yr **9.** 29.3% **11.** 17.3 years
13. 7.3% **15.** 9822 SFr **17.** 2021 **19.** $786.63 **21.** $7985.16 **23.** 15.7 yr **25.** a–B, b–D, c–G, d–A, e–F, f–E, g–H, h–C
27. (a) $200 **(b)** $8 per year **(c)** 4% **(d)** 30 yr **(e)** 30 yr **(f)** $A'(t)$ is a constant multiple of $A(t)$ since $A'(t) = rA(t)$.

29. **31.** .06

Exercises 5.3, page 277

1. 20%, 4% **3.** 30%, 30% **5.** 60%, 300% **7.** −25%, −10% **9.** 12.5% **11. (a)** $f(1) = 3.56, $f'(1) = .4$ dollars per year
(b) $f'(1)/f(1) \approx .11$ or 11% **(c)** $f(6) = 5.06, $f'(6) = .45$ dollars per year $f'(6)/f(6) \approx .09$ or 9% **13.** $p/(140 - p)$, $E(80) = \frac{4}{3}$,
elastic **15.** $2p^2/(116 - p^2)$, $E(6) = .9$, inelastic **17.** $p - 2$, elastic **19. (a)** inelastic **(b)** raised **21. (a)** elastic **(b)** increase
23. (a) 2 **(b)** yes **29. (a)** $p < 2$ **(b)** $p < 2$

Exercises 5.4, page 286

1. (a) $f'(x) = 10e^{-2x} > 0$, $f(x)$ increasing; $f''(x) = -20e^{-2x} < 0$, $f(x)$ concave down **(b)** As x becomes large, $e^{-2x} = \dfrac{1}{e^{2x}}$ approaches 0.

(c)

3. $y' = 2e^{-x} = 2 - (2 - 2e^{-x}) = 2 - y$
5. $y' = 30e^{-10x} = 30 - (30 - 30e^{-10x}) = 30 - 10y = 10(3 - y)$, $f(0) = 3(1 - 1) = 0$
7. 4.8 h **9. (a)** 2500 **(b)** 500 people/day **(c)** day 12 **(d)** day 6 and day 14
(e) at time 10 days **(f)** $f'(t) = .00004f(t)(10,000 - f(t))$ **(g)** 1000 people per day
11. (a) 47,510 **(b)** 15,000 people/day **(c)** $t \approx 2$ **(d)** on day 6.
(e) $f'(t) = .3(50,000 - f(t))$ **(f)** 7500 people per day

13. (a) **(b)** 30 units **(c)** 25 units per hour **(d)** 9 hr **(e)** 65.3 units after 2 hr **(f)** 4 hr

Chapter 5: Review Exercises, page 289

1. $29.92e^{-0.2x}$ **2.** $P'(t) = .0533P(t)$ **3.** $5488.12 **4.** $t \approx 11$ years **5.** .058 **6.** $t \approx 3850$ years old **7. (a)** $17e^{0.018t}$
(b) 20.4 million **(c)** 2011 **8.** 7.85% **9. (a)** $36,693 **(b)** The alternative investment is superior by $3859. **10.** $t \approx 547$ minutes
11. .02; 60,000 people per year; 5 million people **12.** 500,000; 400,000 bacteria per hour **13.** a–F, b–D, c–A, d–G, e–H, f–C, g–B, h–E
14. (a) 25 grams **(b)** 9 years **(c)** $t \approx 3$ years **(d)** −15 grams/year **(e)** 6 years **15.** 6% **16.** $55.35 per year **17.** 400%

18. $E(p) = \dfrac{2}{\dfrac{100}{p^2} - 1}$; $E(5) = \frac{2}{3} < 1$, inelastic **19.** 3%, decrease **20.** −75% **21.** increase **22.** $E(p) = pb$, $E\left(\dfrac{1}{b}\right) = 1$

23. $100(1 - e^{-0.083t})$ **24.** $f(t) = \dfrac{55}{1 + 46.98e^{-0.23t}}$ **25. (a)** 400° **(b)** decreasing at a rate of 100°/sec **(c)** 17 sec **(d)** 2 sec

26. $k = .05$, $P' = (.05)(15000) = 750$ bacteria per day.

CHAPTER 6

Exercises 6.1, page 299

1. $\frac{1}{2}x^2 + C$ **3.** $\frac{1}{3}e^{3x} + C$ **5.** $3x + C$ **7.** $x^4 + C$ **9.** $7x + C$ **11.** $\frac{x^2}{2c} + C$ **13.** $2\ln|x| + \frac{x^2}{4} + C$ **15.** $\frac{2}{5}x^{5/2} + C$

17. $\frac{1}{2}x^2 - \frac{2}{3}x^3 + \frac{1}{3}\ln|x| + C$ **19.** $-\frac{3}{2}e^{-2x} + C$ **21.** $ex + C$ **23.** $-e^{2x} - 2x + C$ **25.** $-\frac{5}{2}$ **27.** $\frac{1}{2}$ **29.** $-\frac{1}{5}$ **31.** -1 **33.** $\frac{1}{15}$

35. 3 **37.** $\frac{2}{5}t^{5/2} + C$ **39.** C **41.** $-\frac{5}{2}e^{-0.2x} + \frac{5}{2}$ **43.** $\frac{x^2}{2} + 3$ **45.** $\frac{2}{3}x^{3/2} + x - \frac{28}{3}$ **47.** $2\ln|x| + 2$

49. Testing all three functions reveals that (b) is the only one that works.

51.

53. $\frac{1}{4}$ **55.** (a) $-16t^2 + 96t + 256$ (b) 8 sec (c) 400 ft
57. $P(t) = 60t + t^2 - \frac{1}{12}t^3$ **59.** $20 - 25e^{-0.4t}\,°C$
61. $-95 + 1.3x + .03x^2 - .0006x^3$ **63.** $5875(e^{0.016t} - 1)$
65. $C(x) = 25x^2 + 1000x + 10{,}000$
67. $F(x) = \frac{1}{2}e^{2x} - e^{-x} + \frac{1}{6}x^3$

Exercises 6.2, page 307

1. $F(x) = x^2 - \frac{3}{4}x, \frac{1}{4}$ **3.** $F(t) = 2t^{3/2} + 2t^2, 44$ **5.** $F(x) = \frac{3}{x}, -\frac{3}{2}$ **7.** $F(x) = -x^{-5} + x^{-2}, \frac{7}{32}$ **9.** $F(t) = e^{3t} + \frac{t^2}{2}, \frac{1}{2} - \frac{1}{e^3}$

11. $F(x) = 2\ln|x|, 2\ln 2 = \ln 4$ **13.** $F(x) = -e^{-x} - \frac{1}{1.5}e^{-1.5x}, \frac{5}{3} - e^{-1} - \frac{2}{3}e^{-1.5}$ **15.** $\int_0^4 f(x)\,dx = \int_0^1 f(x)\,dx + \int_1^4 f(x)\,dx = 8.5$

17. $2\int_1^3 f(x)\,dx - 3\int_1^3 g(x)\,dx = 9$ **19.** $\int_1^2 (-2x^3 + 4x^2 + 21)\,dx = \frac{137}{6}$ **21.** $\int_{-1}^1 (x^3 + x^2)\,dx = \frac{2}{3}$

23. $f(3) - f(1) = \int_1^3 f'(x)\,dx = -2$ **25.** $f(1) - f(-1) = \int_{-1}^1 f'(t)\,dt = \frac{e^2 - e^{-2}}{2}$ **27.** $\int_0^2 f(x)\,dx = \int_0^1 f(x)\,dx + \int_1^2 f(x)\,dx = \frac{5}{2}$

29. $\int_{-1}^0 (1+t)\,dt + \int_0^1 (1-t)\,dt = 1$ **31.** $s(4) - s(2) = \int_2^4 (-32t)\,dt = -192$. The rock moved 192 ft downward in the interval $2 \le t \le 4$.

33. (a) $s(3) - s(1) = \int_1^3 v(t)\,dt = 22$. (b) In the interval of time $1 \le t \le 3$, the ball moved 22 ft in the upward direction. Thus, at time $t = 3$, the ball is 22 ft higher than its position at $t = 1$. (c) $s(5) - s(1) = \int_1^5 v(t)\,dt = -84$. In the interval of time $1 \le t \le 5$, the ball moved 84 ft in the downward direction. Thus, at time $t = 5$, the ball is 84 ft lower than its position at $t = 1$.

35. (a) $C(3) - C(1) = \int_1^3 C'(x)\,dx = \int_1^3 (.1x^2 - x + 12)\,dx = 20.87$ dollars. (b) $C(3) = C(1) + (C(3) - C(1)) = 15 + 20.87 = 35.87$

dollars. **37.** $\int_0^{10} (700e^{0.07t} + 1000)\,dt = 20{,}137.5$ dollars

39. $P(10) - P(0) = \int_0^{10} \left(\frac{7}{300}e^{t/25} - \frac{e^{t/16}}{80}\right)\,dt = \frac{1}{60}(-23 + 35e^{2/5} - 12e^{5/8}) \approx .11325$ million or 113,250 people.

(b) $P(40) - P(10) = \int_{10}^{40} \left(\frac{7}{300}e^{t/25} - \frac{e^{t/16}}{80}\right)\,dt = \frac{7}{12}e^{2/5}(-1 + e^{6/5}) + \frac{1}{5}(e^{5/8} - e^{5/2}) \approx -.043812$ million or $-43{,}812$ people. Between 2000 and 2010, the population increased by 113,250 people. Between 2010 and 2040, the population decreased by 43,812 people, due to emigration.
41. $P(t) = 337.023 - 137.023e^{0.03t}$ thousand dollars **43.** $-\int_0^2 (t + \frac{1}{2})\,dt = -3$ grams per minute. 3 grams of salt were eliminated the first two minutes.

Exercises 6.3, page 317

1. (a) $3 \times 2 = 6$, (b) $\int_1^4 2\,dx = 6$ **3.** (a) $\frac{1}{2} \times 2 \times 2 = 2$, (b) $\int_{-2}^0 -x\,dx = 2$ **5.** (a) $\frac{1}{2} + \frac{1}{2} = 1$,

(b) $\int_0^1 (1 - x)\,dx + \int_1^2 (x - 1)\,dx = 1$ **7.** $\int_1^2 \frac{1}{x}\,dx$ **9.** $\int_1^2 \ln x\,dx$ **11.** $\int_1^3 (x + \frac{1}{x})\,dx$ **13.** $\ln 2$ **15.** $4 + \ln 3$

17.

19. 10 **21.** $-\frac{5}{2} + 4e^{1/2}$ **23.** $\frac{33}{5}$ **25.** $\frac{1}{4}b^4 = 4, b = 2$ **27.** .5; .25, .75, 1.25, 1.75
29. .6; 1.3, 1.9, 2.5, 3.1, 3.7 **31.** 8.625 **33.** 15.12 **35.** .077278 **37.** 40 **39.** 15
41. 5.625; 4.5 **43.** 1.61321; error $= .04241$
45. $A \approx (20)(106) + (40)(101) + (40)(100) + (40)(113) + (20)(113) = 16{,}940$ sq. ft.
49. 1.494 **51.** 9.6

Exercises 6.4, page 329

1. $\int_1^2 f(x)\,dx + \int_3^4 -f(x)\,dx$ **3.**

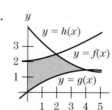
$y = h(x)$
$y = f(x)$
$y = g(x)$

5. Positive   **7.** 4   **9.** $\frac{22}{3}$ **11.** $3\ln 3 - 2$ **13.** $\frac{64}{3}$ **15.** $\frac{52}{3}$
17. $e^2 - e - \frac{1}{2}$ **19.** $\frac{1}{6}$ **21.** $\frac{32}{3}$ **23.** $\frac{1}{2}$ **25.** $\frac{1}{24}$ **27. (a)** $\frac{9}{2}$ **(b)** $\frac{19}{3}$
(c) $\frac{79}{6}$ **29.** $\frac{3}{2}$ **31. (a)** 30 ft   **(b)**

$y = v(t)$

33. (a) $1185.75   **(b)** The area under the marginal cost curve from $x = 2$ to $x = 8$.   **35.** The increase in profits resulting from increasing the production level from 44 to 48 units   **37. (a)** $368/15 \approx 24.5$ **(b)** The amount the temperature falls during the first 2 h
39. 2088 million m³   **41.** $7500   **43.** A is the difference between the two heights after 10 seconds.   **45. (a)** $\frac{15}{2}$ ft
(b) $\int_0^{1/2} v(t)\,dt - \int_{1/2}^1 v(t)\,dt + \int_1^3 v(t)\,dt = \frac{91}{12}$ ft   **47.** $A \approx 3.9100$ **49.** $A \approx 2.2676$

Exercises 6.5, page 338

1. 3   **3.** $50(1 - e^{-2})$ **5.** $\frac{3}{4}\ln 3$ **7.** 55°   **9.** ≈ 82 g   **11.** $20   **13.** $404.72   **15.** $200   **17.** $25   **19.** Intersection (100, 10), consumers' surplus = $100, producers' surplus = $250   **21.** $3236.68   **23.** $75,426   **25.** 13.35 years   **27.** $\frac{26}{3}\pi$ **29.** $\frac{32}{3}\pi$ **31.** $\frac{31\pi}{5}$ **33.** 8π

35. $\frac{13\pi}{3}$ **37.** $n = 4, b = 10, f(x) = x^3$ **39.** $n = 3, b = 7, f(x) = x + e^x$ **41.** The sum is approximated by $\int_0^3 (3 - x)^2\,dx = 9$.
43. (a) $\frac{1000}{3r}(e^{3r} - 1)$ **(b)** 4.5%   **45. (a)** $\frac{1000}{r}(e^{6r} - 1)$ **(b)** 5%

Chapter 6: Review Exercises, page 344

1. $9x + C$ **2.** $\frac{1}{3}x^3 - \frac{3}{2}x^2 + 2x + C$ **3.** $\frac{2}{3}(x + 1)^{3/2} + C$ **4.** $2\ln|x + 4| + C$ **5.** $2(\frac{1}{4}x^4 + x^3 - x) + C$ **6.** $\frac{5}{6}(x + 3)^{6/5} + C$
7. $-2e^{-x/2} + C$ **8.** $10\sqrt{x - 7} + C$ **9.** $\frac{3}{5}x^5 - x^4 + C$ **10.** $\frac{1}{16}(2x + 3)^8 + C$ **11.** $-\frac{2}{3}(4 - x)^{3/2} + C$ **12.** $5\ln|x| - \frac{x^2}{10} + C$
13. $\frac{8}{3}$ **14.** $\frac{3}{64}$ **15.** $\frac{14}{3}\sqrt{2}$ **16.** $2\ln\frac{16}{5}$ **17.** $\frac{15}{16}$ **18.** $\frac{104}{9}$ **19.** $\frac{3}{4}$ **20.** $3(e - 1)$ **21.** $\frac{1}{3}\ln 4$ **22.** $\frac{1}{2}(e^6 - e^{-6})$ **23.** $\frac{1}{2}$ **24.** $\frac{7}{6}$
25. $\frac{80}{81}$ **26.** $2 + e - \frac{3}{e}$ **27.** $\frac{5}{32}$ **28.** $\frac{76}{3}$ **29.** $\frac{1}{3}$ **30.** 4   **31.** $\frac{1}{2}$ **32.** $\approx .370$ **33.** $\frac{28}{3}$ **34.** $\frac{39}{16} - \ln 4$ **35.** $\frac{e}{2} - 1$ **36.** $\frac{253}{12}$ **37.** 8
38. 3   **39.** $\frac{1}{3}(x - 5)^3 - 7$ **40.** $\frac{6}{5} - \frac{1}{5}e^{-5x}$ **41. (a)** $2t^2 + C$ **(b)** Ce^{4t} **(c)** $\frac{1}{4}e^{4t} + C$ **43.** $.02x^2 + 150x + 500$ dollars
44. Loss of $3000   **45.** The total quantity of drug (in cubic centimeters) injected during the first 4 min   **46. (a)** 20.4 meters
(b)

$v(t)$

47. 25   **48.** 17   **49.** .68571; .69315   **50.** ≈ 3.17333 **51.** $433.33   **52.** $\approx 3688.69 **53.** 15
54. $3(1 - e^{-1})$ **55.** .26; 1.96   **56.** $\frac{8\pi}{15}$ **57. (a)** $f(t) = Q - \frac{Q}{A}t$ **(b)** $\frac{Q}{2}$ **58. (a)** $Q - \frac{rt^2}{2}$ **(b)** $r = \frac{2Q}{A^2}$,
(c) $\frac{2}{3}Q$ **59. (a)** The area under the curve $y = \frac{1}{1 + t^2}$ from $t = 0$ to $t = 3$.   **(b)** $\frac{1}{1 + x^2}$
60. (a) $h(0)$ is the area under one quarter of the unit circle. $h(1)$ is the area under one half of the unit circle.
(b) $h'(x) = \sqrt{1 - x^2}$ **61.** $5000\int_0^3 e^{-0.1t}\,dt \approx 13,000$ **62.** $e - 1$ **63.** $\frac{15}{4}$ **64.** 4   **65.** True
66. (a) $(20 - 4t_1)\Delta t$ **(b)** Let $R(t) =$ amount of water added up to time t. Then, $R'(t) = r(t)$, and so, $\int_0^5 r(t)\,dt = R(5) - R(0) =$ total amount of water added from $t = 0$ to $t = 5$.   **67.** 65,687 km³
68. $4,708.71   **69.** $f(x) = x^3 - x^2 + x$ **70.** $a = \sqrt[3]{6}$

CHAPTER 7

Exercises 7.1, page 352

1. $f(5, 0) = 25$, $f(5, -2) = 51$, $f(a, b) = a^2 - 3ab - b^2$ **3.** $g(2, 3, 4) = -2$, $g(7, 46, 44) = \frac{7}{2}$
5. $f(2 + h, 3) = 3h + 6$, $f(2, 3) = 6$, $f(2 + h, 3) - f(2, 3) = 3h$ **7.** $C(x, y, z) = 6xy + 10xz + 10yz$ **9.** $f(8, 1) = 40$, $f(1, 27) = 180$,
$f(8, 27) = 360$ **11.** $\approx $50. $50 invested at 5% continuously compounded interest will yield $100 in 13.8 years   **13. (a)** $1875   **(b)** $2250; yes
15.

$C = 0$
$C = 1$
$C = 2$

17.

$(0, 0)$

19. $f(x, y) = y - 3x$ **21.** They correspond to the points having the same altitude above sea level.   **23. (d)**   **25. (c)**

Exercises 7.2, page 361

1. $5y, 5x$ **3.** $4xe^y, 2x^2 e^y$ **5.** $\dfrac{1}{y} - \dfrac{y}{x^2}; \dfrac{-x}{y^2} + \dfrac{1}{x}$ **7.** $4(2x - y + 5), -2(2x - y + 5)$ **9.** $e^{x^2 y^2}(1 + 2x^2 y^2); 2x^3 y e^{x^2 y^2}$

11. $\dfrac{2y}{(x + y)^2}, -\dfrac{2x}{(x + y)^2}$ **13.** $\dfrac{3}{2}\sqrt{\dfrac{K}{L}}$ **15.** $\dfrac{2xy}{z}, \dfrac{x^2}{z}, -\dfrac{1 + x^2 y}{z^2}$ **17.** $z e^{yz}, xz^2 e^{yz}, x(yz + 1)e^{yz}$ **19.** $1, 3$ **21.** $\dfrac{\partial f}{\partial y} = 2xy, \dfrac{\partial f}{\partial y}(2, -1) = -4.$

If x is kept constant at 2 and y is allowed to vary near -1, then $f(x, y)$ changes at a rate -4 times the change in y.

23. $\dfrac{\partial f}{\partial x} = 3x^2 y + 2y^2, \dfrac{\partial^2 f}{\partial x^2} = 6xy, \dfrac{\partial f}{\partial y} = x^3 + 4xy, \dfrac{\partial^2 f}{\partial y^2} = 4x, \dfrac{\partial^2 f}{\partial y\, \partial x} = \dfrac{\partial^2 f}{\partial x\, \partial y} = 3x^2 + 4y$

25. (a) Marginal productivity of labor $= 480$; of capital $= 40$ **(b)** $480h$ **(c)** Production decreases by 240 units.
27. If the price of a bus ride increases and the price of a train ticket remains constant, fewer people will ride the bus. An increase in train ticket prices coupled with constant bus fare should cause more people to ride the bus.   **29.** If the average price of audio files increases and the average price of an MP3 player remains constant, people will purchase fewer audio files. An increase in average MP3 player prices coupled with constant audio files prices should cause a decline in the number of MP3 players purchased.

31. $\dfrac{\partial V}{\partial P}(20, 300) = -.06, \dfrac{\partial V}{\partial T}(20, 300) = .004$ **33.** $\dfrac{\partial f}{\partial r} > 0, \dfrac{\partial f}{\partial m} > 0, \dfrac{\partial f}{\partial p} < 0$ **35.** $\dfrac{\partial^2 f}{\partial x^2} = -\dfrac{45}{4}x^{-5/4}y^{1/4}$; marginal productivity of labor is decreasing.

Exercises 7.3, page 368

1. $(-2, 1)$ **3.** $(26, 11)$ **5.** $(-1/5, 2/5)$ **7.** $(1, -3), (-1, -3)$ **9.** $(1/2, 0)$ **11.** $(0, 1/2), (-1, 3/2), (-1/2, 3/4)$
13. $(\sqrt{5}, 1), (\sqrt{5}, -1), (-\sqrt{5}, 1), (-\sqrt{5}, -1)$ **15.** $(-1, 3/2), (1, -3/2)$ **17.** $(\frac{1}{3}, \frac{4}{3})$ **19.** Relative minimum; neither relative maximum nor relative minimum.   **21.** Relative maximum; neither relative maximum nor relative minimum; relative maximum.   **23.** Neither relative maximum nor relative minimum.   **25.** $(-5/9, 1/9)$ relative maximum   **27.** $(-3/5, 1/5)$ neither a maximum nor minimum   **29.** Relative minimum at $(-1, -1)$ and $(1, 1)$, neither a maximum nor a minimum at $(0, 0)$ **31.** $(-1/3, 1/3)$ neither a maximum nor a minimum   **33.** Relative maximum at $(-3/7, 1/7)$ **35.** Relative maximum at $(0, 0)$, neither a maximum nor a minimum at $(-2, 4)$ and $(2, -4)$ **37.** $(0, 0)$ Relative minimum   **39.** $(-1, -4)$ relative max   **41.** $(0, -1)$ relative min   **43.** $(-1, 2)$ relative max; $(1, 2)$ neither max nor min.   **45.** $(\frac{1}{4}, 2)$ min; $(\frac{1}{4}, -2)$ neither max nor min.   **47.** $(\frac{1}{2}, \frac{1}{6}, \frac{1}{2})$ **49.** 14 in. $\times$ 14 in. $\times$ 28 in.   **51.** $x = 120, y = 80$

Exercises 7.4, page 376

1. 58 at $x = 6, y = 2, \lambda = 12$ **3.** 13 at $x = 8, y = -3, \lambda = 13$ **5.** $x = \frac{1}{2}, y = 2$ **7.** $x = 2/3, y = 1/3$ **9.** $x = -1/4, y = 1/4$
11. $x = -2/3, y = 3/2$ **13.** $x = -2, y = -1$ **15.** $x = 3/5, y = 1/5, z = 1/5$ **17.** $5, 5$ **19.** Base 10 in., height 5 in.
21. $F(x, y, \lambda) = 4xy + \lambda(1 - x^2 - y^2); \sqrt{2} \times \sqrt{2}$ **23.** $F(x, y, \lambda) = 3x + 4y + \lambda(18{,}000 - 9x^2 - 4y^2); x = 20, y = 60$
25. (a) $F(x, y, \lambda) = 96x + 162y + \lambda(3456 - 64x^{3/4}y^{1/4}); x = 81, y = 16$ **(b)** $\lambda = 3$ **27.** $x = 12, y = 2, z = 4$ **29.** $x = 2, y = 3, z = 1$
31. $F(x, y, z, \lambda) = 3xy + 2xz + 2yz + \lambda(12 - xyz); x = 2, y = 2, z = 3$ **33.** $F(x, y, z, \lambda) = xy + 2xz + 2yz + \lambda(32 - xyz);$
$x = y = 4, z = 2t$

Exercises 7.5, page 383

1. $E = 6.7$ **3.** $E = (2A + B - 6)^2 + (5A + B - 10)^2 + (9A + B - 15)^2$ **5.** $y = 4.5x - 3$ **7.** $y = -2x + 11.5$ **9.** $y = -1.4x + 8.5$
11. (a) $y = 272.8x + 5712.7$ **(b)** 10,077.5   **(c)** 2023   **13. (a)** $y = .157x + 4.986$ **(b)** \$6.24 per hour   **(c)** 2032
15. (a) $y = -4.24x + 22.01$ **(b)** $y = 8.442$ degrees Celsius

Exercises 7.6, page 388

1. $e^2 - 2e + 1$ **3.** $2 - e^2/2 + e^{-2}/2$ **5.** $309\frac{3}{8}$ **7.** $\frac{5}{3}$ **9.** $\frac{38}{3}$ **11.** $e^{-5} + e^{-2} - e^{-3} - e^{-4}$ **13.** $9\frac{1}{3}$

Chapter 7: Review Exercises, page 392

1. $2, \frac{5}{6}, 0$ **2.** $f(-1, 0, 1) = 1, f(1, 3, 3) = e, f(5, -2, 2) = \frac{25}{e}$ **3.** ≈ 19.94. Ten dollars increases to 20 dollars in 11.5 years.

4. $f(1, 2, 3) = 8$ **5.** $6x + y, x + 10y$ **6.** $\dfrac{\partial f}{\partial x} = 3, \dfrac{\partial f}{\partial y} = -2y^3$ **7.** $\dfrac{1}{y}e^{x/y}, -\dfrac{x}{y^2}e^{x/y}$ **8.** $\dfrac{\partial f}{\partial x} = \dfrac{-2y}{(x - 2y)^2}, \dfrac{\partial f}{\partial y} = \dfrac{2x}{(x - 2y)^2}$

9. $3x^2, -z^2, -2yz$ **10.** $\dfrac{\partial f}{\partial x} = y - \lambda, \dfrac{\partial f}{\partial y} = x - \lambda, \dfrac{\partial f}{\partial \lambda} = 5 - x - y$ **11.** $6, 1$ **12.** $\dfrac{\partial f}{\partial y} = z; \dfrac{\partial f}{\partial y}(2, 3, 4) = 4$

13. $20x^3 - 12xy, 6y^2, -6x^2, -6x^2$ **14.** $\dfrac{\partial^2 f}{\partial x^2} = 12x + 2y, \dfrac{\partial^2 f}{\partial y^2} = -2, \dfrac{\partial^2 f}{\partial x\, \partial y} = 2x, \dfrac{\partial^2 f}{\partial x\, \partial y}(1, 2) = 2, \dfrac{\partial^2 f}{\partial x^2}(1, 2) = 16$

15. $-201, 5.5$. At the level $p = 25, t = 10{,}000$, an increase in price of \$1 will result in a loss in sales of approximately 201 calculators, and an increase in advertising of \$1 will result in the sale of approximately 5.5 additional calculators.   **16.** Increases with increased unemployment and decreases with increased social services and police force size.   **17.** $(3, 2)$ **18.** $(2, -1)$ **19.** $(0, 1), (-2, 1)$

20. $(-11, 2), (5, -2)$ **21.** Relative minimum at $(2, 3)$ **22.** Relative minimum at $(1, 3)$.   **23.** Relative minimum at $(1, 4)$; neither max nor min at $(-1, 4)$.   **24.** Minimum value at $(3, -1, 0)$.   **25.** $20; x = 3, y = -1$ **26.** $x = 7, y = 3$ **27.** $x = \frac{1}{2}, y = \frac{3}{2}, z = 2$ **28.** $x = y = z = 10$ **29.** $F(x, y, \lambda) = xy + \lambda(40 - 2x - y); x = 10, y = 20$ **30.** $x = 10.25$ ft, $y = 20.5$ ft   **31.** $y = \frac{5}{2}x - \frac{5}{3}$ **32.** $y = \frac{3}{2}x - \frac{1}{2}$ **33.** $y = -2x + 1$ **34.** $\frac{32}{3}$ **35.** 5160   **36.** 80   **37.** 40   **38.**

CHAPTER 8

Exercises 8.1, page 397

1. $\frac{\pi}{6}, \frac{2\pi}{3}, \frac{7\pi}{4}$ **3.** $\frac{5\pi}{2}$ radians, $-\frac{7\pi}{6}$ radians, $-\frac{\pi}{2}$ radians   **5.** 4π radians   **7.** $\frac{7\pi}{2}$ radians   **9.** -3π radians   **11.** $\frac{2\pi}{3}$ radians

13.

15.

17.

Exercises 8.2, page 402

1. $\sin t = \frac{1}{2}, \cos t = \frac{\sqrt{3}}{2}$ **3.** $\cos t = \frac{2}{\sqrt{13}}, \cos t = \frac{3}{\sqrt{13}}$ **5.** $\sin t = \frac{12}{13}, \cos t = \frac{5}{13}$ **7.** $\sin t = \frac{1}{\sqrt{5}}, \cos t = -\frac{2}{\sqrt{5}}$ **9.** $\sin t = \frac{\sqrt{2}}{2},$

$\cos t = -\frac{\sqrt{2}}{2}$ **11.** $\sin t = -.8, \cos t = -.6$ **13.** $.4$ **15.** 3.6 **17.** 10.9 **19.** $b = 1.3, c = 2.7$ **21.** $\frac{\pi}{6}$ **23.** $\frac{3\pi}{4}$

25. $\frac{5\pi}{8}$ **27.** $\frac{\pi}{4}$ **29.** $\frac{\pi}{3}$ **31.** $-\frac{\pi}{6}$ **33.** $\frac{\pi}{4}$ **35.** Here, $\cos t$ decreases from 1 to -1.   **37.** $0, 0, 1, -1$ **39.** $.2, .98, .98, -.2$

41. (a)   **(b)** $46°$ **(f)** October 27 and April 27
(c) $45°$ coldest, $73°$ warmest
(d) January 26
(e) July 27

Exercises 8.3, page 409

1. $4\cos 4t$ **3.** $4\cos t$ **5.** $-6\sin 3t$ **7.** $1 - \pi \sin \pi t$ **9.** $-\cos(\pi - t)$ **11.** $-3\cos^2 t \sin t$ **13.** $\frac{\cos\sqrt{x-1}}{2\sqrt{x-1}}$ **15.** $\frac{\cos(x-1)}{2\sqrt{\sin(x-1)}}$

17. $-8\sin t(1 + \cos t)^7$ **19.** $-6x^2 \cos x^3 \sin x^3$ **21.** $e^x(\sin x + \cos x)$ **23.** $2\cos(2x)\cos(3x) - 3\sin(2x)\sin(3x)$ **25.** $\cos^{-2} t$

27. $-\dfrac{\sin t}{\cos t}$ **29.** $\dfrac{\cos(\ln t)}{t}$ **31.** -3 **33.** $y = 2$ **35.** $\frac{1}{2}\sin 2x + C$ **37.** $-\frac{7}{2}\sin\dfrac{x}{7} + C$ **39.** $\sin x + \cos x + C$

41. $\cos x + \sin 3x + C$ **43.** $-\frac{1}{4}\cos(4x + 1) + C$ **45.** 1 **47. (a)** $\max = 120$ at $0, \dfrac{\pi}{3}$; $\min = 80$ at $\dfrac{\pi}{6}, \dfrac{\pi}{2}$ **(b)** 57 **49.** 0

51. (a) $69°$ **(b)** increasing $1.6°/$wk   **(c)** weeks 6 and 44   **(d)** weeks 28 and 48   **(e)** week 25, week 51   **(f)** week 12, week 38

Exercises 8.4, page 414

1. $\sec t = \dfrac{\text{hypotenuse}}{\text{adjacent}}$ **3.** $\tan t = \frac{5}{12}$, $\sec t = \frac{13}{12}$ **5.** $\tan t = -\frac{1}{2}$, $\sec t = -\dfrac{\sqrt{5}}{2}$ **7.** $\tan t = -1$, $\sec t = -\sqrt{2}$ **9.** $\tan t = \frac{4}{3}$, $\sec t = -\frac{5}{3}$

11. $75\tan(.7) \approx 63$ ft   **13.** $\tan t\sec t$ **15.** $-\csc^2 t$ **17.** $4\sec^2(4t)$ **19.** $-3\sec^2(\pi - x)$ **21.** $4(2x + 1)\ \sec^2(x^2 + x + 3)$

23. $\dfrac{\sec^2 \sqrt{x}}{2\sqrt{x}}$ **25.** $\tan x + x\sec^2 x$ **27.** $2\tan x\sec^2 x$ **29.** $6[1 + \tan(2t)]^2\sec^2(2t)$ **31.** $\sec t$ **33. (a)** $y - 1 = 2\left(x - \dfrac{\pi}{4}\right)$

(b)

35. $\frac{1}{3}\tan 3x + C$ **37.** 2 **39.** $\tan x + C$

Chapter 8: Review Exercises, page 418

1. $\dfrac{3\pi}{2}$ **2.** $t = -\dfrac{7\pi}{2}$ **3.** $-\dfrac{3\pi}{4}$ **4.**

   5.

   6.

7. $\frac{4}{5}, \frac{3}{5}, \frac{4}{3}$ **8.** $\sin t = .8$; $\cos t = -.6$; $\tan t = -\frac{4}{3}$ **9.** $-.8, -.6, \frac{4}{3}$ **10.** $\sin t = -\frac{4}{5}$; $\cos t = \frac{3}{5}$; $\tan t = -\frac{4}{3}$ **11.** $\pm\dfrac{2\sqrt{6}}{5}$

12. $\sin t = \pm\dfrac{\sqrt{5}}{3}$ **13.** $\frac{\pi}{4}, \frac{5\pi}{4}, -\frac{3\pi}{4}, -\frac{7\pi}{4}$ **14.** $\frac{3\pi}{4}, \frac{7\pi}{4}, -\frac{\pi}{4}, -\frac{5\pi}{4}$ **15.** Negative   **16.** Positive   **17.** 16.3 ft   **18.** $t = 60\tan 53° \approx 79.62$ feet

19. $3\cos t$ **20.** $3\cos 3t$ **21.** $(\cos\sqrt{t})\cdot\frac{1}{2}t^{-1/2}$ **22.** $-3t^2\sin t^3$ **23.** $x^3\cos x + 3x^2\sin x$ **24.** $5\sin 5x\sin 2x - 2\cos 5x\cos 2x$

25. $-\dfrac{2\sin(3x)\sin(2x) + 3\cos(2x)\cos(3x)}{\sin^2(3x)}$ **26.** $\dfrac{-x^3\sin x - 3x^2(\cos x - 1)}{x^6}$ **27.** $-12\cos^2(4x)\sin(4x)$ **28.** $6\tan^2 2x\sec^2 2x$

29. $[\sec^2(x^4 + x^2)](4x^3 + 2x)$ **30.** $-2e^{-2x}\sec^2(e^{-2x})$ **31.** $\cos(\tan x)\sec^2 x$ **32.** $\sec^2(\sin x)\cos x$ **33.** $\sin x\sec^2 x + \sin x$

34. $\dfrac{\cos x}{x} - \sin x(\ln x)$ **35.** $\cot x$ **36.** $-\tan x$ **37.** $4e^{3x}\sin^3 x\cos x + 3e^{3x}\sin^4 x$ **38.** $12e^{3x}\cos(e^{3x})\sin^3(e^{3x})$

39. $\dfrac{\tan(3t)\cos t - 3\sin t\sec^2(3t)}{\tan^2(3t)}$ **40.** $\dfrac{2\cos t\sec^2 2t + \sin t\tan 2t}{\cos^2 t}$ **41.** $e^{\tan t}\sec^2 t$ **42.** $e^t(\sec^2 t + \tan t)$ **43.** $2(\cos^2 t - \sin^2 t)$

44. $y'' = -12\sin 2t - 4\cos 2t = -4y$ **45.** $\dfrac{\partial f}{\partial s} = \cos s\cos(2t)$, $\dfrac{\partial f}{\partial t} = -2\sin s\sin(2t)$ **46.** $\dfrac{\partial z}{\partial w} = t\cos wt$; $\dfrac{\partial z}{\partial t} = w\cos wt$

47. $\dfrac{\partial f}{\partial s} = t^2\cos(st)$, $\dfrac{\partial f}{\partial t} = \sin(st) + st\cos(st)$ **48.** $\dfrac{d}{dt}\sin(s + t) = \cos(s + t)$, $\dfrac{d}{dt}(\sin s\cos t + \cos s\sin t) = \cos s\cos t - \sin s\sin t$. Thus,

$\cos(s + t) = \cos s\cos t - \sin s\sin t$ **49.** $y - 1 = 2\left(t - \dfrac{\pi}{4}\right)$ **50.**

   51.

52. $\frac{1}{3} + \pi$ **53.** 4   **54.** 3   **55.** $\frac{\pi^2}{2} - 2$ **56. (a)** $V(0) = 2.95$, $\left(\frac{1}{320}\right) = 3$, $V\left(\frac{1}{160}\right) = 3.05$, $V\left(\frac{1}{80}\right) = 2.95$ **(b)** $V\left(\frac{1}{160}\right) = 3.05$

57. (a) $V'(t) = 8\pi \cos\left(160\pi t - \frac{\pi}{2}\right)$ **(b)** 8π L/min   **(c)** 16 L/min   **59.** $\cos(\pi - x) + C$ **60.** $\sin 3x + \cos 2x + C$ **61.** 0

62. $-\frac{1}{2}\sin(6 - 2x) + C$ **63.** $\frac{\pi^2}{2}$ **64.** 0   **65.** $2\tan\frac{x}{2} + C$ **66.** $\tan 2x + C$ **67.** $\sqrt{2} - 1$ **68.** $2 - \sqrt{2}$ **69.** $\sqrt{2}$ **70.** 1   **71.** 1

72. $\frac{\pi}{2}$ **73.** $1000 + \frac{400}{3\pi}$ **74.** $-\frac{2}{\pi}$ **75.** $\tan x - x + C$ **76.** $\frac{1}{3}\tan 3x - x + C$ **77.** $\tan x + C$ **78.** $x + \tan x + C$ **79.** $1 - \frac{\pi}{4}$

80. 2

CHAPTER 9

Exercises 9.1, page 426

1. $\frac{1}{6}(x^2 + 4)^6 + C$ **3.** $2\sqrt{x^2 + x + 3} + C$ **5.** $e^{(x^3 - 1)} + C$ **7.** $-\frac{1}{3}(4 - x^2)^{3/2} + C$ **9.** $(2x + 1)^{1/2} + C$ **11.** $\frac{1}{2}e^{x^2} + C$

13. $\frac{1}{2}(\ln 2x)^2 + C$ **15.** $\frac{1}{5}\ln|x^5 + 1| + C$ **17.** $-(2 - 12x + 2x^2)^{-1} + C$ **19.** $\frac{1}{4}(\ln x)^2 + C = (\ln\sqrt{x})^2 + C$ **21.** $\frac{1}{3}\ln|x^3 - 3x^2 + 1| + C$

23. $-4e^{-x^2} + C$ **25.** $\frac{1}{2}\ln|\ln x^2| + C$ **27.** $-\frac{1}{10}(x^2 - 6x)^5 + C$ **29.** $\frac{1}{6}(1 + e^x)^6 + C$ **31.** $\frac{1}{2}\ln(1 + 2e^x) + C$ **33.** $\ln|1 - e^{-x}| + C$

35. $-\ln(e^{-x} + 1) + C$ **37.** $f(x) = (x^2 + 9)^{1/2} + 3$ **39.** $2e^{\sqrt{x+5}} + C$ **41.** $\frac{1}{2}\tan x^2 + C$ **43.** $\frac{1}{2}(\sin x)^2 + C$ **45.** $2\sin\sqrt{x} + C$

47. $-\frac{1}{4}\cos^4 x + C$ **49.** $-\frac{2}{3}\sqrt{2 - \sin 3x} + C$ **51.** $\ln|\sin x - \cos x| + C$ **53.** $\frac{1}{2}(x^2 + 5)^2 + C = \frac{1}{2}x^4 + 5x^2 + \frac{25}{2} + C; \frac{1}{2}x^4 + 5x^2 + C_1$

Exercises 9.2, page 430

1. $\frac{1}{5}xe^{5x} - \frac{1}{25}e^{5x} + C$ **3.** $\frac{x}{5}(x + 7)^5 - \frac{1}{30}(x + 7)^6 + C$ **5.** $-xe^{-x} - e^{-x} + C$ **7.** $2x(x + 1)^{1/2} - \frac{4}{3}(x + 1)^{3/2} + C$

9. $(1 - 3x)\left(\frac{1}{2}e^{2x}\right) + \frac{3}{4}e^{2x} + C$ **11.** $-2xe^{-3x} - \frac{2}{3}e^{-3x} + C$ **13.** $\frac{2}{3}x(x + 1)^{3/2} - \frac{4}{15}(x + 1)^{5/2} + C$ **15.** $\frac{1}{3}x^{3/2}\ln x - \frac{2}{9}x^{3/2} + C$

17. $x\sin x + \cos x + C$ **19.** $\frac{x^2}{2}\ln 5x - \frac{1}{4}x^2 + C$ **21.** $4x\ln x - 4x + C$ **23.** $-e^{-x}(x^2 + 2x + 2) + C$

25. $\frac{1}{5}x(x + 5)^5 - \frac{1}{30}(x + 5)^6 + C$ **27.** $\frac{1}{10}(x^2 + 5)^5 + C$ **29.** $3(3x + 1)e^{x/3} - 27e^{x/3} + C$ **31.** $\frac{1}{2}\tan(x^2 + 1) + C$

33. $\frac{1}{2}xe^{2x} - \frac{1}{4}e^{2x} + \frac{1}{3}x^3 + C$ **35.** $\frac{1}{2}e^{x^2} - x^2 + C$ **37.** $2x\sqrt{x + 9} - \frac{4}{3}(x + 9)^{3/2} + 38$ **39.** $\frac{e^x}{x + 1} + C$

Exercises 9.3, page 434

1. $\frac{1}{15}$ **3.** 312   **5.** $\frac{8}{3}$ **7.** $\frac{64}{3}$ **9.** 0   **11.** $\frac{1}{2}\ln\left(\frac{4}{3}\right)$ **13.** $\frac{1}{3}(e^{27} - e)$ **15.** $\frac{1}{2}$ **17.** 0   **19.** $\frac{1}{\pi}$ **21.** $\frac{\pi}{2}$ **23.** $\frac{9\pi}{2}$ **25.** $\frac{16}{3}$

Exercises 9.4, page 441

1. $\Delta x = .4$; 3, 3.4, 3.8, 4.2, 4.6, 5   **3.** $\Delta x = .5$; $-.75, -.25, .25, .75$

5.

7. $(n = 2)$ 40, $(n = 4)$ 41, exact: $41\frac{1}{3}$ **9.** .63107, exact: .63212   **11.** .09375, exact: .08333   **13.** 1.03740, exact: .8   **15.** $M = 72$, $T = 90$, $S = 78$, exact: 78   **17.** $M = 44.96248$, $T = 72.19005$, $S = 54.03834$, exact: 53.59815   **19.** $M = 573.41797$, $T = 612.10806$, $S = 586.31466$, exact: 586.26358   **21.** 3.24124   **23.** 1.61347   **25.** 25,750 ft^2 **27.** 2150 ft

29. (a) $f''(x) = x^2 + 6$

(b) $A = 10$
(c) .0333
(d) .0244, satisfies the bound in (c)
(e) quartered

37. Midpoint rule: 2.361749156. Trapezoidal rule: 2.474422799. Simpson's rule: 2.399307037. Exact value: 2.397895273. Error using midpoint rule: .036146117. Error using Trapezoidal rule: .076527526. Error using Simpson's rule: .001411764.   **39.** Midpoint rule: .9989755866. Trapezoidal rule: 1.00205197. Simpson's rule: 1.000001048. Exact value: 1. Error using midpoint rule: .0010244134. Error using Trapezoidal rule: .00205197. Error using Simpson's rule: .000001048.   **41.** A bound on the error is .0008333.

Exercises 9.5, page 448

1. $147,656 3. $35,797 5. $182,937 7. (a) $\int_0^2 (30 + 5t)e^{-0.10t}\,dt$ (b) $63.1 million 9. (a) $240\pi \int_0^5 te^{-0.65t}\,dt$ (b) 1,490,493

11. $\approx 1,400,000$;

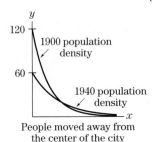

13. (a) $80\pi te^{-0.5t}\Delta t$ thousand people (b) $P'(t)$
(c) the number of people who live between 5 mi and $5 + \Delta t$ mi from the city center
(d) $P'(t) = 80\pi te^{-0.5t}$ (e) $P(b) - P(a) = \int_a^b P'(t)\,dt = \int_a^b 80\pi te^{-0.5t}\,dt$

Exercises 9.6, page 452

1. 0 3. No limit 5. $\frac{1}{4}$ 7. 2 9. 0 11. 5 13. $\frac{1}{2}$ 15. 2 17. 1 19. Area under the graph from 1 to b is $\frac{5}{14}(14b + 18)^{1/5} - \frac{5}{7}$. This has no limit, as $b \to \infty$. 21. $\frac{1}{2}$ 23. $\frac{1}{6}$ 25. Divergent 27. $\frac{2}{3}$ 29. 1 31. $2e$ 33. Divergent 35. $\frac{1}{2}$ 37. 2 39. $\frac{1}{4}$ 41. 2 43. $\frac{1}{6}$ 49. $\dfrac{K}{r}$

Chapter 9: Review Exercises, page 459

1. $-\frac{1}{6}\cos(3x^2) + C$ 2. $\frac{1}{3}(2x + 1)^{3/2} + C$ 3. $-\frac{1}{36}(1 - 3x^2)^6 + C$ 4. $\frac{1}{6}(\ln x)^6 + C$ 5. $\frac{1}{3}[\ln x]^3 + C$ 6. $\frac{1}{2}(4x + 3)^{1/2} + C$
7. $-\frac{1}{3}(4 - x^2)^{3/2} + C$ 8. $-\frac{1}{3}x\cos 3x + \frac{1}{9}\sin 3x + C$ 9. $-\frac{1}{3}e^{-x^3} + C$ 10. $\frac{1}{4}(\ln(x^2 + 1))^2 + C$
11. $\frac{1}{3}x^2\sin(3x) - \frac{2}{27}\sin(3x) + \frac{2}{9}x\cos(3x) + C$ 12. $\frac{1}{2}(\ln(\ln x))^2 + C$ 13. $2(x\ln x - x) + C$ 14. $\frac{2}{3}x(x + 1)^{3/2} - \frac{4}{15}(x + 1)^{5/2} + C$
15. $\frac{2}{3}x(3x - 1)^{1/2} - \frac{4}{27}(3x - 1)^{3/2} + C$ 16. $\frac{x^3}{3}\ln x^2 - \frac{2}{9}x^3 + C$ 17. $\frac{1}{4}\left[\dfrac{x}{(1 - x)^4}\right] - \frac{1}{12}\left[\dfrac{1}{(1 - x)^3}\right] + C$ 18. $\dfrac{x^2}{2}\left[(\ln x)^2 - \ln x + \frac{1}{2}\right] + C$

19. $f(x) = x$, $g(x) = e^{2x}$ 20. Integration by parts: $f(x) = x - 3$, $g(x) = e^{-x}$ 21. $u = \sqrt{x + 1}$ 22. Substitution: $u = x^3 - 1$
23. $u = x^4 - x^2 + 4$ 24. Integration by parts: $f(x) = \ln\sqrt{5 - x} = \frac{1}{2}\ln(5 - x)$, $g(x) = 1$ 25. $f(x) = (3x - 1)^2$, $g(x) = e^{-x}$; then integrate by parts again. 26. Substitution: $u = 3 - x^2$ 27. $f(x) = 500 - 4x$, $g(x) = e^{-x/2}$ 28. Integration by parts: $f(x) = \ln x$, $g(x) = x^{5/2}$ 29. $f(x) = \ln(x + 2)$, $g(x) = \sqrt{x + 2}$ 30. Repeated integration by parts. Starting with $f(x) = (x + 1)^2$, $g(x) = e^{3x}$.
31. $u = x^2 + 6x$ 32. Substitution $u = \sin x$ 33. $u = x^2 - 9$ 34. Integration by parts: $f(x) = 3 - x$, $g(x) = \sin 3x$
35. $u = x^3 - 6x$ 36. Substitution: $u = \ln x$ 37. $\frac{3}{8}$ 38. $\dfrac{-\pi}{16}$ 39. $1 - e^{-2}$ 40. $\frac{1}{4}[(\ln 5)^2 - (\ln 4)^2]$ 41. $\frac{3}{4}e^{-2} - \frac{5}{4}e^{-4}$
42. $-\sqrt{2}\ln 2 - 2\sqrt{2} + 4$ 43. $M = 3.93782$, $T = 4.13839$, $S = 4.00468$ 44. Midpoint rule: ≈ 103.81310: Trapezoidal rule: ≈ 104.63148: Simpson's rule: ≈ 104.08589. 45. $M = 12.84089$, $T = 13.20137$, $S = 12.96105$ 46. Midpoint rule: ≈ 1.57746: Trapezoidal rule: ≈ 1.55747: Simpson's rule: ≈ 1.57080. 47. $\frac{1}{3}e^6$. 48. Diverges 49. Divergent 50. $\frac{1}{3}$ 51. $2^{7/4}$ 52. $\frac{2}{25}$
53. $2e^{-1}$ 54. $\dfrac{1}{k^2}$ 55. $137,668 56. ≈ 11.335 thousand dollars 57. (a) $M(t_1)\Delta t + \cdots + M(t_n)\Delta t \approx \int_0^2 M(t)\,dt$
(b) $M(t_1)e^{-0.1t_1}\Delta t + \cdots + M(t_n)e^{-0.1t_n}\Delta t \approx \int_0^2 M(t)e^{-0.1t}\,dt$ 58. $80,000 + \int_0^\infty 50,000e^{-rt}\,dt$.

CHAPTER 10

Exercises 10.1, page 466

3. $y' = 10e^{2t}$, $y'' = 20e^{2t}$, $y'' - 3y' + 2y = 20e^{2t} - 3(10e^{2t}(+2(5e^{2t}) = 0$, $y(0) = 5$, $y'(0) = 10$. 5. Second order. 7. Yes
9. $y = 5$ 11. $f(0) = 4$, $f'(0) = 5$ 13. 20 feet per second per second 15. (a) Decreasing at $2500 per year (b) $y' = .05(y - 200,000)$
(c) The rate of change of the savings account balance is proportional to the difference between the balance at the end of t years and $200,000. 17. The number of people who have heard the news broadcast after t hours is increasing at a rate that is proportional to the difference between that number and 200,000. At the beginning of the broadcast, there are 10 people tuned in. 19. $y' = k(C - y)$, $k > 0$
21. $y' = k(P_b - y)$, $y(0) = P_0$, where k is a positive constant. 25. No, it will come very close to 5000 but will not attain or exceed this value. 27. $y = 0$ and $y = 1$ 29. All solution curves with $y(0) > 1$ will decrease and tend to 1. A typical solution curve is shown in Fig. 8.

31. (a)

(b) $.2(10 - f(5)) = -36.78794412$,
using nDeriv $f'(5) = -36.78794436$

Exercises 10.2, page 473

1. $y = \sqrt[3]{15t - \frac{3}{2}t^2 + C}$ 3. $y = \ln|e^t + C|$ 5. $y = (-\frac{2}{3}t^{3/2} + C)^{-1}$ or $y = 0$ 7. $y = (e^{t^3} + C)^{1/3}$ 9. $y = (\sqrt{t} + C)^2$ or $y = 0$

11. $y = \frac{-1}{t^3 + C}$ or $y = 0$ 13. $y = \ln(\frac{1}{2}e^{t^2} + C)$ 15. $y = \pm\sqrt{(\ln t)^2 + C}$ 17. $y = \frac{1}{t - t\ln t + C} + 3$ or $y = 3$

19. $y = \frac{1}{2}\ln(2t^2 - 2t + e^6)$ 21. $y = \sqrt[3]{3t \sin t + 3 \cos t + 5}$ 23. $y = \sqrt[3]{\cos t + 1}$ 25. $y = -\sqrt{2t + 2 \ln t + 7}$ 27. $y = \frac{2}{5} + \frac{3}{5}e^{(5/2)t^2}$

29. $y = (3x^{1/2}\ln x - 6x^{1/2} + 14)^{2/3}$ 31. $y = A(p + 3)^{-1/2}, A > 0$ 33. $y' = k(1 - y)$, where $y = p(t)$, $y(0) = 0$; $y = 1 - e^{-kt}$

35. $\frac{dV}{dt} = kV^{2/3}$, where $k < 0$; $V = (3 - \frac{1}{8}t)^3$, $V = 0$ when $t = 24$ weeks.   37. $y = be^{Ce^{-at}}$, C any number   39. **(a)** The population will decrease and approach 5000.   **(b)** The population will increase and approach 5000.

(c)   y (in thousands)

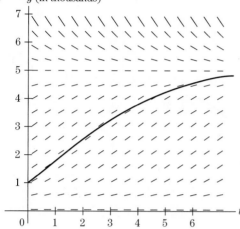

The solution curve represents the number of fish in the pond, if we start with an initial population of 1000 fish.

Exercises 10.3, page 478

1. e^{-2t} 3. $e^{-\frac{1}{2t^2}}$ 5. $\frac{1}{10 + t}$ 7. $y = 1 + Ce^{-t}$ 9. $y = 2 + Ce^{t^2}$ 11. $y = 35 + Ce^{-0.5t}$ 13. $y = \frac{C}{10 + t}$ 15. $y = \frac{C - t}{1 + t}$

17. $y = 1 + Ce^{-\frac{1}{12}t^2}$ 19. $y = 2 - \frac{1}{2}e^t + Ce^{-t}$ 21. $y = \frac{1}{2} + \frac{1}{2}e^{-2t}$ 23. $y = 10 + 10t$ 25. $y = \frac{1}{3}e^{2t} - \frac{4}{3}e^{-t}$ 27. $y = 1 - e^{-\sin 2t}$

29. **(a)** $y'(0) = -40$, $y(t)$ is decreasing at $t = 0$.   **(b)** $y = \frac{50 + 10t + 5t^2}{1 + t}$

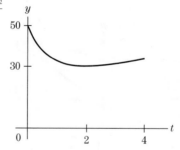

Exercises 10.4, page 483

1. **(a)** \$4200 per year   **(b)** \$40,000   **(c)** $t = \frac{1}{.06}\ln(\frac{80}{41}) \approx 11.1$ years   3. **(a)** $y' = .05y + 3600$ **(b)** $f(t) = -72,000 + 72,000e^{0.05t}$, $f(25) = 179,305$ dollars.   5. At retirement, Kelly has \$41,239 in her savings account, and John has \$31,139 in his savings account.

7. Approximately \$14,214 per year.   9. **(a)** $y' = .031y - A$, $y(0) = 251,010$ **(b)** $A = $12,852$ per year, or \$1071 per month

(c) \$134,556   11. **(a)** $py' + (p + 1)y = 0$ **(b)** $f(p) = 100\frac{e^{-p+1}}{p}$ 13. $f(t) = 10 + 340e^{-0.1t}$ 15. **(a)** $T = 70$ **(b)** $k = .5$

(c) $f(t) = 70 + 28e^{-0.5t}$ **(d)** Approximately 1 hour and 15 minutes   17. $y' = .45y + e^{0.03t} + 2$ 19. **(a)** $k = \frac{2}{7} \approx .286$

(b) $\frac{220}{7} \approx 31.43$ grams per liter per hour. Replacing the dialysate with a fresh solution after 4 hours triples the rate at which waste products are filtered from the body.   21. **(a)** $y' - .04y = -500t - 2000$, $y(0) = 100,000$ **(b)** $f(t) = 362,500 - 262,500e^{0.04t} + 12,500t$

(c)

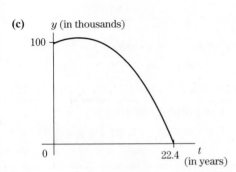

The account is depleted in approximately 22.4 years.

23. (a) 6 years **(b)** $y' - .04y = 3000 - 500t$, $y(0) = 10,000$
25. (a) $y' + .35y = t$ **(b)** $f(t) = \frac{1}{(.35)^2}(.35t - 1 + e^{-0.35t})$, $f(8) \approx 15.2$
milligrams

Exercises 10.5, page 492

1.

3.

5.

7.

9.

11.

13.

15.

17.

19.

21.

23.

25.

27.

29.

31.

33.

35.

37. $y' = ky(H - y)$

39.

Exercises 10.6, page 498

1. (a) Carrying capacity $= 1$, intrinsic rate $= 1$ **(b)**

(c), (d)

3. (a) Carrying capacity $= 100$, intrinsic rate $= 1$ **(b)**

(c), (d)

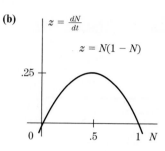

5. $\frac{dN}{dt} = \frac{r}{1000}N(1000 - N)$, $N(0) = 600$;

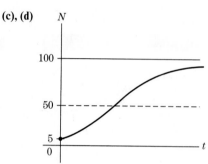

7. $y' = k(100 - y)$, $k > 0$

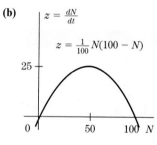

9. $y' = ky(M - y)$, $k > 0$

The reaction is proceeding
fastest when $y = M/2$.

11. $y' = k(c - y)$, $k > 0$

The graph is always concave down.

13. $y' = ky^2$, $k < 0$

15. $y' = k(E - y)$, $k > 0$

17. (a) $\frac{dN}{dt} = \frac{.4}{1000}N(1000 - N) - 75$

(b)

(c) It is sustainable but the fish
population will never reach the
carrying capacity of the pond.

19. (a) $y' = .05y + 10{,}000$, $y(0) = 0$
(b) $y = 200{,}000(e^{0.05t} - 1)$, \$56,805
21. (a) $y' = .05 - .2y$, $y(0) = 6.25$
(b) $y' = .13 - .2y$, $y(0) = 6.25$
23. $y' = -.14y$

25.

Exercises 10.7, page 504

1. 3 **3.** Increasing **5.** $f(1) \approx -\frac{9}{4}$ **7.** $f(2) \approx \frac{27}{8}$

9. Euler's method yields $f(1) \approx .37011$;

solution: $f(t) = \dfrac{1}{\frac{1}{2}t^2 + t + 1}$; $f(1) = .4$; error $= .02989$

11. (a) $y' = .1(1 - y), y(0) = 0$ **(b)** .271

(c) $y = 1 - e^{-0.1t}$, $y(3) = 1 - e^{-0.3} \approx .25918$

(d) error $\approx .01182$

13. (a) C **(b)** A **(c)** E

(d) B **(e)** D

15. $y_i = 1, .75, .6960, .7602, .9093, 1.1342, 1.4397,$
$1.8403, 2.3588; y = 1, .8324, .7948, .8544, .9963, 1.2182,$
$1.5271, 1.9383, 2.4752,$ greatest difference $= .1164$

Chapter 10: Review Exercises, page 510

1. $y = \sqrt[3]{3t^4 - 3t^3 + 6t + C}$ **2.** $-1 + Ae^{\frac{t^2}{2}+t}$ **3.** $y = Ate^{-3t}$ **4.** $\pm\frac{2}{3}t^{3/2} + C$ **5.** $y = \frac{3}{7}t + 3$ **6.** $-\ln(-te^t + e^t)$

7. $y = \sqrt{4t^3 - t^2 + 49}$ **8.** $\frac{5}{8} + \frac{3}{8}e^{-8t}$ **9.** $y(t) = \frac{1}{7}(t-1)^5 + \dfrac{C}{(t-1)^2}$ **10.** $\frac{2}{3}(1+t)^2 + C\sqrt{1+t}$ **11.** $y = -1 - x + e^x$

12. (a) $\frac{dP}{dt} = 1.2 - .33P$ **(b)** $P = \frac{1}{11}(40 - 29e^{-0.33t})$ **13.** Decreasing **14.** $y = 1$

15.

16.

17.

18.

19.

20.

21.

22.

23.

24.

25. (a) $N' = .015N - 3000$

(b) There is a constant solution $N = 200,000$, but it is unstable. It is unlikely a city would have such a constant population.

26. (a) $(10 - \frac{1}{4}y)$ represents the amount of unreacted substance A and $(15 - \frac{3}{4}y)$ that of B.

(b) $k > 0$ since C is increasing.

(c)

27. 13.863 years   **28. (a)** $M \geq \$200,000$ **(b)** $M \approx \$139,761.16$

29. $f(t) = 2t, f(2) = 4$ **30.** $y = |t + 1|, y(1) = 2$

31.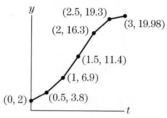

32. $y(1) \approx .34337$

CHAPTER 11

Exercises 11.1, page 518

1. $x - \frac{1}{6}x^3$.   **3.** $5 + 10x + 10x^2 + \frac{20}{3}x^3$.   **5.** $1 + 2x - 2x^2 + 4x^3$.   **7.** $x + 3x^2 + \frac{9}{2}x^3$.   **9.** $p_4(x) = 1 + x + \frac{1}{2}x^2 + \frac{1}{6}x^3 + \frac{1}{24}x^4$, $e^{0.01} \approx p_4(.01) = 1.01005$.

11.

$p_1(x) = 1 + x$; $p_2(x) = 1 + x + x^2$; $p_3(x) = 1 + x + x^2 + x^3$.

13. $p_n(x) = 1 + x + \frac{1}{2}x^2 + \frac{1}{3!}x^3 + \cdots + \frac{1}{n!}x^n$.   **15.** $p_2(x) = x^2$, area $\approx .0417$.   **17.** $1 + (x - 4) + (x - 4)^2 + (x - 4)^3$.

19. $p_3(x) = -1 + \frac{1}{2}(x - \pi)^2, p_4(x) = -1 + \frac{1}{2}(x - \pi)^2 - \frac{1}{24}(x - \pi)^4$.   **21.** 3.04958.   **23.** $p_1(x) = 19 + 33(x - 2)$, $p_2(x) = 19 + 33(x - 2) + 24(x - 2)^2, p_3(x) = 19 + 33(x - 2) + 24(x - 2)^2 + 8(x - 2)^3$, $p_n(x) = 19 + 33(x - 2) + 24(x - 2)^2 + 8(x - 2)^3 + (x - 2)^4, n \geq 4$.   **25.** $f''(0) = -5, f'''(0) = 7$.   **27. (a)** 1;

(b) $|R_3(.12)| \leq \frac{1}{4!}(.12)^4 = 8.64 \times 10^{-6}$.   **29. (a)** $R_2(x) = \dfrac{f'''(c)}{3!}(x - 9)^3$, where c is between 9 and x;

(b) $f^{(3)}(c) = \frac{3}{8}c^{-5/2} \leq \frac{3}{8}9^{-5/2} = \frac{1}{648}$;   **(c)** $|R_2(x)| \leq \frac{1}{648} \cdot \frac{1}{3!}(.3)^3 = \frac{1}{144} \times 10^{-3} < 7 \times 10^{-6}$.

31.
When $b = .55$, the difference is approximately .11; when $b = -.68$, the difference is approximately .087.

33. When $b = 1.85$, the difference is approximately .2552. When $b = 3$, the difference is approximately 3.7105.

Exercises 11.2, page 524

1. Let $f(x) = x^2 - 5$, $x_0 = 2$; then $x_1 = 2.25$, $x_2 \approx 2.2361$, $x_3 \approx 2.23607$.   **3.** Let $f(x) = x^3 - 6$, $x_0 = 2$; then $x_1 \approx 1.8333$, $x_2 \approx 1.81726$, $x_3 \approx 1.81712$.   **5.** If $x_0 = 2$, then $x_3 \approx 2.79130$.   **7.** $x_3 \approx .63707$.
9. $x_0 = -1$, $x_1 = -.8$, $x_2 \approx -.77143$, $x_3 \approx -.77092$.

11. .703.   **13.** 8.21% per month.
15. 1% per month.

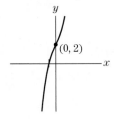

17. $x_1 \approx 3.5$, $x_2 \approx 3.0$.

19. $-\frac{5}{4}$.   **21.** $x_0 > 0$.   **23.** x_1 will be the exact root.
25. $x_0 = 1$, $x_1 = -2$, $x_2 = 4$, $x_3 = -8$.
27. 4; **31.**   **29.** (a) $\sqrt{2}$ (b) $-\sqrt{2}$ (c) 0.

Exercises 11.3, page 532

1. $\frac{6}{5}$.   **3.** $\frac{9}{10}$.   **5.** 3.   **7.** $\frac{25}{124}$.   **9.** $\frac{21}{10}$.   **11.** Divergent $\left(r = -\frac{8}{5}\right)$.   **13.** 25.   **15.** $\frac{3}{11}$.   **17.** $\frac{2}{9}$.   **19.** $\frac{4007}{999}$.
21. $.99\overline{9}\ldots = (.9)\frac{1}{1 - .1} = \frac{9}{10} \cdot \frac{10}{9} = 1$.   **23.** \$190 bill.   **25. (a)** $100 + 100(1.01)^{-1} + 100(1.01)^{-2} + \cdots = \sum_{k=1}^{\infty} 100(1.01)^{-k}$
(b) \$10,100.   **27.** \$1,655,629.   **29.** 20 mg.   **31.** 5 mg.   **33. (a)** 2.5;   **(b)** yes, 3.   **35.** 6.   **37.** $\frac{1}{24}$.   **39.** $\frac{15}{8}$.   **43.** 3.
45. $\frac{1}{4} + \frac{1}{8} + \frac{1}{16} + \frac{1}{32} + \frac{1}{64}; \frac{1}{2}$.

Exercises 11.4, page 538

1. Divergent.   **3.** Convergent.   **5.** Divergent.   **7.** Convergent.   **9.** Convergent.   **11.** Convergent   **13.** Convergent.
15. Divergent.   **17.** $\sum_{k=0}^{\infty} \frac{3}{9 + k^2}$.   **21.** Convergent.   **23.** Convergent.   **25.** Convergent.   **27.** No.   **29.**
31. 15.

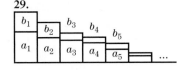

Exercises 11.5, page 545

1. $\frac{1}{3} - \frac{2}{9}x + \frac{2^2}{3^3}x^2 - \frac{2^3}{3^4}x^3 + \cdots$.   **3.** $1 + \frac{1}{2}x - \frac{1}{2^2 \cdot 2!}x^2 + \frac{1 \cdot 3}{2^3 \cdot 3!}x^3 - \frac{1 \cdot 3 \cdot 5}{2^4 \cdot 4!}x^4 + \cdots$.   **5.** $1 + 3x + 3^2x^2 + 3^3x^3 + \cdots$.

7. $1 - x^2 + x^4 - x^6 + \cdots$.   **9.** $1 - 2x + 3x^2 - 4x^3 + 5x^4 - \cdots$.   **11.** $5 + \frac{5}{3}x + \frac{5}{3^2 \cdot 2!}x^2 + \frac{5}{3^3 \cdot 3!}x^3 + \cdots$.

13. $x - \frac{1}{2!}x^2 + \frac{1}{3!}x^3 - \frac{1}{4!}x^4 + \cdots$.   **15.** $x - \frac{1}{2}x^2 + \frac{1}{3}x^3 - \frac{1}{4}x^4 + \cdots$.   **17.** $1 - \frac{3^2}{2!}x^2 + \frac{3^4}{4!}x^4 - \frac{3^6}{6!}x^6 + \cdots$.

19. $3x - \frac{(3x)^3}{3!} + \frac{(3x)^5}{5!} - \cdots$.   **21.** $x + x^3 + \frac{1}{2!}x^5 + \frac{1}{3!}x^7 + \cdots$.   **23.** $1 + \frac{1}{2!}x^2 + \frac{1}{4!}x^4 + \frac{1}{6!}x^6 + \cdots$.

25. $1 + \frac{1}{2}x + \frac{1 \cdot 3}{2 \cdot 4}x^2 + \frac{1 \cdot 3 \cdot 5}{2 \cdot 4 \cdot 6}x^3 + \cdots$.   **27.** $x - \frac{1}{2 \cdot 3}x^3 + \frac{1 \cdot 3}{2 \cdot 4 \cdot 5}x^5 - \frac{1 \cdot 3 \cdot 5}{2 \cdot 4 \cdot 6 \cdot 7}x^7 + \cdots$.   **31.** 48.   **33.** 0.

35. $\left[x - \frac{1}{3}x^3 + \frac{1}{5 \cdot 2!}x^5 - \frac{1}{7 \cdot 3!}x^7 + \cdots\right] + C$.   **37.** $x - \frac{1}{4}x^4 + \frac{1}{7}x^7 - \frac{1}{10}x^{10} + \cdots + C$.   **39.** $1 - \frac{1}{3} + \frac{1}{5 \cdot 2!} - \frac{1}{7 \cdot 3!} + \cdots$.

Chapter 11: Review Exercises, page 549

1. $x + \dfrac{3}{2}x^2$. **2.** $1 + 3x + \dfrac{3}{2}x^2 - \dfrac{1}{2}x^3 + \dfrac{3}{8}x^4$. **3.** $8 - 7x^2 + x^3$. **4.** $p_n(x) = 1 + \dfrac{x}{2} + \dfrac{1}{2^2}x^2 + \dfrac{1}{2^3}x^3 + \cdots + \dfrac{1}{2^n}x^n$.

5. $9 + 6(x-3) + (x-3)^2$. **6.** $p_3(x) = e^2 + e^2(x-2) + \dfrac{e^2}{2!}(x-2)^2 + \dfrac{e^2}{3!}(x-2)^3$. **7.** $p_2(t) = 2t^2; \dfrac{1}{12}$. **8.** $p_2(x) = x; \tan(.1) \approx .1$.

9. (a) $3 + \dfrac{1}{6}(x-9) - \dfrac{1}{216}(x-9)^2$. (b) 2.949583. (c) 2.949576. **10.** (a) $p_3(x) = -x - \dfrac{1}{2}x^2 - \dfrac{1}{3}x^3$, $\ln(1.3) \approx p_3(-.3) = .264$.

(b) $x_0 = 0$, $x_1 = .3$, $x_2 \approx .2631$. **11.** 3.5619. **12.** $x_0 = 0$, $x_1 = \frac{1}{3}$, $x_2 = .2832$, $x_3 = .2812$, $x_4 = .2812$. **13.** $\frac{4}{7}$. **14.** 25. **15.** $\frac{1}{7}$.

16. Divergent **17.** 1. **18.** Convergent to $\dfrac{1}{m+1}$ if $m > 1$. Diverges if $m \le 1$. **19.** e^2. **20.** $e^{\frac{1}{3}}$. **21.** $\frac{9}{2}$. **22.** $\frac{21}{4}$. **23.** Convergent.
24. Converges **25.** Divergent. **26.** Convergent. **27.** $p > 1$ **28.** Convergent when $|p| > 1$. **29.** $1 - x^3 + x^6 - x^9 + x^{12} - \cdots$.

30. $\ln(1 + x^3) = x^3 - \dfrac{1}{2}x^6 + \dfrac{1}{3}x^9 - \dfrac{1}{4}x^{12} + \cdots$, $|x| < 1$. **31.** $1 + 6x + 27x^2 + 108x^3 + \cdots$. **32.** $1 + \dfrac{1}{2!}x + \dfrac{1}{3!}x^2 + \dfrac{1}{4!}x^3 + \cdots$

for all x. **33.** (a) $1 - \dfrac{2^2}{2!}x^2 + \dfrac{2^4}{4!}x^4 - \dfrac{2^6}{6!}x^6 + \cdots$. (b) $x^2 - \dfrac{2^3}{4!}x^4 + \dfrac{2^5}{6!}x^6 - \dfrac{2^7}{8!}x^8 + \cdots$.

34. (a) $\cos 3x = 1 - \dfrac{3^2}{2!}x^2 + \dfrac{3^4}{4!}x^4 - \dfrac{3^6}{6!}x^6 + \cdots$ for all x. (b) $p_4(x) = 1 - \dfrac{3}{2}x^2 + \dfrac{7}{8}x^4$. **35.** $1 + 2x + 2x^2 + 2x^3 + \cdots$.

36. $\dfrac{1}{2} + \dfrac{1}{4 \cdot 2^2} + \dfrac{1}{3!3 \cdot 2^3} + \dfrac{1}{4!4 \cdot 2^4} + \cdots$. **37.** (a) x^2 (b) 0 (c) $.3095$. **38.** $p_4(x) = x + \dfrac{1}{3!}x^3$.

39. (a) $f'(x) = 2x + 4x^3 + 6x^5 + \cdots$. (b) $f'(x) = \dfrac{2x}{(1-x^2)^2}$. **40.** (a) $\dfrac{1}{2}x^2 - \dfrac{1}{2}x^4 + \dfrac{2}{3}x^6 - x^8 + \dfrac{8}{5}x^{10} - \cdots + C$.

(b) $\dfrac{1}{4}\ln(1 + 2x^2) + C$. **41.** $\$566{,}666{,}667$. **42.** $\$365.625$ million. **43.** $\$120{,}066.66$. **44.** $\approx \$49{,}103.30$. **45.** $\$3{,}285{,}603.18$.

CHAPTER 12

Exercises 12.1, page 557

1. $E(X) = \frac{4}{5}$; $\text{Var}(X) = .16$; standard deviation $= .4$. **3.** (a) $\text{Var}(X) = 1$ (b) $\text{Var}(X) = 4$ (c) $\text{Var}(X) = 16$.
5. (a)

Accidents	0	1	2	3
Probability	.21	.5	.25	.04

(b) $E(X) \approx 1.12$ (c) Average of 1.12 accidents per week during the year.
7. (a) 25%, (b) $100c^2\%$. **9.** $E(X) = \$90{,}000$; the grower should spend the $\$5000$.

Exercises 12.2, page 564

1. $f(x) \ge 0$ for $0 \le x \le 6$; $\int_0^6 \frac{1}{18}x\, dx = 1$. **3.** $f(x) \ge 0$; $\int_1^5 \frac{1}{4}\, dx = 1$. **5.** $f(x) \ge 0$ for $0 \le x \le 1$; $\int_0^1 5x^4\, dx = 1$. **7.** $\frac{1}{4}$. **9.** $\frac{1}{15}$.
11. **12.** **13.** y

$y = \frac{1}{8}x$

15. $\frac{1}{12}$. **17.** $\frac{1}{2}$. **19.** $\frac{3}{4}$. **21.** $f(x) = \frac{1}{4}(x-1)^{-1/2}$. **23.** $F(x) = \frac{1}{5}x - \frac{2}{5}$. **25.** (a) $\frac{19}{63}$;
(b) $f(x) = (x^3 - 1)/63$; (c) $\frac{19}{63}$. **31.** $2 - \sqrt{2} \approx .59$. **33.** 1.8. **35.** $.6$.
37. (a) $f(x) \ge 0$ for $x \ge 1$; $\int_1^\infty \frac{4}{x^5}\, dx = 1$; (b) $f(x) = 1 - x^{-4}$; (c) $\frac{15}{16}, \frac{1}{16}$.

Exercises 12.3, page 569

1. $E(X) = 4$; $\text{Var}(X) = 2$. **3.** $E(X) = 3$; $\text{Var}(X) = \frac{4}{3}$. **5.** $E(X) = \frac{5}{6}$; $\text{Var}(X) = \frac{5}{252}$. **7.** $E(X) = \frac{2}{5}$; $\text{Var}(X) = \frac{1}{25}$.
9. (a) $F(x) = 10x^3 - 15x^4 + 6x^5$; (b) $\frac{53}{512}$; (c) $\frac{1}{2}$; on the average about half of the newspaper's space is devoted to advertising;
(d) $\frac{1}{28}$. **11.** (a) 2; the average useful life of the component is 200 h; (b) $\frac{1}{2}$. **13.** 8 min. **15.** (a) $F(x) = \frac{1}{6}x^2 - \frac{1}{54}x^3 - 1$;
(b) $\frac{23}{27}$; (c) 412.5 worker-hours; (d) $.5344$. **17.** $E(X) = \frac{4}{3}$; $\text{Var}(X) = \frac{2}{9}$. **19.** $3\sqrt{2}$. **21.** $\frac{3\sqrt{2}}{2}$ hundred hours. **23.** $\frac{5}{6}$ min.
25. *Hint*: Compute $\int_A^B x f(x)\, dx$ using integration by parts.

Exercises 12.4, page 577

1. $E(X) = \frac{1}{3}$; $\text{Var}(X) = \frac{1}{9}$. **3.** $E(X) = 5$; $\text{Var}(X) = 25$. **5.** $e^{-1} - e^{-2}$. **7.** $1 - e^{-2/3}$. **9.** e^{-3}. **11.** $1 - e^{-2}$. **13.** (a) $e^{-1/3}$;
(b) $r(t) = e^{-t/72}$. **15.** $\mu = 4$; $\sigma = 1$. **17.** $\mu = 0$; $\sigma = 3$. **23.** (a) $.4032$; (b) $.4013$; (c) $.8351$; (d) $.9772$. **25.** (a) 47.72%;
(b) 47.72%. **27.** $.1587$. **29.** The Capital Beltway. **31.** 1.22%. **33.** *Hint*: First show that $\Pr(a \le X \le a+b) = e^{-ka}(1 - e^{-kb})$.
35. 62.13 weeks.

Exercises 12.5, page 584

1. .0504; .0216; .0081.   **3.** .4724; .3543; .1329; .0332; .0062; .0009; .0001.

5. (a) .0000454;   **(b)** .0027694;   **(c)** .9972306.
7. (a) .2231302;   **(b)** .3765321;
(c) .0656425.   **9.** 22,105.

11. .4; .24; .144; .0864; .05184; .031104.

13. (a) $\left(\frac{3}{4}\right)^n\left(\frac{1}{4}\right)$;   **(b)** .4219;   **(c)** 3.   **15.** $1 - p^n$.   **17.** .04073.   **19.** $\lambda = 2$.   **23.** .88705837.   **25. (a)** No;   **(b)** .9979061.

Chapter 12: Review Exercises, page 588

1. (a) $\frac{1}{8}, \frac{19}{64}$;   **(b)** $E(X) = \frac{3}{2}$, $\text{Var}(X) = \frac{3}{20}$.   **2. (a)** .96, 1;   **(b)** $E(X) = \frac{11}{3}$; $\text{Var}(X) = \frac{1}{18}$.   **3.** $f(x) = 1 - e^{A-x}$.   **4.**   I. $f(x) \geq 0$;
II. $\int_A^\infty \frac{ka^k}{x^{k+1}} dx = 1$ **5.** $c_2 = \frac{1}{2}, c_4 = \frac{1}{4}$.   **6.** I. $\frac{1}{2k^3}x^2 e^{-\frac{x}{k}} \geq 0$;   II. $\int_0^\infty \frac{1}{2k^3}x^2 e^{-\frac{x}{k}} dx = 1$.   **7. (a)** 5.01;   **(b)** 100.   **8. (a)** 2.13
(b) ≈ 85 tests.   **9. (a)** .96;   **(b)** 1.8 thousand gallons.   **(c)** $f(x) = 1 - x/2, 0 \leq x \leq 2$.   **10.** $16.67 on each service contract sold.
11. (a) $E(X) = 22.5$; $\text{Var}(X) = 2.0833$ **(b)** 21.5.   **12. (a)** $f(x) = \frac{x}{8}, 3 \leq x \leq 5$;   **(b)** $a = \sqrt{21}$.   **13. (a)** $\frac{1}{300}$,   **(b)** $\frac{3}{8}$,
(c) $17,222.   **14. (a)** $F(x) = \Pr(3 \leq X \leq x)$;

(b) $\Pr(5 \leq X \leq 7)$ **(c)** $\int_5^7 f(x)\, dx$.   **15.** $\frac{2}{\pi}$.   **17.** $54.88.

18. $E(Y) \approx 28.55 > 25$. Buy it for the first year.   **19.** $k \approx .35$.
20. $E(X) = 20$ thousand hours; machine should be purchased.
21. .4938.   **22.** About 24.   **23.** 69.15%.   **24.** About .62% are eligible.
25. $a \approx .25$.   **26.** $t = 525$.   **27. (b)** .6826
28. (a) $\Pr(-2 \leq x \leq 2) = 2A(2) = .954$;   **(b)** .954.
29. (a) $\Pr\left(-\frac{1}{k} \leq X \leq \frac{3}{k}\right) = \Pr\left(0 \leq X \leq \frac{3}{k}\right) \geq \frac{3}{4}$ **(b)** $1 - e^{-3} \approx .9502$.
31. .1953668.   **32.** $\approx .051$.   **33.** 4.   **34.** $\left(\frac{2}{9}\right)\left(\frac{7}{9}\right)^n$.   **35.** $\frac{2}{7}$.
36. .4705.

Index of Applications

Index